Uterine Function

Molecular and Cellular Aspects

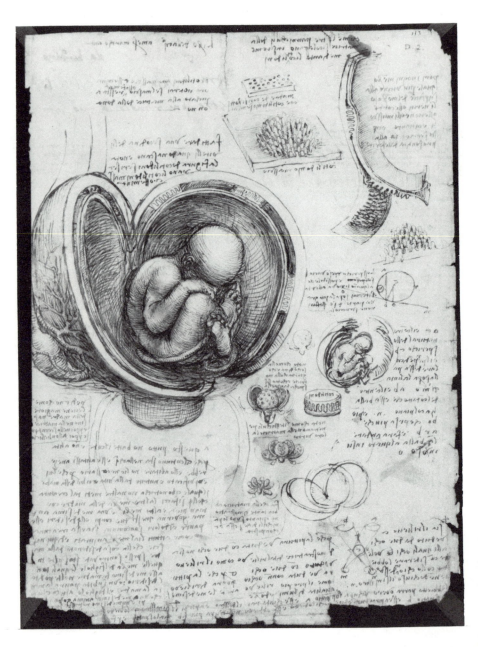

"Embryo in the womb" by Leonardo da Vinci (1452–1519). From the collection at Windsor Castle, Royal Library. © 1989 Her Majesty Queen Elizabeth II.

Uterine Function

Molecular and Cellular Aspects

Edited by

Mary E. Carsten and Jordan D. Miller

School of Medicine
University of California, Los Angeles
Los Angeles, California

Plenum Press • *New York and London*

Library of Congress Cataloging in Publication Data

Uterine function: molecular and cellular aspects / edited by Mary E. Carsten and Jordan D. Miller.
    p.   cm.
    Includes bibliographical references.
    ISBN 0-306-43446-6
    1. Myometrium—Physiology. 2. Uterus—Physiology. 3. Molecular biology. 4. Cytology. I. Carsten, Mary E. II. Miller, Jordan D.
    [DNLM: 1. Uterus—cytology. 2. Uterus—physiology. WP 400 U892]
QP262.U85   1990
612.6′2—dc20
DNLM/DLC   90-6882
for Library of Congress   CIP

© 1990 Plenum Press, New York
A Division of Plenum Publishing Corporation
233 Spring Street, New York, N.Y. 10013

Printed in the United States of America

This book is dedicated to the memory of

My husband, Don Marlin
—M.E.C.

My father, William Miller
—J.D.M.

Contributors

Marlane J. Angle • Departments of Biochemistry and Obstetrics–Gynecology and The Cecil H. and Ida Green Center for Reproductive Biology Sciences, University of Texas Southwestern Medical Center, Dallas, Texas 75235

Kate Bárány • Department of Physiology and Biophysics, College of Medicine, University of Illinois at Chicago, Chicago, Illinois 60612

Michael Bárány • Department of Biological Chemistry, College of Medicine, University of Illinois at Chicago, Chicago, Illinois 60612

Charles R. Brinkman III • Department of Obstetrics and Gynecology, School of Medicine, University of California Los Angeles, Harbor/UCLA Medical Center, Torrance, California 90509

Karen A. Broderick • Division of Neonatology, Department of Pediatrics, St. Agnes Hospital, Baltimore, Maryland 21229

Ray Broderick • Department of Pharmacology and Experimental Therapeutics, University of Maryland, School of Medicine, Baltimore, Maryland 21201

Robert C. Burghardt • Department of Veterinary Anatomy, College of Veterinary Medicine, Texas A&M University, College Station, Texas 77843

Mary E. Carsten • Departments of Obstetrics–Gynecology and Anesthesiology, School of Medicine, University of California Los Angeles, Los Angeles, California 90024

M. Linette Casey • The Cecil H. and Ida Green Center for Reproductive Biology Sciences, and Departments of Biochemistry and Obstetrics–Gynecology, University of Texas Southwestern Medical Center, Dallas, Texas 75235

Gautam Chaudhuri • Departments of Obstetrics–Gynecology and Pharmacology, School of Medicine, University of California Los Angeles, Los Angeles, California 90024

Harold A. Coleman • Department of Physiology, Monash University, Clayton, Victoria 3168, Australia

Kenneth A. Conklin • Department of Anesthesiology, School of Medicine, University of California Los Angeles, Los Angeles, California 90024

M. Joan Dawson • Department of Physiology and Biophysics, College of Medicine, University of Illinois at Urbana-Champaign, Urbana, Illinois 61801

Jack Diamond • Division of Pharmacology and Toxicology, Faculty of Pharmaceutical Sciences, University of British Columbia, Vancouver, British Columbia V6T 1W5, Canada

Charles A. Ducsay • Division of Perinatal Biology, Departments of Physiology and Pediatrics, School of Medicine, Loma Linda University, Loma Linda, California 92350

Jan Fahrenkrug • Department of Clinical Chemistry, Bispebjerg Hospital, University of Copenhagen, Copenhagen, Denmark

William H. Fletcher • Department of Anatomy, School of Medicine, Loma Linda University, and Jerry L. Pettis Memorial Veterans Medical Center, Loma Linda, California 92357

Robert H. Hayashi • Department of Obstetrics and Gynecology, University of Michigan Medical School, Ann Arbor, Michigan 48109

Elwood V. Jensen • Ben May Institute, University of Chicago, Chicago, Illinois 60637

John M. Johnston • Departments of Biochemistry and Obstetrics–Gynecology and The Cecil H. and Ida Green Center for Reproductive Biology Sciences, University of Texas Southwestern Medical Center, Dallas, Texas 75235

Rosemary D. Leake • Department of Pediatrics, School of Medicine, University of California Los Angeles, Harbor/UCLA Medical Center, Torrance, California 90509

Paul C. MacDonald • The Cecil H. and Ida Green Center for Reproductive Biology Sciences, and Departments of Biochemistry and Obstetrics–Gynecology, University of Texas Southwestern Medical Center, Dallas, Texas 75235

Steven B. Marston • Department of Cardiac Medicine, The National Heart and Lung Institute, London SW3 6LY, United Kingdom

Jordan D. Miller • Department of Anesthesiology, School of Medicine, University of California Los Angeles, Los Angeles, California 90024

Bent Ottesen • Departments of Obstetrics and Gynecology, Hvidovre and Herlev Hospitals, University of Copenhagen, Copenhagen, Denmark

Helena C. Parkington • Department of Physiology, Monash University, Clayton, Victoria 3168, Australia

Kevin Pritchard • Department of Cardiac Medicine, The National Heart and Lung Institute, London SW3 6LY, United Kingdom

Jyothi Raman • Department of Physiology and Biophysics, College of Medicine, University of Illinois at Urbana-Champaign, Urbana, Illinois 61801

Melvyn S. Soloff • Department of Biochemistry, Medical College of Ohio, Toledo, Ohio 43699

Preface

The frontispiece, Leonardo da Vinci's drawing of the embryo in the womb, was chosen as a starting point for this book. It was Leonardo who in his notebooks and drawings combined artistic composition and accurate recording of the anatomy of the human body. Leonardo studied human anatomy in order to execute artistic drawings. His aim was to clarify form and function of human organs including reproductive organs. He followed up his extensive research with graphic representation and thereby initiated record keeping as a basis of scientific investigation. His records, accurate three-dimensional drawings, allowed others to reproduce his findings and to test for correctness. Results could be updated and refined. Only after these steps can abnormalities be ascertained and defined as pathology.

Though Leonardo was both artist and scientist, it is assumed that his anatomic drawings were used to improve his art, and thus scientific endeavor was at the service of his art. Anatomy, the offspring of science and art, is an integration of the two and became an accepted branch of the natural sciences. Although art and science continued to interact throughout the Renaissance, art was often placed in the service of science. In the course of history that followed, art and science increasingly followed separate ways.

The integration of art and scientific and mathematical principles enabled the systematic exploration of nature developed in the Renaissance. Thereafter, great advances in the natural sciences took place at an ever-accelerating pace. Compare the progress made in scientific knowledge in the last 500 years with that in the last 50 years or the last 5 years.

Our purpose in assembling this book was to summarize what is known about the molecular and cellular basis of uterine function and to point out what is still unknown. We wanted to bring the discipline up to date, to delineate gaps in our knowledge, and to encourage further research. Each chapter was written so as to be self-contained and should be understandable by researchers who are not experts in the particular area. In order to accomplish this, we have asked outstanding scholars, each in his or her area of expertise, to write the various chapters and to summarize the "state of the art." Thus, we have brought together diverse areas of research: smooth muscle contraction, microanatomy, structure and function of proteins, second messenger function, hormonal effects on the uterus, regulation of uterine blood flow, factors in initiating labor, all involved in uterine function. The first half of the book deals with the contractile mechanisms of the myometrium; the second half discusses hormonal, nervous, and pharmacological regulation of myometrial contraction. It is the ultimate goal to allow methods and concepts used by one area to be adapted to other areas, thus, as in adapting art to science, aiding more rapid progress.

We thank the contributors for their prompt attendance to writing and completing their chapters and for their cheerful compliance with editorial requests. We also thank our colleagues Dr. Michael Berridge, Dr. Simone Harbon, and Dr. C. Y. Kao for invaluable discussions. We gratefully acknowledge the assistance of Suzanne Gonzalez-Drake in typing and editing the manuscripts.

We are grateful to Mary Phillips Born, Senior Editor at Plenum Publishing Corporation, for guidance and advice and for the rapid publication of this volume.

Mary E. Carsten
Jordan D. Miller

Los Angeles

Contents

5. Calcium Control Mechanisms in the Myometrial Cell and the Role of the Phosphoinositide Cycle

Mary E. Carsten and Jordan D. Miller

6. Calcium Channels: Role in Myometrial Contractility and Pharmacological Applications of Calcium Entry Blockers

Charles A. Ducsay

7. The Role of Membrane Potential in the Control of Uterine Motility

Helena C. Parkington and Harold A. Coleman

8. β-Adrenoceptors, Cyclic AMP, and Cyclic GMP in Control of Uterine Motility

Jack Diamond

9. Physiological Roles of Gap Junctional Communication in Reproduction

Robert C. Burghardt and William H. Fletcher

10. Molecular Mechanisms of Steroid Hormone Action in the Uterus

Elwood V. Jensen

11. Oxytocin in the Initiation of Labor

Rosemary D. Leake

12. Oxytocin Receptors in the Uterus

Melvyn S. Soloff

13. Regulatory Peptides and Uterine Function

Bent Ottesen and Jan Fahrenkrug

14. Biosynthesis and Function of Eicosanoids in the Uterus

Gautam Chaudhuri

15. Pharmacological Application of Prostaglandins, Their Analogues, and Their Inhibitors in Obstetrics

Robert H. Hayashi

16. Fetal Tissues and Autacoid Biosynthesis in Relation to the Initiation of Parturition and Implantation

Marlane J. Angle and John M. Johnston

17. Endocrinology of Pregnancy and Parturition

M. Linette Casey and Paul C. MacDonald

18. Circulation in the Pregnant Uterus

Charles R. Brinkman III

19. Effects of Obstetric Analgesia and Anesthesia on Uterine Activity and Uteroplacental Blood Flow

Kenneth A. Conklin

Uterine Function

Molecular and Cellular Aspects

Ultrastructure and Calcium Stores in the Myometrium

Ray Broderick and Karen A. Broderick

1. Introduction

This chapter is designed to provide the nonspecialist with an introduction to and an overview of uterine smooth muscle ultrastructure, particularly as it pertains to the development and transmission of force. A detailed description of intracellular calcium storage sites has been included along with a discussion of the release and recycling of activatable calcium in smooth muscle. Because of the paucity of information available in these areas with respect to uterine muscle, much of what is conveyed herein is a generalized summary of data from other visceral muscles as well as vascular smooth muscle, with specific reference to uterine muscle whenever possible. This is not meant to contend that all smooth muscles are alike; however, most points of ultrastructure are consistent among the muscle types, and, in fact, all salient features discussed are demonstrated with electron micrographs of uterine smooth muscle prepared specifically for this text. Although direct electron-optical quantitation of intracellular calcium stores and demonstration of calcium release/recycling has been made only in vascular muscle, it is hypothesized, based on the similarities in ultrastructure and physiological responses, that this information is applicable to uterine as well as other visceral smooth muscles. Resolution of these questions in myometrial muscle has lagged somewhat, partly because of a concentration of effort on its "wealthier" cousin vascular smooth muscle, and further research is required.

Ray Broderick • Department of Pharmacology and Experimental Therapeutics, University of Maryland, School of Medicine, Baltimore, Maryland 21201.   *Karen A. Broderick* • Division of Neonatology, Department of Pediatrics, St. Agnes Hospital, Baltimore, Maryland 21229.

2. Myometrium

The myometrium is composed of two distinct smooth muscle cell layers that comprise a major part of the rat uterine wall (Fig. 1). The outer longitudinal muscle layer consists of distinct bundles of smooth muscle cells that are oriented along the long axis of the uterus. The cells of the inner circular layer, that lying between the longitudinal muscle layer and the endometrium, are arranged concentrically around the longitudinal axis of the uterus. These cells are more diffuse in arrangement than those of the longitudinal muscle layer, with no apparent bundle formation. Functional and structural studies have indicated, however, that the two muscle layers are continuous.[1] In larger, and especially in simplex uteri (human), these muscle coats are usually not clear-cut circular and longitudinal layers; rather they consist of more or less interwoven trabeculae that are longitudinal on the outer uterine surface but gradually become more circular as they spiral deeper and deeper into the uterine wall.[2] Contraction of the myometrium will both shorten the uterus and decrease the size of the lumen much in the same way that vascular smooth muscle constricts blood vessels.

Uterine smooth muscle is embedded in a connective tissue matrix (Fig. 2). Smooth muscle cells have been shown to synthesize collagen,[3] elastin,[4] glycoproteins,[4] and proteoglycans.[5] These connective tissue elements contribute to the distribution of force development during contraction. In addition to connective tissue, the space between the smooth muscle cells is occupied by fibroblasts, blood and lymphatic vessels, and nerves. The volume of the extracellular space in uterine smooth muscle is relatively high, as it is in vascular smooth muscle, and estimates range between 37% and 57% of the total myometrial volume.[6] Spaces do not vary between pregnant and nonpregnant[7] or estrogen- and progesterone-dominated uterus,[8] and the volumes of marker distribution indicate constant intracellular concentrations of sodium, potassium, and chloride.[6]

The cells of the myometrium have a shape that is typical of most visceral smooth muscle in that they are elongated, fusiform with tapering ends (Fig. 2). They range from 5 to 10 μm in diameter and from 300 to 600 μm in length.[9,10] Uterine smooth muscle cells from 17-day pregnant rats average over 400 μm in length,[11] whereas in estrogen-dominated rat uteri the muscle cells average only 80 to 85 μm in length.[12] Cell length, however, is an uncertain parameter since these cells can shorten and elongate over a wide range. Mammalian visceral muscle cells are relatively small with a volume of only 2300 to 3500 μm^3 (vascular smooth muscle can be less than 25% of this). Approximately 80% to 90% of the cell volume is occupied by myofilaments (contractile apparatus), intermediate filaments, and dense bodies, with the remaining space taken up mostly by the nucleus, mitochondria, sarcoplasmic reticulum, and microtubules. The small total volume occupied by the smooth muscle cell and the presence of an irregular cell surface produce a high surface-to-volume ratio, which is estimated[13] to be between 1.5 and 2.7 μm^{-1}. This would mean, for example, that there is 1.5 μm^2 of cell surface for every cubic micrometer of cell volume. Specializations of the cell membrane in the form of

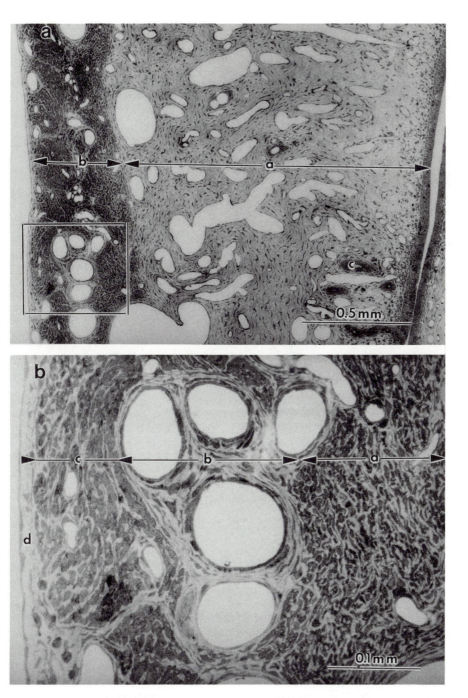

Figure 1. (**a**) Light micrograph of a cross section through the wall of a nonpregnant rat uterus that was perfusion fixed with glutaraldehyde. a, endometrium; b, myometrium; c, uterine gland. (**b**) Inset from panel a. Cross section through the myometrium. a, circular muscle layer; b, vascular layer; c, longitudinal muscle layer; d, perimetrium.

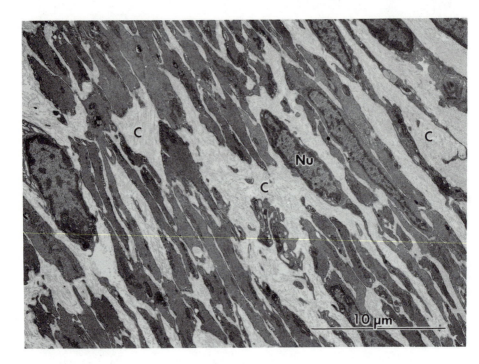

Figure 2. Electron micrograph of rat longitudinal uterine smooth muscle showing a relatively dense connective tissue matrix and the characteristic shape of visceral smooth muscle cells. C, collagen; Nu, nucleus.

inpocketings known as caveolae (see below) increase the surface area by about 70%.

3. Smooth Muscle Ultrastructure

3.1. Sarcolemma

In general, the term *sarcolemma* refers to the muscle cell membrane, the basal lamina or basement membrane, and the immediately adjacent connective tissue. In conventional transmission electron micrographs, the plasma membrane of smooth muscle shows the typical trilaminar appearance common to all eukaryotic cells and measures approximately 8 nm in thickness.[14] The extracellular surface of uterine, as all smooth muscle, plasma membrane is covered by a thin (10 nm), morphologically ill-defined, coat referred to as the basal lamina (Fig. 3), which is separated from the membrane by an electron-lucent layer called the lamina rara or cell surface coat.[15] As a rule, a basal lamina always intervenes between collagen fibrils and the cell

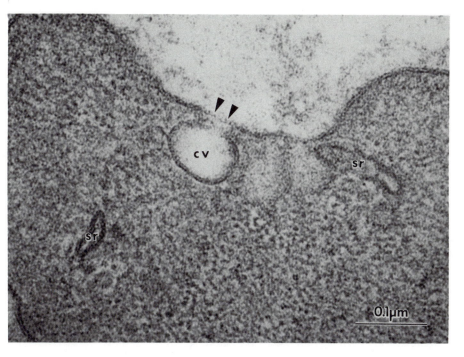

Figure 3. High-magnification electron micrograph of rat uterine smooth muscle showing the plasma membrane with its ill-defined basal lamina. Arrowheads indicate a continuous basal lamina across the opening of a caveola (cv). sr, sarcoplasmic reticulum.

membrane; however, in myometrial smooth muscle the basal lamina is discontinuous and bundles of collagen fibrils can be found directly apposed to the cell membrane (Fig. 4). Both the composition and function of the basal lamina remain largely unknown; however, the basal lamina (basement membrane) of other tissues is known to be made up of collagen, glycoproteins, and glycosaminoglycans.[16] A fine fibrillar texture can be seen in tangential sections of the basal lamina and very small fibers (diameter 3–5 nm) have been demonstrated.[17] Larger fibers (diameter 11 nm) are embedded in the basal lamina and extend into the adjacent extracellular space. In addition, a delicate network of fibrils, extending from the basal lamina and attaching to the surface membrane, has been shown in frozen deep-etched and rotary-shadowed preparations.[18] It is probable that this series of fibrils, extending through the basal lamina, form the external connection from the cell membrane to the surrounding connective tissue or other smooth muscle cells necessary for the transmission of force.

Freeze-fracture replicas of the plasma membrane of uterine smooth muscle have demonstrated intramembrane particles, about 9 nm in diameter, that are more numerous on the P-face (protoplasmic-side leaflet) than on the E-face (external-side

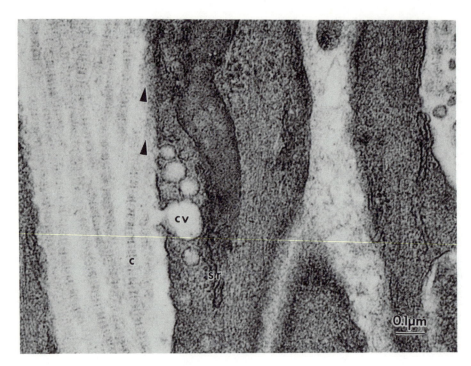

Figure 4. Electron micrograph of longitudinal rat uterine smooth muscle. Arrowheads indicate bundles of collagen fibrils directly apposed to the cell membrane. c, collagen; cv, caveolae; sr, sarcoplasmic reticulum.

leaflet).[19] This occurrence has also been observed in a number of other smooth muscles.[13,20] Devine and Rayns[20] have estimated that there are approximately 450 particles/μm^2 on the P-face and about 300 particles/μm^2 on the E-face in both visceral and vascular smooth muscle. Although the density of these particles is higher in the region of the membrane occupied by caveolae,[21] their exact function remains obscure. It has been postulated[19] that the particles are structural and functional proteins that may be sites for binding and transport.

3.1.1. Membrane Specializations

Several specializations of the plasma membrane occur in smooth muscle cells. These include (1) the presence of numerous vesicular inpocketings of the membrane known as *caveolae,* (2) the presence of the membrane of *electron-dense bands* that appear to act as attachment points for actin filaments, (3) the appearance of *intermediate junctions,* coupled dense bands between adjacent cells for mechanical coupling, and (4) the formation of cell-to-cell communications in the form of *gap junctions.* The gap junctions form in the basal lamina in regions of close apposition between the outer leaflets of the plasma membranes of neighboring cells and act as low-resistance pathways for the rapid spread of electrical signal throughout the tissue. A detailed description of myometrial gap junctions and their role in parturition is provided in Chapter 9.

3.1.2. Caveolae

Surface caveolae are flask-shaped inpocketings of the cell membrane that measure about 70 nm across and 120 nm long (the diameter of the narrowest part of the neck is about 35 nm).[13] Numerous caveolae occupy the surface of uterine smooth muscle (Fig. 5). Gabella[22] has calculated, both serologically and using freeze-fracture,[23] that taenia coli smooth muscle has about 32 caveolae/μm^2 cell surface, and each muscle cell has about 170,000 caveolae. It has been estimated that caveolae occupy about 54% of the cell surface in taenia coli[24] and about 45% in the rabbit pulmonary artery.[25] Each caveola is made up of approximately 23,000 nm^2 of plasma membrane, which increases the cell surface by about 70% compared with the geometrical surface of the cell (excluding all inpocketings).[13] The caveolae are not randomly distributed over the cell surface but rather are grouped in rows parallel

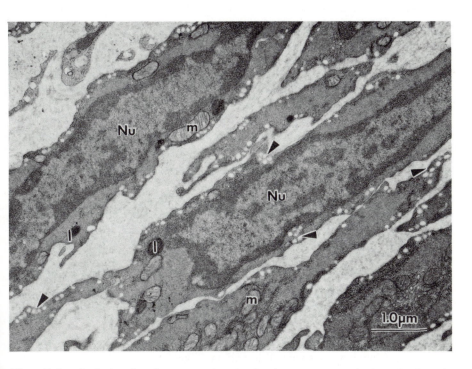

Figure 5. Longitudinal section of rat myometrium showing the numerous caveolae (arrowheads) at the membrane of the muscle cells. 1, lysosomes; m, mitochondria; Nu, nucleus.

to the long axis of the cell with dense bands intervening between.[26–29] In addition, caveolae have a characteristic association with sarcoplasmic reticulum and mitochondria as demonstrated in uterine smooth muscle (Fig. 6); however, no physiological significance has been found for this juxtaposition nor for the longitudinal arrangement of the caveolae in the membrane.

The function of caveolae in smooth muscle is not known. It has been suggested that they are sites for ion transport and binding.[30–32] Although the basal lamina is continuous across the opening (Fig. 3), caveolae of smooth muscle do communicate with the extracellular space as evidenced by their uptake of tracers such a ferritin, colloidal lanthanum, or peroxidase.[33] However, the vesicles have never been seen to be internalized and there has been no other evidence of pinocytotic or endocytotic function.[19] In this respect, it is important not to confuse caveolae with clathrin-containing coated pits, which can serve as receptors for the internalization of, among other molecules, lipoproteins.[34]

3.1.3. Dense Bands

Bands of electron-dense material, elongated along the longitudinal axis of the cell, encrust the cytoplasmic side of the uterine smooth muscle cell membrane (Fig. 7). These dense bands are 0.2 to 0.4 μm wide, up to 2 μm or more in length, and are interposed between longitudinal rows of caveolae.[17] It has been estimated that they occupy between 30 and 50% of the cell profile in the middle portions of the cell (at the level of the nucleus) and up to 100% of the cell perimeter toward the ends of the cell where the profile is smaller.[13] Although they appear as thickenings of the membrane, the presence of dense bands does not alter the usual trilaminar appearance of the cell membrane.

It appears that dense bands are involved in the transmission of force from the contractile apparatus to the cell membrane. The dense bands of smooth muscle contain the actin-binding protein, α-actinin,[35] which is also a major component of the Z bands of striated muscle,[36] thus suggesting a common functional role between the two. Dense bands have been shown to be associated with bundles of actin filaments that penetrate the bands at a very acute angle to the membrane.[13] Actin filaments entering the dense bands appear to be attached to the cell membrane by the intracellular protein vinculin[37]; however, it is not clear if this is the sole point of attachment. Because of the presence of α-actinin and the probable functional similarity, dense bands are considered by some to be membrane-associated dense bodies (see below). The two are considered separately here in order to stress the importance of dense bands in the transmission of force across the membrane.

3.1.4. Intermediate Junctions

Intermediate junctions are cell-to-cell junctions formed by the coupling of two dense bands in closely apposed cells as demonstrated in uterine muscle (Fig. 8). The junctions are symmetrical and extend the entire length of the two dense bands measuring up to 0.4 μm in width and 1 to 2 μm in length.[13] The intercellular gap

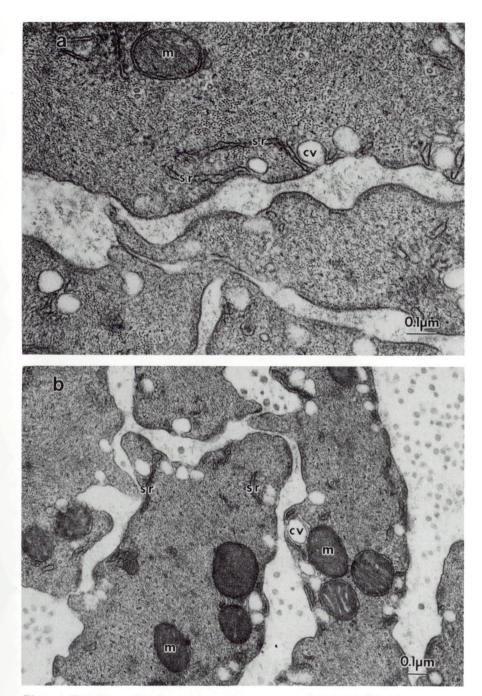

Figure 6. Transverse section of rat uterine smooth muscle. (a) The characteristic association between caveolae (cv) and sarcoplasmic reticulum (sr). (b) The close association often found between caveolae and mitochondria (m).

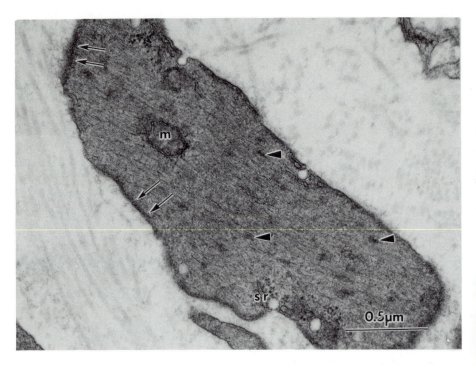

Figure 7. Longitudinal section of rat uterine smooth muscle demonstrating the dense bands (arrows) associated with the cell membrane and the dense bodies (arrowheads) found in the cytoplasm. sr, sarcoplasmic reticulum; m, mitochondria.

measures 40 to 60 nm, and Gabella[13] has described a single, ill-defined layer of electron-dense material, having a faint longitudinal periodicity and being continuous with the basal lamina of the two cells, running along the middle of it (Fig. 8). In all respects, the coupled dense bands appear identical to the uncoupled bands described above.[13]

The intermediate junctions provide mechanical coupling between adjacent cells by linking the contractile apparatus of the two through the apposing dense bands. The association of actin filaments with the intermediate junctions distinguishes them from other types of adherent junctions, i.e., desmosomes, which are most often associated with intermediate filaments. Moreover, desmosomes have a wider intercellular gap and have a multilayered appearance in which the membranes are occupied by characteristic intramembrane particles, probably representing "transmembrane linkers."[17,38] Transmembrane linkers have not yet been observed in smooth muscle. In addition, the association with actin filaments also serves to distinguish intermediate junctions from gap junctions. The cytoplasmic surface of gap junction membranes are relatively free of filaments (appear electron-opaque) while freeze-fracture replicas of myometrial smooth muscle have shown

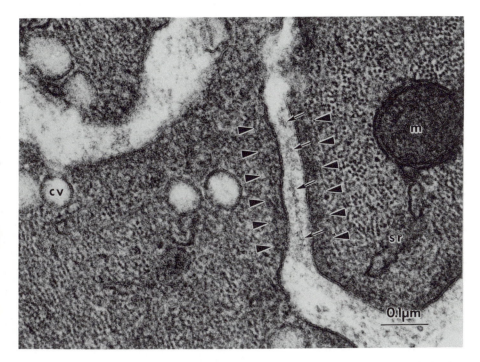

Figure 8. High-magnification electron micrograph of a transverse section of rat uterine muscle showing an intermediate junction (arrowheads) between two apposing smooth muscle cells. Note electron-dense material located in the gap between the cells. cv, caveolae; sr, sarcoplasmic reticulum; m, mitochondria.

gap junctions to be made up of aggregates of membrane particles packed in a hexagonal pattern on the P-face.[39]

3.2. Contractile Apparatus

A large percentage of the uterine smooth muscle cell volume is occupied by filamentous protein (Fig. 9) of which three types have been described: thick, thin, and intermediate. Because of the structural similarities with the well-characterized filaments of striated muscle, thick and thin filaments of smooth muscle are regarded as actin and myosin filaments.[17,40]

3.2.1. Thick Filaments

In situ smooth muscle myosin is found in the filamentous form regardless of whether the muscle is contracted or relaxed, or of the extent of myosin phosphorylation.[41] Myosin filaments are approximately 15 nm in diameter, as seen in uterine smooth muscle (Fig. 9), and are more heterogeneous in diameter and longer in

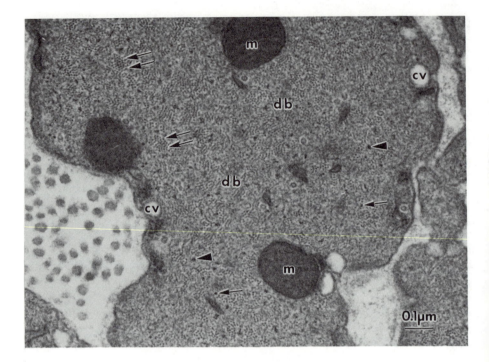

Figure 9. Transverse action of rat uterine smooth muscle demonstrating components of the contractile apparatus. Actin thin filament bundles (double arrow), intermediate filaments (arrow), myosin thick filaments (arrowhead), and dense bodies (db) are shown. m, mitochondria; cv, caveolae.

length (2.2 μm) than those of striated muscle.[42] At both ends the filaments taper over about 14 nm, but it is not clear whether the filaments are bipolar with a central bare zone as in striated muscle [in skeletal muscle myosin filaments, the cross-bridges have the same polarity on one-half of the length of the filament and the opposite polarity on the other half with a short bare area (no cross-bridges) occurring between the two portions]. X-ray diffraction studies[43,44] have shown that *in situ* smooth muscle myosin filaments have a periodicity (14.3 nm) that is suggestive of a cross-bridge repeat identical to that of striated muscle[45] and is also indicative of a longitudinal arrangement of the thick filaments.

Smooth muscle myosin (470 kDa) is immunologically different from myosin found in striated muscle[46]; however, at the ultrastructural level, the different types are indistinguishable. Myosin is composed of two globular head groups joined to a 150-nm-long tail. Associated with each globular head are two light chains: a 20-kDa chain and a 17-kDa chain. The 20-kDa light chain is involved in the calcium-dependent regulation of smooth muscle contraction.[47–50] Biochemically, a characteristic feature of smooth muscles is that myosin and other contractile proteins are highly soluble and can be extracted at low ionic strength.[51] In addition, the shape of isolated smooth muscle myosin molecules, unlike skeletal muscle myosin, is sensitive to the state of phosphorylation of the 20-kDa light chain as well as to the magnesium content of the buffer.[52–54] By altering these parameters, large-scale folding of the molecule can be induced thus producing two populations with respect to sedimentation velocity in the ultracentrifuge: an extended (6 S) form and a folded (10 S) form. Certainly, this extensive folding would not occur when the molecules are joined in the filamentous, *in situ,* state but it has been postulated[55] that small-scale, analogous, changes in the head group region could play a role in contractile regulation (see also Chapter 3).

3.2.2. Thin Filaments

Smooth muscle thin filaments are about 7 nm in diameter, are circular in transverse section, and form a uniform population as demonstrated in uterine smooth muscle (Fig. 9). Thin filaments in uterine muscle are generally organized as orderly packed bundles that have a lattice-like appearance (Fig. 9) commonly dubbed "action cables."[56] This packing of thin filaments is often hexagonal with an 11- to 12-nm interfilament spacing.[57] The length of the bundles has not been determined, but they are known to extend for several micrometers along the cell length, branching and dividing as they go.[17]

Actin is the major component of thin filaments in smooth as well as striated muscle. Monomers of actin (42.5 kDa) organize as a double-stranded helix to form filamentous actin. Smooth muscle F-actin can activate skeletal muscle ATPase and, except for a few amino acids at the N-terminus, has a similar sequence to other actins (see Chapter 4). In addition to actin, smooth muscle thin filaments, as do skeletal muscle, contain tropomyosin.[58] The weight ratio of actin to tropomyosin in uterine artery smooth muscle (3.6) is similar to that found in skeletal muscles (4.2)[59]; however, there are differences in the amino acid composition of the tropomyosin found in the two muscle types.[60] Interestingly, smooth muscle tropomyosin retains the binding site for the regulatory protein troponin, although troponin is absent from smooth muscle. Other proteins present include caldesmon and filamin (see Chapter 4).

3.2.3. Intermediate Filaments

Intermediate (10 nm) filaments are commonly found in uterine smooth muscle (Fig. 9) as well as a variety of other cells. They are not directly involved in the contractile process, but their association with dense bands and dense bodies[42,57] suggests that they might form a cytoskeleton that provides support for the myofilaments.[61] Morphologically, the intermediate filaments are very similar to neurofilaments, glial filaments, and epidermal keratin filaments and, in addition to smooth muscle, are found in cardiac muscle, endothelial cells, fibroblasts, and macrophages, as well as in developing skeletal muscle.[13] The intermediate filaments of visceral smooth muscle, such as uterine, are composed primarily of the protein desmin,[62] as opposed to those of vascular smooth muscle, which may either be composed of or additionally incorporate the protein vimentin.[63,64]

3.2.4. Dense Bodies

Dense bodies are electron dense structures scattered throughout the cytoplasm of uterine, as well as other, smooth muscle cells (Fig. 7, 9). As mentioned previously, they are closely related to the electron-dense bands of the inner aspect of the plasma membrane and are considered by some to be common structures. Dense bodies contain the actin-binding protein α-actinin: Actin filaments, as well as intermediate filaments, have been shown to insert into the dense bodies.[65] This would suggest that smooth muscle dense bodies are the structural and functional equivalent of the Z bands in striated muscle acting as the cytoplasmic anchorage for the contractile apparatus. The actin filaments associated with the dense bodies are polarized. When the actin filaments are labeled with the S-1 subfragment of myosin, the arrowheads formed by the label always point away from the dense body.[65] The filaments, therefore, have opposite polarity on either side of a given dense body which is also analogous to that observed at the Z lines of skeletal muscle.

3.2.5. Arrangement and Distribution of Filaments

The most striking structural feature of the contractile apparatus of smooth muscle is that there are no striking structural features. There is no apparent lateral register between the filaments in smooth muscle so that the transverse striation and regular repeat characteristic of skeletal and cardiac muscle does not exist. There is some apparent alignment of the myosin filaments into small groups of three to five and because of the association of these groups with actin filaments attached to dense bodies[37] and dense bands [13,61] it has been suggested that these constitute small contractile units.[65] However, it is correct to say that the pattern of myofilaments found in smooth muscle does not constitute well-defined myofibrils and sarcomeres. In uterine smooth muscle, myosin and actin filaments do appear to run parallel to one another as well as to the long axis of the cell (Fig. 10). Although not proven experimentally, it is generally accepted that the myofilaments slide past one another as the cell contracts. Still to be determined is whether the myofilaments remain parallel to the cell axis or change their orientation during isotonic contraction.[22,66,67]

Another striking feature of smooth muscle is the relative content of the contractile proteins. The numerical ratio of actin to myosin filaments is approximately 13 : 1, compared to 2 : 1 in vertebrate striated muscle.[42] The much larger number of thin than thick filaments, as compared to striated muscle, is mostly accounted for by the relatively low concentration of myosin in smooth muscle.[68] The concentration of actin can vary by a factor of two between smooth muscles of a given species and the concentration in vascular muscle is almost twice that of visceral muscle while the concentration of myosin is fairly uniform among smooth muscles. This produces a difference in the actin/myosin weight ratio between vascular and visceral smooth

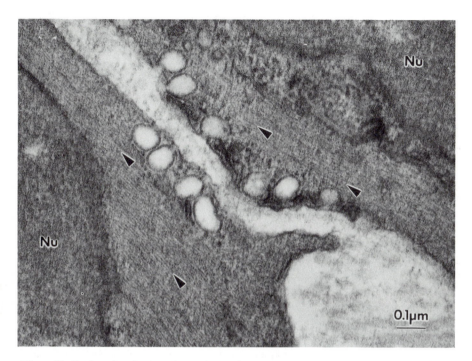

Figure 10. Section of uterine smooth muscle showing longitudinal arrangement of contractile filaments. Note filaments (arrowheads) running parallel to cell nuclei (Nu).

muscle (3.0 versus 1.8, respectively), both of which are much greater than that found in skeletal muscle (0.54).[59] Interestingly, smooth muscle, in spite of its much lower myosin content, can develop as much force per cell cross-section (2.3 × 10^5 N/m^2) as skeletal muscles (1.5–3 × 10^5 N/m^2).[69] In vascular smooth muscle, it has been reported that there are 72 thick filaments and 14.3 dense bodies and bands per square micrometer of sarcoplasm[42]; in selected regions of regular thick filament lattice, the packing density of thick filaments is 160/μm^2.[70]

3.2.6. Transmission of Force

Transmitting the force of contraction of individual smooth muscle cells throughout the myometrium and eventually resulting in the constriction of the uterine lumen must, of course, begin with force transmission from the contractile apparatus to the cell membrane in each cell. The force generated by cross-bridge cycling in the contractile apparatus translates into a shortening of the cell primarily due to the insertion of the actin thin filaments into the membrane-associated dense bands (see above). Since the dense bands are scattered throughout the entire length of the cell, the pull produced by the filaments is distributed over the lateral walls of the cell resulting in shortening in all directions.[17] In addition to the actin filaments, it appears that intermediate filaments anchored to the dense bands may also contribute to force transmission, although the exact contribution remains unclear.

Another important question that remains unresolved is how force is transmitted from the dense bands, which are associated with the inner aspect of the membrane, across the plasma membrane. The transmembrane "linkers" (structural protein linking the inner and outer aspects of the membrane) described in other cell types have not been identified in smooth muscle. In fact, freeze-fracture studies[13,20,21] have demonstrated that in the area of the membrane most often associated with dense bands, and thus the area where transmembrane linkers would most likely be found, there are actually fewer particles (proteins) observed than in other areas of the membrane. It may turn out that the proteins involved in the transmission of force across the membrane are not resolved by freeze-fracture or may not actually be intramembrane proteins.

Finally, the force originating with the shortening of the cell's contractile apparatus will be transmitted out of the cell to surrounding cells and connective tissue. It appears clear that the intermediate junctions provide the primary cell-to-cell adhesions in smooth muscle (see above) and it is very likely that they allow transmission of pull from one cell to another. As is the case with the dense bands that comprise them, intermediate junctions are not preferentially located at the terminal parts of the cells (as are the intercalated disks of cardiac cells) and thus as the cell shortens in all directions the pull is translated in all directions. The force generated at non-coupled dense bands (not forming intermediate junctions) will be transmitted across the membrane to the adjoining connective tissue. It has been argued that structural links exist between the cell membrane of visceral (and vascular) smooth muscle and collagen fibrils.[71,72] The adhesion of the cell membrane to the stroma is apparently mediated by the glycoprotein fibronectin. This protein has an affinity for collagen,[73] glycosaminoglycans,[74] and actin[13] while also showing an affinity for the surface of many cell types.[75,76] The linking of smooth muscle cells through intermediate junctions or adhesion to the stroma serves to make the myometrium a mechanical syncytium.

3.3. Intracellular Structures

3.3.1. Sarcoplasmic Reticulum

All smooth muscle cells have a well-developed endoplasmic reticulum (sarcoplasmic reticulum) that consists of a network of tubules and flattened cisternae within the cytoplasm of the cell[13,77,78] as demonstrated in uterine muscle (Fig. 11). The sarcoplasmic reticulum is a true intracellular system excluding extracellular markets such as ferritin, horseradish peroxidase, and colloidal lanthanum[33] and having an ionic composition, as demonstrated by electron probe microanalysis,[79] similar to that of the cytoplasm. As is the case in most cells, smooth muscle contains both agranular (smooth) and granular (rough; studded with ribosomes) reticulum that appear to be continuous. Cisternae (membrane-enclosed sacs) of rough sarcoplasmic reticulum are present near the nuclear poles, in the cone-shaped

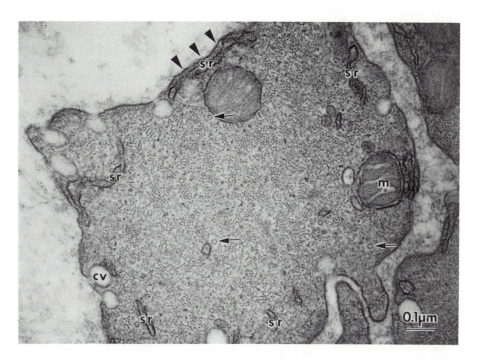

Figure 11. Transverse section of rat uterine smooth muscle showing well-developed sarcoplasmic reticulum (sr) within the cytoplasm. Note surface coupling (arrowheads) where sarcoplasmic reticulum is closely apposed to the surface membrane. Arrows identify microtubules. m, mitochondria; cv, caveolae.

region where the cells' organelles tend to be concentrated (Fig. 12), or scattered among the myofilaments. Close apposition to the cell surface (junctional sacroplasmic reticulum) is most frequently made by smooth sarcoplasmic reticulum (Fig. 11). The sarcoplasmic reticulum occupies from 1.5 to 7.5% of the smooth muscle cell volume (excluding mitochondria and nucleus) compared to 9 to 13% in frog skeletal muscle.[33] The most extensive smooth muscle sarcoplasmic reticulum is found in large elastic arteries[80] and in estrogen-treated or pregnant uteri.[19] The greater volume of sacroplasmic reticulum in these tissues is largely accounted for by an increased prominence of rough sarcoplasmic reticulum and probably reflects an accelerated state of protein synthesis.[81,82]

Smooth sarcoplasmic reticulum is the organelle primarily responsible for the physiological regulation of cytoplasmic calcium in smooth muscle (this is addressed in detail below). Regions where the tubules of sarcoplasmic reticulum approach the surface membrane are called *surface couplings*. These are important structural and functional specializations between the plasma membrane and the sacroplasmic reticulum where a 10- to 18-nm gap separates the two structures (Fig. 11). The region separating the cytoplasmic leaflet of the plasma membrane and the apposing sarcoplasmic reticulum is spanned by quasi-periodic bridging structures.[18] It has been suggested that these bridging structures may be analogous to the densely staining

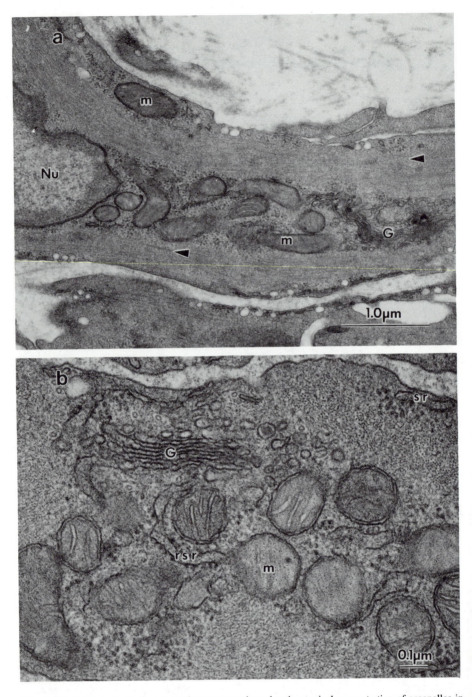

Figure 12. (a) Longitudinal section of rat myometrium showing typical concentration of organelles in the cone-shaped region at the nuclear pole. Nu, nucleus; m, mitochondria; G, Golgi apparatus; arrowheads mark longitudinal filaments. (b) High-magnification electron micrograph of organelles. rsr, rough sarcoplasmic reticulum.

structures in striated muscle dubbed "feet"[83] that connect the T tubules to the terminal cisternae. In smooth muscle, these are probably the sites where depolarization, action potentials, drugs, and transmitters release calcium stored in the junctional sacroplasmic reticulum.[84]

3.3.2. Mitochondria and Other Organelles

Mitochondria occupy approximately 3 to 9% of the smooth muscle cell volume.[13] In uterine, as in other smooth muscle, some mitochondria are clustered in the regions near the nuclear poles (Fig. 12), as are most organelles, while others are scattered throughout the cytoplasm with some preference for the region immediately beneath the cell membrane (Fig. 12). In longitudinal section (Fig. 12), mitochondria can be seen to be elongated and parallel to the myofilaments. They are closely associated with the cisternae and tubules of the sacroplasmic reticulum, and several investigators[85–87] have reported an association with caveolae such as that demonstrated here in uterine smooth muscle (Fig. 6b); however, the electron-dense bridging structures found between sarcoplasmic reticulum and plasma membrane have not been observed. As in other cells, the mitochondria are the site of oxidative metabolism in smooth muscle. Their limited role in the regulation of cytoplasmic calcium is discussed below.

An extensive Golgi apparatus is seen in the perinuclear region of uterine smooth muscle (Fig. 12). Although the exact role of these stacked lamellae in smooth muscle has not been elucidated, it is safe to assume that the Golgi is involved in the processing and packaging of secreted proteins as is the case in other cells[88]; however, additional investigation is needed. Lysosomes, residual bodies, and multivesicular bodies are occasionally found. Lysosomes usually appear as electron-dense bodies within the cytoplasm of uterine smooth muscle (Fig. 5), and their presence seems to be dependent upon estrogen. Lysosomes are particularly prominent in the postpartum uterus during involution of uterine muscle.[19] Their role in the degradative process has been well documented in other cells[89]; yet again, additional investigation is needed for smooth muscle. Microtubules are consistently present in uterine muscle (Fig. 11), but there is no suggestion as to a specific role in this tissue.

4. Cellular Calcium Stores

4.1. Sarcoplasmic Reticulum

4.1.1. Calcium Uptake

Ultrastructural methods have been used to demonstrate the ability of smooth muscle sarcoplasmic reticulum to sequester calcium. Since calcium and its analogue strontium were known to be accumulated by fragmented sarcoplasmic reticulum via the same transport mechanism, it was possible to electron-optically demonstrate the accumulation of the more electron-opaque strontium (higher atomic number) in the sarcoplasmic reticulum of intact smooth muscle cells. Somlyo and Somlyo[90] found deposits of strontium, indicative of uptake, in the sarcoplasmic reticulum of smooth muscles incubated in calcium-free, strontium-containing solutions that were strontium loaded either during spontaneous firing of action potentials or in nonspontaneously active smooth muscle during high-potassium-induced depolarization. Studies using the calcium-precipitating agents, pyroantimonate or oxalate, demonstrated precipitates, indicative of calcium accumulation, in the cisternae of sarcoplasmic reticulum in intact uterine[91] and vascular[92,93] muscle as well as saponin-permeabilized smooth muscle.[94] For alternative methods see Chapter 5.

The application of electron probe microanalysis to the area of smooth muscle, first made by Somlyo and Somlyo in the middle 1970s, not only confirmed the existence of localized regions of high calcium concentration in the cytoplasm of vascular smooth muscle cells[95] but subsequently led to the quantitation of the calcium content of sarcoplasmic reticulum. The calcium concentration in the junctional sacroplasmic reticulum of relaxed guinea pig portal vein smooth muscle is approximately 28 mmoles/kg dry wt[96] and in the central sarcoplasmic reticulum (nonjunctional) of rabbit main pulmonary artery it is approximately 18 mmoles/kg dry wt.[79] Because the diameter of the electron probe used in these studies was somewhat larger than the diameter of the sarcoplasmic reticulum lumen, these values underestimate the actual calcium content. Following correction for the overlap with adjacent cytoplasm, the calcium concentration in smooth muscle sarcoplasmic reticulum is approximately 30–50 mmoles/kg dry wt. Although this is significantly higher than the calcium concentration in other subcellular regions of smooth muscle, it is significantly less than the concentration of calcium in the terminal cisternae of skeletal muscle (approximately 120 mmoles/kg dry wt)[97] that contain the calcium-binding protein calsequestrin. However, when compared to the calcium content of the longitudinal reticulum of skeletal muscle (approximately 8 mmoles/kg dry wt), which is largely devoid of calcium-binding protein, smooth muscle sarcoplasmic reticulum calcium is very significantly higher, suggesting that a large fraction of the calcium, within the sarcoplasmic reticulum lumen of smooth muscle, is associated with calcium-binding proteins.[55]

The existence of a calcium-binding protein within the lumen of smooth muscle sarcoplasmic reticulum has been suggested on the basis of electron microscopy of saponin-permeabilized smooth muscles that showed protein content within the sarcoplasmic reticulum in both conventionally fixed and rotary-shadowed freeze-etched specimens.[18] In addition, a low-affinity calcium-binding protein, similar to the cardiac form of calsequestrin, was recently localized to the sarcoplasmic reticulum of smooth muscle.[98] With the availability of increasingly specific antibodies, it is probable that the proteins contained in smooth muscle sarcoplasmic reticulum will soon be identified.

4.1.2. Calcium Release and Recycling

From a morphological point of view it was perfectly rational to assume that the sarcoplasmic reticulum of smooth muscle was the intracellular source of activator calcium because of the striking similarities that exist with the sarcoplasmic reticulum of striated muscle, which was known to contain activatable calcium. In smooth muscle, the junctional sarcoplasmic reticulum lies directly beneath the plasma membrane, and the two are connected by periodic "feet" or bridging structures similar to those that connect the T tubules to the terminal cisternae in striated muscle.[99] In addition, the distance across the junctional gap between sarcoplasmic reticulum and plasma membrane in smooth muscle (12–20 nm) is very similar to the corresponding dimension at the surface couplings in cardiac[100] and at the triad in skeletal muscle,[83] thus suggesting a similar pathway for excitation–contraction coupling. Although the sarcoplasmic reticulum was known to store calcium in smooth muscle (see above), its sparsity in this tissue (1.5 to 7.5%) compared to striated muscle (9% in frog twitch muscle) raised questions about its relevance as a physiologically significant source of activator calcium.

Several lines of evidence strongly suggest that the calcium stored in the sarcoplasmic reticulum can be released upon stimulation and is sufficient in quantity to activate a maximal contraction of smooth muscle. First, it has been known for some time that smooth muscle contains an intracellular store of agonist-releasable calcium. This comes from studies that demonstrated agonist-induced contractions in smooth muscle following the removal of extracellular calcium.[33,85,101] When uterine smooth muscle is placed in a calcium-free buffer, there is a single transient phasic contraction to agonists such as acetylcholine, carbachol, angiotensin, or oxytocin which cannot be repeated until intracellular calcium stores are replenished.[102–106] Daniel's group recently demonstrated, in isolated smooth muscle cells from rat myometrium, that a maximal transient phasic contraction could be induced by carbachol after extracellular calcium had been removed with EGTA or following blockade of the plasmalemmal calcium channels with nitrendipine but that the inducibility of this contraction declined exponentially with time in the calcium-free buffer.[107] Although the amplitude of contraction following extracellular calcium removal did correlate with the volume of sarcoplasmic reticulum in different smooth muscles,[28] it is likely that, in addition to the volume of sarcoplasmic reticulum, other factors such as the ease of excitation–contraction coupling in calcium-free solutions and/or the rate of passive calcium leak from the sarcoplasmic reticulum also contribute to the variations in maintaining contractility in the absence of extracellular calcium.[108]

Thus, sarcoplasmic reticulum began to be implicated as the probable source of activator calcium. In smooth muscle it was demonstrated that caffeine, a known releaser of calcium from the sarcoplasmic reticulum of skeletal muscle,[109] induced a contraction of similar amplitude and time course as physiological agonists such as acetylcholine or norepinephrine.[110–113] These findings were supported by ^{45}Ca flux studies showing a transient increase in ^{45}Ca efflux from ^{45}Ca-loaded smooth muscle following agonist stimulation in both calcium-free and calcium-containing solutions.[110,101] The time course and magnitude of the agonist-induced increase in ^{45}Ca efflux were similar to those induced by caffeine (similar results have been obtained using uterine microsomal fractions; see Chapter 5). However, this still provided only indirect evidence of physiologically significant release of calcium from sarcoplasmic reticulum.

Direct evidence that excitatory stimuli release calcium from smooth muscle sarcoplasmic reticulum was established through electron probe microanalysis of cryosections from rapidly frozen vascular muscle. The first of such evidence demonstrated that calcium could be activated from its storage sites in junctional (membrane apposed) sarcoplasmic reticulum.[96] Paired strips of guinea pig portal vein were incubated in a calcium-free/lanthanum-containing buffer (lanthanum substitutes for calcium on all extracellular calcium binding sites and also inhibits calcium efflux but does not enter the cell)[114–116] and frozen either at the peak of a norepinephrine contraction or in the relaxed state. Significant differences in the number of high-calcium-containing regions of junctional sarcoplasmic reticulum were detected by electron probe microanalysis (from 4.7/cell cross section in the relaxed to 1.4/cell cross section in the contracted muscle). Furthermore, if the agonist was rapidly removed at the peak of contraction, relaxation of the strip could be immediately followed by another norepinephrine-induced contraction; in other words, calcium could be recycled within the cell between the cytoplasm and the sarcoplasmic reticulum.

Electron probe microanalysis of rabbit main pulmonary artery,[79] a tonic smooth muscle that contains a significant quantity of central (as opposed to junctional) sarcoplasmic reticulum, demonstrated that calcium can also be released from internal regions of the sarcoplasmic reticulum that are not connected to the surface membrane. In these experiments, strips of main pulmonary artery, incubated in a 1.2 mM (normal) calcium-containing buffer, were frozen either at the peak of a norepinephrine contraction or in the relaxed state. Due to the inability to resolve fine ultrastructure in frozen-dried cryosections, the central sarcoplasmic reticulum was identified through its high phosphorous content. The number and calcium content of the high-calcium/high-phosphorous regions of the cell interior (at least 200 nm from the plasma membrane) was significantly reduced in the contracted muscle from a range of 42 to 49 mmoles calcium/kg dry wt in the relaxed to a range of 19 to 32 mmoles calcium/kg dry wt following stimulation. These results confirm that the sarcoplasmic reticulum is the source of calcium release as demonstrated in several smooth muscles both in calcium-containing and in calcium-free solutions.[110,114,117] In addition, the fact that calcium is stored in and mobilized from nonjunctional sarcoplasmic reticulum in smooth muscle may explain the discrepancy between the volume of intracellular calcium stores in smooth and striated muscles. Although the total volume of sarcoplasmic reticulum in mammalian smooth muscle (1.5–7.5%) is significantly less than the total volume found in frog skeletal muscle (9%), calcium storage in the sarcoplasmic reticulum of the frog muscle is largely confined to the terminal cisternae[97] which constitute only about 3–4% of the cell volume.[118,119] It remains to be determined, however, whether all nonjunctional sarcoplasmic reticulum of smooth muscle can store or release calcium, and whether differences exist between smooth and rough sarcoplasmic reticulum in calcium storage.

4.2. Mitochondria

Smooth muscle mitochondria do not play a significant role in the physiologic regulation of cytoplasmic calcium. This has largely been established through electron probe microanalysis of rapidly frozen, intact cells, which has demonstrated that in normal smooth muscle, *in situ* mitochondrial calcium concentration ranges from 0 to 3 mmoles/kg dry wt.[95,120,121] Since mitochondria occupy only 5% of the smooth muscle cell volume, the total calcium stored in the mitochondria (0.05 × ~2 = 0.1 mmole/kg dry cell) is less than 10% of the amount by which cytoplasmic calcium is increased during a maximal sustained contraction.[79,96] In addition, mitochondrial calcium concentration remains unchanged both in intact tissue during a maximal, prolonged contraction[122] and in saponin-permeabilized cells exposed to 10^{-6} M calcium.[123] This is not surprising in light of the fact that mitochondria, isolated from vascular and uterine smooth muscle, have an apparent K_m for calcium of 10–17 μM.[124,125] This K_m is far above the submicromolar to low micromolar cytoplasmic free calcium concentration present during maximal contraction of smooth muscle.[126,127] None of this is compatible with significant rates of mitochondrial calcium transport at physiologic cytoplasmic calcium concentrations. However, this is not to insinuate that the mitochondrial calcium pool is at equilibrium; in fact, mitochondrial calcium is tightly regulated through the combined actions of the active calcium uptake system (calcium-uniporter) and both sodium-dependent (sodium–calcium exchange) and sodium-independent calcium export systems.[128] It now appears that the main function of these systems for transporting calcium into and out of the mitochondria may be the regulation of calcium-dependent enzymes in the mitochondrial matrix. The normal mitochondrial calcium concentration is in a range that is compatible with the idea that the free calcium concentration in the mitochondrial matrix regulates the activity of some of the important mitochondrial enzymes.[129] Furthermore, the calcium content of normal mitochondria can be mobilized *in situ:* mitochondrial calcium is significantly decreased (from 2.0 to 0.5 mmole/kg dry wt) in calcium-depleted smooth muscle.[127]

Despite a relatively low affinity for calcium, mitochondria have a very large capacity for accumulating divalent cations.[121] It has been known for some time that when faced with high external calcium, calcium uptake by the mitochondria (rate dependent on the calcium concentration in the cytosol) will exceed calcium release (rate varies very little), thus producing a net uptake. Longer-term accumulation of calcium is aided by the simultaneous accumulation of phosphate, which penetrates into the mitochondria on an independent carrier system to precipitate the excess calcium that has been taken up.[130] This serves the purpose of permitting the storage of large amounts of calcium in the matrix essentially without increasing its ionic concentration to levels that would prevent the correct regulation of the intra-mitochondrial enzymes. In fact, the ability of mitochondria to store large amounts of precipitated calcium phosphate salts, presumably in the form of amorphous hydroxyapatite, is a safety device of formidable importance to injured cells, which very frequently experience conspicuous increases of the level of cytosolic calcium. Excessively high cytoplasmic calcium levels can have a devastating effect on cellular integrity.[131,132]

In theory, calcium accumulated by mitochondria should be slowly released back into the cytosol when the cellular crises is resolved. This would be initiated when the cytoplasmic calcium had been reduced to levels insufficient to activate the uptake pathway, thus resulting in a net efflux of calcium from the mitochondria through the slow but steady efflux pathways. The rate of calcium release would be compatible with the capacity of the plasma membrane exporting systems thereby preventing a secondary cytosolic calcium overload. However, it must be understood that the ability of mitochondria to buffer calcium is finite and if the noxious condition is not resolved a point of no return will eventually be reached at which the amount of calcium in the mitochondria will interfere with their proper functioning.[129] The effects of mitochondrial calcium loading on smooth muscle cell viability has been difficult to assess due to the inability to calcium-load *in situ* mitochondria in intact, undamaged cells.

Recently, massive uptake of calcium and phosphorus was induced by the rapid washout of high intracellular sodium, from sodium-loaded vascular smooth muscle.[121] The average mitochondrial calcium concentration, measured by electron probe microanalysis, increased over 100-fold from control values to 329 mmoles/kg dry wt at peak loading. Since mitochondria constitute 5% of the smooth muscle cell volume, this would mean that the total cellular calcium at this time was over 20 mmoles/kg dry wt, which is vastly in excess of the normal (approximately 3 mmoles/kg dry wt) value.[96] The sequestration of calcium by the mitochondria was dependent on an increase in cytoplasmic calcium that accompanied the washout of sodium (probably due to reverse mode operation of the plasmalemmal sodium–calcium exchanger) but interestingly, the increase in cytoplasmic calcium was large enough to produce only an 80% maximal contraction of the tissue. This would indicate that the mitochondrial calcium accumulation was occurring at cytoplasmic free calcium concentrations far below the K_m (10–17 μM) of the calcium uniporter and this is the first demonstration of significant calcium uptake by *in situ* mitochondria in the absence of markedly elevated cytosolic free calcium. It may be that the massive calcium uptake was triggered by a sodium-efflux-induced reversal of mitochondrial sodium–calcium exchange (normally exports calcium). Most significantly perhaps is the fact that the mitochondrial calcium loading was completely reversible with time and did not appear to be injurious to the cell. The complete removal of over 300 mmoles calcium/kg dry wt in less than 2 hr (Fig. 13) was accompanied by no evidence of permanent cell damage, either by the appearance of electron-lucent, hyperpermeable cells or by any reduction of the maximal contractile response of the tissue. These results would imply that even severe mitochondrial calcium overloading may not cause irreversible mitochondrial uncoupling and cell death in smooth muscle. This perhaps can be attributed to the ability of smooth muscle to function glycolytically.[133]

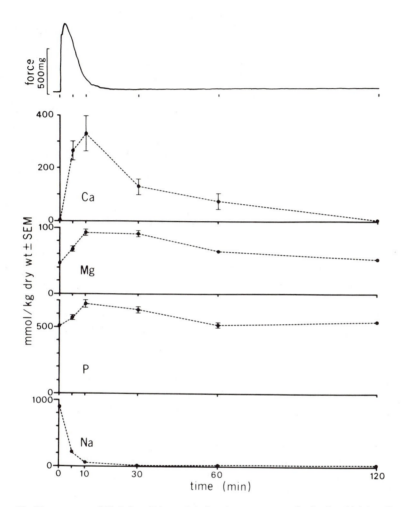

Figure 13. Time course of Na^+-free K^+ wash-induced contracture and mitochondrial ion fluxes in rabbit portal vein. Tissue was Na^+ loaded for 3 hr at 37°C (1 mM ouabain) and washed, beginning at time 0, with a Na^+-free K^+–Li^+ solution at 37°C (1 mM ouabain). Tension tracing represents a typical isometric contraction ($t_{1,2}$ relaxation = 6.5 ± 1.5 min, maximum force developed by 2.8 ± 0.5 min, n = 5) induced by the Na^+-free wash of Na^+-loaded portal vein. Mitochondrial ionic concentrations are expressed as mean ± S.E.M.; each point represents electron probe microanalysis of 20–80 mitochondria from at least two animals. (From Broderick, R., and Somlyo, A. P., 1987, Calcium and magnesium transport by in situ mitochondria: Electron probe analysis of vascular smooth muscle, *Circ. Res.* **61**:523–530, by permission of the American Heart Association, Inc.)

4.3. Sarcolemma

Perhaps in the strictest sense, the sarcolemma would not be classified as an intracellular calcium storage site. However, both the bidirectional transport of calcium across the plasma membrane and membrane-bound calcium have been purported to play a role in cellular calcium homeostasis.

Several reports have indicated that excitatory agents such as high potassium or norepinephrine can increase calcium influx across the surface membrane of smooth muscle, which normally has a low permeability to calcium and other divalent cations.[134] Although it appears evident that the sarcoplasmic reticulum is the primary source and sink of agonist-releasable calcium in smooth muscle (see above), calcium influx across the plasma membrane may play some role in tension maintenance. This is illustrated by the fact that tension development in calcium-free solutions is transient while in the presence of extracellular calcium there is a tonic (maintained) component of tension that is accompanied by a sustained elevation of cytoplasmic free calcium.[135,136] The tonic component of uterine muscle contraction is also eliminated when extracellular calcium is removed or its entry blocked[107] and a single class of high-affinity plasmalemmal calcium-channel antagonist (dihydropyridine) binding sites have been identified in myometrial smooth muscle.[137] The exact role of extracellular calcium influx in the tonic component of smooth muscle contraction, however, remains to be resolved.

The removal of calcium from smooth muscle is mediated primarily through the membrane-bound calmodulin-stimulated calcium ATPase,[138-141] which serves to maintain normal steady-state cellular calcium. In addition, in some smooth muscles, i.e., uterine,[142] a plasma membrane sodium–calcium exchange system may play a role in the maintenance of cellular calcium levels[143] but this remains a controversial issue[84] as does the importance of these exporting systems in the cycling of activatable calcium. The fact that the relaxation of smooth muscle is not inhibited when calcium efflux is blocked by lanthanum[96,117] suggests that calcium accumulation by the sarcoplasmic reticulum plays the dominant role in relaxation. It is possible that the major role of the membrane calcium-exporting systems is the removal of small amounts of calcium that enter the cell during action potentials or graded depolarizations that, while insufficient to activate contraction, could eventually overload the sarcoplasmic reticulum.[84]

The importance of plasma membrane-bound calcium in smooth muscle calcium homeostasis remains to be determined. Although high- and low-affinity calcium-binding sites have been described in the plasma membrane of uterine and other smooth muscle,[139,140] electron probe microanalysis studies have indicated that not more than 1 mmole calcium/kg dry wt is bound to the surface membrane in guinea pig portal vein.[84] Whether any of this calcium can be mobilized has not been determined. However, considering the small amount of calcium bound to the membrane, it is most likely that the physiologically significant role of this calcium is in the regulation of cell membrane permeability[144] and/or in the regulation of membrane ion channels.[145]

5. Conclusions

Much of what we know about the ultrastructure of myometrial smooth muscle has largely been derived by inference from work done on other smooth muscles. Although this chapter is a generalized summary of data from both visceral and vascular muscle, it does represent a step forward for uterine research, since all of the most salient ultrastructural features, previously described in other smooth muscle, have now been demonstrated in the myometrium. However, much work is still needed before we achieve the detailed level of understanding that we now have of other visceral muscles and vascular smooth muscle.

Although there have been many recent advances in the elucidation of the pathways and kinetics of calcium metabolism/homeostasis in uterine muscle, definitive, *in situ,* demonstration of the intracellular sources of activatable calcium has not been made. This question awaits the application of electron probe microanalysis, which awaits available funding.

ACKNOWLEDGMENTS. This work was supported by a grant from the Frank C. Bressler Research Fund. The authors wish to thank Mr. Rick Wierwille for his expert technical assistance.

References

1. Osa, T., and Katase, T., 1975, Physiological comparison of the longitudinal and circular muscles of the pregnant rat uterus, *Jpn. J. Physiol.* **25**:153–164.
2. Massman, H. W., 1980, Comparative morphology of the endometrium, in: *The Endometrium* (F. A. Kimball, ed.), Spectrum Publications, New York, pp. 3–23.
3. Ross, R., and Klebanoff, S. J., 1971, The smooth muscle cell. I. In vivo synthesis of connective tissue proteins, *J. Cell Biol.* **50**:159–171.
4. Ross, R., 1971, The smooth muscle cell. II. Growth of smooth muscle in culture and formation of elastic fibers, *J. Cell Biol.* **50**:172–186.
5. Wight, T. N., and Ross, R., 1975, Glycosaminoglycans in primate arteries II. Synthesis and secretion by arterial smooth muscle cells in culture, *J. Cell Biol.* **67**:675–686.
6. Law, R. O., 1982, Techniques and application of extracellular space determination in mammalian tissues, *Experientia* **38**:411–420.
7. Casteels, R. J., and Kuriyama, H., 1965, Membrane potential and ionic content in pregnant and nonpregnant rat myometrium, *J. Physiol. (London)* **177**:263–287.
8. Jones, A. W., 1968, Influence of oestrogen and progesterone on electrolyte accumulation in the rabbit myometrium, *J. Physiol. (London)* **197**:19–20.
9. Csapo, A. I., 1962, Smooth muscle as a contractile unit, *Physiol. Rev.* **42**(Suppl. 5):7–33.
10. Finn, C. A., and Porter, D. G., 1975, *The Uterus,* Elek Science, London.
11. Cole, W. C., Garfield, R. E., and Kirkaldy, S. J., 1985, Gap junctions and direct intracellular communication between rat uterine smooth muscle cells, *Am. J. Physiol.* **249**:C20–C31.
12. Kyozuka, M., Crankshaw, J., Berezin, I., Collins, S. M., and Daniels, E. E., 1987, Calcium and contractions of isolated smooth muscle cells from rat myometrium, *Can. J. Physiol. Pharmacol.* **65**:1966–1975.
13. Gabella, G., 1981, Structure of smooth muscles, in: *Smooth Muscle: An Assessment of Current*

Knowledge (E. Bulbring, A. F., Brading, A. W. Jones, and T. Tomita, eds.), Arnold, London, pp. 1–46.

14. Singer, S. J., and Nicolson, G. L., 1972, The fluid mosaic model of the structure of cell membranes, *Science* **175:**720.

15. Luft, J. H., 1976, The structure and properties of the cell surface coat, *Int. Rev. Cytol.* **45:**291–382.

16. Kefalides, N. A., Alper, R., and Clark, C. C., 1979, Biochemistry and metabolism of basement membranes, *Int. Rev. Cytol.* **61:**167–228.

17. Gabella, G., 1984, Structural apparatus for force transmission in smooth muscle, *Physiol. Rev.* **64:** 455–477.

18. Somlyo, A. V., and Franzini-Armstrong, C., 1985, New views of smooth muscle structure using freezing, deep-etching and rotary shadowing, *Experientia* **41:**841–856.

19. Garfield, R. E., and Somlyo, A. P., 1985, Structure of smooth muscle, in: *Calcium and Contractility* (A. K. Grover and E. E. Daniel, eds.), Humana Press, Clifton, N.J., pp. 1–36.

20. Devine, C. E., and Rayns, D. G., 1975, Freeze-fracture studies of membrane systems in vertebrate muscle. II. Smooth muscle, *J. Ultrastruct. Res.* **51:**293–306.

21. Wells, G. S., and Wolowyk, M. W., 1971, Freeze-etch observations on membrane structure in the smooth muscle of the guinea-pig taenia coli, *J. Physiol. (London)* **218:**11P–13P.

22. Gabella, G., 1976, Quantitative morphological study of smooth muscle cells of the guinea-pig taenia coli. Structural changes in smooth muscle cells during isotonic contraction, *Cell Tissue Res.* **170:**161–201.

23. Gabella, G., and Blundell, D., 1978, Effect of stretch and contraction on caveolae of smooth muscle cells, *Cell Tissue Res.* **190:**255–271.

24. Goodford, P. J., 1966, The extracellular space of the smooth muscle of the guinea-pig taenia coli, *J. Physiol. (London)* **186:**1–10.

25. Verity, M. A., and Bevan, J. A., 1966, Plurivesicular nerve endings in the pulmonary artery, *Nature* **211:**537–538.

26. Prosser, C. L., Burnstock, G., and Kahn, J., 1960, Conduction in smooth muscle: Comparative structural properties, *Am. J. Physiol.* **199:**545–552.

27. Devine, C. E., Simpson, F. O., and Bertaud, W. S., 1971, Surface features of smooth muscle cells from the mesenteric artery and vas deferens, *J. Cell Sci.* **8:**427–443.

28. Muggli, R., and Baumgartner, H. R., 1972, Pattern of membrane invaginations at the surface of smooth muscle cells of rabbit arteries, *Experientia* **28:**1212–1214.

29. Orci, L., and Perrelet, A., 1973, Membrane-associated particles: Increases at site of pinocytosis demonstrated by freeze-etching, *Science* **181:**868–869.

30. Garfield, R. E., and Daniel, E. E., 1977, Relation of membrane vesicles to volume control and sodium transport in smooth muscle: Studies on sodium rich tissues, *J. Mechanochem. Cell Motil.* **4:**157–176.

31. Garfield, R. E., and Daniels, E. E., 1977, Relation of membrane vesicles to volume control and sodium transport in smooth muscle: Effect of metabolic and transport inhibition on fresh tissue, *J. Mechanochem. Cell Motil.* **4:**115–157.

32. Goodford, P. J., and Wolowyk, M. W., 1972, Localization of cation interactions in the smooth muscle of the guinea-pig taenia coli, *J. Physiol. (London)* **224:**521–535.

33. Devine, C. E., Somlyo, A. V., and Somlyo, A. P., 1972, Sarcoplasmic reticulum and excitation–contraction coupling in mammalian smooth muscle, *J. Cell Biol.* **52:**690–718.

34. Brown, M. S., and Goldstein, J. L., 1976, Receptor-mediated control of cholesterol metabolism, *Science* **191:**150–154.

35. Schollmeyer, J. E., Furcht, L. T., Goll, D. E., Robson, R. M., and Stromer, M. E., 1976, Localization of contactile proteins in smooth muscle cells and in normal and transformed fibroblasts, in: *Cell Mobility,* Volume A (R. Goldman, T. Pollard, and J. Rosenbaum, eds.), Cold Spring Harbor Laboratory, Cold Spring Harbor, N.Y., pp. 361–388.

36. Schollmeyer, J. E., Goll, D. E., Robson, R. M., and Stromer, M. H., 1973, Localization of α-actinin and tropomyosin in different muscles, *J. Muscle Res. Cell Motil.* **59:**306a.

37. Geiger, B., 1979, A 130K protein from chicken gizzard: Its localization at termini of microfilament bundles in cultured chicken cells, *Cell* **18:**193–205.
38. Staehelin, L. A., and Hull, B. E., 1978, Junctions between living cells, *Sci. Am.* **238:**140–152.
39. Ikeda, M., Shibata, Y., and Yamamoto, T., 1987, Rapid formation of myometrial gap junctions during parturition in the unilaterally implanted rat uterus, *Cell Tissue Res.* **248:**297–303.
40. Shoenberg, C. F., and Stewart, M. J., 1980, Filament formation in muscle homogenates, *J. Muscle Res. Cell Motil.* **1:** 117–126.
41. Somlyo, A. V., Butler, T. M., Bond, M., and Somlyo, A. P., 1981, Myosin filaments have phosphorylated light chains in relaxed smooth muscle, *Nature* **294:**567–570.
42. Ashton, F. T.,Somlyo, A. V., and Somlyo, A. P., 1975, The contractile apparatus of vascular smooth muscle: Intermediate high voltage stereo electron microscopy, *J. Mol. Biol.* **98:**17–29.
43. Lowy, J., Poulsen, F. R., and Vibert, P. J., 1970, Myosin filaments in vertebrate smooth muscle, *Nature* **225:**1053–1054.
44. Shoenberg, C. F., and Haselgrove, J. C., 1974, Filaments and ribbons in vertebrate smooth muscle, *Nature* **249:**152–154.
45. Spudich, J. A., Huxley, H. E., and Finch, J. T., 1972, Regulation of skeletal muscle contraction. II. Structural studies of the interaction of the tropomyosin–troponin complex with actin, *J. Mol. Biol.* **72:**619–632.
46. Groschel-Stewart, U., Schreiber, J., Mahl-Meister, C., and Weber, K., 1976, Production of specific antibodies to contractile proteins, and their use in immunofluorescence microscopy. I. Antibodies to smooth and striated chicken myosin, *Histochemie* **46:**229–236.
47. Adelstein, R. S., and Eisenberg, E., 1980, Regulation and kinetics of the actin–myosin–ATP interaction, *Annu. Rev. Biochem.* **49:**921–956.
48. Kamm, K. E., and Stull, J. T., 1985, The function of myosin and myosin light chain kinase phosphorylation in smooth muscle, *Annu. Rev. Pharmacol. Toxicol.* **25:**593–620.
49. Hartshorne, D. J., 1987, Biochemistry of the contractile process in smooth muscle, in: *Physiology of the Gastrointestinal Tract,* (L. R. Johnson, ed.), Raven Press, New York, pp. 423–482.
50. Sommerville, L. E., and Hartshorne, D. J., 1987, Intracellular calcium and smooth muscle contraction, *Cell Calcium* **7**(5–6):353–364.
51. Murphy, R. A., Driska, S. P., and Cohen, D. M., 1977, Variations in actin to myosin ratios and cellular force generation in vertebrate smooth muscle, in: *Excitation–Contraction Coupling in Smooth Muscle* (R. Casteels, T. Godfraind, and J. C. Ruegg, eds.), Elsevier/North-Holland, Amsterdam, pp. 417–424.
52. Smith, R. C., Cande, W. Z., Craig, R., Tooth, P. J., Scholey, J. M., and Kendrik-Jones, J., 1983, Regulation of myosin filament assembly by light-chain phosphorylation, *Philos. Trans. R. Soc. London Ser. B* **302:**73–82.
53. Suzuki, H., Kamata, T., Onishi, H., and Watanabe, S., 1982, ATP induced reversible change in the conformation of chicken gizzard myosin and HMM, *J. Biochem.* **91:**1699–1705.
54. Trybus, K. M., Huiatt, T. W., and Lowey, S., 1982, A bent monomeric conformation of myosin from smooth muscle, *Proc. Natl. Acad. Sci. USA* **79:**6151–6155.
55. Somlyo, A. P., and Somlyo, A. V., 1986, Smooth muscle structure and function, in: *The Heart and the Cardiovascular System* (H. A. Fozzard, ed.), Raven Press, New York, pp. 845–864.
56. Rice, R. V., Moses, J. A., McManus, G. M., Brady, A. C., and Blasik, L. M., 1970, The organization of contractile filaments in a mammalian smooth muscle, *J. Cell Biol.* **47:**183–196.
57. Somlyo, A. P., Devine, C. E., Somlyo, A. V., and Rice, R. V., 1973, Filament organization in vertebrate smooth muscle, *Philos. Trans. R. Soc. London Ser. B* **265:**223–229.
58. Ebashi, S., and Kodama, A., 1966, Interaction of troponin with F-actin in the presence of tropomyosin, *J. Biochem.* **59:**425–426.
59. Cohen, D. M., and Murphy, R. A., 1979, Cellular thin filament protein contents and force generation in porcine arteries and veins, *Circ. Res.* **45:**661–665.
60. Carsten, M. E., 1968, Tropomyosin from smooth muscle of the uterus, *Biochemistry* **7:**960–967.
61. Cooke, P., 1976, A filamentous cytoskeleton in vertebrate smooth muscle fibers, *J. Cell Biol.* **48:**539–556.

62. Bennett, G. S., Fellini, S. A., Croop, J. M., Otto, J. J., Bryans, J., and Holtzer, H., 1976, Differences among 100 A-filament subunits from different cell types, *Proc. Natl. Acad. Sci. USA* **73:**4364–4368.

63. Frank, E. D., and Warren, L., 1981, Aortic smooth muscle cells contain vimentin instead of desmin, *Proc. Natl. Acad. Sci. USA* **78:**3020–3024.

64. Gabbiani, G., Schmid, E., Winter, S., Chaponnier, C., Dechastonay, C., Vandekerckhove, J., Weber, K., and Franke, W. W., 1981, Vascular smooth muscle cells differ from other smooth muscle cells: Predominance of vimentin filaments and a specific α-type actin, *Proc. Natl. Acad. Sci. USA* **78:**298–302.

65. Bond, M., and Somlyo, A. V., 1982, Dense bodies and actin polarity in vertebrate smooth muscle, *J. Cell Biol.* **95:**403–413.

66. Fay, F. S., and Delise, C. M., 1973, Contraction of isolated smooth muscle cells. Structural changes, *Proc. Natl. Acad. Sci. USA* **70:**641–645.

67. Fisher, B. A., and Bagby, R. M., 1977, Reorientation of myofilaments during contraction of a vertebrate smooth muscle, *Am. J. Physiol.* **232:**C5–C14.

68. Murphy, R. A., Herlihy, J. T., and Megerman, J., 1974, Force-generating capacity and contractile protein content of arterial smooth muscle, *J. Gen. Physiol.* **64:**691–705.

69. Driska, S. P., and Murphy, R. A., 1978, Estimation of cellular force generation in an arterial smooth muscle with a high actin:myosin ratio, *Blood Vessels* **15:**26–32.

70. Devine, C. E., and Somlyo, A. P., 1971, Thick filaments in vascular smooth muscle, *J. Cell Biol.* **49:**636–649.

71. Gabella, G., 1979, Smooth muscle cell junctions and structural aspects of contraction, *Br. Med. Bull.* **35:**213–218.

72. Dingemans, K. P., Janses, N., and Becker, A. E., 1981, Ultrastructure of the normal aortic media, *Virchows Arch. A.* **392:**199–216.

73. Engvall, E. Ruoslahti, E., and Miller, E. J., 1978, Affinity of fibronectin to collagen of different genetic types and to fibrinogen, *J. Exp. Med.* **147:**1584–1595.

74. Perkins, M. E., Ji, T. H., and Hynes, R. O., 1979, Cross-linking of fibronectin to sulfated proteoglycans at the cell surface, *Cell* **16:**941–952.

75. Schlessinger, J., Barak, L. S., Hammes, G. G., Yamada, K. M., Pastan, I., Webb, W. W., and Elson, E. L., 1977, Mobility and distribution of a cell surface glycoprotein and its interaction with other membrane components, *Proc. Natl. Acad. Sci. USA* **74:**2909–2913.

76. Gold, L. I., and Pearlstein, E., 1980, Fibronectin–collagen binding and requirement during cellular adhesion, *Biochem. J.* **186:**551–559.

77. Somlyo, A. V., 1980, Ultrastructure of vascular smooth muscle, in: *Handbook of Physiology: The Cardiovascular System: Vascular Muscle* (D. F. Bohr, A. P. Somlyo, and H. V. Sparks, eds.), American Physiological Society, Bethesda, pp. 33–67.

78. Forbes, M. S., 1982, Ultrastructure of vascular smooth muscle cells in mammalian heart, in: *The Coronary Artery* (S. Kalsner, ed.), Oxford University Press, London, pp. 3–58.

79. Kowarski, D., Shuman, H., Somlyo, A. P., and Somlyo, A. V., 1985, Calcium release by norepinephrine from central sarcoplasmic reticulum in rabbit main pulmonary artery smooth muscle, *J. Physiol. (London)* **366:**153–175.

80. Somlyo, A. P., and Somlyo, A. V., 1975, Ultrastructure of smooth muscle, in: *Methods in Pharmacology,* Volume 3 (E. E. Daniels and D. M. Paton, eds.), Plenum Press, New York, pp. 3–45.

81. Bergman, R. A., 1968, Uterine smooth muscle fibers in castrate and oestrogen treated rats, *J. Cell Biol.* **36:**636–648.

82. Ross, R., and Klebanoff, S. J., 1971, The smooth muscle cell. I. *In vivo* synthesis of connective tissue proteins, *J. Cell Biol.* **50:**159–171.

83. Franzini-Armstrong, C., 1970, Studies of the triad. I. Structure of the junction in frog twitch fibers, *J. Cell Biol.* **47:**488–499.

84. Somlyo, A. P., 1985, Excitation–contraction coupling and the ultrastructure of smooth muscle, *Circ. Res.* **57:**497–507.

85. Somlyo, A. P., Devine, C. E., Somlyo, A. V., and North, S. R., 1971, Sarcoplasmic reticulum and the temperature-dependent contraction of smooth muscle in calcium-free solutions, *J. Cell Biol.* **51**:722–741.

86. Somlyo, A. P., Somlyo, A. V., Devine, C. E., Peters, P. D., and Hall, T. A., 1974, Electron microscopy and electron probe analysis of mitochondrial cation accumulation in smooth muscle, *J. Cell Biol.* **61**:723–742.

87. Makita, T., and Kiwaki, S., 1978, Connection of microtubules, caveolae, mitochondria, and sarcoplasmic reticulum in the taenia coli of guinea pigs, *Arch. Histol. Jpn.* **41**:167–171.

88. Beams, H. W., and Kessel, R. G., 1968, The Golgi apparatus: Structure and function, *Int. Rev. Cytol.* **23**:209–276.

89. Dingle, J., Dean, R., and Sly, W. (eds.), 1984, *Lysosomes in Biology and Pathology*, Volume 7, Elsevier, Amsterdam.

90. Somlyo, A. V., and Somlyo, A. P., 1971, Strontium accumulation by sarcoplasmic reticulum and mitochondria in vascular smooth muscle, *Science* **174**:955–958.

91. Rubanyi, G., Balogh, I., Kovach, A. G., Somogyi, E., and Sotonyi, P., 1980, Ultrastructure and localization of calcium in uterine smooth muscle, *Acta Morphol. Acad. Sci. Hung.* **28**:269–278.

92. Popescu, L. M., and Diculescu, I., 1975, Calcium in smooth muscle sarcoplasmic reticulum *in situ*: Conventional and X-ray analytical electron microscopy, *J. Cell Biol.* **67**:911–918.

93. Heumann, H.-G., 1976, The subcellular localization of calcium in vertebrate smooth muscle: Calcium containing and calcium accumulating structures in muscle cells of mouse intestine, *Cell Tissue Res.* **169**:221–231.

94. Somlyo, A. P., Somlyo, A. V., Shuman, H., and Endo, M., 1982, Calcium and monovalent ions in smooth muscle, *Fed. Proc.* **41**:2883–2890.

95. Somlyo, A. P., Somlyo, A. V., and Shuman, H., 1979, Electron probe analysis of vascular smooth muscle: Composition of mitochondria, nuclei and cytoplasm, *J. Cell Biol.* **81**:316–335.

96. Bond, M., Kitazawz, T., Somlyo, A. P., and Somlyo, A. V., 1984, Release and recycling of calcium by the sarcoplasmic reticulum in guinea pig portal vein smooth muscle, *J. Physiol. (London)* **355**:677–695.

97. Somlyo, A. V., Gonzalez-Serratos, H., Shuman, H., McClellan, G., and Somlyo, A. P., 1981, Calcium release and ionic changes in the sarcoplasmic reticulum of tetanized muscle: An electron probe study, *J. Cell Biol.* **90**:577–594.

98. Wuytack, F., Raemaekers, L., Verbist, J., Jones, L. R., and Casteels, R., 1987, Smooth-muscle endoplasmic reticulum contains a cardiac-like form of calsequestrin, *Biochim. Biophys. Acta* **899**:151–158.

99. Somlyo, A. V., 1979, Bridging structures spanning the junctional gap at the triad of skeletal muscle, *J. Cell Biol.* **80**:743–750.

100. Sommer, J. R., and Johnson, E. A., 1979, Ultrastructure of cardiac muscle, in: *Handbook of Physiology: The Cardiovascular System*, Volume I (R. M. Berne, ed.), American Physiological Society, Bethesda, pp. 113–186.

101. Droogmans, G., Raemaekers, L., and Casteels, R., 1977, Electro- and pharmacomechanical coupling in the smooth muscle cell of the rabbit ear artery, *J. Gen. Physiol.* **70**:129–148.

102. Saki, K., Yamaguchi, T., and Uchita, M., 1981, Oxytocin-induced calcium-free contraction of rat uterine smooth muscle: Effects of divalent cations and drugs, *Arch. Int. Pharmacodyn.* **250**:40–50.

103. Kyozuka, M., 1983, Contraction of rat uterine smooth muscle in Ca-free high K solution, *Biomed. Res.* **4**:523–532.

104. Lalanne, C., Mironneau, C., Mironneau, J., and Savineau, J. P., 1984, Contraction of rat uterine smooth muscle induced by acetylcholine and angiotensin II in Ca-free medium, *Br. J. Pharmacol.* **81**:317–326.

105. Ashoori, F., Taki, A., and Tomita, T., 1985, The response of non-pregnant rat myometrium to oxytocin in Ca-free solution, *Br. J. Pharmacol.* **84**:175–183.

106. Mironneau, J., Mironneau, C., and Savineau, J. P., 1984, Maintained contraction of rat uterine smooth muscle incubated in Ca-free solution, *Br. J. Pharmacol.* **82**:735–743.

107. Kyozuka, M., Crankshaw, J., Berezin, I., Collins, S. M., and Daniel, E. E., 1987, Calcium and

contraction of isolated smooth muscle cells from rat myometrium, *Can. J. Physiol. Pharmacol.* **65:**1966–1975.

108. Johansson, B., and Somlyo, A. P., 1980, Electrophysiology and excitation–contraction coupling, in: *Handbook of Physiology: Vascular Smooth Muscle,* Volume II (D. F. Bohr, A. P. Somlyo, and H. V. Sparks, eds.), American Physiological Society, Bethesda, pp. 301–324.

109. Weber, A., and Herz, R., 1968, The relationship between caffeine contracture of intact muscle and the effect of caffeine on reticulum, *J. Physiol. (London)* **52:**750–759.

110. Deth, R., and Casteels, R., 1977, A study of releasable Ca fractions in smooth muscle cells of rabbit aorta, *J. Gen. Physiol.* **69:**401–416.

111. Deth, R. C., and Lynch, C. J., 1981, Mobilization of a common source of smooth muscle calcium in norepinephrine and methylxanthines, *Am. J. Physiol.* **240:**C239–C247.

112. Itoh, T., Kuriyama, H., and Suzuki, H., 1981, Excitation–contraction coupling in smooth muscle cells of the guinea pig mesenteric artery, *J. Physiol. (London)* **321:**513–535.

113. Leijten, P. A., and van Breemen, C., 1984, The effects of caffeine on the noradrenaline-sensitive calcium store in rabbit aorta, *J. Physiol. (London)* **357:**327–339.

114. Deth, R., and van Breemen, C., 1977, Agonist induced release of intracellular Ca in the rabbit aorta, *J. Membr. Biol.* **30:**363–380.

115. Freeman, D. J., and Daniel, E. E., 1973, Calcium movement in vascular smooth muscle and its detection using lanthanum as a tool, *Can. J. Physiol. Pharmac.* **51:**900–913.

116. Weiss, G. B., 1977, Approaches to delineation of differing calcium binding sites in smooth muscle, in: *Excitation–Contraction Coupling in Smooth Muscle* (R. Casteels, ed.), Elsevier/North-Holland, Amsterdam, pp. 32–48.

117. Deth, R., and van Breemen, C., 1974, Relative contributions of calcium influx and cellular calcium release during drug induced activation of the rabbit aorta, *Pfluegers Arch.* **348:**13–22.

118. Eisenberg, B. R., 1983, Quantitative ultrastructure of mammalian skeletal muscle, in: *Handbook of Physiology: Skeletal Muscle* (L. D. Peachey, R. H. Adrian, and S. R. Geiger, eds.), American Physiological Society, Bethesda, pp. 73–112.

119. Yoshioka, T., and Somlyo, A. P., 1984, The calcium and magnesium contents and volume of the terminal cisternae in caffeine-treated skeletal muscle, *J. Cell Biol.* **99:**558–568.

120. Bond, M., Wasserman, A. J., Kowarski, D., Somlyo, A. V., and Somlyo, A. P., 1985, The range of mitochondrial calcium in smooth muscle, *Biophys. J.* **47:**414a.

121. Broderick, R., and Somlyo, A. P., 1987, Calcium and magnesium transport by *in situ* mitochondria: Electron probe analysis of vascular smooth muscle, *Circ. Res.* **61:**523–530.

122. Bond, M., Shuman, H., Somlyo, A. P., and Somlyo, A. V., 1984, Total cytoplasmic calcium in relaxed and maximally contracted rabbit portal vein smooth muscle, *J. Physiol. (London)* **357:** 185–201.

123. Somlyo, A. P., Somlyo, A. V., Shuman, H., and Endo, M., 1982, Calcium and monovalent ions in smooth muscle, *Fed. Proc.* **41:**2883–2890.

124. Vallieres, J., Scarpa, A., and Somlyo, A. P., 1975, Subcellular fractions of smooth muscle: Isolation, substrate utilization and Ca transport by main pulmonary artery and mesenteric vein mitochondria, *Arch. Biochem. Biophys.* **170:**659–669.

125. Wikstrom, M., Ahonen, P., and Luukkaine, T., 1975, The role of mitochondria in uterine contraction, *FEBS Lett.* **56:**120–123.

126. Morgan, K. G., Morgan, J. P., and DeFeo, T. T., 1984, Determination of absolute ionized calcium concentrations in vascular smooth muscle using aequorin, *Fed. Proc.* **43:**767a.

127. Rembold, C. M., and Murphy, R. A., 1986, Myoplasmic calcium, myosin phosphorylation, and regulation of the crossbridge cycle in swine arterial smooth muscle, *Circ. Res.* **58:**803–815.

128. Crompton, M., Moser, R., Ludi, H., and Carafoli, E., 1978, The interrelations between the transport of sodium and calcium in mitochondria of various mammalian tissues, *Eur. J. Biochem.* **82:**25–31.

129. Denton, R. M., McCormack, J. G., and Edgell, N. J., 1980, Role of calcium ions in the regulation of intramitochondrial metabolism, *Biochem. J.* **190:**107–117.

130. Rossi, C. S., and Lehninger, A. L., 1963, Stoichiometric relationship between accumulation of ions by mitochondria and the energy-coupling sites in the respiratory chain, *Biochem. Z.* **338**:698–713.

131. Rudge, M. F., and Duncan, C. J., 1984, Comparative studies on the role of calcium in triggering subcellular damage in cardiac muscle, *Comp. Biochem. Physiol.* **77**:459–468.

132. Zimmerman, A. N., Daems, W., Hulsmann, W. C., Snyder, J., Wise, E., and Durrer, D., 1967, Morphological changes in heart muscle caused by successive perfusion with calcium-free and calcium containing solution (calcium paradox), *Cardiovasc. Res.* **1**:201–209.

133. Paul, R. J., 1980, Chemical energetics of vascular smooth muscle, in: *Handbook of Physiology: The Cardiovascular System* (D. F. Bohr, A. P. Somlyo, and H. V. Sparks, eds.), American Physiological Society, Bethesda, pp. 201–236.

134. Fleckenstein, A., 1983, *Calcium Antagonism in Heart and Smooth Muscle: Experimental Facts and Therapeutic Prospects*, Wiley, New York.

135. Morgan, J. P., and Morgan, K. G., 1984, Stimulus-specific patterns of intracellular calcium levels in smooth muscle of the ferret portal vein, *J. Physiol. (London)* **351**:155–167.

136. Himpens, B., and Somlyo, A. P., 1988, Free-calcium transients during depolarization and pharmacomechanical coupling in guinea-pig smooth muscle, *J. Physiol. (London)* **395**:507–530.

137. Grover, A. K., and Oakes, P. J., 1985, Calcium channel antagonist binding and pharmacology in rat uterine smooth muscle, *Life Sci.* **37**:2187–2192.

138. Grover, A. K., Kwan, C. Y., Crankshaw, D. J., Garfield, R. E., and Daniel, E. E., 1980, Characteristics of Ca transport and binding by rat myometrium plasma membrane subfractions, *Am. J. Physiol.* **239**:C66–C74.

139. Grover, A. K., Kwan, C. Y., and Daniel, E. E., 1982, High affinity pH-dependent passive Ca binding by myometrial plasma membrane vesicles, *Am. J. Physiol.* **241**:C61–C67.

140. Grover, A. K., Kwan, C. Y., and Daniel, E. E., 1982, Ca-concentration dependence of Ca-uptake by rat myometrium plasma membrane enriched fraction, *Am. J. Physiol.* **242**:C278–C287.

141. De Schutter, G., Wuytack, F., Verbist, J., and Casteels, R., 1984, Tissue levels and purification by affinity chromatography of the calmodulin-stimulated Ca-transport ATPase in pig antrum smooth muscle, *Biochim. Biophys. Acta* **773**:1–10.

142. Grover, A. K., Kwan, C. Y., and Daniel, E. E., 1981, Na–Ca exchange in rat myometrium membrane vesicles highly enriched in plasma membranes, *Am. J. Physiol.* **240**:C175.

143. Sheu, S.-S., and Blaustein, M., 1986, Sodium/calcium exchange and regulation of cell calcium and contractility in cardiac muscle, with a note about vascular smooth muscle, in: *The Heart and the Cardiovascular System* (H. M. Fozzard, R. B., Jennings, E. Haber, A. M. Katz, and H. E. Morgan, eds.), Raven Press, New York, pp. 509–536.

144. Hille, B., 1984, *Ionic Channels of Excitable Membranes*, Sinauer Associates, Sunderland, Mass.

145. Nestler, E. J., Walaas, S. I., Greengard, P., 1984, Neuronal phosphoproteins: Physiological and clinical implications, *Science* **225**:1357–1364.

2

Uterine Metabolism and Energetics

M. Joan Dawson and Jyothi Raman

1. Introduction

Exciting progress has been made over the past decade in understanding the cellular biology of the uterus. The internal structure of smooth muscle cells, regulation of actomyosin ATPase, ion-transport mechanisms, and communication between cells have all been areas of great advances. Studies of hormonal control of reproductive tissue function, and studies of the effects of reproductive hormones on gene expression have been on the leading edge of the present surge in molecular biology. However, the subject of this chapter, metabolism and energetics of uterine tissues, has received rather little attention. It continues to be true that most of what is believed about endometrial and myometrial metabolism and energetics is inferred from studies of more tractable tissues.

Ultimately, the genetic code and the systemic control systems that direct cellular function must work upon, and are therefore limited by, a bioenergetic system that converts a limited amount of chemical fuel into mechanical and osmotic work. The design of uterine cells represents an intricate compromise between their various functional assignments and the conflicting requirement to maintain the energy stores needed to carry out these roles. As the structure and function of endometrium and myometrium change throughout reproductive life, the mechanisms that maintain metabolic and bioenergetic health must work under a constantly changing set of conditions. These design limitations on the uterine machine must be understood because they dictate the physicochemical framework within which uterine cells function.

What is sometimes called the golden age in research on cellular metabolism and energetics occurred from the 1930s to the 1950s, during which time the basic stoichiometries of metabolic pathways and bioenergetic cycles were elucidated,

M. Joan Dawson and Jyothi Raman • Department of Physiology and Biophysics, College of Medicine, University of Illinois at Urbana-Champaign, Urbana, Illinois 61801.

using classical biochemical techniques. As powerful as these techniques are for the purpose, their invasive nature prevents them from yielding unambiguous information concerning how the cellular metabolic pathways are controlled in living systems. After a delay of decades, noninvasive applications of biophysical techniques such as nuclear magnetic resonance spectroscopy and fluorescence spectroscopy to living systems are once again making the relation between cellular metabolism and function a promising area to explore.

In the succeeding pages, we shall summarize what is now known about the mechanism by which fuel is provided to uterine cellular machinery, how this fuel supply is regulated, and how the regulatory mechanisms adapt to changing requirements throughout the reproductive cycle. We shall also explore some of the currently active areas of research and the possibilities inherent in new experimental techniques. Myometrium, rather than endometrium, will be emphasized; this reflects the expertise of the authors and of the body of information currently available. The principles that relate function and metabolism apply equally to both tissue types, and the need for bioenergetic control systems that adapt to changing conditions is at least as critical for endometrial as for myometrial tissue. It was recognized long ago[1] that "Oestrus and menstruation are the visible effects of metabolic changes in the endometrium."

2. Myometrial Metabolism and Energetics: An Overview

The basic mechanisms for the supply and utilization of energy in uterine tissues are similar to those in all living systems. Comparison of myometrial bioenergetics with that of other contractile tissues helps to elucidate the relation between metabolism and function.

The major biochemical events associated with contraction of any muscle type are summarized in Fig. 1. The sequence of events is well known, and with differences only in detail, is the basis of energy supply and energy transduction processes in most biological systems. The prominence of phosphorus-containing compounds in cellular energetics was first noted by Lipmann.[2] Reliance on phosphates for catalysis of exchange reactions is unique to biology and indicates shared thermodynamic and physicochemical organization among bioenergetic systems.[3] However, the reasons for the peculiar importance of phosphates still remain speculative. The role of ATP as the "energy coinage" of the cell emerged during the 1930s, as did discussion concerning how the concentration of this compound could be regulated so that the metabolic energy sources (reactions 2–4) are adequate to meet cellular energy demands (reaction 1).

2.1. Contraction

Myometrial contraction involves the same catalytic proteins, actin and myosin (first demonstrated by Csapo[4]), and a similar cycle of cross-bridge activity as in

Mechanochemical energy conversion

1. $H_2O + ATP \xrightarrow{\text{\textit{Actomyosin ATPase}}} ADP + Pi + H^+$ ($\Delta G \approx -60$ kJ mol^{-1})

Anaerobic recovery:

2. $H^+ + PCr + ADP \xrightleftharpoons{\text{\textit{Creatine kinase}}} ATP + Cr$ ($\Delta G \approx 0$ kJ mol^{-1})

3. $C_6H_{12}O_6 + 2ADP + 2Pi \rightarrow\rightarrow\rightarrow 2(C_3H_5O_3)^- + 2H^+ + 2H_2O + 2ATP$
   glucose   lactate

Aerobic recovery:

4. $C_6H_{12}O_6 + 36ADP + 36Pi + 6O_2 \rightarrow\rightarrow\rightarrow 6CO_2 + 42H_2O + 36ATP$
   (ΔG for glucose oxidation ≈ -126 kJ mol^{-1})

Figure 1. The major biochemical reactions occurring in a contracting muscle, details of which can be found in any comprehensive biochemistry textbook. The exact stoichiometries are shown for metabolites and water, but it is not possible to show exact stoichiometries for specific ionic species because these alter with changing intracellular conditions. The stoichiometry of H_2O in reaction 4 includes 36 molecules of water obtained from the condensation of ADP with Pi; in the steady state, an equal amount is absorbed in ATP hydrolysis. Therefore, there is a *net* release of 6 molecules of H_2O as a result of 1 molecule of glucose broken down. ΔG for ATP hydrolysis (reaction 1) is determined from the data shown in Table II. Reaction 2 is always near equilibrium and thus has a $\Delta G \approx 0$ (see Section 5.2). Reaction sequences 3 and 4 summarize the overall effect of metabolic recovery, involving several known intermediate steps. Because lactic acid is a strong acid (pK 3.9), it is invariably ionized under physiological conditions. ΔG for *in vivo* oxidation of glucose is determined by indirect calorimetry in human subjects. PCr, phosphocreatine; Pi, inorganic phosphate; Cr, creatine.

other muscle types (see Chapters 1, 3, and 4). However, the physical arrangement of the proteins and the activation and control of the ATPase differ markedly between striated and smooth muscle types.[5] Important to bioenergetic studies is the fact that smooth muscle contraction is graded and variable at the cellular level, unlike the skeletal muscle twitch, which is an "all or none" event. In addition, whereas in skeletal muscle the chemical energy utilization during contraction is almost exclusively associated directly with mechanical work production, in smooth muscles the osmotic and electrical work done by ion-transport processes accounts for a large and variable fraction of the total energy expenditure. It is therefore difficult to determine the fractional cost of the various energy-requiring processes.

The rate of smooth muscle contraction is remarkably slow, either when measured as the rate of isometric force production or as the rate of shortening. The economy of contraction (force $\times$ duration/energy expended) of isometric contraction is 100-fold greater than in skeletal muscle.[6] Further, the longer a contraction is maintained, the more economical it becomes so that force can be maintained for long periods of time with little expenditure of energy. Bozler[6] showed in studies of pregnant myometrium that the initial rate of energy expenditure is about the same as in skeletal muscle, but that this declines during a sustained isometric contraction. On the other hand, while the economy of smooth muscle contraction is high, the efficiency of working contractions (force × distance shortened/energy expended) has recently been estimated to be about 50-fold lower in smooth than in skeletal muscles.[7,8]

Another bioenergetic difference between contractions of striated and smooth muscle is the clear dissociation, in smooth muscle, between factors affecting isometric force development and factors affecting maximum rate of shortening, although these are closely related steps in mechanochemical energy conversion. At present, none of these bioenergetic differences between smooth and skeletal muscle contraction is fully understood. The explanations must be sought in terms of the thermodynamics as well as the activation and kinetics of contraction. Studies are now beginning, in both skeletal and smooth muscle, of the effects of changes in concentration of ATP and its hydrolysis products on specific steps in the cross-bridge cycle. Such studies may well help to elucidate the fundamental mechanism of energy transduction by these ATP-requiring processes. This point will be considered again in Section 5.3.

2.2. Metabolic Recovery

Reactions 2–4 of Fig. 1 summarize the means by which the ATP hydrolyzed as a result of mechanical or chemical work is replenished via energy-yielding hydrolytic and oxidative reactions. In the absence of these recovery sequences, the limited supply of ATP in the myometrium would be depleted after fewer than ten contractions.[9]

The energy reserve closest to ATP is provided by phosphocreatine through the creatine kinase reaction (reaction 2), which has been shown to be in equilibrium in skeletal muscle at rest and to remain near equilibrium during normal contractions.[10] Phosphocreatine and creatine kinase, the only enzyme with which it is known to interact, are present in nerve, muscle, and epithelial cell types, but there is no documentation of their presence in other cells. This reversible transphosphorylation is most often thought of as a buffering reaction, stabilizing the supply of ATP. Figure 2 shows that the creatine kinase reaction can, indeed, serve very effectively to buffer [ATP]. As the intracellular pH becomes more acidic, as it will under conditions of stress or maximal activity (described below), the effectiveness of the phosphocreatine/creatine kinase buffer system is enhanced. In skeletal muscle the supply of phosphocreatine is large (five times the supply of ATP itself), and the presence of this buffer clearly preserves the muscle from ATP depletion, and consequent rigor. In smooth muscles, phosphocreatine is low in concentration (about 10% that in skeletal muscle), and its effectiveness as a buffer is thereby decreased. The concentrations of phosphocreatine and ATP have been shown to covary during contractions or anaerobiosis of some smooth muscles[12,13]; this is entirely consistent with the creatine kinase equilibrium, given the low initial concentration of the phosphocreatine buffer. It is noteworthy that smooth muscles do not exhibit rigor mortis, and thus that constancy of [ATP] does not have the same urgency as in skeletal muscle. These points will be taken up in more detail in Section 5.2.

Depleted phosphocreatine and ATP are replenished by glycolysis and oxidative

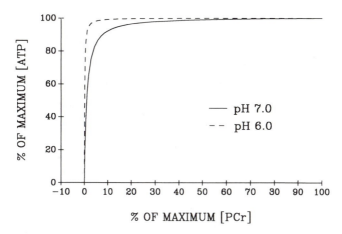

% OF MAXIMUM [PCr]

Figure 2. Variations in [ATP] with changing [PCr] calculated from the creatine kinase equilibrium in a closed system ($K = 10^9$ with H^+ included as a reactant). The x and y axes are normalized so the calculation is valid for all types of muscle. We assumed that total adenosine and total creatine remain constant as their phosphorylated forms are varied, as has been shown in skeletal muscle (Dawson *et al.*, 1978).[11] The solid line shows the result when the pH is held constant at 7.0 and the dashed line shows constant pH = 6.0. Calculations by B. B. Roman.

phosphorylation (reaction sequences 3 and 4, Fig. 1). The same glycolytic and oxidative enzymes have long been known to be present in both smooth and skeletal muscle.[14–16] Both of these pathways are "activated," i.e., increased in rate, as a result of mechanical activity. However, the increase in the rate of oxidative phosphorylation connected with mechanical activity is extremely low, and was utterly missed by early investigators. In 1923, Lovatt-Evans[17] concluded that there was a *decrease* rather than an increase of oxygen uptake during myometrial contraction. On the other hand, early investigators were greatly impressed by the fact that although the absolute amount of lactate produced by smooth muscles is low, it is very high in proportion to the amount of oxygen consumed.[18] Thus, even the earliest studies of smooth muscle metabolism showed that the metabolic response to functional demand differs greatly from that of striated muscle types. Most extraordinary is that even in the presence of an ample supply of oxygen, lactic acid formation is plentiful in smooth muscles. This is often expressed by indicating that smooth muscle lacks or has a weak "Pasteur" effect, i.e., that the presence of oxygen does not cause the automatic switch from anaerobic to fully aerobic metabolism. Pasteur first demonstrated this in yeast and this is characteristic of skeletal muscle and many other mammalian tissues. These points will be taken up again in Section 5.4.

2.3. Kinetics, Thermodynamics, and Cellular Compartmentation

Understanding the relation between function and metabolism in the myometrium requires knowing which variables are responsible for changes in metabolic rate in response to changing demand. In principle, these pathways are governed by the same factors that determine the rates of catalyzed reactions in solutions contained in test tubes. All reactions tend to go toward equilibrium at a rate that is determined by the chemical driving force (i.e., how far the reaction is removed from equilibrium, measured by the free-energy change) and by the catalysis rate (activity) of the enzymes involved. If a reaction is at equilibrium, the forward and reverse flux rates are equal and the free-energy change for the reaction is zero. If an external agent causes the concentration of one of the reactants to change, the flux rates will be adjusted by the law of mass action so that the reaction proceeds in the direction that tends to restore equilibrium. It is unique to biology that the intrinsic rate of the catalyzing enzyme also may be a variable; for example, the activity may be altered by changes in the concentrations of reactants (cooperativity) and by changes in regulators (allosteric effectors) that alter the enzyme activity. The regulated reactions act like switches, turning fluxes through particular metabolic pathways on and off. Such switches necessarily cost metabolic energy and can be expected to occur only where they confer survival value to an organism.

In many cases, the concentrations of compounds necessary to alter enzyme activity *in vitro* by allosteric or cooperative effects are orders of magnitude different from those that are thought to occur *in vivo*.[19] These and other apparent contradictions between the functions of living cells and of isolated enzyme systems have led to great interest in the possibility of microcompartmentation within cells.[20] In smooth muscle cells, the glycolytic and oxidative reaction sequences are postulated to be compartmented; in some vascular tissues, oxidative metabolism appears to support isometric force production while glycolysis supports the Na^+/K^+-ATPase.[21]

While the subjects of microcompartmentation and kinetic control by allosteric effectors are being much investigated at present, some of the most basic properties of chemical reactions receive little attention from biochemists. While standard free-energy changes for biochemical reactions are relatively easy to measure and have been available for some time,[22] they yield virtually no information about the thermodynamics of the reaction *in vivo*. What is needed is a quantitative determination of free-energy changes for bioenergetic reactions as they occur in living cells. This is only beginning to become available. For example, the free-energy change for the hydrolysis of ATP within smooth muscles (shown in Fig. 1) has only recently been estimated. Further, it is not yet possible to say with certainty what is the intracellular chemical driving force for specific steps within the glycolytic and oxidative pathways of metabolism. There is not even general agreement concerning which steps in the pathways are equilibrium steps and which represent metabolic control points. These uncertainties arise because it has been impossible to measure reliably the intracellular concentrations of reactants and ions that affect their equilibria. Prominent among the uncertainties has been the *in vivo* concentrations of the phosphorus-containing metabolites and ions, notably Mg^{2+}, that affect their reactivity. It is necessary to know the *free* concentrations of metabolites (i.e., not bound to macromolecules or otherwise sequestered), because it is that pool that is available to take part in biochemical reactions or to allosterically affect rate-limiting enzymes.

It has long been believed that the concentrations of bioenergetically important phosphorus-containing metabolites in smooth muscle, including myometrium, are much lower than in striated muscle, and that they are present in vastly different proportions. It was long ago suggested that these quantities are hormonally regulated in the myometrium and vary with the physiological state of the tissue.[23] Modern noninvasive techniques are refining and extending the earlier observations. These points will be taken up again in Section 4.

2.4. Summary

ATP is the fuel for myometrial contraction, and rephosphorylation of ADP occurs via the creatine kinase reaction and the reactions of glycolysis and oxidative phosphorylation. The myometrium shares these reaction pathways with skeletal muscle, which is best understood, and with other muscle types. The concentrations of bioenergetically important phosphorus-containing compounds differ markedly between myometrium and skeletal muscle. The character of the actomyosin ATPase and the relationships between rates of oxidative phosphorylation and glycolysis also vary between smooth and striated muscle types. Function and metabolism of the myometrium are under hormonal control, a fact that must be taken into account when theories of cellular control are constructed.

The factors that influence the concentrations of bioenergetically important compounds in the myometrium, and the mechanisms controlling the rates of energy-yielding and energy-utilizing pathways, remain to be clarified. In addition to analyses of allosteric effects on rate-limiting enzymes and the discovery of possible physical substrates for microcompartmentation, this requires assessment of the magnitude *in vivo* of forces driving specific biochemical reactions.

3. Methods—Past, Present, and Possible

New knowledge of cellular metabolism has always followed the development of new technologies. Of particular importance to understanding the thermodynamic framework in which cellular function and metabolism operate is the development of noninvasive methods with which to gather reliable data concerning the *in vivo* cellular milieu.

3.1. Noninvasive Methods

Noninvasive methods of studying the relation between metabolism and function have always been attractive to biologists. Over 100 years ago, T. H. Huxley[24] longed for a means of "making out the molecular structure of living tissues comparable to that which the spectroscope affords to the inquirer into the nature of the heavenly bodies" and complained that "the constituents of our own bodies are more remote from our ken than those of Sirius in this respect." Chemical methods of analyzing tissue composition can yield specific pieces of information with great sensitivity; however, they inevitably suffer from the destruction of intracellular compartmentation and physical and chemical milieu. At best, this results in uncertainties over whether conclusions reached on *in vitro* systems accurately reflect *in vivo* function, and at worst (as in [ADP] in skeletal muscle, Section 5.1) have resulted in incorrect conclusions that have delayed understanding of *in vivo* biochemistry for several decades.

3.2. Heat Production

The first biophysical technique to be used for the study of living tissue was the measurement of heat production, particularly in nerve and muscle. A. V. Hill and his colleagues began such studies in the early 1920s, by showing that an active skeletal muscle produces great quantities of heat, much of which is produced after the contraction is over, during the subsequent period of metabolic recovery. They were able to dissect the heat produced into various identifiable components, thought to be associated with specific events during contraction or metabolic recovery, and to observe how each of these components changed with differing experimental conditions. Such analyses provided a framework in which to place information about molecular events. However, while a change in the rate of heat production is a very sensitive sign of an underlying biochemical reaction, it is utterly nonspecific. One can tell with great accuracy and precision the time course with which something is happening, but can only infer from circumstances or other knowledge what that something may be.

In smooth muscle basal metabolism and heat production are high and variable as a result of spontaneous activity. The time course of the events of contraction and metabolic recovery are not clearly separable as they are in skeletal muscle, but greatly overlap. These and other complexities have hindered the use of heat measurements in bioenergetic studies of smooth muscle. Important work by Bozler[25] and others[26] has shown that the heat production of smooth muscle can be separated into the same components as in skeletal muscle, and thus points to qualitative, but not quantitative, similarities between the two muscle types in the molecular steps associated with the contraction/recovery cycle.

3.3. Oxygen Consumption

Studies of oxygen consumption by skeletal muscle, using the Beckman oxygen electrode, have yielded information concerning both the time course and extent of oxidative metabolism under various experimental conditions. In all cases, the time course of oxygen consumption coincides with production of oxidative-recovery heat, clearly indicating that these are measures of the same biochemical events. These measures are also highly correlated in smooth muscle. Both recovery heat production and oxygen consumption associated with contraction of smooth muscle are slow, low-intensity events, and are therefore difficult to measure reliably. Nevertheless, combined with measurements of lactic acid production, measurement of oxygen consumption yields information concerning the total metabolic cost of contractile or ion-pumping processes, and leads to useful hypotheses concerning the control of metabolism in smooth muscle.[21]

3.4. Fluorescence

Fluorescence is finding increasingly widespread use in biological studies.[27] The method involves light-excitation of electrons into a higher-energy state and detection of the emitted radiation as they return to their ground or low-energy state. The parameters that are measured using this spectroscopic technique are: excitation and emission spectra and the intensity, lifetime, and polarization of the fluorescence. Some biologically important molecules (amino acids, flavins, vitamin A, chlorophyll, and NADH) are naturally fluorescent, but more often, a fluorescent indicator or probe is introduced into the sample. In order to be useful in studies of living cells, some measurable parameter associated with the probe must depend upon the intracellular milieu. Biological applications of this technique include ligand binding (including H^+) to macromolecules, environmental probes (including polarity and conformational changes), molecular dynamics (molecular tumbling and O_2 quenching), measurement of distances between fluorophores (molecular groups giving rise to the fluorescence), and various assays using fluorescent antibodies and chelating agents.

The chief limitation of fluorescence studies in biological systems is the dearth of natural fluorophores and of physiologically sensitive fluorescent probes or labeling reagents. However, a whole new area of research involving the designing of site-specific probes has made this field one of the rapidly developing areas of biological spectroscopy of this decade. Fluorescently labeled antibodies have been used in smooth muscle to great advantage for a number of years to determine the content and distribution of structural, contractile, and regulatory proteins.[28] More recently, Ca^{2+} probes, such as Fura 2, are being used to understand Ca^{2+} dynamics during smooth muscle contraction[29–31] and a pH-sensitive probe has been used to measure myometrial intracellular pH.[32]

3.5. Nuclear Magnetic Resonance Spectroscopy

This technique will be emphasized, both because it is now being applied successfully to studies of uterine energetics and because of its potential for even greater usefulness in basic and clinical studies. NMR relies upon detection of a resonance signal arising from susceptible atomic nuclei when placed in a magnetic field and excited with a radio frequency pulse. Different species of nuclei have greatly different resonance frequencies and nuclei of the same species have slightly different frequencies depending upon their magnetic environments and thus upon the molecules within which they are located. These differences in resonance frequencies allow NMR to detect the presence of particular compounds within a sample and can yield information about their environment as well. Only those atomic nuclei in mobile molecules give rise to sharp NMR signals; thus, in general, NMR detects small molecules that are free in solution. Standard *in vivo* NMR techniques do not usually detect signals from macromolecules, although there is a large field of protein and nucleic acid structural studies by two-dimensional NMR.

Smooth muscles have not been much studied by NMR, largely because of their relative inaccessibility and because of the low concentrations and turnover rates for the metabolites of interest. However, not only can methods available at present be applied productively to studies of uterine metabolism, but after over a dozen years of development, the possibilities for using NMR in noninvasive studies of living systems continue to expand. Emergence of interest in application of NMR to studies of reproductive tissue is exemplified by two recent conferences on the subject.[33,34]

There have been two parallel developments in the application of NMR to the study of living tissues, both of which began in the early 1970s. NMR imaging produces a three-dimensional image, similar to that obtained by computerized X-ray tomography.[35] More sensitive but less spatially localized methods can be used to study tissue metabolism and ion distributions noninvasively. Localized spectroscopic studies using ^{31}P, ^{13}C, and ^1H are now being used to observe regional differences in chemical composition as well as variations in tissue metabolism and enzyme regulation. These and other nuclei are also used to monitor ion transport and distribution (e.g., Na^+, K^+, Mg^{2+}, Ca^{2+}, Cl^-, H^+) within cells, tissues, and organs. The two branches of NMR are now being combined, increasingly successfully, to obtain spatially resolved metabolic information, measurements of blood flow, and images in which spatial contrast is sensitive to various metabolic parameters. A large number of books, monographs, and reviews of various complexity and orientation are now available on virtually all aspects of the subject. Since the field changes rapidly, the reader is advised to survey the most recent publications for the newest developments in the field. Reviews have recently been published concerning present and possible applications of NMR spectroscopy,[36] and imaging,[37] to reproductive tissues.

The biggest drawback to application of NMR spectroscopy for studies of metabolism is relative insensitivity; in general, compounds in concentrations less than 0.1 mmole/liter cannot be detected directly. This means that many intermediates of the glycolytic and oxidative pathways cannot be observed. The important "second messengers" such as inositol phosphates and cyclic nucleotides are even lower in concentration and their direct detection in smooth muscle by NMR is correspondingly more hopeless. On the other hand, it is to be expected that studies in which lactate (which has been detected in smooth muscle by ^1H NMR[38]) is measured together with phosphorus metabolites and ion distributions will greatly expand our knowledge of the regulation of smooth muscle metabolism. ^{13}C studies of myometrium, already begun,[39] should also yield useful information concerning the details of regulation of specific metabolic pathways. Finally, simultaneous NMR spectroscopy measurements of intracellular metabolites, together with fluorescence measurements of oxygen concentration, NAD/NADH ratios, and other accessible metabolic parameters, should prove an extraordinarily powerful method of studying the relation between function and metabolism in myometrium and other tissues.

3.5.1. ^{31}P NMR Spectroscopy of Myometrium

^{31}P NMR spectroscopy can be used to determine the concentrations of phosphorus-containing compounds involved in tissue bioenergetics and in membrane synthesis, modulation, and degradation. It also yields information about the intracellular chemical milieu, including intracellular pH and free [Mg^{2+}].

Figure 3 shows the first published ^{31}P NMR spectrum of myometrium[40]: a study of four nonpregnant uteri isolated from a rat, and superfused at 18°C with oxygenated Krebs' solution, within an NMR sample tube placed in a 4.7-tesla (47 kgauss) magnet. The resonances are identified on the basis of their positions, on the frequency (x) axis, expressed as parts per million shift from the resonance frequency of phosphocreatine. The three peaks on the right are from nucleoside phosphates. The individual nucleotides are indistinguishable in the spectra as they resonate quite close to each other. The peak at −16.5 ppm arises from the middle phosphorus, and therefore solely from the β-phosphorus of nucleotide triphosphate. The other two nucleoside phosphate peaks represent phosphorus at the ribose (α) and the terminal (γ-NTP or β-NDP) end of the molecule. They therefore contain contributions from both di- and triphosphates. Pyridine nucleotides, both NAD and NADP in their oxidized and reduced forms, resonate as a shoulder on the α peak. The peak at −9.6 ppm is in the position of uridine diphosphoglucose, a precursor of glycogen synthesis. Peaks from phosphocreatine and inorganic phosphate are clearly seen. The position of the inorganic phosphate peak is pH-dependent (Section 3.5.3). In this

Figure 3. ^{31}P NMR spectrum of four isolated rat uteri (nonpregnant) obtained at 81 MHz, 1198 pulses of 30 μs duration at 2-sec intervals. The tissue was superfused at 18°C with 25 mM bicarbonate Krebs' solution, bubbled with 95% O_2, 5% CO_2. PME, phosphomonoesters; Pi, inorganic phosphate; EPE, phosphodiesters, PCr, phosphocreatine; PN, pyridine nucleotides; UDPG, uridine diphosphoglucose. The stepped plot above the spectrum represents the integrals of the major peaks. Adapted from Dawson and Wray.[40]

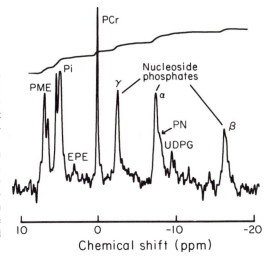

spectrum, the inorganic phosphate peak is split, with the left component representing inorganic phosphate in the bathing medium (pH = 7.4), and the right arising from intracellular inorganic phosphate (pH = 7.1).

The leftmost peak, in the phosphomonoester region of the spectrum, is in the position of phosphoethanolamine, and the shoulder on the right of this peak is in the position of phosphocholine. Both of these metabolites have been shown to be present in myometrium; they are precursors of membrane phospholipid (phosphatidylcholine and phosphatidylethanolamine) synthesis. Three very small peaks are apparent in the phosphodiester region of the spectrum, two of which have been identified[41] in human breast cancers as glycerol-phosphorylethanolamine and glycerol-phosphorylcholine, and the third may be glycerol-phosphorylserine. These compounds are the products of phospholipid degradation via phospholipases. Other phosphates that one might hope to observe in smooth muscles are adenosine and inosine monophosphate resonating closely together at about +6.3 ppm.

Spectra that are both qualitatively and quantitatively very similar to that shown in Fig. 3 have been obtained from isolated myometrium of rabbit,[42] rat,[43] and human.[44,45] In general, ^{31}P NMR spectra from each of the reproductive tissues are easily distinguished from one another, on the basis of the positions and intensities of the observed peaks. Thus, isolated human placenta,[36,44,46,47] ovary from rat[48] or human,[44] and human uterine tube[44] and endometrium[45] all show tissue-specific spectral characteristics. Similar spectra have also been obtained from a localized volume within the uterus of an anesthetized rabbit,[49] using appropriate technology for focusing on the selected volume of interest. There is no technical or safety impediment to obtaining such a spectrum from human uterus, using an intrauterine receiver coil similar in size and shape to some intrauterine devices.

3.5.2. Quantitation of Phosphorus Metabolites

Under ideal conditions, the areas of the peaks in an NMR spectrum are directly proportional to the concentrations of metabolites present; however, this is not always true in intact tissues. The relation between peak area and concentration must be separately determined for each experimental condition. The method used must be considered when evaluating results. In studies using the ^{31}P nucleus, a particular point to watch for is the phenomenon of *saturation,* i.e., a decrease in the proportion of nuclei that can respond to excitation.

Because the signal obtained after a single radio frequency pulse is weak, NMR spectra are often averaged over tens or even hundreds of pulses. It is usual to use a shorter interval between pulses than that required for the system to return to thermodynamic equilibrium via T_1, or spin–lattice relaxation. This practice increases the signal-to-noise ratio because more pulses are applied in a given time, but the response to each individual pulse is decreased by a factor related to the T_1 of the compound being observed. Since different metabolites have different values of T_1, their peak areas can only be compared if the relative saturation is determined in separate experiments. An additional problem is that the observed T_1-relaxation time of a particular nucleus depends, in ways not fully understood, upon its chemical environment, *and* upon the chemical exchange processes it undergoes. Therefore, as experimental conditions are varied, changes in peak intensity could be due either to changes in metabolite concentration or to changes in T_1-relaxation time, or both.

3.5.3. Intracellular pH and Free [Mg²⁺]

The precise resonance positions of charged species are dependent upon the ionic bonds they form in solution. For example, inorganic phosphate is found within cells almost entirely as $H_2PO_4^-$ and HPO_4^{2-}, and the pK that determines the relative proportions of these two species is close to physiological pH. The two ionic forms of inorganic phosphate have slightly different resonance frequencies, and because they exchange rapidly, a single peak appears in an intermediate position depending upon their relative concentrations. Thus, after appropriate calibration, the position of the inorganic phosphate peak can be used to determine intracellular pH. An analogous situation pertains to the β-phosphorus of nucleoside triphosphate. After appropriate calibration, the position of the β-nucleoside triphosphate peak can be used to determine intracellular free [Mg²⁺]. The general form of the equation is

$$pX = pK + \log_{10} (\delta - \delta_1)/(\delta_2 - \delta)$$

where X is the concentration of the ion, K is the equilibrium constant, δ_1 and δ_2 are the resonance positions of the bound and unbound forms of the metabolite, respectively, and δ is the observed resonance frequency of the metabolite in the spectrum of tissue.

A number of careful studies have been made of the accuracy and precision with which intracellular pH can be determined on the basis of the resonance position of inorganic phosphate in the spectrum of living tissues, and of the agreement of the pH determination by NMR and by other methods such as microelectrodes or the distribution of weak acids or bases. In general, if appropriate care is taken, the accuracy of NMR determinations can be as good or better than those using other techniques; agreement between NMR and microelectrode studies is usually within less than 0.1 pH unit.[50]

The situation is not as certain for measurements of intracellular [Mg²⁺]. As shown in Fig. 3, the β-nucleoside triphosphate peak is broad, and uncertainties in measurement of its position are large in comparison to the expected frequency shift with [Mg²⁺]. There are two calibration curves in the literature,[42,51] which differ greatly from each other. The effects of possible intracellular changes in the concentrations of interfering factors (such as K⁺) that affect the calibration curve have been studied only sporadically. Finally, where comparison has been possible, e.g., skeletal and cardiac muscle[52] NMR determinations of intracellular free [Mg²⁺] have been considerably higher (>1.5 mM) than those of recent microelectrode studies[53] (<1 mM). Nevertheless, a number of studies have reported differences among tissues in the position of the β-phosphate peak; these are most simply interpreted as arising from differences in free intracellular $[Mg^{2+}]$.[54] The fact that NMR spectroscopy is virtually the only way of determining free intracellular $[Mg^{2+}]$ in intact tissues must be taken into consideration when the present and possible usefulness of the method is evaluated.

3.5.4. Determination of Flux Rates for Equilibrium Reactions

NMR offers an increasing number of "labeling" methods by which unidirectional flux rates, as well as net reaction rates, may be determined. A simple example is the introduction of a substrate enriched with ^{13}C whose progress through a metabolic pathway can be followed in a manner analogous to familiar experiments using the radioactive ^{14}C isotope. ^{13}C NMR has the advantages that (1) the path is followed noninvasively during a single experiment; (2) the chemical identity of the labeled carbon is readily apparent; and (3) when a product is formed by condensation of two ^{13}C-labeled precursors, the proximity of the two labels within a single compound can be recognized in the spectrum.

There is a family of NMR techniques (magnetization-transfer) that detects the exchange of magnetically labeled nuclei and this allows estimates of rate constants for reactions that are at or near equilibrium. These can be either chemical exchanges, or physical exchanges between membrane-bound compartments. Magnetization-transfer techniques have been applied to studies of myometrium[55] as well as to other tissues. A magnetization-transfer experiment involving the γ phosphate of NTP is shown in Fig. 4. The γ phosphate is continuously irradiated and thus cannot respond to the repeated excitation pulses. Its resonance is therefore absent from the spectrum; this can be thought of as a kind of magnetic label. When the labeled γ-phosphorus nucleus exchanges with the phosphorus of phosphocreatine through the creatine kinase reaction, the magnetic perturbation is carried with it, resulting in a diminution of the proportion of phosphocreatine nuclei that respond to the excitation. Thus, the intensity of the phosphocreatine peak is decreased as well. The rate of exchange of phosphorus between the two compounds is then calculated on the assumption that the diminution of the peak labeled by exchange (phosphocreatine in this case) represents the steady-state gain of label by exchange, and loss of label through magnetic relaxation. Unfortunately, the technique stretches the sensitivity of NMR to its limits and it has therefore been used only to measure fluxes undergone by compounds that are present in living tissue in relatively large concentrations.

4. Bioenergetically Important Metabolites in the Uterus

It is clearly impossible to discuss the relations between function (ATP consumption) and metabolism (ATP generation) without knowledge of the concentrations of this metabolite, its breakdown products, and other substances that affect the

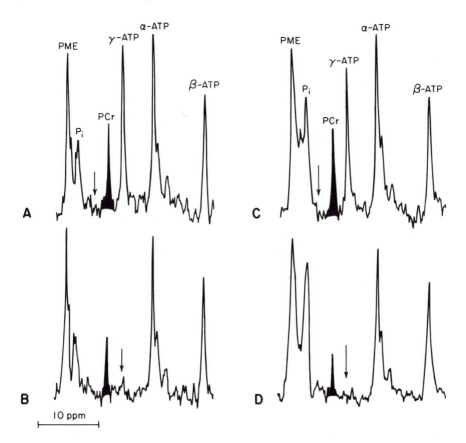

Figure 4. [31]P steady-state magnetization-transfer experiments in immature rat uteri at 36°C. (A) Control uteri; (B) γ-ATP-saturated spectra of control uteri. (C) Control uteri; (D) γ-ATP-saturated spectra of uteri 4 hr after estradiol injection. The arrows indicate the frequencies at which the saturation pulse was applied. Reprinted with permission from Degani *et al.*[55]

free-energy change for ATP hydrolysis. Much is already known at the molecular level about how these metabolites interact with specific proteins. Phosphorus-containing metabolites have been identified as key allosteric effectors on the rate-limiting enzymes of glycolysis, and as the determinants of the rate of oxidative phosphorylation. It has long been known, from studies of isolated enzymes, that release of products of ATP hydrolysis is an integral part of the energy-transduction step in the actomyosin ATPase cycle. Evidence is also accumulating that rate-limiting steps in ATP-dependent ion transport involve interaction of the pump with phosphorus metabolites. Thus, variations in [ATP], [Pi], or [ADP] could well affect muscle activation, metabolism, and mechanics.

In this section, we will review the information available concerning the concentrations of phosphorus-containing metabolites in the uterus and how they vary with physiological status. We will also compare the concentrations of phosphorus metabolites in the myometrium with those in other contractile tissues, and consider how these could be related to functional differences. We will then apply the same considerations to intracellular pH and free $[Mg^{2+}]$. Finally, the effects of alterations in these quantities as a result of disease will be considered.

4.1. Concentrations of Phosphorus Metabolites

There is wide variation in values reported for concentrations of phosphorus metabolites in the myometrium, and the question arises as to which values are most likely to reflect the actual *in vivo* concentrations. A compilation of data available over a decade ago[56] still represents almost all of the information available from chemical analysis of tissue extracts. As described in that review, it was not possible to determine whether the differences observed were due to species variation or to differences in the techniques used in various laboratories. The question of whether concentrations of phosphorus metabolites change as a result of pregnancy was also unresolved, with studies of human[57] and rabbit[23] suggesting an increase in phosphocreatine in the pregnant myometrium, and studies of rats[58] failing to show such an effect.

Table I compares the concentrations of phosphorus metabolites in myometrium of the rat, as determined by ^{31}P NMR and by chemical methods. There are clear differences between the estimates obtained by the two techniques. In nonpregnant myometrium, while the nucleoside triphosphate and the sum of all the phosphorus

Table I. Concentration of Phosphorus Metabolites in the Rat Uterus

Metabolite	Condition	Magnetic resonance spectroscopy[a]	Chemical analysis[b]
Phosphocreatine	Nonpregnant	3.0 ± 0.3 (4)[c]	1.2 ± 0.4 (22)
	Pregnant	4.2 ± 0.5 (5)	0.5 ± 0.4 (7)
	Postpartum	2.3 ± 0.6 (4)	
Pi	Nonpregnant	1.5 ± 0.3 (4)	6.0 ± 0.3 (22)
	Pregnant	1.6 ± 0.2 (5)	5.9 ± 0.9 (7)
	Postpartum	2.5 ± 0.3 (4)	
ATP	Nonpregnant	2.4 ± 0.4 (4)	2.5 ± 0.5 (22)
	Pregnant	3.1 ± 0.4 (5)	1.1 ± 0.2 (7)
	Postpartum	2.3 ± 0.3 (4)	
Phosphomonoester	Nonpregnant	6.3 ± 1.1 (5)	≈1.1[d]
	Pregnant	6.6 ± 0.5 (4)	
	Postpartum	9.8 ± 1.3 (4)	

[a]Data from Dawson and Wray.[59]
[b]Data from Walaas and Walaas[58] unless otherwise indicated.
[c]Concentrations of all metabolites are in mmoles/kg wet wt ± S.E.M. Number of samples in each case is given by *n* value in parentheses.
[d]Data from Volfin *et al.*[60]

metabolites are similar when determined by either method, the phosphocreatine and phosphomonoester are higher, and the inorganic phosphate is lower when determined by ^{31}P NMR. This is but one example of the general tendency for noninvasive NMR studies to yield higher levels of labile metabolites such as ATP, the phosphomonoesters, and phosphocreatine than those obtained by conventional chemical methods. The most likely explanation for this discrepancy is that, even with the best methods of sampling, freezing, and extracting mammalian tissues for chemical analysis, some artifactual hydrolysis of the high-energy phosphates occurs.[61]

In the case of the pregnant rat myometrium, chemical analysis showed virtually no phosphocreatine present, a condition under which the thereby unbuffered stores of ATP could be expected to be artifactually hydrolyzed as well. For this and other reasons,[44,59] we conclude that the ^{31}P measurements of phosphorus metabolite concentrations in isolated uterus are reasonably reliable indications of the *in vivo* values, and that discrepancies with earlier measurements are largely due to known sources of experimental error associated with conventional chemical methods.

4.2. Alterations during Pregnancy and Parturition

Marked differences are found in the levels of phosphorus metabolites between pregnant, nonpregnant, and postpartum rat uterus studied by ^{31}P NMR (Table I) and also between nonpregnant and term-pregnant human myometrium (Figs. 5 and 6). The phosphocreatine increases as a result of pregnancy in both rat and human myometrium, and by approximately the same extent. In the study of human myometrium, an increase in uridine diphosphoglucose (a precursor of glycogen synthesis) could also be discerned (Fig. 5); a similar increase would not have been detected in the rat study, because it was done using less sensitive NMR techniques. The changes observed in the NMR spectrum as a result of pregnancy all tend to increase the stores of immediately available energy-yielding metabolites, and thus to prepare the myometrium for the effort of expelling the fetus against the resistance of the birth canal.

In the rat uterus,[59] the phosphocreatine fell by a third following birth and remained at this low value throughout the first 3 weeks postpartum. The inorganic phosphate practically doubled following parturition and was still elevated 3 weeks later. Intracellular pH became markedly acidic, falling by 0.3 pH unit. The phosphomonoester, tentatively identified as phosphoethanolamine, a precursor of membrane phospholipids, increased dramatically. This last finding may be related to extensive membrane modulation and degradation occurring during uterine involution. The time courses of these changes coincide; the rat uterus halves in size during the first 24–36 hr postpartum[62] and the largest change in the phosphomonoester peak is seen during this same period.

Further studies showed that cesarean-sectioned uteri, which undergo far less muscular work than do uteri during vaginal birth, undergo the same changes in phosphorus metabolites and intracellular pH as do normal uteri.[59] These results

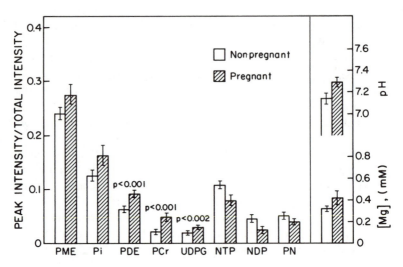

Figure 5. Bar graph of ^{31}P NMR peak intensities from normal pregnant ($n = 6$) and nonpregnant ($n = 12$) human myometria. Relative peak intensities are equivalent to relative concentrations of metabolites. Intracellular pH and free $[Mg^{2+}]$ were determined from peak positions as described in Section 3.5.3. Myometria were superfused with oxygenated De Jalon's solution at 4°C and spectra were averaged over a 1-hr period. Error bars represent ±S.E.M. and p values are for the unpaired t-test. Abbreviations as for Fig. 3. PDE, phosphodiesters; NTP, nucleoside triphosphates; NDP, nucleoside diphosphates. Adapted from Noyszewski *et al.*[44]

support the hypothesis (Section 4.5) that phosphorus metabolite levels in the myometrium are under hormonal control.[23,43,59]

4.3. Intracellular pH in the Myometrium

Measurements of intracellular pH of smooth muscles are difficult and few have been accomplished. The information that is available, much of it from NMR spectroscopy, has recently been reviewed,[63] and suggests that intracellular pH of smooth muscles is similar to that of other mammalian tissues studied under similar conditions. There are more data available for myometrium than for any other smooth muscle type. In both isolated nonpregnant and pregnant rat uterus[59] maintained under physiological conditions at 37°C, pH is equal to 7.1. As shown in Fig. 5, studies of human myometrium at 4°C and in the absence of CO_2 yield a pH of 7.15 in nonpregnant and 7.3 in pregnant uterus, with no significant difference between the two groups. When corrected for dependence upon temperature, intracellular pH is about 7.1 in human myometrium at 37°C, as it is in other muscle types.[50,64] Active pH regulation by the uterus has been demonstrated in ^{31}P NMR studies[65] and more recently in fluorescence studies[32]; in each of these studies, extracellular pH

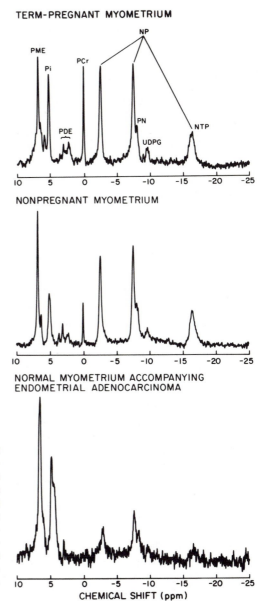

Figure 6. [31]P spectra from term-pregnant (top) and nonpregnant (middle) normal myometrium and apparently normal myometrium accompanying endometrial adenocarcinoma (bottom). Isolated myometria (100–500 mg) were superfused with oxygenated De Jalon's solution at 4°C. The *y* axes have been adjusted so that for any given peak, equal signal intensities in the three spectra represent equal concentrations of metabolites. Abbreviations as for Fig. 3. Reprinted with permission from Noyszewski *et al.*[44]

changes in either the acid or alkaline direction caused much smaller changes in intracellular pH.

We conclude that myometrial intracellular pH is maintained below that required for transmembrane equilibrium. Normal pregnancy does not greatly affect the electrochemical gradient for H^+ across the cell membrane, and intracellular pH in the myometrium is similar to that in most other vertebrate cell types.

4.4. Myometrial Free Magnesium

The importance of this measurement should not be underestimated. Magnesium must be bound to ATP for the latter to exert most of its biological activity. Mg^{2+} is necessary for the catalytic activity of many enzymes and, as a competitor to Ca^{2+}, it affects cell activation processes. Thus, alterations in binding of Mg^{2+} to ATP, or in intracellular free $[Mg^{2+}]$, could be important physiological or pathological variables. In vertebrate cells, free intracellular $[Mg^{2+}]$ appears to be carefully buffered at a value much lower than its thermodynamic equilibrium with the extracellular $[Mg^{2+}]$.[53]

The uterus contains less total Mg^{2+} (6.4 mmoles/kg fat-free wet tissue) than does skeletal muscle (8.4 mmoles/kg), but this difference could be explained by differences in intracellular water in proportion to total fat-free wet weight.[66] Since cells contain a number of known and probably also unknown binding sites for Mg^{2+}, it is not possible to relate quantitatively the total Mg^{2+} content to free intracellular $[Mg^{2+}]$. Until recently, there has been no method available for rigorous determination of free intracellular $[Mg^{2+}]$ in the myometrium.

NMR studies of skeletal and cardiac muscle have invariably concluded that more than 90% of the ATP present is Mg^{2+}-bound.[61] The values for free intracellular $[Mg^{2+}]$ calculated from these estimates depend upon the equilibrium constant assumed and vary between 0.6 and 1.5 mM. NMR studies of smooth muscle show that the position of the β-nucleoside phosphate peak is shifted in the direction of Mg^{2+} unbinding as compared to skeletal or cardiac muscle, and that about 60% of the nucleotide in smooth muscle is Mg^{2+}-bound.[36,42,44,59] This corresponds to a free intracellular $[Mg^{2+}]$ of 0.3–0.6 mM. In one especially painstaking study of rabbit bladder,[42] the [ADP] was raised with metabolic inhibitors, and the free intracellular $[Mg^{2+}]$ was separately determined from the positions of the β-phosphates of ATP and ADP. The results of these two independent measurements agreed closely, showing a free intracellular $[Mg^{2+}]$ of 0.4 mM.

There is little or no effect of pregnancy on free intracellular $[Mg^{2+}]$ in rat or human.[44,59] On the other hand, plasma $[Mg^{2+}]$ tends to fall during pregnancy.[67,68] While Mg^{2+} is lost from the myometrium when it is incubated in Mg^{2+}-free solution,[69] the relation between plasma Mg^{2+} and myometrial free $[Mg^{2+}]$ has yet to be explored.

4.5. Hormonal Control of Phosphorus Metabolite Levels

Throughout pregnancy, there is enormous growth of the uterus, largely as a result of hypertrophy of the muscular layers.[70] Pronounced biochemical changes also occur during this period, e.g., increases in concentrations of contractile proteins, creatine kinase, and glycogen. As described in Section 4.2, the concentrations of high-energy phosphorus metabolites also increase. Studies of the factors causing these biochemical changes of pregnancy have often been done using, as a model, the immature rat exposed to pregnancy hormones. It has been found that exposing immature animals to estrogen mimics the biochemical effects of pregnancy in many cases. Degani and colleagues[43] found that injection of immature animals with estrogen and progesterone caused time-dependent changes in uterine phosphorus metabolite levels, with phosphocreatine initially decreasing and then increasing above its initial level. Binding of nucleotides to Mg^{2+} also underwent a time-dependent decrease, but returned to initial levels. No changes in intracellular pH could be observed.

Since estrogen affects other pregnancy hormones, we have begun investigating the possibility that some of the biochemical changes of pregnancy may directly result from other hormonal effects. Relaxin, for example, is necessary for normal birth, and profoundly affects distensibility of the cervix.[71] Treatment of pregnant rats with monoclonal antibodies to relaxin does not result in a change in placental phosphorus metabolite levels, but there is an unexpected increase in the [phosphocreatine] and a large increase in the [phosphocreatine]/[inorganic phosphate] in the myometrium.[72]

4.6. Effects of Disease

Almost nothing is known about alterations in myometrial metabolism or energetics as a result of gynecological diseases or disorders of pregnancy. This is an important area to explore, both because noninvasive measurements of bioenergetic status could have diagnostic or prognostic value, and because diseases of the reproductive tissues are among the more important and least understood areas of human pathology. Our own investigations of reproductive tissue disorders have mainly concerned placenta[36,44,45] and more recently endometrium.[45,73] We have found that the ^{31}P NMR spectrum of the human placenta is a function of gestational age, and that common disorders of pregnancy result in specific spectral abnormalities. For example,[44] placentas of diabetics show an alkaline shift in intracellular pH that may correlate with the severity of the disease. Molar pregnancies show abnormalities in the phosphomonoester region of the spectrum and growth retardation is associated with abnormal levels of phosphodiesters. Much additional investigations will be required to determine the significance and molecular mechanisms of the observed changes, although we have offered some hypotheses.[36,44]

Results on a small number of human myometrial[44] and endometrial samples[73] suggest that in these tissues there may also be characteristic abnormalities associated with diseases. Both carcinomas and fibroids show abnormally high phosphomonoester peak intensities. This result is consistent with a number of reports that phosphomonoester peaks are elevated in benign and malignant neoplasms. Particularly relevant is the finding that phosphomonoester peak intensities are elevated in breast cancers and are higher in malignant than benign breast tumors.[41]

Perhaps more interesting is the possibility that biochemical changes may be demonstrable in histologically normal tissue associated with a known disease. For example, apparently normal tissue from patients with endometrial adenocarcinoma displays the spectral characteristics of the disease, which are quite different from those of normal myometrial tissues (Fig. 6). On the other hand, spectra from myometrial tissues obtained from patients with cervical cancer did not differ from normal.[44] Preliminary results indicate that abnormalities also may be observed by magnetic resonance spectroscopy in histologically normal endometrial samples near the site of the disease.[73]

5. In Vivo Control of Function and Metabolism in the Myometrium

In this section, we shall summarize ways in which present knowledge of the intracellular milieu should be useful to studies of the control of myometrial function; much of the discussion will be centered upon comparison of myometrial function and biochemistry with those of other muscle types. In addition, we shall discuss some of the important gaps in our knowledge, and ways in which these gaps might be filled.

5.1. Comparison of Phosphorus Metabolites of Myometrium and Striated Muscles

A great deal is known about differences in activation and contractile characteristics between myometrial and striated muscle, and it is appropriate to examine how differences in metabolism and bioenergetics, including concentrations of phosphorus metabolites, might be related to the known functional differences. Table II summarizes estimates of phosphorus metabolite concentrations, intracellular pH and free $[Mg^{2+}]$, and calculated free-energy change for ATP hydrolysis in rat pregnant and nonpregnant myometrium, human skeletal muscle, and ferret cardiac muscle. These data are from ^{31}P NMR spectroscopy studies, as referenced in Table II.

Table II shows that, as estimated by NMR, the concentration of high-energy phosphates in myometrium is only about one-sixth that in human skeletal muscle, a result that is in remarkably good agreement with the early study of Csapo and Gergely,[9] who found about one seventh as much ATP and phosphocreatine in uterine strips as in skeletal muscle. The value for cardiac muscle is intermediate between those for skeletal and uterine muscle. The obvious interpretation of these findings is that stores of high-energy phosphates are related to the maximum mechanical power production, which is also highest in skeletal muscle, intermediate in myocardium, and lowest in myometrium. Some caution is necessary in interpreting data such as these shown in Table II, because metabolite concentrations are determined per gram wet or dry weight of tissue, whereas it is the *intracellular* concentrations of metabolites, or even concentrations within intracellular compartments, that determine the driving force and rate of reactions. The amount of

Table II. Concentrations of Energetically Important Metabolites in Various Muscle Types

	Skeletal[a]	Cardiac[b]	Uterine[c]	
			Pregnant	Nonpregnant
ATP (mmoles/kg wet wt)	5.11 ± 0.12	4.4	3.1 ± 0.4	2.4 ± 0.4
PCr[d]	28.52 ± 0.43	12.76 ± 2.1	4.2 ± 0.5	3.0 ± 0.3
Pi	4.27 ± 0.17	3.57 ± 0.62	1.6 ± 0.2	1.5 ± 0.3
PCr + Pi	32.79 ± 0.41	16.33	5.8	4.5
PCr + Cr	28.6 ± 0.28	14.0		3.2[e]
pH (37°C)	7.13 ± 0.02	6.98 ± 0.03	7.08 ± 0.03	7.09 ± 0.07
Mg^{2+} (mmoles/liter)	>1.5	>1.5	0.4 ± 0.2	0.3 ± 0.05
ADP (μmoles/kg wet wt)	2.5	3.1		0.8
ΔG (kJ/mole)	−70.3	−61.5		−68.6

[a] Data from Edwards *et al.*[74]
[b] Data from Allen *et al.*[75]
[c] Data from Dawson and Wray[59] unless otherwise indicated.
[d] PCr, phosphocreatine; Pi, inorganic phosphate; Cr, creatine.
[e] Data from Volfin *et al.*[60]

intracellular water in any tissue is difficult to determine and precise comparisons of intracellular constituents in myometrial and striated muscles are not possible. Nevertheless, available evidence[68] indicates that intracellular water accounts for a high proportion of the volume of skeletal muscles (85–90%) and a much lower proportion of the myometrium (40–60% and unaltered by pregnancy). When corrected for the differences in intracellular water content, the actual [ATP] may be the same in all the muscle types shown in Table II. The [phosphocreatine], on the other hand, represents an unambiguous difference among these striated and smooth muscle types.

There is also no clear evidence for differences between striated and smooth muscles in concentrations of the products of ATP hydrolysis, which, even more than ATP, are often cited as possible regulators of glycolysis, oxidative phosphorylation, and contraction. Literature values for inorganic phosphate have varied drastically, and over the years there has been a tendency for estimates to decline. The most recent studies yield control values for inorganic phosphate of 0.5 to 5 mM in all types of muscle, with uncertainties in the measurement precluding precise comparisons. It may be more informative to compare the lower limits for physiological concentrations of inorganic phosphate in various muscle types. Since phosphorylated creatine cannot exceed total creatine, inorganic phosphate cannot be less than the difference between [phosphocreatine + inorganic phosphate] and [total creatine]. This value is equal, within experimental uncertainties, for skeletal, cardiac, and myometrial muscle.

It is worth noting that the concentration of inorganic phosphate, as determined by ^{31}P NMR, appears to vary greatly among smooth muscles. For example, inorganic phosphate is almost unobservable in well-oxygenated preparations of vascular muscle.[76,77] whereas it is a prominent peak in spectra of myometrium, irrespective of species of origin, or perfusion methods. The placenta, with its rich mixture of cell types, shows the highest intensity for the inorganic phosphate peak of any tissue studied by ^{31}P NMR spectroscopy.[44]

The intracellular [ADP] requires particular explanation. Chemical analyses of [ADP] in extracts of skeletal muscle or myometrium yield approximately 0.5 and 1.5 mM, respectively. However, it has long been known that virtually all of the ADP present in skeletal muscle extracts can be accounted for by known intracellular binding sites,[61] particularly F-actin. Thus, the extracted ADP has no clear relationship to the free intracellular ADP, which is available to take part in biochemical reactions, or to affect the activity of rate-limiting enzymes. In NMR studies it has been common to estimate the free intracellular [ADP] from the creatine kinase equilibrium. A recent study[78] has presented evidence that the creatine kinase reaction is in equilibrium in arterial muscle, as it is in skeletal and cardiac muscle. Assuming the same is true for myometrium, this leads to a calculated [ADP] in the micromolar range in the nonpregnant human or rat myometrium, similar to that in striated muscle. Further, it leads to calculated values for the free-energy change for ATP hydrolysis that are similar to those in other muscle types. In other words, the amount of mechanical or osmotic work that can be obtained per mole of ATP hydrolyzed appears to be similar in the myometrium to that in striated muscle.

The data shown in Table II thus lead us to a useful conclusion: Although skeletal, cardiac, and smooth muscles differ dramatically in function, they normally have approximately the same intracellular concentrations of ATP and its breakdown products, and a similar driving force for the production of work via hydrolysis of ATP. Functional differences among muscle types are thus not clearly related to differences in resting or control values for these phosphorus metabolite levels. There are, however, two clear bioenergetic differences between myometrium and striated muscle. The first is in [phosphocreatine], which is neither a reactant nor a known regulator of enzymes involved in glycolysis, oxidative phosphorylation, or contraction. Nevertheless, through its participation in the creatine kinase reaction, phosphocreatine could perform a regulatory role by altering the effect of metabolic stress on levels of other phosphorus metabolites. The second difference from striated muscle is the apparently lower level of free intracellular $[Mg^{2+}]$ and Mg^{2+} binding to ATP in the myometrium. This characteristic, which the myometrium shares with other smooth muscles, could have numerous consequences.

5.2. Creatine Kinase

The role of the creatine kinase reaction in muscle has classically been described as that of a buffer[79] to the supply of ATP. More recently, additional functions have been postulated, such as a shuttle for high-energy phosphates between mitochondria and cytoplasm[80] and as a regulator of cellular activity through maintenance of equilibrium among reactants.[81] The first and last of these postulated roles depend upon the activity of creatine kinase being great enough to maintain equilibrium or near equilibrium under all conditions. The second arose originally as an attempt to explain the differences in isoforms of creatine kinase between mitochondria and cytoplasm,[80] and was later elaborated to explain apparent *disequilibrium* among the reactants, especially in the heart.[82] The available evidence at present suggests that the creatine kinase reaction is at or near equilibrium in smooth,[78] cardiac,[83] and skeletal[84] muscle.

The level of creatine kinase in the myometrium is responsive to estrogen treatment, both *in vitro* and *in vivo*.[85] Its synthesis is induced by a wide variety of estrogens, is dependent of RNA synthesis, and is linearly related to the concentration of nucleus-bound estrogen.[86] Consistent with these findings, magnetization-transfer NMR studies have found that the forward rate constant (phosphocreatine to ATP) increases from 0.20 to 0.33 mmole/kg wet wt per sec in immature uteri treated with estrogen.[55] This is to be compared with a rate of 16 mM/sec in rat gastrocnemius muscle.[87] As shown in Table III, creatine kinase activity doubles during pregnancy. This increase is more closely associated with uterine cell growth and division than with a direct effect of estrogens themselves.[88] It is interesting that creatine kinase activity of the endometrium is also variable, being highest in the proliferative phase of the menstrual cycle.[88] All of this suggests that creatine kinase activity may be one of the best indices available of the level of activity of cells that normally contain this enzyme.

5.3. Energetics of Contraction

Metabolic control of contraction could be at either or both of two levels: the activation of contraction, or the turnover of the actomyosin ATPase itself. In both of these areas, smooth muscle presents specific characteristics, peculiarly different from those of striated muscles.

1. *Activation of contraction.* One of the most prominent differences between smooth and skeletal muscle is that while skeletal muscle has a single excitatory input, smooth muscles are characterized by numerous excitatory, inhibitory, and modulatory inputs via neurotransmitters, hormones, and paracrine substances.

Table III. Enzyme Activities (μmoles/min per g wet wt)

	Nonpregnant	Term-pregnant	Skeletal	
			Fast	Slow
Creatine kinase (human)[a]	11	18	200	300
Phosphofructokinase (murine)[b]	3	3	58	10
Cytochrome oxidase (murine)[b]	1.5	2	3	12
Hydroxyacyl dehydrogenase (murine)[b]	11.4	8.5	6	22
Lactate dehydrogenase (murine)[b]	160	120	500	290

[a] Data from Luh and Henkel.[104]
[b] Data from Orlander.[105]

Many of these chemical messengers act by way of a GTP-dependent intracellular messenger system. A biochemical correlate of the complexity of smooth muscle activation is the relatively high concentration of guanosine triphosphate, which accounts for about 10% of the total nucleotides in the myometrium.[89] In contrast, skeletal muscle has uniquely low concentrations of nucleotides other than adenosine phosphate.

Differences between smooth and striated muscles in free intracellular $[Mg^{2+}]$ also have predictable consequences in the activation of contraction. It is generally believed that one of the primary mechanisms for smooth muscle activation is Ca^{2+}-mediated phosphorylation of myosin. *In vitro,* phosphorylated myosin utilizes ATP at a greater rate, and *in vivo,* phosphorylation of myosin precedes smooth muscle contraction.[90] Questions remain as to whether there may also be a direct effect of Ca^{2+} on an actin-linked protein, analogous to troponin in skeletal muscle.[91] Since the actin-linked Ca^{2+}-binding sites in smooth muscle also bind Mg^{2+} in the millimolar range, the precise value for free intracellular $[Mg^{2+}]$ in smooth muscle is relevant to this question. It has recently been found[92] that at 0.7 mM free Mg^{2+}, unloaded shortening velocity (presumed to be proportional to the rate of crossbridge cycling) is dependent upon both myosin phosphorylation and free $[Ca^{2+}]$, whereas at 2.2 mM free Mg^{2+} there is no Ca^{2+}-dependence beyond that for maximum phosphorylation of myosin. If myometrial free intracellular $[Mg^{2+}]$ is less than 0.5 mM (Section 4.4; Table II), Ca^{2+} may well have a regulatory role in this tissue beyond its effect on myosin phosphorylation.

2. *The actomyosin ATPase.* Skeletal muscular fatigue is closely related to phosphorus metabolite levels.[11] When normal muscle is activated maximally, force and power production decline concurrently with a decline in [phosphocreatine] and a rise in [ADP] and [inorganic phosphate]. The great responsiveness of mechanical activity in skeletal muscle to changes in phosphorus metabolite concentrations serves to protect the supply of ATP and thus to prevent rigor. Particular attention is being focused at present on the role of inorganic phosphate,[93] and specifically on the role of the acid form of inorganic phosphate, $H_2PO_4^-$,[94,95] as the possible mediator of force fatigue. The effect of this metabolite might result from the reversible release of inorganic phosphate from the actomyosin complex during the development of the major force-producing state.[93]

Numerous studies in various smooth muscles have shown that force development, rate of rise and relaxation of contraction, and rate of shortening are responsive to changes in phosphorus metabolite levels. For example,[96] ATP augments and inorganic phosphate inhibits Ca^{2+}-dependent contractions of the skinned rabbit mesenteric artery. However, there appears to be no closely analogous fatigue mechanism in smooth muscles; indeed, such a mechanism would seem to have little value. Phosphorus metabolite levels either do not change or change very little in intact smooth muscles as a result of physiologically induced contractions, although they may change markedly under conditions of metabolic stress. In a study of vascular smooth muscle,[97] it was shown that ATP utilization in tonically contracting porcine carotid artery closely matches concurrent ATP synthesis. There have been numerous demonstrations of large changes in phosphorus metabolite levels in intact smooth muscles, but these have generally occurred under conditions of substrate deprivation or inhibition of either glycolysis or oxidative phosphorylation. In all such studies, the relationship between contraction and metabolite levels is variable, leading to the conclusion that factors in addition to phosphorus metabolite levels affect the contractile response to metabolic stress. Importantly, the effects of short periods of metabolic stress on phosphorus metabolite levels and contractile properties of smooth muscle are completely reversible, indicating that the loss of contractile activity may be due to a protective mechanism that reduces energy utilization before cell damage occurs.

We conclude that the decline in force development associated with fatigue in skeletal muscle does not have a direct correlate in smooth muscles, but that there may well be protective mechanisms, as yet not clearly elucidated, that effectively limit energy utilization in smooth muscles. Thus, short periods of anoxia, hypoglycemia, or ischemia, as may occur in labor, do not necessarily cause irreversible damage.

In some smooth muscles, metabolic control is exerted by changes in blood and tissue P_{CO_2}, mediated via its effect on intracellular pH. For example, CO_2 is the main regulator of blood flow to the brain, with a fall in intracellular pH of vascular smooth muscle leading to pronounced vasodilation.[98,99] Blood flow to the heart, skeletal muscle, pulmonary artery, and ductus arteriosus are all altered by physiological changes in CO_2.[63] However, while this effect is easily demonstrated in vascular muscle, and has obvious physiological consequences, the results in other smooth muscles have been variable. There has been little study of the effects of physiological or pathological variables on intracellular pH of the myometrium, and in studies in which myometrial pH was altered,[32,63,65] effects on contractility were not monitored.

5.4. Economy and Efficiency of Contraction

In principle, comparison of mechanical and thermodynamic characteristics of striated and smooth muscles must make important contributions to the basic problem of the nature of muscular contraction. The obvious differences among the various muscle types provide a framework for interpretation of molecular events. In practice, however, uncertainties in the structure of the contractile apparatus in smooth muscles severely limit the utility of studies of mechanical and thermodynamic characteristics of whole smooth muscles or muscle strips. It has long been clear that smooth muscles produce as much, or even more, force per cross-sectional area than do skeletal muscles, in spite of their much lower actomyosin ATPase activity.[4] It is also incontrovertible that smooth muscle contracts and relaxes more slowly and produces force more economically than does striated muscle.[6] These characteristics can be interpreted in terms of a similar actomyosin ATPase cycle in smooth as in striated muscle, with the quantitative differences that smooth muscle cross-bridges cycle more slowly and occupy the force-producing states for longer periods than those of skeletal muscle.

It should be emphasized, however, that interpretation of *in vivo* mechanical and thermodynamic characteristics of smooth muscle in terms of any molecular model remains speculative because of the large uncertainties in orientation of the contractile apparatus in smooth muscle. Even the simplest requirements for interpretation of mechanical and thermodynamic measurements are not yet in place. It is not clear how many contractile units are arranged in parallel (affecting force per cross-sectional area) and whether this may change during contraction. It is also not clear how much internal work and shortening occurs, although the very low efficiency with which smooth muscles accomplish working contractions (Section 2.1) suggests that this may be considerable. We can hope for rapid progress in this area as a result of the numerous studies of mechanical characteristics of single smooth muscle cells, the results of which supplement molecular studies of the actomyosin ATPase. The recent demonstration[100] of corkscrew-like shortening in single smooth muscle cells, together with spontaneous reextension after contraction, could go a long way toward explaining the slow shortening velocity and low efficiency of smooth muscle contraction.

Given the fundamental problems in interpreting thermodynamic data, it is perhaps not surprising that the question of what causes the economy of smooth muscle contraction to increase with duration has been hotly debated. It has been proposed,[101] and strongly contested,[7] that there are two classes of cross-bridges in smooth muscle. The first type consists of phosphorylated cross-bridges that have a relatively fast cycling rate. The second is a "latch-bridge" that results from dephosphorylation of the cross-bridges and maintains force with a very low cycling rate. Such a system would ensure an increase in economy of force maintenance and a decrease in efficiency of shortening (due to the internal load of the slowly cycling "latch-bridges"). Studies aimed at showing the presence of this internal load[7] or a decrease in economy of working contractions[8] have not been successful, and thus tend to refute the hypothesis. It is perhaps worth bearing in mind that in skeletal muscle, too, myosin phosphorylation, economy of contraction, and apparent cross-bridge cycling rate are functions of duration of contraction. Thus, the problem does not belong to the smooth muscle community alone.

5.5. Metabolism

There is a marked and unexplained variation among mammalian tissues in the amount of lactic acid they produce under basal conditions in the presence of an ample supply of oxygen. In resting skeletal muscle, it is negligible, while in brain it represents about 15% of the glucose consumed.[102] The magnitude of aerobic lactate production in myometrium is indicated by the fact that in rat uterus, 64–80% (depending on the phase of the estrous cycle) of [^{14}C]glucose taken up is converted to lactate, while only 8.5–11% is oxidized to CO_2.[103]

The profound difference between skeletal and smooth muscles in the reliance on aerobic generation of lactate is not reflected by the patterns of glycolytic and oxidative enzymes. As shown in Table III, myometrial content of oxidative and glycolytic enzymes is somewhat lower (uncorrected for differences in wet weight) than in fast or slow skeletal muscle, but the relative capacity for aerobic versus anaerobic metabolism is intermediate between striated fast and slow muscles. Reports as to whether there is an adaptive change in enzymatic profile during pregnancy differ. In human myometrium, the activities of mitochondrial enzymes decrease slightly whereas lactate dehydrogenase increases during pregnancy, while in the rat both glycolytic and mitochondrial capacities tend to increase.[106] In the mouse, no metabolic adaptation to pregnancy was observed.[105] On the other hand, there is general agreement that myometrial glycogen content is under hormonal (mainly estrogen) control, and that it increases during pregnancy and decreases during parturition.[107]

In vascular muscles, which have been most studied, the high lactate production may be explained by the fact that aerobic glycolysis preferentially supplies the energy required for the Na^+/K^+-ATPase.[8,108,109] A similar phenomenon can be demonstrated in myometrium. Incubation of rat uterus in K^+-rich but Na^+-free buffer reduced lactate production, while in Na^+-rich buffer, myometrial lactate production markedly rose on addition of 10 mM K^+, an effect which could be inhibited by ouabain.[110] Tight coupling between lactate production and Na-pump activity has also been observed in the myometrium of pregnant rat and rabbit.[111] Glucose uptake is stimulated by insulin in uterine muscle,[112] an effect which is associated with Na-pump stimulation. It has been suggested that the preferential use of specific metabolic pathways to supply ATP for particular energy-requiring reactions results from physical compartmentation[8] or from differential dependence on [ATP].[113] It is perhaps worth remembering that there is an intrinsic interdependence between glycolysis and the Na^+/K^+-ATPase in that for every molecule of ATP generated by lactic acid production, a proton is also liberated (see Fig. 1), and smooth muscles rely heavily on Na^+/H^+ exchange for pH regulation.[63] Thus, the relative dependence upon glycolysis versus oxidative phosphorylation has consequences for ion balance beyond merely the availability of substrate for ATP-dependent ion pumps.

The evidence linking glycolysis to the activity of membrane pumps is complemented by the apparent links between contractile activity and oxygen uptake. Many years ago, Bulbring[114] showed that oxygen uptake is directly proportional to force developed in taenia coli. Paul and colleagues[8] have shown that in at least some vascular muscles, mechanical activity is supported entirely by oxidative metabolism, although this is apparently not universal for vascular muscle. The question has not been addressed in myometrium.

It has been difficult to analyze the role of oxidative metabolism in smooth muscles. Studies of isolated mitochondria from smooth muscles are relatively recent; however, they appear to show that the phosphorylation potential and respiratory control ratios are similar to those of skeletal muscle and other tissues.[115] The data shown in Table II (Section 5.1) indicate that myometrial free [ADP] is in the micromolar range and thus near the mitochondrial K_m for this substrate. Small changes in [ADP] could therefore serve to regulate oxidative phosphorylation in smooth muscle, as they are thought to do in skeletal muscle. The myometrial [Pi] is well above the mitochondrial K_m for this substrate, suggesting that in myometrium, as in other types of muscle,[116] inorganic phosphate could not have much direct effect on the rate of oxidative phosphorylation.

6. Summary and Conclusions

In this chapter we have taken a holistic and comparative approach to evaluation of present knowledge of uterine metabolism. We hope that this complements the efforts of numerous investigators who are making substantial progress in understanding molecular mechanisms of metabolism and function of smooth muscle and epithelial tissues. Certainly, the tabulation and evaluation of recently available data concerning the intracellular milieu of myometrium should be valuable to those who wish to determine which of the molecular mechanisms that are possible in isolated subsystems could actually occur *in vivo*.

We are still far from understanding the cellular function of the uterus. When all of the genetic and molecular mechanisms are understood, it will still be necessary to explain how each subset of functions interacts with the others to form the complexity of the whole. Luckily, there are a few organizing principles that should help us. First, physicochemical principles cannot be circumvented. Second, cellular biochemistry and molecular mechanisms are conserved, from species to species and from tissue to tissue. Just as evolution has conserved useful chemical structures and put them to myriad uses in different organisms and tissues, the integration of cellular metabolism and function is conserved, so that only a few of the possible means by which the molecular building blocks can be put together are actually in evidence in biological systems. Thus, the similarities between uterine and other tissue types serves to indicate the principles that dictate tissue function in general.

Conversely, the differences between myometrium and other tissues, particularly other muscle types, should help us to deduce the organization of myometrial cellular function. Evolution sees to it that there is no superfluous cellular machinery. The myometrium is a myometrium because its function dictates its structure and the choices of available biochemical building blocks. Thus, comparisons of intracellular milieu, metabolism, and functions can lead us to organizational principles related to specific tissue functions.

Finally, the energy available for growth and function is limited. Therefore, we can assume that the metabolism of the uterus has evolved in such a way that its functions are supported by the most effective metabolic system possible. Therefore, holistic studies of metabolism in functioning tissue should yield fruitful understanding of the cellular mechanisms of the cellular functions themselves.

References

1. Holmes, E., (Editor and Author), 1937, in: *The Metabolism of Living Tissue,* Cambridge University Press, London, p. 180.
2. Lipmann, F., 1941, Metabolic generation and utilization of phosphate bond energy, *Adv. Enzymol.* **1**:99–162.
3. Westheimer, F. H. 1987, Why nature chose phosphates, *Science* **235**:1173–1178.
4. Csapo, A., 1948, Actomyosin content of the uterus, *Nature* **162**:218–219.
5. Marston, S. B., 1983, Myosin and actomyosin ATPase: Kinetics, in: *Biochemistry of Smooth Muscle* Volume 2 (N. L. Stephens, ed.), CRC Press, Boca Raton, pp. 167–192.
6. Bozler, E., 1977, Thermodynamics of smooth muscle contraction, in: *The Biochemistry of Smooth Muscle* (N. L. Stephens, ed.), University Park Press, Baltimore, pp. 3–13.
7. Butler, T. M., Siegman, M. J., and Mooers, S. U., 1987, Slowing of the crossbridge cycling rate in mammalian smooth muscle occurs without evidence of an increase in internal load, in: *Regulation and Contraction of Smooth Muscle* (M. J. Siegman, A. P., Somlyo, and N. L. Stephens, eds.), Liss, New York, pp. 289–301.
8. Paul, R. J., Strauss, J. D., and Krisanda, J. M., 1987, The effects of calcium on smooth muscle mechanics and energetics, in: *Regulation and Contraction of Smooth Muscle* (M. J. Siegman, A. P. Somlyo, and N. L. Stephens, eds.), Liss, New York, pp. 319–332.
9. Csapo, A., and Gergely, J., 1950, Energetics of uterine muscle contraction, *Nature* **166**:1078–1079.
10. Gadian, D. G., Radda, G. K., Brown, T. R., Chance, E. M., Dawson, M. J., and Wilkie, D. R., 1981, The activity of creatine kinase in frog skeletal muscle studied by saturation-transfer nuclear magnetic resonance, *Biochem. J.* **194**:215–228.
11. Dawson, M. J., Gadian, D. G., and Wilkie, D. R., 1978, Muscular fatigue investigated by phosphorus nuclear magnetic resonance, *Nature* **274**:861–866.
12. Nakayama, S., Seo, Y., Takai, A., Tomita, T., and Watari, H., 1988, Phosphorus compounds studied by ^{31}P nuclear magnetic resonance spectroscopy in the taenia of guinea-pig caecum, *J. Physiol. (London)* **402**:565–578.
13. Lundholm, L., Pettersson, G., Andersson, R. G. G., and Mohme-Lundholm, E., 1983, Regulation of the carbohydrate metabolism of smooth muscle; some current problems, in *Biochemistry of Smooth Muscle,* Volume 2 (N. L. Stephens, ed.), CRC Press, Boca Raton, pp. 85–128.
14. Hollman, S., 1949, Uber die anaerobe glycolyse in der uterus muskulatur, *Z. Physiol. Chem.* **284**:89–128.
15. Zemplenyi, T., 1968, Enzyme linking energy-rich bonds with muscular contraction, in: *Enzyme Biochemistry of the Arterial Wall* (T. Zemplenyi, ed.), Lloyd–Luke, London, pp. 45–50.
16. Kirk, E. J. (ed.), 1969, *Enzymes of Arterial Wall,* Academic Press, New York.
17. Lovatt-Evans, C. L., 1923, Studies on the physiology of plain muscle. II. The oxygen usage of plain muscle and its relation to tonus, *J. Physiol. (London)* **58**:22–32.
18. Lovatt-Evans, C. L., 1925, Studies on the physiology of plain muscle; the lactic acid content of plain muscle under various conditions (Section IV), *Biochem. J.* **19**:1115–1127.
19. Wilkie, D. R., 1983, The control of glycolysis in living muscle studied by nuclear magnetic resonance and other techniques, *Biochem. Soc. Trans.* **2**:244–246.
20. Jones, D. P., (Editor and Author), 1988, *Microcompartmentation,* CRC Press, Boca Raton.
21. Paul, R. J., 1983, Functional compartmentation of oxidative and glycolytic metabolism in vascular smooth muscle, *Am. J. Physiol.* **244**:C399–C409.
22. Lehninger, A. L., (Editor and Author), 1965, *Bioenergetics,* Benjamin, New York.
23. Menkes, J. H., and Csapo, A., 1952, Changes in adenosine triphosphate and creatine phosphate content of the rabbit uterus throughout sexual maturation and after ovulation, *Endocrinology* **50**:37–50.
24. Huxley, T. H., 1885, Presidential address, *Proc. R. Soc.* **39**:294.
25. Bozler, E., 1930, The heat production of smooth muscle, *J. Physiol. (London)* **69**:442–462.

26. Gibbs, C. L., 1983, Smooth muscle heat production, in: *Biochemistry of Smooth Muscle,* Volume 2 (N. L. Stephens, ed.), CRC Press, Boca Raton, pp. 127–136.

27. Campbell, I. D., and Dwek, R. A., 1984, (Editors and Authors), in: *Biological Spectroscopy,* Benjamin–Cummings, Menlo Park, pp. 91–125.

28. Bagby, R. M., 1983, Organization of contractile/cytoskeletal elements, in: *Biochemistry of Smooth Muscle,* Volume 1 (N. L. Stephens, ed.), CRC Press, Boca Raton, pp. 4–84.

29. Himpens, B., and Somlyo, A. P., 1988, Free-calcium and force transients during depolarization and pharmacochemical coupling in guinea-pig smooth muscle, *J. Physiol. (London)* **395:**507–530.

30. Yagi, S., Becker, D. L., and Fay, F. S., Relationship between force and Ca^{2+} concentration in smooth muscle as revealed by measurements on single cells, *Proc. Natl. Acad. Sci. USA* **85:**4109–4113.

31. Ishii, N., Simpson, W. M., and Ashley, C. C., 1989, Free calcium at rest during "catch" in single smooth muscle cells, *Science* **243:**1367–1368.

32. Baro, I., Eisner, D. A., Raimbach, S. J., and Wray, S., 1989, The measurement of intracellular pH in single, isolated smooth muscle cells, *Proc. Physiol. Soc.* University College, London, 7P.

33. McCarthy, S., 1987, MRI of the uterus, in: *Magnetic Resonance of the Reproductive System* (S. McCarthy and F. Haseltine, eds.), Slack, Thorofare, N.J., pp. 143–160.

34. Miller, R. K., and Thiede, H. A. (eds.), 1989, 11th Rochester Trophoblast Conference, *Troph. Res.* in press.

35. Buddinger, T. H., and Lauterbur, P. C., 1984, Nuclear magnetic resonance technology for medical studies, *Science* **226:**288–298.

36. Dawson, M. J., McFarlane, D. K., McFarlin, B. L., Noyszewski, E. A., and Trupin, S. R., 1988, The biochemistry of female reproductive tissues studied by phosphorus nuclear magnetic resonance spectroscopy: Effects of pregnancy, hormonal manipulation, and disease, *Biol. Reprod.* **38:**31–38.

37. Mattison, D. R., and Angtuaco, T., 1988, Magnetic resonance imaging in prenatal diagnosis, *Clin. Obstet. Gynecol.* **31:**353–389.

38. Yoshizaki, K., Radda, G. K., Inubushi, T., and Chance, B., 1987, ^{1}H and ^{31}P-NMR studies on smooth muscle of bullfrog stomach, *Biochim. Biophys. Acta* **928:**36–44.

39. Degani, H., and Kaye, A. M., 1988, ^{13}C-NMR studies of glucose metabolism in rat uteri, Presented to the 6th Annual Meeting of the Society of Magnetic Resonance in Medicine, New York, p. 1008.

40. Dawson, M. J., and Wray, S., 1983, ^{31}P nuclear magnetic resonance studies of isolated rat uterus, *J. Physiol. (London)* **336:**19P–20P.

41. Degani, H., 1987, NMR spectroscopy studies of the reproductive organs and associated malignancies, in: *Magnetic Resonance of the Reproductive System* (F. Haseltine and S. McCarthy, eds.), Slack, Thorofare, N.J., pp. 81–96.

42. Kushmerick, M. J., Dillon, P. F., Meyer, R. A., Brown, T. R., Krisanda, J. M., and Sweeney, H. L., 1986, ^{31}P NMR spectroscopy, chemical analysis and free Mg^{2+} of rabbit bladder and uterine smooth muscle, *J. Biol. Chem.* **261:**14420–14429.

43. Degani, H., Shaer, A., Victor, T. A., and Kaye, A. M., 1984, Estrogen-induced changes in high-energy phosphate metabolism in rat uterus, ^{31}P NMR studies, *Biochemistry* **23:**2572–2577.

44. Noyszewski, E. A., Raman, J., Trupin, S. R., McFarlin, B. L., and Dawson, M. J., 1989, [31]Phosphorus nuclear magnetic resonance examination of female reproductive tissues, *Am. J. Obstet. Gynecol.* **161:**282–288.

45. Raman, J., Dawson, M. J., and Trupin, S. R., 1990, ^{31}P magnetic resonance spectroscopic evaluations of gynecologic diseases and pregnancy disorders, *Troph. Res.* in press.

46. Kay, H. H., Gordon, J. D., Ribeiro, A. A., and Spicer, L. D., 1990, ^{31}P nuclear magnetic resonance (NMR) spectroscopy of human placenta, *Troph. Res.* in press.

47. Malek, A., Miller, R. K., Mattison, D. R., Ceckler, T., Panigel, M., Bryant, R., and Neth, L., 1990, Continuous measurement of ATP using ^{31}P NMR spectroscopy in dually perfused human placenta, *Troph. Res.* in press.

48. Haseltine, F. P., Arias-Mendoza, F., Kaye, A. M., and Degani, H., 1986, ^{31}P NMR studies of

adenosine-stimulated ATP synthesis in perfused luteinized ovaries, *Magn. Reson. Med.* **3:**796–800.

49. Haseltine, F. P., 1987, Introduction, in: *Magnetic Resonance of the Reproductive System* (S. McCarthy and F. Haseltine, eds.), Slack, Thorofare, N.J., pp. ix–xiii.
50. Elliott, A. C., Roman, B. B., and Dawson, M. J., 1989, The temperature dependence of intracellular pH in isolated frog skeletal muscle, submitted.
51. Gupta, R. K., and Moore, R. D., 1980, ^{31}P NMR studies of intracellular free Mg^{2+} in intact frog skeletal muscle, *J. Biol. Chem.* **225:**3987–3993.
52. Dawson, M. J., 1983, Nuclear magnetic resonance, in: *Cardiac Metabolism* (A. J. Drake-Holland and M. I. M. Noble, eds.), Wiley, New York, pp. 309–337.
53. Alvarez-Leefmans, F. J., Gamino, S. M., Giraldez, F., and Gonzalez-Serratos, H., 1986, Intracellular free magnesium in frog skeletal muscle fibres measured with ion-selective micro-electrodes, *J. Physiol. (London)* **378:**461–483.
54. Gupta, R. K., and Gupta, P., 1987, ^{31}P NMR measurements of intracellular free magnesium in cells and organisms, in: *NMR Spectroscopy of Cells and Organisms*, (R. K. Gupta, ed.), CRC Press, Boca Raton, pp. 33–44.
55. Degani, H., Victor, R. A., and Kaye, A. M., 1988, Effects of 17-β estradiol on high energy phosphate concentrations and the flux catalysis by creatine kinase in immature rat uteri: ^{31}P nuclear magnetic resonance, *Endocrinology* **122:**1631–1638.
56. Hamoir, G., 1977, Biochemistry of the myometrium, in: *Biology of the Uterus* (R. Wynn, ed.), Plenum Press, New York, pp. 377–414.
57. Cretius, K., 1957, Der kreatin phosphat gehalt menschlicher skelett- und uterus muskulatur, *Z. Geburtshilfe Gynaekol.* **149:**113–122.
58. Walaas, O., and Walaas, E., 1950, The metabolism of uterine muscle studied with radioactive phosphorus P^{32}, *Acta Physiol. Scand.* **21:**18–26.
59. Dawson, M. J., and Wray, S., 1985, The effects of pregnancy and parturition on phosphorus metabolites in rat uterus studied by ^{31}P nuclear magnetic resonance, *J. Physiol. (London)* **368:**19–31.
60. Volfin, P., Clauser, H., and Gautheron, D., 1957, Influence of oestradiol and progesterone injections on the acid-soluble phosphate fractions of the rat uterus, *Biochim. Biophys. Acta* **24:**137–140.
61. Dawson, M. J., and Wilkie, D. R., 1985, Muscle and brain metabolism studied by ^{31}P nuclear magnetic resonance, in: *Recent Advances in Physiology* Volume 10 (P. F. Baker, ed.), Churchill Livingstone, Edinburgh, pp. 247–276.
62. Harkness, M. L. R., and Harkness, R. D., 1954, The collagen content of the reproductive tract of the rat during pregnancy and lactation, *J. Physiol. (London)* **123:**492–500.
63. Wray, S., 1988, Smooth muscle intracellular pH: Measurement, regulation and function, *Am. J. Physiol.* **254:**C213–C225.
64. Dawson, M. J., and Elliot, A. C., 1984, Intracellular pH of isolated frog muscle as a function of temperature: A ^{31}P nuclear magnetic resonance study, *J. Physiol. (London)* **360:**59P.
65. Wray, S., 1986, A ^{31}phosphorus nuclear magnetic resonance study of the regulation of intracellular pH following changes in extracellular pH in rat uterine smooth muscle, *J. Physiol. (London)* **373:**81P.
66. Widdowson, E. M., and Dickerson, J. W. T., 1964, Chemical composition of the body, in: *Mineral Metabolism*, Volume 2, Part A (C. L. Comar and F. Bronner, eds.), Academic Press, New York.
67. Pitkin, R. M., Reynolds, W. A., Williams, G. A., and Hargis, G. K., 1979, Calcium metabolism in normal pregnancy: A longitudinal study, *Am. J. Obstet. Gynecol.* **133:**781–790.
68. Law, R. O., 1982, Techniques and applications of extracellular space determination in mammalian tissues, *Experientia* **38:**411–421.
69. Moawad, A. H., and Daniel, E. E., 1971, Total contents and net movements of magnesium in the rat uterus, *Am. J. Physiol.* **220:**75–82.

70. Reynolds, S. R. M., 1965, Structural changes and uterine enlargement, in: *The Physiology of the Uterus,* Hafner, New York, pp. 235–244.

71. Downing, S. J., and Sherwood, O. D., 1985, The physiological role of relaxin in the pregnant rat. III. The influence of relaxin on cervical extensibility, *Endocrinology* **116:**1215–1220.

72. Raman, J., Noyszewski, E. A., Hwang, J. J., Sherwood, O. D., and Dawson, M. J., unpublished results.

73. Raman, J., Apte, D. V., Dawson, M. J., Bobowski, S. J., and Trupin, S. R., 1989, Biochemical evaluation of endometrial biopsies by magnetic resonance spectroscopy, *Am. J. Obstet. Gynecol.* (in press).

74. Edwards, R. H. T., Dawson, M. J., Wilkie, D. R., Gordon, R. E., and Shaw, D., 1982, Clinical use of nuclear magnetic resonance in the investigation of myopathy, *Lancet* **1:**725–731.

75. Allen, D. G., Morris, P. G., Orchard, C. H., and Pirolo, J. S., 1985, A nuclear magnetic resonance study of metabolism in the ferret heart during hypoxia and inhibition of glycolysis, *J. Physiol. (London)* **361:**185–204.

76. Dawson, M. J., Spurway, N. C., and Wray, S., 1985, A ^{31}P nuclear magnetic resonance (NMR) study of isolated rabbit arterial smooth muscle, *J. Physiol. (London)* **365:**72P.

77. Spurway, N. C., and Wray, S., 1987, A phosphorus nuclear magnetic resonance study of metabolites and intracellular pH in rabbit vascular smooth muscle, *J. Physiol. (London)* **393:**57–71.

78. Fischer, M. J., and Dillon, P. F., 1988, Direct determination of ADP in hypoxic porcine carotid artery using ^{31}P-NMR, *NMR Biomed.* **1:**1–6.

79. Carlson, F. D., and Wilkie, D. R., 1974, Chemical and energetic changes during contraction, in: *Muscle Physiology* (F. D. Carlson and D. R. Wilkie, eds.), Prentice–Hall, Englewood Cliffs, N.J., pp. 87–104.

80. Jacobus, W. E., and Lehninger, A. L., 1973, Creatine kinase of rat heart mitochondria, *J. Biol. Chem.* **248:**4803–4810.

81. McGilvery, R. W., and Murray, T. W., 1974, Calculated equilibria of phosphocreatine and adenosine phosphates during utilization of high energy phosphate by muscle, *J. Biol. Chem.* **249:**5845–5850.

82. Gudbjarnason, S., Mathes, P., and Ravens, K. G., 1970, Functional compartmentation of ATP and creatine phosphate in heart muscle, *J. Mol. Cell. Cardiol.* **1:**325–339.

83. Kingsley-Hickman, P. B., Sako, E. Y., Mohanakrishnan, P., Robitaille, P. M. L., From, A. H. L., Foker, J. E., and Ugurbil, K., 1987, ^{31}P NMR studies of ATP synthesis and hydrolysis kinetics in the intact myocardium, *Biochemistry* **26:**7501–7510.

84. Hseih, P. S., and Balaban, R. S., 1988, Saturation and inversion transfer studies of creatine kinase kinetics in rabbit skeletal muscle *in vivo, Magn. Reson. Med.* **7:**56–64.

85. Katzenellenbogen, B. S., and Leake, R., 1974, Distribution of the oestrogen induced protein and of total protein between endometrial and myometrial fractions of the immature rat uterus, *J. Endocrinol.* **63:**439–449.

86. Katzenellenbogen, B. S., and Gorski, J., 1975, Estrogen actions on synthesis of macromolecules in target cells, in: *Biochemical Actions of Hormones,* Volume 3, (G. Litwack, ed.), Academic Press, New York, pp. 187–243.

87. Shoubridge, E. A., Bland, J. L., and Radda, G. K., 1984, Regulation of creatine kinase during steady-state isometric twitch contraction in rat skeletal muscle, *Biochim. Biophys. Acta* **805:**72–78.

88. Bell, S. C., Halmer, J., and Heald, P. J., 1980, Induced protein and deciduoma formation in rat uterus, *Biol. Reprod.* **23:**935–940.

89. Oliver, J. M., and Kellie, A. E., 1970, The effects of oestradiol on the acid soluble nucleotides of rat uterus, *J. Biochem.* **119:**187–191.

90. Carsten, M. E., and Miller, J. D., 1987, A new look at uterine muscle contraction. *Am. J. Obstet. Gynecol.* **157:**1303–1315.

91. Ebashi, S., Mikawa, T., Kuwayama, H., Suzuki, M., Ikemoto, H., Ishizaki, Y., and Koga, R.,

1987, Ca^{2+} regulation in smooth muscle; dissociation of myosin light chain kinase activity from activation of actin–myosin interaction, in: *Regulation and Contraction of Smooth Muscle* (M. J. Siegman, A. P. Somlyo, and N. L. Stephens, eds.), Liss, New York, pp. 109–117.

92. Barsotti, R. J., Ikebe, M., and Hartshorne, D. J., 1987, Effects of Ca^{2+}, Mg^{2+} and myosin phosphorylation on skinned smooth muscle fibres, *Am. J. Physiol.* **252:**C543–C554.

93. Hibberd, M. G., Dantzig, J. A., Trentham, D. R., and Goldman, Y. E., 1985, Phosphate release and force generation in skeletal muscle fibres, *Science* **228:**1317–1319.

94. Nosek, T. M., Fender, K. Y., and Godt, R. E., 1987, It is diprotonated inorganic phosphate that depresses force in skinned skeletal muscle fibres, *Science* **236:**191–193.

95. Dawson, M. J., 1988, The relation between muscle contraction and metabolism: Studies by ^{31}P nuclear magnetic resonance spectroscopy, in: *Molecular Mechanism of Muscle Contraction* (H. Sugi and G. H. Pollack, eds.), Plenum Press, New York, pp. 433–448.

96. Itoh, T., Kanmura, Y., and Hirosi, K., 1986, Inorganic phosphate regulates the contraction–relaxation cycle in skinned muscles of the rabbit mesenteric artery, *J. Physiol.* (*London*) **376:**231–252.

97. Krisanda, J. M., and Paul, R. J., 1983, High energy phosphate and metabolite content during isometric contraction in porcine carotid artery, *Am. J. Physiol.* **244:**C385–C390.

98. Keatinge, W. R., and Harman, M. C. (eds.), 1980, Local regulation of blood vessels by chemical agents and by intravascular pressure and flow, in: *Local Mechanisms Controlling Blood Vessels*, Academic Press, New York, pp. 73–87.

99. Kontos, H. A., 1981, Regulation of the cerebral circulation, *Annu. Rev. Physiol.* **43:**397–407.

100. Warshaw, D. M., McBride, W. J., and Work, S. S., 1987, Corkscrew-like shortening in single smooth muscle cells, *Science* **236:**1457–1459.

101. Murphy, R. A., Ratz, P. H., and Hai, C. M., 1986, Determinants of the latch state in vascular smooth muscle, in: *Regulation and Contraction of Smooth Muscle* (M. J. Siegman, A. P., Somlyo, and N. L. Stephens, eds.), Liss, New York, pp. 411–413.

102. McIlwain, H., and Bachelard, H. S., 1985, Metabolic, ionic and electrical phenomena in separated cerebral tissues, in: *Biochemistry and the Central Nervous System* (H. McIlwain and H. S. Bachelard, eds.), Churchill Livingstone, Edinburgh, pp. 54–83.

103. Yochim, J. M., and Saldarini, R. J., 1969, Glucose utilization by the myometrium during early pseudopregnancy in the rat, *J. Reprod. Fertil.* **20:**481–489.

104. Luh, W., and Henkel, E., 1965, Gehalt und verteilung von phosphotransferase in meseschlicher skeletl-und uterus muskulatur, *Z. Geburtshilfe Gynaekol.* **163:**279–288.

105. Orlander, J., 1984, Enzyme activities of energy metabolism in skeletal muscle, and in the uterus, during late pregnancy in the mouse, *Acta Physiol. Scand.* **120:**305–310.

106. Geyer, H., Muller, V., and Afting, E.-G., 1977, Enzyme activities in different cell compartments of the involuting rat myometrium, *Eur. J. Biochem.* **79:**483–490.

107. Chew, C. S., and Rinard, G. A., 1979, Glycogen levels in the rat myometrium at the end of pregnancy and immediately postpartum, *Biol. Reprod.* **20:**1111–1114.

108. Paul, R. J., Bauer, M., and Pease, W., 1979, Vascular smooth muscle: Aerobic glycolysis linked to sodium and potassium transport processes, *Science* **206:**1414–1416.

109. Lynch, R. M., and Paul, R. J., 1983, Compartmentation of glycolytic and glycogenolytic metabolism in vascular smooth muscle, *Science* **222:**1344–1346.

110. Kroeger, E. A., 1977, Regulation of metabolism by cyclic adenosine 3′-5′-monophosphate and ion pumping in smooth muscle, in: *The Biochemistry of Smooth Muscle* (N. L. Stephens, ed.), University Park Press, Baltimore, pp. 315–327.

111. Rubanyi, G., Toth, A., and Kovach, A. G. B., 1982, Distinct effect of contraction and ion transport on NADH fluorescence and lactate production in uterine smooth muscle, *Acta Physiol. Acad. Sci. Hung.* **59:**45–58.

112. Hopkinson, L., and Kearly, M., 1955, The interaction of insulin, oestrone and progesterone on the metabolism of isolated rat uterus, *J. Physiol.* (*London*) **128:**113–121.

113. Lundholm, L., Pettersson, G., Andersson, R. G. G., and Mohme-Lundholm, E., 1983, Regula-

tion of the carbohydrate metabolism in smooth muscle; some current problems, in: *Biochemistry of Smooth Muscle,* 2nd ed. (N. L. Stephens, ed.), CRC Press, Boca Raton, pp. 85–108.

114. Bulbring, E., 1953, Measurements of oxygen consumption in smooth muscle, *J. Physiol. (London)* **122:**111–134.

115. Stephens, N. L., and Wrogemann, K., 1970, Oxidative phosphorylation in smooth muscle, *Am. J. Physiol.* **219:**1796–1801.

116. Chance, B., Leigh, J. S., Kent, J., McCully, K., Nioka, S., Clark, B. J., Maris, N. M., and Graham, T., 1986, Multiple controls of oxidative metabolism in living tissues as studied by phosphorus magnetic resonance, *Proc. Natl. Acad. Sci. USA* **83:**9458–9462.

3

Myosin Light Chain Phosphorylation in Uterine Smooth Muscle

Kate Bárány and Michael Bárány

1. Introduction

The uterus is a unique organ capable of increasing its size enormously during pregnancy. The strong contractions of the myometrium bring about the expulsion of the fetus. An understanding of uterine contractility has been the goal of many workers in the last half century.[1] Only in this decade has it been recognized that phosphorylation of a small subunit in the head part of the myosin molecule plays a key role in uterine contractility.[2−5]

In this review, we summarize work from our laboratory and others on myosin light chain phosphorylation–dephosphorylation in relation to contraction–relaxation in the uterus. A broader description of the biochemistry and physiology of the uterus may be found in other chapters of this book.

2. Myosin

2.1. Historical Notes

Frog skeletal myosin was first observed by Kühne in 1859,[6] and his "final" method of myosin preparation was described in his book of 1864.[7] In spite of the extensive work done on the characterization of myosin from various muscles, it was not until 1938 that Mehl extracted the first smooth muscle myosin from beef intestine.[8] Actually, the isolated myosin preparations of the initial workers was an

Kate Bárány • Department of Physiology and Biophysics, College of Medicine, University of Illinois at Chicago, Chicago, Illinois 60612.   *Michael Bárány* • Department of Biological Chemistry, College of Medicine, University of Illinois at Chicago, Chicago, Illinois 60612.

actomyosin, containing variable amounts of actin depending on the time of extraction with a concentrated salt solution.

Functional studies of human uterus actomyosin began with the work of Csapo.[9] One of us (M. B.), who shared a laboratory with Csapo, witnessed his experiments in 1946. Csapo arrived in the late afternoon from the gynecological clinic carrying a surgically dissected frozen human uterus. The tissue was ground in a porcelain mortar with sand in the presence of 3 volumes of 0.5 M KCl containing ATP, and left to stand at 4°C. During the next evening the suspension was centrifuged for several hours in the cold room at 600g, the greatest centrifugal force available in the Nobel Prize winner Albert Szent-Györgyi's Institute. The supernatant was analyzed for actomyosin and myosin by a viscometric method, and it was shown that during pregnancy the actomyosin content of the uterus increases.

Muscle biochemists did not pay much attention to smooth muscle myosin because of its low ATPase activity. However, in 1959 Needham and Williams reported that tryptic treatment increased the Ca^{2+}-ATPase activity of uterine actomyosin 4- to 8-fold,[10] and subsequently these authors noted a 3.5-fold increase in this ATPase of uterine myosin.[11] We followed up this phenomenon using chicken gizzard myosin that was homogeneous in the analytical ultracentrifuge.[12] In addition to trypsin, low concentrations of urea, guanidine·HCl, or ethylene glycol activated gizzard myosin ATPase. The fact that structurally dissimilar agents could activate this ATPase suggested that activation was caused by a conformational change in myosin. This was confirmed in experiments that demonstrated increased accessibility of sulfhydryl groups of iodoacetate in the urea-modified enzyme; 12 cysteine residues per mole of gizzard myosin reacted with iodoacetate in the presence of 1.5 M urea in contrast to the nonreactivity of the sulfhydryl groups in the absence of urea. Urea, guanidine·HCl, or ethylene glycol also activated uterine myosin; the maximal activation (5- to 7-fold) was observed with 2.5 M urea.[12]

From 1953 on, several investigators reported that myosin contains low-molecular-weight, noncovalently bound protein components.[13−16] Initially, these components were viewed as contaminants and a decade passed till these components, the light chains, were established as essential subunits of myosin.[17−20] Perry and collaborators found that one type of light chain, called P-light chain, can be phosphorylated (reviewed by Perry[21]). The discovery that phosphorylation of the 20-kDa light chain of smooth muscle myosin increases its actin- and Mg^{2+}-activated ATPase activity[22−24] and that this phosphorylation is involved in smooth muscle contraction[25,26] opened a new chapter in smooth muscle biochemistry.

2.2. Structure and Function

Smooth muscle myosin is a large molecule with a length of approximately 160 nm comprising a long tail and two heads (Fig. 1). It contains two heavy chains, each with a molecular weight of approximately 200 kDa. From the tail end, the two chains are wound around each other. The chains then separate to form the two heads. Limited proteolysis of myosin separates the heads from the tail; the head is

TAIL HEAD

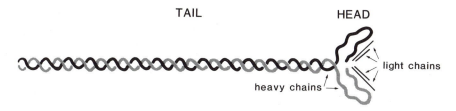

Figure 1. Schematic representation of smooth muscle myosin molecule.

called myosin subfragment 1. Each head contains a 20-kDa and a 17-kDa light chain. Thus, the molecular weight of myosin is almost a half million.

In the smooth muscle cell the tails form the backbone of the thick filaments from which the heads protrude to build the cross-bridges. The thick filaments in smooth muscle are longer than those in skeletal muscle (2.2 versus 1.6 μm).[27]

The head of myosin carries all of the biological functions. The ATPase and actin-binding sites of myosin are located in the heavy chain part of the head. The two light chains are attached by noncovalent forces to the heavy chain in each head. The 20-kDa light chain can be phosphorylated and dephosphorylated by specific enzymes. Phosphorylation of the 20-kDa light chain changes the conformation of the myosin molecule into an elongated form that favors filament formation, whereas dephosphorylation of this light chain results in a folded myosin.[28,29] The role of the 17-kDa light chain in myosin is unknown. (For review, see Bailin.[30])

2.3. Isolation of Myosin

Myosin from smooth muscle has been prepared employing techniques modified from those used for isolation of myosin from skeletal muscle. The most common problem is proteolysis, which is minimized by the addition of several protease inhibitors. In our laboratories, the extraction solutions contain 0.15 mM phenylmethylsulfonyl fluoride, 1 mg/liter leupeptin, 2 mg/liter trypsin inhibitor, and 70 μM streptomycin. Another problem is to remove the large quantities of tropomyosin and actin; this may be solved by column chromatographic procedures.[31] Column-purified human uterine myosin was prepared by Cavaille and collaborators.[32]

2.4. Myosin–Actin Interaction

Uterine myosin readily combines with actin at both high and low ionic strength, and the actin-binding ability of uterine myosin is approximately the same as that of skeletal muscle myosin.[33] The actin-binding ability of myosin is the *in vitro* correlate of the *in vivo* force production by the muscle. The maximum force developed per cross-sectional area of smooth muscle approximates that of skeletal muscle. Such a high force generation would not be expected, because smooth muscle contains only about one-fourth of the myosin present in skeletal muscle.[34] However, the mass ratio of actin to myosin is about three times higher in smooth muscle than in skeletal muscle, and this correlates with the higher ratio of actin filaments to myosin filaments in smooth muscle (13–15 to 1) as measured by electron microscopy[27,35] (see Chapter 1). An additional factor in the force generation may be the length of the thick filament, which is longer in smooth muscle than in skeletal muscle.[27]

Uterine myosin, as any other myosin, hydrolyzes ATP in the presence of high $[Ca^{2+}]$ (10 mM), or in the absence of divalent cations at high $[K^+]$ (0.5–1.0 M). However, these ATPase activities of smooth muscle myosins are much lower than those of skeletal muscle myosin.

Purified smooth muscle myosin, like skeletal muscle myosin, hydrolyzes MgATP very slowly in the absence of actin. However, unlike that of skeletal muscle myosin, the Mg^{2+}-ATPase activity of smooth muscle myosin is not activated by actin unless its 20-kDa light chain is phosphorylated (reviewed by Small and Sobieszek[35]). Even so, the specific actin- and Mg^{2+}-activated ATPase activity of phosphorylated smooth muscle myosin[36] remains well below that of its skeletal muscle myosin counterpart, 20–50 versus 500–1000 nmoles ATP/mg actomyosin per min at 25°C.[33] It should be added that actin in smooth muscle thin filaments binds both tropomyosin and caldesmon, and these regulatory proteins may be involved in modulating the Mg^{2+}-ATPase activity of phosphorylated myosin *in vivo*[37] (see Chapter 4).

Dillon and collaborators[26] analyzed the interaction of myosin with actin during smooth muscle contraction. Light chain phosphorylation was correlated with V_0, the intrinsic shortening velocity at zero external load; thus, the actin- and Mg^{2+}-activated ATPase activity of the phosphorylated myosin determines the rate of cross-bridge cycling. However, on continued stimulation the phosphate content of light chain substantially decreased, while force (reflecting the number of attached cross-bridges) was maintained. This mechanical state, termed *latch,* corresponds to the noncycling attached cross-bridge (latch-bridge), which is responsible for the high economy of force maintenance in smooth muscle (reviewed by Aksoy and Murphy[38]). We have also observed force maintenance in drug-challenged uterine strips while light chain phosphorylation decreased (Fig. 6).

3. Phosphorylation–Dephosphorylation

Phosphorylation and dephosphorylation of proteins by enzymes, the protein kinases and the protein phosphatases, is one of the major mechanisms for regulating cellular functions. Smooth muscle myosin light chain kinase (MLCK), activated by Ca^{2+}·calmodulin complex, transfers the terminal phosphate group of ATP to serine (and/or to threonine) hydroxyl groups of light chains (LC) according to the following reaction:

$$\text{LC-OH} + \text{MgATP}^{2-} \xrightarrow{\text{MLCK}} \text{LC-O-PO}_3^{2-} + \text{MgADP}^- + \text{H}^+ \qquad (1)$$

Dephosphorylation is brought about by the hydrolytic enzyme smooth muscle myosin light chain phosphatase (MLCP) according to the following reaction:

$$\text{LC-O-PO}_3^{2-} + \text{H}_2\text{O} \xrightarrow{\text{MLCP}} \text{LC-OH} + \text{HPO}_4^{2-} \qquad (2)$$

Highly purified MLCK has been prepared from avian gizzard and characterized extensively (reviewed by Edelman *et al.*[39]). Homogeneous MLCK was also isolated from porcine myometrium.[40] The molecular mass of the enzyme, 130 kDa, is in the same range as that of MLCK from chicken and turkey gizzard. Porcine myometrium contains 111 mg MLCK/kg, 0.85 μM, a concentration comparable to that in turkey gizzard, 1.2 μM.

Porcine myometrium MLCK has the following K_m values: 20 μM for light chain, 23 μM for ATP, 0.3 μM for Ca^{2+}, and 0.6 nM for calmodulin; its V_{max} is 16.1 μmoles/mg per min. This is about 22 times more than the estimated myosin content of myometrium, 20 g/kg.[41]

Porcine myometrium MLCK may be phosphorylated by the catalytic subunit of cAMP-dependent protein kinase, resulting in a four- to seven-fold decrease in the affinity of the enzyme to calmodulin but without affecting the V_{max}.[40] This type of phosphorylation was described first by Adelstein and Hathaway[42] for turkey gizzard MLCK. Phosphorylation of MLCK may be involved in the mechanism of relaxation of the myometrium by β-adrenergic stimulants, which increase the cAMP content of the uterus, as shown by Polacek and Daniel.[43] However, these authors pointed out that the level of cAMP is not the primary determinant of uterine contractility.

The importance of MLCK in the function of myometrium is further evidenced by the finding that the enzyme is present in cultures of human myometrial cells.[44] Moreover, the specific activity of MLCK in homogenates of the cultured cells was similar to that of normal myometrial tissue.

Protein kinase C is another serine/threonine phosphorylating enzyme that may be involved in smooth muscle myosin light chain phosphorylation. In the presence of phospholipids, Ca^{2+} activates protein kinase C; the enzymatic activity is further enhanced by diacylglycerol. Phorbol esters and other tumor promoters with structural similarities to diacylglycerol activate protein kinase C (see Chapter 5). Protein kinase C-dependent phosphorylation of smooth muscle myosin[45] and smooth muscle heavy meromyosin[46] has also been reported. In rat uterus, phorbol esters block contractions produced by neurotransmitters (oxotremorine, bradykinin, serotonin), but do not alter contractions produced by potassium depolarization.[47] In rat thoracic aorta and rabbit iris sphincter smooth muscle, phorbol esters induce contraction.[48,49] The opposite effects of phorbol esters on the contractility of various smooth muscles are not understood.

Only limited information is available about the properties of MLCP (reviewed by Kamm and Stull[50]). Enzymes that preferentially dephosphorylate light chain have been isolated from bovine aorta[51,52] and turkey gizzard.[53]

4. Isoforms of the 20-kDa Light Chain

High-resolution two-dimensional gel electrophoresis of smooth muscle proteins resolved the light chain into four spots and ^{32}P-labeling of the muscles revealed that three of the four spots contain phosphorylated components.[54–56] Figure 2 shows the gel electrophoretic analysis of light chain from rat uteri frozen in various functional states.[57] The first row shows the Coomassie blue-stained protein spots, with the light chain spots being indicated. Four light chain spots are visible in uterine strips contracted with oxytocin (frame 1), relaxed spontaneously (frame 2), or relaxed with isoproterenol (frame 3). However, in the uterine strip treated with EGTA for a prolonged time, only three light chain spots appear (frame 4); this is caused by the depletion of the intracellular Ca^{2+}, the activator of MLCK. The second row shows the corresponding autoradiograms; three phosphorylated spots appear in frames 1, 2, and 3, but no phosphorylation is apparent in frame 4.

The nature of the four light chain spots has been debated. It has been suggested that they are contaminants[58] or that they result from artifactual charge modification,[54,59] but experiments that contradict these suggestions have also been described.[60–62] We have investigated the possibility that the four uterine light chain spots are due either to contamination or to artifactual charge modification during two-dimensional gel electrophoresis. The results are shown in Fig. 3. Purified uterine light chain exhibits four spots on the Coomassie blue-stained gels (top panel of Fig. 3) just as there are four light chain spots on the stained gel representing the total uterine proteins (frames 1–3 of top row of Fig. 2). The purified uterine light chain spots were transferred by electroblotting to a nitrocellulose membrane and incubated with an antiserum against purified light chain. This immunoblot shows that all the four light chain spots reacted with the antibody (middle panel of Fig. 3). This provides evidence that the spots do not represent contaminants. Artifactual charge modification may be detected by two-dimensional isoelectric focusing through the appearance of off-diagonal spots. The purified light chain was isoelectrofocused in both first and second dimensions (bottom panel of Fig. 3); the absence of off-diagonal spots excludes the possibility that any of the spots were produced by artifactual charge modification.

Uterine light chain contains two kinds of phosphoamino acids, phosphoserine (Ser-P) and phosphothreonine (Thr-P), as shown in Fig. 4. Electrophoresis of total light chain hydrolysates at pH 3.5 revealed the presence of Ser-P and Thr-P but no tyrosine-P (not shown). Electrophoresis at pH 1.9, which separates Ser-P from Thr-P, showed that not only the total light chain, but also each radioactive light chain spot contained both phosphoamino acids (Fig. 4).

Although each individual light chain spot contained both Ser-P and Thr-P, the molar content of the combined phosphorylated amino acids in the individual light chain spots varied greatly (Table I). Thus, in untreated muscles or in muscles treated with relaxing agents, the incorporation varied from 1.16 to 2.03 mole [^{32}P]phosphate per mole light chain in spot 1, from 0.23 to 0.41 in spot 2, from 0.40 to 0.59 in spot 3, and from 0.01 to 0.04 in spot 4. In muscles contracted with carbachol or oxytocin, the incorporation was 2.21–2.43 in spot 1, 1.19–1.66 in spot 2, 0.89–0.93 in spot 3, and 0.03–0.05 in spot 4. These data indicate that spot 1 contains diphosphorylated light chain, spot 3 contains monophosphorylated light chain, and spot 4 contains nonphosphorylated light chain. The phosphate content of light chain in spot 2 suggests that this spot contains a mixture of diphosphorylated, monophosphorylated, and nonphosphorylated light chain.

An important point is the analysis of the stain distribution in light chain that is not phosphorylated. This is shown in frame 4 of Fig. 2: three spots are visible with the major one corresponding to spot 4 and the two minor ones to spots 3 and 2. This finding is in agreement with the data of Table I, which suggests that in relaxed muscles both spots 2 and 3, in addition to spot 4, contain considerable amounts of nonphosphorylated light chain. The three nonphosphorylated light chain in spots 2, 3, and 4, which have the same molecular weight but different isoelectric points, represent light chain isoforms. Subtle differences in their amino acid compositions result in different net charges and consequently in different isoelectric points.

Based on these data and theoretical considerations,[56] the following scheme has been suggested for the composition of the individual light chain spots:

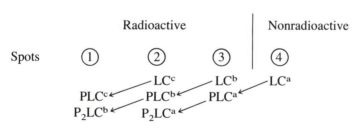

where LC^a, LC^b, and LC^c are the nonphosphorylated light chain isoforms, PLC is monophosphorylated light chain, and P$_2$LC is diphosphorylated light chain (reproduced from Ref. 56 with the permission of the publisher).

The polymorphism of myosin is well established. The various forms may reflect different biological activities. Changes in myosin may accompany developmental and functional adaptation of muscle (see the book by Pette,[63] and the review by Swynghedauw[64]). It has been shown by two-dimensional gel electrophoresis that the 17-kDa myosin light chain of human and monkey uterus exist in two isoelectric forms, the more acidic one becoming progressively predominant at the end of pregnancy.[32] It remains to be explored whether changes in the 20-kDa light chain, involved in the contractile activity of the muscle, also take place during pregnancy.

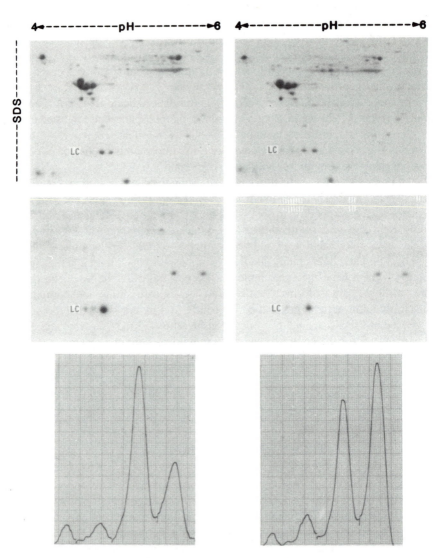

Figure 2. Two-dimensional gel electrophoretic analysis of myosin light chain phosphorylation in uterine smooth muscle. Top row: gel staining profiles of the uterine proteins; middle row: corresponding autoradiograms; bottom row: densitometric scans of the 20-kDa light chain on the stained gels. First frame: uterine strip contracted with oxytocin for 1 min; second frame: spontaneously relaxed uterine strip; third frame: uterine strip relaxed with isoproterenol for 1 min; fourth frame: uterine strip incubated with 1 mM EGTA for 120 min. LC, 20-kDa light chain. For details, see Ref. 57. (Reproduced from ref. 57 with the permission of the publisher.)

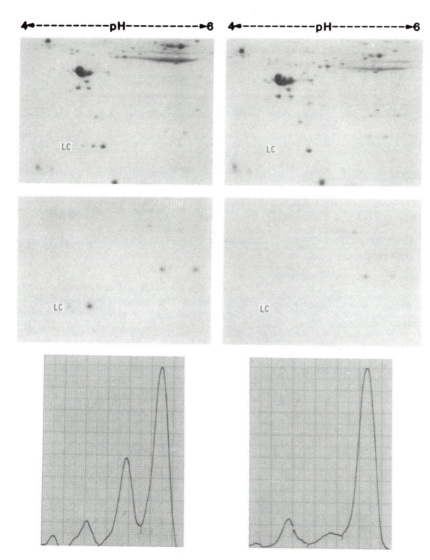

Figure 2. (*Continued*)

4.5 - - - - -pH- - - - -5.2

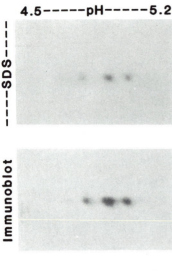

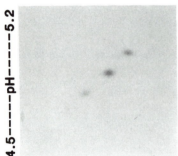

Figure 3. Electrophoretic and immunological analysis of the purified 20-kDa myosin light chain. Top panel: Coomassie blue-stained electrophoretogram, isoelectrofocused in the first dimension and electrophoresed on SDS polyacrylamide gel in the second dimension. Middle panel: immunoblot of the four spots from two-dimensional gel. Bottom panel: Coomassie blue-stained electrophoretogram, isoelectrofocused in both first and second dimensions. For details see ref. 62. (Reproduced from ref. 62 with the permission of the publisher.)

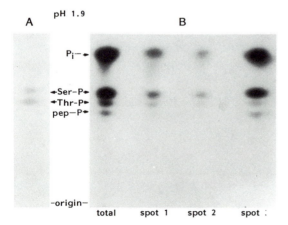

Figure 4. Identification of phosphoamino acid residues in the 20-kDa myosin light chain. ^{32}P-labeled uterine strips frozen at the relaxed state were analyzed at pH 1.9. (A) The ninhydrin stain patterns of the unlabeled marker phosphoamino acids; (B) the autoradiograms obtained with the hydrolysates of the total light chain, and each individual radioactive spot. pep-P, phosphopeptide. For details see ref. 57. (Reproduced from ref. 57 with the permission of the publisher.)

Table I. The Incorporation of [^{32}P]phosphate into the 20-kDa Myosin Light Chain in Uterine Muscles[a]

Treatment of muscle (1 min)	Moles [^{32}P]phosphate/mole light chain in spots				
	1	2	3	4	n
None	1.37	0.38	0.55	0.04	3
Relaxing agents					
Isoproterenol	1.85	0.23	0.49	0.01	4
MIX	1.16	0.26	0.40	0.03	2
EGTA	2.03	0.41	0.59	0.02	2
Contractile agents					
Carbachol	2.43	1.66	0.93	0.03	6
Oxytocin	2.21	1.19	0.89	0.05	5

[a] The light chain spots were separated by two-dimensional gel electrophoresis. The incorporation of [^{32}P]phosphate was calculated from the percentage stain distribution of the same spots from which the radioactivity was determined, from the known amount of protein applied to the gels, and from the specific radioactivity of [^{32}P]phosphocreatine of the treated uterine muscles. For details see Bárány *et al.*[56] (Reproduced from Ref. 56 with the permission of the publisher.)

5. Quantitation of Light Chain Phosphorylation

Since light chain may be di-, mono-, and nonphosphorylated, the actual level of light chain phosphorylation, expressed as moles of phosphate per mole of light chain, requires radioactivity measurements to be determined with microgram amounts of protein. In our laboratory two uterine strips are incubated at 37°C in 50 ml physiological salt solution containing 1.5–2.0 mCi carrier-free [^{32}P]orthophosphate for 1 hr.[65] The strips are washed with phosphate-free physiological salt solution 15 times to remove any ^{32}P from the extracellular space of the muscle. After a challenge with various agents, the strips are frozen, powdered, extracted with 3% perchloric acid, and the suspension is centrifuged. The specific radioactivity of phosphocreatine, which is equal to that of the γ-[^{32}P]phosphate of ATP,[5] is determined from the supernatant. The incorporation of [^{32}P]phosphate into light chain is determined from the residue, after dissolution in SDS and separation of light chain by two-dimensional gel electrophoresis. The phosphate content of light chain is calculated from the counts measured in light chain spots excised from the gel, from the myosin content of the protein applied to the gel, and from the specific radioactivity of [^{32}P]phosphate in phosphocreatine.[5]

Hathaway and Haeberle[56] developed a radioimmunoblotting method for measuring light chain phosphorylation in intact uterine muscle, based on labeling the light chain–antibody complex with [^{125}I]-protein A. The radioimmunoassay method was linear over a light chain protein range of 0.1–0.5 μg. The sensitivity of the light chain detection could be enhanced by using [^{125}I]-IgG Fab fragments allowing detection of 20 ng light chain.[66]

To circumvent the use of radioactivity, we determined the relationship between the actual and the apparent phosphorylation as measured by the two-dimensional gel electrophoretic technique.[56] The percentage distribution of the staining intensities of the four light chain spots on two-dimensional gel electrophoretogram is measured densitometrically. The sum of the fractional staining intensities in the spots exhibiting radioactivity (spots 1, 2, and 3) gives the apparent phosphorylation. The relationship between the actual phosphorylation (P) and the apparent phosphorylation (S) was determined by plotting the data pairs and fitting a straight line according to the least squares method:

$$P = (S - 0.29)/0.52$$

Relative staining intensities of phosphorylated and unphosphorylated light chain spots, separated by two-dimensional gel electrophoresis or isoelectric focusing, were measured in several laboratories and used for quantitation of light chain phosphorylation (see reviews of Aksoy and Murphy[38] and Kamm and Stull[50]). The sensitivity of light chain detection was increased by using protein silver stain instead of Coomassie blue.[59,67]

6. Light Chain Phosphorylation in Uterine Smooth Muscle

6.1. Spontaneous Activity

The uterus is a complex organ that is under myogenic, neural, and hormonal control. The innervation is severed when the uterus is removed from the body. To circumvent the large variations of hormonal effects, estrogen-primed rats were utilized in the following studies.

The horn consisting of myometrium and endometrium is the functional unit for uterine activity. This system was characterized by measuring its mechanical properties in physiological salt solution. Figure 5 shows the length–force relationships between length and spontaneous force (A), and length and passive force (B). The curve is asymmetric for the spontaneous force; a half-maximal force ($0.5F_0$) was attained at $0.65L_0$ (L_0, resting length) on the ascending limb and at $1.75L_0$ on the descending limb. Extrapolating the data from the descending limb suggests that stretching the horn to $2.5L_0$ would prevent spontaneous force development. As the length, starting from the slack length, was increased, the passive force increased first slowly, then rapidly.

Janis and collaborators[4] were the first to show that the spontaneous activity of the uterus was accompanied by cyclic phosphorylation and dephosphorylation of light chain. Subsequent work[5,56] quantitated the phosphate content of light chain. During spontaneous force development, phosphorylation increased from 0.35 mole phosphate/mole light chain to 0.8 mole/mole, while during spontaneous relaxation the phosphate content decreased to 0.35 mole/mole.

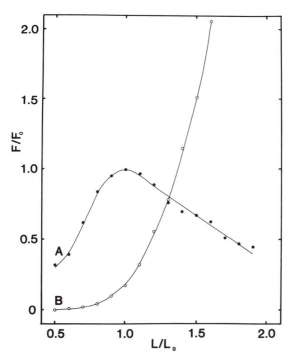

Figure 5. Length–force relationship in rat uterus. Measurements were performed in a physiological salt solution containing in mM: 118 NaCl, 5.7 KCl, 1.2 MgCl$_2$, 2.5 CaCl$_2$, 0.03 CaNa$_2$EDTA, 11 glucose, 12.5 NaHCO$_3$, while bubbled with 95% O$_2$–5% CO$_2$ at 37°C, pH 7.4. Abscissa: length values normalized as a fraction of L_0, at which length maximal spontaneous force was attained. Ordinate: force values normalized as a fraction of F_0, the maximal spontaneous force. (●) Length–spontaneous force relationship (A); (○) length–passive force relationship (B). For details see ref. 56. (Reproduced from ref. 56 with the permission of the publisher.)

6.2. Drug-Induced Contraction

Smooth muscle contraction is initiated by an increased [Ca^{2+}] in the sarcoplasm which may result from the entry of extracellular Ca^{2+} and/or the release of Ca^{2+} from internal sources. Drugs may elicit changes in intracellular [Ca^{2+}] by promoting/inhibiting Ca^{2+} influx/efflux through different pathways. We studied light chain phosphorylation during carbachol-, oxytocin-, and K$^+$-induced contractions (Fig. 6).[68] These three different stimulations immediately increased the force to 100–110% of the maximal spontaneous force, F_0. However, in the subsequent 30 min the force developments showed different patterns (top row of Fig. 6). The tonic component of the carbachol-induced contraction gradually decreased and the rhythmic activity returned usually with a higher frequency and lower amplitude. The tonic component of the oxytocin-induced contraction remained at $0.7F_0$ after 30 min. The amplitude of the superimposed rhythmic activity was approximately

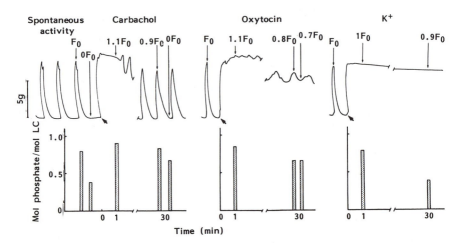

Figure 6. Effect of drugs on isometric force and myosin light chain phosphorylation in rat uterus. First row: representative force records of spontaneously active, carbachol-, oxytocin-, or K^+-challenged uteri. F_0 is the maximal force of the spontaneously active uterine strip. Second row: corresponding light chain phosphorylation. For details see ref. 68. (Reproduced from ref. 68 with the permission of the publisher.)

$0.1F_0$. When the uterus was depolarized with 100 mM K^+, the force was maintained at $0.9F_0$ for 30 min and no rhythmic activity was observed.

The bottom row of Fig. 6 shows the level of light chain phosphorylation 1 and 30 min after the addition of the stimulants. In all cases, the initial force developments were accompanied by high-level phosphorylation, 0.8–1.0 mole phosphate/mole light chain. However, after 30 min the responses were different: In carbachol-treated uteri 0.83 mole phosphate/mole light chain was measured in the contracted state and 0.67 mole/mole in the relaxed state, which is much higher than the 0.35 mole/mole in the untreated relaxed muscle. In oxytocin-contracted uteri no significant difference was found in the phosphorylation between uterine strips frozen at the maximum $(0.8F_0)$ or at the minimum force values $(0.7F_0)$ of their superimposed activities; the phosphorylation was 0.66 mole phosphate/mole light chain in both cases. In K^+-challenged uterine horns, the phosphorylation decreased to 0.39 mole phosphate/mole light chain in 30 min while isometric force was essentially maintained. These data indicate that the initial phase of contractile activity is related to light chain phosphorylation, but in the subsequent events no direct relationship was established between the extent of force and light chain phosphorylation. Hence, phosphorylated light chain is not necessary for the maintenance of force in uterine muscle. Similar results were obtained for arterial,[54,55,69,70] tracheal,[71,72] and esophageal smooth muscles.[73]

Haeberle and collaborators[74] have studied the regulation of isometric force maintenance and isotonic shortening velocity by light chain phosphorylation in K^+-depolarized rat uterine muscle in the presence of 10 mM extracellular Ca^{2+}. Light chain phosphorylation peaked at the initial state of contraction (0.46 mole phosphate/mole light chain) and then declined to a steady-state value (0.28 mole phosphate/mole light chain). From data obtained by varying the extracellular $[Ca^{2+}]$, the authors concluded that in addition to light chain phosphorylation, other factors are also involved in the regulation of uterine smooth muscle contraction.

6.3. Drug-Induced Relaxation

It is generally believed that a decrease in intracellular $[Ca^{2+}]$ or an increase in intracellular $[cAMP]$ is coupled with relaxation. Ca^{2+} decrease in the sarcoplasm may be accomplished by inhibiting the Ca^{2+} influx by Ca^{2+}-channel blockers such as verapamil or D600, or by stimulating the Ca^{2+} efflux with isoproterenol.[75] The cellular cAMP level depends on the relative activities of adenylate cyclase, which synthesizes cAMP, and phosphodiesterase, which hydrolyzes cAMP. Adenylate cyclase is activated by β-adrenergic agonists, such as isoproterenol, whereas phosphodiesterase is inhibited by papaverine or theophylline.[76]

The spontaneous contractile activity of uterus immediately stops after addition of isoproterenol, 3-isobutyl-1-methylxanthine (MIX), verapamil, or EGTA.[56] Light chain phosphorylation in a spontaneously relaxed muscle decreases from 0.35 mole phosphate/mole light chain to 0.2 mole/mole or less upon addition of isoproterenol, MIX, verapamil, or EGTA.

The effects of isoproterenol, MIX, verapamil, and EGTA have been studied on carbachol-, oxytocin-, and K^+-induced contractions of uterus. Figure 7 shows representative polygraph records: to spontaneously active horns, a contraction-inducing agent was added for 1 min, then the bath was exchanged for a solution containing the same contraction-inducing agent and one of the relaxing agents. The top panel of Fig. 7 shows that addition of isoproterenol or MIX to carbachol-contracted uteri decreased the tonic component of the contraction while spontaneous activity was partially regained, whereas addition of verapamil or EGTA brought the force down to zero. The middle panel shows that addition of isoproterenol or MIX to oxytocin-contracted uteri decreased the tonic component of the contraction while spontaneous activity was superimposed, whereas addition of verapamil or EGTA brought the force down to zero. The lower panel shows that addition of any relaxing agent to K^+-challenged uteri reduced the force, though most of the time approximately $0.3F_0$ was sustained for 10 min after the addition of isoproterenol, MIX, or EGTA. No spontaneous activity was noted in the presence of K^+, with or without muscle relaxants.

Figure 8 shows the light chain phosphorylation values determined in parallel experiments. All relaxing agents induced a time-dependent light chain dephosphorylation. The effect was most pronounced with EGTA, which after 10 min reduced the phosphate content from 0.8–0.9 mole phosphate/mole light chain to 0.15–0.25 mole/mole. The extent of light chain dephosphorylation in the presence of verapamil was similar to that of EGTA, whereas with isoproterenol and MIX the phosphate content of light chain was reduced to 0.3–0.6 mole/mole.

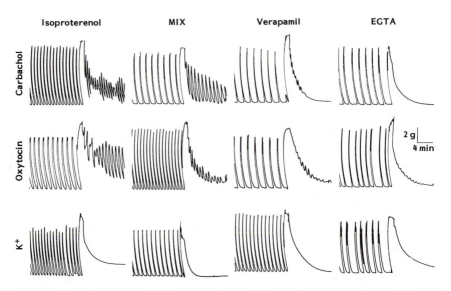

Figure 7. Effect of isoproterenol, MIX, verapamil, or EGTA on carbachol-, oxytocin-, or K^+-induced contractions of uteri. To spontaneously active horns, the contraction-inducing agents were added for 1 min, then the bath was exchanged for solutions containing the same contraction-inducing agent and isoproterenol (10^{-6} M), MIX (10^{-4} M), verapamil (2×10^{-4} M), or EGTA (10^{-3} M in Ca^{2+}-free physiological salt solution), respectively. For details see ref. 56. (Reproduced from ref. 56 with the permission of the publisher.)

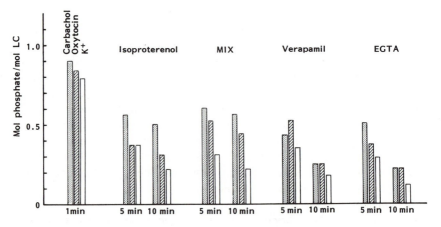

Figure 8. Level of myosin light chain phosphorylation after addition of isoproterenol, MIX, verapamil, or EGTA to carbachol-, oxytocin-, or K^+-contracted uteri. Level of light chain phosphorylation was determined at 1 min after a contraction-inducing agent was added, at 5 and 10 min after a relaxing agent was added in the presence of the contraction-inducing agent when the force was at its lowest level. Corresponding polygraph records are shown in Fig. 7. LC, 20-kDa light chain. For details see ref. 56. (Reproduced from ref. 56 with the permission of the publisher.)

Relaxation of uterus may be brought about by the hormone relaxin.[77] Nishikori and collaborators showed that relaxin suppresses rat uterine contractile activity by decreasing light chain phosphorylation. Furthermore, they found that relaxin decreases MLCK activity by decreasing the affinity of the enzyme to myosin, Ca^{2+}, and calmodulin.

Following the observation of Adelstein and Hathaway[42] that phosphorylation of MLCK by cAMP-dependent protein kinase decreases its activity, it is generally accepted that agents that increase intracellular cAMP relax smooth muscles, including uterus, by inhibiting MLCK. However, recently, Dokhac and collaborators[58] showed that the prostaglandin E_2-induced increase in cAMP results in an increase in light chain phosphorylation and an accompanying increase in contraction of myometrial strips from estrogen-primed rat uterus. This is an example of an increase in cAMP that is not coupled with relaxation, and it confirms Polacek and Daniel's prediction[43] that besides cAMP, other factors are involved in regulating the contractile state (see Chapter 8).

The enzyme that dephosphorylates light chain in muscle is MLCP. To study the effect of phosphatases on relaxation of uterine muscle, Haeberle and collaborators[78] skinned the uterus chemically by glycerine treatment. This removed the muscle membrane and consequently the soluble components of muscle while the contractile system was left intact. Addition of a purified phosphatase to the bath of the contracted muscle containing saturating Ca^{2+} and calmodulin produced complete relaxation of the muscle. The phosphatase-induced relaxation could be reversed by adding to the bath purified chicken gizzard MLCK. Changes in steady-state isometric force were associated with parallel changes in light chain phosphorylation over a range of phosphorylation extending from 0.01 to 0.97 mole phosphate/mole light chain. The authors suggest that the contraction–relaxation cycle of glycerinated uterine muscle is correlated with light chain phosphorylation–dephosphorylation.

6.4. Stretch Activation

Smooth muscle has the unique property that it may be stretched considerably as the internal volume of the organ it surrounds is increased. The uterus is exposed to chronic stretch during pregnancy and stretching may play a role in parturition.

Stretching of a rat uterine strip elicits light chain phosphorylation.[65] Figure 9 compares light chain phosphorylation in a carbachol-contracted uterine strip developing 10 g of active force with that in stretched strip producing no active tension (cf. Frame 1 and 2 in Fig. 9). The staining intensity patterns of these two differently stimulated muscles were very similar; the phosphate contents were 0.79 and 0.77 mole phosphate/mole light chain, respectively. Furthermore, in EGTA-relaxed strip the light chain phosphorylation was 0.37 mole phosphate/mole light chain, which increased to 0.77 mole/mole, when the 1-min EGTA treatment was followed by stretch (cf. Frame 3 with 4 in Fig. 9). These data indicate that solely stretching the

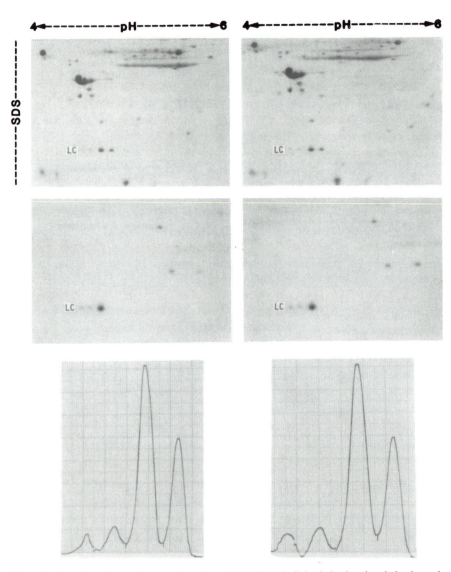

Figure 9. Two-dimensional gel electrophoretic analysis of myosin light chain phosphorylation in uterine smooth muscle. Top row: gel staining profiles of the uterine proteins; middle row: corresponding autoradiograms; bottom row: densitometric scans of the 20-kDa light chain on the stained gels. First frame: carbachol-contracted strip; second frame: stretched strip; third frame: EGTA-relaxed strip; fourth frame: EGTA-treated and stretched strip. LC, 20-kDa light chain. For details see ref. 65. (Reproduced from ref. 65 with the permission of the publisher.)

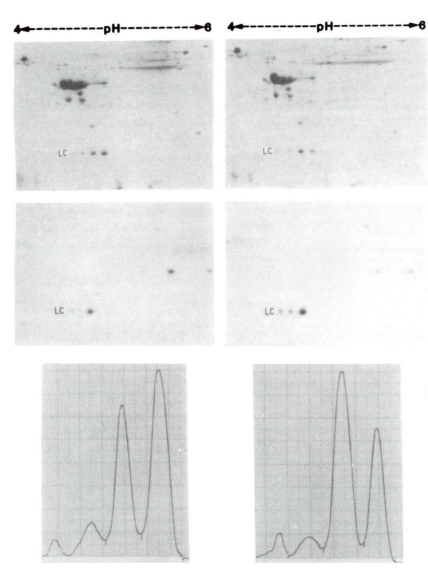

Figure 9. (*Continued*)

uterus in the absence of a stimulating agent, or even in the presence of a relaxing agent, can induce maximal light chain phosphorylation.

In the first studies of light chain phosphorylation during uterine muscle contraction,[4,5] it was established that there is an association between light chain phosphorylation and contraction. In a subsequent work,[65] we have shown that light chain phosphorylation can be separated from contraction. Namely, addition of carbachol or oxytocin to uterine strips at resting length elicited both maximal light chain phosphorylation and maximal active force development, but addition of these agents to stretched strips did not bring about active force development while light chain phosphorylation remained maximal. Thus, interaction between myosin and actin is not necessary for light chain phosphorylation.

In an attempt to elucidate the mechanism of Ca^{2+} mobilization during stretch, the inhibitory potency of the Ca^{2+}-channel blocker, verapamil, was compared in cases of three different stimulations (stretching, oxytocin, or K^+). Figure 10 shows

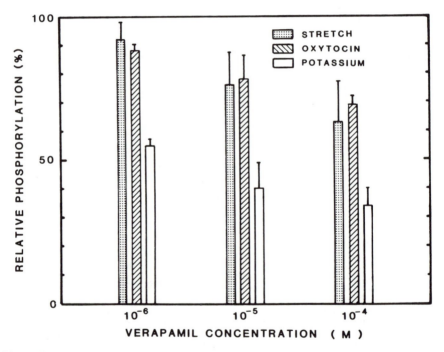

Figure 10. Comparison of relative phosphorylation of myosin light chain in stretched, oxytocin-contracted or potassium-contracted uterine strips in the presence of verapamil. The uterine strips were exposed to 1, 10, or 100 μM verapamil for 5 min, then stretched, or challenged with 1 μM oxytocin or with 100 mM K^+ for 1 min, and frozen. The extent of phosphorylation upon stimulation in the presence of verapamil was compared with that in the absence of verapamil. The latter values were: 0.83, 0.85, and 0.79 mole phosphate/mole light chain for stretched, oxytocin-, and K^+-stimulated strips, respectively. Each bar represents the mean ± S.D. For details see ref. 65. (Reproduced from ref. 65 with the permission of the publisher.)

that the higher the verapamil concentration, the lower is the phosphorylation. There was no significant difference between the relative phosphorylation values of stretched or oxytocin-stimulated strips, but the K^+-induced phosphorylation values were consistently lower at any verapamil concentration. These results indicate that verapamil is more effective against K^+-induced light chain phosphorylation than against stretch- or oxytocin-induced phosphorylation. Accordingly, Ca^{2+} is mobilized from extracellular space for K^+-induced phosphorylation, and mainly from intracellular sources of stretch- or oxytocin-induced phosphorylation.

Stretching smooth muscle under different experimental conditions led to the conclusion that light chain phosphorylation can occur without force development; however, force cannot develop without prior light chain phosphorylation. Stretching a 1-min EGTA-treated uterine strip elicited maximal phosphorylation, but stretching of a 50-min EGTA-treated strip did not cause phosphorylation. It is assumed that during the prolonged treatment, EGTA was not only chelating the extracellular Ca^{2+} but was also causing depletion of intracellular Ca^{2+}. Figure 11, top row, shows that the spontaneous activity of a strip was stopped with EGTA between 20 and 21 min. The strip was stretched at 21 min, and upon quick release of the stretch, active force immediately developed. After removal of EGTA by washes, the spontaneous activity was restored (between 40 and 50 min). Then the strip was treated with EGTA for 50 min, stretched and quick released; no force developed. The bottom row of Fig. 11 shows the light chain phosphorylation values determined in

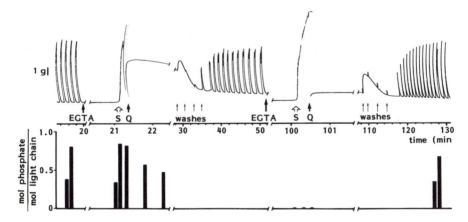

Figure 11. Stretch-release-induced force development and phosphorylation level of the 20-kDa light chain in uterus. Top row: force records of the same uterine strip; bottom row: phosphorylation from different strips, corresponding to the force records; each bar represents the mean value of two to five experiments. For EGTA treatment, the bath contained 1 mM EGTA in Ca^{2+}-free physiological salt solution. Washes were carried out with physiological salt solution. The phosphate content of light chain in spontaneously active strips was determined both in the relaxing phase and when force peaked. S, stretching, during which the polygraph sensitivity was decreased tenfold. Q, quick release of the stretch. During stretch and quick release the time scale was increased tenfold. For details see ref. 65. (Reproduced from ref. 65 with the permission of the publisher.)

parallel experiments; the values were 0.35 mole phosphate/mole light chain in the 1-min EGTA-treated strip, 0.83 upon its stretch, 0.81 upon its quick release, 0.56 and 0.47 during force development. In the 50-min EGTA-treated strip the phosphate content was less than 0.05 mole/mole and did not change during stretching or quick release. These processes were reversible; washes with physiological salt solution restored spontaneous activity and cyclic phosphorylation. The data show that the stretched strip must contain phosphorylated light chain to produce force upon stretch release. Consequently, light chain phosphorylation is a prerequisite for stretch-release-induced force.

Stretch-induced light chain phosphorylation is a novel mechanochemical coupling. The external work performed on the uterine muscle is transformed into a chemical activation of the contractile protein, myosin.

7. Concluding Remarks

The data presented demonstrate the role of myosin light chain phosphorylation in contraction of uterine muscle. The major events that lead to contraction are outlined in Fig. 12. The intracellular $[Ca^{2+}]$ can be increased from the extracellular space through voltage-dependent Ca^{2+} channels, by release of bound Ca^{2+} from the inner face of the plasma membrane, or from the sarcoplasmic reticulum (reviewed by van Breemen *et al.*[79]). The Ca^{2+} release from the internal membranes is initiated by the binding of an agonist (AGO) to a surface receptor. It has been suggested that subsequent to receptor activation, inositol-1,4,5-trisphosphate (IP_3), liberated from the plasma membrane, may increase the Ca^{2+} permeability of the sarcoplasmic reticulum[80] (see Chapter 5). The intracellular free Ca^{2+} combines with calmodulin to form the Ca^{2+}·calmodulin complex, which binds to MLCK to activate the enzyme. Phosphorylation of myosin·LC from MgATP in the presence of actin yields actomyosin·LCP, the contractile protein complex that generates tension and/or shortens the uterine muscle. When muscle stimulation terminates, Ca^{2+} is removed from the intracellular fluid by the Ca^{2+} pumps of the sarcoplasmic reticulum and the cell membrane, and to a lesser extent by the Na^+–Ca^{2+} exchange carrier.

By lowering the free intracellular $[Ca^{2+}]$, MLCK will be partially inactivated, and MLCP dephosphorylates light chain to a certain extent. In a spontaneously active uterine muscle this is the relaxed state. The spontaneous activity of uterus may be stopped by various relaxing agents, accompanied by further light chain dephosphorylation. When the uterus is contracted by drugs, the same relaxing agents may elicit partial relaxation and dephosphorylation even in the presence of the drugs (Figs. 7, 8).

We now turn to describe possible mechanisms for the physiological role of light chain phosphorylation in uterine muscle contraction. Kinetic studies on the phosphorylation in contracting muscle clearly show that light chain phosphorylation precedes tension development. The orientation of myosin- and actin-containing

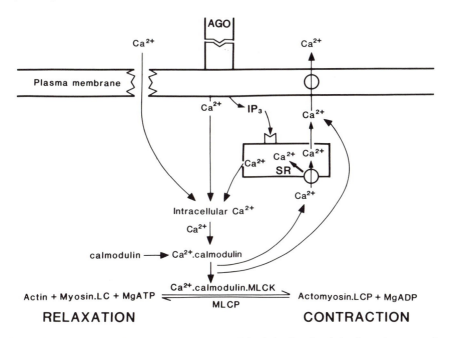

Figure 12. A scheme for the involvement of myosin light chain phosphorylation in uterine contraction. The intracellular [Ca^{2+}] can be increased via voltage- or receptor-operated channels. Agonist binding to a receptor may initiate Ca^{2+} release from internal membranes. The intracellular free Ca^{2+} combines with calmodulin, subsequently the Ca^{2+}·calmodulin complex binds to MLCK, which in this activated form phosphorylates the myosin light chain. This leads to the interaction between myosin and actin that constitutes muscle contraction. Light chain, 20-kDa myosin light chain; LCP, phosphorylated myosin light chain; MLCK, myosin light chain kinase; MLCP, myosin light chain phosphatase; SR, sarcoplasmic reticulum; IP_3, inositol-1,4,5-triphosphate; AGO, agonist.

filaments in smooth muscle differs from that in skeletal muscle. Electron micrographs of skeletal muscle demonstrate the orderly arrangement of thick and thin filaments running parallel to each other in the longitudinal section. In contrast, such an orderly arrangement is not apparent in longitudinal section of the myometrium[81]; instead, the thick and thin filaments display a less regular array in bundles. For tension production, actin and myosin filaments must combine with each other; this is achieved at the molecular level through the combination of the complementary amino acid residues that form part of the binding sites of actin and myosin molecules.

It is postulated that light chain phosphorylation plays a major role in the orientation of actin and myosin filaments close to each other in uterine muscle. This is illustrated in the left panel of Fig. 13. Two kinds of sites are shown in the cross-bridges: the actin binding site (which includes the ATPase site) and the phosphorylation site. One phosphate group, which is virtually ionized at physiological pH, carries two negative charges (the exact number is 1.8 charges/mole phosphoserine).[82] Assuming three myosin molecules per cross-bridge, full phosphoryla-

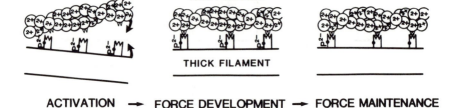

THIN FILAMENT

THICK FILAMENT

ACTIVATION → FORCE DEVELOPMENT → FORCE MAINTENANCE

Figure 13. A hypothesis for the role of myosin light chain phosphorylation in smooth muscle contraction. In the relaxed muscle the myosin-containing thick filaments and the actin-containing thin filaments are not in ordered, parallel position. In the activation phase of contraction, phosphorylation of light chain facilitates the alignment of the filaments through electrostatic forces (left panel). Consequently, the combination of myosin with actin can take place at the specific binding sites developing force (middle panel). The phosphorylation site of the light chain is not involved in the myosin–actin bond and, therefore, it is accessible for both MLCK and myosin light chain phosphatase. Subsequently, myosin remains combined to actin, force is maintained, while some light chain becomes dephosphorylated (right panel). Note, in addition to the 2+ charged groups, the specific myosin binding sites (enlarged out of proportion) are shown on some but not on all actin molecules. The specific actin binding sites (W) and phosphorylation sites (•) are indicated on the cross-bridges. Only one phosphorylated residue with 2− charges is shown on selected cross-bridges, but when fully phosphorylated, a cross-bridge bears 11 negative charges. For details, see text.

tion of light chain will result in 11 negative charges per cross-bridge.[82] Such a dense charge distribution will produce a considerable attractive force to positively charged groups on the surface of the actin filament. The bound bivalent cation (Ca^{2+} or Mg^{2+}) of actin[83,84] is a good candidate for participation in the generation of electrostatic forces. Alternatively, clustered lysine and/or arginine residues, which carry positive charges at physiological pH, may be involved in orienting the phosphate residues of light chain to near juxtaposition with actin filaments. Accordingly, the primary role of light chain phosphorylation during the activation phase of uterine contraction is the orientation of actin and myosin filaments in smooth muscle to an arrangement similar to that in skeletal muscle.

When the binding sites of actin and myosin reach close proximity, the complementary amino acid surfaces of these proteins lock into each other (middle panel of Fig. 13). Short-range hydrophobic interactions contribute greatly to this binding and ionic interactions also do so to a lesser extent. The muscle is now ready to develop force. This requires high actin- and Mg^{2+}-activated ATPase activity of myosin which is facilitated by the phosphorylation of light chain.

Force maintenance in smooth muscle, including uterus muscle, does not require a major energy cost. The biochemical equivalent of this is a low actomyosin ATPase and a reduced level of light chain phosphorylation. Since the phosphorylation site of light chain in the myosin head does not participate in the myosin–actin linkage, the magnitude of force is maintained while the light chain is partially dephosphorylated (right panel of Fig. 13).

Relaxation of contracted uterine muscle is brought about by ATP, which dissociates actomyosin into actin and myosin, in a manner similar to that in skeletal muscle. This is not shown in Fig. 13.

In summary, myosin light chain phosphorylation is a prerequisite of uterine muscle contraction. Its unique role is the activation of the contractile apparatus. In the regulation of force maintenance, other factors are also involved. The mechanisms of uterine muscle contraction is most likely the same as that of other smooth muscles.

ACKNOWLEDGMENTS. We express our appreciation to our present and past collaborators, Drs. John T. Barron, Sándor Csabina, and Ferenc Erdödi, for valuable discussions. We thank Donna M. Lattyak for careful typing of the manuscript and Kathryn Peregrine for dedicated assistance. The authors are supported by a grant, AR34602, from the National Institutes of Health.

References

1. Needham, D. M., 1971, *Machina Carnis,* Cambridge University Press, London.
2. Huszar, G., and Bailey, P., 1979, Relationship between actin–myosin interaction and myosin light-chain phosphorylation in human placental smooth muscle, *Am. J. Obstet. Gynecol.* **135:**718–726.
3. Lebowitz, E. A., and Cooke, R., 1979, Phosphorylation of uterine smooth muscle myosin permits actin-activation, *J. Biochem.* **85:**1489–1494.
4. Janis, R. A., Moats-Staats, B. M., and Gualtieri, R. T., 1980, Protein phosphorylation during spontaneous contraction of smooth muscle, *Biochem. Biophys. Res. Commun.* **96:**265–270.
5. Janis, R. A., Bárány, K., Bárány, M., and Sarmiento, G., 1981, Association between myosin light chain phosphorylation and contraction of rat uterine smooth muscle, *Mol. Physiol.* **1:**3–11.
6. Kühne, W., 1859, Untersuchungen über Bewegungen und Veränderungen der contractilen Substanzen, *Arch. Anat. Physiol. Wiss. Med.* p. 748. Quoted from Needham, D. M., 1971, *Machina Carnis,* Cambridge University Press, London, p. 674.
7. Kühne, W., 1864, *Untersuchungen über das Protoplasma und die Contractilität,* W. Engelmann, Leipzig. Quoted from Needham, D. M., 1971, *Machina Carnis,* Cambridge University Press, London, p. 674.
8. Mehl, J. W., 1938, Studies on the proteins of smooth muscle, I, *J. Biol. Chem.* **123:**lxxxiii.
9. Csapo, A., 1948, Actomyosin content of the uterus, *Nature* **162:**218–219.
10. Needham, D. M., and Williams, J. M., 1959, Some properties of uterus actomyosin and myofilaments, *Biochem. J.* **73:**171–181.
11. Needham, D. M., and Williams, J. M., 1963, Proteins of the uterine contractile mechanism, *Biochem. J.* **89:**552–561.
12. Bárány, M., Bárány, K., Gaetjens, E., and Bailin, G., 1966, Chicken gizzard myosin, *Arch. Biochem. Biophys.* **113:**205–221.
13. Tsao, T. C., 1953, Fragmentation of myosin molecule, *Biochim. Biophys. Acta* **11:**368–382.
14. Locker, R. H., 1956, The dissociation of myosin by heat coagulation, *Biochim. Biophys. Acta* **20:**514–521.
15. Kominz, D. R., Carroll, W. R., Smith, E. N., and Mitchell, E. R., 1959, A subunit of myosin, *Arch. Biochem. Biophys.* **79:**191–199.
16. Wetlaufer, D. B., and Edsall, J. T., 1960, Sedimentation of myosin in urea solutions, *Biochim. Biophys. Acta* **43:**132–134.
17. Dreizen, P., Hartshorne, D. J., and Stracher, A., 1966, The subunit structure of myosin, I. Polydispersity in 5 M guanidine, *J. Biol. Chem.* **241:**443–448.

18. Gershman, L. C., Dreizen, P., and Stracher, A., 1966, Subunit structure of myosin. II. Heavy and light alkali components, *Proc. Natl. Acad. Sci. USA* **56:**966–973.
19. Oppenheimer, H., Bárány, K., Hamoir, G., and Fenton, J., 1967, Succinylation of myosin, *Arch. Biochem. Biophys.* **120:**108–118.
20. Bárány, K., and Oppenheimer H., 1967, Succinylated meromyosins, *Nature* **213:**626–627.
21. Perry, S. V., 1979, The regulation of contractile activity in muscle, *Biochem. Soc. Trans.* **7:**596–617.
22. Sobieszek, A., 1977, Vertebrate smooth muscle myosin. Enzymatic and structural properties, in: *The Biochemistry of Smooth Muscle* (N. L. Stephens, ed.), University Park Press, Baltimore, pp. 413–443.
23. Gorecka, A., Aksoy, M. O, and Hartshorne, D. J., 1976, The effect of phosphorylation of gizzard myosin on actin activation, *Biochem. Biophys. Res. Commun.* **71:**325–331.
24. Chacko, S., Conti, M. A., and Adelstein, R. S., 1977, Effect of phosphorylation of smooth muscle myosin on actin-activation and on Ca^{2+} regulation, *Proc. Natl. Acad. Sci. USA* **74:**129–133.
25. Barron, J. T., Bárány, M., and Bárány, K., 1979, Phosphorylation of the 20,000-dalton light chain of myosin of intact arterial smooth muscle in rest and in contraction, *J. Biol. Chem.* **254:**4954–4956.
26. Dillon, P. F., Aksoy, M. O., Driska, S. P., and Murphy, R. A., 1981, Myosin phosphorylation and the cross-bridge cycle in arterial smooth muscle, *Science* **211:**495–497.
27. Ashton, F. T., Somlyo, A. V., and Somlyo, A. P., 1975, The contractile apparatus of vascular smooth muscle: Intermediate high voltage stereo electron microscopy, *J. Mol. Biol.* **98:**17–29.
28. Trybus, K. M., Huiatt, T. W., and Lowey, S., 1982, A bent monomeric conformation of myosin from smooth muscle, *Proc. Natl. Acad. Sci. USA* **79:**6151–6155.
29. Ikebe, M., Hinkins, S., and Hartshorne, D. J., 1983, Correlation of enzymatic properties and conformation of smooth muscle myosin, *Biochemistry* **22:**4580–4587.
30. Bailin, G., 1986, Structure and function of smooth muscle myosin, *Biochem. Arch.* **2:**229–236.
31. Sobieszek, A., and Bremel, R. D., 1975, Preparation and properties of vertebrate smooth-muscle myofibrils and actomyosin, *Eur. J. Biochem.* **55:**49–60.
32. Cavaille, F., Janmot, C., Ropert, S., and d'Albis, A., 1986, Isoforms of myosin and actin in human, monkey, and rat myometrium. Comparison of pregnant and non-pregnant uterus proteins, *Eur. J. Biochem.* **160:**507–513.
33. Bárány, M., 1967, ATPase activity of myosin correlated with speed of muscle shortening, *J. Gen. Physiol.* **50:**197–218.
34. Murphy, R. A., 1979, Filament organization and contractile function in vertebrate smooth muscle, *Annu. Rev. Physiol.* **41:**737–748.
35. Small, J. V., and Sobieszek, A., 1983, Contractile and structural proteins of smooth muscle, in: *Biochemistry of Smooth Muscle*, Volume I (N. L. Stephens, ed.), CRC Press, Boca Raton, pp. 85–140.
36. Marston, S. B., 1983, Myosin and actomyosin ATPase: Kinetics, in: *Biochemistry of Smooth Muscle*, Volume I (N. L. Stephens, ed.), CRC Press, Boca Raton, pp. 167–191.
37. Horiuchi, K. Y., Miyata, H., and Chacko, S., 1986, Modulation of smooth muscle actomyosin ATPase by thin filament associated proteins, *Biochem. Biophys. Res. Commun.* **136:**962–968.
38. Aksoy, M. O., and Murphy, R. A., 1983, Regulation of the dynamic properties of smooth muscle: Ca^{++}-stimulated cross-bridge phosphorylation, in: *Biochemistry of Smooth Muscle*, Volume I (N. L. Stephens, ed.), CRC Press, Boca Raton, pp. 141–166.
39. Edelman, A. M., Blumenthal, D. K., and Krebs, E. G., 1987, Protein serine/threonine kinases, *Annu. Rev. Biochem.* **56:**567–613.
40. Higashi, K., Fukunaga, K., Matsui, K., Maeyama, M., and Miyamoto, E., 1983, Purification and characterization of myosin light-chain kinase from porcine myometrium and its phosphorylation and modulation by cyclic AMP-dependent protein kinase, *Biochim. Biophys. Acta* **747:**232–240.
41. Cohen, D. M., and Murphy, R. A., 1978, Differences in cellular contractile protein contents among porcine smooth muscles, *J. Gen. Physiol.* **72:**369–380.

42. Adelstein, R. S., and Hathaway, D. R., 1979, Role of calcium and cyclic adenosine 3′:5′ monophosphate in regulating smooth muscle contraction, *Am. J. Cardiol.* **44**:783–787.

43. Polacek, I., and Daniel, E. E., 1971, Effect of α- and β-adrenergic stimulation on the uterine motility and adenosine 3′,5′-monophosphate level, *Can. J. Physiol. Pharmacol.* **49**:988–998.

44. Richardson, M. R., Taylor, D. A., Casey, M. L., MacDonald, P. C., and Stull, J. T., 1987, Biochemical markers of contraction in human myometrial smooth muscle cells in culture, *In Vitro Cell. Dev. Biol.* **23**:21–28.

45. Endo, T., Naka, M., and Hidaka, H., 1982, Ca^{2+}-phospholipid dependent phosphorylation of smooth muscle myosin, *Biochem. Biophys. Res. Commun.* **105**:942–948.

46. Nishikawa, M., Hidaka, H., and Adelstein, R. S., 1983, Phosphorylation of smooth muscle heavy meromyosin by calcium-activated, phospholipid-dependent protein kinase. The effect on actin-activated MgATPase activity, *J. Biol. Chem.* **258**:14069–14072.

47. Baraban, J. M., Gould, R. J., Peroutka, S. J., and Snyder, S. H., 1985, Phorbol ester effects on neurotransmission: Interaction with neurotransmitters and calcium in smooth muscle, *Proc. Natl. Acad. Sci. USA* **82**:604–607.

48. Sawamura, M., Kobayashi, Y., Nara, Y., Hattori, K., and Yamori, Y., 1987, Effect of extracellular calcium on vascular contraction induced by phorbol ester, *Biochem. Biophys. Res. Commun.* **145**:494–501.

49. Howe, P. H., and Abdel-Latif, A. A., 1987, Phorbol ester-induced protein-phosphorylation and contraction in sphincter smooth muscle of rabbit iris, *FEBS Lett.* **215**:279–284.

50. Kamm, K. E., and Stull, J. T., 1985, The function of myosin and myosin light chain kinase phosphorylation in smooth muscle, *Annu. Rev. Pharmacol. Toxicol.* **25**:593–620.

51. Werth, D. K., Haeberle, J. R., and Hathaway, D. R., 1982, Purification of a myosin phosphate from bovine aortic smooth muscle, *J. Biol. Chem.* **257**:7306–7309.

52. Di Salvo, J., Gifford, D., Bialojan, C., and Rüegg, J. C., 1983, An aortic spontaneously active phosphatase dephosphorylates myosin and inhibits actin–myosin interaction, *Biochem. Biophys. Res. Commun.* **111**:906–911.

53. Pato, M. D., and Kerc, E., 1985, Purification and characterization of a smooth muscle myosin phosphatase from turkey gizzards, *J. Biol. Chem.* **260**:12359–12366.

54. Driska, S. P., Aksoy, M. O., and Murphy, R. A., 1981, Myosin light chain phosphorylation associated with contraction in arterial smooth muscle, *Am. J. Physiol.* **240**:C222–C233.

55. Bárány, K., Ledvora, R. F., Mougios, V., and Bárány, M., 1985, Stretch-induced myosin light chain phosphorylation and stretch-release-induced tension development in arterial smooth muscle, *J. Biol. Chem.* **260**:7126–7130.

56. Bárány, K., Csabina, S., and Bárány, M., 1985, The phosphorylation of the 20,000 dalton myosin light chain in rat uterus, in: *Advances in Protein Phosphatases,* Volume II (W. Merlevede and J. DiSalvo, eds.), Leuven University Press, Leuven, Belgium, pp. 37–58.

57. Csabina, S., Mougios, V., Bárány, M., and Bárány, K., 1986, Characterization of the phosphorylatable myosin light chain in rat uterus, *Biochim. Biophys. Acta* **871**:311–315.

58. Dokhac, L., D'Albis, A., Janmot, C., and Harborn, S., 1986, Myosin light chain phosphorylation in intact rat uterine smooth muscle. Role of calcium and cyclic AMP, *J. Muscle Res. Cell Motil.* **7**:259–268.

59. Haeberle, J. R., Hott, J. W., and Hathaway, D. R., 1984, Pseudophosphorylation of the smooth muscle 20,000 dalton myosin light chain. An artifact due to protein modification, *Biochim. Biophys. Acta* **790**:78–86.

60. Gagelmann, M., Ruegg, J. C., and Di Salvo, J., 1984, Phosphorylation of the myosin light chains and satellite proteins in detergent-skinned arterial smooth muscle, *Biochem. Biophys. Res. Commun.* **120**:933–938.

61. Mougios, V., and Bárány, M., 1986, Isoforms of the phosphorylatable myosin light chain in arterial smooth muscle, *Biochim. Biophys. Acta* **872**:305–308.

62. Bárány, K., Csabina, S., de Lanerolle, P., and Bárány, M., 1987, Evidence for isoforms of the phosphorylatable myosin light chain in rat uterus, *Biochim. Biophys. Acta* **911**:369–371.

63. Pette, D., 1980, *Plasticity of Muscle,* de Gruyter, Berlin.

64. Swynghedauw, B., 1986, Developmental and functional adaptation of contractile proteins in cardiac and skeletal muscles, *Physiol. Rev.* **66:**710–771.

65. Csabina, S., Bárány, M., and Bárány, K., 1986, Stretch-induced myosin light chain phosphorylation in rat uterus, *Arch. Biochem. Biophys.* **249:**374–381.

66. Hathaway, D. R., and Haeberle, J. R., 1985, A radioimmunoblotting method for measuring myosin light chain phosphorylation levels in smooth muscle, *Am. J. Physiol.* **249:**C345–C351.

67. Silver, P. J., and Stull, J. T., 1982, Quantitation of myosin light chain phosphorylation in small tissue samples, *J. Biol. Chem.* **257:**6137–6144.

68. Csabina, S. Bárány, M., and Bárány, K., 1987, Comparison of myosin light chain phosphorylation in uterine and arterial smooth muscles, *Comp. Biochem. Physiol.* **87B:**271–277.

69. Bárány, K., Ledvora, R. F., and Bárány, M., 1985, The phosphorylation of the 20,000 dalton myosin light chain in intact arterial muscle, in: *Calmodulin Antagonists and Cellular Physiology* (H. Hidaka and D. J. Hartshorne, eds.), Academic Press, New York, pp. 199–223.

70. Aksoy, M. O., Mras, S., Kamm, K. E., and Murphy, R. A., 1983, Ca^{2+}, cAMP, and changes in myosin phosphorylation during contraction of smooth muscle, *Am. J. Physiol.* **245:**C255–C270.

71. Silver, P. J., and Stull, J. T., 1982, Regulation of myosin light chain and phosphorylase phosphorylation in tracheal smooth muscle, *J. Biol. Chem.* **257:**6145–6150.

72. Gerthoffer, W. T., and Murphy, R. A., 1983, Myosin phosphorylation and regulation of the cross-bridge cycle in tracheal smooth muscle, *Am. J. Physiol.* **244:**C182–C187.

73. Weisbrodt, N. W., and Murphy, R. A., 1985, Myosin phosphorylation and contraction of feline esophageal smooth muscle, *Am. J. Physiol.* **249:**C9–C14.

74. Haeberle, J. R., Hott, J. W., and Hathaway, D. R., 1985, Regulation of isometric force and isotonic shortening velocity by phosphorylation of the 20,000 dalton myosin light chain of rat uterine smooth muscle, *Pfluegers Arch.* **403:**215–219.

75. Kroeger, E. A., Marshall, J. M., and Bianchi, C. P., 1975, Effect of isoproterenol and D600 on calcium movements in rat myometrium, *J. Pharmacol. Exp. Ther.* **193:**309–316.

76. Huszar, G., 1981, Biology and biochemistry of myometrial contractility and cervical maturation, *Semin. Perinatol.* **5:**216–234.

77. Nishikori, K., Weisbrodt, N. W., Sherwood, O. D., and Sanborn, B. M., 1983, Effects of relaxin on rat uterine myosin light chain kinase activity and myosin light chain phosphorylation, *J. Biol. Chem.* **258:**2468–2474.

78. Haeberle, J. R., Hathaway, D. R., and DePaoli-Roach, A. A., 1985, Dephosphorylation of myosin by the catalytic subunit of a type-2 phosphatase produces relaxation of chemically skinned uterine smooth muscle, *J. Biol. Chem.* **260:**9965–9968.

79. van Breemen, C., Leijten, P., Yamamoto, H., Aaronson, P., and Cauvin, C., 1986, Calcium activation of vascular smooth muscle, *Hypertension* **8:**II-89–II-95.

80. Carsten, M. E., Miller, J. D., 1985, Ca^{2+} release by inositol trisphosphate from Ca^{2+}-transporting microsomes derived from uterine sarcoplasmic reticulum, *Biochem. Biophys. Res. Commun.* **130:**1027–1031.

81. Riemer, R. K., and Roberts, J. M., 1986, Activation of uterine smooth muscle contraction: Implications for eicosanoid action and interactions, *Semin. Perinatol.* **10:**276–287.

82. Bárány, K., Bárány, M., Gillis, J. M., and Kushmerick, M. J., 1979, Phosphorylation–dephosphorylation of the 18,000-dalton light chain of myosin during the contraction–relaxation cycle of frog muscle, *J. Biol. Chem.* **254:**3617–3623.

83. Gershman, L. C., Selden, L. A., and Estes, J. E., 1986, High affinity binding of divalent cation to actin monomer is much stronger than previously reported, *Biochem. Biophys. Res. Commun.* **135:**607–614.

84. Bond, M., Shuman, H., Somlyo, A. P., and Somlyo, A. V., 1984, Total cytoplasmic calcium in relaxed and maximally contracted rabbit portal vein smooth muscle, *J. Physiol. (London)* **357:**185–201.

4

Thin Filament Control of Uterine Smooth Muscle

Steven B. Marston and Kevin Pritchard

1. Introduction

The contractile apparatus of uterine smooth muscle consists of thick (myosin) filaments and thin (actin-based) filaments. There has been little ultrastructural work published on uterine smooth muscle; but such as has been done[1] (see Chapter 1) suggests the thick and thin filaments are arranged in overlapping arrays in much the same way as in other smooth muscles[2] and that contraction is produced by the same myosin cross-bridge mechanism.[3]

Thin filaments have three functions in the contractile apparatus.

1. Force generation by interaction with myosin
2. Control of force generation by interaction with Ca^{2+}
3. Transmission of force to the ends of the cell to produce contraction of the muscle as a whole

This chapter will consider the mechanisms responsible for these functions.

2. Location of Thin Filaments in the Contractile Matrix

The structure of all thin filaments is basically the same (Fig. 1): a double helix of actin in which the monomers repeat every 5.9 nm and the helix completes one turn every 74 nm.[4,5] Tropomyosin is present in the thin filaments in a molar ratio of 1 : 7[6–8] and is probably located along the grooves between the two actin helices as it is in skeletal muscle thin filaments.[5,9] Other protein components of uterine smooth

Steven B. Marston and Kevin Pritchard • Department of Cardiac Medicine, The National Heart and Lung Institute, London SW3 6LY, United Kingdom.

Figure 1. Negatively stained electron microscope image of isolated smooth muscle thin filaments. The double-stranded helix structures of polymerized actin can be clearly seen. Tropomyosin and caldesmon, although present in the filaments (Fig. 2), cannot be seen in this image. × 200,000. From ref. 24 with permission.

muscle thin filaments include caldesmon, calcium-binding protein, and filamin (Fig. 2). These filaments are arranged in parallel tracts along the cell length in relaxed smooth muscle with a spacing of about 11 nm between filaments[10]; sometimes these tracts are organized into regular arrays of up to 100 filaments. Uterine smooth muscle contains 6.8 times as much actin as myosin in molar terms in common with visceral smooth muscle (arterial smooth muscle contains an even higher actin/myosin ratio)[11]; consequently, most actin molecules are not interacting with myosin at any one moment during contraction. Where thin filaments are in

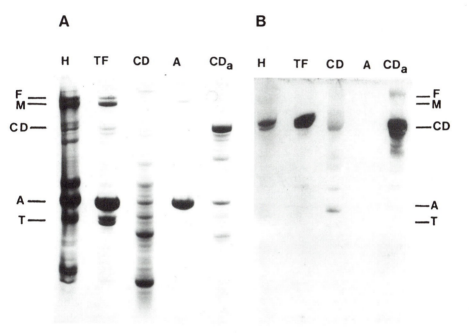

Figure 2. Uterine smooth muscle proteins separated on 4%–30% gradient polyacrylamide, 0.1% SDS gel electrophoresis. (**A**) Stained with Coomassie blue. (**B**) Electroblotted onto nitrocellulose and labeled with anti-chicken gizzard caldesmon antibody and ^{125}I protein A.

H, washed homogenate of rabbit uterus. The main components include filamin (F), myosin (M), caldesmon (CD) (two bands), actin (A), tropomyosin (T). In the immunoblot, only the paired CD bands are labeled.

TF, native thin filaments isolated from the washed homogenate. The main bands are actin, tropomyosin, and caldesmon present at $1 : 0.38 \pm 0.04 : 0.093 \pm 0.017$ (w/w) by densitometry.[8] Some filamin and myosin are also present in this preparation.

CD, caldesmon isolated from uterus.

A, purified uterine polymeric actin isolated from the native thin filaments. From ref. 8 with permission.

CD_a, caldesmon isolated from sheep aorta.

By comparison with CD_a, and with the CD band in the uterus native thin filaments, uterine caldesmon has been significantly degraded during its preparation. The immunoblot identifies CD immunoreactivity from whole caldesmon and from a 40,000-molecular-weight fragment. Despite the proteolysis, this caldesmon preparation was an active inhibitor (see Fig. 5).

association with myosin they tend to be seen as rosettes of up to 15 filaments surrounding one thick filament or in random arrangements.[12,13]

Thin filaments are attached at one end to amorphous patches known as dense bodies, which are presumed to provide an anchorage for transmission of forces generated by the cross-bridge mechanism. The smooth muscle cell surface contains numerous "membrane-associated" dense bodies that contain the actin-binding protein vinculin.[14] There are also dense bodies in the cytoplasm, which contain the actin-binding protein α-actinin.[15,16] Numerous actin filaments insert into both types of dense bodies and when myosin molecules are attached to these filaments it is observed that they are correctly orientated such that myosin filaments would be pulled toward the dense bodies.[15,17,19] Immunological work has established that the actin filaments can be found in two spatially distinct domains: one in which actin is associated with myosin and caldesmon, which makes up 80% of the cell area and presumably represents the contractile apparatus; and one in which actin is associated with filamin, which may represent the attachment sites of the thin filaments.[18] The way in which thin filament–dense body complexes and myosin filaments are organized into a contractile apparatus is still poorly understood.

3. Protein Components of the Thin Filament

3.1. Actin

Actin forms the backbone of the thin filament and makes up 60% of its mass in vertebrate smooth muscles. It has a molecular weight of 42,000 and is made up of 375 or 374 amino acids.

Actin is expressed in a tissue-specific manner by a multigene family. A total of six actin isoforms have been identified: uterine smooth muscle contains the α and γ forms which are specific to smooth muscles and also some of the "cytoplasmic" β form of actin. Mature uterus contains roughly equal quantities of α and γ isoforms; it has been suggested that the ratio of these two isoforms is related to whether the muscle is predominantly phasic (like uterus) or tonic (e.g., vascular smooth muscle).[20] The isoform distribution changes during pregnancy (γ-actin increased early in pregnancy)[21] and it has been observed that α-actin mRNA is induced by estrogens.[22]

Actin is a highly conserved protein: α and γ isoforms differ from each other by only three amino acids, all within the first five N-terminal amino acids, and from skeletal actin by only eight and six amino acids, respectively.[23,24] So far no significant functional differences between the various isoforms have been observed.

The isolation of uterine actin has presented considerable difficulties. Traditionally, monomeric actin has been obtained from muscle acetone powders extracted at low ionic strength.[25] The application of the technique to uterus produces a monomeric actin preparation that is not pure and usually fails to polymerize into filaments.[26] This problem appears to be due to proteolysis, which is particularly active in uterine tissue[27,28] and to the presence of contaminating proteins such as gelsolin[29] that inhibit polymerization. A number of procedures that circumvent these problems have been published,[26,28,30,31] but have not yet come into general use.

The skeletal muscle actin monomer may be used as a model for the very similar uterine actin. Its overall dimensions are 6.7 × 4.1 × 3.7 nm and from X-ray crystallography it is known to be formed into a large and a small domain.[32] The actin monomer has specific binding sites for ATP and divalent cations. Smooth muscle actin monomer is relatively unstable and irreversible denaturation occurs unless both of these sites are occupied.[33,34]

Monomeric actin polymerizes into filaments by first forming a nucleus of three or four actins followed by rapid addition of monomers to one end of the nucleus.[35] The equilibrium is in favor of the polymeric form when salt concentrations approach the physiological. The actin monomers form into a right-handed double helix with a subunit repeat of 5.9 nm and 15.5 subunits per turn of the helix (Fig. 3). The large domain of each monomer is in contact with three other monomers. The structure of polymerized actin is identical to the actin helix in the thin filament *in vivo*.[9]

Since the formation of uterine actin filaments from monomers has proven difficult *in vitro,* the alternative method of isolating filaments directly has been developed. The starting material is crude actomyosin extract; thin filaments are separated from myosin by high-speed sedimentation under conditions where myosin is soluble and does not bind to actin.[36,37] Subsequent resedimentation in 0.8 M KCl results in sedimentation of pure polymeric actin.[38–40] We have found that this procedure can be satisfactorily used for isolating rabbit uterine polymeric actin (Fig. 2).[8]

3.2. Tropomyosin

Tropomyosin is a major component of uterine thin filaments (Fig. 2); a stoichiometry of one tropomyosin dimer per seven actin monomers has been found in whole uterine smooth muscle[11,20] and in thin filament preparations from enteric,[37,38] vascular,[40] and uterine smooth muscle.[8]

Figure 3. Structural model for the smooth muscle thin filament. Two complete turns of the actin double helix are shown. The pitch is 74 nm and the structure repeats every 37 nm. A tropomyosin molecule (dimer) is wrapped around the helix in each groove. Each tropomyosin dimer is 42 nm long and spans one half turn of the helix. Caldesmon is proposed also to be located in each groove, alongside tropomyosin. Caldesmon is shown as a black line; triangles indicate the beginning and end of the caldesmon molecule. Each caldesmon is 75 nm long. Adapted from ref. 24, with permission.

In smooth muscle, tropomyosin monomer is present in two isoforms each with 284 amino acids that are distinct from striated muscle and nonmuscle isoforms. The two tropomyosin isoforms (termed β and γ by Sanders and Smillie[41]) are present in approximately equal quantities in uterus. These isoforms are closely related to skeletal muscle β- and α-tropomyosin, respectively; indeed, they are derived from the same two genes by differential splicing.[42,43]

Uterine tropomyosin has not been widely investigated[36,44] but it is probably identical to other smooth muscle tropomyosins,[20] and it may be prepared by standard methods developed for isolating skeletal tropomyosin.[44,45] Tropomyosin is normally a dimer made up of two peptides, each of which is virtually pure α helix, coiled around each other (Fig. 3). The amino acid sequence of smooth muscle tropomyosin indicates that the coiled form should be a stable conformation[41] and studies on thermal stability confirm this.[46] The tropomyosin dimer is estimated to be 41–42 nm long with a helical pitch of 13.7 nm.[47,48]

Tropomyosin is normally incorporated into the thin filaments and has a high affinity for actin *in vitro*. Binding of smooth muscle tropomyosin to actin at 25°C requires at least 45 mM KCl (maximal at 129 mM KCl) and Mg^{2+} (optimal at 2 mM) in the medium. Under optimal conditions, binding is highly cooperative; half-maximal binding is observed at 0.5 μM tropomyosin dimer, and binding saturates with a stoichiometry of one tropomyosin dimer bound for 6.5–7.5 actins,[48,49] which is compatible with the amount of tropomyosin present in native thin filaments. Dissociation of tropomyosin occurs at low temperatures when [KCl] is less than 50 mM, or at any temperature when [KCl] is more than 500 mM.

The tropomyosin dimer has a sevenfold-repeated sequence of amino acids that are presumed to be actin-binding sites.[41] Tropomyosin binds in the grooves of the actin double helix in such a way that it is in contact with seven actins in one of the strands[50] (Fig. 3). Adjacent tropomyosin molecules overlap slightly: the part of the polypeptide chain involved in the overlap zone (C-terminus 254–284) is significantly different in smooth compared with skeletal tropomyosin[41,43] and this could be responsible for many of the differences between the tropomyosins' influence on actin-myosin interaction (see Section 4.3).

Skeletal muscle tropomyosin has certain sites where it interacts with regulatory proteins, notably troponin T. One of these sites is at the C-terminus, which is greatly altered in smooth muscle tropomyosin, and the other is around Cys-190, which also shows differences. However, troponin T fragments still bind to the Cys-190 region of smooth muscle tropomyosin, indicating the retention of a functional site here.[51] Smooth muscle thin filaments do not contain troponin, but they do contain caldesmon, which has similarities to troponin T[52] and thus may bind at that site.

3.3. Caldesmon

Caldesmon is the third most abundant protein component of smooth muscle thin filaments after actin and tropomyosin (Fig. 2). It represents 6% of the total mass of the thin filament, one-tenth the actin mass, and in molar terms is present in a ratio of 1 per 14 to 20 actins.[7,8,40] This protein has only been recognized within the last 5 years; partly because it is very sensitive to proteolytic digestion and partly because it tends to be removed during thin filament preparation procedures using techniques intended for skeletal muscle proteins.[40] Since its discovery, a large number of techniques have been developed for its purification.[53–57] Isolation techniques rely on one or more of caldesmon's distinctive properties: it binds actin, it binds calmodulin, it is stable at 90°C, and it is stable and soluble at pH 3. There is little doubt that all these procedures yield the same protein,[7] but it is by no means certain that the properties are always the same. Therefore, in considering findings about caldesmon, due attention must be paid to how it was obtained.

According to SDS gel electrophoresis (Fig. 2), caldesmon has a molecular weight between 120,000 and 150,000 and may in some tissues exist as two isoforms of slightly different molecular weight.[8,53] Since the apparent molecular weight of a single sample of caldesmon can be different under different electrophoresis conditions,[7] this approach cannot produce an accurate estimate of molecular weight. The native molecular weight of gizzard caldesmon has been reported to be 93,000 ± 4000 based on sedimentation equilibrium measurements.[58] Bryan *et al.* sequenced the gizzard caldesmon gene, demonstrating that the actual molecular weight is 87,000.[59] Data on molar ratios of caldesmon to actin have been amended accordingly. Thus the molecular ratios presented in this chapter differ from those cited in the literature which used a higher molecular weight. Such studies as have been made on uterine caldesmon[8] indicate that it does not differ significantly from caldesmon in any other smooth muscle.

Caldesmon is an extremely elongated molecule. This was first recognized from gel filtration experiments, where caldesmon eluted in the position of globular proteins around 600,000 molecular weight[56,60,61] and has been confirmed by electron microscope studies.[18,62] Single caldesmon molecules visualized by rotary shadowing appear as extremely thin random chains up to 150 nm long, frequently with a slight thickening at one end.[62] Circular dichroism measurements show caldesmon to have no regular secondary structure[62] and calculations comparing the length to the molecular weight predict that most of the caldesmon molecule must be an extended polypeptide chain. Pure caldesmon has a marked tendency to self-aggregation. At salt concentrations above 200 mM, reversible formation of end-to-end dimers is observed.[57] Caldesmon appears to contain one or two active-SH groups (presumably cysteine); and aggregates formed by S–S cross-linking between caldesmons readily occur in media that do not contain high levels of SH-protecting reagents.[61,62]

Caldesmon is able to bind to all the components of the thin filament. Binding to actin and actin + tropomyosin has been studied in some detail.[63] Caldesmon binds to two types of sites on polymeric actin: it binds to "tight" sites with affinities in the 10^8–10^7 M^{-1} range with a stoichiometry of 1 caldesmon to every 20 actins (estimated from Lowry protein assay against an actin standard and assuming a molecular weight of 87,000) and also to "weak" sites with an affinity about 100-fold less and a stoichiometry of about 1 per 4 actins. Only the tight sites are likely to be physiologically relevant since the stoichiometry is the same as that found in native thin filaments, and binding at tight sites is directly correlated to regulation of the filament activity. Binding affinity of caldesmon to the tight sites is dependent on salt concentration, temperature, and tropomyosin. The effect of added tropomyosin on caldesmon affinity for actin probably occurs because caldesmon also binds to tropomyosin, with an affinity in the 10^5–10^6 M^{-1} range.[63]

The location of caldesmon bound in the native thin filament has been deduced from electron microscopy studies (Fig. 1). Thin filaments, whether single, in paracrystals, or in bundles induced by various cross-linking agents, have so far proven indistinguishable from actin + tropomyosin; consequently, the caldesmon cannot be a compact structure since a molecule of that size would be visible.[64] This idea was confirmed by examining thin filaments that had been aggregated into parallel bundles by polyclonal antibody.[65,66] These bundles had no periodic structures and specific labeling of the antibody with gold particles showed it to be positioned all along the filaments; thus, caldesmon is probably extended along the whole surface of the thin filament (Fig. 3). Since the caldesmon molecule may be as long as 150 nm, a single caldesmon molecule could extend over about 20 actins in one strand of the actin helix; thus, this structural arrangement can account for the observed binding stoichiometry. A detailed analysis of the structure of thin filaments has recently been published.[67]

Besides binding actin and tropomyosin, caldesmon also binds to calcium-binding proteins; indeed, caldesmon was first isolated on the basis of its binding to a calmodulin affinity column.[53] The binding of brain calmodulin has been studied in some detail, but it is likely that caldesmon can bind to a number of similar proteins (e.g., skeletal muscle troponin C).[68]

Ca^{2+}·calmodulin can bind to caldesmon, but calmodulin does not. Binding affinity is about 10^6 M^{-1} and is largely independent of temperature (range 25–37°C) and [KCl].[63,66]

The sites where caldesmon binds to actin and Ca^{2+}·calmodulin are retained in quite small regions of the molecule obtained by careful proteolytic digestion. The C-terminal 40,000-molecular-weight peptide appears to contain these binding sites unimpaired while binding activity can still be detected in a smaller 20,000-molecular-weight peptide derived from the larger subunit.[69,70] It is suggested, but not proven, that the actin- and calmodulin-binding domain of caldesmon corresponds to the thickened "head" region seen at the end of the caldesmon molecule. In contrast, tropomyosin-binding ability is not found in the 40,000-molecular-weight fragment, but it is found in the other half of caldesmon, a 100,000-molecular-weight fragment, suggesting that tropomyosin is in contact with the "tail" of caldesmon.[71] The fragment molecular weights quoted are estimates from gel electrophoresis and hence differ from values of native caldesmon or as obtained by sedimentation.

3.4. Native Thin Filament

The native thin filament preparation is obtained from smooth muscles by homogenizing and washing muscle in a low-salt, low-Ca^{2+} buffer with 0.5% Triton X-100. An actin-rich actomyosin is extracted from the homogenate in high-ATP solution and the actin, plus its complement of actin-binding components, is separated from soluble components, such as myosin, by high-speed sedimentation.[40] Thin filament preparations contain actin, tropomyosin, and caldesmon in a similar ratio to that present in whole tissue; consequently, it is believed they correspond to native thin filaments, mainly from the "contractile domain"[18] since they contain only small quantities of filamin (Fig. 2).

The ratio of protein components in uterine thin filaments is actin: tropomyosin: caldesmon $1:0.38:0.093$[8] w/w by densitometry (Fig. 2); this is the same as thin filaments from other smooth muscles[7] and also corresponds to the binding ratios of pure tropomyosin and caldesmon (see Section 3.3). Thin filaments are found to be rather variable in length (Fig. 1). The length distribution is roughly reciprocal (frequency proportional to l/length) with a median of 1 μm; the longest filaments observed have been 4 μm[40] It is not known whether the filaments are longer, or even shorter *in vivo*. Figure 3 indicates the most probable structure, which has been deduced from electron microscope studies of native and reconstituted thin filaments.

4. Function of the Thin Filaments in Uterine Muscle Contraction

In smooth muscle, as in striated muscle, contractility is believed to be produced by the interaction of thick-filament cross-bridges (myosin heads) with actin in the thin filaments. Force and movement are generated at the expense of MgATP hydrolysis at the active site of myosin. When investigating the interaction of purified actin and myosin, only MgATPase activity can be measured, but studies with intact muscle preparations have shown that this correlates reasonably well with contractility.[72,73] Contractility is controlled by the Ca^{2+} concentration in the vicinity of the contractile apparatus.[74,75] The stimulation of contraction of Ca^{2+} is based on a reversal of the inhibitory effects of regulatory protein components in both the thick filaments and the thin filaments. This release of inhibitory control is accomplished via the mediation of Ca^{2+}-binding proteins. Ca^{2+} control of thick filaments involves Ca^{2+}·calmodulin activation of myosin light chain kinase, which phosphorylates the regulatory (inhibitory) light chain of the myosin head, allowing actin to activate myosin MgATPase (see Chapter 3).

4.1. Native Thin Filaments plus Myosin

The interaction of actin with phosphorylated myosin is, however, controlled by the thin filament regulatory system involving caldesmon.[76,77] Ca^{2+} regulation of myosin MgATPase is an intrinsic property of thin filaments extracted intact from smooth muscle (Fig. 4).

The Ca^{2+} sensitivity is particularly high for smooth muscle thin filaments activating smooth muscle heavy meromyosin or myosin (phosphorylated) (Fig. 4b) while with skeletal muscle myosin, Ca^{2+} sensitivity is rarely better than threefold (Fig. 4a). At high $[Ca^{2+}]$ (i.e., $> 10^{-5}$ M), thin filaments activate myosin maximally, with a K_m quite similar to that of actin-tropomyosin, but at low (i.e., $< 10^{-8}$ M) $[Ca^{2+}]$, activation is very low and sometimes zero. The Ca^{2+}-dependent regulators of thin filaments evidently work by inhibiting activation in the absence of

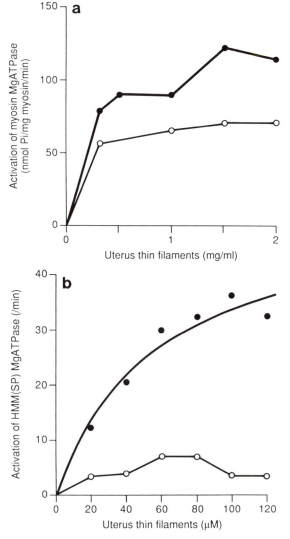

Figure 4. Uterine thin filament activation of myosin MgATPase. (a) Uterine thin filaments with skeletal muscle myosin. Conditions: 23°C, 40 mM KCl, 5 mM PIPES, pH 7.0, 5 mM $MgCl_2$, 1 mM dithiothreitol (DTT), 5 mM Na-azide, 0.25 mg/ml skeletal myosin, 0–2 mg/ml thin filaments (0–27 μM actin monomer), 0.1 mM $CaCl_2$ (●) or 1 mM EGTA (○). Background ATPase and myosin MgATPase (9 nmoles/mg per min) have been subtracted.[8]

(b) Uterine thin filaments with smooth muscle (aorta) thiophosphorylated heavy meromyosin [HMM(SP)]. Conditions: 37°C, 5 mM PIPES K_2, pH 7.1, 2.5 mM $MgCl_2$, 1 mM DTT, 4 μM HMM(SP), 0–150 μM (actin monomer concentration) thin filaments (0–12 mg/ml), 0.1 mM $CaCl_2$ (●) or 1 mM EGTA (○). Background ATPase and myosin MgATPase (1.2 min^{-1}) were subtracted.

Maximal activation of skeletal myosin by uterine thin filaments was 130 nmoles/mg myosin per min (equivalent to a turnover of 30 min^{-1}). Maximal Ca^{2+} sensitivity was 1.7-fold, at 23°C. With smooth muscle myosin at 37°C, maximal activation was 52 min^{-1} and estimated K_m for activation was 56 μM. The Ca^{2+} sensitivity was at least 10-fold. It is commonly found that Ca^{2+} sensitivity is greatest with assay mixtures made up wholly of smooth muscle proteins. From ref. 8 with permission.

Ca^{2+}. This form of regulation is thus analogous to the troponin system in striated muscles, and is quite the opposite of the so-called "leiotonin" mechanism of thin filament control once proposed by Ebashi.[78]

Early work on Ca^{2+}-regulatory factors in uterine muscle provided evidence for troponin[79] and for troponin-I and troponin-C like proteins.[80] Both these reports suggested they were proteins of 20,000–40,000 molecular weight. However, all Ca^{2+}-regulated thin filament preparations contain caldesmon and the evidence that this is the regulatory protein is quite compelling.[7,40] Caldesmon is easily split into 40,000- and 20,000-molecular-weight active components by endogenous proteolysis,[69] so it is not unreasonable to believe that the early reports had isolated caldesmon fragments.

Our goal is to understand the mechanism of Ca^{2+} regulation in thin filaments on the basis of the properties of their component proteins. This has required *in vitro* studies in which synthetic thin filaments have been reconstituted from the purified individual proteins. The simplest system is F-actin plus myosin; if the other constituents are added one by one, we can reconstitute systems with progressively better resemblance to the native thin filaments.

4.2. Actin plus Myosin

Actin activates myosin MgATPase activity. The amount of activation is a specific property of the myosin; skeletal muscle myosins generally have a high actin-activated MgATPase rate, while smooth muscle myosin is activated the least. This low actin activation of smooth muscle myosin is directly related to the characteristically slow rates of contraction in smooth muscles[81,82] [see Marston (1983)[82] for a full account of smooth muscle actomyosin ATPase]. Uterine and other smooth muscle actins activate myosin and its proteolytic fragments, heavy meromyosin or subfragment-1, from either smooth or striated muscles; their properties differ little from those of skeletal muscle F-actin[40,83] (Fig. 5). Actin is a highly conserved molecule, and the small differences in amino acid sequence (Section 3.1) evidently play no larger role in the actin–myosin interaction.

4.3. Actin–Tropomyosin plus Myosin

When tropomyosin is bound to actin filaments, the properties of the actin are modified. The influence of tropomyosin depends on the tropomyosin type, and is further variable according to experimental conditions. All types of tropomyosin appear to increase the affinity of actin for myosin·ADP·Pi, as determined from the K_m in reactions using heavy meromyosin.[84] Under most conditions, smooth muscle tropomyosins also increase actin activation, a phenomenon known as "potentiation"—although in other conditions, inhibition is observed instead. Potentiation is favored at ionic strengths greater than 0.1, at temperatures greater than 25°C, and at small actin/myosin ratios. Under conditions we generally use for our assays (e.g., Figs. 4–6), a potentiation of about 50% is the norm, although as much as 300% has

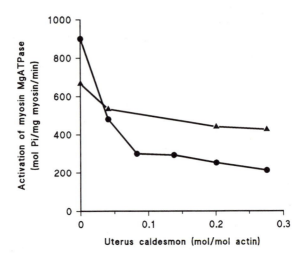

Figure 5. Caldesmon inhibition of actin and actin–tropomyosin activation of skeletal muscle myosin. Conditions: 35°C, 60 mM KCl, 5 mM PIPES, pH 7.0, 5 mM $MgCl_2$, 5 mM Na-azide, 1 mM DTT; 0.125 mg/ml skeletal myosin, 0.5 mg/ml uterine actin (▲) or 0.5 mg/ml (22 μM) uterine actin plus 0.125 mg/ml aorta tropomyosin (●) with 0–0.13 mg/ml (1.2 μM) uterine caldesmon.

The presence of tropomyosin on actin filaments potentiated activation of myosin MgATPase by one-third. Caldesmon was a potent inhibitor of actin–tropomyosin activation. It was very much less effective as an inhibitor of actin activation. Adapted from ref. 8 with permission.

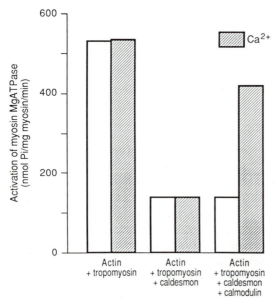

Figure 6. Ca^{2+} regulation of actin–tropomyosin–caldesmon by the addition of calmodulin. Conditions: 35°C, 70 mM KCl, 5 mM PIPES, pH 7.1, 5 mM $MgCl_2$, 5 mM Na-azide, 1 mM DTT, 0.125 mg/ml skeletal myosin, 0.5 mg/ml uterine actin, and 0.125 mg/ml aorta tropomyosin; 0.1 mM $CaCl_2$ (hatched bars) or 1 mM EGTA (white bars); 0.15 mg/ml uterine caldesmon, 0.4 mg/ml bovine brain calmodulin.

Actin–tropomyosin activation of myosin MgATPase activity was completely insensitive to $[Ca^{2+}]$ changes. Caldesmon inhibition of activation was also Ca^{2+} insensitive. Addition of calmodulin reversed caldesmon inhibition in the presence, but not in the absence, of Ca^{2+}. From ref. 8 with permission.

been observed. Inhibition due to smooth muscle tropomyosin can be observed at low temperatures, or at very high actin/myosin ratios at low ionic strength [see Marston and Smith (1985)[24] for a full analysis]. Skeletal muscle tropomyosin is substantially different since potentiation is only seen under certain rather specific conditions and it is usually inhibitory by up to 70%. The way in which tropomyosin changes actin activation should not be considered regulatory, since none of these effects are Ca^{2+}-regulated; however, tropomyosin clearly does play some role in the actin–myosin interaction which could potentially be under the control of additional regulatory factors.

4.4. Actin–Tropomyosin–Caldesmon plus Myosin

Actin, tropomyosin, and caldesmon make up at least 80% of the protein in thin filaments.[8,40] When caldesmon is added to actin + tropomyosin, it always inhibits activation of myosin MgATPase (Fig. 5).[54,63,85,88] Maximal inhibition is usually > 80% and can be > 95% and requires an unusually low stoichiometry of caldesmon : actin. When caldesmon binding is measured together with inhibition, it is found that inhibition is directly proportional to the fractional occupation of sites on the filament. These sites have an affinity of about 10^7 M^{-1} and saturate with 0.048 caldesmon bound per actin.[63] Since the native thin filaments contain caldesmon in precisely the same stoichiometry, it is highly likely that caldesmon inhibition is involved in regulating the thin filament. Under many conditions, tropomyosin must be bound on actin for caldesmon inhibition to be observed[63]: in the absence of tropomyosin, caldesmon binds but has minimal effect. The tropomyosin requirement is, however, not absolute; at low ionic strengths or temperatures, caldesmon can inhibit equally well with or without tropomyosin.[86]

The central role of caldesmon in this system is virtually certain, since caldesmon is known to be the only protein present in native thin filaments in the correct stoichiometry and because it has been shown that a specific anticaldesmon antibody, which reverses caldesmon inhibition, can antagonize Ca^{2+} regulation of native thin filaments.[87] Inhibition by caldesmon is completely independent of the $[Ca^{2+}]$ (Fig. 6). The system lacks any mechanism for the release of inhibition at high $[Ca^{2+}]$. It is conceivable that native caldesmon has regulatory Ca^{2+}-binding sites that are destroyed during purification, but it is more likely that actin, tropomyosin, and caldesmon cannot suffice to reconstitute a Ca^{2+}-regulated thin filament. Native thin filaments seem to contain an additional, Ca^{2+}-sensitizing protein that is lost when caldesmon is purified. In many experiments, brain calmodulin has been used as a model Ca^{2+}-binding regulatory protein.

4.5. Actin–Tropomyosin–Caldesmon with Calmodulin plus Myosin

Under suitable conditions, Ca^{2+}·calmodulin reverses the inhibitory effect of caldesmon on actin–tropomyosin activation of myosin MgATPase activity.[54,63,85,88] Figure 6 illustrates such an experiment, using uterine proteins.[8] Ca^{2+}·calmodulin binds to caldesmon with an affinity of about 10^6 M^{-1},[63,68] which is little affected by changes in temperature or ionic strength. This binding abolishes caldesmon inhibition. In the absence of Ca^{2+}, calmodulin has no effect, since it does not bind to caldesmon. Thus, the two proteins, calmodulin and caldesmon, can form a mechanism for the Ca^{2+} regulation of thin filaments. Ca^{2+}·calmodulin is effective at releasing caldesmon inhibition at [KCl] > 70 mM, at 37°C (as little as 2 μM excess free Ca^{2+}·calmodulin may be required under optimum conditions) but extremely large (> 60 μM) concentrations of Ca^{2+}·calmodulin are required at 25°C. This behavior is a consequence of different types of interaction of calmodulin with caldesmon; at the high temperatures and ionic strengths, Ca^{2+}·calmodulin binds directly to actin–tropomyosin–caldesmon, forming a quaternary complex in which caldesmon   inhibition   is   reversed.[63]   However,   at   lower   temperatures Ca^{2+}·calmodulin binding to caldesmon is competitive with caldesmon binding to actin–tropomyosin, and consequently very high Ca^{2+}·calmodulin concentrations are required to displace caldesmon from actin–tropomyosin.

4.6. The Ca^{2+}-Binding Component of Native Thin Filaments

We are not certain whether calmodulin is involved in the intrinsic Ca^{2+} regulation of native thin filaments. In contrast to the reconstituted system,[63] they are Ca^{2+} regulated under most conditions.[40] Ca^{2+} sensitivity is not affected by changing [KCl] in the range 10–150 mM but varies with temperature, being best at temperatures below 25°C and considerably reduced at 37°C (using skeletal muscle myosin). This behavior is opposite to that described above for reconstituted systems, containing Ca^{2+}·calmodulin. Furthermore, there is no evidence that, in high [Ca^{2+}], caldesmon dissociates from native thin filaments, or that in the absence of Ca^{2+}, there is a dissociation of the Ca^{2+}-binding protein. Calmodulin has been used experimentally since it is readily available, and its use has enabled us to elucidate the basic mechanism of control, but we believe that the calcium-binding component of the thin filament is a different protein. This putative calcium-binding protein (CBP) has not been identified, but we have studied mixtures of calcium-binding proteins extracted from smooth muscle and found that they contain a protein activity that closely mimics the properties expected of the thin-filament calcium-binding protein (e.g., the release of caldesmon inhibition at 25°C[66]). It is to be hoped that this calcium-binding protein activity can be purified and characterized in the near future.

5. A Consensus Model for the Ca^{2+} Regulatory Mechanism in Smooth Muscle Thin Filaments

It is now appropriate to gather together the structural and biochemical data discussed in the previous sections into a concise model of the regulatory mechanism (Fig. 7).

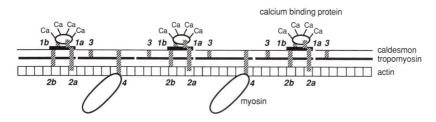

Figure 7. A structural and mechanistic model of the Ca^{2+}-regulated thin filament of smooth muscles. The model combines the structural data, information on the domain structure of caldesmon and intermolecular binding contacts, and the regulatory properties of the intact and reconstituted thin filament systems. It is based mainly on data obtained with aorta smooth muscle, but it is compatible with results obtained with uterine thin filaments. The figure represents one strand of the actin helix (Figs 1, 3) as a line of actin monomers. It is suggested that each unit of 14 actins binds four tropomyosin monomers, one caldesmon, and one molecule of a calcium-binding protein. The tropomyosin shown is the dimer. The locations of interaction sites are derived from studies of caldesmon's domain structure. Five different types of interaction between caldesmon (CD) and other proteins have been identified.

1a, CD–CBP inactivates 2a when Ca^{2+} is bound.

1b, CD–CBP binding contact independent of Ca^{2+}.

2a, CD–actin binding with inhibition of up to 14 actins.

2b, CD–actin binding contact only, Ca^{2+}-independent.

3, CD–tropomyosin, located in CD tail. Binding contact which is required to propagate 2a at physiological temperatures and ionic strengths.

4, CD–myosin, binding contact only, may be involved in latch-bridges.

1. Caldesmon is bound to actin with a high affinity and a stoichiometry of 1 per 10 to 20 actin monomers.

2. In the presence of tropomyosin the bound caldesmon inhibits actin activation of myosin MgATPase by up to 95%. Neither the binding nor the inhibition is Ca^{2+}-sensitive.

3. A Ca^{2+}-binding protein binds to caldesmon. In the presence of Ca^{2+}, the $Ca^{2+} \cdot CBP$ interacts with caldesmon and releases its inhibition of the actomyosin MgATPase activity. This interaction may be associated with weakening of the caldesmon–actin and caldesmon–tropomyosin links. Figure 7 illustrates the likely structure of a regulated unit of the thin filament and the protein–protein interactions involved.

6. Physiological Role for Ca^{2+} Regulation of Thin Filaments

There is a wealth of evidence that sarcoplasmic $[Ca^{2+}]$ controls uterine and other smooth muscle contraction.[74,89,90] It is also well established that physiological $[Ca^{2+}]$ can activate both the myosin filaments (via myosin light chain kinase) and the actin filaments (via caldesmon and a calcium-binding protein)[24,77] (see Chapters 1 and 3). Measurements relating myosin phosphorylation to contractility have been made in uterine muscle[75,90,91] and to date they seem to indicate that activation of myosin by phosphorylation always correlates with contractility. On the face of it, this finding leaves little role for thin filament regulation, except perhaps as a modulator of muscle response to phosphorylation. In other smooth muscles, notably vascular smooth muscles, which have been investigated in depth,[89,92] it has been observed that myosin phosphorylation correlates with the rise of tension following stimulation, but subsequent sustained contractures are maintained while myosin phosphorylation levels fall toward the relaxed level. This response (called "latch") may, therefore, be under the control of a second regulatory system, which could be the thin filament.

Recent work indicates that Ca^{2+} regulation of the thin filament can control contractility:

1. The contraction of actomyosin gels is inhibited by caldesmon.[93]
2. The movement of myosin along immobilized actin filaments is inhibited by caldesmon.[94]
3. Caldesmon can partly relax desensitized skeletal muscle fibers.[95]

Also, it has recently been suggested that thin filaments can bind to myosin via a high-affinity second site that is distinct from the ATPase site and probably involves caldesmon.[87] This binding could conceivably be analogous to the latch state of intact muscle.

Our current working hypothesis is that Ca^{2+} control of the thin filaments occurs in intact uterine smooth muscle and has the potential for regulating contractility but more direct evidence is needed.

7. Conclusion

The actin-based thin filaments form one half of the contractile mechanism of uterine smooth muscles. Their interaction with myosin results in contraction at the expense of MgATP hydrolysis. This function is a property of all actin filaments and is essentially the same in all muscle systems; the uniquely slow contraction velocity of smooth muscle is a property of the myosin, not the actin. The unique properties of smooth muscle thin filaments, of which uterine thin filaments are a typical example, are related either to the way in which the filaments are organized into contractile elements or its Ca^{2+}-dependent regulatory system.

We still have only a poor idea of how actin filaments are organized other than that they produce contraction via the sliding filament mechanism and that a number of actin-associated proteins are believed to be involved (e.g., filamin, vinculin, α-actinin, talin). In contrast, recent work on the caldesmon system has shown us that the thin filaments are Ca^{2+}-regulated and we can describe the regulatory mechanism in some detail. There can now be little doubt that Ca^{2+} does control contractility in part via the receptors on the thin filaments and we are now eager to explore how it does so in concert with myosin regulation and its physiological role.

ACKNOWLEDGMENT. We are grateful to the British Heart Foundation for funding our work.

References

1. Haeberle, J. R., Coolican, S. A., Evan, A., and Hathaway, D. R., 1985, The effects of a calcium dependent protease on the ultrastructure and contractile mechanism of skinned uterine smooth muscle. *J. Muscle Res.* **6:**347–363.
2. Bagby, R. M., 1983, Organisation of the contractile/cytoskeletal elements, in: *The Biochemistry of Smooth Muscle* (N. L. Stephens, ed.), CRC Press, Boca Raton, pp. 1–84.
3. Murphy, R. A., 1979, Filament organisation and contractile function in smooth muscle, *Annu. Rev. Physiol.* **41:**737–748.
4. Hanson, J., and Lowy, J., 1963, The structure of F-actin filaments isolated from muscle, *J. Mol. Biol.* **6:**46.
5. Vibert, P. J., Haselgrove, J. C., Lowy, J., and Poulsen, F. R., 1972, Structural changes in actin-containing filaments of muscle, *J. Mol. Biol.* **71:**757–767.
6. Cohen, P. M., and Murphy, R. A. 1978, Differences in cellular contractile protein contents among porcine smooth muscles, *J. Gen. Physiol.* **72:**369–380.
7. Marston, S. B., and Lehman, W., 1985, Caldesmon is a Ca^{2+}-regulatory component of native smooth muscle thin filaments, *J.* **231:**517–522.
8. Marston, S. B., 1989, Ca^{2+}-dependent regulation of uterus smooth muscle thin filaments by caldesmon, *Am. J. Obstet. Gynecol.* **160:**252–257.
9. Egelman, E. H., 1985, The structure of F-actin, *J. Muscle Res. Cell Motil.* **6:**129–151.
10. Ashton, F. T., Somlyo, A. V., and Somlyo, A. P., 1975, The contractile apparatus of vascular smooth muscle: Intermediate high voltage stereoelectron microscopy, *J. Mol. Biol.* **98:**17–29.
11. Cohen, D. M., and Murphy, R. A., 1979, Cellular thin filament protein contents and force production in porcine arteries and veins, *Circ. Res.* **45:**661–665.
12. Rice, R. V., Moses, J. A., McManus, G. M., Brady, A. C., and Blasik, L. M., 1970, The organisation of contractile filaments in mammalian smooth muscle, *J. Cell Biol.* **47:**183–196.
13. Somlyo, A. V., Butler, T. M., Bond, M., and Somlyo, A. P., 1981, Myosin filaments have non-phosphorylated light chains in relaxed smooth muscle, *Nature* **294:**567–570.
14. Small, J. V., 1985, Geometry of actin–membrane attachments in smooth muscle cells: The localisation of vinculin and alpha-actinin, *EMBO J.* **4:**45–49.
15. Bond, M., and Somlyo, A. V., 1982, Dense bodies and actin polarity in vertebrate smooth muscle, *J. Cell Biol.* **95:**403–413.
16. Fay, F. S., Fujiwara, K., Rees, D. D., and Fogarty, K. E., 1983, Distribution of alpha actinin in single isolated smooth muscle cells, *J. Cell Biol.* **96:**783–795.
17. Gabella, G., 1973, Fine structure of smooth muscle, *Philos. Trans. R. Soc. London Ser. B* **265:**7–16.
18. Furst, D. O., Cross, R. A., DeMey, R. A., and Small, J. V., 1986, Caldesmon is a long, flexible molecule localised in actomyosin domains of smooth muscle, *EMBO J.* **5:**251–257.
19. Begg, D. A., Rodenwald, R., and Rubhun, L. I., 1979, The visualisation of actin filament polarity in thin sections. Evidence for the uniform polarity of membrane associated filaments, *J. Cell Biol.* **79:**846–852.
20. Fatigati, V., and Murphy, R. A., 1984, Actin and tropomyosin variants in smooth muscles, *J. Biol. Chem.* **259:**14383–14388.
21. Cavaille, F., Janmot, C., Ropert, S., and d'Albis, A., 1986, Isoforms of myosin and actin in human, monkey, and rat myometrium, *Eur. J. Biochem.* **160:**507–513.
22. Hsu, C.-Y. J., and Frankel, F. R., 1987, Effect of oestrogen on the expression of mRNAs of different actin isoforms in immature rat uterus, *J. Biol. Chem.* **262:**9594–9600.
23. Vandekerckhove, J., and Weber, K., 1979, The complete amino-acid sequence of actins from bovine

aorta, bovine heart, bovine fast skeletal muscle and rabbit slow skeletal muscle, *Differentiation* **14:**123–133.

24. Marston, S. B., and Smith, C. W. J., 1985, The thin filaments of smooth muscles, *J. Muscle Res. Cell Motil.* **6:**669–708.

25. Spudich, J. A., and Watt, S., 1971, The regulation of rabbit skeletal muscle contraction, *J. Biol. Chem.* **246:**4866–4871.

26. Carsten, M. E., 1965, A study of uterine actin, *Biochemistry* **4:**1049–1054.

27. Haeberle, J. R., Coolican, S. A., Evan, A., and Hathaway, D. R., 1985. The effects of a calcium dependent protease on the ultrastructure and contractile mechanism of skinned uterine smooth muscle, *J. Muscle Res. Cell Motil.* **6:**347–363.

28. Elce, J. S., Elbrecht, A. S. V., Middelstadt, M. V., McIntyre, E. J., and Anderson, P. J., 1981, Actin from pig and rat uterus, *Biochem. J.* **193:**891–898.

29. Strzelecka-Golaszewska, H., Hinssen, H., and Sobieszek, A., 1984, Influence of an actin-modulating protein from smooth muscle on actin–myosin interaction. *FEBS Lett.* **177:**209–216.

30. Calvadore, J. C., Axelrud-Calvadore, C., Berta, P., Harricane, M. C., and Haiech, J., 1985, Preparation and characterisation of bovine aortic actin, *Biochem. J.* **228:**433–441.

31. Pardee, J. D., and Spudich, J. A., 1982, Purification of muscle actin, *Methods Enzymol.* **82B:**164–179.

32. Suck, D., Kabsch, W., and Mannherz, H., 1981, Three dimensional structure of a complex of skeletal muscle actin and bovine pancreatic DNAse I at 6A resolution, *Proc. Natl. Acad. Sci. USA* **78:**4319–4323.

33. Suzuki, K., Yamaguchi, M., and Sekine, T., 1978, Two forms of chicken gizzard F-actin depending on preparations with and without added calcium, *J. Biochem.* **83:**869–878.

34. Strzelecka-Golaszewska, H., Prochniewiczs, E., Zmorzynski, S., and Drabikowski, W., 1980, Chicken gizzard actin: Polymerisation and stability, *Eur. J. Biochem.* **104:**41–52.

35. Oosawa, F., and Kasai, M., 1962, A theory of linear and helical aggregation of macromolecules, *J. Mol. Biol.* **4:**10–21.

36. Needham, D. M., and Williams, J. M., 1963, Proteins of the uterine contractile mechanism, *Biochem. J.* **89:**552.

37. Sobieszek, A., and Bremel, R. D., 1975, Preparation and properties of vertebrate smooth muscle myofibrils and actomyosin, *Eur. J. Biochem.* **55:**49–60.

38. Driska, S. P., and Hartshorne, D. J., 1975, The contractile properties of smooth muscle. Properties and components of a calcium sensitive actomyosin from chicken gizzard, *Arch. Biochem. Biophys.* **167:**203–212.

39. Small, J. V., and Sobieszek, A., 1983, Contractile and structural proteins of smooth muscle, in: *The Biochemistry of Smooth Muscle* (N. L. Stephens, ed.), CRC Press, Boca Raton, pp. 85–140.

40. Marston, S. B., and Smith, C. W. J., 1984, Purification and properties of Ca^{2+}-regulated thin filaments and F-actin from sheep aorta smooth muscle, *J. Muscle Res. Cell Motil.* **5:**559–575.

41. Sanders, C., and Smillie, L. B., 1985, Amino acid sequence of chicken gizzard beta tropomyosin: Comparison of the chicken gizzard, rabbit skeletal and equine platelet tropomyosins, *J. Biol. Chem.* **260:**7264–7275.

42. Ruiz-Opazo, N., Weinberger, J., and Nadal-Guinard, B., 1985, One smooth and two striated, skeletal and cardiac alpha-tropomyosin isoforms are encoded by the same gene, *Nature* **315:**67–70.

43. Helfman, D. M., Cheley, S., Kuismanen, E., Finn, L. A., and Yamawaki-Kataoka, Y., 1986, Nonmuscle and muscle tropomyosin isoforms are expressed from a single gene by alternative RNA splicing and polyadenylation, *Mol. Cell. Biol.* **6:**351–354.

44. Carsten, M. E., 1968, Tropomyosin from smooth muscle of the uterus, *Biochemistry* **7:**960–967.

45. Bailey, K., 1948, Tropomyosin, a new asymmetric protein component of the muscle fibril, *Biochem. J.* **43:**271–279.

46. Woods, E. F., 1976, The conformational stabilities of tropomyosins, *Aust. J. Biol. Sci.* **29:**405–418.

47. McLachan, A. D., and Stewart, M., 1976, The 14-fold periodicity in alpha tropomyosin and the interaction with actin, *J. Mol. Biol.* **103**:271–298.

48. Dabrowska, R., Nowak, E., and Drabikowski, W., 1980, Comparative studies of chicken gizzard and rabbit skeletal tropomyosin, *Comp. Biochem. Physiol.* **65B**:75–83.

49. Sanders, C., and Smillie, L. B., 1984, Chicken gizzard tropomyosin: Head–tail assembly and interaction with F-actin and troponin, *Can. J. Biochem. Cell. Biol.* **62**:443–448.

50. Taylor, K. A., and Amos, L. A., 1981, A new model for the geometry of the binding of myosin crossbridges to actin filaments. *J. Mol. Biol.* **147**:297–324.

51. Pearlstone, J. R., and Smillie, L. B., 1982, Binding of troponin-T fragments to several types of tropomyosin, *J. Biol. Chem.* **257**:10587–10592.

52. Lim, S. S., Tu, Z., and Lemanski, L. F., 1984, Anti-troponin-T monoclonal antibody crossreacts with all muscle types, *J. Muscle Res. Cell Motil.* **5**:515–526.

53. Sobue, K., Muramoto, Y., Fujita, M., and Kakiuchi, S., 1981, Purification of a calmodulin binding protein from chicken gizzard that interacts with F-actin, *Proc. Natl. Acad. Sci. USA* **78**:5652–5655.

54. Smith, C. W. J., and Marston, S. B., 1985, Disassembly and reconstitution of the Ca^{2+}-sensitive thin filaments of vascular smooth muscle, *FEBS Lett.* **184**:115–119.

55. Ngai, P. K., Carruthers, C. A., and Walsh, M. P., 1984, Isolation of the native form of chicken gizzard myosin light chain kinase, *Biochem. J.* **218**:863–870.

56. Bretscher, A., 1984, Smooth muscle caldesmon: Rapid purification and F-actin cross-linking properties, *J. Biol. Chem.* **259**:12873–12880.

57. Cross, R. A., Cross, K. E., and Small, J. V., 1987, Salt dependent dimerisation of caldesmon, *FEBS Lett.* **219**: 306–310.

58. Graceffa, P., Wang, C.-L.A., and Stafford, W. F., 1988, Caldesmon. Molecular weight and subunit composition by analytical ultracentrifugation, *J. Biol. Chem.* **263**:14196–14202.

59. Bryan, J., Imai, M., Lee, R., Moore, P., Cook, R. G., and Lin, W. G., 1989, Cloning and expression of a smooth muscle caldesmon. *J. Biol. Chem.* **264**:13873–13879.

60. Smith, C. W. J., 1985, Thin filament linked regulation of vascular smooth muscle contraction, Ph.D. thesis, University of London.

61. Sobue, K., Morimoto, K., Kanda, K., and Kakiuchi, S., 1985, Crosslinking in actin filaments is caused by caldesmon aggregates but not by its dimers, *FEBS Lett.* **182**:201–204.

62. Lynch, W. P., Riseman, V. M., and Bretscher, A., 1987, Smooth muscle caldesmon is an extended flexible monomeric protein in solution that can readily undergo reversible intra- and intermolecular sulphydryl cross-linking *J. Biol. Chem.* **262**:7429–7437.

63. Smith, C. W. J., Pritchard, K., and Marston, S., 1987, The mechanism of Ca^{2+}-regulation of vascular smooth muscle thin filaments by caldesmon and calmodulin, *J. Biol. Chem.* **262**:116–122.

64. Moody, C. J., Lehman, W., and Marston, S. B., 1986, The structural basis for caldesmon regulation of vascular smooth muscle contraction, *J. Muscle Res. Cell Motil.* **7**:373.

65. Lehman, W., and Marston, S. B., 1986, Caldesmon association with smooth muscle thin filaments, *Biophys. J.* **49**:67a.

66. Marston, S. B., Lehman, W., Moody, C., and Pritchard, K., 1988, Caldesmon and calcium regulation in smooth muscles, in: *Calcium and Calcium Binding Proteins* (C. Gerday, R. Gilles, and L. Bolis, eds.), Springer-Verlag, Berlin, pp. 69–81.

67. Lehman, W., Craig, R., Lui, J., and Moody, C., 1989, Caldesmon and the structure of smooth muscle thin filaments: Immunolocalization of caldesmon on thin filaments. *J. Muscle Res. Cell Motil.* **10**:101–112.

68. Pritchard, K., and Marston, S., 1989, Ca^{2+}-calmodulin binding to caldesmon and the caldesmon–actin–tropomyosin complex, *Biochem. J.* **257**:839–843.

69. Szpacenko, A., and Dabrowska, R., 1986, Functional domains of caldesmon, *FEBS Lett.* **202**:182–186.

70. Fujii, T., Imai, M., Rosenfeld, G. C., and Bryan, J., 1987, Domain mapping of chicken gizzard caldesmon, *J. Biol. Chem.* **262**:2757–2763.

71. Riseman, V. M., and Bretscher, A., 1986, Proteolytic dissection of caldesmon into fragments that retain F-actin, Ca^{2+}-calmodulin and tropomyosin-binding activities, *J. Cell Biol.* **103:**536a.
72. Paul, R. J., Gluck, E., and Ruegg, J. C., 1976, Cross-bridge ATP utilization in arterial smooth muscle, *Pfluegers Arch.* **361:**297–299.
73. Siegman, M. J., Butler, T. M., Mooers, S. V., and Davies, R. E., 1980, Chemical energetics of force development, force maintenance and relaxation in mammalian smooth muscle, *J. Gen. Physiol.* **76:**609.
74. Filo, R. S., Bohr, D. F., and Ruegg, J. C., 1965, Glycerinated smooth and skeletal muscle: Calcium and magnesium dependence, *Science* **147:**1581–1583.
75. Haeberle, J. R., Holt, J. W., and Hathaway, D. R., 1985, Regulation of isometric force and isotonic shortening by phosphorylation of the 20,000 dalton myosin light chain of rat uterine smooth muscle, *Pfluegers Arch.* **403:**215–219.
76. Marston, S. B., Trevett, R. M., and Walters, M., 1980, Calcium ion regulated thin filaments from vascular smooth muscle, *Biochem. J.* **1985:**355–365.
77. Marston, S. B., 1982, The regulation of smooth muscle contractile proteins, *Prog. Biophys. Mol. Biol.* **41:**1–41.
78. Ebashi, S., 1980, the Croonian Lecture 1979: Regulation of muscle contraction, *Proc. R. Soc. London Ser. B* **207:**259–286.
79. Carsten, M. E., 1971, Uterine smooth muscle: Troponin, *Arch. Biochem. Biophys.* **147:**353–357.
80. Grand, R. J. A., Perry, S. V., and Weeks, R. A., 1977, The troponin like components of smooth muscle, in: *Excitation–Contraction Coupling in Smooth Muscles* (R. Casteels, T. Godfraind, and J. C. Ruegg, eds.), Elsevier/North-Holland, Amsterdam, pp. 335–341.
81. Marston, S. B., and Taylor, E. W., 1980, Comparison of the myosin and actomyosin ATPase mechanisms of the four types of vertebrate muscles, *J. Mol. Biol.* **139:**573–600.
82. Marston, S. B., 1983, Myosin and actomyosin ATPase: Kinetics, in: *Biochemistry of Smooth Muscle,* Volume 1 (N. L. Stephens, ed.), CRC Press, Boca Raton, pp. 167–191.
83. Strzelecka-Golaszewska, H., and Sobieszek, A., 1981, Activation of smooth muscle myosin by smooth and skeletal muscle myosins, *FEBS Lett.* **134:**197–201.
84. Sobieszek, A., 1982, Steady state kinetics of the actin activation of skeletal muscle heavy meromyosin subfragments. Effects of skeletal, smooth and non-muscle tropomyosins, *J. Mol. Biol.* **157:**275–286.
85. Sobue, K., Morimoto, K., Inui, M., Kanda, K., and Kakiuchi, S., 1982, Control of actin–myosin interaction of gizzard smooth muscle by calmodulin and caldesmon linked flip–flop mechanism, *Biomed. Res.* **3:**188–196.
86. Chalovich, J. M., Cornelius, P., and Benson, C. E., 1987, Caldesmon inhibits skeletal actomyosin subfragment-1 ATPase activity and the binding of subfragment-1 to actin, *J. Biol. Chem.* **262:**5711–5716.
87. Marston, S. B., Pritchard, K., Redwood, C., and Taggart, M., 1988, Ca^{2+}-regulation of the thin filaments: Biochemical mechanism and physiological role, *Trans. Biochem. Soc.* **16:**494–497.
88. Dabrowska, R., Goch, A., Galazkiewicz, B., and Osinska, H., 1985, The influence of caldesmon on ATPase activity of the skeletal muscle actomyosin and bundling of actin filaments, *Biochim. Biophys. Acta* **842:**70–75.
89. Kamm, K. E., and Stull, J. T., 1985, The function of myosin light chain kinase phosphorylation in smooth muscle, *Annu. Rev. Pharmacol.* **25:**593–620.
90. Haeberle, J. R., Hathaway, D. R., and DePaoli-Roach, A. A., 1985, Dephosphorylation of myosin by the catalytic subunit of a type-2 phosphatase produces relaxation of chemically skinned uterine smooth muscle, *J. Biol. Chem.* **260:**9965–9968.
91. Janis, R. A., Barany, K., Barany, M., and Sarmiento, G. J., 1981, Association between myosin light chain phosphorylation and contraction of rat uterine smooth muscle, *Mol. Physiol.* **1:**3–11.
92. Murphy, R. A., 1982, Myosin phosphorylation and crossbridge regulation in arterial smooth muscle. A state-of-the-art review, *Hypertension* **4** (Suppl. II):3–7.
93. Nomura, M., and Sobue, K., 1987, Caldesmon regulates the three dimensional contraction (myosin

dependent contraction of the actin binding protein induced actin gel), *Biochem. Biophys. Res. Commun.* **144:**936–943.

94. Sellers, J. R., and Shirinsky, V. P., 1987, Caldesmon inhibits movement of myosin coated beads on nitella actin cables, *J. Cell Biol.* **105:**194a.

95. Taggart, M., and Marston, S., 1988, Caldesmon relaxes desensitised rabbit fibres, *FEBS Lett.* **242:**171–174.

5

Calcium Control Mechanisms in the Myometrial Cell and the Role of the Phosphoinositide Cycle

Mary E. Carsten and Jordan D. Miller

1. From Excitation to Contraction

Smooth muscle contraction follows excitation at the plasma membrane. The link between these two events is the rise in intracellular calcium. Calcium can enter through plasma membrane channels activated either by depolarization of the plasma membrane or by specific binding of a hormone to its receptor. Hormone–receptor binding can be followed by the breakdown of plasma membrane phosphoinositides with the generation of a series of messengers. These serve multiple functions, including the release of calcium from intracellular stores into the cytoplasm. In this chapter we first discuss the biochemical reactions that cause changes in intracellular calcium concentration. This is followed by a review of the phosphoinositide system and its role in the release of second messengers that initiate myometrial contraction.

2. Calcium as Activator of Contraction

The necessity of calcium for smooth muscle contraction has long been recognized. The molecular mechanisms underlying the involvement of calcium in smooth muscle contraction have only recently been discovered (see Chapters 3 and 4), and our knowledge is still incomplete. However, we can say with certainty that calcium couples stimulus to contraction. This role of calcium is obligatory, whether the

Mary E. Carsten • Departments of Obstetrics–Gynecology and Anesthesiology, School of Medicine, University of California Los Angeles, Los Angeles, California 90024.   *Jordan D. Miller* • Department of Anesthesiology, School of Medicine, University of California Los Angeles, Los Angeles, California 90024.

stimulus is of hormonal nature or voltage induced, and irrespective of whether the calcium originates from extracellular or intracellular sources.

2.1. Intracellular Calcium

Estimates of the calcium requirement for contraction in smooth muscle were based on the calcium sensitivity of skinned fibers[1] including those from pregnant rat myometrium.[2] The threshold for contraction was about 0.2 μM calcium with a peak contraction at 1–6 μM. Whether these values apply to the *in vivo* situation depends on the accuracy of the many assumptions made about the intracellular environment including binding constants, free cation concentrations, ATP concentration, possible soluble cofactors, and so forth. Therefore, attempts at direct measurement were made in living cells. The intracellular concentration of calcium is most directly measured with calcium-selective electrodes in nonmobile large cells such as nerves. The electrode, which requires cells larger than 10–20 μm, has detection limits of 10^{-8} M Ca^{2+}, response times of 1 sec or more and is subject to local transients.[3] The smooth muscle cell is difficult to analyze because of its small size, motion, and low resting levels of calcium. Thus, much preliminary work was done in larger cells. Newer techniques have allowed progress using the smooth muscle cell.

Aequorin, a bioluminescent protein, emits a blue light in the presence of free calcium. It can be injected or more recently loaded into cells by temporarily making the cells permeable. This procedure does not seem to change the cellular sensitivity to contractile agents,[4,5] and muscle strips can be observed both at rest and during a contraction. Under these conditions, resting intracellular free calcium is 0.1 μM and, during contraction with potassium, 0.4 μM. With phenylephrine there is an initial peak of 1.3 μM, which reduces to 0.2 μM during the steady-state contraction.[6] A shortcoming of the aequorin method is that the signal is not linearly related to the free calcium. Thus, two areas, one of high and one of low calcium, generate a larger signal than one area with the average of the calcium concentrations.[4]

There are now a number of fluorescent dyes that can be used for the measurement of free calcium. The wavelength at which excitation occurs or the intensity of the emission of the dye changes with the concentration of free calcium. These dyes are best used in single cells; they are loaded as permeable esters and then made impermeant by the action of cellular esterases. The fluorescent intracellular dye quin-2 requires a high concentration of dye, which buffers the free calcium, and the results are not definitive. Fura-2, with greater sensitivity and ionic specificity than quin-2, is expected to define more clearly the range of calcium values seen in intact cells.[7] The most recent work has observed the regional changes and rapid oscillations[8] in free calcium; however, since the stored calcium in the sarcoplasmic reticulum is released to the cytoplasm, and since the dyes penetrate to some extent into the sarcoplasmic reticulum, the resting values are high, and the values at contraction will be low.[9] For a more detailed discussion of the methodology and problems, see the review by Cobbold and Rink.[10] In spite of some shortcomings, the application of these methods to uterine cells would be highly desirable.

2.2. Sources of Calcium and Calcium Transport

For contraction to occur, the intracellular calcium concentration must be changed from 10^{-7} M to 10^{-6} M. This may occur from either or both of two sources. Calcium can enter the cell through the plasma membrane or be released from membrane-bound intracellular storage areas. Relaxation would occur when the calcium is removed from the cytoplasm across these same membranes. The source of calcium thus may be extracellular or intracellular. In the skeletal muscle cell, intracellular calcium stores are primarily responsible for contraction. In contrast, calcium entering the cell through the plasma membrane may be an important source of calcium for contraction in the smooth muscle cell because of the relatively slow contraction and high surface-to-volume ratio. The uterine muscle cell ranges in diameter from 5 to 10 μm and from 300 to 600 μm in length, with the largest size being present toward the end of gestation. The average volume of the myometrial smooth muscle cell has been estimated at 21,000 μm^3 and the surface area at 23,000 μm^2, thus yielding a high surface-to-volume ratio of 1.1, as compared to 0.1 for skeletal muscle (for review see Garfield and Somlyo[11]).

The concentration of calcium in the interstitial fluid is 10^{-3} M; that inside the smooth muscle cell is 10^{-6} M in contraction and 10^{-7} M in relaxation; i.e., there is a 10,000-fold concentration gradient. The electrochemical gradient from outside to inside the cell is maintained by the plasma membrane. This gradient is maintained because of the low calcium permeability of the plasma membrane. Electron microscopy[12,13] and electron probe analysis[14,15] (see Chapter 1) have established that the intracellular calcium storage area at physiological calcium levels is the sarcoplasmic reticulum. The sarcoplasmic reticulum is a system of intracellular membranes that form channels or tubules. It is very abundant in skeletal muscle, where it mediates rapid release of calcium. The smooth muscle cell has been shown to have a sarcoplasmic reticulum similar to that of skeletal muscle. The sarcoplasmic reticulum represents 2–7.5% of the cell volume in smooth muscles, with pregnant myometrium being in the higher range[14] (see Chapter 1). The intracellular stored calcium recycles and is adequate to produce contractions in the absence of extracellular calcium.[16] It is presently thought that mitochondria play no role in the regulation of cytoplasmic calcium in smooth muscles except at very high and thus nonphysiological levels.[14]

Contraction is terminated by a reduction in the intracellular free calcium concentration. To maintain calcium homeostasis, so-called calcium pumps move calcium against its concentration gradient across the plasma membrane and/or the membranes of the sarcoplasmic reticulum. This calcium transport requires energy. The first demonstration of calcium sequestration in smooth muscle (uterus) microsomal preparations was in 1969.[17] In these preparations the calcium pump requires the splitting of ATP in the presence of calcium and magnesium. The pump is thus an enzyme, the Ca,Mg-ATPase. For the purpose of this chapter we use the terms *calcium pump, Ca,Mg-ATPase,* and *transport ATPase* interchangeably. There are two different Ca,Mg-ATPases, one located in the plasma membrane and the other in the membranes of the sarcoplasmic reticulum, and they have different properties.[18–20]

3. Uterine Muscle-Specific Problems

Uterine smooth muscle can be studied in intact muscle strips, permeabilized cells, or isolated cellular components. Intact muscle strips are discussed in Chapter 7 on electrophysiology. Permeabilized cells are useful for studying effects of compounds normally produced inside the cell; leakiness of the plasma membrane allows their application outside the cell and observation of their intracellular action. Isolated cellular components offer the best prospects for studying biochemical reactions, short of isolated enzyme systems. Thus, in order to characterize the different ATPases involved in calcium transport, one must separate plasma membrane and sarcoplasmic reticulum from one another.

3.1. Microsomal Preparations

Unlike skeletal muscle, smooth muscle cells are held together by tough collagen fibers, and even treatment with collagenase will not completely free the cells, at least not in our hands. Because of the small cell size, the uterine smooth muscle cell has relatively more plasma membrane and less sarcoplasmic reticulum than skeletal muscle cells. Hence, it is more difficult to obtain preparations of sarcoplasmic reticulum free of plasma membrane.

The most frequently used procedure for isolation of plasma membrane and sarcoplasmic reticulum is by dissecting and homogenizing the myometrium followed by centrifugation. In order to obtain good preparations, it is essential that the uteri be removed quickly (even in the slaughterhouse this can be done within 15 min after sacrificing the cow), immediately dissected, rinsed, and placed in ice-cold buffer. The muscle tissue is then minced in a meat grinder and homogenized twice in a Waring blender. It is essential that during these operations the tissue be kept ice cold.

Initially, heavier, larger particles are discarded by stepwise centrifugation at increasing force. We carry out this differential centrifugation at 2500 g for 20 min, 15,000 g for 20 min, and 40,000 g for 90 min. The use of two low-speed spins rather than one increases the yield and purity. The final pellet is resuspended and placed on a sucrose density gradient, where it is centrifuged to equilibrium,[21] thereby achieving a separation. Indeed, the use of a 35, 45, and 55% sucrose density gradient made possible the first demonstration of ATP-dependent calcium uptake into microsomal vesicles derived from the myometrium.[17] The main protein layer was at 35% sucrose, and electron microscopy confirmed the vesicular nature of the preparation. Properties associated with sarcoplasmic reticulum including oxalate stimulation of ATP-dependent calcium uptake were identified.[22] The protein can be stored at 4°C and used the following day.[23] However, the separation of plasma membrane and sarcoplasmic reticulum is often incomplete because of the small density differential between the separated microsomal fractions. Aggregation of microsomes may occur, changing the density of components. With an alternative sucrose gradient of 24, 28, 33, and 45%, purer fractions are obtained, although yield is sacrificed.[21] The sarcoplasmic reticulum (density 1.11) is concentrated at the 0/24% and 24/28% sucrose interfaces, whereas plasma membrane (density 1.16) is concentrated at the 28/33% and 33/45% sucrose interfaces.[21] The degree of purity of this sarcoplasmic reticulum preparation was established by oxalate stimulation of ATP-dependent calcium uptake[22] and measurement of plasma membrane markers, such as 5'-nucleotidase, ouabain inhibition of Na,K-ATPase, ouabain inhibition of p-nitrophenylphosphatase, [³H]ouabain binding, and [¹²⁵I]wheat germ agglutinin binding (for discussion see Section 3.3).

In other procedures the density of specific components is artificially changed. Thus, the density of the plasma membrane fraction can be increased by treatment with digitonin, which reacts with cholesterol,[18] so that the plasma membrane sediments at higher density than normal. However, digitonin treatment may increase the permeability of the membrane and thus make the preparation less useful for further studies including calcium uptake. Alternatively, the density of the sarcoplasmic reticulum fraction can be increased by calcium uptake in the presence of large amounts of ATP, calcium, and oxalate.[24] The calcium-loaded sarcoplasmic reticulum vesicles are then centrifuged into the pellet while the supernatant contains plasma membrane and inside-out and leaky sarcoplasmic reticulum vesicles. There was a sevenfold increase in the calcium content of the pelleted material as compared to the original calcium-loaded sarcoplasmic reticulum vesicles. Depending on the specificity of the oxalate stimulation (see below), there could be contamination with inside-out plasma membrane vesicles.

Isolation procedures based on affinity chromatography usually suffer from poor yield and technical difficulties with passage of particulate preparations through columns. In our hands, wheat germ agglutinin affinity purification resulted in only 8% of the uterine microsomal protein being recovered as plasma membrane.[21] In other smooth muscles, calmodulin affinity chromatography was used to purify the Ca,Mg-ATPase of plasma membrane.[25] In view of the small yields obtained in these procedures, one must be aware of the possibility of purifying a contaminant such as the erythrocyte Ca,Mg-ATPase, as one has to expect contamination with blood. A criterion to use is that the yield of the purified material must be greater than the amount of that material originally present in any possible contaminant.

3.2. Criteria for Purity

In order to evaluate the properties of these preparations, one first has to estimate the purity of the isolated fractions. This has posed some problems inasmuch as there is no generally agreed on marker for sarcoplasmic reticulum. Furthermore, it has been shown in several laboratories that one marker for plasma membrane, for instance, may appear in several microsomal fractions.[21,26] If this were a result of contamination of one fraction with another, the degree of contamination computed from markers should be the same for all markers. For example, if the plasma membrane marker 5'-nucleotidase shows 10% plasma membrane contamination of a fraction, then the alternative plasma membrane marker, wheat germ agglutinin binding, should also show 10% contamination in that fraction. This has not been the case.[21,26] Thus, the specificity of these markers is questionable,[27] and their appearance in various fractions may well be an inherent property of each.

3.3. Specific Markers

The most frequently used enzyme markers for plasma membrane are 5'-nucleotidase, K^+-stimulated ouabain-sensitive phosphatase, phosphodiesterase,[26] ouabain-sensitive p-nitrophenylphosphatase, and Na,K-ATPase.[21] In plasma membrane preparations of a variety of smooth muscles, ATP-dependent calcium uptake[18] as well as ATPase activity[28] is stimulated by calmodulin and inhibited by vanadate, with the plasma membrane showing a greater sensitivity to vanadate than the sarcoplasmic reticulum.[18,29]

ATP-dependent calcium uptake in ileum and myometrial smooth muscle plasma membrane preparations showed specificity for ATP, and other nucleotide triphosphates were less effective or ineffective.[19,29,30] This is in contrast to sarcoplasmic reticulum preparations from striated muscles (which showed no nucleotide specificity),[30] but this may be tissue, not organelle, specific, since sarcoplasmic reticulum from gastric smooth muscle uses ATP as the preferred substrate.[19]

For identification of sarcoplasmic reticulum one uses the oxalate stimulation of ATP-dependent calcium uptake[26] by analogy with skeletal muscle. The basis for oxalate stimulation is the greater oxalate permeability of the sarcoplasmic reticulum membranes compared to the oxalate permeability of the plasma membrane.[19] When the concentration of calcium oxalate inside the microsomal vesicles exceeds the solubility product of calcium oxalate, crystals precipitate. The solubility product of calcium oxalate is approximately $2.6 \times 10^{-9} M^2$; this is an approximation, since it may be 100 times larger in the presence of ATP and magnesium, and is modified by the pH inside the vesicle. In the presence of oxalate, the free calcium concentration inside the vesicles is kept low, the calcium gradient is increased, and calcium uptake is potentiated. However, the increase in oxalate-stimulated calcium uptake in sarcoplasmic reticulum from smooth muscle is much less than in that from skeletal muscle.[31] We observed only a three- to four-fold increase in the amount of calcium taken up in the presence of oxalate as compared to its absence. Since phosphate penetrates the plasma membrane vesicles more readily, it is not surprising that phosphate was found to be better than oxalate as a calcium-trapping anion in plasma membrane. It potentiates ATP-dependent calcium uptake into plasma membrane fraction.[19,30] Some questions concerning the use of the oxalate effect as a marker have recently been raised on the basis of lack of association of this marker with other enzyme activities used as markers, and of inconsistent results.[32] A more reasonable interpretation of these findings would be that there was insufficient separation of plasma membrane and sarcoplasmic reticulum; indeed, no sarcoplasmic reticulum fraction was identified.[32]

Another marker used for smooth muscle sarcoplasmic reticulum is NADH cytochrome C reductase (rotenone insensitive). Since this enzyme is also found in outer mitochondrial membrane,[19] its validity as a marker is questionable.

Antibodies to the plasma membrane Ca,Mg-ATPase inhibited the Ca,Mg-ATPase and calcium uptake in a plasma-membrane-enriched fraction from smooth muscle, whereas the sarcoplasmic-reticulum-enriched fraction was only slightly inhibited, and oxalate-stimulated calcium uptake was not inhibited.[28] For meaningful results, absolute purity of the plasma membrane Ca,Mg-ATPase used as antigen must be assured, and the source must be known with certainty. The source of the antigen cannot be known, since red cells are present in any tissue, and the erythrocyte antigen generated identical antibodies; the argument is thus circular.

Another marker used for identification of plasma membrane is specific binding. We have had good success with binding of [^{125}I]wheat germ agglutinin and of [^{3}H]ouabain.[21] The use of specific hormone receptor binding as a marker is based on the premise that hormones and agonists bind exclusively to the plasma membrane. This is not entirely true, since some hormones have been shown to cross the plasma membrane and others originate within cells. For example, prostaglandin E_2 was bound to plasma membrane and sarcoplasmic reticulum fractions in accord with the presence of receptors in both plasma membrane and sarcoplasmic reticulum.[33] The ideal marker would be one applied to the whole tissue that could be followed through the purification procedure by electron microscopy, immunology, radioactivity, and so forth. But even that might encounter difficulties such as impermeability of the tissue or cell to the marker or displacement of the marker during the purification procedure. We made an attempt to apply a similar technique by labeling uterine muscle strips with [^{125}I]wheat germ agglutinin and obtained greater specificity than when [^{125}I]wheat germ agglutinin was used on the homogenate.[21]

3.4. Molecular Parameters

There are differences in some molecular properties of plasma membrane and sarcoplasmic reticulum. For example, the lipid composition is different. The cholesterol/phospholipid ratio is higher in the plasma membrane than in the sarcoplasmic reticulum.[19] However, this is only a quantitative difference and hence not of much value in identifying the origin of isolated fractions.

The plasma membrane ATPase of smooth muscles has a molecular weight of 130,000–140,000, whereas that of the sarcoplasmic reticulum is 100,000–110,000.[20,34] In uterine microsomal preparations, the sarcoplasmic reticulum ATPase was shown to have a molecular weight of 100,000–110,000.[35] Furthermore, the tryptic digestion patterns of the plasma membrane ATPase and the sarcoplasmic reticulum ATPase of smooth muscle are different from one another, and the tryptic digestion patterns of the sarcoplasmic reticulum ATPase closely resemble those of skeletal and cardiac sarcoplasmic reticulum, though they are immunologically distinct.[36]

4. Calcium Pumps and Relaxation

Two pathways are available to the uterine muscle cell to lower the intracellular calcium concentration to produce relaxation. It can transport calcium either through the plasma membrane to the interstitial fluid or across the membranes of the sarcoplasmic reticulum for intracellular storage. In both cases calcium is moved against a large concentration gradient.

4.1. Plasma Membrane

Two mechanisms for removal of calcium from the cytoplasm through the plasma membrane have been proposed. They operate against a concentration gradient and are driven by different sources of energy.

4.1.1. The Ca,Mg-ATPase

The plasma membrane Ca,Mg-ATPase constitutes the primary mechanism mediating calcium efflux from the uterine muscle cell. Isolated plasma membrane vesicles from rat myometrium of varying degrees of purity have been shown to elicit ATP-dependent calcium uptake from reaction media containing submicromolar concentrations of calcium.[37−39] The plasma membrane fraction was extensively characterized and found to be relatively pure.[37] However, in other experiments two supposed plasma membrane fractions were separated[38]; since one of these showed oxalate-stimulated calcium uptake, it appears that this fraction was sarcoplasmic reticulum or was heavily contaminated with sarcoplasmic reticulum.

Unfortunately, in these studies the rat was used as a model. The rat may not be the best model for the human. For example, the sensitivity of rodents to cardiac glycosides as well as [^{3}H]ouabain binding and ouabain inhibition of the plasma membrane Na,K-ATPase is much lower than that in the human (two orders of magnitude for the I_{50} value).[40] Another species-related difference is found in the 5'-nucleotidase activity, which is phenomenally high in the rat[26,37,38] and much lower in rabbit,[41] cow, and human myometrium[21] and pig stomach.[19] Thus, one must emphasize that biochemical properties vary from species to species and tissue to tissue and that one cannot assume that information obtained from one is applicable to another.

In preparations of sarcolemmal sheets from human myometrium a Ca,Mg-ATPase was identified.[42] This plasma membrane Ca,Mg-ATPase is stimulated by calmodulin and inhibited by vanadate. A K_m for calcium binding of 0.25 μM was calculated, which is of the same order of magnitude as the K_d of calcium binding calculated for sarcoplasmic reticulum.[43] Care must be taken in interpreting results when Ca,Mg-ATPase activity is measured rather than calcium transport, since other highly active Ca,Mg-ATPases that are not the transport ATPase are present in the cell membrane, and they have very different properties.[29]

4.1.2. Na$^+$–Ca^{2+} Exchange

Another mechanism to be considered for moving calcium across the plasma membrane is Na$^+$–Ca^{2+} exchange. An antiporter present in the plasma membrane carries Na$^+$ in one direction and Ca^{2+} in the other. It is an important and well-accepted mechanism in cardiac muscle, where it does not require ATP for energy, is concentration dependent, and is electrogenic, since three Na^+ are exchanged for one Ca^{2+} (for review see Reeves[44]). It may work in either direction, but under physiological conditions, where Na^+ outside is within 30–50 mM of normal range, it functions only to remove Ca^{2+}.[45] The Na^+–Ca^{2+} exchange has been suggested to be present in smooth muscle based on observations in intact cells[45−49] and in microsomal preparations[50] including those originating from rat uterine plasma membrane.[29,51,52] Increased calcium uptake in sodium-preloaded (plasma membrane-containing) microsomes from bovine myometrium has also been observed in our laboratory (unpublished). In general, Na^+–Ca^{2+} exchange is of low magnitude, varies with the smooth muscle studied, and its physiological significance is unknown at this time. However, the observed low magnitude of Na^+–Ca^{2+} exchange in smooth muscle may be related to technical difficulties. The proof of the existence of Na^+–Ca^{2+} exchange requires careful control of conditions. The calcium gradient must be such that the opening of a channel will not carry calcium in the same direction, and the accumulated calcium must be releasable when a calcium ionophore is added. For Na^+–Ca^{2+} exchange to occur, the Na^+ gradient must be maintained during exchange. Lack of an adequate Na^+ gradient may account for the low values obtained.[50−52] Sodium loading and Na^+ permeability must be followed, since absolute values for Na^+ permeability differ between tissues and possibly between species. Net changes in the concentration of the two cations must be demonstrated, which is difficult when using radioactive tracers, since the degree of replacement of hot for cold cation may vary and confuse the interpretation of results.

4.2. Sarcoplasmic Reticulum

The second pathway by which the uterine smooth muscle cell lowers the intracellular calcium concentration is by storing the calcium inside the cell, in the sarcoplasmic reticulum. This process, like the plasma membrane calcium pump, requires energy in the form of ATP. It is calculated that the calcium gradient across the sarcoplasmic reticulum is similar to that across the plasma membrane; thus, the concentration of calcium on the inside of the sarcoplasmic reticulum is in the millimolar range. Whether the calcium is free in solution, complexed with insoluble anions, or bound to soluble or insoluble proteins remains to be determined. The use of oxalate *in vitro,* to precipitate calcium and thus increase the amount of calcium stored, is clearly not physiological. It is not known if phosphate is a physiological substitute; it has not been demonstrated *in vivo.* Various calcium-binding proteins are present in skeletal muscle sarcoplasmic reticulum, but their presence in smooth muscle sarcoplasmic reticulum has not been demonstrated conclusively. We have found calcium-binding proteins in our sarcoplasmic reticulum preparation that may be comparable to calsequestrin and high-affinity calcium-binding proteins of skeletal muscle sarcoplasmic reticulum. Recently a protein with properties like cardiac calsequestrin has been identified in gastric smooth muscle sarcoplasmic reticulum.[53] The quantities isolated so far are too small to allow full characterization.

The reactions and mechanisms of ATP-dependent calcium uptake into the sarcoplasmic reticulum are best studied in microsomal preparations. In our laboratory these originate from bovine uterus obtained at the slaughterhouse from freshly killed cows. The relevance of bovine studies to the human was checked by repeating our experiments on human uteri obtained at cesarean hysterectomy.[54] Since the latter procedure is only infrequently performed, most of our work was done on uteri from pregnant cows.

4.2.1. Measurement of ATP-Dependent Calcium Uptake

Calcium uptake is demonstrated in a buffered solution (pH 7.0) in the absence and presence of ATP (usually 2 mM) and calcium (20 μM). The difference between the two measurements is the ATP-dependent calcium uptake. There are essentially three methods for determination of calcium uptake. (1a) With the filtration method, either the calcium taken up into the sarcoplasmic reticulum vesicles or the calcium released can be measured. The incubation medium contains ^{45}Ca (500,000 cpm/ml). Samples are taken out at specified time intervals and filtered through 0.45-μm pore size Millipore™ filters with prefilters. Aliquots of the filtrate are counted in a liquid scintillation system. The calcium taken up is computed from the difference in counts in the incubation medium and the filtrate, using appropriate controls not containing ATP.[23] Low protein concentrations (<0.5 mg/ml) or EGTA buffers with a high total calcium but low free calcium cannot be used with this technique, since the difference in total calcium would be too small to measure accurately. (1b) A minor modification of this method is to filter the aliquot and, after two brief washings, directly count the filters for ^{45}Ca. No prefilters are used.[55] The advantage of this technique is that smaller amounts of protein and EGTA buffering of calcium can be used. (2) In the centrifugation method, the calcium taken up is determined. Aliquots of sarcoplasmic reticulum preparations are incubated in the presence and absence of ATP. Centrifugation is used to separate the microsomes from the incubation medium. After centrifuging at 165,000 g (av.) for 30 min, washing, and recentrifuging, calcium is determined in each pellet by atomic absorption spectroscopy in the presence of trichloroacetic acid, lanthanum, and cesium. ATP-dependent calcium uptake again is the difference between calcium uptake in the presence and absence of ATP.[23]

Each of these methods has its advantages and disadvantages. The filtration methods are suitable for kinetic measurements and also when the objective is to measure calcium release. The centrifugation atomic absorption method, developed in our laboratory,[23] gives highly accurate results but is more time consuming. We found this method superior because of the considerable amount of exchange taking place between added ^{45}Ca and intrinsic ^{40}Ca.[23] This can introduce large errors in the calculation of calcium uptake in method 1a because of low calcium uptake (in contrast to skeletal muscle), a large intrinsic exchangeable calcium pool, low total calcium concentration in the medium, and possible dependence of the size of the exchangeable calcium pool on added ATP or pharmacological agents. This will be particularly pertinent when comparing ATP-dependent calcium uptake in different fractions of one preparation. With reference to method 1b, commercial filters contain variable amounts of calcium; this makes measurement of nanomolar amounts of calcium inaccurate.[22] The centrifugation method, for example, enabled us to show that cyclic AMP does not increase ATP-dependent calcium uptake into sarcoplasmic reticulum preparations.[56]

(3) Spectrophotometric methods depend on a shift in the absorption spectrum of a dye alone as compared to the calcium–dye complex. Most frequently metallochromic indicators for calcium are used. These are substances that undergo color changes when the concentration of free calcium in solution changes. With changes in the free calcium concentration a shift in wavelength of the absorbance maximum and in the extent of molar absorbance will occur. The most frequently used metallochromic indicators are murexide and arsenazo III. For calcium concentrations varying from 25 to 250 μM, murexide is used; at lower concentrations, 1–25 μM, arsenazo III is used. The latter has sensitivity two orders of magnitude greater than murexide,[57] which brings it into the range for measuring ATP-dependent calcium uptake into smooth muscle sarcoplasmic reticulum. These indicators may not be entirely specific for calcium; selection of an appropriate wavelength is essential to exclude reactions with interfering ions[57] and variability associated with pH changes.[58]Since these methods are dependent on the free calcium in the medium, changes in free calcium unrelated to uptake such as that produced by the metabolism of ATP to ADP will affect the results and must be taken into account. Under our standard conditions the calculated free calcium rises from 7 to 10 μM, and the free magnesium from 0.39 to 0.69 mM in 8 min.[59] A dual-wavelength spectrophotometer is essential for accurate measurements. The methods allow measurement of kinetics.

4.2.2. Calcium Transport

The ATP-dependent uptake of calcium from solution requires intact SH groups[17] and is temperature dependent.[23] The amount of ATP-dependent calcium uptake depends on the calcium concentration in the medium. When the free calcium concentration is varied over a 100-fold range with EGTA, increased calcium uptake with increased free calcium is observed.[22]

To demonstrate that the calcium taken up in the presence of ATP is actually transported into the microsomal vesicles, one needs to show that the calcium taken up cannot be removed simply by washing or by treatment with EGTA. EGTA complexes calcium only on the outside of the sarcoplasmic reticulum vesicles and releases accumulated calcium slowly. In contrast, calcium ionophores move calcium across membranes down the concentration gradient; they are ideal tools to identify intravesicular calcium. The calcium ionophore X-537A rapidly released calcium ($t_{1/2} = 12$ min), whereas calcium release with EGTA was extremely slow ($t_{1/2} = 93$ min), as the vesicles are relatively impermeable to calcium (see Fig. 1).[24] We have used calcium-oxalate-loaded sarcoplasmic reticulum vesicles both to further document the source of the vesicles as sarcoplasmic reticulum and to show that

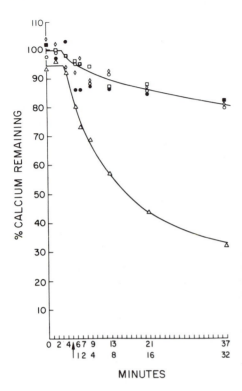

Figure 1. Calcium release from microsomes derived from pregnant uterine sarcoplasmic reticulum. Microsomes were loaded with calcium in the presence of oxalate at pH 7.5. Additions for release: None ($\Diamond$), 2 mM EGTA ($\Box$), 2 mM EGTA and 2×10^{-5} M X-537A ($\triangle$), 2 mM EGTA and 50 μg/ml PGE$_2$ ($\bigcirc$), 2 mM EGTA and 50 μg/ml PGI$_2$ ($\bullet$). Addition of EGTA at zero time; X-537A, PGE$_2$, PGI$_2$ after 5 min (at arrow). (From Carsten and Miller,[24] with permission.)

the calcium is inside the vesicle. Oxalate will only precipitate calcium in a closed vesicle where the product of the free calcium concentration times the oxalate concentration exceeds the solubility product of calcium oxalate. In our preparations oxalate increased the rate of calcium accumulation by a factor of three. This is in contrast to sarcoplasmic reticulum from skeletal or cardiac muscle, where it is over 50-fold.[41] The presence of calcium oxalate crystals inside sarcoplasmic reticulum vesicles has been demonstrated in other smooth muscle preparations by electron microscopy.[60]

4.2.3. The Ca,Mg-ATPase

The Ca,Mg-ATPase that is coupled to calcium uptake is only a small fraction of the total Mg-ATPase, in contrast to skeletal muscle preparations where the Ca,Mg-ATPase constitutes the major portion of the total ATPase. Therefore, the Ca,Mg-ATPase activity is difficult to measure in smooth muscle sarcoplasmic reticulum. It is measured as the difference between ATP split in the presence of calcium and magnesium and the ATP split in the presence of magnesium alone, i.e., in the presence of sufficient EGTA to bind essentially all calcium present as contaminant, usually 1 mM EGTA.[23] This ATPase was shown to be temperature dependent[23] and to have a molecular weight of 100,000 to 110,000[35]; it was not stimulated by calmodulin or lanthanum or inhibited by vanadate.[61]

The structure of the Ca,Mg-ATPase has become clearer as both immunologic and DNA probes have been used. Based on the highly conserved sequence around the phosphorylation site, cDNA probes have been constructed to identify many aspartylphosphate cation transferases including the Ca,Mg-ATPase, Na,K-ATPase, H,K-ATPase, and their isoforms.[62] The sarcoplasmic reticulum Ca,Mg-ATPase of fast skeletal muscle differs from the Ca,Mg-ATPase of other muscle types, whereas the slow skeletal, cardiac, and smooth muscle Ca,Mg-ATPases have many similarities.[36,63] Differences between uterine and cardiac muscle Ca,Mg-ATPases have been shown at the 3' end of the mRNA,[64] and the C terminal four amino acids of slow skeletal are replaced by 50 amino acids in the gastric[62] and uterine smooth muscle enzyme.[65] This is a rapidly expanding area of investigation, and no full review is intended.

4.2.4. Mechanism of Calcium Transport

The mechanism of calcium transport in uterine sarcoplasmic reticulum as well as in other smooth muscles has only been partly elucidated. The mechanism in skeletal muscle may serve as a model. It is helpful for our understanding of the calcium transport mechanism to look at partial reactions; some of these have been demonstrated in myometrium.

An obligatory early step in the transport of calcium across the internal membranes of smooth muscle cells is the binding of calcium to the Ca,Mg-ATPase. Calcium binding can be conveniently measured by atomic absorption spectroscopy. An accepted method for examination of ligand–receptor interaction is Scatchard analysis.[66] Fitting of the binding data is often difficult, and deciding whether to use a model with one or two types of specific binding sites has been subjective. With the advent of the computer age, weighted least-square curve fitting combined with the use of an exact mathematical model of binding allows objective measurement of goodness of fit for several models using Scatchard analysis.[67,68] Thus, it has been possible to use statistical analysis to justify a more complex model and to determine calcium affinity constants with great accuracy. By this method, calcium transporting microsomes in the absence of ATP were shown to have two binding sites and nonspecific binding. The association constants were $K_1 = 8.4 \pm 0.4 \times 10^6$ M$^-$, $K_2 = 0.11 \pm 0.02 \times 10^6$ M$^-$. The maximum number of binding sites was $r_1 = 3.9 \pm 0.4 \times 10^{-6}$, $r_2 = 10.5 \pm 0.5 \times 10^{-6}$ mol/g, and nonspecific binding of 0.021. In the presence of magnesium the number of low-affinity binding sites was reduced to about half, whereas receptor affinity and the number of high-affinity sites were unchanged (see Fig. 2).[43] At the present time it cannot be stated with certainty that the measured constants refer to the enzyme, though the high-affinity site is in the correct range for the Ca,Mg-ATPase.

Because the transport of calcium across the sarcoplasmic reticulum membrane is coupled with the enzymatic hydrolysis of ATP, one hypothesizes that there is an

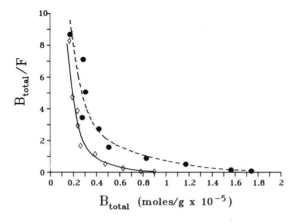

B_{total} (moles/g x 10^{-5})

Figure 2. Scatchard plot of calcium binding to mirosomes derived from pregnant uterine sarcoplasmic reticulum in the presence ($\Diamond$) and absence ($\bullet$) of 5 mM free magnesium. Each point represents the average of five experiments. Analysis by weighted least-squares curve fitting. (From Miller and Carsten,[43] with permission.)

enzyme–substrate intermediate. Therefore, the next step in the calcium transport is the formation of a high-energy intermediate in the presence of ATP and calcium. This has been demonstrated by showing a phosphorylated ATPase in the sarcoplasmic reticulum preparation.[35] The proposed reaction scheme is as follows:

$$Ca_{out} + E + ATP \rightleftharpoons Ca \cdot E \cdot ATP \rightleftharpoons Ca \cdot E \sim P + ADP$$
$$Ca \cdot E \sim P \rightleftharpoons E + P + Ca_{in}$$

where E is the Ca,Mg-ATPase enzyme, Ca_{out} calcium outside the sarcoplasmic reticulum vesicle, Ca_{in} calcium inside the sarcoplasmic reticulum vesicle, and P inorganic phosphate. $Ca \cdot E \cdot ATP$ is the enzyme substrate complex going to $Ca \cdot E \sim P$, the high-energy intermediate of the enzyme, while splitting off ADP.

Only part of this reaction scheme has been demonstrated at this time. The calcium binding that we observed in the absence of ATP clearly shows that calcium binding can precede ATP binding. The formation of the phosphoenzyme complex was demonstrated using [^{32}P]ATP, separating the protein from the reaction mixture, and measuring the radioactivity. The yield of calcium-stimulated phosphoprotein depended strictly on the ATP concentration. Calcium and magnesium were both required for the reaction to proceed.

The phosphoprotein was extensively characterized. It was established that the protein was indeed the sarcoplasmic reticulum ATPase and not a Na,K-ATPase, as addition of ouabain or omission of KCl had no effect on the phosphorylation. Absence or presence of sodium azide also did not affect the formation of the phosphoprotein, thus excluding mitochondrial origin of the ATPase. The ^{32}P-labeled microsomes were further characterized by sodium dodecylsulfate (SDS) polyacrylamide gel electrophoresis. The protein bands were stained, and SDS gel slices were counted in a liquid scintillation system. There were at least ten protein bands demonstrated in the SDS electrophoresis of the preparation; one corresponded with a large peak of calcium-dependent phosphorylation (see Fig. 3). The molecular weight of this phosphoprotein was estimated to be approximately 100,000–110,000, again confirming it as the sarcoplasmic reticulum ATPase. Over 95% of the counts in the 100,000 mol. wt. peak were lost with a 10-min incubation in the presence of hydroxylamine. Lability of the counts in the presence of hydroxylamine is a property of an acylphosphate bond and was found in skeletal muscle Ca,Mg-ATPase. The formation of the acylphosphate intermediate results from the phosphate group attaching to the β-carboxyl group of an aspartyl residue in the sarcoplasmic reticulum ATPase active site. This is in contrast to the ester linkage to the hydroxyl group of serine, typical of protein kinase action.

A phosphorylatable protein of molecular weight 22,000–25,000 called phospholamban has been found in cardiac and slow skeletal muscle and may influence the rate of calcium uptake by the sarcoplasmic reticulum. In bovine uterine smooth muscle sarcoplasmic reticulum, a 25,000 molecular weight protein is phosphorylated only in the presence of calcium and is relatively resistant to hydroxylamine. This protein is likely to be phospholamban, although no further characterization has been performed.[35] Recently others have found a phospholamban-like protein in some but not all smooth muscle sarcoplasmic reticulum preparations, but its physiological significance is uncertain. The protein is phosphorylated by a calcium-dependent protein kinase; cAMP-, or cGMP-dependent protein kinase will also cause

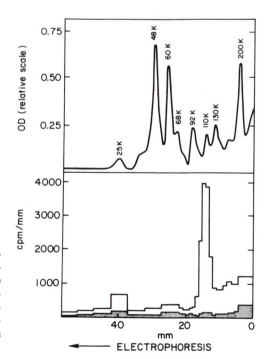

Figure 3. Sodium dodecyl sulfate polyacrylamide gel electrophoresis of microsomal proteins. Top, scan of Coomassie blue-stained proteins; bottom, ^{32}P distribution in the presence of calcium (open histogram) and same after treatment with hydroxylamine (stippled histogram). (From Carsten and Miller,[35] with permission).

phosphorylation and an increase in calcium uptake (though less in smooth muscle than in cardiac).[69]

4.2.5. Role of Magnesium

Magnesium may have many actions at the sarcoplasmic reticulum, and the effects on calcium transport may explain its relaxing effect at the cellular level. Magnesium ATP is thought to be the substrate for the Ca,Mg-ATPase. Magnesium is obligatory for ATPase activity as well as for calcium uptake. Furthermore, phosphorylation of the Ca,Mg-ATPase does not occur in the absence of magnesium.[35] Magnesium can be measured by atomic absorption spectroscopy in a manner similar to that used for calcium. We have shown that magnesium is bound specifically to sarcoplasmic reticulum preparations derived from bovine uterus. The association constant is $K_a = 2.9 \pm 0.3 \times 10^3$ M$^-$ ($K_d = 3.4 \pm 0.3 \times 10^{-4}$ M); the maximum number of binding sites is 38 ± 5 μmol/g protein.[59] Though the association constant is low compared to that for calcium, it must be remembered that the free intracellular magnesium is in the millimolar range >1000 times higher than the free calcium concentration. When magnesium was measured in the same sample as calcium, we found that bound calcium decreased and bound magnesium increased with an increase in free magnesium in the absence of ATP. Is magnesium the counterion for calcium, so that when calcium is taken up by the sarcoplasmic reticulum, magnesium is released? In the presence of ATP and an ATP-regenerating system, no change in bound magnesium is demonstrable with calcium uptake proceeding for as long as 32 min.[59] The ATP concentration must be kept constant; otherwise the free magnesium would increase as the ATP is hydrolyzed, because ATP binds more magnesium than its hydrolytic products. The conclusion is that magnesium is not the counterion for calcium. Though it was not demonstrated in this study, it is probable that either hydrogen or potassium ion serves this role.

4.2.6. Role of cAMP

Conflicting results have been published on the role of cAMP on the calcium uptake by the sarcoplasmic reticulum. No effect of cAMP was found in preparations of rat uterine microsomes,[70] but increased calcium uptake and increased phosphorylation of protein were observed without a change in the Ca,Mg-ATPase activity.[71,72] Another group found increased calcium uptake and increased Ca,Mg-ATPase activity.[73] Using the centrifugation method for the measurement of calcium, we were able to show that cAMP does not increase ATP-dependent calcium uptake into sarcoplasmic reticulum preparations, although it decreases base-line calcium content of the sarcoplasmic reticulum.[56] When the ^{45}Ca filtration method is used (see Section 4.2.1), the decreased base-line calcium in the presence of cAMP and calcium exchange markedly change the results. Taking these factors into account and recalculating, the data from Nishikori[71] likewise demonstrate no change in the uptake of calcium in the presence or absence of cAMP.

5. Rise of Intracellular Calcium and Contraction

In raising intracellular calcium for contraction, the uterine smooth muscle cell has two sources. These are calcium entry from the extracellular fluid and calcium release from internal structures. It appears that both are operative.

5.1. Calcium Entry from the Extracellular Fluid

A well-established mechanism for calcium entry into the cell is calcium flux through channels. In the myometrium as in other smooth muscle, the flux of calcium occurs through channels in the plasma membrane. These channels can be voltage or receptor activated. In the latter case, hormone or messenger binding is obligatory. Calcium channels are discussed in detail in Chapters 6 and 7.

5.2. Calcium Release Mechanisms from the Sarcoplasmic Reticulum

It has been much more difficult to demonstrate calcium release than calcium uptake in sarcoplasmic reticulum preparations; thus, less is known about calcium release. Mechanisms postulated include (1) reversal of the calcium pump, (2) ionophore action (part of ATPase or other), and (3) opening of calcium channels, either receptor or depolarization operated.

5.2.1. Reversal of the Calcium Pump

It has been proposed that reversal of ATP-dependent calcium uptake and of the ATPase reaction takes place. This hypothesis is attractive, since the energy expended in calcium uptake would be at least partially regenerated. In gastric smooth muscle microsomes the exchange of the γ phosphate of ATP with orthophosphate occurs and is taken as evidence that the reaction is reversible.[74] Quantitative measurement of the reverse reaction has been difficult even in skeletal muscle, and the physiological role in calcium release is at best questionable.[75-77] *In vitro* optimal reversal requires low calcium and high ADP and orthophosphate concentrations; the latter two are not present physiologically.[78] Reversal of the pump is a relatively slow process (for review see Martonosi[79]) and is the major argument against reversal playing a role in skeletal muscle, where calcium release is very rapid. This argument may not hold for smooth muscle.

5.2.2. Calcium Release by Ionophore

When the Ca,Mg-ATPase is added to artificial lipid membranes, the rate of diffusion of calcium goes up markedly. Though this effect may be interpreted as an ionophoretic action of the enzyme, it is nonspecific since the permeability to many other compounds is also increased (for review see Martonosi[79]). The phosphorylated Ca,Mg-ATPase can also mediate rapid calcium release without the synthesis of

ATP, but the conditions required (low magnesium, less than micromolar calcium, and relatively high ATP concentrations) may never be present *in vivo*.[80] Prostaglandins (PG) and phosphatidic acid have been shown to act as ionophores in artificial systems,[81,82] but whether they work in this manner physiologically has not been shown.

5.2.3. Calcium Channels

Other proposed mechanisms for calcium release from the sarcoplasmic reticulum include the opening of channels in the membrane. This hypothesis is consistent with the rapid release of calcium and is well accepted.[76,79] yet there is no direct evidence for this hypothesis, as it is very difficult to distinguish experimentally between a carrier and a fluid-filled channel. In skeletal muscle sarcoplasmic reticulum there seems to be activation of a channel to open after binding of calcium (calcium-induced calcium release), addition of caffeine, or depolarization. Inhibition of the channel by ruthenium red (or other polyamines) or local anesthetics such as procaine prevents calcium release.[76,79] As is true for much of skeletal muscle physiology, the relevance to smooth muscle[83] and in particular to myometrium is unknown.

5.3. Hormonal Effects on Intracellular Calcium Storage

Smooth muscle contractile hormones such as acetylcholine, norepinephrine, and, for the uterus, oxytocin and prostaglandins are known to raise the intracellular calcium concentration.

5.3.1. Prostaglandin and Oxytocin

Prostaglandins applied externally to smooth muscle cells were shown to cause a rise in intracellular calcium. Prostaglandin $F_{2\alpha}$ caused a dose-dependent elevation of calcium in vascular smooth muscle cells in calcium-free solution (2 mM EGTA).[84] In rat uterine horns PGE_1 caused sustained contractions in calcium-free medium, obtained by incubation in 3 mM EDTA.[85] Under these conditions entry of extracellular calcium is highly unlikely. In partially purified microsomal preparations from bovine uterus smooth muscle, it was shown that prostaglandin E_2 or $F_{2\alpha}$ or oxytocin decreases ATP-dependent calcium uptake and that prostaglandin E_2 releases calcium.[23,86] Prostaglandin E_2 was found to be more potent than $PGF_{2\alpha}$ in inhibiting ATP-dependent calcium uptake, consistent with the *in vivo* pharmacological action of these prostaglandins. Microsomes derived from pregnant uterus were more sensitive to oxytocin than were microsomes from nonpregnant uterus. Receptors for prostaglandins have been demonstrated in sarcoplasmic reticulum membranes[33]; thus, it is quite possible that prostaglandins, which are synthesized within the cell, have a direct intracellular effect.

Oxytocin, on the other hand, is more likely to have its initial effect on the plasma membrane, although direct intracellular effects are possible. Some contamination of the crude microsomes with plasma membrane might also account for the observations. The effects of oxytocin on inhibition of the plasma membrane ATPase[87,88] (this is discussed in Section 5.4), opening of calcium channels, or generation of a second messenger (as discussed in Section 7.3.2) could influence the interpretation of the results.

5.3.2. The α-Adrenergic Agonists

In vitro α_1 agonists produce a greater strength of contraction of the myometrium than potassium depolarization.[89,90] These observations in the myometrium are consistent with the α_1 receptor mediating release of internally bound calcium. In contrast to vascular smooth muscle, where both α_1 and α_2 agonists cause contraction, in the myometrium only α_1 agonists cause contraction. The present evidence is that α_2 agonists have no contractile action even though the actual number of α_2 receptors is greater than that of α_1 receptors in rabbit[91] and human [92,93] myometrium. Whether the α_2 receptor modulates relaxation by inhibiting cAMP synthesis is debatable.[90] Observations in the myometrium are consistent with the α_1 receptor mediating release of internally bound calcium through inositol trisphosphate release[94] as the mechanism (see Section 7.3).

5.3.3. Messenger Calcium

An alternative hypothesis for the rise of intracellular calcium on stimulation of the cell is called calcium-induced calcium release. The calcium influx caused by the opening of channels in the plasma membrane does not activate the myofilaments directly. The small amount of calcium that enters the cell could cause a large amount of calcium to be released from the sarcoplasmic reticulum.[95] Another hypothesis is that on receptor occupation calcium from a small labile store on the inside of the plasma membrane is released. This calcium could trigger calcium release from the sarcoplasmic reticulum.[96]

5.4. Inhibition of Calcium Efflux

Inhibition of calcium efflux through the plasma membrane would also raise the intracellular calcium concentration. Inhibition of the Ca,Mg-ATPase activity of the plasma membrane by oxytocin has been shown.[87,88,97] However,[98,99] comparable levels of oxytocin failed to prevent calcium extrusion; whether this represents incomplete inhibition of the Ca,Mg-ATPase or an alternative mechanism of calcium extrusion such as Na^+-Ca^{2+} exchange or problems of methodology is not clear. Research on both calcium influx and efflux uses a method of measuring the rate of accumulation of ^{45}Ca in the presence of a stimulus. The extracellular calcium bound to the cell is washed off with La^{3+}, and total tissue ^{45}Ca is counted. However, there have been serious problems with this methodology. Early work failed to show an increase in calcium after potassium depolarization,[100,101] in which calcium influx is thought to be the major mechanism responsible for contraction. The effects of lanthanum are not as simple as initially thought, and this may change the interpretation of results.[99] Care in the interpretation is also mandatory because efflux of calcium from the cell does not necessarily indicate the source or the extent of the increase in the intracellular free calcium. Calcium influx, efflux, intracellular storage, and free calcium are all interconnected; thus, it is difficult to distinguish direct from indirect effects.

6. Pharmacomechanical Coupling

Strength of smooth muscle contraction in response to hormonal stimuli is not solely dependent on the extent of membrane depolarization.[102] In vitro greater strength of contraction is seen in the presence of agonists such as acetylcholine and epinephrine as compared with potassium depolarization.[89] Prostaglandins can also contract KCl-depolarized uterus, i.e., independent of a change in membrane potential. When uterus is maximally contracted by potassium, cumulative additions of PGE$_1$ induced additional concentration-dependent contraction.[85] Though some of the contraction is explainable on the basis of receptor-operated calcium channels, there remains a portion independent of external calcium[89] (see Chapter 7). This is called pharmacomechanical coupling and may be an important mechanism in uterine contraction.

Explanation of the phenomenon has usually entailed the following scenario. The first step, and the one conferring specificity, is the binding of a hormone to its receptor at the plasma membrane. The next step is the generation of a signal to the inside of the cell. The signal is carried by a so-called second messenger, which transposes nonspecifically to the inside of the cell. In smooth muscle the second messenger initiates various biochemical reaction that lead to muscle contraction, including calcium release from internal stores, enzyme activation, and phosphorylation reactions. The recent identification of a physiological signal for release of intracellular calcium has added to our understanding of pharmacomechanical coupling.

7. The Phosphoinositide System and Release of Second Messengers

There is mounting evidence that binding of agonists to receptors hydrolyzes phosphoinositides in the plasma membrane, which in turn leads to calcium mobilization.[103,104]

7.1. Occurrence, Structure, and Metabolism

The phosphoinositides are minor components of plasma membranes and make up approximately 3–5% of the total phospholipids.[105] The molecule consists of a glycerol backbone with two fatty acids attached in positions 1 and 2; the one attached in position 2 is always arachidonic acid. The 3 position attaches inositol through a phosphate group. This compound is called phosphatidylinositol. Approximately 10–20% of the phosphatidylinositol is further phosphorylated at positions 4 and 5 of inositol (the structures are shown in Fig. 4). On agonist binding to a specific receptor, phosphoinositidase C (PIC), previously called phospholipase C, is activated. This enzyme is a specific phosphodiesterase, distinct from the ones responsible for the hydrolysis of cyclic nucleotides. Phosphoinositidase C hydrolyzes phosphatidylinositol-4,5-bisphosphate to diacylglycerol and D-*myo*-inositol-1,4,5,-trisphosphate (InsP$_3$).[103] At least five different enzymes have been identified, with remarkable structural heterogeneity.[106] A phosphoinositidase C was recently isolated from guinea pig uterus, and at least two enzymes were identified; the only one purified was shown to be a 62,000-mol. wt. enzyme.[107] Though somewhat depen-

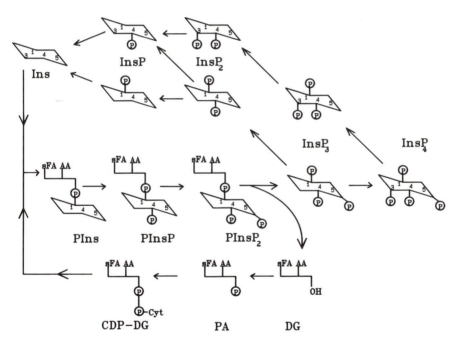

Figure 4. Pathway of phosphoinositide metabolism. Center row: Phosphatidylinositol (PIns), phosphatidylinositol-4-phosphate (PInsP); phosphatidylinositol-4,5-bisphosphate (PInsP$_2$). Clockwise: Diacylglycerol (DG); phosphatidic acid (PA); cytidine diphosphate diacylglycerol (CDP-DG); saturated fatty acid (sFA); arachidonic acid (AA). Counterclockwise: Inositol-1,4,5-trisphosphate (InsP$_3$); inositol-1,4-bisphosphate (InsP$_2$); inositol-1-monophosphate or inositol-4-monophosphate (InsP); and inositol (Ins). Alternately from inositol-1,4,5-trisphosphate (InsP$_3$) to inositol-1,3,4,5-tetrakisphosphate (InsP$_4$) to inositol-1,3,4-trisphosphate (InsP$_3$); inositol-3,4-bisphosphate (InsP$_2$); inositol-4-monophosphate (InsP); and inositol (Ins). Inositol is recycled.

dent on cofactors and assay conditions, phosphatidylinositol-4,5-bisphosphate is the preferred substrate. The activation of phosphoinositidase C requires calcium levels normally found inside the resting cell, $0.1-1$ μM, and does not depend on a rise in calcium concentration.[108]

The $InsP_3$ diffuses into the cytoplasm and is broken down sequentially to inositol bisphosphate ($InsP_2$), inositol monophosphate (InsP), and free myo-inositol by specific phosphomonoesterases.[109] The dephosphorylation of $InsP_3$ requires Mg^{2+}[110] and physiological concentrations of calcium.[111] The InsP phosphomonoesterase is inhibited by lithium,[112] whereas the $InsP_3$ 5′-phosphomonoesterase is not inhibited.[109] Phosphatidylinositol is resynthesized from myo-inositol and indirectly from diacylglycerol. The diacylglycerol must first be phosphorylated to phosphatidic acid; then phosphatidic acid and CTP react to form cytosine monophosphate (CMP)-phosphatidic acid.[105,113] The resynthesis occurs in the endoplasmic reticulum and in the plasma membrane.[113] In the mouse uterus it was shown to be stimulated by priming with estrogen.[114] A rapid equilibrium is established between the mono- and polyphosphoinositides in the plasma membrane by two very active specific kinases. A phosphatidylinositol 4-kinase has been purified from bovine uterus; it has a molecular weight of 55,000 and is specific for the 4 position.[115] This completes the cycle of phosphoinositide metabolism.

7.2. Inhibitors of the Phosphoinositide Cycle

Inhibition of the cycle has been claimed for several compounds. Lithium, as mentioned above, is thought to inhibit the step from inositol monophosphate to inositol. This would prevent further synthesis of PIns and would allow the accumulation of InsP, $InsP_2$ and $InsP_3$. The hydrolysis of $InsP_3$ is inhibited in the presence of high concentrations of 2,3-diphosphoglycerate.[110] This compound also competes for the $InsP_3$ specific binding site on the endoplasmic reticulum.[116] Neomycin was found to inhibit both the formation and action of $InsP_3$. Originally it was felt that this was specific and resulted from binding of the antibiotic to the polyphosphorylated inositol or phosphatidylinositol.[117] Recent evidence indicates that the effect is nonspecific and related to the binding of neomycin to several polyphosphorylated compounds including ATP. Thus, neomycin can block both the breakdown of $PInsP_2$ to $InsP_3$ and the action of $InsP_3$ on calcium release, but the fact that neomycin acts as an inhibitor is not evidence of the involvement of inositides in that process.[118] Adding an antibody to bacterial phosphoinositidase C inhibited contractions by ileal muscle strips in a muscle bath in response to histamine or acetylcholine, though not to prostaglandin.[119] The specificity and reproducibility of this approach require confirmation.

7.3. Biological Function of the Phosphoinositides

The presence of the phosphoinositides in the plasma membrane has been known for a long time. The role of phosphoinositides in signal transduction in many systems is now established. Though not universal, it seems to be a common system involved in the release of calcium to the cytoplasm by agonist stimulation.

7.3.1. Receptors and Phosphoinositide Turnover

Historically, phosphoinositides were linked to a receptor mechanism in the 1950s and 1960s. It was found that agonists caused changes in PIns and InsP$_2$ content and ^{32}P labeling of phosphoinositides in neural tissues. For a more extensive historical review use Abdel Latif.[105] In smooth muscle, acetylcholine induced the breakdown of phosphatidylinositol-4,5-bisphosphate (PInsP$_2$) labeled with ^{32}P accompanied by an increase in ^{32}P labeling of phosphatidic acid (PA) and phosphatidylinositol (PIns).[120] Though many products of phosphoinositide breakdown were known and even used to measure the stimulated breakdown, the action was felt to depend on the loss of the phosphoinositide rather than on the effects of any of the products.

This view changed when it was shown that the stimulated breakdown of phosphatidylinositol-4,5-bisphosphate generated InsP$_3$ in less than 5 sec. This finding, in blowfly salivary gland, suggested that InsP$_3$ might act as a second messenger.[121] The formation of InsP$_2$, InsP, and inositol occurs later, thus indicating that they are produced sequentially from InsP$_3$. Contraction as produced by a calcium agonist[122] or by K$^+$[123,124] does not increase InsP$_3$ production (though this is not a universal observation). Likewise, blocking calcium entry and contraction does not interfere with InsP$_3$ production.[122] The above observations are strong arguments for InsP$_3$ as the mediator rather than the consequence of contraction. Studies in smooth muscles have shown increased phosphoinositide metabolism on stimulation by acetylcholine,[120] carbachol, or norepinephrine (α_1-adrenergic)[123] in the rabbit iris; carbachol[124,125] or acetylcholine in the trachea[126,127]; acetylcholine, histamine, substance P, or norepinephrine in visceral smooth muscle[122]; and angiotensin and vasopressin in vascular smooth muscle.[128,129] In both dose–response and timed studies using carbachol in rabbit iris, the increase in InsP$_3$ paralleled the muscle contraction and the increase in myosin light chain (MLC) phosphorylation.[108]

The effects of prostaglandins remain controversial; in many tissues PGF$_{2\alpha}$-stimulated InsP$_3$ release was demonstrated,[130–134] including in rat aorta,[135] but this was not present in trachea.[126]

7.3.2. Phosphoinositide Turnover in Uterine Muscle

To date only a few studies have been done on agonist-stimulated phosphoinositide turnover in the uterus. In myometrial strips from immature estrogen-pretreated guinea pigs, oxytocin and carbachol evoke rapid release of InsP$_3$, followed by release of InsP$_2$ and InsP$_1$. The response depended on the agonist concentration. The increase in InsP$_3$ occurred within 30 sec and reached a plateau at 3 min. Effects were maximal at 0.2 µM oxytocin and 100 µM carbachol, with half-maximal effects at 0.03 µM and 15 µM, respectively. Most experiments were carried out with 0.2 µM oxytocin and 50 µM carbachol. The increase in InsP$_3$ was not enhanced by raising cytosolic free calcium. Maximal activation of contraction required a tenfold lower concentration of agonist than did maximal release of InsP$_3$, and InsP$_3$ was increased more by oxytocin than by carbachol at equal maximum contraction.[136] Thus, a simple relationship between the $InsP_3$ generated and contraction is not found. The use of lithium to prevent phosphatase activity somewhat confuses the results, since lithium leads to accumulation of $InsP_2$, $InsP_1$, and $Ins(1,3,4)P_3$, an inactive isomer of $InsP_3$.

In human gestational myometrium, formation of $InsP_1$, $InsP_2$, and $InsP_3$ has been shown to be dose dependent, with half-maximal response (of $InsP_1$) at 2×10^{-8} M oxytocin. Oxytocin was ineffective in nonpregnant myometrium. Vasopressin induced $InsP_3$ formation and was more effective than oxytocin. It was effective in nonpregnant human myometrium, though less than in pregnant[137]; this was similar to findings in decidual cells.[138] The $InsP_3$ was also increased by α_1-adrenergic agonists,[137] angiotensin II,[139] and platelet-activating factor.[137] However, carbachol was ineffective,[137] contrary to the findings in animals.[136,140]

Prostaglandins, on the other hand, showed only small (at 1 μM $PGF_{2\alpha}$[140]; at 30 μM PGE_2 or $PGF_{2\alpha}$) (S. Harbon, personal communication) or no effects (10 μM PGE_2, $PGF_{2\alpha}$[137]) on $InsP_3$ release, whereas in the same preparations oxytocin stimulation released considerably more $InsP_3$.[137,140]

At this time care must be used in drawing conclusions from experiments on the synthesis of $InsP_3$ after agonist stimulation, since in many cases measurement is made after relatively long incubation times, 5–30 min, no distinction is made between the isomers of $InsP_3$ or all InsP are measured, and inhibitors such as lithium are used. Comparison with contraction is very useful but frequently overlooked.

7.3.3. Calcium Mobilization on Agonist Stimulation

Receptors linked to phosphoinositide breakdown are found in many tissues and are related to a calcium-mediated event. A correlation of phosphatidylinositol hydrolysis and calcium flux was first demonstrated[141] in the blowfly salivary gland. Inositol trisphosphate acts to mobilize intracellular calcium from a nonmitochondrial pool; this occurs at $InsP_3$ concentrations well below those calculated to exist (30 μM) in the stimulated cell.[142] When agonist is applied to the outside of the plasma membrane of intact cells, the free calcium inside the cell rises. Agonist-induced phosphoinositide breakdown and mobilization of intracellular calcium have been shown for vasopressin[129,143] and angiotensin in cultured vascular smooth muscle cells.[128,143] A diagram of the agonist stimulation of calcium mobilization is shown in Fig. 5.

For $InsP_3$ to be the messenger, it must cause calcium release when directly applied to the cell. Since extracellular $InsP_3$ is inactive, either the plasma membrane must be made permeable to $InsP_3$ or broken cell preparations must be used. Smooth muscle strips or cells are allowed to accumulate calcium with or without ^{45}Ca. Then the cell is made permeable by detergent treatment, most frequently saponin. Rapid release of calcium by $InsP_3$ from an intracellular store of various smooth muscles was shown by this method (porcine coronary artery cells,[144] rabbit main pulmonary artery,[145] guinea pig stomach cells,[146] tracheal muscle strips or single cells,[126] and vascular smooth muscle cells[147]). Calcium release is shown not to be influenced

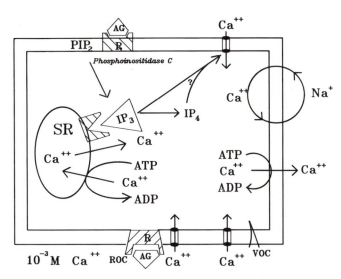

Figure 5. Calcium cycle in the uterine smooth muscle cell. Agonist (AG) receptor (R) binding activates phosphoinositidase C to hydrolyze phosphatidylinositol-4,5-bisphosphate (PIP_2), to inositol-1,4,5-trisphosphate ($InsP_3$) and diacylglycerol (not shown). SR, sarcoplasmic reticulum; ROC, receptor-operated channel; VOC, voltage-operated channel.

by inhibitors of mitochondrial oxidative phosphorylation such as sodium azide or oligomycin.

With permeabilized gastric muscle cells in culture it has been shown that $InsP_3$ and agonist both cause similar contraction and increase in free calcium.[146] The contractile response is concentration dependent and correlates with the increase in free calcium.[146] Repeated applications of $InsP_3$ elicited repeated dose-dependent contractions, and single additions of $InsP_3$ evoked sustained contractions, implying that tonic contractions can occur in response to intracellular calcium release mediated by $InsP_3$. The amount of calcium released was thought to be sufficient to produce maximum contraction.[145] More recently, by use of permeabilized cells, it has been possible to compare the time of onset of contraction evoked by an agonist or $InsP_3$. The molecule to be tested is "caged," added to the medium, and then activated with a laser, allowing one accurately to time the onset of action. Under these conditions $InsP_3$ latency is 0.5 sec compared to an α agonist, where the latency is 1.5 sec. The time difference is expected to be long enough to allow for release of $InsP_3$ by the agonist.[148] Application of $InsP_3$ to saponin-skinned pregnant uterine muscle lead to a contractile response. The prior exposure to a calcium ionophore prevented the response.[148a] Additionally, it has been found that heparin at <1 μg/ml (approximately 0.2 μM) specifically and competitively inhibits calcium release by $InsP_3$. It does not inhibit GTP- or caffeine-induced release.[149]

Though the above experiments are suggestive of the effect of $InsP_3$ being on the sarcoplasmic reticulum, several alternative sites of action are possible. With saponin-treated cells one cannot distinguish between release of calcium bound to the outside of the plasma membrane and that from the inside of the cell. Nor does this procedure tell us the exact location of the calcium release site inside the cell. To identify the location of calcium release, microsomal preparations derived from the sarcoplasmic reticulum have to be used. In addition, there are problems inherent in the saponin treatment. Because of its high cholesterol content,[19] the plasma membrane permeability is altered by saponin but, depending on temperature, concentration, and time of exposure, saponin can affect membranes inside the cell such as the sarcoplasmic reticulum. Also, changes in some enzyme activities have been reported.[150] Although most studies were done in the presence of inhibitors of mitochondrial oxidative phosphorylation, a question as to the identity of the calcium pool remains.

7.3.4. InsP₃-Induced Calcium Mobilization in the Uterus

In our laboratory we showed directly that inositol trisphosphate rapidly releases calcium from a microsomal fraction of bovine uterine myometrium.[151] This sarcoplasmic reticulum fraction (described above) has been well characterized in our laboratory in order to distinguish it from other cell fractions.[43] The microsomes were allowed to take up calcium and ^{45}Ca from solution. At 8 min, 1 ml was filtered to determine the ^{45}Ca concentration in the vesicles; 2 mM EGTA was immediately added to bring down the free calcium concentration from 7×10^{-6} to $<10^{-8}$ M and to remove bound calcium from the outside of the vesicles. Calcium released after this must then originate from inside the vesicles. Addition of InsP₃ caused further release of calcium (see Fig. 6). This calcium release is concentration dependent and, at the highest concentration used, amounts to 40% of calcium releasable by an ionophore. With this experimental setup we showed that the released calcium actually comes from inside the vesicles, and specifically from the sarcoplasmic reticulum.

7.3.5. Mechanism of InsP₃ Action

Calcium release is relatively specific for Ins(1,4,5)P₃. In various systems some activity is found for Ins(2,4,5)P₃ (one-sixth of the activity of InsP₃),[152] Ins(1,3,4)P₃ (less than $\frac{1}{27}$ the activity), and Ins(1,3,4,5)P₄ (less than $\frac{1}{60}$ the activity of InsP₃); Ins(1,4)P₂ and Ins(4,5)P₂ are inactive (to 60 times the concentration of InsP₃).[144,153,154] Theoretically InsP₃ could act through any proposed mechanism for regulating cell calcium. These include (1) calcium channels, (2) ionophoretic action, and (3) Ca,Mg-ATPases. In saponin-treated vascular smooth muscle cells it was first demonstrated that InsP₃-induced calcium efflux was not appreciably delayed at lowered temperature. This suggested that calcium release was mediated by a channel rather than by a carrier, as it was assumed that an increase in the viscosity of the lipid bilayer would not affect a channel mechanism.[147] Incorporation of smooth muscle, but not skeletal or cardiac, sarcoplasmic reticulum into an artificial lipid bilayer conferred InsP₃-activatable calcium channel activity to the membrane.[83]

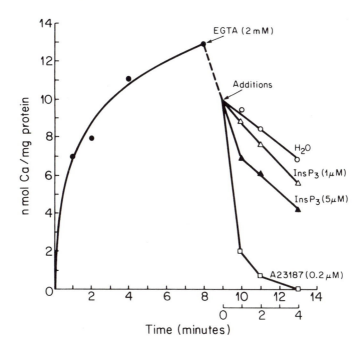

Figure 6. Calcium release by inositol trisphosphate from microsomal vesicles, derived from bovine myometrium. Calcium loading in the presence of ^{45}Ca and ATP at 37°; 2mM EGTA added at 8 min; 1 min later (time 0) additions of H_2O, inositol trisphophate ($InsP_3$), or ionophore A23187. (From Carsten and Miller,[151] with permission.)

Specific binding of $InsP_3$ has been demonstrated and analyzed by Scatchard plots. In the endoplasmic reticulum fraction of adrenal cortex a K_d of 1.7×10^{-9} M was calculated for the high-affinity site, and receptor concentration was 104 fmol/mg protein.[116] Similar values for K_d were obtained in other tissues (microsomes from anterior pituitary and liver). It was suggested that the specific binding site is the physiological receptor for $InsP_3$, since the affinity of the $InsP_3$ phosphatase for IP_3 is much lower. A low-affinity site with K_d of 16 μM was found in earlier work, and this is equal to the K_m of the phosphatase.[155] Interestingly, there is approximately a 1000-fold difference between the nanomolar binding affinity of $InsP_3$ and the micromolar concentration needed for calcium mobilization.[116] No studies on the uterus have been done to date.

Other suggested sites of action for $InsP_3$ include inhibition of the plasma membrane calcium extrusion ATPase, at least in vascular smooth muscle.[156] However, in isolated adipocyte plasma membrane vesicles, $InsP_3$ had no effect on the Ca,Mg-ATPase activity, calcium uptake, or release. Similarly, no effect was observed on the endoplasmic reticulum calcium uptake (Ca,Mg-ATPase), though the expected release of calcium from the endoplasmic reticulum occurred. It is concluded that $InsP_3$ acts exclusively at the endo (sarco) plasmic reticulum to promote calcium release.[157]

7.3.6. Unresolved Problems with the InsP$_3$ Model

The data presented above do not conclusively prove the hypothesis of InsP$_3$ action. First, the concentration of agonist necessary to cause contraction in a given tissue is frequently less than one-tenth the concentration that produces a comparable effect on InsP$_3$ release.[108,122,129,136,141] This is not necessarily a fatal flaw: (1) a small change in InsP$_3$ might produce a maximal effect, and (2) one would anticipate that the InsP$_3$ would rise just prior to the onset of the contraction, seconds to 1–2 min; whether physiologically it remains elevated during a 30-min stimulus or not would have little bearing on the initial mechanism. Thus, problems with the methodology of InsP$_3$ measurement such as waiting 10–30 min, adding lithium to increase the InsP, and not distinguishing between the Ins(1,4,5)P$_3$ and other inositol phosphates may be responsible. A second problem is that for a series of contractile agents the potencies for contraction and for the generation of InsP$_3$ are not necessarily related. Even worse, noncontractile agents such as prostaglandin F$_{1\beta}$ cause large increases in the levels of InsP$_3$.[135] Again, problems with methodology may be responsible for these findings, but the model presented may be simplistic, and other factors may override the effect of InsP$_3$ on calcium release.

7.4. Other Inositol Phosphates

Many other inositol phosphates have been shown to increase in response to agonist stimulation. The functions and importance have been discussed, but no definitive conclusions can be drawn. The most likely candidates as mediators of cellular action are discussed below.

7.4.1. Inositol Tetrakisphosphate

A recently described product of PInsP$_2$ metabolism is inositol tetrakisphosphate [Ins(1,3,4,5)P$_4$; InsP$_4$].[158] This finding was confirmed in studies of the longitudinal muscle of the guinea pig small intestine; stimulation with carbachol or histamine produced InsP$_3$ and InsP$_4$. The mechanical response occurs with the very early rise in InsP$_3$.[159] An Ins(1,4,5)P$_3$ kinase has been identified that phosphorylates Ins(1,4,5)P$_3$ to Ins(1,3,4,5)P$_4$.[160] ATP and magnesium are required,[161] and the kinase, in aorta, is stimulated by calcium and calmodulin.[162] Ins(1,3,4,5)P$_4$ is further metabolized to the inactive Ins(1,3,4)P$_3$ by a 5'-phosphomonoesterase in aorta.[163] This explains the large rise in InsP$_3$ seen late (after 10 or more minutes) after stimulation. The rise in InsP$_3$ results from the accumulation of the slowly metabolized and inactive Ins(1,3,4)P$_3$. Since this isomer of Ins(1,4,5)P$_3$ appears more slowly than Ins(1,4,5)P$_3$ and is more slowly metabolized, it accumulates in large amount. This is further accentuated in the presence of lithium, which inhibits the breakdown of the Ins(1,3,4)P$_3$.[164,165]

Recent evidence suggests that inositol tetrakisphosphate may be involved in calcium entry from outside the cell. A prerequisite for calcium entry is the prior discharge of calcium from the intracellular store by $InsP_3$. To that extent $InsP_3$ is required for calcium entry,[166] at least in sea urchin eggs, but calcium alone cannot substitute for $InsP_3$.[167] In binding experiments to membranes it was found that a specific binding site is present for $InsP_4$ that has a much lower affinity for $InsP_3$, thus implying a unique role for the $InsP_4$.[168] Observed action of $InsP_3$ alone in patch-clamp experiments[169] must be interpreted to have occurred through metabolism of $InsP_3$ to $InsP_4$. The stimulus to calcium entry by $InsP_4$ may be the mechanism for what has previously been described as a receptor-operated channel.

7.4.2. Cyclic Inositol Trisphosphate

In addition to the above-described inositol phosphates, cyclic inositol phosphates have been isolated by some investigators.[170,171] The cyclic compounds are unstable in acid and so would not be seen using the routine purification methods. However, using proper techniques, other investigators have failed to find cyclic inositides on stimulation.[172,173] Cyclic inositol trisphosphate [cycIns(1–2,4,5)P_3; cycInsP_3] formed by the action of phosphoinositidase C is dephosphorylated stepwise in a manner similar to $InsP_3$.[174] The last step is the breakdown of the cycIns(1–2)P to Ins(1)P by a specific enzyme that opens the ring as shown in the human placenta.[175] The cycInsP_3 is degraded much more slowly than $InsP_3$[176]; hence, it builds up in agonist-stimulated tissues. Its activity varies with the system used, being more active than $InsP_3$ in the photoreceptor, somewhat less active in releasing calcium from permeabilized platelets,[177] and inactive in Swiss 3T3 cells.[178] The physiological function of the cyclic inositol phosphates in mammalian tissue is largely unknown (M. J. Berridge, personal communication).

7.5. Diacylglycerol

Diacylglycerol is the second major product directly released on hydrolysis of phosphatidylinositol-4,5-bisphosphate.[179] It can act directly by stimulating protein kinase C. Diacylglycerol can be phosphorylated to phosphatidic acid by a diacylglycerol kinase or sequentially hydrolyzed by a diacylglycerol lipase and monoacylglycerol lipase, releasing arachidonic acid, the obligatory precursor for formation of prostaglandins of the 2 series (see Chapters 14 and 16 for review) or reconverted to phosphatidylinositol.[179]

7.5.1. Protein Kinase C

Protein kinase C is a ubiquitous enzyme that phosphorylates proteins; at least 20 endogenous substrates have been identified in smooth muscle.[180] Protein kinase C is really a heterogeneous group of enzymes, frequently with more than one type in a given cell.[181] All require phosphatidylserine and calcium for activation, as well as diacylglycerol containing one unsaturated fatty acid.[182] Diacylglycerol increases the affinity of the enzyme for calcium; thus, full activity can occur without an increase in the calcium concentration above that present in the resting state.[182,183] For a comparative review see Kikkawa and Nishizuka.[183] Protein kinase C has been found in fetal membranes[184] (see Chapter 16), various smooth muscles including uterus,[185] and in many systems that use InsP$_3$ as a second messenger. Protein kinase C occurs mainly in soluble form and is translocated to membranes in the presence of increased calcium, as in stimulated cells.[186,187] *In vitro* activation of protein kinase C can occur by phorbol esters that have a diacylglycerol-like structure.[188]

The actions of protein kinase C have been used to explain two events: the first is the activation of systems that end the InsP$_3$-mediated events; the second is to explain long sustained contractions (latch-bridge mechanism). The effects of protein kinase C on contraction can be summarized as follows. The acute action of agonists that cause contraction through InsP$_3$ mechanisms are antagonized by protein kinase C activation, whereas contraction dependent on external calcium is potentiated. Exposure for much longer time periods (20–60 min) causes contraction in the absence of any other stimulus.

The diverse results obtained by activating protein kinase C *in vitro* relate to the multitude of sites that can be phosphorylated. The observed effect will depend on the relative importance and actions of the phosphorylated proteins. Interference with contraction as seen in many smooth muscles including rat uterus[185] can be explained by the *in vitro* finding that activation of protein kinase C interferes with the InsP$_3$ mechanism. This is supported by the finding that phosphoinositidase-C-induced phosphoinositide hydrolysis is inhibited either through phosphorylation of the receptor[189] or through phosphorylation of,[107] or action on, the phosphoinositidase C.[190–192] The InsP$_3$ degradation is enhanced by the protein-kinase-C-phosphorylated inositol trisphosphate 5′-phosphomonoesterase.[193,194] Protein kinase C was also shown to stimulate the ATP-dependent calcium pump of the plasma membrane,[195] which would attenuate InsP$_3$-initiated calcium mobilization.

Furthermore, direct effects on the contractile proteins have been described and may play a role in the relaxant effect. In turkey gizzard and tracheal smooth muscle, it was demonstrated that phosphorylation of myosin light chain occurred at sites different from that produced by myosin light chain kinase. Contractile agents cause phosphorylation at the myosin light chain kinase sites but not the protein kinase C site.[196] Phosphorylation at the latter site decreases the rate of phosphorylation of myosin light chain by myosin light chain kinase and thereby slows contraction or favors the relaxed state.[197,198] Phosphorylation of myosin light chain kinase by protein kinase C also slows contraction by decreasing myosin light chain kinase activity. This occurs through a decrease in affinity of myosin light chain kinase for calmodulin when myosin light chain kinase is in the phosphorylated state.[199,200] Correspondingly, protein kinase C in the presence of phorbol ester and phosphatidylserine inhibited calcium-activated tension in skinned vascular smooth muscle fibers.[201]

Protein kinase C activation causes contraction in situations where the contraction stems from enhanced calcium entry as in the case of KCl,[202] calcium ionophore, or calcium channel agonist.[203,204] The mechanism is not evident but may be increased entry of calcium or potentiation of the calcium effect. The observation that potentiation is seen even in skinned cells[205] requires an effect other than on calcium entry. Slow-onset contractions brought about by the activation of protein kinase C as in tracheal[206] and vascular smooth muscle[207–209] may well be accounted for by the late phosphorylation of contractile proteins including myosin light chain.[108] The quantitative correlation of phorbol-ester-induced contraction in skinned vascular smooth muscle with myosin light chain phosphorylation has, however, been questioned.[205]

It was suggested that protein kinase C may be involved in the latch state in vascular smooth muscle[205] (see Chapters 2 and 3) and that protein kinase C may have a role in tonic contraction of bronchial,[210] aorta,[211,212] and iris[108] smooth muscle. A sustained calcium-dependent contractile response was observed in the rabbit ear artery following phorbol ester treatment.[203] In tracheal smooth muscle, activation of protein kinase C was associated with phosphorylation of multiple proteins such as caldesmon, desmin, and synemin.[186,213]

7.5.2. Phosphatidic Acid

The production of phosphatidic acid from diacylglycerol is tissue-[124] and possibly agonist-dependent.[214] In some cases the production is quite large, and the concentration increases with time and to much greater extent than is seen with diacylglycerol.[124] This has led to suggestions that this compound is at least a potentiator of the calcium-releasing effects of $InsP_3$[215] and may well be related to late effects of agonists. Arguments against the universality of the action relate to the fact that phosphatidic acid labeling was slower than calcium release[216] or that phosphatidic acid formation does not necessarily accompany $InsP_3$ formation.[214] The explanation may be in the fact that there are different pathways for metabolism of diacylglycerol. Whether phosphatidic acid has any other functions is not known at present, although some suggestions have been made.[215]

7.5.3. Prostaglandin Biosynthesis

Since arachidonic acid is usually present in diacylglycerol, prostaglandins can be a product of phosphoinositide hydrolysis. This could be the origin of prostaglandin biosynthesis inside the cell that occurs after hormonal stimulation and serves to augment the effects of these agonists.[217] Support for this comes from the observation that oxytocin action is often diminished in the presence of a prostaglandin synthesis inhibitor.[218–220] *In vitro* studies on human pregnant myometrium showed oxytocin to increase formation of PGE_2, $PGF_{2\alpha}$, and prostacyclin; another uterotonic agent, ergotamine, increased formation of PGE_2 and $PGF_{2\alpha}$.[221]

Prostaglandin production is limited by the availability of substrate and thus would be expected to increase as arachidonic acid is produced. The pathway for arachidonic acid release may be either from diacylglycerol produced by the action of phosphoinositidase C or through stimulation of phospholipase A_2, which is calcium sensitive. The relative importance of the two mechanisms has not been quantified. To complicate the picture, phospholipase A_2 is activated by protein kinase C.[222,223] Activation of both protein kinase C and phospholipase A_2 was suggested in timed experiments; histamine induced increased phosphoinositide hydrolysis and release of thromboxane and 6-keto-PGF_1.[224] Norepinephrine and α_1-adrenergic agonists stimulated arachidonic acid release to the same extent as did protein kinase C activators. The arachidonic acid release was blocked by an inhibitor of phospholipase A_2 but not by an inhibitor of diacylglycerol lipase.[222,223] In summary, the prostaglandins are intimately involved in the phosphoinositide system, being produced from diacylglycerol and indirectly by diacylglycerol activation of protein kinase C.

Other products of arachidonic acid metabolism may also play an important role, including the lipooxygenase products. Recent work has found that lipoxin A, one such product, activates protein kinase C at low and thus potentially physiological concentrations.[225]

7.6. The Regulatory Role of Guanine Nucleotides

Recent studies have revealed that guanine nucleotides might be involved in signal transmission leading to the release of intracellular calcium.

7.6.1. Guanine Nucleotide Binding Proteins

In most cellular systems there is no direct contact between the receptors on the outside of the plasma membrane and the effector enzymes (activators) at the inside. They communicate with each other through a pair of guanine nucleotide binding proteins (G proteins). Most information on G proteins relates to the adenylate cyclase system (see also Chapter 8). One of the G proteins, G_s, enhances the activity of adenylate cyclase (when a stimulatory receptor is activated), and the other, G_i, inhibits (when an inhibitory receptor is activated). Each G protein consists of three subunits. The primary structure of the subunits in the adenylate cyclase system has been largely worked out, and the mode of action is emerging.[226–228] The presence of guanosine 5'-diphosphate (GDP) on the G_s protein inhibits the enzyme. Hormone–receptor interaction stimulates the dissociation of G_s protein–GDP and the binding of guanosine 5'-triphosphate (GTP) to the G_s protein. Binding of GTP to G_s protein is accompanied by its dissociation into subunits and a conformational change and activation of the enzyme (possibly also causing the dissociation of the hormone from the receptor[226]). GTP slowly hydrolyzes to GDP, and the G_s protein again inhibits the enzyme action. Less well worked out is the mechanism of inhibition produced by the G_i protein, which prevents activation of the enzyme by the G_s protein.

The G proteins involved in the phosphoinositide cycle are still unknown. Evidence to date linking a stimulatory G protein in the signal transduction between a hormone receptor and phosphoinositidase C is based on the following. GTP is essential for the action of phosphoinositidase C in the presence of physiological cation concentrations,[111,229,230] and agonist effects are potentiated.[111,230] Nonhydrolyzable analogues of GTP can substitute for GTP.[111,230-232] Other nucleotides such as ATP, CTP, GDP, and GMP cannot substitute for GTP,[229] and GDPβS is inhibitory.[231]

In the myometrium the generation of InsP$_3$ induced by oxytocin or carbachol appears to occur through activation of G proteins, since fluoride ion in the presence of aluminum activates the enzyme.[233] However, unlike the effect on the adenylate cyclase, cholera toxin could not activate phosphoinositidase C, so a different G$_s$ protein must be present. Pertussis toxin, which blocks the effects of inhibitors of adenylate cyclase, did not change the effect of either carbachol or oxytocin on InsP$_3$ generation. This would make it unlikely that a standard G$_i$ protein was present, at least in the guinea pig.[233] These results differ from those reported in rat myometrium where cholera toxin[140] or pertussis toxin[140,234] decreased oxytocin-stimulated InsP$_3$ formation. Figure 7 shows a diagram of the emerging evidence of G protein involvement in InsP$_3$ generation.

7.6.2. Guanine Nucleotide Involvement in InsP₃ Action

The InsP$_3$-induced calcium release from sarcoplasmic reticulum was observed in permeabilized vascular smooth muscle to require GTP or its nonhydrolyzable analogue.[235] Though no specific tests for the presence of a guanine nucleotide binding protein were performed, the description is highly suggestive of such a system being present. This effect, however, must be distinguished from that discussed below.

7.6.3. GTP-Induced Calcium Release

GTP can directly mobilize calcium from intracellular stores and induce calcium efflux from cells. Like InsP$_3$-mediated calcium efflux, GTP-mediated calcium efflux is not affected by calcium channel blockers.[236] GTP-induced calcium mobilization has been observed in permeabilized cells,[237-239] in permeabilized smooth muscle cells,[240] and in isolated microsomal fractions.[236,237,241,242] The mechanism for this specific action of GTP is not well understood at present. Nonhydrolyzable GTP analogues are ineffective or may inhibit the effect of GTP.[236,237,242] In contrast to InsP$_3$-induced calcium release, GTP-induced calcium release is slow[236] and temperature dependent[236,238] and in some instances requires the presence of a viscosity-increasing agent.[236,238,242] The latter might be a substitute for serum albumin present under physiological conditions.[236] Since GTP released 70% of the available cellular calcium pool in contrast to InsP$_3$, which released 30%, an additional calcium pool must be tapped,[240] and these pools have indeed been partially separated.[241] In electron micrographs of microsomal vesicles taken after GTP addi-

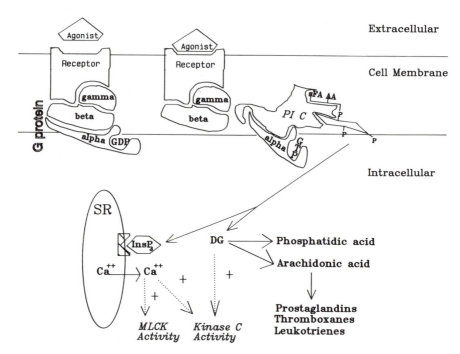

Figure 7. Suggested scheme for activation of contraction in the uterine smooth muscle cell. At rest the receptor is not occupied and binds to a G protein consisting of α, β, and γ subunits with bound GDP. When an appropriate agonist binds to receptor, GDP dissociates, and GTP binds in its place; the α subunit with bound GTP dissociates from the $\beta\gamma$ complex and activates phosphoinositidase C (PI C). The phosphoinositidase C hydrolyzes phosphatidylinositol-4,5-bisphosphate to inositol-1,4,5-trisphosphate (InsP$_3$) and diacylglycerol (DG), both intracellular messengers. InsP$_3$ causes release of calcium from the sarcoplasmic reticulum (SR). The released intracellular free calcium activates the myosin light chain kinase (MLCK) to phosphorylate the myosin light chain. Phosphorylated myosin then interacts with actin, resulting in contraction. Calcium and diacylglycerol activate protein kinase C (kinase C), which phosphorylates various proteins including MLCK. Diacylglycerol (DG) is further metabolized to phosphatidic acid and/or arachidonic acid. The latter can give rise to prostaglandins, thromboxanes, and leukotrienes.

tion in the presence of a viscosity-increasing agent, fusion of microsomal vesicles has been observed[242] and may be the crucial step in GTP-induced calcium release. To date no studies of GTP action on the uterus are available.

8. Summary and Conclusions

The obligatory role of calcium in smooth muscle contraction has long been known. The calcium arises at least in part from intracellular stores. Hormonal control of the rise in free intracellular calcium has long been suspected. The link from hormone binding to calcium mobilization has now been identified in the action of phosphoinositidase C, which hydrolyzes membrane phosphoinositides, leading to the liberation of InsP$_3$ and diacylglycerol. It is presently thought that InsP$_3$ releases calcium from the sarcoplasmic reticulum, which causes a phasic contraction in smooth muscle through mobilization of intracellular calcium. Its phosphorylation product, InsP$_4$, then initiates influx of calcium. Calcium together with diacylglycerol activates protein kinase C. The actions of protein kinase C remain unclear, and arguments for its causing contraction and relaxation have both been supported by observations. Coming back to our theme, it is obvious that work on the uterus is sorely lacking. There are few studies on the phosphoinositide system in the myometrium, and it is hoped that this review will stimulate interest.

References

1. Filo, R. S., Bohr, D. F., and Ruegg, J. C., 1965, Glycerinated skeletal and smooth muscle: Calcium and magnesium dependence, *Science* **147:**1581–1583.
2. Savineau, J. P., Mironneau, J., and Mironneau, C., 1988, Contractile properties of chemically skinned fibers from pregnant rat myometrium: Existence of an internal Ca-store, *Pfluegers Arch.* **411:**296–303.
3. Ammann, D., 1985, Ca^{2+}-selective microelectrodes, *Cell Calcium* **6:**39–55.
4. Morgan, J. P., and Morgan, K. G., 1984, Calcium and cardiovascular function, *Am. J. Med.* **5:**33–46.
5. Morgan, J. P., and Morgan, K. G., 1984, Alteration of cytoplasmic ionized calcium levels in smooth muscle by vasodilators in the ferret, *J. Physiol.* (*London*) **357:**539–551.
6. DeFeo, T. T., and Morgan, K. G., 1986, A comparison of two different indicators: Quin 2 and aequorin in isolated single cells and intact strips of ferret portal vein, *Pfluegers Arch.* **406:**427–429.
7. Grynkiewicz, G., Poenie, M., and Tsien, R. Y., 1985, A new generation of Ca^{2+} indicators with greatly improved fluorescence properties, *J. Biochem. Chem.* **260:**3440–3450.
8. Berridge, M. J., and Galione, A., 1988, Cytosolic calcium oscillators, *FASEB J.* **2:**3074–3082.
9. Williams, D. A., Becker, P. L., and Fay, F. S., 1987, Regional changes in calcium underlying contraction of single smooth muscle cells, *Science* **235:**1644–1647.
10. Cobbold, P. H., and Rink, T. J., 1987, Fluorescence and bioluminescence measurement of cytoplasmic free calcium, *Biochem. J.* **248:**313–328.
11. Garfield, R. E., and Somlyo, A. P., 1985, Structure of smooth muscle, in: *Calcium and Contractility* (A. K. Grover, ed.), Futura Press, pp. 1–36.
12. Gabella, G., 1971, Caveolae intracellulares and sarcoplasmic reticulum in smooth muscle, *J. Cell Sci.* **8:**602–609.
13. Devine, C. E., Somlyo, A. V., and Somlyo, A. P., 1972, Sarcoplasmic reticulum and excitation–contraction coupling in mammalian smooth muscles, *J. Cell Biol.* **52:**690–718.
14. Somlyo, A. P., 1985, Excitation–contraction coupling and the ultrastructure of smooth muscle, *Circ. Res.* **57:**497–507.
15. Somlyo, A. P., 1985, Cell calcium measurement with electron probe and electron energy loss analysis, *Cell Calcium* **6:**197–212.
16. Bond, M., Kitazawa, T., Somlyo, A. P., and Somlyo, A. V., 1984, Release and recycle of calcium by the sarcoplasmic reticulum in guinea-pig portal vein smooth muscle, *J. Physiol.* (*London*) **355:**677–695.
17. Carsten, M. E., 1969, Role of calcium binding by sarcoplasmic reticulum in the contraction and relaxation of uterine smooth muscle, *J. Gen. Physiol.* **53:**414–426.

18. Wibo, M., Morel, N., and Godfraind, T., 1981, Differentiation of Ca²⁺ pumps linked to plasma membrane and endoplasmic reticulum in the microsomal fraction from intestinal smooth muscle, *Biochim. Biophys. Acta* **649:**651–660.

19. Raeymaekers, L., Wuytack, F., Eggermont, J., De Schutter, G., and Casteels, R., 1983, Isolation of a plasma-membrane fraction from gastric smooth muscle. Comparison of the calcium uptake with that in endoplasmic reticulum, *Biochem. J.* **210:**315–322.

20. Wuytack, F., Raeymaekers, L., Verbist, J., De Smedt, H., and Casteels, R., 1984, Evidence for the presence in smooth muscle of two types of Ca²⁺-transport ATPase, *Biochem. J.* **224:**445–451.

21. Carsten, M. E., and Miller, J. D., 1980, Characterization of cell membrane and sarcoplasmic reticulum from bovine uterine smooth muscle, *Arch. Biochem. Biophys.* **204:**404–412.

22. Carsten, M. E., and Miller, J. D., 1977, Purification and characterization of microsomal fractions from smooth muscle, in: *Excitation–Contraction in Smooth Muscle* (R. Casteels, T. Godfraind, and J. C. Ruegg, eds.), Elsevier/North-Holland Biomedical Press, Amsterdam, pp. 155–163.

23. Carsten, M. E., and Miller, J. D., 1977, Effects of prostaglandins and oxytocin on calcium release from a uterine microsomal fraction, *J. Biol. Chem.* **252:**1576–1581.

24. Carsten, M. E., and Miller, J. D., 1984, Calcium-loaded microsomes from uterine smooth muscle, *Gynecol. Obstet. Invest.* **17:**78–83.

25. Wuytack, F., De Schutter, G., and Casteels, R., 1981, Purification of (Ca²⁺ + Mg²⁺)-ATPase from smooth muscle by calmodulin affinity chromatography, *FEBS Lett.* **129:**297–300.

26. Janis, R. A., Crankshaw, D. J., and Daniel, E. E., 1977, Control of intracellular Ca²⁺ activity in rat myometrium, *Am. J. Physiol.* **232:**C50–C58.

27. Carsten, M. E., and Miller, J. D., 1985, Problems specific to smooth muscle fractionation, *INSERM* **124:**293–298.

28. Verbist, J., Wuytack, F., Raeymaekers, L., and Casteels, R., 1985, Inhibitory antibodies to plasmalemmal Ca²⁺-transporting ATPases, *Biochem. J.* **231:**737–742.

29. Enyedi, A., Minami, J., Caride, A. J., and Penniston, J.T., 1988, Characteristics of the Ca²⁺ pump and Ca²⁺-ATPase in the plasma membrane of rat myometrium, *Biochem. J.* **252:**215–220.

30. Sharma, R. V., Butters, C. A., McEldoon, J. P., and Bhalla, R. C., 1987, Characterization of Ca²⁺ uptake in plasma membrane vesicles isolated from guinea pig ileum smooth muscle, *Cell Calcium* **8:**65–77.

31. Hasselbach, W., 1964, Relaxing factor and the relaxation of muscle, in: *Progress in Biophysics,* Vol. 14, (J. A. V. Butler and H. E. Huxley, eds.), Pergamon Press, Oxford, pp. 167–222.

32. Kwan, C. Y., Grover, A. K., Triggle, C.R., and Daniel, E., 1983, On the oxalate stimulation of ATP-dependent calcium accumulation by smooth muscle subcellular membranes, *Biochem. Int.* **6:**713–722.

33. Carsten, M. E., and Miller, J. D., 1981, Prostaglandin E₂ receptor in the myometrium: Distribution in subcellular fractions, *Arch. Biochem. Biophys.* **212:**700–704.

34. Raeymaekers, L., Wuytack, F., and Casteels, R., 1985, Subcellular fractionation of pig stomach smooth muscle. A study of the distribution of the (Ca²⁺ + Mg²⁺)-ATPase activity in plasmalemma and endoplasmic reticulum, *Biochim. Biophys. Acta* **815:**441–454.

35. Carsten, M. E., and Miller, J. D., 1984, Properties of a phosphorylated intermediate of the Ca,Mg-activated ATPase of microsomal vesicles from uterine smooth muscle, *Arch. Biochem. Biophys.* **232:**616–623.

36. Chiesi, M., Gasser, J., and Carafoli, E., 1984, Properties of the Ca-pumping ATPase of sarcoplasmic reticulum from vascular smooth muscle, *Biochem. Biophys. Res. Commun.* **124:**797–806.

37. Matlib, M. A., Crankshaw, J., Garfield, R. E., Crankshaw, D. J., Kwan, C.-Y., Branda, L. A., and Daniel, E. E., 1979, Characterization of membrane fractions and isolation of purified plasma membranes from rat myometrium, *J. Biol. Chem.* **254:**1834–1840.

38. Grover, A. K., Kwan, C.-Y., Crankshaw, J., Crankshaw, D. J., Garfield, R. E., and Daniel, E. E., 1980, Characteristics of calcium transport and binding by rat myometrium plasma membrane subfractions, *Am. J. Physiol.* **239:**C66–C74.

39. Grover, A. K., Kwan, C.-Y., and Daniel, E. E., 1982, Ca²⁺ dependence of calcium uptake by rat myometrium plasma membrane-enriched fraction, *Am. J. Physiol.* **242**:C278–C282.
40. Gupta, R. S., Chopra, A., and Stetsko, D. K., 1986, Cellular basis for the species differences in sensitivity to cardiac glycosides (digitalis), *J. Cell Physiol.* **127**:197–206.
41. Vallieres, J., Fortier, M., Somlyo, A. V., and Somlyo, A. P., 1978, Isolation of plasma membranes from rabbit myometrium, *Int. J. Biochem.* **9**:487–498.
42. Popescu, L. M., and Ignat, P., 1983, Calmodulin-dependent Ca²⁺-pump ATPase of human smooth muscle sarcolemma, *Cell Calcium* **4**:219–235.
43. Miller, J. D., and Carsten, M. E., 1987, The binding of Ca²⁺ to Ca²⁺-transporting microsomes derived from bovine uterine sarcoplasmic reticulum, *Biochem. Biophys. Res. Commun.* **147**:13–17.
44. Reeves, J. P., 1985, The sarcolemmal sodium–calcium exchange system, *Curr. Top. Membr. Transp.* **25**:77–127.
45. Smith, J. B., Zheng, T., and Smith, L., 1989, Relationship between cytosolic free Ca²⁺ and Na⁺–Ca²⁺ exchange in aortic muscle cells, *Am. J. Physiol.* **256**:C147–C154.
46. Hirata, M., Itoh, T., and Kuriyama, H., 1981, Effects of external cations on calcium efflux from single cells of the guinea-pig taenia coli and porcine coronary artery, *J. Physiol.* **310**:321–336.
47. Aickin, C. C., Brading, A. F., and Walmsley, D., 1987, An investigation of sodium–calcium exchange in the smooth muscle of guinea-pig ureter, *J. Physiol.* (*London*) **391**:325–346.
48. Pritchard, K., and Ashley, C. C., 1986, Na⁺/Ca²⁺ exchange in isolated smooth muscle cells demonstrated by the fluorescent calcium indicator fura-2, *FEBS Lett.* **195**:23–27.
49. Ashida, T., and Blaustein, M. P., 1987, Regulation of cell calcium and contractility in mammalian arterial smooth muscle: The role of sodium–calcium exchange, *J. Physiol.* (*London*) **392**:617–635.
50. Morel, N., and Godfraind, T., 1984, Sodium/calcium exchange in smooth muscle microsomal fractions, *Biochem. J.* **218**:421–427.
51. Grover, A. K., Kwan, C.-Y., Rangachari, P. K., and Daniel, E. E., 1983, Na–Ca exchange in a smooth muscle plasma membrane-enriched fraction, *Am. J. Physiol.* **244**:C158–165.
52. Grover, A. K., and Kwan, C.-Y., 1987, Sodium–calcium exchange in rat myometrium, *Biochem. Arch.* **3**:353–362.
53. Wuytack, F., Raeymaekers, L., Verbist, J., Jones, L. R., and Casteels, R., 1987, Smooth-muscle endoplasmic reticulum contains a cardiac-like form of calsequestrin, *Biochim. Biophys. Acta* **899**:151–158.
54. Carsten, M. E., 1979, Calcium accumulation by human uterine microsomal preparations: Effects of progesterone and oxytocin, *Am. J. Obstet. Gynecol.* **133**:598–601.
55. Godfraind, T., Sturbois, X., and Verbeke, N., 1976, Calcium incorporation by smooth muscle microsomes, *Biochim. Biophys. Acta* **455**:254–268.
56. Carsten, M. E., and Miller, J. D., 1983, Regulation of myometrial contractions, in: *Initiation of Parturition: Prevention of Prematurity, Report of the Fourth Ross Conference on Obstetric Research* (P.C. MacDonald and J. Porter, eds.), Ross Laboratories, Columbus, OH, pp. 166–171.
57. Scarpa, A., 1979, Measurement of calcium ion concentrations with metallochromic indicators, in: *Detection and Measurement of Free Ca²⁺ in Cells* (C. C. Ashley and A. K. Campbell, eds.), Elsevier/North-Holland, New York, pp. 85–115.
58. Brown, H. M., and Rydqvist, B., 1981, Arsenazo III–Ca²⁺ effect of pH, ionic strength, and arsenazo III concentration on equilibrium binding evaluated with Ca²⁺ ion-sensitive electrodes and absorbance measurements, *Biophys. J.* **36**:117–137.
59. Miller, J. D., and Carsten, M. E., 1984, Magnesium binding to uterine microsomes and the effect of calcium uptake, *INSERM* **124**:265–272.
60. Raeymaekers, L., Agostini, B., and Hasselbach, W., 1980, Electron cytochemistry of oxalate-stimulated calcium uptake in microsomes from the smooth muscle of the pig stomach, *Histochemistry* **65**:121–129.
61. Carsten, M. E., and Miller, J. D., 1985, Modulation of phosphoprotein formation in bovine uterine microsomes, *Soc. Gynecol. Invest.* **32**:377.

62. Gunteski-Hamblin, A.-M., Greeb, J., and Shull, G. E., 1988, A novel Ca^{2+} pump expressed in brain, kidney, and stomach is encoded by alternative transcript of the slow-twitch muscle sarcoplasmic reticulum Ca-ATPase gene, *J. Biol. Chem.* **263:**15032–15040.

63. Wuytack, F., Kanmura, Y., Eggermont, J. A., Raeymaekers, L., Verbist, J., Hartweg, D., Gietzen, K., and Casteels, R., 1989, Smooth muscle expresses a cardiac/slow muscle isoform of the Ca^{2+}-transport ATPase in its endoplasmic reticulum, *Biochem. J.* **257:**117–123.

64. de la Bastie, D., Wisnewsky, C., Schwartz, K., and Lompre, A. M., 1988, $(Ca^{2+} + Mg^{2+})$-dependent ATPase mRNA from smooth muscle sarcoplasmic reticulum differs from that in cardiac and fast skeletal muscles, *FEBS Lett.* **229:**45–48.

65. Lytton, J., Zarain-Herzberg, A., Periasamy, M., and MacLennan, D. H., 1989, Molecular cloning of the mammalian smooth muscle sarco(endo)plasmic reticulum Ca^{2+}-ATPase, *J. Biol. Chem.* **264:**7059–7065.

66. Scatchard, G., 1949, The attractions of proteins for small molecule and ions, *Ann. N.Y. Acad. Sci.* **51:**660–671.

67. Munson, P. J., and Rodbard, D., 1980, Ligand: A versatile computerized approach for characterization of ligand-binding systems, *Anal. Biochem.* **107:**220–239.

68. Teicher, M., 1982, Adaptation of Munson's and Rodbard's LIGAND program to the Apple II+ microcomputer, MED-58:LIGAND, Biomedical computing Technology Information Center, Vanderbilt Medical Center, Nashville, TN.

69. Raeymaekers, L., Hofmann, F., and Casteels, R., 1988, Cyclic GMP-dependent protein kinase phosphorylates phospholamban in isolated sarcoplasmic reticulum from cardiac and smooth muscle, *Biochem. J.* **252:**269–273.

70. Batra, S. C., and Daniel, E. E., 1971, Effect of multivalent cations and drugs on calcium uptake by the rat myometrial microsomes, *Comp. Biochem. Physiol.* **38:**285–300.

71. Nishikori, K., Takenaka, T., and Maeno, H., 1977, Stimulation of microsomal calcium uptake and protein phosphorylation by adenosine cyclic 3′,5′-monophosphate in rat uterus, *Mol. Pharmacol.* **13:**671–678.

72. Nishikori, K., and Maeno, H., 1979, Close relationship between adenosine 3′:5′-monophosphate-dependent endogenous phosphorylation of a specific protein and stimulation of calcium uptake in rat uterine microsomes, *J. Biol. Chem.* **254:**6099–6106.

73. Krall, J. F., Swensen, J. L., and Korenman, S. G., 1976, Hormonal control of uterine contraction. Characterization of cyclic AMP-dependent membrane properties in the myometrium, *Biochim. Biophys. Acta* **448:**578–588.

74. Raeymaekers, L., and Hasselbach, W., 1981, Ca^{2+} uptake, Ca^{2+}-ATPase activity, phosphoprotein formation and phosphate turnover in a microsomal fraction of smooth muscle, *Eur. J. Biochem.* **116:**373–378.

75. Feher, J. J., and Briggs, F. N., 1983, Determinants of calcium loading at steady state in sarcoplasmic reticulum, *Biochim. Biophys. Acta* **727:**389–402.

76. Kim, D. H., Ohnishi, S. T., and Ikemoto, N., 1983, Kinetic studies of calcium release from sarcoplasmic reticulum *in vitro*, *J. Biol. Chem.* **258:**9662–9668.

77. Palade, P., Mitchell, R. D., and Fleischer, S., 1983, Spontaneous calcium release from sarcoplasmic reticulum, *J. Biol. Chem.* **258:**8098–8107.

78. Inesi, G., 1985, Mechanism of calcium transport, *Annu. Rev. Physiol.* **47:**573–601.

79. Martonosi, A. N., 1984, Mechanisms of Ca^{2+} release from sarcoplasmic reticulum of skeletal muscle, *Pharmacol. Rev.* **64:**1240–1320.

80. Chiesi, M., and Wen, Y. S., 1983, A phosphorylated conformational state of the $(Ca^{2+}-Mg^{2+})$-ATPase of fast skeletal muscle sarcoplasmic reticulum can mediate rapid Ca^{2+} release, *J. Biol. Chem.* **258:**6078–6085.

81. Carsten, M. E., and Miller, J. D., 1978, Comparison of calcium association constants and ionophoretic properties of some prostaglandins and ionophores, *Arch. Biochem. Biophys.* **185:** 282–283.

82. Serhan, C., Anderson, P., Goodman, E., Dunham, P., and Weissman, G., 1981, Phosphatidate and oxidized fatty acids are calcium ionophores, *J. Biol. Chem.* **256:**2736–2741.

83. Ehrlich, B. E., and Watras, J., 1988, Inositol 1,4,5-trisphosphate activates a channel from smooth muscle sarcoplasmic reticulum, *Nature* **336:**583–586.

84. Fukuo, K., Morimoto, S., Koh, E., Yukawa, S., Tsuchiya, H., Imanaka, S., Yamamoto, H., Onishi, T., and Kumahara, Y., 1986, Effects of prostaglandins on the cytosolic free calcium concentration in vascular smooth muscle cells, *Biochem. Biophys. Res. Commun.* **136:**247–252.

85. Villar, A., D'Ocon, M. P., and Anselmi, E., 1985, Calcium requirement of uterine contraction induced by PGE$_1$: Importance of intracellular calcium stores, *Prostaglandins* **30:**491–497.

86. Carsten, M. E., 1974, Prostaglandins and oxytocin: Their effects on uterine smooth muscle, *Prostaglandins* **5:**33–40.

87. Soloff, S., and Sweet, P., 1982, Oxytocin inhibition of (Ca^{2+} + Mg^{2+})-ATPase activity in rat myometrial plasma membranes, *J. Biol. Chem.* **257:**10687–10693.

88. Popescu, L. M., Nutu, O., and Panoiu, C., 1985, Oxytocin contracts the human uterus at term by inhibiting the myometrial Ca^{2+}-extrusion pump, *Biosci. Rep.* **2:**21–28.

89. Edman, K. A. P., and Schild, H. O., 1963, Calcium and the stimulant and inhibitory effects of adrenaline in depolarized smooth muscle, *J. Physiol. (London)* **169:**404–411.

90. Riemer, R. K., Goldfien, A., and Roberts, J. M., 1987, Rabbit myometrial adrenergic sensitivity is increased by estrogen but is independent of changes in alpha adrenoceptor concentration, *J. Pharmacol. Exp. Ther.* **240:**44–49.

91. Hoffman, B. B., Lavin, T. N., Lefkowitz, R. J., and Ruffolo, R. R., 1981, Alpha adrenergic receptor subtypes in rabbit uterus: Mediation of myometrial contraction and regulation by estrogens, *J. Pharmacol. Exp. Ther.* **219:**290–295.

92. Jacobs, M. M., Hayashida, D., and Roberts, J. M., 1985, Human myometrial adrenergic receptors during pregnancy: Identification of the α-adrenergic receptor by [^{3}H]dihydroergocryptine binding, *Am. J. Obstet. Gynecol.* **152:**680–684.

93. Bottari, S. P., Vokaer, A., Kaivez, E., Lescrainier, J. P., and Vauquelin, G., 1985, Regulation of alpha- and beta-adrenergic receptor subclasses by gonadal steriods in human myometrium, *Acta Physiol. Hung.* **65:**335–346.

94. Reynolds, E. E., and Dubyak, G. R., 1985, Activation of calcium mobilization and calcium influx by alpha$_1$-adrenergic receptors in a smooth muscle cell line, *Biochem. Biophys. Res. Commun.* **130:**627–632.

95. Fabiato, A., 1983, Calcium-induced release of calcium from the cardiac sarcoplasmic reticulum, *Am. J. Physiol.* **245:**C1–C14.

96. Leijten, P., Cauvin, C., Lodge, N., Saida, K., and van Breemen, C., 1985, Ca^{2+} sources mobilized by α$_1$-receptor activation in vascular smooth muscle, *Clin. Sci.* **68**(Suppl. 10):47s–50s.

97. Akerman, K. E. O., and Wikstrom, M. K. F., 1979, (Ca^{2+} + Mg^{2+})-stimulated ATPase activity of rabbit myometrium plasma membrane is blocked by oxytocin, *FEBS Lett.* **97:**283–285.

98. Ashoori, F., Takai, A., and Tomita, T., 1985, The response of non-pregnant rat myometrium to oxytocin in Ca-free solution, *Br. J. Pharmacol.* **84:**175–183.

99. Batra, S., 1986, Effect of oxytocin on calcium influx and efflux in the rat myometrium, *Eur. J. Pharmacol.* **120:**57–61.

100. van Breemen, C., and Daniel, E. E., 1966, The influence of high potassium depolarization and acetylcholine on calcium exchange in the rat uterus, *J. Gen. Physiol.* **49:**1299–1317.

101. Hodgson, B. J., and Daniel, E. E., 1973, Studies concerning the source of calcium for contraction of rat myometrium, *Can. J. Physiol. Pharmacol.* **51:**914–932.

102. Somlyo, A. V., and Somlyo, A. P., 1968, Electromechanical and pharmacomechanical coupling in vascular smooth muscle, *J. Pharmacol. Exp. Ther.* **159:**129–145.

103. Berridge, M. J., 1984, Inositol trisphosphate and diacylglycerol as second messengers, *Biochem. J.* **220:**345–360.

104. Hokin, L. E., 1985, Receptors and phosphoinositide-generated second messengers, *Annu. Rev. Biochem.* **54:**205–235.

105. Abdel-Latif, A. A., 1986, Calcium-mobilizing receptors, polyphosphoinositides, and the generation of second messengers, *Pharmacol. Rev.* **38:**227–272.

106. Rhee, S. G., Suh, P. G., Ryu, S. H., and Lee, S. Y., 1989, Studies of inositol phospholipid-specific phospholipase C, *Science* **244:**546–550.
107. Bennett, C. F., and Crooke, S. T., 1987, Purification and characterization of a phosphoinositide-specific phospholipase C from guinea pig uterus, *J. Biol. Chem.* **262:**13789–13797.
108. Howe, P. H., Akhtar, R. A., Naderi, S., and Abdel-Latif, A. A., 1986, Correlative studies on the effect of carbachol on *myo*-inositol trisphosphate accumulation, myosin light chain phosphorylation and contraction in sphincter smooth muscle of rabbit iris, *J. Pharmacol. Exp. Ther.* **239:**574–583.
109. Storey, D. J., Shears, S. B., Kirk, C. J., and Michell, R. H., 1984, Stepwise enzymatic dephosphorylation of inositol 1,4,5-trisphosphate to inositol in liver, *Nature* **312:**374–376.
110. Downes, C. P., Mussat, M. C., and Michell, R. H., 1982, The inositol trisphosphate phosphomonoesterase of the human erythrocyte membrane, *Biochem. J.* **203:**169–177.
111. Sasaguri, T., Hirata, M., and Kuriyama, H., 1985, Dependence on Ca^{2+} of the activities of phosphatidylinositol 4,5-bisphosphate phosphodiesterase and inositol 1,4,5-trisphosphate phosphate in smooth muscles of the porcine coronary artery, *Biochem. J.* **231:**497–503.
112. Hallcher, L. M., and Sherman, W. R., 1980, The effects of lithium and other agents on the activity of *myo*-inositol-1-phosphatase from bovine brain, *J. Biol. Chem.* **255:**10896–10901.
113. Imai, A., and Gershengorn, M. C., 1987, Independent phosphatidylinositol synthesis in pituitary plasma membrane and endoplasmic reticulum, *Nature* **325:**726–728.
114. Grove, R. I., and Korach, K. S., 1987, Estrogen stimulation of phosphatidylinositol metabolism in mouse uterine tissue, *Endocrinology* **121:**1083–1088.
115. Porter, F. D., Li, Y.-S., and Deuel, T. F., 1988, Purification and characterization of a phosphatidylinositol 4-kinase from bovine uteri, *J. Biol. Chem.* **263:**8989–8995.
116. Guillemette, G., Balla, T., Baukal, A. J., Spat, A., and Catt, K. J., 1987, Intracellular receptors for inositol 1,4,5-trisphosphate in angiotensin II target tissues, *J. Biol. Chem.* **262:**1010–1015.
117. Schacht, J., 1976, Inhibition by neomycin of polyphosphoinositide turnover in subcellular fractions of guinea-pig cerebral cortex *in vitro*, *J. Neurochem.* **27:**1119–1124.
118. Prentki, M., Deeney, J. T., Matschinsky, F. M., and Joseph, S. K., 1986, Neomycin: A specific drug to study the inositol–phospholipid signalling system? *FEBS Lett.* **197:**285–288.
119. Moraru, I. I., Popescu, L. M., Vidulescu, C., and Tzigaret, C., 1987, Antibodies against phospholipase C inhibit smooth muscle contraction induced by acetylcholine and histamine, *Eur. J. Pharmacol.* **138:**427–431.
120. Abdel-Latif, A. A., Akhtar, R. A., and Hawthorne, J. N., 1977, Acetylcholine increases the breakdown of triphosphoinositide of rabbit iris muscle prelabelled with [^{32}P]phosphate, *Biochem. J.* **162:**61–73.
121. Berridge, M. J., 1983, Rapid accumulation of inositol trisphosphate reveals that agonists hydrolyse polyphosphoinositides instead of phosphatidylinositol, *Biochem. J.* **212:**849–858.
122. Best, L., Brooks, K. J., and Bolton, T. B., 1985, Relationship between stimulated inositol lipid hydrolysis and contractility in guinea-pig visceral longitudinal smooth muscle, *Biochem. Pharmacol.* **34:**2297–2301.
123. Akhtar, R. A., and Abdel-Latif, A. A., 1984, Carbachol causes rapid phosphodiesteratic cleavage of phosphatidylinositol 4,5-bisphosphate and accumulation of inositol phosphates in rabbit iris smooth muscle: Prazosin inhibits noradrenaline- and ionophore A23187-stimulated accumulation of inositol phosphates, *Biochem. J.* **224:**291–300.
124. Takuwa, Y., Takuwa, N., and Rasmussen, H., 1986, Carbachol induces a rapid and sustained hydrolysis of polyphosphoinositide in bovine tracheal smooth muscle measurements of the mass of polyphosphoinositides, 1,2-diacylglycerol, and phosphatidic acid, *J. Biol. Chem.* **261:**14670–14675.
125. Baron, C. B., Cunningham, M., Strauss, J. F., and Coburn, R. F., 1984, Pharmacomechanical coupling in smooth muscle may involve phosphatidylinositol metabolism, *Proc. Natl. Acad. Sci. U.S.A.* **81:**6899–6903.
126. Hashimoto, T., Hirata, M., and Ito, Y., 1985, A role for inositol 1,4,5-trisphosphate in the initiation of agonist-induced contractions of dog tracheal smooth muscle, *J. Pharmacol.* **86:**191–199.

127. Duncan, R. A., Krzanowski, J. J., Davis, J. S., Polson, J. B., Coffey, R. G., Shimoda, T., and Szentivanyi, A., 1987, Polyphosphoinositide metabolism in canine tracheal smooth muscle (CTSM) in response to a cholinergic stimulus, *Biochem. Pharmacol.* **36**:307–310.

128. Smith, J. B., Smith, L., Brown, E. R., Barnes, D., Sabir, M. A., Davis, J. S., and Farese, R. V., 1984, Angiotensin II rapidly increased phosphatidate–phosphoinositide synthesis and phosphoinositide hydrolysis and mobilizes intracellular calcium in cultured arterial muscle cells, *Proc. Natl. Acad. Sci. U.S.A.* **81**:7812–7816.

129. Doyle, V. M., and Ruegg, U. T., 1985, Vasopressin induced production of inositol trisphosphate and calcium efflux in a smooth muscle cell line, *Biochem. Biophys. Res. Commun.* **131**: 469–476.

130. Orlicky, D. J., Silio, M., Williams, C., Gordon, J., and Gerschenson, L. E., 1986, Regulation of inositol phosphate levels by prostaglandins in cultured endometrial cells, *J. Cell. Physiol.* **128**:105–112.

131. Macphee, C. H., Drummond, A. H., Otto, A. M., and Jimenez de Asua, L. L., 1984, Prostaglandin $F_{2\alpha}$ stimulates phosphatidylinositol turnover and increases the cellular content of 1,2-diacylglycerol in confluent resting Swiss 3T3 cells, *J. Cell. Physiol.* **119**:35–40.

132. Sasaki, T., 1985, Formation of inositol phosphates and calcium mobilization in Swiss 3T3 cells in response to prostaglandin $F_{2\alpha}$, *Prostaglandins Leukotrienes Med.* **19**:153–159.

133. Mene, P., Dubyak, G. R., Scarpa, A., and Dunn, M. J., 1987, Stimulation of cytosolic free calcium and inositol phosphates by prostaglandins in cultured rat mesangial cells, *Biochem. Biophys. Res. Commun.* **142**:579–586.

134. Davis, J. S., Weakland, L. L., Weiland, D. A., Farese, R. V., and West, L. A., 1987, Prostaglandin $F_{2\alpha}$ stimulates phosphatidylinositol 4,5-bisphosphate hydrolysis and mobilizes intracellular Ca^{2+} in bovine luteal cells, *Proc. Natl. Acad. Sci. U.S.A.* **84**:3728–3732.

135. Suba, E. A., and Roth, B. L., 1987, Prostaglandins activate phosphoinositide metabolism in rat aorta, *Eur. J. Pharmacol.* **136**:325–332.

136. Marc, S., Leiber, D., and Harbon, S., 1986, Carbachol and oxytocin stimulate the generation of inositol phosphates in the guinea pig myometrium, *FEBS Lett.* **201**:9–14.

137. Schrey, M. P., Cornford, P. A., Read, A. M., and Steer, P. J., 1988, A role for phosphoinositide hydrolysis in human uterine smooth muscle during parturition, *Am. J. Obstet. Gynecol.* **159**:964–970.

138. Schrey, M. P., Read, A. M., and Steer, P. J., 1986, Oxytocin and vasopressin stimulate inositol phosphate production in human gestational myometrium and decidua cells, *Biosci. Rep.* **6**:613–619.

139. Varol, F. G., Hadjiconstantinou, M., Zuspan, F. P., and Neff, N. H., 1989, Angiotensin II stimulates phosphoinositide turnover in the rat myometrium, *Eur. J. Pharmacol.* **162**:37–41.

140. Ruzycky, A. L., and Crankshaw, D. J., 1988, Role of inositol phospholipid hydrolysis in the initiation of agonist-induced contractions of rat uterus: Effects of domination by 17β-estradiol and progesterone, *Can. J. Physiol. Pharmacol.* **66**:10–17.

141. Fain, J. N., and Berridge, M. J., 1979, Relationship between hormonal activation of phosphatidylinositol hydrolysis, fluid secretion, and calcium flux in the blowfly salivary gland, *Biochem. J.* **178**:45–58.

142. Streb, H., Irvine, R. F., Berridge, M. J., and Schulz, I., 1983, Release of Ca^{2+} from a nonmitochondrial intracellular store in pancreatic acinar cells by inositol 1,4,5-trisphosphate, *Nature* **306**:67–69.

143. Nabika, T., Velletri, P. A., Lovenberg, W., and Beaven, M. A., 1985, Increase in cytosolic calcium and phosphoinositide metabolism induced by angiotensin II and [Arg]vasopressin in vascular smooth muscle cells, *J. Biol. Chem.* **260**:4661–4670.

144. Suematsu, E., Hirata, M., Hashimoto, T., and Kuriyama, H., 1984, Inositol 1,4,5-trisphosphate releases Ca^{2+} from intracellular store sites in skinned single cells of porcine coronary artery, *Biochem. Biophys. Res. Commun.* **120**:481–485.

145. Somlyo, A. V., Bond, M., Somlyo, A. P., and Scarpa, S. M., 1985, Inositol trisphosphate-

induced calcium release and contraction in vascular smooth muscle, *Proc. Natl. Acad. Sci. U.S.A.* **82:**5231–5235.

146. Bitar, K. N., Bradford, P. G., Putney, J. W., and Makhlouf, G. M., 1986, Stoichiometry of contraction and Ca^{2+} mobilization by inositol 1,4,5-trisphosphate in isolated gastric smooth muscle cells, *J. Biol. Chem.* **261:**16591–16596.

147. Smith, J. B., Smith, L., and Higgins, B. L., 1985, Temperature and nucleotide dependence of calcium release by *myo*-inositol 1,4,5-trisphosphate in cultured vascular smooth muscle cells, *J. Biol. Chem.* **260:**14413–14416.

148. Somlyo, A. P., Walker, J. W., Goldman, Y. E., Trentham, D. R., Kobayashi, S., Kitazawa, T., and Somlyo, A. V., 1988, Inositol trisphosphate, calcium and muscle contraction, *Phil. Trans. R. Soc. Lond.* [*Biol.*] **320:**399–414.

148a. Kanmura, Y., Missiaen, L., and Casteels, R., 1988, Properties of intracellular calcium stores in pregnant rat myometrium, *Brit. J. Pharmacol.* **95:**284–290.

149. Ghosh, T. K., Eis, P. S., Mullaney, J. M., Ebert, C. L., and Gill, D. L., 1988, Competitive, reversible, and potent antagonism of inositol 1,4,5-trisphosphate-activated calcium release by heparin, *J. Biol. Chem.* **263:**11075–11079.

150. Kwan, C. Y., and Lee, R. M. K. W., 1984, Interaction of saponin with microsomal membrane fraction isolated from the smooth muscle of rat vas deferens, *Mol. Physiol.* **5:**105–114.

151. Carsten, M. E., and Miller, J. D., 1985, Ca^{2+} release by inositol trisphosphate from Ca^{2+}-transporting microsomes derived from uterine sarcoplasmic reticulum, *Biochem. Biophys. Res. Commun.* **130:**1027–1031.

152. Burgess, G. M., Irvine, R. F., Berridge, M. J., McKinney, J. S., and Putney, J. W., 1984, Actions of inositol phosphates on Ca^{2+} pools in guinea-pig hepatocytes, *Biochem. J.* **224:**741–746.

153. Irvine, R. F., Letcher, A. J., Lander, D. J., and Berridge, M. J., 1986, Specificity of inositol phosphate-stimulated Ca^{2+} mobilization from Swiss-mouse 3T3 cells, *Biochem. J.* **240:**301–304.

154. Irvine, R. F., Brown, K. D., and Berridge, M. J., 1984, Specificity of inositol trisphosphate-induced calcium release from permeabilized Swiss-mouse 3T3 cells, *Biochem. J.* **221:**269–272.

155. Spat, A., Fabiato, A., and Rubin, R. P., 1986, Binding of inositol trisphosphate by a liver microsomal fraction, *Biochem. J.* **233:**929–932.

156. Popescu, L. M., Hinescu, M. E., Musat, S., Ionescu, M., and Pistritzu, F., 1986, Inositol trisphosphate and the contraction of vascular smooth muscle cells, *Eur. J. Pharmacol.* **123:**167–169.

157. Delfert, D. M., Hill, S., Pershadsingh, H. A., Sherman, W. R., and McDonald, J. M., 1986, *myo*-Inositol 1,4,5-trisphosphate mobilizes Ca^{2+} from isolated adipocyte endoplasmic reticulum but not from plasma membranes, *Biochem. J.* **236:**37–44.

158. Batty, I. R., Nahorski, S. R., and Irvine, R. F., 1985, Rapid formation of inositol 1,3,4,5-tetrakisphosphate following muscarinic receptor stimulation of rat cerebral cortical slices, *Biochem. J.* **232:**211–215.

159. Bielkiewicz-Vollrath, B., Carpenter, J. R., Schulz, R., and Cook, D. A., 1987, Early production of 1,4,5-inositol trisphosphate and 1,3,4,5-inositol tetrakisphosphate by histamine and carbachol in ileal smooth muscle, *Mol. Pharmacol.* **31:**513–522.

160. Irvine, R. F., Letcher, A. J., Heslop, J. P., and Berridge, M. J., 1986, The inositol tris/tetrakisphosphate pathway—demonstration of Ins(1,4,5)P_3 3-kinase activity in animal tissues, *Nature* **320:**531–534.

161. Hawkins, P. T., Stephens, L., and Downes, C. P., 1986, Rapid formation of inositol 1,3,4,5-tetrakisphosphate and inositol 1,3,4-trisphosphate in rat parotid glands may both result indirectly from receptor-stimulated release of inositol 1,4,5-trisphosphate from phosphatidylinositol 4,5-bisphosphate, *Biochem. J.* **238:**507–516.

162. Yamaguchi, K., Hirata, M., and Kuriyama, H., 1988, Purification and characterization of inositol 1,4,5-trisphosphate 3-kinase from pig aortic smooth muscle, *Biochem. J.* **251:**129–134.

163. Rossier, M. F., Capponi, A. M., and Vallotton, M. B., 1987, Metabolism of inositol 1,4,5-trisphosphate in permeabilized rat aortic smooth-muscle cells, *Biochem. J.* **245:**305–307.

164. Burgess, G. M., McKinney, J. S., Irvine, R. F., and Putney, J. W., 1985, Inositol 1,4,5-trisphosphate and inositol 1,3,4-trisphosphate formation in Ca^{2+}-mobilizing-hormone-activated cells, *Biochem. J.* **232**:237–243.

165. Hansen, C. A., Mah, S., and Williamson, J. R., 1986, Formation and metabolism of inositol 1,3,4,5-tetrakisphosphate in liver, *J. Biol. Chem.* **261**:8100–8103.

166. Irvine, R. F., and Moor, R. M., 1986, Micro-injection of inositol 1,3,4,5-tetrakisphosphate activates sea urchin eggs by a mechanism dependent on external Ca^{2+}, *Biochem. J.* **240**:917–920.

167. Irvine, R. F., and Moor, R. M., 1987, Inositol(1,3,4,5)tetrakisphosphate-induced activation of sea urchin eggs requires the presence of inositol trisphosphate, *Biochem. Biophys. Res. Commun.* **146**:284–290.

168. Bradford, P. G., and Irvine, R. F., 1987, Specific binding sites for [³H]inositol (1,3,4,5)tetrakisphosphate on membranes of HL-60 cells, *Biochem. Biophys. Res. Commun.* **149**:680–685.

169. Kuno, M., and Gardner, P., 1987, Ion channels activated by inositol 1,4,5-trisphosphate in plasma membrane of human T-lymphocytes, *Nature* **326**:301–304.

170. Wilson, D. B., Bross, T. E., Sherman, W. R., Berger, R. A., and Majerus, P. W., 1985, Inositol cyclic phosphates are produced by cleavage of phosphatidylphosphoinositols (polyphosphoinositides) with purified sheep seminal vesicle phospholipase C enzymes, *Proc. Natl. Acad. Sci. U.S.A.* **82**:4013–4017.

171. Sekar, M. C., Dixon, J. F., and Hokin, L. E., 1987, The formation of inositol 1,2-cyclic 4,5-trisphosphate and inositol 1,2-cyclic 4-bisphosphate on stimulation of mouse pancreatic minilobules with carbamylcholine, *J. Biol. Chem.* **262**:340–344.

172. Hawkins, P. T., Berrie, C. P., Morris, A. J., and Downes, C. P., 1987, Inositol 1,2-cyclic 4,5-trisphosphate is not a product of muscarinic receptor-stimulated phosphatidylinositol 4,5-bisphosphate hydrolysis in rat parotid glands, *Biochem. J.* **243**:211–218.

173. Dean, N. M., and Moyer, J. D., 1987, Separation of multiple isomers of inositol phosphates formed in GH₃ cells, *Biochem. J.* **242**:361–366.

174. Inhorn, R. C., Bansal, V. S., and Majerus, P. W., 1987, Pathway for inositol 1,3,4-trisphosphate and 1,4-bisphosphate metabolism, *Proc. Natl. Acad. Sci. U.S.A.* **84**:2170–2174.

175. Ross, T. S., and Majerus, P. W., 1986, Isolation of D-*myo*-inositol 1 : 2-cyclic phosphate 2-inositolphosphohydrolase from human placenta, *J. Biol. Chem.* **261**:11119–11123.

176. Connolly, T. M., Wilson, D. B., Bross, T. E., and Majerus, P. W., 1986, Isolation and characterization of the inositol cyclic phosphate products of phosphoinositide cleavage by phospholipase C, *J. Biol. Chem.* **261**:122–126.

177. Wilson, D. B., Connolly, T. M., Bross, T. E., Majerus, P. W., Sherman, W. R., Tyler, A. N., Rubin, L. J., and Brown, J. E., 1985, Isolation and characterization of the inositol cyclic phosphate products of polyphosphoinositide cleavage by phospholipase C, *J. Biol. Chem.* **260**:13496–13501.

178. Willcocks, A. L., Strupish, J., Irvine, R. F., and Nahorski, S. R., 1989, Inositol 1 : 2-cyclic 4,5-trisphosphate is only a weak agonist at inositol 1,4,5-trisphosphate receptors, *Biochem. J.* **257**:297–300.

178. Berridge, M. J., 1987, Inositol trisphosphate and diacylglycerol: Two interacting second messengers, *Annu. Rev. Biochem.* **56**:159–193.

180. Park, S., and Rasmussen, H., 1986, Carbachol-induced protein phosphorylation changes in bovine tracheal smooth muscle, *J. Biol. Chem.* **261**:15734–15739.

181. Nishizuka, Y., 1988, The molecular heterogeneity of protein kinase C and its implications for cellular regulation, *Nature* **334**:661–665.

182. Takai, Y., Kishimoto, A., Kikawa, U., Mori, T., and Nishizuka, Y., 1979, Unsaturated diacylglycerol as a possible messenger for the activation of calcium-activated, phospholipid-dependent protein kinase system, *Biochem. Biophys. Res. Commun.* **91**:1218–1224.

183. Kikkawa, U., and Nishizuka, Y., 1986, The role of protein kinase C in transmembrane signalling, *Annu. Rev. Cell. Biol.* **2**:149–178.

184. Okazaki, T., Ban, C., and Johnston, J. M., 1984, The identification and characterization of protein kinase C activity in fatal membranes, *Arch. Biochem. Biophys.* **229:**27–32.
185. Baraban, J. M., Gould, R. J., Peroutka, S. J., and Snyder, S. H., 1985, Phorbol ester effects on neurotransmission: Interaction with neurotransmitters and calcium in smooth muscle, *Proc. Natl. Acad. Sci. U.S.A.* **82:**604–607.
186. Wolf, M., LeVine, H., May, W. S., Cuatrecasas, P., and Sahyoun, N., 1985, A model for intracellular translocation of protein kinase C involving synergism between Ca^{2+} and phorbol esters, *Nature* **317:**546–549.
187. May, W. S., Sahyoun, N., Wolf, M., and Cuatrecasas, P., 1985, Role of intracellular calcium mobilization in the regulation of protein kinase C-mediated membrane processes, *Nature* **317:**549–551.
188. Castagna, M., Takai, Y., Kaibuchi, K., Sano, K., Kikkawa, U., and Nishizuka, Y., 1982, Direct activation of calcium-activated, phospholipid-dependent protein kinase by tumor-promoting phorbol esters, *J. Biol. Chem.* **257:**7847–7851.
189. Leeb-Lundberg, L. M., Cotecchia, S., Lomasney, J. W., DeBernardis, J. F., Lefkowitz, R. J., and Caron, M. G., 1985, Phorbol esters promote α_1-adrenergic receptor phosphorylation and receptor uncoupling from inositol phospholipid metabolism, *Proc. Natl. Acad. Sci. U.S.A.* **82:**5651–5655.
190. McMillan, M., Chernow, B., and Roth, B. L., 1986, Phorbol esters inhibit alpha₁-adrenergic receptor-stimulated phosphoinositide hydrolysis and contraction in rat aorta: Evidence for a link between vascular contraction and phosphoinositide turnover, *Biochem. Biophys. Res. Commun.* **134:**970–974.
191. Griendling, K. K., Rittenhouse, S. E., Brock, T. A., Ekstein, L. S., Gimbrone, M. A., and Alexander, R. W., 1986, Sustained diacylglycerol formation from inositol phospholipids in angiotensin II-stimulated vascular smooth muscle cells, *J. Biol. Chem.* **261:**5901–5906.
192. Brock, T. A., Rittenhouse, S. E., Powers, C. W., Ekstein, L. S., Gimbrone, M. A., and Alexander, R. W., 1985, Phorbol ester and 1-oleoyl-2-acetylgycerol inhibit angiotensin activation of phospholipase C in cultured vascular smooth muscle cells, *J. Biol. Chem.* **260:**14158–14162.
193. Connolly, T. M., Lawing, W. J., and Majerus, P. W., 1986, Protein kinase C phosphorylated human platelet inositol trisphosphate 5′-phosphomonoesterase, increasing the phosphatase activity, *Cell* **46:**951–958.
194. Molina y Vedia, L. M., and Lapetina, E. G., 1986, Phorbol 12,13-dibutyrate and 1-oleyl-2-acetyldiacylglycerol stimulate inositol trisphosphate dephosphorylation in human platelets, *J. Biol. Chem.* **261:**10493–10495.
195. Lagast, H., Pozzan, T., Waldvogel, F. A., and Lew, P. D., 1984, Phorbol myristate acetate stimulates ATP-dependent calcium transport by the plasma membrane of neutrophils, *J. Clin. Invest.* **73:**878–883.
196. Colburn, J. C., Michnoff, C. H., Shu, L.-C., Slaughter, C. A., Kamm, K. E., and Stull, J. T., 1988, Sites phosphorylated in myosin light chain in contracting smooth muscle, *J. Biol. Chem.* **263:**19166–19173.
197. Nishikawa, M., Hidaka, H., and Adelstein, R. S., 1983, Phosphorylation of smooth muscle heavy meromyosin by calcium-activated, phospholipid-dependent protein kinase, *J. Biol. Chem.* **258:**14069–14072.
198. Nishikawa, Y., Sellers, J. R., Adelstein, R. S., and Hidaka, H., 1984, Protein kinase C modulates *in vitro* phosphorylation of the smooth muscle heavy meromyosin by myosin light chain kinase, *J. Biol. Chem.* **259:**8808–8814.
199. Ikebe, M., Inagaki, M., Kanamaru, K., and Hidaka, H., 1985, Phosphorylation of smooth muscle myosin light chain kinase by Ca^{2+}-activated phospholipid-dependent protein kinase, *J. Biol. Chem.* **260:**4547–4550.
200. Nishikawa, M., Shirakawa, S., and Adelstein, R. S., 1985, Phosphorylation of smooth muscle myosin light chain kinase by protein kinase C, *J. Biol. Chem.* **260:**8978–8983.
201. Inagaki, M., Yokokura, H., Itoh, T., Kanmura, Y., Kuriyama, H., and Hidaka, H., 1987, Purified

rabbit brain protein kinase C relaxes skinned vascular smooth muscle and phosphorylates myosin light chain, *Arch. Biochem. Biophys.* **254**:136–141.

202. Menkes, H., Baraban, J. M., and Snyder, S. H., 1986, Protein kinase C regulates smooth muscle tension in guinea-pig trachea and ileum, *Eur. J. Pharmacol.* **122**:19–27.

203. Forder, J., Scriabine, A., and Rasmussen, H., 1985, Plasma membrane calcium flux, protein kinase C activation and smooth muscle contraction, *J. Pharmacol. Exp. Ther.* **235**:267–273.

204. Litten, R. Z., Suba, E. A., and Roth, B. L., 1987, Effects of a phorbol ester on rat aortic contraction and calcium influx in the presence and absence of BAY k 8644, *Eur. J. Pharmacol.* **144**:185–191.

205. Chatterjee, M., and Tejada, M., 1986, Phorbol ester-induced contraction in chemically skinned vascular smooth muscle, *Am. J. Physiol.* **251**:C356–C361.

206. Park, S., and Rasmussen, H., 1985, Activation of tracheal smooth muscle contraction: Synergism between Ca^{2+} and activators of protein kinase C, *Proc. Natl. Acad. Sci. U.S.A.* **82**:8835–8839.

207. Gleason, M. M., and Flaim, S. F., 1986, Phorbol ester contracts rabbit thoracic aorta by increasing intracellular calcium and by activating calcium influx, *Biochem. Biophys. Res. Commun.* **138**:1362–1369.

208. Rasmussen, H., Forder, J., Kojima, I., and Scriabine, A., 1984, TPA-induced contraction of isolated rabbit vascular smooth muscle, *Biochem. Biophys. Res. Commun.* **122**:776–784.

209. Laher, I., and Bevan, J. A., 1987, Protein kinase C activation selectively augments a stretch-induced, calcium-dependent tone in vascular smooth muscle, *J. Pharmacol. Exp. Ther.* **242**:566–572.

210. Dale, M. M., and Obianime, W., 1985, Phorbol myristate acetate causes in guinea-pig lung parenchymal strip a maintained spasm which is relatively resistant to isoprenaline, *FEBS Lett.* **190**:6–10.

211. Danthuluri, N. R., and Deth, R. C., 1984, Phorbol ester-induced contraction of arterial smooth muscle and inhibition of α-adrenergic response, *Biochem. Biophys. Res. Commun.* **125**:1103–1109.

212. Nakaki, T., Roth, B. L., Chuang, D.-M., and Costa, E., 1985, Phasic and tonic components in 5-HT_2 receptor-mediated rat aorta contraction: Participation of Ca^{++} channels and phospholipase C, *J. Pharmacol. Exp. Ther.* **234**:442–446.

213. Umekawa, H., and Hidaka, H., 1985, Phosphorylation of caldesmon by protein kinase C, *Biochem. Biophys. Res. Commun.* **132**:56–62.

214. Mallows, R. S. E., and Bolton, T. B., 1987, Relationship between stimulated phosphatidic acid production and inositol lipid hydrolysis in intestinal longitudinal smooth muscle from guinea pig, *Biochem. J.* **244**:763–768.

215. Brass, L. F., and Laposata, M., 1987, Diacylglycerol causes Ca release from the platelet dense tubular system: Comparisons with Ca release caused by inositol 1,4,5-triphosphate, *Biochem. Biophys. Res. Commun.* **142**:7–14.

216. Campbell, M. D., Deth, R. C., Payne, R. A., and Honeyman, T. W., 1985, Phosphoinositide hydrolysis is correlated with agonist-induced calcium flux and contraction in the rabbit aorta, *Eur. J. Pharmacol.* **116**:129–136.

217. Brummer, H. C., 1972, Further studies on the interaction between prostaglandins and syntocinon on the isolated pregnant human myometrium, *J. Obstet. Gynaecol. Br. Commonw.* **79**:526–5.

218. Garrioch, D. B., 1978, The effect of indomethacin on spontaneous activity in the isolated human myometrium and on the response to oxytocin and prostaglandin, *Br. J. Obstet. Gynaecol.* **85**:47–52.

219. Chan, W. Y., 1983, Uterine and placental prostaglandins and their modulation of oxytocin sensitivity and contractility in the parturient uterus, *Biol. Reprod.* **29**:680–688.

220. Riemer, R. K., Goldfien, A. C., Goldfien, A., and Roberts, J. M., 1986, Rabbit uterine oxytocin receptors and *in vitro* contractile response: Abrupt changes at term and the role of eicosanoids, *Endocrinology* **119**:699–709.

221. Hensby, C. N., Seed, M. P., Williams, K. I., and Antipolis, S., 1986, Effects of oxytocic drugs on prostaglandin synthesis by human pregnant myometrium, *Br. J. Pharmacol.* [*Suppl.*] **86**:806P.

222. Ho, A. K., and Klein, D. C., 1987, Activation of α_1-adrenoreceptors, protein kinase C, or treatment with intracellular free Ca^{2+} elevating agents increases pineal phospholipase A_2 activity, *J. Biol. Chem.* **262**:11764–11770.

223. Parker, J., Daniel, L. W., and Waite, M., 1987, Evidence of protein kinase C involvement in phorbol diester-stimulated arachidonic acid release and prostaglandin synthesis, *J. Biol. Chem.* **262**:5385–5393.

224. Resink, T. J., Grigorian, G. Y., Moldabaeva, A. K., Danilov, S. M., and Buhler, F. R., 1987, Histamine-induced vein endothelial cells. Association with thromboxane and prostacyclin release, *Biochem. Biophys. Res. Commun.* **144**:438–446.

225. Hansson, A., Serhan, C. N., Haeggstrom, J., Ingelman-Sundberg, M., and Samuelsson, B., 1986, Activation of protein kinase C by lipoxin A and other eicosanoids. Intracellular action of oxygenation products of arachidonic acid, *Biochem. Biophys. Res. Commun.* **134**:1215–1222.

226. Gilman, A. G., 1987, G proteins: Transducers of receptor-generated signals, *Annu. Rev. Biochem.* **56**:615–649.

227. Spiegel, A. M., 1987, Signal transduction by guanine nucleotide binding proteins, *Mol. Cell. Endocrinol.* **49**:1–16.

228. Bockaert, J., Homburger, V., and Rouot, B., 1987, GTP binding proteins: A key role in cellular communication, *Biochimie* **69**:329–338.

229. Uhing, R. J., Jiang, H., Prpic, V., and Exton, J. H., 1985, Regulation of a liver plasma membrane phosphoinositide phosphodiesterase by guanine nucleotides and calcium, *FEBS Lett.* **188**:317–320.

230. Fulle, H.-J., Hoer, D., Lache, W., Rosenthal, W., Schultz, G., and Oberdisse, E., 1987, *In vitro* synthesis of ^{32}P-labelled phosphatidylinositol 4,5-bisphosphate and its hydrolysis by smooth muscle membrane-bound phospholipase C, *Biochem. Biophys. Res. Commun.* **145**:673–679.

231. Cockcroft, S., 1986, The dependence on Ca^{2+} of the guanine-nucleotide-activated polyphosphoinositide phosphodiesterase in neutrophil plasma membranes, *Biochem. J.* **240**:503–507.

232. Roth, B. L., 1987, Modulation of phosphatidylinositol-4,5-bisphosphate hydrolysis in rat aorta by guanine nucleotides, calcium and magnesium, *Life Sci.* **41**:629–634.

233. Marc, S., Leiber, D., and Harbon, S., 1988, Fluoroaluminates mimic muscarinic- and oxytocin-receptor-mediated generation of inositol phosphates and contraction in the guinea pig myometrium. Role for a pertussis/cholera toxin-insensitive G protein, *Biochem. J.* **255**:705–713.

234. Anwer, K., Hovington, J. A., and Sanborn, B. M., 1989, Antagonism of contractants and relaxants at the level of intracellular calcium and phosphoinositide turnover in the rat uterus, *Endocrinology* **124**:2995–3002.

235. Saida, K., and van Breemen, C. 1987, GTP requirement for inositol 1,4,5-trisphosphate-induced Ca^{2+} release from sarcoplasmic reticulum in smooth muscle, *Biochem. Biophys. Res. Commun.* **144**:1313–1316.

236. Henne, V., and Soling, H. D., 1986, Guanosine 5'-triphosphate releases calcium from rat liver and guinea pig parotid gland endoplasmic reticulum independently of inositol 1,4,5-trisphosphate, *FEBS Lett.* **202**:267–273.

237. Ueda, T., Chueh, S. H., Noel, M. W., and Gill, D. L., 1986, Influence of inositol 1,4,5-trisphosphate and guanine nucleotides on intracellular calcium release within the N1E-115 neuronal cell line, *J. Biol. Chem.* **261**:3184–3192.

238. Chueh, S.-H., and Gill, D. L., 1986, Inositol 1,4,5-trisphosphate and guanine nucleotides activate calcium release from endoplasmic reticulum via distinct mechanisms, *J. Biol. Chem.* **261**:13883–13886.

239. Wolf, B. A., Florholmen, J., Colca, J. R., and McDaniel, M. L., 1987, GTP mobilization of Ca^{2+} from the endoplasmic reticulum of islets, *Biochem. J.* **242**:137–141.

240. Chueh, S.-H., Mullaney, J. M., Ghosh, T. K., Zachary, A. L., and Gill, D. L., 1987, GTP- and

inositol 1,4, 5-trisphosphate-activated intracellular calcium movements in neuronal and smooth muscle cell lines, *J. Biol. Chem.* **262:**13857–13864.

241. Henne, V., Piiper, A., and Soling, H.-D., 1987, Inositol 1,4,5-trisphosphate and 5′-GTP induce calcium release from different intracellular pools, *FEBS Lett.* **218:**153–158.

242. Dawson, A. P., Hills, G., and Comerford, J. G., 1987, The mechanism of action of GTP on Ca^{2+} efflux from rat liver microsomal vesicles, *Biochem. J.* **244:**87–92.

6

Calcium Channels

Role in Myometrial Contractility and Pharmacological Applications of Calcium Entry Blockers

Charles A. Ducsay

1. Introduction

Calcium plays a major role in a wide range of cellular functions. In the extracellular fluid matrix, calcium is present in millimolar concentrations, whereas free intracellular calcium is up to 10,000 times less plentiful. This compartmentalization is accomplished by (1) limited permeability of the resting plasma membrane to calcium, (2) cellular mechanisms for calcium extrusion, (3) buffering within the cytosol, and (4) sequestration of calcium into intracellular organelles.[1]

Precise control of intracellular concentration in the resting state permits cells to detect and respond to both physiological and pharmacological stimuli. The movement of this ion across the surface membrane of excitable cells can regulate cellular functions ranging from excitation–secretion coupling to excitation–contraction coupling.

This chapter focuses on how cells vary the permeability of their plasma membrane to calcium. The change in permeability and its specificity are accomplished through calcium channels in the plasma membrane. Emphasis is placed on the role of calcium channels in the contractile process in uterine and vascular smooth muscle. In addition, the current trends and future possibilities of altering calcium movement across the uterine sarcolemma are discussed.

Charles A. Ducsay • Division of Perinatal Biology, Departments of Physiology and Pediatrics, School of Medicine, Loma Linda University, Loma Linda, California 92350.

2. Calcium Channels

During the contraction sequence, (1) calcium enters the interior of the smooth muscle cell through the cell membrane and/or (2) calcium is released from intracellular stores in the sarcoplasmic reticulum, (3) contraction occurs, and then (4) reuptake into intracellular stores and outward transport of calcium occur, leading to (5) relaxation. The initial penetration of the cell membrane by extracellular calcium is accomplished through changes in the permeability of calcium channels in the cell membrane.

Although the principal focus of this chapter is on the uterus, studies describing the characterization of calcium channels cover a variety of tissue types ranging from vascular smooth muscle to neural tissue. It appears, however, that both uterine vascular smooth muscle and myometrial tissue contain functional calcium channels that behave in a manner similar to that described for other tissues for which characterization is more complete.

The initial phase of movement of calcium through the membrane appears to be the reversible interaction between a calcium binding site and the calcium ion. The location of the calcium coordination or binding site appears to be on the outer surface of the cell membrane.[2−4]

2.1. General Properties

The presence of calcium channels has been demonstrated on the surface of all known excitable cells.[5,6] A feature common to all of the channels studied is a selectivity for calcium.[6] Despite the fact that the extracellular fluid contains significantly higher concentrations of Na^+ ions, calcium channels are extremely selective in permitting passage of calcium ions while excluding the physiological monovalent (Na^+ and K^+) and divalent (Mg^{2+}) cations. Cations that are not normally present in the extracellular fluid such as Ba^{2+} and Sr^{2+} are, however, passed preferentially over Ca^{2+}. Other cations ($Mg^{2+} << Ni^{2+}$, Mn^{2+}, $Co^{2+} << La^{3+}$, and Cd^{2+}) inhibit the passage of calcium across the channel by blocking the channel pore in the cell membrane.[2]

2.2. Selectivity

An ion channel can separate ions that differ in charge, size, coordination chemistry, and hydration by two distinct mechanisms.[6] Selectivity can be imparted to a channel by (1) selection for the particular molecule by specific binding followed by elution of the bound ligand or (2) selection by rejection of undesired ions by means of a molecular sieving mechanism.[6] In addition, some overlap of the two mechanisms may occur.

Experimental evidence strongly favors selectivity based on affinity, the presence of specific Ca^{2+} binding sites, rather than rejection of ions. With the use of large organic cations as probes, the pore size of the Ca^{2+} channel was estimated to be approximately 6 Å.[7-9] This is three times the diameter of the Ca^{2+} ion and approximately the same size as a hydrated Ca^{2+} ion. The minimal pore diameter is larger than that of K^+ or Na^+ channels.[9,10] Based on these data, the calcium channel would be unable to differentiate between the various ions found in the extracellular fluid. Selectivity solely on the basis of molecular sieving is therefore highly unlikely.[6]

Studies using a double sucrose-gap technique with isolated strips of rat myometrium revealed that the selective permeability of the calcium channel is dependent on the presence of extracellular calcium.[11] If normal action potentials are inhibited by the removal of extracellular calcium, long-lasting action potentials are recorded.[12] The calcium channels become permeable to Na^+ when extracellular calcium concentrations are less than 1 mM, which is below physiological levels.[11] Similar results have been reported in smooth muscle,[13] skeletal muscle,[14] and cardiac muscle[15] preparations.

A key element in the movement of calcium or other cations across the cell membrane is the interaction between the charged ion and the electrical field within the membrane. The force acting on the cation to move it into the cell is countered by reactive groups and steric configurations, i.e., energy barriers, that function to impede the flow of ions through the channel.[6] This energy barrier is structured in such a way as to permit certain types of cations to pass through the channel while excluding others. Based on its functional role within the channel, the barrier has been labeled the selectivity filter.[4,5] This is illustrated in Fig. 1.[5]

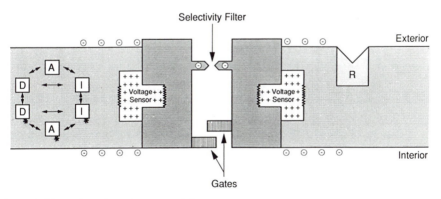

Figure 1. The calcium channel represented diagrammatically as a membrane pore. The selectivity filter is a negatively charged site within the pore that has the capacity to distinguish between different cations. The voltage sensors confer voltage dependence on the channel, while the channel gates are responsible for the open or shut configuration of the channel. The cyclic nature of the channel transitions between the activated (A), deactivated (D), and inactivated (I) states is illustrated. The rate constants governing these transitions may be modified by drug binding to any of these states (*). A receptor site (R) adjacent to the channel may alter channel gating. (Adapted from Triggle.[5])

2.3. Channel Type

Although the mechanisms that confer specificity to the calcium channel are still under scrutiny, the calcium ions that do pass across the "selectivity filter" do so only when the channel is in an open configuration. Calcium channels have a gating mechanism similar to that of the sodium channel.[16,17] This transforms the channel into one of three conformational states: (1) activated, which is an open conformation, (2) resting or deactivated, and (3) inactivated, both of which are closed conformations[16−18] (Fig. 1).

The amount of time a channel is open is dependent on the ability of the channel to remain in the activated state. Reuter[19] has shown that calcium channels open in bursts and that burst lengths increase with membrane depolarization. Within a given time span, the calcium channels in a membrane will fluctuate between open and closed conformations. The probability that a particular channel is in the activated or open conformation is the same as the fraction of the total population that will be in this state.[20]

Calcium channels are divided into two general categories depending on the type of excitatory stimulus necessary to transform the channel into the activated, open state.[7] Activation of the channel can be achieved in response to a change in membrane potential (voltage-dependent channel) or in response to the binding of a ligand to specific channel-associated receptors (ligand-gated or receptor-operated channel)[21] (Fig. 2). Both types of channels appear to be present in a range of excitable cells.[22−24]

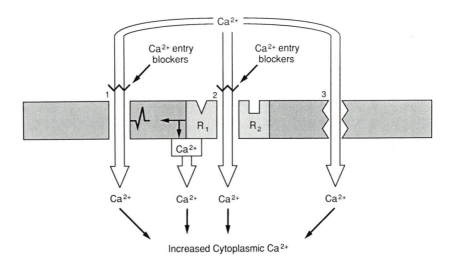

Figure 2. Regulation of intracellular calcium by calcium channels. Cytoplasmic calcium (Ca^{2+}) concentrations are increased by calcium influx through (1) voltage-dependent channels (VDC), (2) receptor-operated channels (ROC), or (3) passive leak channels (LC). Receptor (R_2) occupation by a specific agonist may directly affect channel opening. In addition, receptor (R_1)–agonist binding may also lead to release of Ca^{2+} from intracellular stores and/or depolarization of the membrane and activation of VDC. (Adapted from Rimele and Vanhoutte.[21])

Although receptor-operated channels play a role in the function of smooth muscle as well as in different types of secretory cells, their specific properties remain poorly defined because of a dearth of experimental data.[21] The majority of information characterizing receptor-operated channels comes from research using arterial smooth muscle. The early work of Bolton[22] and van Breemen and co-workers[24] postulated the existence of channels distinct from calcium channels regulated by changes in membrane potential.

In the rabbit aorta D-600 preferentially inhibited calcium influx stimulated by potassium-induced depolarization.[23] In contrast, norepinephrine-stimulated calcium influx was not blocked. In addition, additivity of calcium fluxes stimulated by high potassium and norepinephrine was observed. Although later work using rat aorta was unable to demonstrate additivity in calcium influx,[25] evidence for distinct receptor-operated channels is compelling.[25-27]

Data from Sakai and co-workers[28] suggested the presence of receptor-operated channels in rat myometrium. Oxytocin was found to stimulate myometrium in calcium-free media containing manganese. The existence of multiple types of receptor-operated channels was demonstrated in bovine coronary arteries.[26] Contractions produced by acetylcholine were inhibited by the calcium antagonist diltiazem while serotonin-induced contractions were unaffected. This suggested an interaction between acetylcholine-receptor-operated and voltage-operated calcium channels. (Alternative explanations are possible; see Chapter 5).

Interaction between the two pathways of calcium gating was demonstrated by quantitating calcium fluxes in rat aorta.[25] Calcium influx following activation by potassium was greater than that stimulated by norepinephrine. The effect of norepinephrine was blocked by prazosin, which had no effect on potassium-induced calcium influx. Further, both 3-isobutyl-1-methylxanthine and forskolin inhibited calcium influx stimulated by norepinephrine but were ineffective against potassium-stimulated calcium influx. The strongest evidence for the existence of receptor-operated channels comes from patch-clamp techniques (see Chapter 7), which demonstrated the presence of ATP-sensitive receptor-operated channels independent of voltage.[27] Taken together, the results suggest that receptor- and voltage-operated channels, though they may share common structural similarities, utilize different modes of activation.

Voltage-dependent channels are the most widely distributed and most studied of the calcium channels.[1,2,29,30] Voltage-clamp techniques demonstrated that prior to a detectable calcium flux, the membrane potential must first be reduced to a threshold level.[29-31] The inward calcium current continues to increase as the membrane potential becomes more positive. The overall increase in calcium current is a result of a decrease in electrical resistance (i.e., an increase in ionic conductance) that is indicative of activation of increased numbers of calcium channels.[18] The membrane potential becomes more positive until a peak magnitude is attained. At that point, the current will fall progressively.[29,31,32]

Despite the obvious importance of the initial activation and opening of the calcium channels, it is imperative that this event is ultimately followed by the channels entering an inactive, closed conformation to terminate the calcium influx.

Hurwitz[18] indicated that there are four types of voltage-dependent calcium channels based on the factors that induce the formation of the inactivated state. The various types of channels are characterized by (1) voltage-dependent inactivation, (2) calcium-dependent inactivation, (3) both voltage- and calcium-dependent inactivation, or (4) neither. Voltage-dependent inactivation is a time-dependent process following depolarization of the cell membrane. The rate of inactivation depends on the magnitude of depolarization within a restricted voltage range specific for each cell type.[33-35] In calcium-dependent inactivation, secondary to the increase in intracellular calcium ions, free calcium ions react reversibly with calcium channel binding sites. This leads to an enzymatic dephosphorylation of a channel component, resulting in inactivation of the channel.[18,36,37] In combined voltage- and calcium-dependent inactivation, the binding of increased amounts of intracellular calcium changes the membrane potential and leads to inactivation.[18,37,38] The last channel type appears to be relatively unaffected by calcium concentrations or voltage changes; these channels are slowly or only partially inactivated.[18,39]

Calcium channels have also been differentiated on the basis of their pharmacological sensitivity and on the membrane potential at which they reach the activated state.[1,7] Multiple calcium channel subtypes have been distinguished in a number of tissue types including smooth,[40,41] cardiac,[42,43] and skeletal[44] muscle as well as neurons[45,46] and endocrine tissue.[47]

It appears that any single smooth muscle may contain several different types of voltage-dependent calcium channels.[18] These different channel types include the most common, long-lasting (L)-type channels that are activated by large changes in membrane potential and are acutely sensitive to dihydropyridine calcium antagonists.[1,48] In contrast, transient (T) calcium channels are activated by weak depolarizations and are relatively insensitive to dihydropyridines.[49,50] The third category of channel is the N type, which appears to be similar to the L type with regard to ion permeation but is activated by strong depolarizations and relatively unaffected by dihydropyridines.[50]

2.4. Modulation of Calcium Channels

The entry of calcium into the cell can be modified by several processes. The primary regulator of calcium channels is the membrane potential. Thus, agents that alter membrane potential will influence the opening of channels. In general, it appears that the manner in which hormones and neurotransmitters exert their effects on the regulation of calcium channels is through their action on the membrane potential.[25,26,51-53] Agents may change extracellular or intracellular calcium concentrations, affect receptors that regulate calcium channels, or alter a biochemical reaction involved in the process of channel operation.[18] One agent may work through several of these mechanisms.

Modulation of calcium channels unrelated to membrane potential is a subject of increasing interest. Three of these mechanisms of calcium channel regulation have been investigated, though none of the mechanisms has been fully defined in smooth muscle.

The effects of phosphorylation by a protein kinase (cAMP- or cGMP-dependent or C) on channel opening has been examined.[54] Studies on cultured rat aortic cells suggested that cAMP may reduce the calcium current produced from voltage-dependent calcium channels.[48] Measurement of unidirectional calcium flux in rat aorta also demonstrated decreased calcium entry in response to increased cellular cGMP concentration.[55] In tracheal smooth muscle, cAMP or cGMP suppressed the inward calcium current associated with tetraethylammonium-induced action potentials.[57,58]

A second mechanism involves the G proteins, regulators that have been identified first with receptor-mediated synthesis of cAMP (activation of the cyclase) but have subsequently been implicated in numerous other receptor processes. Since the G proteins might stimulate cyclic nucleotide production, this would indirectly influence the opening of channels by the previously stated mechanism. Direct gating of calcium channels by G protein in cardiac muscle has been demonstrated in membrane patches from cardiac cells.[54,59] Care must be exercised in the extrapolation of data generated in cardiac muscle to smooth muscle, since cAMP has opposite effects on these two muscle types.

A third mechanism involves inositol tetrakisphosphate: This originates from the intracellular second messenger inositol trisphosphate and has been hypothesized to control calcium channels[60] (see Chapter 5).

3. Calcium Entry Blockers

An effective means of controlling contractile activity of smooth muscle is to block the transmembrane influx of calcium. The obvious target for such intervention is the calcium channel (Fig. 2).

A diverse group of compounds have the capacity to alter calcium movement across the cell membrane. These can be divided into two groups: inorganic inhibitors such as Co^{2+}, Mn^{2+}, and La^{3+} and organic compounds.[28]

The first pharmacological agents recognized to alter excitation–contraction coupling were drugs of the diphenylpiperazine group such as cinnarizine. These agents were manufactured in the 1950s before their mode of action was recognized.[61] Cinnarizine, for example, was first viewed as an antihistaminic.[62] Later reports demonstrated that agents such as verapamil and prenylamine exhibited the same inhibitory effect on cardiac muscle as did withdrawal of calcium from the extracellular fluid space.[63] The antivasoconstrictive activity of cinnarizine was also attributed to its inhibition of calcium influx.[64] Since that time, numerous compounds have been recognized to exert similar effects despite differences in structure.[61,65,66] These agents were originally referred to as "calcium antagonists," since their effects on calcium dose–response curves in depolarized arteries were similar to the effects of classical antagonists on the dose–response curves of agonists.[61,64,67] More simply stated, the effects of this group of compounds are reversed by increasing the extracellular calcium concentration in the medium.[50] The antagonist effects are not competitive, since at higher concentrations, as can be seen in Fig. 3, there is a reduction in the maximum contraction achieved. This does not

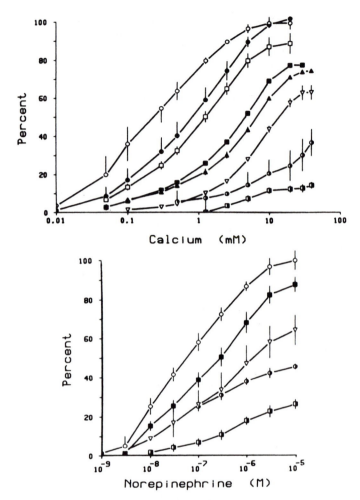

Figure 3. The effects of nifedipine on contraction of rat blood vessel induced by (top) depolarization with 100 mM KCl at varying calcium concentrations or (bottom) norepinephrine. Each graph shows concentration effect curves in the absence (○), and at various concentrations of nifedipine, 3×10^{-10}M (●); 10^{-9}M (□); 2×10^{-9}M (■); 4×10^{-9}M (▲); 10^{-8}M (▽); 10^{-7}M (◑); 10^{-6}M (◨). (From Godfraind [160] with permission.)

differ from calcium antagonist effects on norepinephrine contractions. These pharmacological agents are now more commonly referred to as calcium entry blockers.

3.1. Localization of Ligand Binding Sites

The advent of radiolabeled compounds led to the demonstration of high-affinity calcium entry blocker binding sites that appeared to be part of the calcium channel in membrane preparations from smooth muscle.[68-71] A number of these utilized preparations in which there was incomplete separation of the various membrane types (i.e., plasmalemma, endoplasmic/sarcoplasmic reticulum).[70] Binding in rat myometrial muscle has been demonstrated to be limited primarily to the plasma membrane.[69]

In general, the density of dihydropyridine binding sites in nonvascular smooth muscle is approximately 0.8 pmol/mg protein, whereas that for vascular smooth muscle is 0.2 pmol/mg protein.[72] Recently, the presence of reversible high-affinity dihydropyridine binding sites in the plasma membrane of nonpregnant human myometrium was confirmed.[71] They had similar binding properties to those identified in rabbit[70] and rat myometrium.[69] In human myometrial preparations, ^{3}H-labeled nitrendipine binding sites were greatest in the fraction with the highest concentration of plasmalemma markers and not with endoplasmic reticulum or mitochondrial markers.[71] It becomes apparent that in the myometrium, the plasmalemma is the organelle with the highest concentration of dihydropyridine binding sites and, therefore, the highest density of calcium channels. The binding sites are localized on the inside of the membrane.[18]

3.2. Characterization of Ligand Binding Sites

Affinities for 1,4-dihydropyridines in depolarized preparations are similar in various smooth muscle types (K_d approximately 10^{-10} M) and correlate with the concentration necessary to produce functional blockade.[72] The 1,4-dihydropyridine receptors are susceptible to phospholipases, heat, and proteases.[73-75] Further purification on lectin columns demonstrated that the receptors are glycoproteins.[72]

Studies by Hess *et al.*[76] revealed that organic agents appear to modify the duration and frequency of channel opening rather than actually plugging the channel. Thus, it is envisioned that the channel is stabilized by antagonists that increase the likelihood of the channel gates being in the inactivated or closed position. Likewise, the excitatory dihydropyridine Bay K 8644 exerts its effect by increasing the time in which the channel remains in the open state.

The degree of binding of one type of organic blocking agent may be influenced by the introduction of other similar agents. For example, by interacting with a site that is allosterically linked to the dihydropyridine binding site, diltiazem can increase while verapamil can decrease the binding of dihydropyridines to specific membrane sites.[77]

3.3. Alternative Sites of Action

Although the main body of evidence suggests that the principal site of action of calcium entry blockers is the plasma membrane, a direct intracellular effect cannot be discounted. Several reports have indicated that calcium entry blockers do indeed cross the cell membrane and enter the cell and may exert a direct effect on the contractile mechanism.[78,79] Metzger and co-workers[79] demonstrated the ability of nicardipine to inhibit the contractile activity in preparations of muscle fibers that were stripped of their membranes. Additional work has shown that these pharmacological agents appear to bind to calmodulin[80] and may further exert their effects by inhibiting the activation of key enzymes (i.e., phosphodiesterase) by the calcium–calmodulin complex.[81]

Sakamoto and Huszar[81] demonstrated the direct inhibitory effect of nitrendipine on actomyosin ATPase activity in human myometrium and placenta at high concentrations. This inhibition is consistent with the inhibition of myosin light chain phosphorylation.[82] Since these studies were conducted in the absence of cell membranes and hence calcium channels, two possible steps in the process of kinase activation were implicated where nitrendipine might exert its inhibitory effects. Nitrendipine may either block the formation of the calcium–calmodulin complex by blocking the calcium sites on the calmodulin molecule, or it may inhibit the activation of the myosin light chain kinase by the calcium–calmodulin complex.[81]

Nitrendipine was shown to inhibit binding of the calcium–calmodulin complex to the myosin light chain complex, and this inhibition was not overcome by increasing calcium concentration in the medium.[84] It appears that calcium entry blockers do not affect the active site of the myosin light chain kinase. Their effects occur at the calmodulin regulatory site, since removal of the site by limited proteolytic digestion leads to a diminution of inhibition of myosin light chain kinase by nitrendipine.[84]

An obvious question that arises is whether the intracellular effects observed will have any significance in intact muscle. On the basis of the concentrations of dihydropyridine derivatives utilized in numerous studies,[85–87] it becomes apparent that myometrial contractility is eliminated at much lower concentrations than are required to exert direct effects on myosin light chain phosphorylation. Inhibition of contractility occurs at concentrations of 10^{-7} to 10^{-9} M,[87] whereas levels of 10^{-4} M were required to elicit 15–20% inhibition of actomyosin ATPase.[81]

3.4. Pharmacology

A vast array of agents have the capacity to block calcium influx. A simple grouping of these agents is offered by Spedding[65] (Table I). The compounds found in groups 1 and 2 demonstrate highly specific and potent effects on calcium channels, whereas the compounds in group 3 are less specific and affect the activity of both calcium and sodium channels.[18,88] In general, compounds listed in groups 1 and 3 appear to have preferential effects on vascular smooth muscle, whereas those listed in group 2 exert similar effectiveness in both cardiac and vascular smooth muscle.[66,88]

3.5. Pharmacokinetics

The pharmacokinetics of the three most clinically utilized calcium antagonists are discussed below.

Table I. Common Categories of Calcium Entry Blockers

Group 1	Dihydropyridines: nifedipine, nitrendipine, nimodipine, nisoldipine, nicardipine
Group 2a	Verapamil derivatives: verapamil, gallopamil (D-600), tiapamil
Group 2b	Benzothiazepines: diltiazem, KB944
Group 3	Diphenylalkylamines: flunarizine, cinnarizine, lidoflazine

3.5.1. Absorption/Metabolism

Verapamil, diltiazem, and nifedipine are extensively metabolized as evidenced by recovery of very low percentages of the parent compounds in the urine.[89]

3.5.1a. Verapamil. Absorption of verapamil following oral administration is moderate, but only a small percentage is available because of first-pass metabolism by the liver.[90,91] Reported bioavailability is approximately 22%.[91] The initiation of clinical effects occurs 1 to 2 hr following oral administration, whereas intravenous administration results in clinical effects within 10 min. The mean elimination half-life was determined to be approximately 4.8 hr.[92]

3.5.1b. Diltiazem. Following oral administration, peak plasma concentrations are observed within 3 to 4 hr, and plasma half-life was reported to be between 3 and 6 hr.[89,93] As for verapamil, the principal route of metabolism is through the liver. Bioavailability has been shown to be approximately 50%.

3.5.1c. Nifedipine. The most complete studies of nifedipine absorption and metabolism are from Raemsch and Sommer.[94] Nifedipine is almost completely absorbed from the gastrointestinal tract following oral or sublingual administration. Because of presystemic metabolism, bioavailability ranges from 56 to 77%.[94,95] Nifedipine has a much shorter apparent half-life of elimination, estimated at 1.25 hr.[94]

3.5.2. Protein Binding

It is apparent that only the unbound portion of a drug is available for distribution and binding to active sites or susceptible to metabolism and elimination.[89] Thus, the extent of plasma protein binding of a calcium entry blocker is important.

Of the three agents described above, nifedipine appears to exhibit the highest degree of protein binding, estimated to be as high as 92% depending on initial concentration.[89,96] Verapamil is approximately 90% protein bound, but the degree of binding is not concentration dependent.[97,98] Based on equilibrium dialysis, diltiazem has been demonstrated to be about 80% bound to plasma proteins.[99]

4. Effects of Calcium Entry Blockers on the Female Genital Tract

The three calcium entry blockers described above exert different degrees of effectiveness depending on the tissue type. For example, all three compounds were found to decrease peripheral resistance and to lower blood pressure in anesthetized cats. In addition, all three increased cerebral blood flow, but only the dihydropyridine also increased skeletal muscle blood flow and cardiac output.[100,101] The variable effects of calcium entry blockers again raise the question of whether these agents act solely through blockade of the membrane calcium channels or whether other mechanisms are involved.

Calcium entry blockers appear to have a dramatic effect on both vascular and myometrial tissues in the female genital tract. This raises the possibility of their use as tocolytics.[102–104] However, the safety and efficacy of these pharmacological agents remain to be elucidated. This section examines the effects of calcium entry blockers on the female genital tract with specific reference to the use of these agents during pregnancy.

Each year, almost 6% of all neonates in the United States are born 3 or more weeks before term, and neonatal morbidity and mortality are a frequent consequence of this disorder.[105] A number of different compounds have been utilized in the treatment of premature labor, but all have met with only moderate success. The full potential of calcium entry blockers as tocolytic agents has not been fully explored. The effects of these agents both *in vitro* and *in vivo* are described in the following section.

4.1. In Vitro Studies

It appears that contractile activity of smooth muscle in the female reproductive tract is inhibited by calcium entry blockers.

4.1.1. Fallopian Tube

In isolated strips of outer longitudinal and inner circular smooth muscle layers, it was demonstrated that although these tissues are in close anatomic proximity, they exhibit distinct functional characteristics.[106–108] Stimulatory substances such as prostaglandin (PG) $F_{2\alpha}$ and norepinephrine produce consistent phasic contractions in the circular smooth muscle layer, whereas the same agents lead to primarily tonic-type activity in the longitudinal layer. Contractile activity was inhibited in both muscle layers on removal of extracellular calcium.[108] In contrast, nifedipine was only effective in the circular layer. Nifedipine eliminated only the phasic contractile activity, while the tonic contractions induced in the longitudinal layer remained relatively unaffected.[108,109] These findings suggest that there may be more than one membrane mechanism or channel responsible for the two different types of contractile activity. The fact that the circular layer is affected by calcium entry blockers leads to the speculation that the activity in this muscle layer is regulated by potential-sensitive or voltage-dependent channels that are directly affected by calcium entry blockers. The longitudinal layer would be expected to be regulated primarily by receptor-operated channels, which are less susceptible to the influence of calcium entry blockers.[108]

Arteries in the human fallopian tube respond to α-adrenergic stimulation by norepinephrine. However, contractile activity is only marginally reduced following calcium entry blocker treatment.[109]

4.1.2. Placenta

Umbilical and placental vessels appear to lack autonomic innervation and are therefore believed to be regulated by humoral and locally produced substances.[109–112] The dihydropyridines nifedipine and nitrendipine have been shown to produce concentration-dependent inhibition of agonist-induced contractile activity.[113–115] Based on *in vitro* work, Forman and co-workers[109] suggested that treatment with calcium channel blockers would lead to increased placental perfusion. However, this has not been verified *in vivo,* as discussed in Section 4.2.2.

There appears to be controversy in regard to whether calcium antagonists act on receptor- or voltage-dependent channels. Broughton Pipkin and Chinnery[115] attributed the relaxant effects produced by nitrendipine in PGE_2-stimulated placental vessels to the effect of the calcium entry blocker on receptor-operated channels. This is in contrast to other smooth muscle studies, which demonstrated that calcium entry blockers exert their primary effect on voltage-dependent channels. An explanation for the apparent contradiction is that prostaglandins may alter the cell membrane potential.[116] Thus, the effects may actually be on the voltage-dependent calcium channels. At present, the precise mode of action of calcium entry blockers at the level of the placental vasculature is uncertain.

4.1.3. Myometrium

The myometrium is the most studied of the tissues of the female genital tract in relation to the effects of calcium entry blockers. The rationale is understandable since calcium entry blockers have the potential to inhibit contractions of myometrial smooth muscle. The full potential of this inhibition has only been explored recently. Although a variety of calcium entry blockers are currently available, agents of the dihydropyridine type have received the most attention. In contrast to the information currently available regarding the effects of these agents on the cardiovascular system, there is a paucity of information on the myometrium.

The early landmark work of Csapo[117] examined the role of calcium in the contractile activity of uterine smooth muscle. This revealed that only a small portion of extracellular calcium was responsible for contractile activity of the uterus. If calcium was eliminated from the medium, there was an effective uncoupling of excitation and contraction. Subsequently, it was confirmed that calcium serves as a "coupler" in the contractile process, since the uterus became refractory to external stimuli in a calcium-free environment.[118,119] It was proposed that calcium entry blockers, with their ability to inhibit calcium-dependent muscle function, would effectively block transmembrane calcium influx in the myometrium and inhibit contractile activity.[103,120] In isolated rabbit uteri nicardipine was effective in inhibiting spontaneous contractions at concentrations of 1 μg/ml (approximately 2 μM) in the medium bathing the tissue. Concentrations of 10 μg/ml were capable of abolishing contractile activity induced by oxytocin as well as by electrical stimulation.[103]

The majority of the studies on uterine smooth muscle have focused on the effects of the dihydropyridine subclass of calcium entry blockers. The rationale for this is that the dihydropyridines appear to be the most potent of the major subclasses of calcium entry blockers in their effect on the myometrium. In both isolated rat[121,122] and human[122] myometrium the order of potency is dihydropyridines > verapamil derivatives > benzothiazepines > diphenylpiperazines. It is interesting to note that compounds from all four groups were less potent on human than on rat myometrial tissue. But concentrations of 10^{-7} M nifedipine totally blocked spontaneous and K^+-induced contractions in the human.[123] The reasons for such differences are unclear. A common site of action for inhibition of K^+-induced contractile activity in both rat and human myometrium was proposed to be the voltage-operated or potential-sensitive calcium channel.[121,123]

4.2. In Vivo Studies

Based on the general mechanisms that regulate smooth muscle contractions and the *in vitro* data described above, it seems reasonable to postulate that calcium entry blockers would be of value in the treatment of undesired uterine contractions. The following section describes the effects of *in vivo* administration of calcium entry blockers on both the pregnant and nonpregnant female genital tract.

4.2.1. Nonpregnant

Contractile activity of the human uterus varies during the menstrual cycle but is normally maximal at the onset of menstruation.[124–126] In instances where this contractile activity is exaggerated, primary dysmenorrhea results, with associated discomfort that ranges in degree of severity.[127,128] Uterine ischemia has also been observed in cases of dysmenorrhea.[126]

Specific factors are believed to be responsible for the development of dysmenorrhea. Increased prostaglandin production appears to be the principal contributing factor, since the disorder involves increased endometrial $PGF_{2\alpha}$ output.[129] Whether this is the underlying factor for the observed dysmenorrheic pain is uncertain.[109] Other substances have been implicated in the etiology of this disorder, including vasopressin and neuronally released norepinephrine, since both compounds produce contraction of isolated myometrial strips as well as inducing contractions of intramyometrial arteries.[129,130]

There has been a tremendous amount of attention focused on treatment that would enhance relaxation of the myometrium with subsequent relief of discomfort associated with this disorder. The efficacy of prostaglandin synthetase inhibitors,[109] ethanol,[131] terbutaline,[126] and synthetic analogues of vasopressin have all been examined.[132] Based on the documented relaxant effects of calcium entry blockers on the uterine vasculature and myometrium, recent interest has focused on these pharmacological agents.

Early studies by Ulmsten *et al.*[133] revealed the effectiveness of nifedipine. An oral dose of 20–30 mg significantly reduced the amplitude of contractions as well as uterine tone during the first and second days of menstruation. The only observable side effects were facial flushing with minor alterations in heart rate and blood pressure in these supine subjects.[133] Andersson and Ulmsten[134] demonstrated a rapid reduction of uterine contractions accompanied by relief of pain. The clinical effectiveness of nifedipine in the treatment of dysmenorrhea has also been assessed.[135,136] Although relaxation of the uterus resulted, some patients complained that treatment either induced or exacerbated headaches. Although prostaglandin synthetase inhibitors and hormonal replacement therapy are the most widely accepted treatment regimens,[109] calcium entry blockers present an alternative that deserves further investigation.

4.2.2. Postpartum

Another model that has been used effectively for testing the effect of calcium entry blockers on myometrial contractions is the postpartum uterus. The postpartum uterus provides a unique model for testing pharmacological agents to be used for the inhibition of uterine activity before and after delivery.[137] This depends on accurate measurements of intrauterine pressure, which can be obtained using the micro-pressure transducer technique.[138] Immediately following normal vaginal delivery, intrauterine pressure was recorded. Patients were treated with intravenous oxytocin infusions for 2 hr. Half of the women were also given 30 mg of nifedipine orally 25 to 40 min following the start of the oxytocin infusion. As expected, oxytocin treatment significantly increased intrauterine pressure. Women receiving nifedipine demonstrated a significant reduction in both contraction frequency and amplitude. The onset of uterine relaxation was observed within 15 min, with maximal relaxation occurring approximately 30 min following nifedipine treatment.[139] In the same study, nifedipine also eliminated $PGF_{2\alpha}$-induced contractility. A slight increase in heart rate was noted, but mean arterial pressure was unaffected.[139]

The effectiveness of another dihydropyridine, nicardipine, in the inhibition of uterine contractions has been demonstrated in the postpartum rat.[103] Like nifedipine, nicardipine inhibited contractile activity induced by both oxytocin and $PGF_{2\alpha}$. These effects are directly comparable to those observed with nifedipine treatment in the postpartum human uterus.[139]

Despite the fact that nifedipine reduced uterine contractions in postpartum women with little or no adverse side effects, possible untoward effects on the fetus must remain a concern. The effects of calcium entry blockers on the fetus is a topic specifically examined later in this chapter (Section 4.2.3). The authors did, however, conclude that calcium entry blockers may provide an effective alternative for use in the treatment of uterine hypertony, which sometimes occurs following $PGF_{2\alpha}$-induced labor.[139]

As indicated in Table I, there are a relatively large number of calcium entry blockers of various types. Because of their different molecular structures, the specificity and relative potencies of the pharmacological agents vary according to tissue type. A number of *in vitro* investigations cited in Section 4.1 demonstrated that calcium entry blockers of the dihydropyridine type appear to be the most potent agents affecting vascular and uterine smooth muscle. In an effort to determine the ideal calcium entry blocker, the agent with the greatest specificity for the target tissue is the one of choice. For example, if the goal of treatment is the prevention of or inhibition of myometrial contractions, the ideal calcium entry blocker is one that has its specificity directed at the uterus with little or no effects on the heart or the vasculature.

The relative potencies and selectivities of four calcium entry blockers on the inhibition of myometrial contractions were determined in ovariectomized postpartum rats.[140] Since regular and sustained intrauterine pressure change occurs following removal of the ovaries, this has become a useful model in testing calcium entry blockers as well as other potential tocolytic agents. The selectivities of these agents on the uterus relative to their effects on heart rate and blood pressure were compared with the β-agonist salbutamol. In agreement with *in vitro* studies using both rat[121] and human[123] myometrium, nifedipine was the most effective of the agents examined.

Nifedipine was also demonstrated to have the most potent vasodepressor activity. Tachycardia was observed with all of the calcium entry blockers but was less than that observed with salbutamol. At doses sufficient to induce a reduction of uterine contractile activity comparable to that of salbutamol, both gallopamil and verapamil led to episodes of temporary cessation of heartbeat. In summary, these agents were found to exhibit no selectivity for inhibition of uterine contractions, whereas nifedipine demonstrated moderate selectivity.[140] Diltiazem demonstrated selectivity, but its use is excluded because of its ability to inhibit conduction in the AV node.[141,142]

At present, the most obvious choice for a calcium entry blocker that is effective at the level of the uterus is one of the dihydropyridines (nicardipine, nifedipine, etc.). As previously described, the most important model to test the efficacy and safety of calcium entry blockers is the pregnant subject.

4.2.3. Pregnant

A novel approach to study the effects of calcium entry blockers on the inhibition of labor was utilized by Csapo *et al.*[103] in his final contribution to the study of the uterus. Nicardipine and nifedipine were administered to rats following delivery of the first pup. These calcium entry blockers arrested labor and delayed parturition in a dose–response manner. Despite the delay in delivery, all of the pups were delivered alive. Similar results were obtained in investigations that examined the time interval between delivery of the first and second pup following administration of nifedipine, verapamil, or diltiazem.[143] This study confirmed earlier work that the dihydropyridines are the most effective of the calcium entry blockers when used as tocolytics.[103,140]

The above studies were a preliminary effort in establishing the efficacy of calcium entry blockers in the inhibition of labor. However, they failed to examine the cardiovascular impact of calcium entry blockers administered during the latter part of gestation. Following $PGF_{2\alpha}$-induced premature uterine activity in a chronically instrumented pregnant rabbit preparation, three infusion rates of nicardipine were studied.[144] Along with a dose-related reduction in uterine activity, a significant reduction of both systolic and diastolic blood pressure with a concomitant elevation of pulse rate was observed. The effects of calcium entry blockers appear to mimic those of β-agonists in both human and animal studies.[140,145–147] Whether the same types of serious cardiovascular complications such as tachycardia, arrhythmias, and pulmonary edema will also be as prevalent remains to be elucidated. Because of obvious considerations of size, no physiological measurements were taken in the fetus. The authors did indicate, however, that gross observations of fetal viability at the end of the study did not reveal any instances of fetal demise.[144]

It is apparent that research on the effects of calcium entry blockers on the fetus must be carried out in larger animal species where access to the fetus is more readily achieved. Golichowski and co-workers[148] examined both the tocolytic and hemodynamic effects of nifedipine administration to pregnant ewes. Antepartum administration of nifedipine to ewes led to a significant elevation of fetal heart rate; values for fetal arterial pressure and blood gases were not reported. During labor, nifedipine administration had no further effect on fetal heart rate since values were already elevated.[148] Although it was subsequently shown that dihydropyridines do cross the placenta,[149,150] the observed changes in fetal heart rate may not be the result of action of calcium entry blockers on the fetus. Nifedipine administration produced a 22% decline in maternal peripheral resistance and a significant albeit slight reduction in maternal arterial pressure. Combined, these changes themselves could have profound effects on uteroplacental perfusion and indirectly alter fetal cardiorespiratory parameters.[148]

More recently it was demonstrated that maternal administration of nifedipine to the pregnant ewe resulted in alterations in uterine blood flow and in distribution of blood flow in the fetus.[149] Infusion of nifedipine at a rate of 10 μg/kg per min led to a 21% reduction in uterine blood flow, a significant reduction in maternal arterial pressure, and an increase in heart rate. The decline in uterine blood flow was attributed solely to the maternal hypotensive effect of the calcium entry blocker, since uterine vascular resistance was unchanged. A trend toward fetal hypoxia and acidosis developed, and a significant decline in fetal O_2 saturation was observed. This study also confirmed transplacental passage of nifedipine by measuring levels in the fetus. Fetal-to-maternal plasma concentration ratios reached 0.40 following maternal infusion at a rate of 10 μg/kg per min, resulting in mean fetal plasma concentrations of 110 ng/ml.[149]

Bolus injections of nifedipine in the pregnant pygmy goat produced only a transient decline in maternal arterial pressure immediately following the bolus of calcium entry blocker.[151] In contrast, to the sheep,[148,149] no significant alteration in uterine blood flow was observed. Decreases in uterine blood flow following maternal nifedipine administration occur only when there is a significant reduction in maternal arterial pressure. Nicardipine administered to hypertensive pregnant ewes in doses sufficient to render them normotensive resulted in fetal hypoxemia and acidosis and, in severe cases, fetal demise.[152] It was also determined that angiotensin-II-induced vasoconstriction was reversed by nicardipine in both the systemic and uterine vasculature but not in the placenta.

Ducsay et al.[150,153] were the first to examine the effects of maternal calcium entry blocker administration on the primate fetus. Dihydropyridine calcium channel blockers were effective in significantly reducing spontaneous uterine contractions in pregnant rhesus monkeys. Maternal arterial pressure declined approximately 10% while fetal blood pressure remained unchanged. However, both nifedipine[153] and nicardipine[150] were found to lead to deleterious alterations in fetal arterial pH, Po_2, and Pco_2. These changes are similar to those observed in other animal investigations. The question that remains is what are the factors responsible for these changes?

Studies in the ewe[149] indicated that although substantial quantities of maternally administered nifedipine entered the fetal circulation, the reduction in uterine blood flow was attributable to maternal hypotension. Ducsay et al.[150] showed that in the rhesus monkey, fetal levels of nicardipine never exceeded 6% of maternal plasma concentrations. Transplacental transport of nicardipine in the primate is therefore minimal. Since fetal heart rate and mean arterial pressure remained relatively unchanged, a direct effect on the fetus appears unlikely. It would be reasonable to assume that alterations in fetal oxygenation in response to maternal calcium entry blocker administration are the result of changes in uteroplacental perfusion secondary to changes in maternal arterial pressure as previously described in the ewe[149,152] and the rabbit.[144] Recently, in the rhesus monkey uteroplacental blood flow was measured following maternal nicardipine treatment; this study revealed a significant reduction in placental perfusion.[154] However, the approximate 50% reduction in placental blood flow was not solely attributable to maternal hypotension. Placental vascular resistance was also significantly elevated following calcium entry blocker treatment. The reason for the alteration in placental vascular dynamics was unknown.

Limited human studies utilizing nifedipine to treat premature labor demonstrated a significant tocolytic effect with no apparent deleterious effects on neonatal outcome.[155-158] An additional study using Doppler assessment of fetal and uteroplacental circulation following maternal nifedipine treatment found no alterations in blood flow.[159] It is imperative to note, however, that delivery of a viable infant with an acceptable Apgar score is only a gross indicator of fetal well-being. One problem with comparing clinical and animal data, aside from the obvious species differences, is the dosage and route of calcium entry blocker administration.

A comparison of dosages used in animal and clinical applications reveals that the apparent amounts are comparable. For example, dihydropyridine doses in the rhesus monkey averaged approximately 0.45 mg/kg.[150,153] A common range of tocolytic doses of nifedipine in women is 20–30 mg/dose or approximately 0.4 to 0.6 mg/kg.[154,157,158] Despite the fact that calcium entry blockers are most commonly administered orally, maternal uptake is quite rapid. Detectable plasma concentrations of nifedipine were observed within 6 min following administration with maximum concentrations reached within 1 hr.[94]

5. Summary

Calcium entry blockers have increased our understanding of calcium modulation at the cellular level.

Cell membrane voltage- and receptor-operated calcium channels permit calcium entry into the cell. Thus, they regulate the intracellular calcium concentration, which is up to 10,000 times lower than that in the extracellular fluid.

Calcium entry blockers appear to have specific binding sites in or near the channel and affect calcium movement across the sarcolemmal membrane. Progress has centered around the elucidation of channel types.

Human and animal studies have demonstrated that calcium entry blockers are powerful tocolytic substances. The dihydropyridines appear to be the most specific in their effects on the uterus. However, deleterious effects, specifically a reduction in fetal oxygenation, have generated concern regarding the safety of these agents as tocolytics. Further detailed studies of the hemodynamic and metabolic side effects in the fetus are necessary before the potential of calcium entry blockers can be fully realized. The future use of calcium entry blockers in this area will be dependent on the development of agents with greater specificity for the myometrium and the elimination of maternal and/or fetal side effects. Future areas of research will include development of new calcium channel ligands, identification of calcium channel defects in human disease states, and development of new therapeutic uses.

ACKNOWLEDGMENTS. The author would like to express his gratitude to Drs. Steve Yellon and Ray Gilbert for their expert editorial assistance. Supported in part by NIH Grant HD 22865.

References

1. Greenburg, D. A., 1987, Calcium channels and calcium channel antagonists, *Ann. Neurol.* **21:**317–330.
2. Hagiwara, S., and Byerly, L., 1981, Membrane biophysics of calcium currents, *Fed. Proc.* **40:**2220–2225.

3. Hille, B., 1984, Elementary properties of ions in solution, in: *Ionic Channels of Excitable Membranes*, Sinauer Associates, Sunderland, MA, pp. 249–271.

4. Hille, B., 1984, Selective permeability: Saturation and binding, in: *Ionic Channels of Excitable Membranes*, Sinauer Associates, Sunderland, MA, pp. 272–291.

5. Triggle, D. J., 1982, Biochemical pharmacology of calcium blockers, in: *Calcium Blockers, Mechanisms of Action and Clinical Applications* (S. F. Flaim and R. Zelis, eds.), Urban & Schwarzenberg, Baltimore, Munich, pp. 121–134.

6. Tsien, R. W., Hess, P., McClesky, E. W., and Rosenberg, R. L., 1987, Calcium channels: Mechanisms of selectivity, permeation, and block, *Annu. Rev. Biophys. Biophys. Chem.* **16:**265–290.

7. McClesky, E. W., and Almers, W., 1985, The Ca channel in skeletal muscle is a large pore, *Proc. Natl. Acad. Sci. U.S.A.* **82:**7149–7153.

8. McClesky, E. W., Hess, P., and Tsien, R. W., 1985, Interaction of organic cations with the cardiac Ca channel, *J. Gen. Physiol.* **86:**22a.

9. Hille, B., 1973, Potassium channels in mylenated neurons. Selective permeability to small cations, *J. Gen. Physiol.* **61:**669–686.

10. Coronado, R., and Miller, C., 1982, Conduction and block by organic cations in a K^+-selective channel from sarcoplasmic reticulum incorporated into planar phospholipid bilayers, *J. Gen. Physiol.* **79:**529–547.

11. Jmari, K., Mironneau, C., and Mironneau, J., 1987, Selectivity of calcium channels in rat uterine smooth muscle: Interactions between sodium, calcium and barium ions, *J. Physiol. (Lond.)* **384:**247–261.

12. Mironneau, J., Eugene, D., and Mironneau, C., 1982, Sodium action potentials induced by calcium chelation in rat uterine smooth muscle, *Pflugers Arch.* **352:**197–210.

13. Prosser, C. L., Kreulen, D. L., Weigel, R. J., and Yau, W., 1977, Prolonged action potentials in gastrointestinal muscles induced by calcium chelation, *Am. J. Physiol.* **233:**C19–C24.

14. Potreau, D., and Raymond, G., 1982, Existence of a sodium-induced calcium release mechanism on frog skeletal muscle fibers, *J. Physiol. (Lond.)* **333:**463–480.

15. Miller, D. J., 1979, Are cardiac cells "skinned" by EGTA or EDTA? *Nature* **277:**142–143.

16. Kaas, R. S., and Scheuer, T., 1982, Calcium ions and cardiac electrophysiology, in: *Calcium Blockers: Mechanisms of Action and Clinical Applications* (S. F. Flaim and R. Zelis, eds.), Urban & Schwarzenberg, Baltimore, pp. 3–19.

17. Beeler, G. W., and Reuter, H., 1970, Membrane calcium current in ventricular myocardial fibers, *J. Physiol. (Lond.)* **207:**191–209.

18. Hurwitz, L., 1986, Pharmacology of calcium channels and smooth muscle, *Annu. Rev. Pharmacol. Toxicol.* **26:**225–258.

19. Reuter, H., 1983, Calcium channel modulation by neurotransmitters, enzymes and drugs, *Nature* **301:**569–574.

20. Hille, B., 1984, Counting channels, in: *Ionic Channels of Excitable Membranes*, Sinauer Associates, Sunderland, MA, pp. 205–225.

21. Rimele, T. J., and Vanhoutte, P. M., 1984, Differential effects of calcium entry blockers on vascular smooth muscle, *Int. Angiol.* **3:**17–23.

22. Bolton, T. B., 1979, Mechanisms of action of transmitter and other substances on smooth muscle, *Physiol. Rev.* **59:**606–718.

23. Meisheri, K. D., Hwang, O., and Van Breemen, C., 1981, Evidence for two separate Ca^{2+} pathways in smooth muscle plasmalemma, *J. Membr. Biol.* **59:**19–25.

24. Van Breemen, C., Aaronson, P. I., Cauvin, C. A., Loutzenhiser, R. D., Mangel, A. W., and Saida, K., 1982, The calcium cycle in arterial smooth muscle, in: *Calcium Blockers: Mechanisms of Action and Clinical Applications* (S. F. Flaim and R. Zelis, eds.), Urban & Schwarzenberg, Baltimore, pp. 53–63.

25. Chiu, A. T., McCall, D. E., and Timmermans, P. B. M., 1986, Pharmacological characterization of receptor-operated and potential-operated Ca^{2+} channels in rat aorta, *Eur. J. Pharmacol.* **127:**1–8.

26. Ratz, P. H., and Flaim, S. F., 1985, Acetylcholine- and 5-hydroxtryptamine-stimulated contraction and calcium uptake in bovine coronary arteries: Evidence for two populations of receptor-operated calcium channels, *J. Pharmacol. Exp. Ther.* **234:**641–647.

27. Benham, C. D., and Tsien, R. W., 1987, A novel receptor-operated Ca^{2+}-permeable channel activated by ATP in smooth muscle, *Nature* **328:**275–278.

28. Sakai, K., Yamaguchi, T., Morita, S., and Uchida, M., 1983, Agonist-induced contraction of rat myometrium in Ca^{2+}-free solution containing Mn, *Gen. Pharmacol.* **14:**391–400.

29. Inomata, H., and Kao, C. Y., 1976, Ionic currents in the guinea-pig taenia coli, *J. Physiol. (Lond.)* **255:**347–378.

30. Hagiwara, S., and Ohmori, H., 1982, Studies of Ca channels in rat clonal pituitary cells with patch electrode voltage clamp, *J. Physiol. (Lond.)* **331:**599–635.

31. Brown, A. M., Morimoto, K., Tsuda, Y., and Wilson, D. L., 1981, Calcium current-dependent and voltage-dependent inactivation of calcium channels in *Helix aspersa*, *J. Physiol. (Lond.)* **320:**193–218.

32. Hille, B., 1984, Calcium channels, in: *Ionic Channels of Excitable Membranes*, Sinauer Associates, Sunderland, MA, pp. 76–98.

33. Fukushima, Y., and Hagiwara, S., 1983, Voltage-gated Ca^{2+} channel in mouse myeloma cells, *Proc. Natl. Acad. Sci. U.S.A.* **80:**2240–2242.

34. Fox, A. P., 1981, Voltage-dependent inactivation of a calcium channel, *Proc. Natl. Acad. Sci. U.S.A.* **78:**953–956.

35. Eckert, R., and Chad, J. E., 1984, Inactivation of channels, *Prog. Biophys. Mol. Biol.* **44:**215–267.

36. Chad, J., Eckert, R., and Ewald, D., 1984, Kinetics of calcium dependent inactivation of calcium current in voltage clamped neurons of *Aplysia californica*, *J. Physiol. (Lond.)* **347:**279–300.

37. Eckert, R., and Tillotoson, D., 1981, Calcium mediated inactivation of the calcium conductance in calcium loaded giant neurons of *Aplysia californica*, *J. Physiol. (Lond.)* **314:**265–280.

38. Jmari, K., Mironneau, C., and Mironneau, J., 1986, Inactivation of calcium channel current in rat uterine muscle: Evidence for calcium- and voltage-mediated mechanisms, *J. Physiol. (Lond.)* **380:**111–126.

39. Llinas, R., Steinberg, I. Z., and Walton, K., 1981, Presynaptic calcium currents in squid giant synapse, *Biophys. J.* **33:**289–322.

40. Worley, J. F., Deitmer, J. W., and Nelson, M. T., 1986, Single nisoldipine-sensitive calcium channels in smooth muscle cells isolated from rabbit mesenteric artery, *Proc. Natl. Acad. Sci. U.S.A.* **83:**5746–5750.

41. Sturek, M., and Hermsmeyer, K., 1986, Calcium and sodium channels in spontaneously contracting vascular muscle cells, *Science* **233:**475–478.

42. Bean, B. P., 1985, Two kinds of calcium channels in canine atrial cells, *J. Gen. Physiol.* **86:**1–30.

43. Mitra, R., and Morad, M., 1986, Two types of calcium channels in guinea pig ventricular myocytes, *Proc. Natl. Acad. Sci. U.S.A.* **83:**5340–5344.

44. Cognard, C., Lazdunski, M., and Romey, G., 1986, Different types of Ca^{2+} channels in mammalian skeletal muscle cells in culture, *Proc. Natl. Acad. Sci. U.S.A.* **83:**517–521.

45. Nowycky, M. C., Fox, A. P., and Tsien, R. W., 1985, Three types of neuronal calcium channels with different calcium agonist sensitivity, *Nature* **316:**440–443.

46. Carbone, E., and Lux, H. D., 1984, A low voltage-activated, fully inactivating Ca channel in vertebrate sensory neurons, *Nature* **310:**501–502.

47. Armstrong, C. M., and Matteson, D. R., 1985, Two distinct populations of calcium channels in a clonal line of pituitary cells, *Science* **227:**65–67.

48. Hess, P., Lansman, J. B., and Tsien, R. W., 1986, Calcium channel selectivity for monovalent and divalent cations, *J. Gen. Physiol.* **88:**293–319.

49. Lansman, J. B., Hess, P., and Tsien, R. W., 1986, Blockade of current through single calcium channels by Cd^{2+}, Mg^{2+}, and Ca^{2+}, *J. Gen. Physiol.* **88:**321–347.

50. Godfraind, T., 1986, Calcium antagonism and calcium entry blockade, *Pharmacol. Rev.* **38:**321–416.

51. Kuriyama, H., and Suzuki, H., 1976, Effects of prostaglandin E_2 and oxytocin on the electrical activity of hormone-treated and pregnant rat myometria, *J. Physiol. (Lond.)* **260:**335–349.

52. Reiner, O., and Marshall, J. M., 1976, Action of $PGF_{2\alpha}$ on the uterus of the rat, *Arch. Pharmacol.* **292:**243–250.

53. Carsten, M. E., 1974, Prostaglandins and oxytocin: Their effects on uterine smooth muscle, *Prostaglandins* **5:**33–40.

54. Levitan, I., 1985, Phosphorylation of ion channels, *J. Membr. Biol.* **87:**177–190.

55. Ousterhout, J. M., and Sperelakis, N., 1985, Role of cyclic nucleotides in regulation of slow channel function in vascular smooth muscle, *Biophys. J.* **47:**266a.

56. Godfraind, T., 1986, EDRF and cyclic GMP control gating of receptor operated calcium channels in vascular smooth muscle, *Eur. J. Pharmacol.* **126:**341–343.

57. Richards, I. S., Ousterhout, J. M., Sperelakis, N., and Murlas, C. G., 1987, cAMP suppresses Ca^{2+}-dependent electrical activity of airway smooth muscle induced by TEA, *J. Appl. Physiol.* **62:**175–179.

58. Richards, I. S., Murlas, C., Ousterhout, J. M., and Sperelakis, N., 1987, 8-Bromo-cyclic GMP abolishes TEA-induced slow action potentials in canine tracheal muscle, *Eur. J. Pharmacol.* **128:**299–302.

59. Brown, A. M., and Birnbaumer, L., 1988, Direct G protein gating of ion channels, *Am. J. Physiol.* **254:**H401–H410.

60. Irvine, R. F., 1987, Inositol phosphate and calcium entry, *Nature* **328:**386.

61. Godfraind, T., 1986, Calcium entry blockade and excitation contraction coupling in the cardiovascular system, *Acta Pharmacol. Toxicol.* **58**(Suppl. 2):5–30.

62. van Proosdij-Hartzema, E. G., and de Jongh, D. K., 1959, A new piperazine compound with antihistaminic activity, *Acta Physiol. Pharmacol. Neerl.* **8:**337–342.

63. Fleckenstein, A., 1964, Die Bedeutung der energiereichen Phosphate fur Kontraktilitat und Tonus des Myokards, *Verh. Dtsch. Ges. Inn. Med.* **70:**81–99.

64. Godfraind, T., Kaba, A., and Polster, P., 1968, Differences in sensitivity of arterial smooth muscles to inhibition of their contractile response to depolarization by potassium, *Arch. Int. Pharmacodyn. Ther.* **172:**235–239.

65. Spedding, M., 1985, Calcium antagonist subgroups, *Trends Pharmacol. Sci.* **6:**109–114.

66. Janis, R. A., and Triggle, D. J., 1983, New developments in Ca^{2+} channel antagonists, *J. Med. Chem.* **26:**775–785.

67. Godfraind, T., and Kaba, B., 1969, Blockade or reversal of the contraction induced by calcium and adrenaline in depolarized arterial smooth muscle, *Br. J. Pharmacol.* **35:**549–560.

68. Bolger, G. T., Gengo, P., Klockowski, R., Luchowski, E., Siegel, H., Janis, R. A., Triggle, A. M., and Triggle, D. J., 1983, Characterization of binding of the Ca^{2+} channel antagonist, [^{3}H]nitrendipine, to guinea-pig ileal smooth muscle, *J. Pharmacol. Exp. Ther.* **225:**291–309.

69. Grover, A. K., Kwan, C.-Y., Luchowski, E., Daniel, E. E., and Triggle, D. J., 1984, Subcellular distribution of [^{3}H]nitrendipine binding in smooth muscle, *J. Biol. Chem.* **259:**2223–2226.

70. Miller, W. C., and Moore, J. B., 1984, High affinity binding sites for [^{3}H]nitrendipine in rabbit uterine smooth muscle, *Life Sci.* **34:**1717–1724.

71. Golichowski, A. M., and Tzeng, D. Y., 1985, Binding of the calcium antagonist (^{3}H)nitrendipine to human myometrial plasmalemma, *Biol. Reprod.* **33:**1105–1112.

72. Triggle, D. J., and Janis, R. A., 1987, Calcium channel ligands, *Annu. Rev. Pharmacol. Toxicol.* **27:**347–369.

73. Janis, R. A., and Triggle, D. J., 1984, 1,4-Dihydropyridine Ca^{2+} channel activators: A comparison of binding characteristics with pharmacology, *Drug Dev. Res.* **4:**257–274.

74. Triggle, D. J., and Janis, R. A., 1984, Calcium channel antagonists: New perspectives from radioligand binding assay, in: *Modern Methods in Pharmacology* (N. Back and S. Spector, eds.) Alan R. Liss, New York, pp. 1–28.

75. Glossman, H., Ferry, D. R., Goll, A., Striessnig, J., and Zernig, G., 1985, Calcium channels and calcium channel drugs: Recent biochemical and biophysical findings, *Arzneim. Forsch.* **35:**1917–1935.

76. Hess, P., Lansman, J. B., and Tsien, R. W., 1984, Different modes of Ca channels gating behavior favored by dihydropyridine Ca agonists and antagonists, *Nature* **311**:599–635.

77. Murphy, K. M. M., Gould, R. J., Largent, B. L., and Snyder, S. H., 1983, A unitary mechanism of calcium antagonist drug action, *Proc. Natl. Acad. Sci. U.S.A.* **80**:860–864.

78. Pang, D. C., and Sperelakis, N., 1983, Nifedipine, diltiazem, beperidil and verapamil uptakes into cardiac and smooth muscles, *Eur. J. Pharmacol.* **87**:199–207.

79. Metzger, H., Stern, H. O., Pfilzer, G., and Reugg, J. C., 1982, Calcium antagonists affect calmodulin-dependent contractility of a skinned smooth muscle, *Drug Res.* **32**:1452–1457.

80. Epstein, P. M., Fiss, K., Hachisu, R., and Andrenyak, D. M., 1982, Interaction of calcium antagonists with cyclic AMP phosphodiesterases and calmodulin, *Biochem. Biophys. Res. Commun.* **105**:1142–1146.

81. Sakamoto, H., and Huszar, G., 1986, Pharmacologic levels of nitrendipine do not affect actin–myosin interaction in the human uterus and placenta, *Am. J. Obstet. Gynecol.* **154**:402–407.

82. Huszar, G., and Bailey, B., 1979, Relationship between actin–myosin interaction and myosin light-chain phosphorylation in human placental smooth muscle, *Am. J. Obstet. Gynecol.* **135**:718–726.

83. Csapo, A. I., 1959, Studies on excitation–contraction coupling, *Ann. N.Y. Acad. Sci.* **81**:453–467.

84. Movsesian, M. A., Swain, A. L., and Adelstein, R. S., 1984, Inhibition of turkey gizzard myosin light chain kinase activity by dihydropyridine calcium antagonists, *Biochem. Pharmacol.* **33**:3759–3764.

85. Maigaard, S., Forman, A., and Andersson, K.-E., 1986, Inhibitory effects of nitrendipine on myometrial and vascular smooth muscle in human pregnant uterus and placenta, *Acta Pharmacol. Toxicol.* **59**:1–10.

86. Anderson, C. A., Scriabine, A., Maurer, S. C., and Janis, R. A., 1983, Effects of nitrendipine on rat uterine smooth muscle: Inhibition of contraction and ligand binding studies, *Fed. Proc.* **42**:367.

87. Hollingsworth, M., Edwards, D., and Donnai, P., 1987, Inhibition of contractions of the isolated pregnant human myometrium by calcium entry blockers, *Med. Sci. Res.* **15**:15–16.

88. Fleckenstein, A., 1981, Fundamental actions of calcium antagonists on myocardial and cardiac pacemaker cell membranes, in: *New Perspectives on Calcium Antagonists* (G. B. Weiss, ed.), Williams & Wilkins, Baltimore, pp. 59–81.

89. Hermann, P., and Morselli, P. L., 1985, Pharmacokinetics of diltiazem and other calcium entry blockers, *Acta Pharmacol. Toxicol.* **57**:10–20.

90. Reiter, M. J., Shand, D. G., and Pritchett, L. C., 1982, Comparison of intravenous and oral verapamil dosing *Clin. Pharmacol. Ther.* **32**:711–720.

91. Eichelbaum, M., Somogyi, A., von Unruh, G. E., and Dengler, H. J., 1981, Simultaneous determination of the intravenous and oral pharmakokinetic parameters of D,L-verapamil using stable isotope-labelled verapamil, *Eur. J. Clin. Pharmacol.* **19**:133–137.

92. Mcallister, R. G., Jr., 1982, Clinical pharmakokinetics of calcium channel antagonists, *J. Cardiovasc. Pharmacol.* **4** (Suppl. 3):S340–S345.

93. Ochs, H. R., and Knuchel, M., 1984, Pharmacokinetics and absolute bioavailability of diltiazem in humans, *Klin. Wochenschr.* **62**:303–306.

94. Raemsch, K. D. and Sommer, J., 1983, Pharmacokinetics and metabolism of nifedipine, *Hypertension* **5**(Suppl. II):18–24.

95. Foster, T. S., Hamann, S. R., Richards, V. R., Bryant, P. J., Graves, D. A., and McAllister, R. G., 1983, Nifedipine kinetics and bioavailability after single intravenous and oral doses in normal subjects, *J. Clin. Pharmacol.* **23**:161–170.

96. Schlossman, K., Mendenwald, H., and Rosenkranz, H., 1975, Investigation on the metabolism and protein binding of nifedipine, in: *2nd International Adalat Symposium. New Therapy in Ischemic Heart Disease* (W. Lochner, W. Braasch, and G. Kronenberg, eds.), Springer-Verlag, Berlin, pp. 33–39.

97. Yong, C. L., Kunka, R. L., and Bates, T. R., 1980, Factors affecting the plasma protein binding of verapamil and norverapamil in man, *Res. Commun. Chem. Pathol. Pharmacol.* **30**:329–339.

98. Keefe, D. L., Yee, Y. G., and Kates, R. E., 1981, Verapamil protein binding in patients and in normal subjects, *Clin. Pharmacol. Ther.* **29:**21–26.

99. Bloedow, D. C., Piepho, R. W., Nies, A. S., and Gal, J. 1982, Serum binding of diltiazem in humans, *J. Clin. Pharmacol.* **22:**201–205.

100. Hof, R. P., 1983, Calcium antagonist and the peripheral circulation: Differences and similarities between PY 108-068, nicardipine, verapamil and diltiazem, *Br. J. Pharmacol.* **78:**375–394.

101. Andersson, K. E., 1986, Pharmacodynamic profiles of different calcium channel blockers, *Acta Pharmacol. Toxicol.* **58**(Suppl. 2):31–42.

102. Andersson, K. E., Ingemarsson, I., Ulmsten, U., and Wingerup, L., 1979, Inhibition of prostaglandin-induced uterine activity by nifedipine, *Br. J. Obstet. Gynaecol.* **86:**175–179.

103. Csapo, A. I., Puri, C. P., Tarro, S., and Henzl, M. R., 1982, Deactivation of the uterus during normal and premature labor by the calcium antagonist nicardipine, *Am. J. Obstet. Gynecol.* **142:**483–491.

104. Forman, A., 1979, Calcium entry blockade as a therapeutic principle in the female urogenital tract, *Acta Obstet. Gynecol. Scand. [Suppl.]* **121:**1–26.

105. U.S. Department of Health and Human Services, Public Health Service, 1986, *Vital Statistics of the United States, 1982,* Vol. 1 (*Natality*), National Center for Health Statistics, Hyattsville, MD.

106. Lindblom, B., Hamberger, L., and Wiqvist, N., 1978, Differentiated contractile effects of prostaglandins E and F on the isolated circular and longitudinal smooth muscle of the human oviduct, *Fertil. Steril.* **30:**553–559.

107. Lindblom, B., Hamberger, L., and Ljung, B., 1979, contractile patterns of isolated oviductal smooth muscle under different hormonal conditions, *Fertil. Steril.* **33:**283–287.

108. Forman, A., Andersson, K.-E., and Ulmsten, U., 1983, Effects of calcium and nifedipine on noradrenaline- and $PGF_{2\alpha}$-induced activity of the ampullary–isthmic junction of the human oviduct *in vitro*, *J. Reprod. Ferti.* **67:**343–349.

109. Forman, A., Andersson, K. E., and Maigaard, S., 1986, Effects of calcium channel blockers on the female genital tract, *Acta Pharmacol. Toxicol.* **58**(Suppl. 2):183–192.

110. von Euler, U. S., 1938, The action of adrenaline, acetylcholine and other substances on nerve-free vessels (human placenta), *J. Physiol. (Lond.)* **93:**129–143.

111. Kitson, G. E., and Broughton Pipkin, F., 1981, Effects and interactions of prostaglandins E_1 and E_2 on human chorionic plate arteries, *Am. J. Obstet. Gynecol.* **140:**683–688.

112. Reilly, F. D., and Russel, P. P., 1977, Neurohistochemical evidence supporting an absence of adrenergic and cholinergic innervation in the human placenta, *Anat. Rec.* **188:**277–286.

113. Maigaard, S., Forman, A., and Andersson, K.-E., 1984, Effects of nicardipine on human placental arteries, *Gynecol. Obstet. Invest.* **18:**217–224.

114. Maigaard, S., Forman, A., and Andersson, K.-E., 1986, Inhibitory effects of nitrendipine on myometrial and vascular smooth muscle in human pregnant uterus and placenta, *Acta Pharmacol. Toxicol.* **59:**1–10.

115. Broughton Pipkin, F., and Chinnery, E., 1985, The effects of calcium antagonist, nitrendipine, on responses of isolated strips of human chorionic plate artery to prostaglandin E_2 and E_1, *Br. J. Pharmacol.* **86:**705–709.

116. Reiner, O., and Marshall, J. M., 1976, Action of $PGF_{2\alpha}$ on the uterus of the rat, *Arch. Pharmacol.* **292:**243–250.

117. Csapo, A. I., 1959, Studies on excitation–contraction coupling, *Ann. N.Y. Acad. Sci.* **81:**453–467.

118. Rubanyi, G., and Csapo, A. I., 1976, The "activator-Ca" of the pregnant and postpartum uterus, *Life Sci.* **19:**1239–1252.

119. Rubanyi, G., and Csapo, A. I., 1977, The effect of Ca-transport inhibitors on the excitability of the pregnant and post partum rabbit uterus, *Life Sci.* **20:**289–300.

120. Fleckenstein, A., 1977, Specific pharmacology of calcium in myocardium, cardiac pacemakers and vascular smooth muscle, *Annu. Rev. Pharmacol. Toxicol.* **17:**149–166.

121. Granger, S. E., Hollingsworth, M., and Weston, A. H., 1985, A comparison of several calcium antagonists on uterine, vascular and cardiac muscles from the rat, *Br. J. Pharmacol.* **85:**255–262.

122. Granger, S. E., Hollingsworth, M., Weston, A. H., 1986, Effects of calcium entry blockers on tension development and calcium influx in rat uterus, *Br. J. Pharmacol.* **87:**147–156.

123. Hollingsworth, M., Edwards, D., and Donnai, P., 1987, Inhibition of contractions of the isolated pregnant human myometrium by calcium entry blockers, *Med. Sci. Res.* **15:**15–16.

124. Hendricks, C. H., 1964, A new technique for the study of motility of the non-pregnant human uterus, *J. Obstet. Gynaecol. Br. Commonw.* **71:**712–715.

125. Moawad, A. H., and Bengtsson, L. P., 1967, *In vivo* studies of the motility patterns of the non-pregnant human uterus I: The normal menstrual cycle, *Am J. Obstet. Gynecol.* **98:**1052–1064.

126. Akerlund, M., and Andersson, K.-E., 1976, Effects of terbutaline on human myometrial activity and endometrial blood flow, *Obstet. Gynecol.* **47:**529–535.

127. Ylikorkala, O., and Dawood, M. Y., 1978, New concepts in dysmenorrhea, *Am. J. Obstet. Gynecol.* **130:**833–847.

128. Akerlund, M., 1979, Pathophysiology of dysmenorrhea, *Acta Obstet. Gynecol. Scand.* **87:**21–32.

129. Maigaard, S., Forman, A., and Andersson, K.-E., 1985, Differences in contractile activation between human nonpregnant myometrium and intramyometrial vessels, *Acta Physiol. Scand.* **124:**371–379.

130. Maigaard, S., Forman, A., and Andersson, K.-E., 1985, Different responses to prostaglandin $F_{2\alpha}$ and E_2 in human nonpregnant extra- and intramyometrial arteries, *Prostaglandins* **30:**599–607.

131. Fuchs, A.-R., Coutinho, E. M., Xavier, R., Bates, P. E., and Fuchs, F., 1968, Effects of ethanol on the activity of the nonpregnant uterus and its reactivity to neurohypophyseal hormones, *Am. J. Obstet. Gynecol.* **101:**997–1000.

132. Akerlund, M., Stromberg, P., Forsling, M. L., Melin, P., and Vilhardt, H., 1983, Inhibition of vasopressin effects on the uterus by a synthetic analogue, *Obstet. Gynecol.* **62:**309–312.

133. Ulmsten, U., Andersson, K.-E., and Forman, A., 1978, Relaxing effects of nifedipine on the non-pregnant human uterus *in vitro* and *in vivo*, *Obstet. Gynecol.* **52:**436–441.

134. Andersson, K.-E., and Ulmsten, U., 1978, Effects of nifedipine on myometrial activity and lower abdominal pain in women with primary dysmenorrhea, *Br. J. Obstet. Gynaecol.* **85:**142–148.

135. Sandahl, B., Ulmsten, U., and Andersson, K.-E., 1979, Trial of the calcium antagonist nifedipine in the treatment of primary dysmenorrhea, *Arch. Gynecol.* **227:**147–151.

136. Mondero, N. A., 1983, Nifedipine in the treatment of dysmenorrhea, *J. Am. Osteopath. Assoc.* **82:**704–708.

137. Forman, A., Gandrup, P., Andersson, K.-E., and Ulmsten, U., 1982, Effects of nifedipine on spontaneous and methylergometrine-induced uterine activity post-partum, *Am. J. Obstet. Gynecol.* **144:**442–448.

138. Ulmsten, U., and Andersson, K.-E., 1979, Multichannel intrauterine pressure recording by means of microtransducers, *Acta Obstet. Gynecol. Scand.* **58:**115–120.

139. Forman, A., Gandrup, P., Andersson, K. E., and Ulmsten, U., 1982, Effects of nifedipine on oxytocin- and prostaglandin $F_{2\alpha}$-induced activity in the postpartum uterus, *Am. J. Obstet. Gynecol.* **144:**665–669.

140. Abel, M. H., and Hollingsworth, M., 1985, The potencies and selectivities of four calcium antagonists as inhibitors of uterine contractions in the rat *in vivo*, *Br. J. Pharmacol.* **85:**263–269.

141. Narimatsu, A., and Taira, N., 1976, Effects on atrioventricular conduction of calcium-antagonistic coronary vasodilators, local anesthetics and quinidine injected into the posterior and anterior septal artery of the atrioventricular node preparation of the dog, *Naunyn Schmiedebergs Arch. Pharmacol.* **294:**169–177.

142. Spedding, M., 1982, Differences between the effects of calcium antagonists in the pithed rat preparation, *J. Cardiovasc. Pharmacol.* **4:**973–979.

143. Hahn, D. W., McGuire, J. L., Vanderhoof, M., Ericson, E., and Pasquale, S. A., 1984, Evaluation of drugs for arrest of premature labor in a new animal model, *Am. J. Obstet. Gynecol.* **148:**775–778.

144. Lirette, M., Holbrook, H., and Katz, M., 1985, Effect of nicardipine HCl on prematurely induced uterine activity in the pregnant rabbit, *Obstet. Gynecol.* **65**:31–36.

145. Siimes, A. S. I., and Creasy, R. K., 1979, Cardiac and uterine hemodynamic responses to ritodrine hydrochloride administration in pregnant sheep, *Am. J. Obstet. Gynecol.* **133**:20–28.

146. Bieniarz, J., Ivankovich, A., and Scommenga, A., 1974, Cardiac output during ritodrine treatment on premature labor, *Am. J. Obstet. Gynecol.* **118**:910–920.

147. Katz, M., Robertson, P. A., and Creasy, R. K., 1981, Cardiovascular complications associated with terbutaline treatment for preterm labor, *Am. J. Obstet. Gynecol.* **139**:605–608.

148. Golichowski, A. M., Hathaway, D. R., Fineberg, N., and Peleg, D., 1985, Tocolytic and hemodynamic effects of nifedipine in the ewe, *Am. J. Obstet. Gynecol.* **151**:1134–1140.

149. Harake, B., Gilbert, R. D., Ashwal, S., and Power, G., 1987, Nifedipine: Effects on fetal and maternal hemodynamics in pregnant sheep, *Am. J. Obstet. Gynecol.* **157**:1003–1008.

150. Ducsay, C. A., Thompson, J. S., Wu, A. T., and Novy, M. J., 1987, Effects of calcium entry blocker (nicardipine) tocolysis in rhesus macaques: Fetal plasma concentrations and cardiorespiratory changes, *Am. J. Obstet. Gynecol.* **157**:1482–1486.

151. Veille, J. C., Bissonnette, J. M., and Hohimer, R. J., 1986, The effect of a calcium channel blocker (nifedipine) on uterine blood flow in the pregnant goat, *Am. J. Obstet. Gynecol.* **154**:1160–1163.

152. Parisi, V. M., Salinas, J. K., and Stockman, E. J., 1989, Placental vascular responses to nicardipine in the hypertensive ewe, *Am. J. Obstet. Gynecol.* **161**:1039–1043.

153. Ducsay, C. A., Cook, M. J., and Novy, M. J., 1985, Cardiorespiratory effects of calcium channel-blocker tocolysis in pregnant rhesus monkeys, in: *The Physiological Development of the Fetus and Newborn* (C. T. Jones and P. W. Nathanielsz, eds.), Academic Press, London, pp. 423–428.

154. Ducsay, C. A., McNutt, C. M., Harvey, L. M., and Gilbert, R. D., 1989, Effects of nicardipine on uteroplacental blood flow in the rhesus macaque, in: *Proceedings of the Thirty-Sixth Annual Meeting of the Society for Gynecological Investigation,* San Diego, p. 147.

155. Forman, A., Andersson, K. E., and Ulmsten, U., 1981, Inhibition of myometrial activity by calcium antagonists, *Semin. Perinatol.* **5**:288–294.

156. Ulmsten, U., 1984, Treatment of normotensive and hypertensive patients with preterm labor using oral nifedipine, a calcium antagonist, *Arch. Gynecol.* **236**:69–72.

157. Kaul, A. F., Osathanondh, R., Safon, L. E., Frigoletto, F. D., and Friedman, P. A., 1985, The management of preterm labor with the calcium channel-blocking agent nifedipine combined with the β-mimetic terbutaline, *Drug Intell. Clin. Pharm.* **19**:369–371.

158. Read, M. D., and Welby, D. E., 1986, The use of a calcium antagonist (nifedipine) to suppress preterm labour, *Br. J. Obstet. Gynaecol.* **93**:933–937.

159. Mari, G., Moise, K. J., Kirshon, B., Lee, W., and Cotton, D. B., 1989, Doppler assessment of the fetal and uteroplacental circulation after nifedipine therapy for preterm labor, in: *Proceedings of the Thirty-Sixth Annual Meeting of the Society for Gynecologic Investigation,* p. 284.

160. Godfraind, T., 1983, Action of nifedipine on calcium fluxes and contraction in isolated rat arteries, *J. Pharmacol. Exp. Ther.* **224**:443–450.

7

The Role of Membrane Potential in the Control of Uterine Motility

Helena C. Parkington and Harold A. Coleman

1. Introduction

An increase in the concentration of free calcium in the cytoplasm is essential for the initiation of contraction in smooth muscle. This calcium may be released from stores within the cell, but the most important source of calcium for long-term and repeated contraction is that which enters the cell from the external environment. At the resting membrane potential the influx of calcium is very small but increases dramatically as the membrane becomes depolarized. Thus, the membrane potential plays a crucial role in determining the level of contractility in uterine smooth muscle. The mechanisms involved in the control of calcium influx by membrane potential are considered in this chapter, along with how these processes vary throughout the estrous cycle and pregnancy and how they may be modified by endogenous hormones that are known to influence uterine motility.

2. Origins of Membrane Potential

2.1. Generation of Ionic Gradients

A potential difference exists across the cell membrane. This potential arises as a consequence of an unequal distribution of ions on either side of a membrane, which has a limited permeability to the ions (see Section 3). For example, the concentration of potassium inside the cell is high, while that of sodium is low. The external concentrations of these ions are the reverse. The polarity of the cell, i.e.,

Helena C. Parkington and Harold A. Coleman • Department of Physiology, Monash University, Clayton, Victoria 3168, Australia.

inside negative, is determined by the presence in the cytoplasm of large anions to which the membrane is absolutely impermeable.

In the first instance it is useful to consider the factors responsible for establishing and maintaining concentrations of ions in the cell that differ from their concentrations outside. Active transport systems, usually involving expenditure of energy by the cell through the breakdown of ATP, are among the most important factors that establish and maintain the potential difference across the membrane. Energy stored by an established gradient to one ion can also be harnessed to alter the concentration gradient of a second ion: the ion moving "downhill" along its concentration gradient results in "uphill" movement of the second ion. Such a mechanism is referred to as an antiporter when the ions involved move in opposite directions across the membrane, e.g., sodium–calcium exchange, or as a coporter when the ions move in the same direction, e.g., sodium–potassium–chloride cotransport.

2.1.1. Sodium–Potassium ATPase

The concentrations of sodium and potassium ions on either side of the cell membrane are the most important in determining the value of the resting membrane potential. There is a small but significant leak of sodium into and potassium out of the cell at rest. The importance of an ATPase in maintaining the high concentration of potassium and low concentration of sodium inside the cell was first demonstrated by Skou.[1] Activity of the ATPase pump in mammals is blocked when the temperature drops below about 20°C, by metabolic inhibitors, when potassium is omitted from the perfusing solution, or by the cardiac glycoside ouabain, and the result is an accumulation of sodium within and loss of potassium from the cell.

During each cycle of the pump, two potassium ions are required to bind to the outer surface, and these are internalized while three sodium ions are extruded.[2,3] Thus, the pump is electrogenic; i.e., there is a net transfer of charge from one side of the membrane to the other. The electrogenic nature of the pump can be demonstrated when potassium is readmitted to the solution perfusing sodium-rich myometrium (rendered so by storage at 0°C for several hours). The pump activity is increased, and the membrane rapidly but transiently (20–30 min) hyperpolarizes to levels greater than those observed in normal tissue. This effect is abolished by ouabain.[4] Although stimulation of the pump is capable of hyperpolarizing the membrane, its contribution to the resting potential under normal conditions has yet to be established. Hirst and van Helden[5] calculated that the electrogenic activity of the pump would be expected to contribute less than 1 mV to the resting potential recorded in their preparation of arteriole smooth muscle. The situation in arterioles may not be representative of events in nonvascular smooth muscle; many steroid hormones have been shown to increase the synthesis of new pump protein in a variety of tissues, and many other substances such as steroids and polypeptides can modulate activity of existing pump complexes.[2]

Thus, this pump not only is important in establishing and maintaining the cytoplasmic concentrations of the ions sodium and potassium but may also contribute additional negativity to the inside of the cell as a direct consequence of its electrogenic nature.

2.1.2. Calcium ATPase

The concentration of free calcium in the cytoplasm of smooth muscle at rest is less than 10^{-7} M,[6,7] while its external concentration is of the order of 10^{-3} M. The low concentration of internal calcium is achieved either by extrusion of ions across the plasma membrane, or by their sequestration into internal stores. These processes balance the small leakage of calcium into the cytoplasm that occurs at rest. The presence in plasma membrane of an ATPase that removes calcium from the cytoplasm against a gradient that can be as high as 10^4 was initially demonstrated in erythrocytes[8] (see also Chapters 5 and 12). It is likely that this pump is electrogenic since there is no indication of co- or countertransport involving another ion. However, its contribution to the membrane potential at rest would be expected to be negligible because of the small number of calcium ions that leak into the cell. During activity in smooth muscle, calcium influx into the cell is greatly enhanced (see Section 4). The electrogenic contribution of the pump to the membrane potential in the recovery period following activity has not been studied. Recent advances in raising antibodies to the pump molecule[9–12] and studies incorporating pump proteins into lipid vesicles[12,13] make kinetic studies of the pump possible.

2.1.3. Sodium–Calcium Exchange

The existence of transmembrane sodium–calcium exchange was initially demonstrated in cardiac muscle[14] and squid axon.[15] It was envisaged that sodium influx along its electrochemical gradient powered calcium extrusion, thereby providing a system for maintaining cytoplasmic calcium at low levels. The system is electrogenic, since three sodium ions exchange for one calcium.[16] This exchange system is reversible; i.e., it can operate to extrude calcium (forward mode) or to extrude sodium (reverse mode), depending on the transmembrane gradient of sodium and on membrane potential.[16–18] Depolarization favors the reverse mode, i.e., sodium extrusion and calcium influx, whereas hyperpolarization favors the forward mode.[17] This voltage dependence is probably related to the presence of charged groups on the carrier protein.

The observation that removal of sodium from the external solution results in contraction in many smooth muscles, including myometrium of rats, has been interpreted in terms of calcium accumulation resulting from a cessation of sodium–calcium exchange in forward mode or to a change in its operation to reverse mode.[18a,18b] Its electrogenic contribution to membrane potential in smooth muscle is unknown.

2.1.4. Regulation of Hydrogen Ions

Hydrogen ions are produced in the cytoplasm as a result of cellular activity. Control of their removal has been studied in considerable detail in skeletal and cardiac muscles, and it seems clear that substantial sodium–hydrogen exchange is involved.[19] This antiport mechanism is not thought to be electrogenic. The situation in smooth muscle appears to be more complex. Some dependence on external sodium has been demonstrated,[20,21] but the involvement of carbon dioxide and of bicarbonate and chloride coport systems, again nonelectrogenic, may be more prominent than in the other two muscle types.[19,21a]

2.1.5. Regulation of Internal Chloride

The cytoplasmic concentration of free chloride in smooth muscle is higher than would be predicted by passive distribution.[22] Exchange for bicarbonate appears to be the principal mechanism for the accumulation of most of the internal chloride in vas deferens[23] and ureter.[24] This has not been associated with net transfer of charge. In some vascular smooth muscles a mechanism implicating sodium–potassium–chloride cotransport is thought to be involved in regulating chloride,[25–27] whereby each sodium ion moving downhill into the cell powers the uphill transport of one potassium and two chloride ions into the cell. Thus, not all smooth muscles may regulate their cytoplasmic chloride in the same way, and the mechanisms operating in the uterus are not known.

2.2. Distribution of Ions in Smooth Muscle

The distribution of potassium, sodium, and chloride in smooth muscle has been estimated by three methods. With the ion analysis method the extracellular component is determined from the volume of tissue occupied by large impermeable molecules such as inulin, when used as a marker, and the intracellular content of the ion is calculated by subtraction from the total tissue content of the ion.[28] This method gives no indication of the concentration of an ion that is free in the cytoplasm, since it includes ion contained in intracellular compartments and bound to macromolecules.

A second method, in which the efflux curve of radiolabeled ion has been analyzed, attempted to circumvent this problem. The component of the curve considered to represent efflux of cytoplasmic ion is extrapolated to zero time. Both of these methods are fraught with problems, which have been discussed thoroughly by Brading.[28]

A knowledge of ionic activity, in contradistinction to concentration, is more desirable because ions do not behave like ideal gases. Recently it has become possible to determine directly the activities of sodium and chloride ions within the cytoplasm using ion-sensitive microelectrodes.[23,24,36] This method has only been applied to vas deferens and ureter. The cytoplasmic activity of chloride in these tissues is 40–50 mM[22,24] and is not very different from the value of 50–60 mM obtained from ion analysis.[29–31] On the other hand, ion analysis has yielded values of 20–40 mM for cytoplasmic sodium in a variety of tissues.[29–31] This is very different from the activity of 8 mM measured in ureter with ion-sensitive micro-electrodes.[36] Ion-sensitive electrodes have not been applied to myometrium. The cytoplasmic concentrations of potassium, sodium, and chloride obtained by ion analysis and flux studies for longitudinal myometrium are summarized in Table I.

2.3. Equilibrium Potentials in Smooth Muscle

The existence of a potential difference across the cell membrane means that an ion experiences not only a concentration gradient but also an electrical gradient. The potential that just balances the concentration gradient for a particular ion is called the equilibrium potential (E_x) for that ion. Calculation of E_x requires a knowledge of the activity of the ion, x, on either side of the membrane according to the equation

$$E_x = (RT/zF) \ln ([x]_o/[x]_i) \tag{1}$$

(Nernst equation), where R is the gas constant, T the absolute temperature, F Faraday's constant, z the valency and charge on the ion, and $[]_o$ and $[]_i$ are the activities of the ion outside and inside the cell, respectively. E_K has a value of -80 to -90 mV in smooth muscle. It is clear from equation (1) that the electrical gradient for potassium is inward, while its chemical gradient is outward. Thus, for values of membrane potential more positive than E_K potassium will leave the cell, whereas for values of membrane potential more negative than E_K potassium will enter the cell. It is interesting to note that ion analysis yields a value for E_{Na} of around 20 mV,[35,36] whereas the lower activity of sodium inside the cell as measured by ion-sensitive microelectrode suggests that E_{Na} may be 60–65 mV.[36] The values of the equilibrium potential for potassium, sodium, and chloride in myometrium, determined from ion analysis and flux studies, are included in Table I.

Table I. Ion Concentrations and Equilibrium Potentials for Sodium, Potassium, and Chloride in Myometrium

	K	Na	Cl
Intracellular concentration (mM)			
Rat[32,33]	164	21	40
Cat[34]	164	67	89
Guinea pig[34]	179	30	66
Extracellular concentration (mM)	6[a]	137	134[a]
Equilibrium potential (mV)	-80 to -90	20–30	-20 to -30

[a] Not physiological.

3. Membrane Permeability and Passive Electrical Properties

Under normal conditions the distribution of ions on either side of the cell membrane remains essentially constant, and changes in membrane potential usually occur as a result of changes in the permeability of the membrane.

3.1. Membrane Permeability

3.1.1. Relationship between Membrane Potential and Permeability

The ability of the cell membrane to conduct ions (the permeability of the membrane) is limited in the resting state. Since the concentrations of potassium, sodium, and chloride on either side of the plasma membrane are unequal, a diffusion potential occurs across the membrane, which can be described by

$$E_m = (RT/F) \ln \frac{[K]_o + P_{Na}[Na]_o + P_{Cl}[Cl]_i}{[K]_i + P_{Na}[Na]_i + P_{Cl}[Cl]_o} \qquad (2)$$

(the Goldman–Hodgkin–Katz voltage equation), where P_x is the permeability of the membrane to the ion relative to the potassium permeability. It is clear that the contribution of an ion to the membrane potential depends on its distribution on either side of the membrane and on the permeability of the membrane to it. For example, an increase in the extracellular concentration of either potassium or sodium or a decrease in the concentration of chloride would result in the membrane potential becoming less negative. Likewise, an increase in potassium permeability would result in the membrane potential becoming more negative, whereas an increase in sodium permeability would have the opposite effect.

3.1.2. Ionic Permeabilities in Smooth Muscle

Calculations of permeability have, until recently, relied on measurements of transmembrane ion fluxes, and the problems inherent in this method have been discussed in detail.[28] By this technique the permeability of chloride, relative to potassium, has been estimated to be 0.65 in taenia coli[29,30] and in myometrium of estrogen-treated rats.[33] In the last 5 years Aickin and Brading have made considerable progress in determining ion permeabilities using ion-sensitive microelectrodes. The latter technique suggests that chloride permeability in vas deferens is much lower, around 0.045.[23]

Flux studies have produced wide variability in estimates of relative sodium permeability, and values of around 0.2[29,31] or 0.01[30] have been reported for taenia coli and vas deferens, although values of 1–3 have been reported for rat myometrium![33] In the light of the recent results with ion-sensitive microelectrodes, the apparently high values for ion permeabilities calculated from flux measurements have been reinterpreted in terms of appreciable transport mechanisms. The latter provide alternative pathways for the transmembrane movements of ions, and their operation is considered to contribute to the overestimate of permeability (by definition, a purely passive event).[25,36] Further discussion of the relative contribution of the various ions to membrane potential in myometrium awaits more reliable estimates of the activities of ions in the cytoplasm and the permeability of the membrane to the ions.

Relative permeabilities of 0.05–0.1 for sodium and 0.05 for chloride and concentrations of the ions as given in Table I result in a calculated membrane potential for myometrium in the range of -50 to -60 mV. This range of values is in reasonable agreement with many values of resting potential measured in myometrium and other smooth muscles (see Section 3.2.2).

The concentration of extracellular calcium is 10^{-3} M, and free cytoplasmic calcium is generally accepted as being around 10^{-7} M at rest.[6,7] These concentrations of calcium result in a calculated value of E_{Ca} of $+120$ to $+150$ mV. It is unlikely that calcium contributes significantly to the value of the resting membrane potential directly because of the small number of ions that leak into the cell. From flux studies the permeability of the membrane to calcium appears to be similar to that of sodium,[37] but the influx of calcium under resting conditions (0.02 pmol/cm² per min)[38] is very much less than that of sodium because the concentration of calcium in the extracellular compartment is much lower.

3.2. Passive Electrical Properties

3.2.1. Relationship between the Passive Electrical Properties

The permeability of the membrane is inversely related to membrane resistance (R_m, specific membrane resistance, Ωcm^2). In the simplest case of a spherical cell, the membrane can be regarded as a resistor and capacitor in parallel. When a step of current I is injected into such a cell, the membrane potential, E_m, changes according to $E = IR$. The voltage response is termed an electrotonic potential. The change in voltage is not instantaneous but has an exponential time course (Fig. 1). The rate of change is determined by the membrane resistance and capacitance, and the amplitude of the electrotonic potential is determined by the membrane resistance. The time taken to reach $1 - e^{-1}$ (63%) of the final value is defined as the membrane

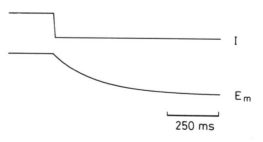

Figure 1. The voltage response of the membrane (E_m) (the electrotonic potential) to injecting a current, I, into a small, idealized, spherical cell. The time constant of the response is 200 msec.

250 ms

time constant, τ_m. This parameter can be determined experimentally and is used as an indicator of the membrane resistance, since they are directly related according to

$$\tau_m = R_m \, C_m \tag{3}$$

(C_m is specific membrane capacitance, $\mu F \cdot cm^{-2}$). Changes in τ_m are usually attributed to changes in R_m because C_m is determined by membrane thickness and dielectric constant, which may be assumed to vary only minimally. C_m has a similar value in a variety of disparate tissues and is considered to be of the nature of a biological constant.[39,40] Thus, an increase in the permeability of the membrane to an ion or ions (a decrease in resistance) can be measured as a decrease in τ_m. (In view of the changes that occur in the uterus during pregnancy, and especially at term, it may be unwise to adhere too tenaciously to the notion of the constancy of the value of C_m in this somewhat unusual case. The possibility that C_m may change during pregnancy has not been investigated experimentally.)

The cells of many tissues, including the myometrium, are not simple spheres, and the situation is complicated further by the fact that the cells are connected to each other via low-resistance pathways (discussed in detail in Section 6.1). As a consequence, not all of the current arising in a cell leaves via the membrane of the same cell. A portion of the current flows into the neighboring cells and leaves via their membranes; i.e., it "sees" a larger area of membrane than that of a single cell, has additional avenues of exit, and hence "sees" a smaller resistance.

This three-dimensional flow of current is difficult to deal with experimentally. The situation can be simplified to one dimension if strips of muscle greater in length than in width or thickness are used, and the entire cross-sectional area of the strip is polarized with large extracellular electrodes placed at one end. The current applied is forced to flow in one dimension only, that is, longitudinally along the tissue. It cannot flow radially in the other two dimensions because all of the cells in that plane are equipotential.[41] When stimulated thus, smooth muscle can be approximated by a one-dimensional cable. The amplitude of the steady-state electrotonic potential decays exponentially with distance along the strip.[42] This decay is described by the length constant (λ), which is defined as the distance at which the amplitude of the electrotonic potential has declined to e^{-1} (37%) of the initial value. This technique has been used in determining λ and τ_m in myometrium. τ_m is calculated according to a method in which the time for the electrotonic potential to reach half of the steady-state value is plotted against the distance from the nearer polarizing electrode and is obtained from the equation[43,44]

$$slope = \tau_m / 2\lambda \tag{4}$$

The length constant is related to membrane resistance (R_m, Ωcm^2) and internal resistivity (R_i, Ωcm) according to the expression

$$\lambda = \sqrt{aR_m / 2R_i} \tag{5}$$

where a is the radius of the cable and seems to be equivalent to the radius of the smooth muscle cell.[41,45] R_i results from the cytoplasmic resistivity in series with the resistance of the intercellular pathways. The length constant describes how well current will spread throughout the tissue and becomes significant in considering the propagation of action potentials through the tissue (see Section 6.2).

A change in the permeability of the membrane has repercussions through the three passive electrical properties discussed in this section. For example, a decrease in the permeability of the membrane will be measured as an increase in τ_m and may not only change the resting membrane potential but will also result in an increase in λ, and as a consequence current will spread further throughout the tissue.

3.2.2. Resting Membrane Potential

The resting potential is generally considered to be the level at which there is minimal change in membrane potential, and presumably at a time during which there is minimal movement of ions across the membrane. Quiescent smooth muscles such as vas deferens[46] and arteries,[47] invariably have large (more negative) resting potentials of -60 to -70 mV. However, some smooth muscles with large resting potentials are spontaneously active because of the occurrence of slow oscillations (see Section 5.1) of large amplitude.[48] The resting potential recorded in uterine smooth muscle is in the range -35 to -65 mV.[32,34,49−58] Circular myometrium of guinea pigs is quiescent and has a large resting potential (around -60 mV),[59] and although the value recorded from circular myometrium of estrogen-treated rats is of similar magnitude, the tissue is spontaneously active.[55]

In only a limited number of studies[56,57] has a definite attempt been made to determine how the resting potential of the myometrium might change throughout pregnancy and, in particular, immediately prior to and during labor in both muscle layers. Events occurring in myometrium of rats have been studied in most detail by Marshall and colleagues, and these[51,57,60] have been combined with the results of others[32,50,56] to produce Fig. 2.

Although values of resting membrane potential in myometrium of species other than rat are available, detailed information is scarce. The data are summarized in Table II. The membrane potential recorded from the longitudinal myometrium of guinea pigs is smaller than that observed in the circular layer,[52] in contrast to findings in rats. The resting potential in circular myometrium of sheep and guinea pigs does not change during labor.[61]

3.2.3. Membrane Time Constant

The time constant of the myometrial cell membrane is of the order of 100–300 msec in rats,[50,64,65] guinea pigs,[59] and sheep,[61] and this is similar to values reported for other visceral smooth muscles. For the myometrium of rats[65] and sheep,[61] τ_m increases up to two- to threefold when the animal goes into labor (Fig. 3). Observations in circular myometrium of guinea pigs are interesting in that there

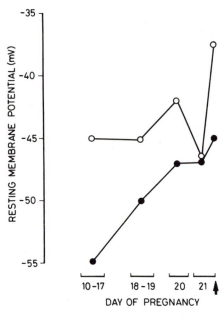

Figure 2. Resting membrane potential in longitudinal (filled circles) and circular (open circles) myometrium of rats from day 10 of pregnancy onwards, including during delivery (arrow). (Data from references 32, 50, 51, 56, 57, and 60 combined.)

are small fluctuations in τ_m during the estrous cycle, τ_m being higher in estrus (330 msec) than in diestrus (210 msec).[45]

3.2.4. Length Constant

Values for λ in the myometrium (1–3 mm)[50,61,64,65] are similar to those observed in other visceral smooth muscles. There is a marked increase in λ during labor in rats[50,65] and sheep,[61] but not in guinea pigs[59] (Fig. 3).

3.3. Effect of Reproductive State on Passive Electrical Properties

From the observations in Fig. 3 it can be seen that there is no consistent pattern in the changes that occur in the passive electrical properties as rats, guinea pigs, and

Table II. Resting Membrane Potential Measured in Myometrium

	Longitudinal	Circular
Cat[34]	−60 mV	
Rabbit[62]	−50 mV	
Sheep[61]		−50 mV
Guinea pig[59,63]	−55 mV	−58 mV

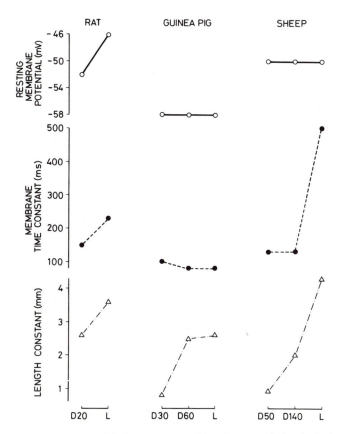

Figure 3. Resting membrane potential, time constant, and length constant in longitudinal myometrium of rats (day 20 of pregnancy and labor, day 21)[65] and circular myometrium of guinea pigs (days 30 and 60 of pregnancy and labor, day 68)[59] and sheep (days 50 and 140 of pregnancy and labor, day 145).[61]

sheep go into labor. There is a decline in the resting potential in longitudinal myometrium of rats,[51,65] and Sims and colleagues[65] have shown that it is associated with an increase in τ_m. This suggests that there is a decrease in potassium permeability. There is also an increase in λ, which might be expected as a consequence of the increase in R_m (equations 3 and 5). However, the increase in λ could not be fully explained by the increase in R_m. Calculations indicated that R_i had decreased by some 33%. Direct measurements confirmed a decrease in R_i of 42% during labor,[65] indicating that passive spread of current throughout the uterus would be facilitated at this time (see also Chapter 9). Since the concentration of estrogen increases markedly during the last 2 days of pregnancy in rats, while that of progesterone decreases,[66] it is interesting to consider how changes in the steroids may contribute to the changes in passive electrical properties as labor approaches.

Progesterone administration to ovariectomized, estrogen-primed rats induces an increase in the permeability of the myometrial cell membrane to potassium,[32] and progesterone injection on day 19 prolongs pregnancy and results in hyperpolarization of the membrane.[56] Thus, the decline in the concentration of progesterone at term could explain the membrane depolarization, the decrease in membrane permeability, and the increase in λ observed in the longitudinal myometrium at this time.

RNA extracted from the cytoplasm of uterine cells of ovariectomized rats injected with estrogen induces additional potassium permeability when injected into *Xenopus* oocytes.[67] The values of membrane potential recorded in the myometrium of nonpregnant rats are most negative following estrogen treatment, and this is consistent with the notion that estrogen increases the permeability of the myometrial cell membrane to potassium. This is in contrast with observations during pregnancy; the membrane is at its most hyperpolarized up to day 16,[50,51] when circulating estrogen levels are very low,[66] and the progressive depolarization in the last few days of pregnancy coincides with an increase in the concentration of circulating estrogen.[66] There is no effect on membrane potential when intact pregnant rats are injected with estrogen.[56] Thus, a direct role for estrogen in the depolarization of longitudinal myometrium at term in this species is not immediately apparent, and any effects may be via interaction with other factors peculiar to pregnancy. This is not surprising since attempts to induce the pattern of uterine activity observed during late pregnancy and labor by injecting various regimens of steroids have failed in rats[50] and sheep (unpublished observations). Estrogen treatment of ovariectomized nonpregnant rats results in an increase in λ but no change in τ_m[50] (in contrast to observations at term; see Sections 3.2.3 and 3.2.4). Since there is no increase in R_m, the increase in λ is most likely caused by either an increase in cell coupling or an increase in the diameter and/or length of the smooth muscle cells. In addition, the increase in λ resulting from estrogen injection of nonpregnant rats is but half to one-third that observed during labor.[65]

Although circulating progesterone may be implicated in determining the resting potential in the longitudinal myometrium of rats, experimental results are less forthcoming for the circular layer. Progressive depolarization of the membrane occurs in the "pregnant" horn of unilaterally pregnant animals. Values recorded in the "nonpregnant" horn remain at the relatively hyperpolarized level characteristic of midpregnancy. The depolarization associated with the onset of labor has been attributed, at least in part, to the stretching effect of the fetus (see Section 5.5.1). In addition, depolarization does not occur following ovariectomy in the absence of estrogen supplementation.[60]

Resting potential and τ_m do not change as guinea pigs go into labor, indicating that there is no change in R_m. Throughout pregnancy there is a gradual increase in λ, which could be explained by a gradual increase in cell-to-cell coupling or an increase in cell diameter (equation 5). Similar changes have been shown to take place during the estrous cycle in this species.[45] There is little change in λ as the animal goes into labor.

In sheep, as in guinea pigs, the resting potential and τ_m do not change throughout pregnancy, although there is a gradual increase in λ (Fig. 3).[61] As the ewe goes into labor there is a large increase in τ_m, but without any change in membrane potential. This suggests that there is a decrease in permeability not only to potassium (which would cause depolarization) but also to other ions, sodium being the most likely candidate (which would tend to hyperpolarize). The increase in R_m contributes to the increase in λ, but there is an additional increase in λ that suggests that cell coupling is increased by about 35%.[61]

4. Changes in Membrane Permeability and Active Responses

4.1. Mechanism of Ion Permeation

Permeation of the membrane by ions is made possible by protein complexes that span the cell membrane and have aqueous pores through which ions can traverse a membrane that is otherwise impermeable to ions.[68,69] Such proteins are called channels. Channels possess several important functional features. They have a selectivity filter such that, under physiological conditions, a channel can be quite specific for a particular ion. They also possess a gate such that no ions can pass through a channel unless the gate is open.[68,69] The gate in some channels can be operated by chemicals (receptor activated channels); the acetylcholine receptor/channel of the neuromuscular junction is among the best studied examples of chemical activation of channel activity.[69]

Many channels contain charged groups that make the gate sensitive to the voltage across the cell membrane,[68,69] and such channels are termed voltage dependent. The channels are in one of three conditions: closed, open, or inactivated (Fig. 4). Transitions usually occur between the three conditions as random events, with each transition having a different probability. The probability of a transition occurring varies depending on the type of channel, on the membrane potential, and, in some instances, on cytoplasmic components, e.g., calcium, second messengers.

Under resting conditions the probability of most channels being open is small; most channels are closed. The effect of depolarization of the membrane on many channels is to increase the frequency of opening. They remain open for a brief but

Figure 4. Channels can be in the open, closed, or inactivated conditions. In both the closed and inactivated conditions, ions cannot pass through the channel. Channels do not progress from one condition to another in an orderly manner, but do so randomly, depending on the probabilities of the various transitions. In general, for a particular type of channel, the probabilities of the various transitions are different. The probabilities also vary with channel type, membrane potential, and, in some instances, cytoplasmic components. For example, some potassium channels do not seem to enter the inactivated condition, whereas calcium channels have a significant probability of entering the inactivated condition at depolarized potentials.

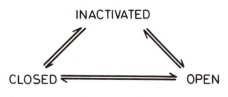

random period. In addition, the mean time that a channel spends in the open configuration is also voltage dependent, but to a lesser extent than that of frequency of opening (for example see Fig. 10). During continued depolarization most types of channel (but not all) eventually make the transition to the inactivated condition. When inactivated, a channel is nonconducting. This condition differs from being closed in that the channel usually requires the membrane to be repolarized (deinactivated) before it can reopen again (reactivated) (Fig. 4). For many channels the probability of leaving the inactivated condition is small at depolarized potentials and becomes much greater at more negative potentials. Not all channels are activated by depolarization; some are activated by hyperpolarization. In addition, the activity of some channels is increased by stretching the cell.

Under physiological conditions, the changes in membrane potential that are necessary for the activation of voltage-dependent channels are frequently achieved by operation of receptor-activated channels, e.g., neurotransmitters and hormones, or stretch. However, it is the activation of voltage-dependent channels that results in the functional response of many excitable cells.

4.2. Excitability

A cell is deemed to be excitable if it can be stimulated to produce an action potential, and the modern view of excitability has been elucidated in studies on the giant axon of squid.[70] An action potential is initiated by depolarization of the membrane, resulting in an increase in the number of voltage-dependent sodium channels in the open state; i.e., the membrane becomes more permeable to sodium. This results in an increased influx of sodium into the cell, which tends to further depolarize the membrane, and which increases even more the number of channels in the open state. With sufficient depolarization, to threshold, the inward current carried by sodium exceeds the outward current and results in a rapid, self-perpetuating (regenerative) depolarization toward the equilibrium potential for sodium—the rising phase of the action potential. The depolarization also increases the frequency of opening of voltage-dependent potassium channels, but with a delay and a slower time course. The increased potassium efflux, together with the decline in the sodium influx because of the channels entering the inactivated state, results in repolarization, termination of the action potential, and reestablishment of resting conditions. Thus, excitability is interpreted in terms of changes in the ease with which ions can pass into or out of the cell along their electrochemical gradients.

In contrast to the squid axon, the inward current underlying the spike of the action potential in smooth muscle is carried by an influx of calcium ions, which may be involved in contraction. Excitability and events underlying the activation of membrane channels in smooth muscle are complex, and action potentials of very diverse time courses can be recorded from different smooth muscles. Nevertheless, such diversity in the form of the action potential may result from a variable mixture of a finite number of channel types rather than resulting from a great diversity of channel types in different smooth muscles. Thus, it is of interest to consider the ionic mechanisms that have been found in smooth muscle cells in general before considering the observations in uterine smooth muscle in detail.

4.2.1. Voltage Clamp

It can be appreciated from the foregoing discussion that many conductances arising from channel activity in excitable cells change with both time and membrane potential.[69] Although a variety of techniques have been used in attempts to elucidate the changes in conductance in excitable cells, most techniques have centered on clamping the membrane at known potentials, thus eliminating that variable, and studying the conductance changes as a function of time.[68,69] This has been difficult to achieve with preparations of smooth muscle because the cells are electrically coupled to form a syncytium, usually of three dimensions (see Sections 3.2.1 and 6.1).

In the earliest attempts to voltage clamp syncytial preparations the double-sucrose-gap method was used.[71] This technique has, in theory, a number of limitations, which have been discussed elsewhere in detail.[72–75] In brief, the main problems involve uncertainty regarding the isopotentiality in the node being studied, various resistances in series with the membrane, leakage pathways in parallel with the membrane, and the possible deleterious effects of sucrose on the cells. It must be remembered that the sucrose-gap technique was developed for preparations that could not easily be voltage clamped by other methods at the time. Despite, and bearing in mind, the theoretical limitations of the technique, which prevent detailed kinetic analyses, in practice it has yielded information that otherwise would not have been feasible to obtain. Of the studies on smooth muscle with this technique, most have involved longitudinal myometrium of rats, and therefore the results are discussed in detail in Section 5.2.

A more recent development of voltage clamp that has been applied to the study of conductance changes in smooth muscle cells involves the use of a switching amplifier to voltage clamp small preparations with a single intracellular microelectrode. This technique has been applied to the study of cerebral arterioles[76] and small bundles of uterine smooth muscle[77] (see Section 5.3.2).

4.2.2. Patch Clamp

A method that has been developed in recent years, the patch-clamp technique,[113] has resulted in a remarkable series of advances in the study of ionic channels in a wide variety of cells. A glass electrode with an internal tip diameter of approximately 1 μm is pushed against the cell membrane, and suction is applied. If the cell membrane is sufficiently clean, then a very tight seal can form between the cell membrane and the glass electrode. This preparation is known as the "cell-attached mode"[113] (Fig. 5A). "Gigaseals" with electrical resistances greater than 1 $G\Omega$ (10^9 Ω) can be attained routinely. With seal resistances in the gigaΩ range, the outside surface of a patch of membrane inside the electrode is effectively isolated from the surrounding environment, both electrically and chemically. It is therefore

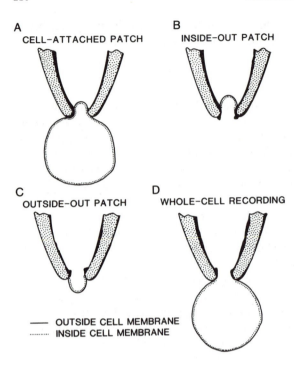

A
CELL-ATTACHED PATCH

B
INSIDE-OUT PATCH

C
OUTSIDE-OUT PATCH

D
WHOLE-CELL RECORDING

—— OUTSIDE CELL MEMBRANE
·········· INSIDE CELL MEMBRANE

Figure 5. Various associations between a glass electrode and cell membrane used in the patch-clamp technique. In each case a tight seal is formed between glass and membrane. **A:** The cell and patch remain intact. **B:** The electrode is pulled away from the cell, and a patch of membrane, cytoplasmic face out, is isolated from the remainder of the cell. **C:** The membrane inside the electrode is disrupted, and when the electrode is pulled away the remnants of membrane adhering to the glass seal over the end of the electrode to form an outside-out patch that is isolated from the remainder of the cell. **D:** The patch of membrane inside the electrode is disrupted, allowing exchange of materials between pipette and cytoplasm. (Modified from Hamill *et al.*[113])

possible to "clamp" the potential across this patch of membrane at various values. The ionic current that flows across the patch of membrane when only a single channel in the membrane opens is usually recorded as a square step of current (for example, see Fig. 10). Different types of channels can be distinguished by the size of the current step, their reversal potential, their sensitivity to various putative blockers, the duration of opening of the channel, and whether they are activated by voltage, by chemical means, or a mixture of both.

The patch of membrane at the tip of the electrode can be removed from the cell in such a way that the cytoplasmic face of the patch of membrane is facing the bath solution (inside-out mode)[113] (Fig. 5B), or the external surface of the patch of membrane is facing the bathing solution (outside-out mode)[113] (Fig. 5C). These procedures enable single-channel currents to be studied under conditions of precisely controlled solutions on both sides of the membrane but have the disadvantage that unknown cytoplasmic factors that could be involved in regulating the channel are washed away.

A commonly used procedure with the gigaseal technique is to disrupt the patch of membrane at the end of the pipette, permitting continuity between the contents of the pipette and the cytoplasm (whole-cell mode)[113] (Fig. 5D). Because of the tight seal between the electrode and the membrane and the low resistance of the electrodes, this recording arrangement is often used to voltage clamp single cells that are small. In contrast to the cell-attached (Fig. 5A) or isolated patch modes (Fig. 5B,C), the whole-cell mode records the currents flowing through channels in the membrane of the whole cell. This means that, typically, such a large number of channels is involved that it is not possible to resolve the currents flowing through individual channels, and it is not possible to be confident of recording the currents flowing through a single population of channels or to be able to identify unequivocally all the different populations of channels in the cell. Furthermore, there is considerable exchange between the contents of the patch pipette and the cytoplasm. This exchange provides the opportunity for regulating the composition of the cytoplasm, but has the disadvantage that components, which may be involved in channel activity, can be washed away.

The patch-clamp technique, when used to record single-channel currents, has the potential to identify unequivocally the various types of channels that occur in a cell's membrane. The technique can also be used to study modulation of channels by second messengers. The kinetics of channel activity can be described in considerable detail. Eventually it should be possible to reconstruct the electrical activity of the cell as a whole and in this way to demonstrate the role and importance of the various channels in determining the response of the cell. In this regard the application of the patch-clamp technique to smooth muscle, and indeed to many preparations, is still very much in its infancy, with most studies being at the level of identifying the various channels. Voltage-clamp studies indicate that a number of channel types remain to be identified. Of the channels that have been identified, there are little quantitative data available describing their behavior, thus precluding any attempts at constructing models to describe them. Perhaps the biggest difficulty is that most data have been obtained under conditions that are far from physiological. This not only makes comparison between studies difficult but can also mean that the physiological significance of the results is not clear. This problem also applies to the voltage-clamp studies in which the patch-clamp technique has been used in the whole-cell mode. Overall, the application of the patch-clamp technique has enabled useful advances to be made in our understanding of the ionic mechanisms in smooth muscle. However, there is considerable scope for further progress with this technique.

When the patch-clamp technique has been applied to smooth muscle cells, in all studies except those on guinea pig myometrium (see Section 5.3.2), the cells have been isolated from the smooth muscle syncytium and the membranes "cleaned" with the use of enzymes. Such isolated cells have high input resistances, i.e., the resistance of the entire membrane of the cell; this means that they can be readily voltage clamped using the whole-cell mode. In studies of single-channel currents the high input resistance of isolated smooth muscle cells can be a problem because of the risk of distortions and inaccuracies that may occur because of the currents flowing through channels of large conductance or through those with long open times.[78] Such studies have usually been made with isolated patches.

4.3. Ion Channels in Smooth Muscle

4.3.1. Calcium Channels

Calcium channels are of prime interest in attempting to understand the control of uterine motility because it is through these that the calcium that is so important for contraction passes into the cell.

At the single-channel level two different types of voltage-dependent calcium channels have been found in all preparations of vascular[79–81] and visceral[82,83] smooth muscle studied so far. Some properties of these channels are compared in Table III. These two calcium channels most likely correspond to the calcium channels termed T (transient) and L (long-lasting) that have been studied in cardiac[84] and neuronal[85] tissues. It should be noted that both types of channel stay open for a brief time, which on average is of the order of milliseconds.[85] The T channels are so named because their openings mainly occur as a burst of activity soon after the membrane potential is changed while the openings of L channels occur more uniformly during the period of change in membrane potential.[82] Attempts have been made to differentiate further the two channels according to their sensitivity to dihydropyridines. Although all reports indicate that the larger-conductance L channel is sensitive to these compounds,[79–82] the situation is not resolved for the smaller conductance channel in smooth muscle; several studies indicate that it is insensitive to dihydropyridines,[80–82] whereas another study reports that the activity of the channel is almost completely inhibited by nisoldipine at a concentration as low as 5 × 10^{-8} M[79] (see also Chapter 6).

In most studies[79–85] barium has been used to substitute for calcium as the permeant ion for several reasons. First, unlike calcium, barium does not inactivate calcium channels. Second, some evidence suggests that barium may pass through the L channels in cardiac[84] and neuronal[85] tissues more readily than does calcium, although both ions seem to permeate the T channels to a similar degree.[84,85] Finally, barium blocks some potassium channels, and this allows calcium channels to be studied without contamination from currents flowing through potassium channels.

Various voltage-clamp studies indicate that there may be two components to the current flowing through calcium channels in whole cells.[76,80–82,86–90] The data are consistent with the stronger evidence obtained at the single-channel level for the

Table III. Characteristics of the Two Calcium Channels Recorded from Smooth Muscle Cells with the Patch-Clamp Technique[a]

	T channel	L channel
Conductance	6–12 pS	15–28 pS
Activation	Transient	Long lasting

[a] Barium (50–110 mM) was present in the patch pipette.

occurrence of two different calcium channels. Additional evidence obtained from voltage-clamp studies of cardiac myocytes and neurons indicates that the L channels are inactivated by cytoplasmic calcium, although the T channels do not appear to be.[91−93]

4.3.2. Potassium Channels

The channels in smooth muscle cells that have been studied in most detail at the single-channel level are those that conduct potassium. The significance of potassium channels lies in the fact that they are the most important channels involved in terminating action potentials and in returning the membrane potential to its resting level. In addition, they are important in establishing the resting membrane potential. Changes in their activity at rest can result in changes in the likelihood of the membrane potential reaching threshold for the initiation of an action potential and hence contraction.

A number of different potassium channels have been identified in smooth muscle, and they can be distinguished from each other by their conductances and distributions of open and closed times at various levels of membrane potential. They can be characterized further by their sensitivity to calcium ions and to various compounds that block potassium channels, two of the most commonly used being tetraethylammonium and 4-aminopyridine (Table IV).

Three different outward currents have been recorded from smooth muscle cells dissociated from rabbit pulmonary artery[99] or rabbit small intestine[98,100] and studied under voltage clamp. Only one of these corresponds to the potassium channels identified in smooth muscle at the single-channel level (the large conductance channel, i in Table IV). The other two currents are likely to be carried by potassium channels not yet identified at the single-channel level.

Table IV. Potassium Channels in Smooth Muscle[a]

	i[94,95]	ii[94]	iii[96,97]
Patch clamp			
Conductance	100–130 pS	50 pS	50 pS
Calcium	Activates	Activates	Insensitive
Tetraethylammonium	Blocks	Insensitive	Blocks
Kinetics	Fast	Fast	Slow
Voltage clamp	i[98]	iv[99]	v[100]
Calcium	Activates	Insensitive	Insenstive
Tetraethylammonium	Blocks	Insensitive	Blocks
4-aminopyridine		Blocks	Insensitive
Kinetics	Fast	Fast	Fast

[a]Characteristics of three potassium channels recorded with physiological concentrations of potassium ion using the patch-clamp technique and of three potassium currents recorded with voltage clamp in smooth muscle preparations.

From the data in Table IV it can be appreciated that, during the upstroke of a calcium action potential, both the depolarization of the membrane and the increase in the cytoplasmic concentration of calcium would activate some potassium channels (i and ii in Table IV), which would, in turn, tend to repolarize the membrane. Channel iii has slow kinetics; i.e., it takes a number of seconds to activate fully and does not seem to inactivate thereafter. This channel may underlie the very slowly activating potassium current that has been recorded with the voltage-clamp technique from *Xenopus* oocytes that had been injected with RNA extracted from the uterus of estrogen-treated rats.[67]

4.3.3. Other Channels

Apart from the channels that may be directly involved in the action potential, other channels can also play a significant role in determining the contractility of smooth muscle by regulating the level of the membrane potential.

It has been known for some time that when many different types of smooth muscle are stretched, the membrane becomes depolarized. Sufficient stretch can result in action potential activity and consequently contraction.[48] A channel has been identified in smooth muscle cells dissociated from the stomach of amphibia which has a conductance of about 33 pico siemens (pS) in normal physiological solution and is equally permeable to both sodium and potassium ions, giving it a reversal potential close to 0 mV.[100a] [The reversal potential is the equivalent of the equilibrium potential (one ion) when several ions are involved.] When the membrane is stretched, the activity of the channel increases, and this would tend to depolarize the membrane. This channel may therefore underly stretch-induced contraction.

A voltage-dependent potassium current (called I_M) has been recorded from dissociated amphibian visceral smooth muscle cells under whole-cell voltage clamp.[101,102] This channel is inhibited by acetylcholine [acting on muscarinic (M) receptors], which results in depolarization of the membrane. Activity of this channel is enhanced by activation of β-adrenoceptors or by analogues of cAMP, and this causes hyperpolarization. Thus, this channel can modify excitability in these cells. If they occur in uterine smooth muscles, the suppression of uterine activity in preterm labor by the administration of β-adrenoceptor agonists could be explained by the activation of these channels.

A chloride conductance, studied in visceral smooth muscles under voltage clamp, appears to be activated by cytoplasmic calcium, e.g., the calcium released from intracellular stores by some agonists.[103] At the single-channel level, a large-conductance (340-pS) chloride channel has been identified in isolated patches from vascular smooth muscle.[104] These channels are insensitive to calcium ions and are only rarely recorded in the cell-attached mode, making their physiological function unclear.[104] Two other chloride channels, with conductances of about 12 pS and >150 pS, have been reported in vascular smooth muscle cells.[105] Activation of a chloride conductance in smooth muscle would tend to depolarize the cell membrane toward threshold for an action potential.

Increasing hyperpolarization of isolated visceral smooth muscle cells results in the increasing activation of a slow inward current that has a reversal potential at -25 mV.[106] The current is thought to be carried by a mixture of sodium and potassium ions[106] and would tend to depolarize the membrane toward threshold for the initiation of action potentials. It is tempting to speculate that this channel may give rise to pacemaker potentials, so common in many visceral smooth muscles, including the myometrium.

4.4. Modulation of Channel Activity

In addition to the channels that are activated solely by voltage, the activity of many channels can be influenced by neurotransmitters and hormones. This action may be direct, for example, the activation of the end-plate channel in striated muscle by acetylcholine.[69] Chemicals may also influence channels indirectly by inducing the production in the cell of modulator substances (second messengers), which in turn either activate the channel or modulate (increase or decrease) activity that is primarily voltage dependent.

Two channels that appear to be activated solely by chemical means have been identified in smooth muscle at the single-channel level.[107,108] Both channels have conductances of 20–25 pS and are permeant to both sodium and potassium ions under physiological conditions. One channel is activated by acetylcholine,[108] and the other channel is activated by ATP.[107] The activation of these channels would tend to depolarize the membrane. Thus, they are likely to be responsible for the depolarization leading to action potential generation during neuroeffector transmission. There is some evidence that suggests that oxytocin and prostaglandin $F_{2\alpha}$ may activate a similar channel in myometrium (see Sections 5.5.2 and 5.5.3).

Cytoplasmic calcium itself can also modify the activity of channels in smooth muscle. This includes enhancement of some potassium and chloride channels. Voltage-dependent calcium channels (L) are at least partially inactivated by calcium ions.

Voltage-dependent calcium channels in smooth muscle may also have their activity modified by other means. For example, in cells from toad stomach, acetylcholine, acting on muscarinic receptors, increases the magnitude of the currents flowing through L-type channels.[90,109] Diacylglycerol analogues known to activate protein kinase C increase the calcium current in this preparation, suggesting that acetylcholine exerts its effect on the calcium channel via diacylglycerol.[90,110] Calcium currents have been reported to be increased by norepinephrine.[89,111,112] Thus, modulator substances can alter the excitability of a cell in a variety of ways. For instance, acetylcholine can increase the probability of reaching threshold for action potentials by inhibiting potassium channels and/or by activating channels that carry a mixture of cations. Once threshold is reached there can be an additional effect of increasing the influx of calcium.

5. Electrical Activity in the Myometrium

In view of the various ionic channels that have been described for smooth muscle, it is hardly surprising that the form of the active response in myometrium is highly variable. Slow changes in potential also occur in the cell membrane and these may have an important influence on excitability in uterine smooth muscle. Such changes serve to bring the membrane potential closer to the threshold for activation of the voltage-dependent channels involved in action potentials.

5.1. Pacemaker Potentials

Rhythmic oscillations in membrane potential have been recorded from uterine smooth muscle[49,114,115] (Fig. 6A). These can take the form of discrete events up to 5 mV in amplitude, several seconds in duration, and arising from the most hyper-polarized level of membrane potential (Fig. 6). It can be appreciated from Fig. 6A that slow oscillations are not associated with changes in tension unless the threshold for an action potential is reached. These are similar to the slow waves recorded from various regions of the gut.[48] Slow changes in membrane potential may take the form of a continuous depolarization[116] similar to the pacemaker potentials in the SA node of the heart. Although all myometrial cells appear capable of generating these slow oscillations, their occurrence at any given time is probably restricted to a few areas

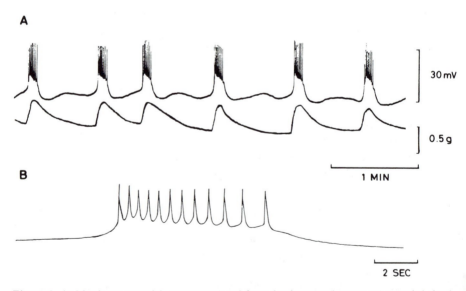

Figure 6. **A:** Membrane potential (upper trace) and force development (lower trace) recorded simultaneously in circular myometrium from a guinea pig on day 31 of pregnancy. Slow oscillations that do not reach threshold for the initiation of action potentials are not associated with contraction. **B:** A burst of spikes from A recorded at a faster chart speed to illustrate prepotentials between spikes within the burst.

within the mass of uterine smooth muscle. In the uterus the pacemaker areas are mobile rather than fixed,[49,117,118] in contrast with the heart.

Slow oscillations are myogenic in nature; that is, they are not generated by extrinsic factors such as nerves or brief changes in circulating hormones but represent an intrinsic property of the smooth muscle cells. The ionic mechanisms underlying the generation of slow oscillations in myometrium have received scant attention. The gradual depolarization observed in longitudinal myometrium of rats appears to be associated with an increase in membrane resistance, as judged from the increase in amplitude of the electrotonic potential recorded at a single site (see Section 3.2.1), and this has been attributed to a decrease in potassium permeability.[50] This interpretation is limited by a lack of knowledge of the current–voltage relationship in the membrane of rat myometrium; in mouse myometrium depolarization alone results in an increase in membrane resistance.[119] In longitudinal myometrium of rats the slow depolarization disappears when external sodium is reduced,[115] suggesting that the slow depolarization results from an increase in membrane permeability involving sodium. It should be pointed out that because of the large electrochemical gradient for sodium and the relatively small permeability of the membrane to it, an increase in sodium permeability sufficient to cause the 5–10 mV depolarization would involve only a small ($<5\%$) increase in the total permeability of the membrane.

More rapid changes in membrane potential have also been identified in myometrium. They directly precede each spike within a burst of action potentials (Fig. 6B) and have been named prepotentials.[50,115,120] The rate of rise of prepotentials dictates the frequency of spikes within a burst. The higher the frequency the greater the rate of rise of the accompanying contraction and possibly the higher the level of tetanus attained. The ionic basis of prepotentials has not been studied.

5.2. Simple Action Potentials

Action potentials occurring in longitudinal myometrium of rats[49,50] and rabbits[62,114] in late pregnancy and in circular myometrium of guinea pigs[59] and sheep[61] have a simple time course consisting of a depolarizing upstroke followed by monotonic repolarization (Fig. 7A). In midgestation a small afterhyperpolarization occurs in longitudinal myometrium of rats.[121]

The ionic mechanisms underlying these action potentials have been elucidated in considerable detail by the Mironneaus and colleagues using the double-sucrose-gap technique to voltage clamp strips of longitudinal myometrium from pregnant rats and, more recently, using a single microelectrode or patch electrode to voltage clamp single cells from this preparation in short-term culture. The inward current mediating the upstroke of the action potential seems to be carried by calcium ions permeating calcium channels similar to those described in other smooth muscle cells (see Section 4.3.1). The channels are voltage dependent such that they first become activated by depolarization to a membrane potential of about -40 mV and are fully activated at membrane potentials more positive than 0 mV.[122,123] The channels are

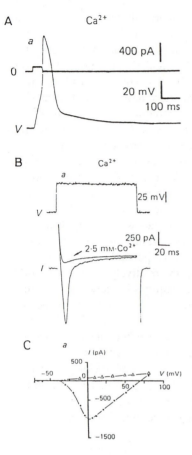

Figure 7. Intracellular recordings from single myometrial cells in short-term culture after dissociation from the longitudinal myometrium of pregnant rats. **A:** An action potential with a simple time course recorded in a solution containing 10 mM calcium. **B:** Calcium currents recorded under voltage clamp in 10 mM calcium (lower trace) and evoked by a depolarizing voltage step (upper trace) in the absence and presence of cobalt (a calcium channel blocker). **C:** Current–voltage relationship for the calcium current in the absence (circles) and presence of cobalt (triangles). (From Amédée *et al.*,[123] with permission.)

also time dependent such that, under voltage clamp, a depolarizing step in the membrane potential sufficient to activate the channels results in an inward current that peaks within about 10 msec and then declines (Fig. 7B). Part of the decline is caused by a voltage-dependent inactivation of the calcium channels. At small depolarizations only a relatively small number of calcium channels enter the inactivated state, and it requires a depolarization to around 0 mV before the maximum number of calcium channels become inactivated.[122,123] There is an additional component of inactivation caused by an accumulation of calcium ions on the inside surface of the membrane.[122,123] Sodium ions can permeate these calcium channels, but only if the external calcium concentration is less than 10^{-6} M.[124] These channels are also permeable to the divalent cations barium and strontium.[123,124] Other divalent cations, particularly cobalt and manganese but also nickel and cadmium, do not readily permeate the channels but bind to sites in the channels so strongly that they block the channel.[124,125] The channels are also largely blocked by the dihydropyridine derivatives nifedipine at 10^{-7} M[123,124] and isradipine at 10^{-8}

M.[125a] The sensitivity of the current to block by dihydropyridines and to inactivation by cytoplasmic calcium have led to the suggestion that the calcium current underlying the action potential is carried predominantly by L-type channels (see Section 4.3.1). It is important to note that the dihydropyridines are potent inhibitors of tension development in uterine smooth muscle.[125b]

In early experiments sodium ions were thought to be at least partially involved in the inward current underlying the upstroke of the spike in longitudinal myometrium of rats.[126–129] These results can be reevaluated in the light of subsequent observations. The contraction accompanying lowering sodium in the solution perfusing this preparation is consistent with stimulation of sodium–calcium exchange in reverse mode, with a consequent increase in cytoplasmic calcium (see Section 2.1.3).[130] This calcium could then partially inactivate the calcium channels as discussed above and result in the depression of the inward calcium current that is observed. Furthermore, in calcium-free solution, sodium ions can pass through the calcium channels (discussed above[124]).

The repolarization phase of the action potential is mediated by an outward current[125,127] carried by potassium ions,[129] and, as with other smooth muscle cells (see Section 4.3.2), it has been suggested that more than one set of channels may be involved in longitudinal myometrium of rats. From studies involving sucrose gap voltage clamp, it appears that one component of the current is partly blocked by tetraethylammonium, and it has been suggested that the current is calcium dependent.[121,131] A more recent study of isolated cells suggests that there is little or no calcium dependence of the outward current in this preparation.[131a] The conclusion of the earlier study[131] relied heavily on analysis of tail currents and could be explained by limitations of the sucrose gap technique.

Another component of the potassium current in longitudinal rat myometrium is most prominent during mid pregnancy. It is quickly activated, voltage dependent, and blocked by both 4-aminopyridine and tetraethylammonium.[121,131] This current is largely inactivated at the resting membrane potential in myometrium of rats in late gestation and can only be maximally activated by depolarization from membrane potentials more negative than -80 mV.[121] In midgestation the resting potential is more negative than in late pregnancy (see Section 3.2.2). As a result, this fast current would be more available and would therefore be activated to a greater extent during the action potential, perhaps explaining the more rapid rate of repolarization of the action potential and the afterhyperpolarization seen at this time.[121] The relationship between this current and the 4-aminopyridine-sensitive outward current recorded from other smooth muscles (see Section 4.3.2) is not clear, since the voltage dependencies appear to differ between the two currents.

5.3. Complex Action Potentials

Not all active responses in uterine smooth muscle take the form of a simple spike action potential. In the circular myometrium of rats,[52,55,60,132] mice,[116] and brush-tailed possums (*Trichosurus vulpecula*)[133] and in longitudinal myometrium of guinea pigs,[63] complex action potentials occur that consist of spikes followed by a sustained plateau of depolarization. Despite the superficial similarities in form, the mechanisms underlying these action potentials in different species and in different reproductive states may not be the same. Of the two preparations studied, free-running voltage recordings alone have been made from circular myometrium of rats. On the other hand, double-sucrose-gap voltage-clamp, single-electrode voltage-clamp, and patch-clamp techniques have been used to study longitudinal myometrium of guinea pigs. Since there is little overlap in the results, the complex action potentials in these two preparations are discussed separately.

5.3.1. Rats

It can be appreciated from Fig. 8 that both the spike and the plateau components of the action potential are most pronounced in circular myometrium from ovariectomized rats that have been injected with estradiol. There is considerable variability in the amplitude of the initial spike, the appearance of oscillations on the plateau component, and the duration of the plateau following estrogen treatment and during pregnancy (Fig. 8).

Following estrogen treatment and during the first 15 days of pregnancy, the level of membrane potential attained during the plateau appears to be consistently around -20 to -30 mV.[52,55,60,134,135] A current step that produces some 10 mV of hyperpolarization when applied between action potentials fails to induce a detectable shift in membrane potential when applied at early times during the plateau,[134,135] and a high membrane conductance is suggested during the early stages of the plateau.[135] Current steps applied later in the plateau result in a prompt return to the resting potential, indicating that voltage- and time-dependent conductances may be involved in the plateau. The amplitude of the spike component is more sensitive to reducing external calcium and to calcium channel blockers verapamil and D-600 than is the plateau.[134,135] As the concentration of calcium in the external medium is increased, the duration of the plateau increases, dictating the duration of the accompanying contraction.[132] The level of the plateau is unaffected by tetraethylammonium even though the latter reduces membrane conductance.[135] This suggests that the membrane is "clamped" at the level of the plateau by a considerable conductance not involving tetraethylammonium-sensitive potassium channels. Replacement of calcium by barium or of sodium by Tris or choline is also without effect on the level of the plateau.[134] The only ion whose external concentration is found to alter its level is chloride; the level increases from around -26 mV in 120 mM chloride to approximately -13 mV within 2 min of reducing chloride to 12 mM.[134]

It is interesting to note that the hormone relaxin abolishes the plateau, leaving the initial spikes intact. This is accompanied by a marked reduction in the duration and amplitude of the contraction.[55] Although the events occurring during the plateau in estrogen-primed myometrium and up to day 16 of pregnancy are far from being understood, it is likely that calcium enters the cell during the initial spike. It is

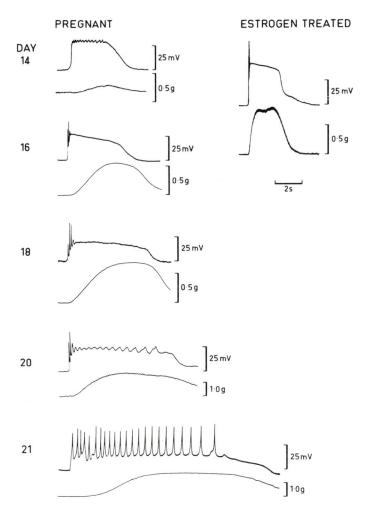

PREGNANT   **ESTROGEN TREATED**

Figure 8. Membrane potential (upper trace in each pair of recordings) and mechanical activity (lower trace) recorded from the circular myometrium of rats throughout pregnancy and following estradiol injection. Membrane potentials were −50 mV on day 14, −56 mV on day 16, −50 mV on day 18, −51 mV on day 20, −47 mV on day 21, and −65 mV following estrogen treatment. Note the decrease in sensitivity of the mechanogram on days 20 and 21. (From Parkington,[52] with permission.)

suggested that this ion is pivotal in the activation of subsequent conductances, as indicated by the calcium dependence of the duration of the plateau. A chloride conductance may be involved,[134] and it is envisaged that the resulting depolarization then activates calcium channels. Not all calcium channels would be inactivated at the level of the plateau (see Section 5.2). This would allow sustained influx of calcium, thereby controlling contraction. Repolarization may be brought about by a decline in the chloride conductance and a more prominent role of a potassium conductance, for which the slowly activated (several seconds[67]) potassium conductance (see Section 4.3.2) seems ideally suited. From day 16 of pregnancy onwards the plateau becomes less pronounced (Fig. 8), and the conductance occurring during the early stages of the plateau decreases. Small oscillations appear on the plateau, and these become larger and more pronounced in the presence of tetraethylammonium as the time of delivery approaches.[135] These results suggest that the conductance that underlies the plateau component declines in late pregnancy and is no longer able to "clamp" the membrane at the depolarized level of the plateau. The calcium and potassium conductances underlying the spike component of the action potential become relatively more significant, so that by the time of delivery the "plateau" appears as a small depolarization on which is superimposed a burst of spikes.

5.3.2. Guinea Pigs

The action potential recorded from longitudinal myometrium of estrogen-treated guinea pigs consists of an initial spike followed by an afterdepolarization of 15–20 mV, which can last for several hundreds of milliseconds.[136,137] This preparation has been studied under voltage clamp in a double-sucrose-gap apparatus.[136,137] During pregnancy the action potential becomes more complex: the afterdepolarization becomes more pronounced and appears as a plateau that can last for many seconds and on which further spikes or oscillations may be superimposed (Fig. 9).[63] During the plateau phase the membrane conductance appears to be high.[63] Some of the ionic currents involved in this complex action potential have been studied in our laboratory using small bundles of uterine smooth muscle cells. Bundles can be obtained that are sufficiently small to be isopotential and can therefore be voltage clamped with a single electrode using a switching amplifier (similar to the method used on arterioles by Hirst and colleagues[76]).

Depolarizing steps activate two inward currents.[77,136,137] A transient inward current is recorded with a time course similar to that recorded in rat myometrium (see Section 5.2), which is blocked by 10^{-6} M nifedipine and does not occur when calcium is omitted from the external medium. This current is most likely responsible for the upstroke of the spike. A slower inward current is recorded in both reproductive states. In estrogen-treated animals it reaches a peak after about 150 msec and inactivates (declines) with a time course of 300–1200 msec, whereas in tissue from pregnant guinea pigs the current is still activating (increasing) at the end of a 1-sec voltage step, suggesting that it may be involved with the plateau component of the action potential.[77] Following the termination of a depolarizing voltage step, the current in both preparations decays with a time constant in the range of 170–200 msec. The ions carrying the current have not been resolved, but the duration of contraction parallels that of the plateau. Both plateau and contraction, as well as the slow inward current, are blocked by 10^{-6} M nifedipine and 1 mM cobalt and are not observed following removal of external calcium, indicating that calcium is involved in carrying or regulating this current.

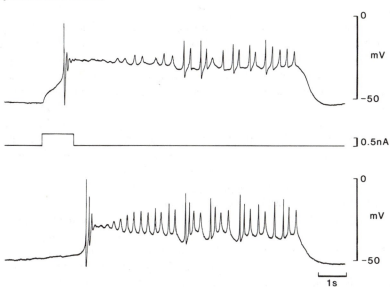

Figure 9. Action potentials with complex time courses recorded with intracellular microelectrodes from the myometrium of a pregnant guinea pig. Upper trace: An action potential evoked by a depolarizing current step. Lower trace: A spontaneous action potential recorded from the same cell. (From Coleman and Parkington,[63] with permission.)

Depolarization of estrogen-treated myometrium activates a transient outward current that is calcium dependent and blocked by tetraethylammonium. The channels conducting the current seem to have relatively fast kinetics since the current develops, peaks, and inactivates (decays) within about 60 msec,[136,137] and they most likely correspond to the calcium-activated channels that have been studied at the single-channel level in the pregnant myometrium (see below). Depolarization from a hyperpolarized potential results in a fast transient outward current that is not completely blocked by tetraethylammonium.[136,137] This may correspond to the outward current recorded from rat myometrium during midpregnancy that is blocked by 4-aminopyridine and is observed only when the membrane is depolarized from a hyperpolarized potential (see Section 5.2). Depolarization to beyond 0 mV results in a large outward current that reaches its maximal value in 2–3 sec.[136,137] The ionic nature of this current is unknown.

Hyperpolarization of the membrane of pregnant myometrium activates an inward current with slow kinetics that is not affected by potassium or calcium channel blockers.[77] This current seems similar to the hyperpolarization-activated current recorded from other visceral smooth muscle cells[106] (see Section 4.3.3). Its modulation, perhaps via biochemical processes within the cell, may be responsible for the slow depolarization that underlies the generation of spontaneous activity observed in longitudinal myometrium.

To elucidate further the ionic currents in uterine smooth muscle, the patch-clamp technique has been used in our laboratory to make single-channel recordings from small bundles of cells similar to those that have been studied with the single-electrode voltage clamp.[63,77,137a] These small bundles have several advantages: (1) Gigaohm seals can be formed without subjecting the cells to enzyme treatment; (2) the cells remain coupled to each other, and this means that even large, long-duration current steps can be recorded in the cell-attached mode without the errors and inaccuracies caused by a large input impedance; and (3) the resting membrane potential can be simultaneously measured with an intracellular microelectrode recording from a nearby cell.[63,77] It has proved possible to make simultaneous recordings for more than an hour with this arrangement, in which single channel currents can be recorded under normal physiological conditions.

At the resting potential, small inward current steps can be recorded. As the membrane potential is made more negative, the probability of a channel being in the open state increases, mainly because of an increase in the frequency of opening. The channels have a conductance of 30–40 pS and an extrapolated reversal potential of -15 to -30 mV. These channels most likely carry a mixture of cations, especially sodium and potassium ions. The voltage dependence of the channels and the averaged response of several hundred steps suggest that they probably correspond to the hyperpolarization-activated currents recorded from these cells under voltage clamp (discussed above).[77,137a]

When the membrane is depolarized, single-channel potassium currents of large amplitude are recorded (Fig. 10).[63] As the membrane is depolarized progressively, the channels open more frequently with an e-fold increase in frequency for every 10–15 mV. On depolarization, the channels tend to stay open longer, with an e-fold increase in the mean open time for every 44–48 mV of depolarization. With physiological potassium gradients, the conductance is 130–170 pS over the membrane potential range of -30 to $+10$ mV.[63] The conductance and voltage dependence of the channels suggest that they correspond to the large-conductance calcium-activated potassium channels studied in other smooth muscles (see Section 4.3.2). When the patch of membrane is repeatedly stepped to a depolarized potential, the averaged response, after leakage current subtraction, gives an outward current that reaches its maximal value within 25 msec. The rise time of the spike in this tissue is 20–30 msec, which means that the channels will be largely activated by the time the spike reaches its peak and may explain why the repolarization phase of the spike (8–15 V/sec) in this tissue is more rapid than its upstroke (5–8 V/sec).[63] Calcium entering during the spike will activate the channels to an even greater degree, thus accentuating the repolarization.

In isolated patches of membrane, single-channel currents are recorded that remain open for many seconds. These are of large conductance, of the order of 420 pS, and seem to be permeant to chloride ions.[63] Channels similar to these have been recorded in patches of membrane isolated from vascular smooth muscle[104] (see Section 4.3.3). The factors that regulate these channels are not clear, and since they have not been recorded in the cell-attached mode, it is difficult to postulate their

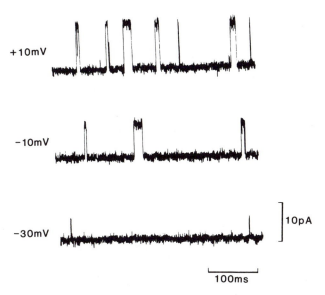

Figure 10. Single-channel currents resulting from the openings of potassium channels. Currents recorded in the cell-attached mode with the patch-clamp technique from myometrium of a pregnant guinea pig at membrane potentials of −30 mV, −10 mV, and 10 mV. As the membrane is progressively depolarized, the current steps become larger and occur with a greater frequency, and the channels tend to remain open for longer times. (From Coleman and Parkington,[63] with permission.)

function, although the possible involvement of chloride during the plateau component in the rat is suggestive of a role for channels such as these.

From our limited knowledge of the ionic mechanisms involved in the complex action potential, a hypothesis can be suggested that attempts to explain our observations in the myometrium of guinea pigs. The upstroke of the initial spike is generated in a manner that is similar to a simple spike (see Section 5.2) and carried by calcium. Its rapid repolarization probably results from activation of large-conductance potassium channels, which are voltage sensitive and calcium activated. Meanwhile, the slow inward current develops, perhaps carried by either chloride ions or calcium. The chloride channels could be activated by the increase in cytoplasmic calcium resulting from the spike (see Section 4.3.3). A combination of calcium or chloride and potassium conductances could clamp the membrane at a depolarized potential; the calcium channels underlying the initial spike would be partially but not totally inactivated. A slow time-dependent inactivation of the inward current underlying the plateau could unclamp the potential from the plateau level, thus allowing sufficient repolarization to partially deinactivate the calcium channels, i.e., make the transition from the inactivated to the closed condition, such that they could reopen (Fig. 4), thus allowing influx of more calcium to activate the potassium channels, resulting in increasing oscillations in calcium and potassium conductances (Fig 9). A voltage dependence of the inward current could result in the eventual rapid repolarization of the membrane and terminate contraction. Given the diversity of channels identified in smooth muscle so far (see Section 4), it would not be surprising if a much larger number of different channel types, including chloride channels, are eventually found to be involved in these complex action potentials.

5.4. Relationship between Membrane Potential and Contraction

An increase in free calcium within the cytoplasm appears to be essential for the initiation of contraction in smooth muscle.[6,139-141] It is useful to consider whether the amount of calcium entering the cell during an action potential, as discussed above, is sufficient in itself to activate the contractile mechanism or whether additional calcium from internal stores is required, so amplifying the transmembrane signal.

The initial experiments in which attempts were made to answer this question involved vascular smooth muscle. In order to evoke a simple spike in mesenteric artery it is necessary that the perfusion solution contain a potassium channel blocker; otherwise the inward calcium current cannot overcome the large outward potassium current (see Section 4.2). In the presence of tetraethylammonium, a depolarizing current step evokes a simple spike that is accompanied by a contraction. However, in the presence of procaine, another potassium channel blocker, the spike is not accompanied by a detectable change in tension.[142] Since procaine is also believed to prevent release of calcium from endoplasmic reticulum, it has been concluded from these experiments that calcium influx during a simple spike alone may be insufficient to initiate contraction and that additional calcium is required to be released from the endoplasmic reticulum.[142] However, procaine is also a local anesthetic and as such may inhibit a variety of ion channels.[138,143,144] In the preparations of mesenteric artery the maximum rate of rise of the spikes in tetraethylammonium and procaine were not measured, and so it is not known whether the amount of calcium entering during the spikes was equivalent in both instances.

The question has been reexamined by a more direct approach. The kinetics of the calcium currents during the spike action potential in smooth muscle cells isolated from urinary bladder have been studied under voltage clamp using patch clamp in whole-cell mode.[145] The amount of calcium entering during a typical spike (70 mV in amplitude and 30 msec in duration) has been calculated by an integration technique, and it is estimated that cytoplasmic calcium increases by 2×10^{-5}M during such a spike. In cerebral arterioles the action potential evoked by a depolarizing current step is more complex, consisting of a transient spike followed by an afterdepolarization[76] reminiscent of the action potential occurring in longitudinal myometrium of estrogen-treated guinea pigs.[136] It has been calculated that calcium entry during depolarization of arteriolar smooth muscle to -35 mV for 1 sec increases intracellular calcium by 10^{-6} M.[76] Although much of the calcium entering the cell during an action potential in smooth muscle may become buffered or removed by transport mechanisms, approximately 2×10^{-6} M has been shown to cause near-maximal contraction in skinned smooth muscle.[142,146] Since some myometrial preparations generate complex action potentials whose duration closely correlates with that of the contraction (see Section 5.3), it is tempting to suggest that transmembrane calcium may be sufficient for contraction in these tissues (for an alternative view see Chapter 5). The contractile response of longitudinal myometrium of guinea pigs to oxytocin in calcium-free solution declines from around 50% of the response in normal solution during midpregnancy to approximately 15% at term.[147] This suggests that there may be a decline in the capacity or the accessibility of the intracellular calcium store at term. There may be an increase in the reliance on transmembrane influx of calcium for contraction. It is interesting that complex action potentials are prominent in this tissue.

5.5. Factors Affecting the Action Potential

5.5.1. Pregnancy

The effect of pregnancy on the shape of the action potential in uterine smooth muscle is perhaps most striking in the circular myometrium of rats, and this preparation has been studied in considerable detail by Marshall and her colleagues. A complex action potential with a prominent plateau is recorded during most of pregnancy in this species. Electrical activity undergoes a progressive transformation on the days preceding delivery, when the complex action potential is replaced by a burst of simple spikes[52,57,60] (Fig. 8). This change in electrical activity is associated with a large increase in the amplitude and duration of the phasic contraction. The contraction associated with a spontaneously occurring complex action potential is small, and many action potentials are recorded in the absence of any change in tension.[51]

It has been considered that calcium influx during the complex action potential might be small and thereby directly responsible for the relatively poor contractile performance.[57] We suggest that the small size of the contraction accompanying the complex action potential in circular myometrium of rats during most of pregnancy may result from limited propagation rather than any inherent property of the action potential or the contractile apparatus, and several experimental observations contribute to the formulation of this view. First, when propagation is enhanced, e.g., following estrogen injection (see Sections 3.3 and 6.2.3), the plateau component of the action potential is responsible for a large contraction.[55] Second, during pregnancy the circular myometrium has the capacity to produce strong contractions when all cells are stimulated simultaneously, e.g., in response to high-potassium solution or to acetylcholine. Contractions thus evoked were found to be equivalent to those generated in longitudinal muscle at all stages of pregnancy.[57] Third, during pregnancy λ is around 3 mm in longitudinal myometrium, and the conduction velocity of action potentials is of the order of 20 cm/sec, whereas in the circular layer λ is 1.3 mm, and conduction velocity is only 0.7 cm/sec.[116] Thus, the apparent asynchrony between membrane potential and tension and the variability in amplitude of contractions suggest that not all cells in a muscle strip are active at the same time during spontaneous activity.

Although the concentration of estrogen in the circulation of rats rises sharply during the final few days of pregnancy,[66] the disappearance of the plateau at this time is unlikely to be a direct consequence of this because estrogen treatment of ovariectomized rats results in a pronounced plateau[55] (Fig. 8). Rather than being directly responsible, estrogen may play a permissive role in the transition from complex to simple action potentials at term.

The transition from complex action potential to a burst of simple spikes appears to be associated with depolarization of the membrane and has been attributed to the stretching effect of the fetuses. This view resulted from two observations. First, the change did not take place in the "nonpregnant" horn of unilaterally pregnant rats. Second, the transition progressed as normal following placental dislocation.[60] It is interesting that the circular myometrium of pregnant rats is more sensitive to stretch than longitudinal muscle; i.e., tonic tension is maintained in response to stretch in the circular but not in the longitudinal layer.[148] It is tempting to suggest that stretch-activated channels (see Section 4.3.3) depolarize the membrane (see Section 5.3.1) and inactivate the channels responsible for conducting the plateau current. Alternatively, a change in the proportion of the various channels or the induction of cellular mediators that alter channel properties may be responsible.

5.5.2. Oxytocin

At term, the myometrium is exquisitely sensitive to the spasmogenic actions of oxytocin (10^{-12}–10^{-10} M). Rabbit uterus *in vivo*[182] and circular myometrium of rats *in vitro*[149] do not respond to oxytocin until late in pregnancy. Preparations of longitudinal myometrium isolated from mice,[119,150] rats,[151,152] and rabbits[62] at term have been used extensively to study the effects of oxytocin on electrical activity. At the lower concentrations, oxytocin has only a small effect on the resting potential, and its most striking effect is on the pattern of contractile activity. In the absence of oxytocin, contractions are variable in amplitude and frequency but become large, uniform in amplitude, and constant in frequency in the presence of low concentrations of oxytocin.[62,150,151] Intracellular recording suggests that oxytocin increases resting membrane conductance,[119] which may represent activation of a channel permeable to sodium and potassium ions, similar to other receptor-activated channels in smooth muscle (see Section 4.4). It also increases the amplitude of the action potential and the number and frequency of action potentials in a burst, all of which result in a larger contraction.[62] Voltage-clamp studies support this evidence in that the inward current underlying the action potential, carried by calcium, is increased.[152]

Higher concentrations of oxytocin cause depolarization, which results in suppression or block of spikes. Tone is elevated and topped with small phasic contractions that occur at an irregular frequency.[150] In contrast to the effects of lower concentrations of oxytocin, the phasic contractions do not always coincide with bursts of spikes recorded intracellularly from individual cells but are more closely correlated with electrical activity recorded extracellularly (in a 1-cm length of tissue).[150] We suggest that the depolarization produced by high concentrations of oxytocin probably allows considerable entry of calcium into the smooth muscle cells, which, as well as causing contraction in the individual cells, may also uncouple intercellular connections, as has been shown to occur in salivary glands[153] (see Chapter 9). Thus, the contractile effort of uncoupled cells would no longer be expected to result in a coordinated phasic response but rather would result in an increase in tone. The repercussions of such actions on myometrial function can be considered. Prior to expulsion of the fetus, the lower levels of oxytocin would result in strong coordinated phasic contractions, as are observed. The passage of the fetus through the cervix would be expected to result in massive release of oxytocin, which would help to provide tonic spasm of the myometrium behind the fetus and reduce loss of blood *post partum*.

5.5.3. Prostaglandin $F_{2\alpha}$

In longitudinal myometrium of pregnant rats the membrane potential response to prostaglandin $F_{2\alpha}$ (PGF) involves an initial hyperpolarization that persists for 5–10 sec, followed by depolarization on which bursts of spikes are superimposed.[154] The depolarization is associated with an increase in tone, and each burst of spikes triggers an additional phasic contraction. The spikes and phasic contractions are abolished by D-600, leaving intact the initial hyperpolarization, the depolarization, and the increase in tone. When external sodium is reduced in the continued presence of D-600, the initial hyperpolarization still occurs, no depolarization is observed, and the increase in tone, though still present, is reduced in amplitude.[154] The persistence of contraction in the presence of D-600 suggests that at least part of the response to PGF involves release of calcium from internal stores. This internally released calcium may also activate calcium-sensitive potassium channels, resulting in hyperpolarization of the membrane. The depolarization is likely to reflect activation of a channel that is permeable to sodium and potassium (see Section 4.4).

As well as having a direct effect on contractility during late pregnancy, PGF has an additional, indirect action in circular myometrium of guinea pigs during early pregnancy.[155–157] In this tissue PGF causes depolarization accompanied by an increase in tone, on which is superimposed complex action potentials, each of which results in a large phasic contraction.[157] Depolarization alone appears to be incapable of evoking an active response or contraction in this tissue. At low concentration, PGF appears to transform a set of calcium channels in the membrane that are fundamentally voltage dependent but require transformation before they can be activated by depolarization. Transformation of these channels does not appear to be achieved by oxytocin or acetylcholine.[157] At higher concentrations PGF also transforms the channels but, in addition, causes depolarization (by the mechanism discussed above), which activates the transformed channels and results in contraction.[157] Intact tissues removed from the animal are spontaneously active and excitable by depolarization, but following removal of the endometrium or perfusion with indomethacin to inhibit PG synthesis, they become inexcitable within about 30 min. This suggests that endogenous PGs of endometrial origin may maintain the excitability of the channels in the intact animal.[157]

5.5.4. Prostaglandin E_1

When studied in the sucrose-gap apparatus, PGE_1 increased the contractile response to depolarizing current steps in longitudinal myometrium of pregnant rats without any change in the inward current.[158] It was concluded that PGE_1 increased contractility solely by releasing calcium from internal stores.[158] Further study is required to elucidate fully the mechanisms involved.

In circular myometrium of guinea pigs during early pregnancy, PGE_1 transforms voltage-dependent calcium channels in a manner similar to PGF^{157} (see Section 5.5.3).

5.5.5. Prostaglandin E_2

The sensitivity of both muscle layers of pregnant mouse myometrium to PGE_2 increases 100-fold between midpregnancy and term.[119] In addition, longitudinal muscle is approximately an order of magnitude more sensitive than the circular layer.

The effects of PGE_2 on the contractile activity resembles those in response to oxytocin in many respects.[119] A small depolarization occurs that is associated with an increase in the frequency of spikes, whose maximum rate of rise is not different from that occurring in control solution. The depolarization does not occur when external sodium is removed and may result from an increase in sodium conductance (see Section 5.1).[119,159]

Prostaglandin E_2 also transforms calcium channels in circular myometrium of guinea pigs.[157]

6. Spread of Excitation

6.1. The Syncytial Nature of Smooth Muscle and Cell-to-Cell Coupling

Coordinated contraction of uterine smooth muscle is essential for successful expulsion of the fetus during labor. In order to achieve this, many smooth muscle cells must contract at more or less the same time. Electrophysiological studies support the notion that most smooth muscle tissues behave as a syncytium with low-resistance pathways between adjacent cells (see Section 3.2). This was first demonstrated in 1961 when strips of frog stomach were polarized at one end by large extracellular electrodes and it was found that λ was of the order of $1-2$ mm.[42] The fact that the smooth muscle cells were so much shorter than λ suggested that the cells were electrically coupled. In subsequent experiments, current was injected intracellularly into intestinal smooth muscle cells and the voltage response recorded in a nearby cell via a second microelectrode 4 μm to 700 μm away. With this arrangement, the electrotonic potential decayed sharply with distance, and λ appeared to be less than one cell length. Furthermore, the input resistance was very small, and the time course of the electrotonic potential was reduced some tenfold,[160,161] which suggested that membrane resistance was very low in smooth muscle. Resolution of these ambiguous results was achieved by Tomita[162] when he compared the observations following extracellular polarization with large electrodes with the responses to intracellular current injection in the same tissue. He concluded that the reduction in λ and the time course of the electrotonic potential could be explained in terms of the syncytial nature of smooth muscle. Because of the low resistance between cells, current injected at a point in a syncytium is presented with a large area of low resistance membrane across which the current is rapidly dissipated. In contrast, polarization of the entire cross-sectional area of a smooth muscle strip with large extracellular electrodes forces current to flow in one dimension only, as discussed (see Section 3.2).

6.1.1. Longitudinal Tissue Resistance

The ease with which current can flow through a tissue can be estimated from measurements of τ_m and λ as discussed in Section 3.2.1. However, the value of R_i derived from these parameters represents the sum of cytoplasmic and junctional resistances (R_c and R_j, respectively). The relative contributions of R_c and R_j to R_i can be measured by applying alternating current to strips of muscle. The extracellular medium is replaced with isotonic sucrose on the assumption that the current applied to the strip will only flow intracellularly, through the cytoplasm and low-resistance intercellular pathways. Alternating current is applied at a range of frequencies. The cytoplasm presents a pure resistance, while the intercellular junction consists of a resistance and capacitance in parallel. At high frequencies the current flows across the junctional capacitance, bypassing R_j and yielding an estimate of R_c. At the lower frequencies junctional capacitance forces current to pass through R_j, and thus the sum of R_c and R_j is measured.[41,65]

R_c remains unchanged during pregnancy and in labor[65] or following steroid treatment.[163] This is consistent with reports that the ionic composition of cytoplasm appears to remain constant irrespective of reproductive state.[32,34] In contrast, R_j decreases during delivery[65] and also following estrogen treatment.[163] Thus, electrophysiological evidence, whether measuring spread of direct current λ (see Section 3.2) or longitudinal resistance to alternating current flow, supports the notion that myometrial cells are electrically coupled to form a functional syncytium.

6.1.2. The Anatomic Correlate for Low-Resistance Pathways

Gap junctions have assumed particular prominence regarding their possible role in electrical and chemical communication between cells[164,165] (see also Chapter 9). The membranes between adjacent cells in the region of a gap junction are separated by a distance of only 2–3 nm, and both membranes and the gap are spanned by large protein oligomers that form channels that provide continuity between the cytoplasms of both cells and are reminiscent of the ion channels discussed above. In the open configuration junctional channels (connexons) are some 2 nm in diameter, are permeable to molecules with a nominal diameter of 1.2 nm, and can be open or closed as a result of movement of the oligomers. In lacrimal gland and cardiac cells the conductance of a single junctional channel appears to be of the order of 50–170 pS.[166–169] It has been demonstrated in salivary gland and cardiac cells that the conductance state of the junctional channels can change such that the total current flowing from one cell to another is reduced when cytoplasmic calcium increases or pH decreases.[153,167,168] Such changes may take place in other tissues.

Whether gap junctions are the morphological correlate for the low-resistance pathways in smooth muscle has not been unequivocally demonstrated. Most longitudinally arranged muscles in the gut contain few such junctions,[170–172] but they are abundant in the circular muscle,[170,172] although passive spread of current occurs in both muscle layers. Gap junctions are scarce in myometrium except during labor. In a careful comparison of passive spread of current and the occurrence of gap junctions in rats, Sims and co-workers found that while R_j decreased by about 40% during labor, the number of junctions increased 11-fold, and their area increased 50-fold in samples from the same animals.[65] This correlation may seem disappointing, but a precise comparison between current spread and gap junctions is difficult because gap junctions are observed in histological sections irrespective of where they occur in the three-dimensional preparation, whereas current spread is measured longitudinally in the one-dimensional cable model of smooth muscle (see section 3.2). In such a preparation only some of the radially situated junctions will contribute to longitudinal spread of current, and this proportion is not known.

6.2. Spread of Activity throughout the Uterus

6.2.1. Propagation of Action Potentials

In order for an action potential to propagate, the current generated in the local region of its initiation must be sufficient to depolarize the membrane in the surrounding area to reach threshold for the initiation of an action potential in the quiescent region. When a propagating action potential reaches a region of increased membrane area, it is possible that the current density may be insufficient to depolarize the larger area of the membrane to threshold and the current will decay passively (see Section 3.2). In a nerve cell this can occur where an axon or dendrite branches or at the junction between an axon or dendrite and the soma.[173]

Propagation of spikes is constant in a uniform cable.[174] However, the myometrium approximates a cable only under the very restricted conditions of radial polarization of a strip by large extracellular electrodes (see Sections 3.2.1 and 6.1). Furthermore, the membrane properties, e.g., membrane resistance, are not necessarily uniform, as evidenced by the occurrence of spontaneous depolarizations in isolated regions (pacemaker activity that may be associated with changes in membrane permeability), and these have a tendency to move around the tissue. In addition, the myometrium is not structurally uniform and spontaneous activity originating in a pacemaker region faces structural heterogeneities at several levels.

First, the myometrium consists of small cells that are connected in a three-dimensional syncytium. Heterogeneities at the most fundamental level result from the requirement of current to spread into neighboring cells in the restricted areas of the intercellular junctions.

Second, the cells are arranged into discrete bundles, which are especially prominent in the longitudinal layer, and these may constitute another level of cable. Bundles give off frequent branches that become part of neighboring bundles, creating further areas of discontinuity in the electrical load encountered by a propagating action potential. Consider the case of a spike in a bundle that gives off a small branch a few cells in diameter that is destined to join another large bundle. As the spike approaches the region of reduced diameter, the current density may initially increase. In such a case the amplitude and maximum rate of rise of the spike will increase because of the increase in density of the current in the narrow region of reduced membrane area (Fig. 11A). The three parameters will return to the original values with distance from the branch point, provided that the membrane properties are not different, because in this case the action potential will provide the same current density as in the original cable. At the junction between the branch and the large bundle these parameters will decrease because of a decrease in density of the membrane current (Fig. 11B). If the diameter of the large bundle is sufficiently large, then propagation will fail.

Third, the muscle layers of the uterus are connected by strands of smooth muscle that pass between the blood vessels and connective tissue, thus providing the physical link between the two muscle layers. The branching provides additional heterogeneity.

The effects of inhomogeneities on spike propagation have been studied comprehensively in relation to the cardiac syncytium in which conduction velocity, conduction failure, and direction of propagation have been explained in terms of tissue geometry.[175,176] That propagation failure occurs in the myometrium is indicated by the spikelike events, usually of small and variable amplitude, that can be recorded.[45,61]

6.2.2. Effect of Junctional Resistance on Propagation

It is commonly assumed that an increase in cell-to-cell coupling, resulting in a reduction in junctional resistance, must necessarily improve propagation and hence coordination of uterine contraction. However, simulations of propagation through electrically coupled cells in cardiac muscle indicate that, particularly in the region of a pacemaker, the successful initiation and propagation of action potentials depend on a balance among a number of conflicting factors.[177] For example, a group of

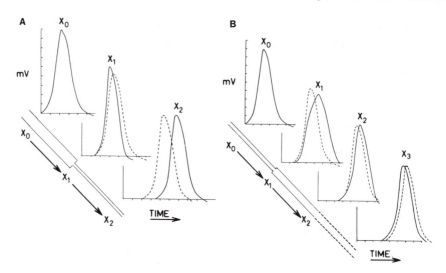

Figure 11. Theoretical considerations of the fate of an action potential propagating along a cable that undergoes a change in diameter (solid lines). The dotted lines represent the situation if no such change has occurred. A: At point X_1, where the diameter decreases, the peak is earlier and higher, and the half-width is reduced. After the action potential passes X_1, it returns to its initial shape in the time domain, but it travels more slowly, and an increased latency is seen. B: At point X_1, where the diameter increases, the peak is later, the amplitude falls (if great enough, there is failure of propagation), and the half-width is wider. At X_2, the action potential returns to its original form but travels faster, and a decreased latency is seen. Further along the cable, at X_3, the action potential occurs at an earlier time because of its increased velocity. (Modified from Goldstein and Rall,[174] with permission. Reproduced from the *Biophysical Journal* by copyright permission of the Biophysical Society.)

cells with pacemaker properties may need to be poorly coupled to nonpacing cells in order to reduce the area of membrane such that the current density is sufficient to bring the membrane potential to threshold for the initiation of spikes within the pacemaker region. Subsequently, this poor coupling can then make it difficult for the pacemaker cells to initiate action potentials in the surrounding cells. Whether a group of cells, i.e., the pacemaker region, can produce action potentials that then successfully propagate may also depend on the size of the pacemaker region.[177]

Consider also the situation where a small bundle joins a larger bundle and thus faces an increased electrical load because of the larger membrane area. The extent of electrical coupling between the cells in the region where the two bundles join is critical: an increase or decrease in this coupling, outside of the critical values, will result in a failure of an action potential to propagate.[178]

6.2.3. Conduction Velocity

Apparent conduction velocity of extracellularly recorded spikes evoked in longitudinal myometrium of whole strips of uterus isolated from rats in late gestation is 9 cm/sec prior to delivery and increases to 11 cm/sec during delivery.[179] Values of 8 and 13 cm/sec, respectively, have been observed in the myometrium of conscious ewes in which the myometrium is stimulated during quiescent periods via indwelling platinum electrodes.[180] Estrogen treatment of nonpregnant rats increases conduction velocity in longitudinal myometrium from 1 cm/sec to 5 cm/sec.[181] These changes in conduction velocity may be partly explained by the better spread of current that occurs under these conditions, as measured by the changes in the one-dimensional length constant, λ (see Section 3.2) and in junctional resistance (see Section 6.1.1).

Spread of activity is much reduced in the circular orientation, and spikes evoked in the myometrium of conscious ewes are recorded at 1 cm but not at 2 cm from the stimulating electrode.[180] A similar pattern is observed in the myometrium of pregnant mice, in which apparent conduction velocity is 20 cm/sec in the longitudinal orientation but only 0.7 cm/sec in the circular direction.[116]

6.2.4. Interaction between Muscle Layers

The wall of most hollow viscera, including the uterus, contains an outer longitudinally arranged muscle layer and an inner circular layer. Electrical coupling between the two muscle layers of the uterus has been studied in rats, mice, and guinea pigs using an L-shaped preparation, each arm consisting of one layer free of the other, while in the angle the layers remain in their normal association. Mechanical activity recorded in both muscle components of rats and mice appeared to occur synchronously.[116,148]

In our laboratory a similar study has been made of coupling between the two layers of the myometrium in guinea pigs during late pregnancy. Circular muscle is quiescent when detached from the longitudinal component, but spontaneous activity originating in the longitudinal layer in an L-shaped preparation results in action potentials and contractions in the circular component also. However, although each spontaneous contraction persists for 3–5 min in longitudinal muscle, the accompanying response in the circular layer is transient. Despite the continuous train of spikes in the longitudinal component, it is associated with only a single spike or brief burst of spikes in the circular layer. It appears that, although electrical activity in the myometrium can and does spread from one muscle layer to the other, the experimental evidence available suggests that the barriers in this region may be considerable.

7. Studies in Vivo

7.1. Pattern of Activity throughout Pregnancy

Phasic contractions of low amplitude occur throughout pregnancy in many species,[183–190] but this has been most extensively studied in sheep and rats. Activity recorded via extracellular electrodes has been used to correlate electromyographical (EMG) activity with changes in intrauterine pressure in conscious ewes from the earliest stages of pregnancy up to and including expulsion of the fetus.[187] During the first 1–2 months of pregnancy, EMG activity is low in amplitude, and activity recorded at various sites on the uterus is not synchronous.[191] This suggests limited propagation of activity in the early stages of pregnancy. Between days 50 and 100 of pregnancy a definite pattern of activity develops that is characterized by a grouping of the EMG spikes into bursts that have durations of around 10 min and a frequency of 1–2/hr.[191] Each burst recorded at various sites on the uterus occurs at approximately the same time and is associated with a small increase in intrauterine pressure. This pattern of activity persists until 30–70 hr prior to delivery (around day 145).[187] At no time is there any consistency in the location at which activity seems to be generated,[180] thus confirming observations *in vitro* that the pacemaker in myometrium is not restricted to a specific region but moves around the organ. The random location of pacemaker activity has also been observed in pigs.[184] In contrast, it has been claimed that synchronous activity originates at the ovarian end of the uterus during labor in rats.[192]

The onset of labor in sheep can be recognized by a decrease in duration and increase in frequency of EMG bursts[180,187,190] (Fig. 12). There is an increase in the amplitude of the EMG activity and in the amplitude of the pressure pulse accompanying each burst, suggesting an increase in coordination of activity as delivery approaches.

It is interesting to consider the relationship between individual EMG spikes recorded at various sites on the uterus. During most of pregnancy, correlation is observed between spikes within a burst recorded by two sets of electrodes separated by 3 cm or less. Surprisingly, this association disappears during labor, at which time activity seems to become chaotic.[180] Two possibilities come to mind that might

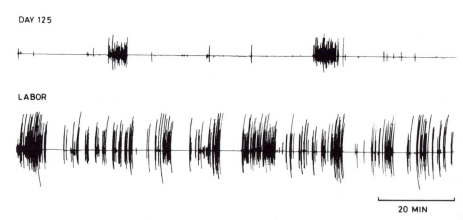

Figure 12. Electromyographic (EMG) activity recorded with extracellular electrodes from the myometrium of a conscious ewe. Upper trace: On day 125 of pregnancy bursts occur approximately once per hour. Lower trace: During labor in the same ewe, bursts of EMG activity occur more frequently but are of shorter duration.

explain this transition. If propagation were increased during labor, as suggested by increased conduction velocity, this, together with the complex anatomic arrangement of uterine smooth muscle, could give rise to apparent chaos. This could result from either propagation failure (see Section 6.2) or spread of activity from a greater number of pacemakers. Since there is an increase in the number of spikes, the latter possibility appears more likely. Alternatively, it is possible that uterine contractions observed during labor in the intact animal may be extrinsically generated (e.g., by circulating hormones), giving rise to activity at each recording site that is independent of propagation (see Section 7.2.2).

7.2. Comparison between Observations in Vivo and in Vitro

7.2.1. Pregnancy

Much is known about the EMG activity of the myometrium in pregnant sheep and how such activity relates to intrauterine pressure. In our laboratory we were in the unique situation of being able to compare spontaneous activity in isolated strips of myometrium using intracellular microelectrodes with EMG activity recorded in the uterus of the same animal immediately prior to its isolation.[61] Recordings *in vitro* suggest that electrical coupling between smooth muscle cells is poor prior to day 90 of pregnancy,[61] and this corroborates the conclusion drawn from observations *in vivo*.[191] From day 100 of pregnancy to term, excluding labor, myometrial strips are quiescent for the initial 30–60 min following isolation, and then bursts of action potentials, each accompanied by a phasic contraction, occur continuously for the remainder of the experimental period.[61] The frequency of the bursts is higher in circular myometrium (0.6/min) than in longitudinal muscle (0.2/min),[193] but in both it is much higher than spontaneous bursts of EMG activity that are observed in intact ewes (1–2/hr). These results suggest that myometrial activity is under tonic inhibition *in vivo* by factor(s) that are washed out *in vitro* within 30–60 min. This raises two questions. First, what is the nature of the tonic inhibition, and second, what factors may be involved in generating the bursts of activity *in vivo*?

In attempts to elucidate these questions, segments of myometrium, fixed to a frame to maintain original dimensions, have been translocated to an omental fold. The frequency of EMG bursts in isolated segments is close to that observed in the intact uterus of the same animal,[194–196] and this is consistent with the view that a circulating factor is responsible for maintaining tonic inhibition of uterine activity *in vivo*. Close examination of individual bursts of EMG activity shows that there is no temporal association between bursts, either between isolated segments and intact uterus or between different isolated segments in the same animal.[194–196] Thus, a circulating factor is unlikely to be responsible for the initiation of the bursts of EMG activity, in contrast to the inhibition. Local production of prostaglandins may be involved, since infusion of the PG synthesis inhibitors indomethacin[193] or 4-aminoantipyrine[197] reduces the frequency of spontaneous bursts of EMG activity by about half. Alternatively, the bursts may be purely myogenic in origin. Clearly, much work remains to be done in this area.

7.2.2. During Labor

A period of quiescence occurs in the circular myometrium of rats on the last day of pregnancy, prior to delivery.[57] The membrane potential is more negative at this time (see Fig. 2), suggesting that the permeability of the membrane may be altered. The precise temporal relationship of the quiescent period to labor has not been established. Strips of circular myometrium isolated from ewes during the early stages of labor remain quiescent for hours in the organ bath.[61] In contrast, in intact ewes, bursts of EMG activity and the associated increase in intrauterine pressure are large in amplitude and of high frequency.[180,187] It is difficult to follow individual spontaneous spikes from one electrode to the next, even when electrode separation is small. This is in contrast to the one-to-one correspondence between spikes recorded by the same array of electrodes immediately before labor. We suggest that the apparent chaos during labor may result from an external factor (in contradistinction to a local factor or of myogenic origin) that generates activity in a muscle that is fundamentally quiescent. Strips of myometrium are again spontaneously active during the later stages of labor, and the duration and frequency of bursts *in vitro* are remarkably similar to those occurring in the intact animal immediately prior to their isolation. It is possible that oxytocin may be responsible for the activity *in vivo* and also in the isolated strips; exposure of isolated myometrium from animals late in pregnancy to this peptide results in activity that persists for hours (in rats, guinea pigs, and sheep).

8. Concluding Remarks

In this chapter we have discussed the control of uterine motility in terms of ion channels in the membrane of myometrial cells. Many of these channels are primarily voltage dependent such that the probability of their being in the open, conducting state is determined by membrane potential. Thus it is clear that membrane potential is important in determining contractility in uterine smooth muscle.

Although calcium channels are the fundamental units regulating calcium influx, their study in isolation is not sufficient for an understanding of uterine function. For instance, other channel types exist that provide conductance pathways for ions such as potassium, sodium, and chloride, which would control membrane potential and thus regulate activity of the calcium channels because of their voltage dependence. Thus, it is the totality of channel activity over the membrane of the whole cell that gives rise to the resting membrane potential, the action potential, and hence contraction. The diversity of ion channels provides scope for subtle (as well as gross) regulation of smooth muscle cell function by a variety of hormones. Additional modulation of activity is possible as a result of induction by hormones of intracellular mediators that may enhance or inhibit channel activity.

Transport mechanisms may also assume importance in regulating uterine motility by providing alternative routes for calcium entry other than conventional ion channels, by contributing to the level of membrane potential, and via involvement in calcium removal from the cytoplasm. The experimental evidence suggests that the relative importance of such mechanisms may vary from one smooth muscle to another. It may also be that the mechanisms of calcium extrusion in myometrium vary with reproductive state.

Events at the level of the organ must be considered since coordination of individual smooth muscle cells is essential for successful expulsion of the fetus during labor. Furthermore, the uterus of the intact animal is exposed to a myriad of local and circulating agents that may modulate channel activity and thus alter the behavior of the smooth muscle. This is exemplified by the differences that occur in the activity recorded *in vitro* compared with that *in vivo*. The complexity of the system therefore requires an integrated approach to its study.

However, in considering the actions of hormones, many of which may be involved in subtle regulation and fine tuning of myometrial activity, it must be borne in mind that, even for agents recognized as major regulators of myometrial activity, such as the steroids, prostaglandins, and oxytocin, surprisingly little is known about their mechanism of action at the cellular level.

Conventional electrophysiological techniques have contributed much to what we understand of the events underlying uterine motility. It is envisaged that further application of the most recent innovations in electrophysiology, e.g., single-electrode voltage-clamp and patch-clamp techniques, will clarify further the nature of the control of membrane potential and calcium entry in myometrium. However, we feel that the greatest advances will only be achieved by an integrative approach involving biochemistry, molecular biology, and electrophysiology applied across the spectrum from molecule to membrane, cell, tissue, and the intact organism.

ACKNOWLEDGMENTS. The authors thank Professor M. E. Holman, Professor G. D. Thorburn, Dr. G. Jenkin, Dr. I. McCance, Dr. R. Harding, and Dr. R. Lang for fruitful discussion and comments on the manuscript. This work was supported by the National Health and Medical Research Council of Australia, grant numbers 850119 and 860865.

References

1. Skou, J. C., 1957, The influence of some cations on the adenosine triphosphatase from peripheral nerves, *Biochim. Biophys. Acta* **23**:394–401.
2. Rossier, B. C., Geering, K., and Kraehenbuhl, J. P., 1987, Regulation of the sodium pump: How and why? *Trends Biochem.* **12**:483–487.
3. Casteels, R., Droogmans, G., and Hendrickx, H., 1973, Active ion transport and resting potential in smooth muscle cells, *Phil. Trans. Roy. Soc. Lond.* **265**:47–56.
4. Taylor, G. S., Paton, D. M., and Daniel, E. E., 1970, Characteristics of electrogenic sodium pumping in rat myometrium, *J. Gen. Physiol.* **56**:360–375.
5. Hirst, G. D. S., and van Helden, D. F., 1982, Ionic basis of the resting potential of submucosal arterioles in the ileum of the guinea-pig, *J. Physiol. (Lond.)* **333**:53–67.

6. Himpens, B., and Somlyo, A. P., 1988, Free-calcium and force transients during depolarization and pharmacomechanical coupling in guinea-pig smooth muscle, *J. Physiol. (Lond.)* **395:**507–530.

7. Pritchard, K., and Ashley, C. C., 1987, Evidence for Na^+/Ca^{++} exchange in isolated smooth muscle cells: A fura-2 study, *Pflugers Arch.* **410:**401–407.

8. Schatzmann, H. J., and Vincenzi, F. F., 1969, Calcium movements across the membrane of human red cells, *J. Physiol. (Lond.)* **201:**369–395.

9. Wuytack, F., Raeymaekers, L., De Schutter, G., and Casteels, R., 1982, Demonstration of the phosphorylated intermediates of the Ca^{2+} transport ATPase in a microsomal fraction and in a $(Ca^{2+} + Mg^{2+})$-ATPase purified from smooth muscle by means of calmodulin affinity chromatography, *Biochim. Biophys. Acta* **693:**45–52.

10. Raeymaekers, L., Wuytack, F., Eggermont, J., De Schutter, G., and Casteels, R., 1983, Isolation of plasma-membrane fraction from gastric smooth muscle. (Comparison of the calcium uptake with that in endoplasmic reticulum), *Biochem. J.* **210:**315–322.

11. Wuytack, F., De Schutter, G., Verbist, J., and Casteels, R., 1983, Antibodies to the calmodulin-binding Ca^{2+}-transport ATPase from smooth muscle, *FEBS Lett.* **154:**191–195.

12. Verbist, J., Wuytack, F., Raeymaekers, L., and Casteels, R., 1985, Inhibitory antibodies to plasmalemmal Ca^{2+}-transporting ATPases. Their use in subcellular localization of $(Ca^{2+} + Mg^{2+})$-dependent ATPase activity in smooth muscle, *Biochem. J.* **231:**737–742.

13. Wuytack, F., De Schutter, G., and Casteels, R., 1981, Purification of $(Ca^{2+} + Mg^{2+})$-ATPase from smooth muscle by calmodulin affinity chromatography, *FEBS Lett.* **129:**297–300.

14. Reuter, H., and Seitz, N., 1968, The dependence of calcium efflux from cardiac muscle on temperature and external ion composition, *J. Physiol. (Lond.)* **195:**451–470.

15. Baker, P. F., Blaustein, M. P., Hodgkin, A. L., and Steinhandt, R. A., 1969, The influence of calcium on sodium efflux in squid axons, *J. Physiol. (Lond.)* **200:**431–458.

16. Rasado-Flores, H., and Blaustein, M. P., 1987, Na/Ca exchange in barnacle muscle cells has a stoichiometry of $3Na^+ : Ca^{2+}$, *Am. J. Physiol.* **252:**C499–C504.

17. Mullins, L. J., and Brindley, F. J., 1975, Sensitivity of calcium efflux from squid axons to changes in membrane potential, *J. Gen. Physiol.* **65:**135–152.

18. Mullins, L. J., 1979, The generation of electric currents in cardiac fibers by Na/Ca exchange, *Am. J. Physiol.* **236:**C103–C110.

18a. Masahashi, T., and Tomita, T., 1983, The contracture produced by sodium removal in the non-pregnant rat myometrium, *J. Physiol. (Lond.)* **334:**351–363.

18b. Matsuzawa, K., Masahashi, T., Kihira, M., and Tomita, T., 1987, Contracture caused by sodium removal in the pregnant rat myometrium, *Jpn. J. Physiol.* **37:**19–31.

19. Aickin, C. C., 1986, Intracellular pH regulation by vertebrate muscle, *Annu. Rev. Physiol.* **48:**349–361.

20. Aickin, C. C., 1985, The effect of Na^+ and HCO_3^- ions on recovery from an acid load in the smooth muscle of guinea-pig ureter, *J. Physiol. (Lond.)* **368:**80P.

21. Weissberg, P. L., Little, P. J., Cragoe, E. J., and Bobik, A., 1987, Na–H antiport in cultured rat aortic smooth muscle: Its role in cytoplasmic pH regulation, *Am. J. Physiol.* **253:**C193–C198.

21a. Wray, S., 1988, Smooth muscle intracellular pH: measurement, regulation, and function, *Am. J. Physiol.* **254:**C213–C225.

22. Aickin, C. C., and Brading, A. F., 1982, Measurement of intracellular chloride in guinea-pig vas deferens by ion analysis, ^{36}Cl efflux and microelectrodes, *J. Physiol. (Lond.)* **326:**139–154.

23. Aickin, C. C., and Brading, A. F., 1983, Towards an estimate of chloride permeability in the smooth muscle of guinea-pig vas deferens, *J. Physiol. (Lond.)* **336:**179–197.

24. Aickin, C. C., and Vermue, N. A., 1983, Microelectrode measurement of intracellular chloride activity in the smooth muscle cells of guinea-pig ureter, *Pflugers Arch.* **397:**25–28.

25. Gerstheimer, F., Muhleisen, M., Nehring, D., and Kreye, V. A., 1987, A chloride–bicarbonate exchanging anion carrier in vascular smooth muscle of the rabbit, *Pflugers Arch.* **409:**60–66.

26. Brading, A. F., 1987, The effect of Na ions and loop diuretics on transmembrane ^{36}Cl fluxes in isolated guinea-pig smooth muscles, *J. Physiol. (Lond).* **394**:86P.
27. Aickin, C. C., 1987, Na, K, Cl co-transport is involved in Cl accumulation in the smooth muscle of isolated guinea-pig vas deferens, *J. Physiol. (Lond.)* **394**:87P.
28. Brading, A. F., 1981, Ionic distribution and mechanisms of transmembrane ion movements in smooth muscle, in: *Smooth muscle: An Assessment of Current Knowledge* (E. Bülbring, A. F. Brading, A. W. Jones, and T. Tomita, eds.), Edward Arnold, London, pp. 65–92.
29. Casteels, R., 1969, Calculation of the membrane potential in smooth muscle cells of the guinea-pig's taenia coli by the Goldman equation, *J. Physiol. (Lond.)* **205**:193–208.
30. Brading, A. F., 1971, Analysis of the effluxes of sodium, potassium and chloride ions from smooth muscle in normal and hypertonic solutions, *J. Physiol. (Lond.)* **214**:393–416.
31. Casteels, R., 1969, Ion content and ion fluxes in the smooth muscle cells of the longitudinal layer of the guinea-pig's vas deferens, *Pflugers Arch.* **313**:95–105.
32. Casteels, R., and Kuriyama, H., 1965, Membrane potential and ionic content in pregnant and non-pregnant rat myometrium, *J. Physiol. (Lond.)* **177**:263–287.
33. Hamon, G., Papadimitriou, A., and Worcel, M., 1976, Ionic fluxes in rat uterine smooth muscle, *J. Physiol. (Lond.)* **254**:229–243.
34. Bülbring, E., Casteels, R., and Kuriyama, H., 1968, Membrane potential and ion content in cat and guinea-pig myometrium and the response to adrenaline and noradrenaline, *Br. J. Pharmacol.* **34**:388–407.
35. Aickin, C. C., Brading, A. F., and Burdyga, V., 1984, Evidence for sodium–calcium exchange in the guinea-pig ureter, *J. Physiol. (Lond.)* **347**:411–430.
36. Aickin, C. C., and Brading, A. F., 1985, Advances in the understanding of transmembrane ionic gradients and permeabilities in smooth muscle obtained by using ion-selective micro-electrodes, *Experientia* **41**:879–887.
37. Loutzenhiser, R., Leyten, P., Saida, K., and van Breemen, C., 1984, Calcium compartments and mobilization during contraction of smooth muscle, in: *Calcium and Contractility: Smooth Muscle* (A. K. Grover and E. E. Daniel, eds.), Humana Press, Clifton, NJ, pp. 61–92.
38. Casteels, R., and van Breemen, C., 1975, Active and passive Ca^{2+} fluxes across cell membranes of the guinea pig taenia coli, *Pflugers Arch.* **359**:197–207.
39. Cole, K. S., 1968, *Membrane, Ions and Impulses,* University of California Press, Berkeley.
40. Woodbury, J. W., White, S. H., Mackey, M. C., Hardy, W. L., and Chang, D. B., 1970, Bioelectrochemistry, in: *Physical Chemistry IXB* (H. Eyring, D. Henderson, and W. Jost, eds.), Academic Press, New York, pp. 904–985.
41. Bennett, M. R., 1972, *Autonomic Neuromuscular Transmission,* Cambridge University Press, Cambridge.
42. Shuba, M. F., 1961, Electrotonus in smooth muscle, *Biofizika* **6**:56–64.
43. Hodgkin, A. L., and Rushton, W. A. H., 1946, The electrical constants of a crustacean nerve fibre, *Proc. R. Soc. Lond. [Biol.]* **133**:444–479.
44. Abe, Y., and Tomita, T., 1968, Cable properties of smooth muscle, *J. Physiol. (Lond.)* **196**:87–100.
45. Parkington, H. C., 1983, Electrical properties of the costo-uterine muscle of the guinea-pig, *J. Physiol. (Lond.)* **355**:15–27.
46. Burnstock, G., and Holman, M. E., 1960, Autonomic nerve–smooth muscle transmission, *Nature* **187**:951–952.
47. Holman, M. E., and Surprenant, A., 1979, Some properties of the excitatory junction potentials recorded from saphenous arteries of rabbits, *J. Physiol. (Lond.)* **287**:337–351.
48. Holman, M. E., 1968, Introduction to electrophysiology of visceral smooth muscle, in: *Handbook of Physiology—Alimentary Canal,* American Physiological Society, Bethesda, pp. 1665–1708.
49. Marshall, J. M., 1959, Effect of estrogen and progesterone on single uterine muscle fibers in the rat, *Am. J. Physiol.* **942**:935–942.

50. Kuriyama, H., and Suzuki, H., 1976, Changes in electrical properties of rat myometrium during gestation and following hormonal treatments, *J. Physiol. (Lond.)* **260:**315–333.
51. Anderson, G. F., Kawarabayashi, T., and Marshall, J. M., 1981, Effect of indomethacin and asprin on uterine activity in pregnant rats: Comparison of circular and longitudinal muscle, *Biol. Reprod.* **24:**359–372.
52. Parkington, H. C., 1983, The electrical and mechanical activity of the circular myometrium during pregnancy, *Proc. Aust. Physiol. Pharmacol. Soc.* **14:**284–292.
53. Kanda, S., and Kuriyama, H., 1980, Specific features of smooth muscle cells recorded from the placental region of the myometrium of pregnant rats, *J. Physiol. (Lond.)* **299:**127–144.
54. Osa, T., and Ogasawara, T., 1984, Changes in the adrenergic effects and membrane activity of the circular muscle of the rat uterus during late postpartum, *Jpn. J. Physiol.* **34:**113–126.
55. Chamley, W. A., and Parkington, H. C., 1984, Relaxin inhibits the plateau component of the action potential in the circular myometrium of the rat, *J. Physiol. (Lond.)* **353:**51–65.
56. Kishikawa, T., 1981, Alterations in the properties of the rat myometrium during gestation and post partum, *Jpn. J. Physiol.* **31:**515–536.
57. Bengtsson, B., Chow, E. M. H., and Marshall, J. M., 1984, Activity of circular muscle of rat uterus at different times in pregnancy, *Am. J. Physiol.* **246:**C216–C223.
58. Ohashi, H., 1970, Effects of changes of ionic environment on the negative after-potential of the spike in rat uterine muscle, *J. Physiol. (Lond.)* **210:**785–797.
59. Parkington, H. C., 1984, Changes in the excitability of the circular myometrium of the guinea-pig throughout pregnancy, *Proc. Aust. Physiol. Pharmacol. Soc.* **15:**96P.
60. Kawarabayashi, T., and Marshall, J. M., 1981, Factors influencing circular muscle activity in the pregnant rat uterus, *Biol Reprod.* **24:**373–379.
61. Parkington, H. C., 1985, Some properties of the circular myometrium of the sheep throughout pregnancy and during labour, *J. Physiol. (Lond.)* **359:**1–15.
62. Kleinhaus, A. L., and Kao, C. Y., 1969, Electrophysiological actions of oxytocin on the rabbit myometrium, *J. Gen. Physiol.* **53:**758–780.
63. Coleman, H. A., and Parkington, H. C., 1987, Single channel Cl$^-$ and K$^+$ currents from cells of uterus not treated with enzymes, *Pflugers Arch.* **410:**560–562.
64. Abe, Y., 1971, Effects of changing the ionic environment on passive and active membrane properties of pregnant rat uterus, *J. Physiol. (Lond.)* **214:**173–190.
65. Sims, S. M., Daniel, E. E., and Garfield, R. E., 1982, Improved electrical coupling in uterine smooth muscle is associated with increased numbers of gap junctions at parturition, *J. Gen. Physiol.* **80:**353–375.
66. Yoshinaga, K., Hawkins, R. A., and Stocker, J. F., 1969, Estrogen secretion by the rat ovary *in vivo* during the estrous cycle and pregnancy, *Endocrinology* **85:**103–112.
67. Boyle, M. B., Azhderian, E. M., MacLusky, N. J., Naftolin, F., and Kaczmarek, L. K., 1987, *Xenopus* oocytes injected with rat uterine RNA express very slowly activating currents, *Science* **235:**1221–1224.
68. Stanfield, P., 1986, Voltage-dependent calcium channels of excitable membranes, *Br. Med. Bull.* **42:**359–367.
69. Hille, B., 1984, *Ionic Channels of Excitable Membranes,* Sinauer, Sunderland, MA.
70. Hodgkin, A. L., and Huxley, A. F., 1952, A quantitative description of membrane current and its application to conduction and excitation in nerve, *J. Physiol. (Lond.)* **117:**500–544.
71. Rougier, O., Vassort, G., and Stämpfli, R., 1968, Voltage clamp experiments on frog atrial heart muscle fibres with the sucrose gap technique, *Pflugers Arch.* **301:**91–108.
72. Johnson, E. A., and Lieberman, M., 1971, Heart: Excitation and contraction, *Annu. Rev. Physiol.* **33:**479–532.
73. Attwell, R., and Cohen, I., 1977, The voltage clamp of multicellular preparations, *Prog. Biophys. Mol. Biol.* **31:**201–245.
74. Ramón, F., Anderson, N. C., Joyner, R. W., and Moore, J. W., 1975, Axon voltage-clamp simulations IV. A multicellular preparation, *Biophys. J.* **15:**55–69.

75. Daniel, E. E., Posey-Daniel, V., Jager, L. P., Berezin, I., and Jury, J., 1987, Structural effects of exposure of smooth muscle in sucrose gap apparatus, *Am J. Physiol.* **252**:C77–C87.
76. Hirst, G. D. S., Silverberg, G. D., and van Helden, D. F., 1986, The action potential and underlying ionic currents in proximal rat middle cerebral arterioles, *J. Physiol. (Lond.)* **371**:389–404.
77. Parkington, H. C., and Coleman, H. A., 1988, Ionic mechanisms underlying action potentials in myometrium, *Clin. Exp. Pharmacol. Physiol.* **15**:657–665.
78. Fischmeister, R., Ayer, R. K., and DeHaan, R. L., 1986, Some limitations of the cell-attached patch clamp technique: A two-electrode analysis, *Pflugers Arch.* **406**:73–82.
79. Worley, J. F., Deitmer, J. W., and Nelson, M. T., 1986, Single nisoldipine-sensitive calcium channels in smooth muscle cells isolated from rabbit mesenteric artery, *Proc. Natl. Acad. Sci. U.S.A.* **83**:5746–5750.
80. Benham, C. D., Hess, P., and Tsien, R. W., 1987, Two types of calcium channels in single smooth muscle cells from rabbit ear artery studied with whole-cell and single-channel recordings, *Circ. Res.* **61**:I.10–I.16.
81. Yatani, A., Seidel, C. L., Allen, J., and Brown, A. M., 1987, Whole-cell and single-channel calcium currents of isolated smooth muscle cells from saphenous vein, *Circ. Res.* **60**:523–533.
82. Yoshino, M., Someya, T., Nishio, A., Yazawa, K., Usuki, T., and Yabu, H., 1989, Multiple types of voltage-dependent Ca channels in mammalian intestinal smooth muscle cells, *Pflugers Arch.* **414**:401–409.
83. Isenberg, G., and Klöckner, U., 1986, Single channel and whole cell calcium currents as studied in isolated smooth muscle cells, in: *Proceedings of the Smooth Muscle Function Symposium (Banff)*, XXX International Congress, IUPS, Vancouver, p. 61.
84. Nilius, B., Hess, P., Lansman, J. B., and Tsien, R. W., 1985, A novel type of cardiac calcium channel in ventricular cells, *Nature* **316**:443–446.
85. Fox, A. P., Nowycky, M. C., and Tsien, R. W., 1987, Single-channel recordings of three types of calcium channels in chick sensory neurones, *J. Physiol. (Lond.)* **394**:173–200.
86. Loirand, G., Pacaud, P., Mironneau, C., and Mironneau, J., 1986, Evidence for two distinct calcium channels in rat vascular smooth muscle cells in short-term primary culture, *Pflugers Arch.* **407**:566–568.
87. Bean, B. P., Sturek, M., Puga, A., and Hermsmeyer, K., 1986, Calcium channels in muscle cells isolated from rat mesenteric arteries: Modulation by dihydropyridine drugs, *Circ. Res.* **59**:229–235.
88. Nakazawa, K., Saito, H., and Matsuki, N., 1988, Fast and slowly inactivating components of Ca-channel current and their sensitivities to nicardipine in isolated smooth muscle cells from rat vas deferens, *Pflugers Arch.* **411**:289–295.
89. Benham, C. D., and Tsien, R. W., 1988, Noradrenaline modulation of calcium channels in single smooth muscle cells from rabbit ear artery, *J. Physiol. (Lond.)* **404**:767–784.
90. Vivaudou, M. B., Clapp, L. H., Walsh, J. V., and Singer, J. J., 1988, Regulation of one type of Ca^{2+} current in smooth muscle cells by diacylglycerol and acetylcholine, *FASEB J.* **2**:2497–2504.
91. Dupont, J.-L., Bossu, J.-L., and Feltz, A., 1986, Effect of internal calcium currents in rat sensory neurons, *Pflugers Arch.* **406**:433–435.
92. Fedulova, S. A., Kostyuk, P. K., and Veselovsky, N. S., 1985, Two types of calcium channels in the somatic membrane of new-born rat dorsal root ganglion neurones, *J. Physiol. (Lond.)* **359**:431–446.
93. Lee, K. S., Marban, E., and Tsien, R. W., 1985, Inactivation of calcium channels in mammalian heart cells: Joint dependence on membrane potential and intracellular calcium, *J. Physiol. (Lond.)* **364**:395–411.
94. Inoue, R., Kitamura, K., and Kuriyama, H., 1985, Two Ca-dependent K-channels classified by the application of tetraethylammonium distribute to smooth muscle membranes of rabbit portal vein, *Pflugers Arch.* **405**:173–179.
95. Benham, C. D., Bolton, T. B., Lang, R. J., and Takewaki, T., 1986, Calcium-activated potassium

channels in single smooth muscle cells of rabbit jejunum and guinea-pig mesenteric artery, *J. Physiol. (Lond.)* **371**:45–67.

96. Benham, C. D., and Bolton, T. B., 1983, Patch-clamp studies of slow potential-sensitive potassium channels in longitudinal muscle cells of rabbit jejunum, *J. Physiol. (Lond.)* **340**:469–486.

97. Bolton, T. B., Lang, R. J., Takewaki, T., and Benham, C. D., 1985, Patch and whole-cell voltage clamp of single mammalian visceral and vascular smooth muscle cells, *Experientia* **41**:887–894.

98. Ohya, Y., Kitamura, K., and Kuriyama, H., 1987, Cellular calcium regulates outward currents in rabbit intestinal smooth muscle cell, *Am. J. Physiol.* **252**:C401–C410.

99. Okabe, K., Kitamura, K., and Kuriyama, H., 1987, Features of 4-aminopyridine sensitive outward current observed in single smooth muscle cells from the rabbit pulmonary artery, *Pflugers Arch.* **409**:561–568.

100. Terada, K., Kitamura, K., and Kuriyama, H., 1987, Different inhibitions of the voltage-dependent K^+ current by Ca^{2+} antagonists in the smooth muscle cell membrane of rabbit small intestine, *Pflugers Arch.* **408**:558–564.

100a. Kirber, M. T., Walsh, J. V., and Singer, J. J., 1988, Stretch-activated ion channels in smooth muscle: a mechanism for the initiation of stretch-induced contraction, *Pflugers Arch.* **412**:339–345.

101. Sims, S. M., Singer, J. J., and Walsh, J. V., 1985, Cholinergic agonists suppress a potassium current in freshly dissociated smooth muscle cells of the toad, *J. Physiol. (Lond.)* **367**:503–529.

102. Sims, S. M., Singer, J. J., and Walsh, J. V., 1987, Antagonistic adrenergic–muscarinic regulation of M current in smooth muscle cells, *Science* **239**:190–193.

103. Byrne, N. G., and Large, W. A., 1987, Action of noradrenaline on single smooth muscle cells freshly dispersed from the rat anococcygeus muscle, *J. Physiol. (Lond.)* **389**:513–525.

104. Soejima, M., and Kokubun, S., 1988, Single anion-selective channel and its ion selectivity in the vascular smooth muscle cell, *Pflugers Arch.* **411**:304–311.

105. Shoemaker, R., Naftel, J., and Farley, J., 1985, Measurement of K^+ and Cl^- channels in rat cultured vascular smooth muscle cells, *Biophys. J.* **47**:465a.

106. Benham, C. D., Bolton, T. B., Denbigh, J. S., and Lang, R. J., 1987, Inward rectification in freshly isolated single smooth muscle cells of the rabbit jejunum, *J. Physiol. (Lond.)* **383**:461–476.

107. Nakazawa, K., and Matsuki, N., 1987, Adenosine triphosphate-activated inward current in isolated smooth muscle cells from rat vas deferens, *Pflugers Arch.* **409**:644–646.

108. Inoue, R., Kitamura, K., and Kuriyama, H., 1987, Acetylcholine activates single sodium channels in smooth muscle cells, *Pflugers Arch.* **410**:69–74.

109. Clapp, L. H., Vivaudou, M. B., Walsh, J. V., and Singer, J. J., 1987, Acetylcholine increases voltage-activated Ca^{2+} current in freshly dissociated smooth muscle cells, *Proc. Natl. Acad. Sci. (U.S.A.)* **84**:2092–2096.

110. Clapp, L. H., Singer, J. J., Vivaudou, M. B., and Walsh, J. V., 1987, A diacylglycerol analogue enhances voltage-activated Ca^{2+} current in freshly dissociated smooth muscle cells from the stomach of the toad, *J. Physiol. (Lond.)* **394**:44P.

111. van Helden, D. F., 1986, Neuro-modulation of the calcium current through α-receptors in guinea-pig mesenteric veins, *Neurosci. Lett.* **23**:S88.

112. Pacaud, P., Loirand, G., Mironneau, C., and Mironneau, J., 1987, Opposing effects of noradrenaline on the two classes of voltage-dependent calcium channels of single vascular smooth muscle cells in short-term primary culture, *Pflugers Arch.* **410**:557–559.

113. Hamill, O. P., Marty, A., Neher, E., Sakmann, B., and Sigworth, F. J., 1981, Improved patch-clamp techniques for high-resolution current recording from cells and cell-free membrane patches, *Pflugers Arch.* **391**:85–100.

114. Kao, C. Y., 1961, Contents and distribution of potassium, sodium and chloride in uterine smooth muscle, *Am. J. Physiol.* **201**:717–723.

115. Reiner, O., and Marshall, J. M., 1975, Action of D-600 on spontaneous and electrically stimulated activity of the parturient rat uterus, *Naunyn Schmeidebergs Arch. Pharmacol.* **290**:21–28.

116. Osa, T., 1974, An interaction between the electrical activities of longitudinal and circular smooth muscles of pregnant mouse uterus, *Jpn. J. Physiol.* **24:**189–203.

117. Marshall, J. M., 1973, The physiology of the myometrium, in: *The Uterus* (H. J. Norris, A. T. Hertig, and M. R. Abel, eds.), Williams & Wilkins, Baltimore, pp. 89–109.

118. Lodge, S., and Sproat, J. E., 1981, Resting membrane potentials of pacemaker and nonpacemaker areas in rat uterus, *Life Sci.* **28:**2251–2256.

119. Suzuki, H., and Kuriyama, H., 1975, Comparison between prostaglandin E_2 and oxytocin actions on pregnant mouse myometrium, *Jpn. J. Physiol.* **25:**345–356.

120. Marshall, J. M., 1974, Effects of neurohypophysial hormones on the myometrium, in: *Handbook of Physiology—Endocrinology Vol. IV. Part 1* (R. O. Greep and E. B. Astwood, eds.), American Physiological Society, Bethesda, pp. 469–492.

121. Mironneau, J., Savineau, J. P., and Mironneau, C., 1981, Fast outward current controlling electrical activity in rat uterine smooth muscle during gestation, *J. Physiol. (Paris)* **77:**851–859.

122. Jmari, K., Mironneau, C., and Mironneau, J., 1986, Inactivation of calcium channel current in rat uterine smooth muscle: Evidence for calcium- and voltage-mediated mechanisms, *J. Physiol. (Lond.)* **380:**111–126.

123. Amédée, T., Mironneau, C., and Mironneau, J., 1987, The calcium channel current of pregnant rat single myometrial cells in short-term primary culture, *J. Physiol. (Lond.)* **392:**253–272.

124. Jmari, K., Mironneau, C., and Mironneau, J., 1987, Selectivity of calcium channels in rat uterine smooth muscle: Interactions between sodium, calcium and barium ions, *J. Physiol. (Lond.)* **384:**247–261.

125. Anderson, N. C., Ramón, F., and Snyder, A., 1971, Studies on calcium and sodium in uterine smooth muscle excitation under current-clamp and voltage-clamp conditions, *J. Gen. Physiol.* **58:**322–339.

125a. Honoré, E., Amédée, T., Martin, C., Dacquet, C., Mironneau, C., and Mironneau, J., 1989, Calcium channel current and its sensitivity to (+)isradipine in cultured pregnant rat myometrial cells, *Pflugers Arch.* **414:**477–483.

125b. Granger, S. E., Hollingsworth, M., and Weston, A. H., 1986, Effects of calcium entry blockers on tension development and calcium influx in rat uterus, *Br. J. Pharmacol.* **87:**147–156.

126. Marshall, J. M., 1963, Behavior of uterine muscle in Na^+-deficient solutions: Effects of oxytocin, *Am. J. Physiol.* **204:**732–738.

127. Anderson, N. C., 1969, Voltage-clamp studies on uterine smooth muscle, *J. Gen. Physiol.* **54:**145–165.

128. Mironneau, J., 1973, Excitation–contraction coupling in voltage clamped uterine smooth muscle, *J. Physiol. (Lond.)* **233:**127–142.

129. Kao, C. Y., and McCullough, J. R., 1975, Ionic currents in the uterine smooth muscle, *J. Physiol. (Lond.)* **246:**1–36.

130. Savineau, J. P., Mironneau, J., and Mironneau, C., 1987, Influence of the sodium gradient on contractile activity in pregnant rat myometrium, *Gen. Physiol. Biophys.* **6:**535–560.

131. Mironneau, J., and Savineau, J. P., 1980, Effects of calcium ions on outward membrane currents in rat uterine smooth muscle, *J. Physiol. (Lond.)* **302:**411–425.

131a. Kao, C. Y., Wakui, M., Wang, S. Y., and Yoshino, M., 1989, The outward current of the isolated rat myometrium, *J. Physiol. (Lond.)* **418:**20P.

132. Osa, T., and Kawarabayashi, T., 1977, Effects of ions and drugs on the plateau potential in the circular muscle of pregnant rat myometrium, *Jpn. J. Physiol.* **27:**111–121.

133. Parkington, H. C., 1982, The electrical activity of the possum uterus, *Proc. Aust. Physiol. Pharmacol. Soc.* **13:**126P.

134. Parkington, H. C., 1984, Factors affecting the action potential of the circular myometrium of the rat, *Proc. Aust. Physiol. Pharmacol. Soc.* **15:**188P.

135. Wilde, D. W., and Marshall, J. M., 1988, Effects of tetraethylammonium and 4-aminopyridine on the plateau potential of circular myometrium from the pregnant rat, *Biol. Reprod.* **38:**836–845.

136. Vassort, G., 1975, Voltage-clamp analysis of transmembrane ionic currents in guinea-pig myo-

metrium: Evidence for an initial potassium activation triggered by calcium influx, *J. Physiol. (Lond.)* **252**:713–734.

137. Vassort, G., 1981, Ionic currents in longitudinal muscle of the uterus, in: *Smooth Muscle: An Assessment of Current Knowledge* (E. Bülbring, A. F. Brading, A. W. Jones, and T. Tomita, eds.), Edward Arnold, London, pp. 353–366.

137a. Coleman, H. A., and Parkington, H. C. 1990, Hyperpolarization-activated channels in myometrium: A patch clamp study, in: *Frontiers in Smooth Muscle Research* (N. Sperelakis and J. D. Wood, eds.), Liss, New York, Wiley, pp. 665–672.

138. Feinstein, M. B., 1966, Inhibition of contraction and calcium exchange-ability in rat uterus by local anesthetics, *J. Pharmacol. Exp. Ther.* **152**:516–524.

139. Aksoy, M. O., Murphy, R. A., and Kamm, K. E., 1982, Role of Ca^{2+} and myosin light chain phosphorylation in regulation of smooth muscle, *Am. J. Physiol.* **242**:C109–C116.

140. Morgan, J. P., and Morgan, K. G., 1984, Stimulus-specific patterns of intracellular calcium levels in smooth muscle of ferret portal vein, *J. Physiol. (Lond.)* **351**:155–167.

141. Williams, D. A., and Fay, F. S., 1986, Calcium transients and resting levels in isolated smooth muscle cells as monitored with quin-2, *Am. J. Physiol.* **250**:C779–C791.

142. Itoh, T., Kuriyama, H., and Suzuki, H., 1981, Excitation–contraction coupling in smooth muscle cells of the guinea-pig mesenteric artery, *J. Physiol. (Lond.)* **321**:513–535.

143. Neher, E., 1983, The charge carried by single-channel currents of rat cultured muscle cells in the presence of local anaesthetics, *J. Physiol. (Lond.)* **339**:663–678.

144. Gage, P. W., 1985, Ion channels and signal transduction in animal cells, *Proc. Aust. Physiol. Pharmacol. Soc.* **16**:61–77.

145. Klöckner, U., and Isenberg, G., 1985, Calcium currents of cesium loaded isolated smooth muscle cells (urinary bladder of the guinea pig), *Pflugers Arch.* **405**:340–348.

146. Savineau, J. P., Mironneau, J., and Mironneau, C., 1988, Contractile properties of chemically skinned fibres from pregnant rat myometrium: Existence of an internal Ca-store, *Pflugers Arch.* **411**:296–303.

147. Coleman, H. A., McShane, P. G., and Parkington, H. C., 1988, Gestational changes in the utilization of intracellularly stored calcium in the myometrium of guinea-pigs, *J. Physiol. (Lond.)*. **399**:13–32.

148. Osa, T., and Katase, T., 1975, Physiological comparison of the longitudinal and circular muscles of the pregnant rat uterus, *Jpn. J. Physiol.* **25**:153–164.

149. Crankshaw, D. J., 1986, The sensitivity of the longitudinal and circular muscle layers of the rat's myometrium to oxytocin *in vitro* during pregnancy, *Can. J. Physiol. Pharmacol.* **65**:773–777.

150. Osa, T., and Taga, F., 1973, Electrophysiological comparison of the action of oxytocin and carbachol on pregnant mouse myometrium, *Jpn. J. Physiol.* **23**:81–96.

151. Edwards, D., Good, D. M., Granger, S. E., Hollingsworth, M., Dobson, A., Small, R. C., and Weston, A. H., 1986, The spasmogenic action of oxytocin in the rat uterus—comparison with other agonists, *Br. J. Pharmacol.* **88**:899–908.

152. Mironneau, J., 1976, Effects of oxytocin on ionic currents underlying rhythmic activity and contraction in uterine smooth muscle, *Pflugers Arch.* **363**:113–118.

153. Bennett, M. V. L., Spray, D. C., and Harris, A. L., 1981, Gap junctions and development, *Trends Neurosci.* **4**:159–163.

154. Reiner, O., and Marshall, J. M., 1976, Action of prostaglandin, $PGF_{2\alpha}$ on the uterus of the pregnant rat, *Naunyn-Schmeidebergs Arch. Pharmacol.* **292**:243–250.

155. Hall, W. J., and Pickles, V. R., 1963, The dual action of menstrual stimulant A2 (prostaglandin E_2), *J. Physiol. (Lond.)* **169**:90P–91P.

156. Clegg, P. C., Hall, W. J., and Pickles, V. R., 1966, The action of ketonic prostaglandins on the guinea-pig myometrium, *J. Physiol. (Lond.)* **183**:123–144.

157. Coleman, H. A., and Parkington, H. C., 1988, Induction of prolonged excitability in myometrium of pregnant guinea-pigs by prostaglandin $F_{2\alpha}$, *J. Physiol. (Lond.)* **399**:33–47.

158. Grosset, A., and Mironneau, J., 1977, An analysis of the actions of prostaglandin E_1 on membrane currents and contraction in uterine smooth muscle. *J. Physiol. (Lond.)* **270**:765–784.

159. Osa, T., Suzuki, H., Katase, T., and Kuriyama, H., 1974, Excitatory action of synthetic pros-taglandin E_2 on the electrical activity of pregnant mouse myometrium in relation to temperature changes and external sodium and calcium concentrations, *Jpn. J. Physiol.* **24:**233–248.

160. Nagai, T., and Prosser, C. L., 1963, Electrical parameters of smooth muscle cells, *Am. J. Physiol.* **204:**915–924.

161. Kuriyama, H., and Tomita, T., 1965, The response of single smooth muscle cells of guinea-pig taenia coli to intracellularly applied currents and their effects on the spontaneous electrical activity, *J. Physiol. (Lond.)* **178:**270–289.

162. Tomita, T., 1966, Electrical responses of smooth muscle to external stimulation in hypertonic solution, *J. Physiol. (Lond.)* **183:**450–468.

163. Bortoff, A., and Gilloteaux, J., 1980, Specific tissue impedances of estrogen- and progesterone-treated rabbit myometrium, *Am. J. Physiol.* **238:**C34–C42.

164. Garfield, R. E., Sims, S. M., Kannan, M. S., and Daniel, E. E., 1978, Possible role of gap junctions in activation of myometrium during parturition, *Am. J. Physiol.* **235:**C168–C179.

165. Puri, C. P., and Garfield, R. E., 1982, Changes in hormone levels and gap junctions in the rat uterus during pregnancy and parturition, *Biol. Reprod.* **27:**967–975.

166. Neyton, J., and Trautman, A., 1985, Single-channel currents of an intercellular junction, *Nature,* **317:**331–335.

167. Veenstra, R. D., and DeHaan, R. L., 1988, Cardiac gap junction channel activity in embryonic chick ventricle cells, *Am. J. Physiol.* **254:**H170–H180.

168. Ramón, F., and Rivera, A., 1986, Gap junction channel modulation—a physiological viewpoint, *Prog. Biophys. Mol. Biol.* **48:**127–153.

169. Burt, J. M., and Spray, D. C., 1988, Single-channel events and gating behavior of the cardiac gap junction channel, *Proc. Natl. Acad. Sci. U.S.A.* **85:**3431–3434.

170. Daniel, E. E., Daniel, V. P., Duchon, G., Garfield, R. E., Nichols, M., Malhotra, S. K., and Oki, M., 1976, Is the nexus necessary for cell-to-cell coupling of smooth muscle? *J. Membr. Biol.* **28:**207–239.

171. Fry, G. N., Devine, C. E., and Burnstock, G., 1977, Freeze-fracture studies of nexuses between smooth muscle cells. Close relationship to sarcoplasmic reticulum, *J. Cell Biol.* **72:**26–34.

172. Gabella, G., and Blundell, D., 1979, Nexuses between the smooth muscle cells of the guinea-pig ileum, *J. Cell Biol.* **82:**239–247.

173. Coleman, H. A., 1987, Multiple sites for the initiation of action potentials in neurons of the inferior mesenteric ganglion of the guinea-pig, *Neuroscience* **20:**357–363.

174. Goldstein, S. S., and Rall, W., 1974, Changes of action potential shape and velocity for changing core conductor geometry, *Biophys. J.* **14:**731–757.

175. Joyner, R. W., 1982, Effects of the discrete pattern of electrical coupling on propagation through an electrical syncytium, *Circ. Res.* **50:**192–200.

176. Spach, M. S., and Kootsey, J. M., 1983, The nature of electrical propagation in cardiac muscle, *Am. J. Physiol.* **244:**H3–H22.

177. Joyner, R. W., and Van Capelle, F. J. L., 1986, Propagation through electrically coupled cells. How a small SA node drives a large atrium, *Biophys. J.* **50:**1157–1164.

178. Joyner, R. W., Veenstra, R., Rawling, D., and Chorro, A., 1984, Propagation through electrically coupled cells. Effects of a resistive barrier, *Biophys. J.* **45:**1017–1025.

179. Miller, S. M., Garfield, R. E., and Daniel, E. E., 1989, Improved propagation in myometrium associated with gap junctions during parturition, *Am. J. Physiol.* **256:**C130–C141.

180. Parkington, H. C., Harding, R., and Sigger, J. N., 1988, Co-ordination of electrical activity in the myometrium of pregnant ewes, *J. Reprod. Fertil.* **82:**697–705.

181. Melton, C. E., and Saldivar, J. T., 1964, Impulse velocity and conduction pathways in the rat myometrium, *Am. J. Physiol.* **207:**279–285.

182. Fuchs, A.-R., 1972, Uterine activity during and after mating in the rabbit, *Fertil. Steril.* **23:**915–923.

183. Porter, D. G., 1971, Quantitative changes in myometrial activity in the guinea-pig during pregnancy, *J. Reprod. Fertil.* **27:**219–226.

184. Taverne, M. A., Naaktegeboren, C., Elshesser, F., Forsling, H. L., van der Weyden, G. C., Ellendorff, F., and Schmidt, D., 1979, Myometrial electrical activity and plasma concentrations of progesterone, estrogens and oxytocin during late pregnancy and parturition in the miniature pig, *Biol. Reprod.* **21:**1125–1134.

185. Germain, G., Gabrol, D., Visser, A., and Sureau, C., 1982, Electrical activity of the pregnant uterus in the cynomolgus monkey, *Am. J. Obstet. Gynecol.* **142:**513–519.

186. Demianczuk, N., Towell, M. E., and Garfield, R. E., 1984, Myometrial electrophysiologic activity and gap junctions in the pregnant rabbit, *Am. J. Obstet. Gynecol.* **149:**485–491.

187. Harding, R., Poore, E. R., Bailey, A., Thorburn, G. D., Jansen, C. A., and Nathanielsz, P. W., 1982, Electromyographic activity of the nonpregnant and pregnant sheep uterus, *Am. J. Obstet. Gynecol.* **142:**448–457.

188. Taverne, M. A., and Scheerboom, J. E., 1985, Myometrial electrical activity during pregnancy and parturition in the pygmy goat, *Res. Vet. Sci.* **30:**120–123.

189. Bell, R., 1983, The prediction of preterm labour by recording spontaneous antenatal uterine activity, *Br. J. Obstet. Gynaecol.* **90:**884–887.

190. Lye, S. J., Sprague, C. L., Mitchell, B. F., and Challis, J. R. G., 1983, Activation of ovine fetal adrenal function by pulsatile or continuous administration of adrenocorticotropin-(1–24). I. Effects on fetal plasma corticosteroids, *Endocrinology* **113:**770–776.

191. Sigger, J. N., Harding, R., and Bailey, A., 1984, Development of myometrial electrical activity during the first half of pregnancy in sheep, *Aust. J. Biol. Sci.* **37:**153–162.

192. Fuchs, A. R., and Poblete, V. F., 1970, Oxytocin and uterine function in pregnant and parturient rats, *Biol. Reprod.* **2:**287–400.

193. Shepherd, V., Parkington, H. C., Jenkin, G., Ralph, M. M., and Thorburn, G. D., 1986, Activity of ovine circular and longitudinal myometrium during pregnancy and parturition, *Proc. Aust. Soc. Reprod. Biol.* **18:**107.

194. Sigger, J. N., Harding, R., and Jenkin, G., 1984, Relationship between electrical activity of the uterus and surgically isolated myometrium in the pregnant and non-pregnant ewe, *J. Reprod. Fertil.* **70:**103–114.

195. Thorburn, G. D., Parkington, H. C., Rice, G., Jenkin, G., Harding, R., Sigger, J. N., Ralph, M. M., Shepherd, V., and Myles, K., 1988, Regulation of electrical activity in the myometrium of the pregnant ewe, in: *The Endocrine Control of the Fetus, Physiologic and Pathophysiologic Aspects* (W. Kunzel and A. Jensen, eds.), Springer-Verlag, Heidelberg, pp. 391–400.

196. Lye, S. J., and Freitag, C.. L., 1988, An *in vivo* model to examine the electromyographic activity of isolated myometrial tissue from pregnant sheep, *J. Reprod. Fertil.* **82:**51–61.

197. Adrianakis, P., Walker, D. W., Ralph, M. M., and Thorburn, G. D., 1987, Effect of 4-aminoantipyrine on prostaglandin (PG) concentrations in uterine and umbilical circulations of pregnant sheep, *Proc. Endocrinol. Soc. Aust.* **30:**115.

8

β-Adrenoceptors, Cyclic AMP, and Cyclic GMP in Control of Uterine Motility

Jack Diamond

1. β-Adrenoceptors and Cyclic AMP in Uterine Relaxation

Adenosine 3′,5′-cyclic monophosphate (cAMP) and the enzymes necessary for its synthesis and degradation have been shown to be present in most mammalian cells, including myometrial cells. Cyclic AMP is considered to be the "second messenger" responsible for mediating the actions of numerous drugs and hormones, and such a role has been suggested for this cyclic nucleotide in the uterine relaxant effects of β-adrenoceptor agonists such as isoproterenol and epinephrine. Some of the evidence for and against a role for cAMP as a mediator of the uterine relaxant effects of β agonists is discussed in the following sections. By way of background, brief descriptions of the roles of adrenergic receptors in control of uterine motility and the effects of estrogen and progesterone on these responses are included. The factors involved in coupling adrenergic receptors to the cAMP system are also discussed.

1.1. Adrenergic Receptors in the Uterus

1.1.1. Function of α- and β-Adrenoceptors in Control of Uterine Motility

The classification of adrenergic receptors into α and β receptors was originally suggested by Ahlquist in 1948.[1] He later postulated that all uteri possess both α and β receptors and that stimulation of α-adrenoceptors promotes contraction of uterine muscle, whereas stimulation of β receptors causes uterine relaxation.[2] Evidence

---

Jack Diamond • Division of Pharmacology and Toxicology, Faculty of Pharmaceutical Sciences, University of British Columbia, Vancouver, British Columbia V6T 1W5, Canada.

supporting this hypothesis was subsequently obtained in a variety of species including rabbit,[3] guinea pig,[4] cat,[5] rat,[6] and human.[7] Since that time β-adrenoceptors have been further subdivided into β_1 and β_2 receptors,[8] and α receptors into α_1 and α_2 subtypes.[9] These classifications are based on the relative responsiveness of these tissues to various agonists and antagonists and, more recently, on the binding of selective radioligands to specific receptors. All four types of adrenergic receptors have been identified in myometrial tissues from a variety of species including humans.[10] In most species the β receptors responsible for uterine inhibition appear to be primarily of the β_2 type,[8,11−13] although the presence of β_1 receptors has also been demonstrated in some cases.[12,13] Both α_1- and α_2-adrenoceptors are also present in the myometrium, but there is evidence in at least some species that only the α_1 receptors mediate contraction.[14] Thus, the generalization can be made that the uterine relaxant effects of catecholamines are exerted primarily through β_2-adrenoceptors, whereas the uterine contractile effects of these agents are mediated via α_1 receptors.

It has been suggested that the α_1 and β_1 receptors are "innervated" receptors (located postsynaptically in close proximity to sympathetic nerve endings and preferentially activated by the sympathetic neurotransmitter norepinephrine), whereas the α_2 and β_2 receptors are "hormonal" receptors (located presynaptically and extrajunctionally and preferentially activated by the circulating hormone epinephrine).[15] Although this is consistent with much of the available evidence, it may be an oversimplification in that there are species differences and exceptions to this general scheme. For a more complete discussion of myometrial adrenergic receptors, the reader is referred to an excellent recent review by Bulbring and Tomita.[16]

1.1.2. Effects of Ovarian Hormones on Adrenergic Receptors in the Uterus

It is well established that the relative preponderance of α- and β-adrenoceptors in myometrial tissues can be altered by changes in concentrations of estrogen and progesterone, such as those occurring during pregnancy or different stages of the estrous cycle.[17] For example, in the cat, β receptors were predominant in uteri from nonpregnant or estrogen-treated animals, whereas α receptors were predominant in uteri from pregnant or progesterone-treated animals.[5,18] The β-adrenoceptor predominance seen in estrogen-treated uteri could be reversed by administration of progesterone or an aqueous extract of pregnant or progesterone-proliferated uteri. Thus, the response to catecholamines was reversed from a primarily inhibitory one in estrogen-dominated cats to an excitatory one in progesterone-dominated cats. Exactly the opposite response is seen in the rabbit, where the estrogen-dominated uterus is contracted by norepinephrine or stimulation of the hypogastric nerve (an α-adrenergic response) but the progesterone-dominated uterus is relaxed by these procedures (a β-adrenergic response).[3]

Responses seen in the rat are virtually the same as those seen in the rabbit. Uterine strips obtained from ovariectomized rats that had been injected with estradiol were contracted by norepinephrine (an α-adrenergic response), whereas uteri from rats injected with progesterone or with estrogen followed by progesterone were relaxed by norepinephrine (a β-adrenergic response).[19] A similar pattern appears to occur in the human, where the contractile effects of the catecholamines have been reported to be more prominent in estrogen-dominated uteri, whereas inhibitory effects were predominant under conditions of high progesterone levels such as those occurring during pregnancy or the luteal phase of the estrous cycle.[10]

It is obvious from the above reports that there are marked species differences with regard to the effects of ovarian hormones on the adrenergic receptor patterns in myometrial tissues. The picture is further complicated by observations that the responses to catecholamines, and the modification of these responses by ovarian hormones, may be quite different in the longitudinal and circular muscle layers of the myometrium.[20,21] In pregnant rats, during the middle stages of gestation, inhibitory β receptors were predominant in longitudinal muscle cells, whereas excitatory α receptors were predominant in the circular muscle layer. These patterns were reversed at term. This would seem to make sense from a teleological standpoint, since these patterns of activity would tend to favor a quiescent uterus during midgestation and expulsion of the fetuses at term. In the earlier study on rat uterus referred to above,[19] no attempt was made to separate the myometrium into longitudinal and circular layers. However, contractions were monitored in strips oriented in the longitudinal direction, and the results are consistent with those observed in longitudinal muscle cells in the more recent studies.

Since the advent of reliable techniques for estimating the numbers of adrenergic receptors in various tissues, attempts have been made to explain the earlier observations by changes in density of specific adrenoceptors. Initial studies on receptor density in rabbit myometrium under different hormonal conditions appeared to be consistent with the observed mechanical changes.[22–24] In these studies, increased numbers of α-adrenoceptors were seen in myometrial preparations following estrogen treatment, and it was suggested that this might be responsible for the enhanced contractile responses to catecholamines that were seen in the estrogen-dominated tissues. However, when specific subtypes of α receptors were measured, it was found that the increase in α-adrenoceptor density seen in rabbit uterus following estrogen treatment was caused by a selective increase in the number of α_2-adrenoceptors.[14] There was no significant change in the concentration of α_1-adrenoceptors under these conditions.

Since contraction in rabbit myometrium was mediated via α_1- and not α_2-adrenoceptors,[14] these results suggest that changes in α-adrenoceptor density might not be responsible for the changes in mechanical response usually observed following estrogen treatment. Interestingly, in this report, estrogen treatment did not change the sensitivity of the uterus to the contractile effects of phenylephrine or norepinephrine. This is consistent with the lack of change in α_1-adrenoceptors observed by these authors, but it does not agree with previous reports in the literature.[3,22] No changes in β-receptor density were found in these studies with estrogen or progesterone treatment. The predominance of β inhibitory responses in progesterone-treated rabbit uteri could not be explained by changes in the concentration or affinity of adrenergic receptors.[22] Similarly, in the study of Bottari *et al.*[10] in human uterus, estrogen treatment increased the number of α_2- and β_1-adrenoceptors but did not change the concentration of α_1 or β_2 receptors. These results cannot explain the ability of estrogen to alter the mechanical responses of the uterus to catecholamines, since the receptors involved in mediating the mechanical responses are believed to be the α_1- and β_2-adrenoceptors.[10] Thus, at the present time, the ability of the ovarian hormones to alter mechanical responses to the catecholamines in myometrial preparations from various species is not readily explained by their effects on the relative numbers of adrenergic receptors in the tissues.

In summary, the excitatory, α-adrenoceptor-mediated responses to cate-cholamines in myometrial tissues from several species (e.g., rat, rabbit, human) are enhanced under conditions of estrogen dominance, whereas the inhibitory, β-adre-noceptor-mediated responses are more prominent in preparations under the influ-ence of progesterone. These altered contractile responses are not easily explained by measurable changes in the concentrations of specific adrenoceptors in the myo-metrium. Interpretation of the literature in this area is complicated not only by species differences but by differences in experimental techniques between studies. Most studies have not taken into account possible differences in responses between circular and longitudinal smooth muscle cells. Such studies may be needed before any definite conclusions can be reached regarding the mechanisms by which the ovarian hormones can affect the responses of myometrial preparations to cate-cholamines. Possible influences of these hormones on the generation of second messengers might play a role, and this is discussed in Section 1.3 below.

1.2. Coupling of Adrenergic Receptors to Adenylate Cyclase

Both β_1- and β_2-adrenoceptors are coupled to adenylate cyclase, the enzyme responsible for the synthesis of the second messenger cAMP from its substrate, ATP.[13] It is generally assumed that cAMP is the mediator of the effects of β-adrenoceptor stimulation. Because of the wide distribution of β-adrenoceptors, the diversity of responses mediated by them, and the therapeutic importance of activa-tion or blockade of these receptors, the β-adrenoceptor–adenylate cyclase complex has been widely studied. It is now recognized that receptor coupling to adenylate cyclase involves intermediate guanine-nucleotide-dependent regulatory proteins known as N proteins or G proteins (for reviews see refs. 25–28). Coupling of β-adrenoceptors to adenylate cyclase is accomplished by a G protein designated as G_s ("s" signifies that the protein is stimulatory to adenylate cyclase). Another protein, G_i ("i" for inhibitory), mediates the effects of hormones known to inhibit the activity of adenylate cyclase. Other G proteins have also been described, including transducin, which couples photon-induced excitation of retinal rod cells to the activation of a cGMP phosphodiesterase, and G_o ("o" for other), a G protein found in brain cells whose function has not yet been elucidated. All of these G proteins appear to be heterotrimers composed of three distinct subunits designated α, β, and γ. Activation of the G proteins is assumed to involve binding of GTP to the α

subunits and dissociation of the α subunits from the βγ complexes. The α subunits of G_s and G_i differ structurally from one another, but both bind GTP, and both have intrinsic GTPase activity enabling them to hydrolyze the GTP to GDP. The β and γ subunits from different G proteins are structurally similar and may be functionally interchangeable. They are not capable of hydrolyzing GTP.

Some of the relationships between these G proteins and the other components of the adrenoceptor–adenylate cyclase complex are illustrated schematically in Fig. 1. When a β-adrenoceptor agonist such as isoproterenol attaches to the stimulatory receptor (R_s), GTP binds to the α subunit of G_s, which then dissociates from the β and γ subunits and stimulates the catalytic unit of adenylate cyclase (AC). The

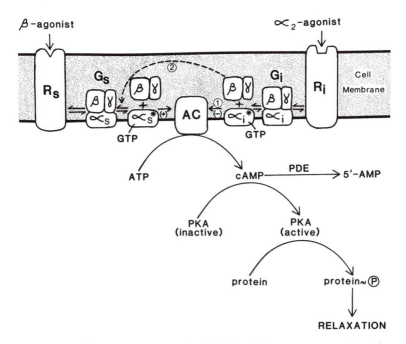

Figure 1. Putative cAMP-dependent pathways involved in mediating responses to adrenergic receptor stimulation in the uterus. Coupling of stimulatory receptors (R_s) to the catalytic unit of adenylate cyclase (AC) is accomplished by means of a stimulatory guanine nucleotide-dependent regulatory protein, or G protein, designated as G_s. When an appropriate agonist combines with R_s, GTP binds to the α subunit of G_s, which then dissociates from the βγ subunits and activates the AC. Coupling of inhibitory receptors (R_i) to AC involves an inhibitory G protein (G_i). Two possible mechanisms by which the G_i could inhibit the activity of AC are indicated by the circled numbers 1 and 2 in the figure. See text for further explanation. Note that the terms stimulatory and inhibitory refer to effects on AC, not on the mechanical activity of the uterus. Stimulation or inhibition of AC results in an increase or decrease, respectively, in the rate of production of AMP. Cyclic AMP is hydrolyzed to 5′-AMP by the action of one or more phosphodiesterases (PDE). An increase in cAMP levels will result in the activation of cAMP-dependent protein kinases (PKA), which, in turn, phosphorylate specific proteins within the cell. Phosphorylation of these proteins presumably mediates the response attributed to activation of R_s, i.e., uterine relaxation.

resulting increase in cAMP levels activates a specific cAMP-dependent protein kinase (PKA), which, in turn, phosphorylates specific proteins within the cell. Presumably, phosphorylation of these proteins is somehow responsible for the observed response (relaxation in the case of the uterus). The identity of these proteins and the precise underlying mechanism for the relaxation have not been elucidated for uterine smooth muscle. When the GTP bound to α_s is hydrolyzed to GDP, α_s reassociates with the $\beta\gamma$ subunit, and G_s no longer activates adenylate cyclase. Thus, as far as coupling of R_s to adenylate cyclase is concerned, the binding of GTP to α_s can be considered as the "on" reaction, and hydrolysis of GTP to GDP as the "off" reaction. When stimulation of adenylate cyclase ceases, cAMP levels are rapidly returned to resting levels by the action of one or more cAMP phosphodiesterases (PDE) present in the cells. Cyclic AMP levels can therefore be increased by agents that increase the activity of adenylate cyclase or inhibit the activity of the phosphodiesterase.

An exotoxin produced by the cholera bacillus can inhibit the GTPase activity of the G_s (via ADP-ribosylation of the α subunit of G_s), thus blocking the off reaction and producing a persistent activation of adenylate cyclase. This may be responsible for the severe diarrhea characteristic of cholera, since cAMP is a potent activator of fluid secretion in the intestine. Nonhydrolyzable GTP derivatives such as GTPγS or GppNHp can also cause persistent activation of adenylate cyclase even in the absence of receptor activation. Because of this ability to activate G_s, both cholera toxin and nonhydrolyzable GTP derivatives have been frequently used as tools in studies designed to elucidate the roles of the G proteins in various physiological processes.

Some analogies can be drawn between the situation described above for activation of G_s and that occurring during activation of the inhibitory guanine-nucleotide-dependent regulatory protein G_i. Stimulation of the inhibitory receptor (R_i) by an agonist such as norepinephrine results in binding of GTP to the α subunit of G_i and dissociation of α_i from its β and γ subunits (Fig. 1). This then results in inhibition of the catalytic activity of adenylate cyclase and/or prevention of its activation by the stimulatory G protein. Two possible mechanisms by which adenylate cyclase activity could be inhibited by G_i are depicted in Fig. 1. One possibility is that the dissociated α_i subunit directly inhibits the catalytic activity of adenylate cyclase (mechanism 1 in Fig. 1). A second possibility is that the $\beta\gamma$ complex from dissociated G_i may interact with G_s to prevent or reverse the activation of adenylate cyclase by the stimulatory G protein.[27] Since the β and γ subunits of G_s and G_i are believed to be the same, the presence of large amounts of $\beta\gamma$ subunits from dissociated G_i would favor the association of α_s with $\beta\gamma$ subunits, thus inactivating α_s (mechanism 2 in Fig. 1). It should be noted that alternative explanations are possible, and there is some controversy regarding the precise mechanism by which G_i inhibits adenylate cyclase.[28]

As was the case with the stimulatory G protein, the activity of G_i can be affected by a bacterial toxin, in this case the toxin produced by *Bordetella pertussis,* the causative agent in whooping cough. In analogy with the effect of cholera toxin on G_s, pertussis toxin can cause ADP-ribosylation of the α subunit of G_i. However, rather than causing a persistent activation of the α subunit, this ribosylation somehow interferes with the activation of the G_i. The net effect is to prevent the inhibition of adenylate cyclase normally caused by stimulation of the inhibitory receptors. The magnitude of the effect seen in a given tissue with pertussis toxin will depend on the degree of tonic inhibition present in the tissue. If a substantial amount of endogenous inhibitory activity is present, pertussis toxin may produce a marked increase in cAMP production. Pertussis toxin has also been a valuable experimental tool for defining the roles of inhibitory G proteins.

The presence of guanine-nucleotide-dependent regulatory proteins coupled to adenylate cyclase has been demonstrated in rat uterine smooth muscle by Krall *et al.*[29] and by Tanfin and Harbon.[30] Possible roles for these G proteins in the regulation of myometrial cAMP levels in the presence of various agonists are further discussed in Section 1.3 below.

1.3. Evidence for and against a Role for cAMP as a Mediator of Uterine Relaxation

As noted above, it is generally assumed that the effects of β-adrenoceptor agonists are mediated by increases in tissue levels of cAMP. Consequently, when it was discovered that β agonists could increase cAMP levels in isolated uterine preparations, it was suggested that cAMP was responsible for the uterine relaxation caused by these agents.[31] Since that time many of the criteria usually used to establish a role for cAMP as the mediator of a particular physiological response have been satisfied, at least to some extent, for the relaxant effects of β agonists on uterine smooth muscle. Early evidence in support of this hypothesis included the following: (1) Epinephrine, norepinephrine, and isoproterenol stimulate adenylate cyclase activity[32,33] and increase cAMP levels[31] in rat uterine preparations. (2) The order of potency for stimulation of adenylate cyclase (isoproterenol > epinephrine > norepinephrine) is the same as that for relaxation of isolated uterine strips.[33] (3) The increase in cAMP and the relaxation caused by the catecholamines are both blocked by the β-adrenoceptor blocker propranolol.[31,33] (4) A dose–response relationship was reported between the concentrations of catecholamines required to elevate cAMP and to relax the uterus.[33] (5) The increase in uterine cAMP levels caused by isoproterenol is correlated temporally with relaxation.[34] (6) A phosphodiesterase inhibitor, papaverine, relaxes pregnant rat myometrium and causes a parallel increase in cAMP levels.[34] (7) Another phosphodiesterase inhibitor, theophylline, potentiates the uterine relaxant effects of low concentrations of epinephrine.[35,36] (8) The dibutyryl derivative of cAMP inhibits spontaneous and oxytocin-induced contractions in isolated rat uteri.[31,34−36] All of these observations are consistent with, and tend to support, the hypothesis that the uterine relaxant effects of β-adrenoceptor agonists are mediated by increases in tissue levels of cAMP.

In one of the early reports by Triner and co-workers cited above,[32] evidence was presented suggesting that both α- and β-adrenoceptors might be coupled to adenylate cyclase in rat myometrium. In this study, adenylate cyclase activity was measured in homogenates of uteri obtained from estrogen-primed rats. Epinephrine, isoproterenol, and norepinephrine all increased cAMP production by 80–90% in these homogenates, and this response was blocked by propranolol. Pretreatment of the homogenates with an α blocker, phentolamine, tended to increase the amount of cAMP production caused by epinephrine and norepinephrine. These two catecholamines are capable of activating both α- and β-adrenoceptors. Finally, in the presence of propranolol, low concentrations of epinephrine decreased adenylate cyclase activity below control levels. These results suggested that stimulation of β-adrenoceptors increased adenylate cyclase activity in the rat uterus, whereas stimulation of α-adrenoceptors inhibited cAMP production in this tissue. Since these experiments were done in tissue homogenates, no direct correlations could be made between the effects of the catecholamines on adenylate cyclase activity and their effects on uterine contractility. Nevertheless, these results are consistent with the known effects of α and β agonists on rat uterine contractility as described above, and it was suggested that changes in adenylate cyclase activity might be responsible for the α (excitatory) effects as well as for the β (inhibitory) effects of the catecholamines.[32] These results are also consistent with the general mechanisms depicted in Fig. 1 for the dual regulation of adenylate cyclase activity by G proteins.

In spite of this presumptive evidence for α-adrenoceptor-mediated inhibition of uterine adenylate cyclase activity, there does not appear to be any direct evidence to support the contention that decreases in myometrial cAMP levels are responsible for the uterine contractile effects of α-adrenoceptor agonists (see for example ref. 37). In fact, the most recent evidence suggests that the two phenomena are mediated by different α-adrenoceptor subtypes. In at least some species, the uterine contractile effects of adrenergic agonists are mediated through α_1-adrenoceptors,[14] whereas the adenylate-cyclase-inhibiting effects of these agents appear to be mediated through α_2-adrenoceptors.[13] Thus, most of the early evidence in the literature seems to favor a role for cAMP elevation as a mediator of the uterine relaxant effects of β agonists but not for adenylate cyclase inhibition as a mediator of the uterine contractile effects of α agonists.

Studies similar to those described above for uterine smooth muscle have also been carried out in a number of other smooth muscle types. As a result of these studies, cAMP has been implicated in the relaxant effects of a variety of agents, both catecholamines and noncatecholamines, in a variety of smooth muscles. The evidence favoring a role for cAMP as a mediator of relaxation in other types of smooth muscle has been extensively reviewed elsewhere.[16,38,39] The evidence for such a role is substantial, and the hypothesis has become widely accepted. Consequently, a number of laboratories have turned their attention to attempting to elucidate the mechanism(s) by which elevation of tissue cAMP could bring about smooth muscle relaxation. Several possible mechanisms have been suggested, including (1) lowering of cytoplasmic calcium by intracellular sequestration,[34,40,41] (2) stimulation of Na,K-ATPase activity resulting in acceleration of Na–Ca exchange and increased calcium efflux from the cytoplasm,[42] and (3) interference with contraction at the level of the contractile proteins by cAMP-dependent phosphorylation of myosin light chain kinase.[43] In each of these proposed mechanisms it is assumed that elevation of cAMP results in activation of a cAMP-dependent protein kinase that, in turn, phosphorylates a specific protein or proteins somehow involved in the regulation of smooth muscle contraction (see Fig. 1).

In spite of the diversity of evidence favoring a role for cAMP as a mediator of smooth muscle relaxation and the general acceptance of the hypothesis, there is a substantial amount of data in the literature that are difficult to reconcile with this hypothesis. The remainder of this section focuses on some of these apparently contradictory observations and possible explanations for them. The discussion is limited to the results of experiments utilizing uterine smooth muscle preparations. For further discussion of the literature relating to the role of cAMP in other types of smooth muscle, the reader is referred to the general reviews cited above.[16,38,39]

It should be noted at the outset that cAMP elevation by itself is not always sufficient to cause smooth muscle relaxation, nor is it required for the smooth-muscle-relaxing effects of a variety of agents.[44] Even for agents assumed to act via cAMP elevation, such as β-adrenoceptor agonists or phosphodiesterase inhibitors, dissociations between cAMP elevation and relaxation have been reported. One of the earliest reports of such a dissociation was that of Polacek and Daniel,[37] who found that when estrogen-primed rat uteri were treated with a relaxant concentration of isoproterenol, the cAMP levels increased and remained elevated even when uterine contractions were restored by addition of propranolol to the muscle baths. It was also found that low concentrations of theophylline and papaverine relaxed the uterus with little or no elevation of cAMP.[45] These authors concluded that the level of cAMP in the uterus is not the primary determinant of uterine motility. Similar conclusions were drawn from subsequent studies on KCl-depolarized rat myometrium. Cyclic AMP levels were significantly increased in uterine strips contracted by KCl depolarization, but the muscles remained contracted indefinitely in spite of the elevated cAMP levels.[46,47] The KCl-contracted muscles could be relaxed by concentrations of papaverine[46] and isoproterenol[46,47] that produced little or no further elevation of cAMP. These results argued against an obligatory role for cAMP in the uterine relaxation caused by papaverine and isoproterenol.

Other more recent reports have also noted apparent dissociations between cAMP elevation and relaxation of rat myometrium by β-adrenoceptor agonists. In one such report,[48] the effects of ritodrine, an agonist with some selectivity for β_2-adrenoceptors, were compared with those of isoproterenol on myometrial preparations obtained from pregnant rats. The effects of these agents on membrane potential, cAMP levels, and force of contraction were studied in preparations of circular and longitudinal muscle excised from rats on the 17th to 19th days or on the 22nd (final) day of gestation. In each case, isoproterenol and ritodrine hyperpolarized the membrane, increased cAMP production, and relaxed the muscles. However, in any given experiment the concentrations of the β- agonists required to elevate cAMP were much greater than those required to relax the myometrium. For example, in longitudinal muscle cells from 17- to 19-day pregnant animals, ritodrine produced significant inhibition of contraction at concentrations of 0.01 μM or higher, where-as cAMP levels were not increased until concentrations of 1.0 or 10 μM were used. In circular muscle cells from midgestational rats, progressively greater relaxation occurred with concentrations of ritodrine from 0.01 to 10 μM, but no cAMP elevation and no significant membrane hyperpolarization were observed with any concentration of the drug. It was concluded that relaxation of rat myometrium by β-adrenoceptor agonists is probably not caused solely by an increased concentration of cAMP.

A similar but less dramatic dissociation between cAMP content and contrac-tility has also been observed in rat myometrium with forskolin, a direct activator of adenylate cyclase that does not act through β-adrenoceptors.[49] A low concentration of forskolin (0.1 μM) reduced the amplitude of spontaneous contractions of term pregnant rat myometrium by about 75% but did not increase cAMP levels. The inhibition of contractile force by that concentration of forskolin was accompanied by a decrease in the frequency of action potentials but no change in the resting membrane potential. A higher concentration of forskolin (1.0 μM), in addition to abolishing spontaneous contractions and action potentials, also hyperpolarized the membrane and increased cAMP levels. Pretreatment with 1.0 mM ouabain reduced the magnitude of the hyperpolarization produced by 1.0 μM forskolin but had no effect on the relaxation or cAMP accumulation caused by the drug. It was concluded that changes in membrane potential are not prerequisite for the inhibitory actions of forskolin.[49] This is consistent with earlier results indicating that hyperpolarization is not required for the β-adrenoceptor-mediated relaxant effects of catecholamines in pregnant rat myometrium.[50] These results also suggest the existence of a cAMP-independent relaxation mechanism for low doses of forskolin. However, the pos-sibility exists that there may be a local increase in cAMP not detected by total tissue measurements of the cyclic nucleotide but sufficient to cause a decrease in tension.

Dissociations between isoproterenol-induced relaxation and cAMP elevation have also been reported in uterine preparations from other species, including the rabbit. For example, a low concentration of isoproterenol (0.02 μM) produced an inhibition of spontaneous contractions in myometrial strips from estrogen-primed rabbits but had no effect on cAMP levels.[51] A higher concentration of isoproterenol (2.0 μM) produced both inhibition of mechanical activity and elevation of cAMP. Pretreatment of the uterine strips with the β-adrenoceptor-blocking drug propranolol prevented the cAMP elevation caused by 2.0 μM isoproterenol but did not prevent the inhibition of mechanical activity caused by the drug. However, the relaxant effect of isoproterenol could be mimicked by addition of the dibutyryl derivative of cAMP. These results were interpreted to mean that isoproterenol could cause relaxa-tion of the rabbit uterus by two mechanisms, one cAMP-dependent and one cAMP-independent.[51]

In a second report from this laboratory,[52] the effects of isoproterenol on elec-trical activity of the membrane as well as on mechanical activity were monitored. Isoproterenol was found to cause hyperpolarization of the membrane and abolition of spike potentials in addition to inhibition of mechanical activity. All of these effects were mimicked by dibutyryl cAMP. Propranolol blocked the membrane-hyperpolarizing effect of isoproterenol but not its effects on spike generation or mechanical activity. These results, together with those in the earlier report,[51] suggest that cAMP elevation and membrane hyperpolarization are related phenomena but that neither one is obligatory for isoproterenol-induced relaxation of the rabbit uterus. Surprisingly, the membrane-hyperpolarizing effect of dibutyryl cAMP was also blocked by propranolol. No explanation was offered for this unexpected finding. Since dibutyryl cAMP appeared to mimic all of the effects of isoproterenol, and since no direct evidence for a cAMP-independent mechanism for isoproterenol-induced relaxation was found in this study, it was suggested that the component of relaxation originally assumed to be cAMP-independent might, in fact, be related to a small, undetectable increase in cAMP in a localized compartment.[52] Why this hypothetical cAMP pool would be resistant to the effects of propranolol is not clear, since propranolol blocked the elevation of cAMP in all measurable compartments.

A similar dissociation between cAMP elevation and relaxation caused by isoproterenol in estrogen-primed rabbit uterus has been noted in a more recent report from another laboratory.[53] In this study, isoproterenol (0.005 to 0.5 μM) produced a dose-dependent inhibition of acetylcholine-induced myometrial contractions but had no effect on tissue levels of cAMP at any of the concentrations tested. This appears to be a clear dissociation between the cAMP-elevating and the uterine-relaxing effects of the β agonist. The effects of isoproterenol in this tissue were also compared with those of the adenylate cyclase activator forskolin. In agreement with the results noted above for the rat myometrium,[49] the lowest dose of forskolin used (0.1 μM) inhibited spontaneous contractions of the rabbit myometrium by about 55% but had no effect on tissue cAMP levels. However, higher concentrations of forskolin further relaxed the muscles and caused significant increases in cAMP levels. There was a statistically significant correlation between cAMP accumulation and mechanical inhibition for concentrations of forskolin between 0.2 and 5.0 μM. It was concluded that although cAMP might mediate relaxation in the myometrium under certain conditions, it is not an obligatory mediator, and β-adrenoceptor agonists can affect muscle tension via cAMP-independent as well as cAMP-dependent mechanisms.[53]

A very interesting series of papers concerning the role of cAMP in uterine motility has come from the laboratory of Dr. Simone Harbon.[30,54–58] Over a period of years, and using progressively more sophisticated techniques, Dr. Harbon and her associates have studied the interactions between catecholamines and prostaglandins of the E series (PGE_1 and PGE_2) on cAMP levels and tension in rat myometrium. In early experiments it was demonstrated that epinephrine, PGE_1, and PGE_2 all produced similar degrees of cAMP elevation in uterine strips from estrogen-primed rats.[54,55] However, epinephrine relaxed the muscles, whereas PGE_1 and PGE_2 contracted them. There appeared to be a good correlation between the cAMP-elevating and the uterine-relaxing effects of epinephrine, thus supporting a role for cAMP in uterine relaxation. On the other hand, the fact that the uterine strips remained contracted during exposure to the prostaglandins in spite of the elevated cAMP levels was at odds with this concept. It was also found that prostaglandin-contracted muscles could be relaxed by concentrations of epinephrine that produced no further increases in cAMP beyond those caused by the prostaglandins themselves.

In order to explain these observations within the context of the cAMP hypothesis, it was necessary to suggest that cAMP might be compartmentalized within the myometrial cells and that the prostaglandins increased cAMP levels in a compartment that was different from that affected by epinephrine and that was not involved in regulation of contractility.[55] However, in subsequent reports from the same laboratory, evidence was presented indicating that epinephrine and PGE_1 increased cAMP levels in the same intracellular compartment.[56,57] This conclusion was based on observations that epinephrine and PGE_1 produced the same degree of saturation of intracellular cAMP binding sites[56] and the same magnitude of activation of cAMP-dependent protein kinase[57] for a given amount of cAMP elevation. To rule out the possibility that the catecholamines and the prostaglandins might exert different effects on adenylate cyclase from endothelial versus myometrial cells, the above experiments were performed on myometrial preparations from which the endothelium had been removed. It was concluded that there was no difference between the cAMP formed as a result of PGE_1 or epinephrine administration and that cAMP compartmentalization could not explain the differential effects of the two agents on uterine tension.[56] It was further suggested that factors other than cAMP elevation must be important for the uterine-relaxing effects of the catecholamines.

The role of cAMP in uterine relaxation was reevaluated, and the hypothesis modified somewhat, in a later report from Dr. Harbon's laboratory.[58] In this report, the effects of isoproterenol on myometrial tension and cAMP levels were compared with those of PGE_2 and forskolin, alone or in combination. Uterine relaxation seen following forskolin or forskolin plus PGE_2 was accompanied by much larger elevations of cAMP than those seen with equirelaxant concentrations of isoproterenol. For example, complete inhibition of carbachol-induced contractions by 20 μM isoproterenol was accompanied by an increase in cAMP levels to 12.6 pmol/mg protein (from a control level of 5 pmol/mg protein). Elevation of cAMP to the same level (12.4 pmol/mg protein) by 0.1 μM forskolin was accompanied by little or no relaxation. Elevation of cAMP to 115.6 pmol/mg protein by 3.0 μM forskolin was accompanied by an 80% inhibition of the carbachol-induced contractions, i.e., approximately ten times as much cAMP elevation as that required for a similar relaxation caused by isoproterenol.

As a result of these and other observations, it was proposed that relaxation induced by β-adrenoceptor stimulation is the result of the combined effects of both a cAMP-dependent and a cAMP-independent process.[58] It was assumed that, with the moderate elevations of cAMP produced by the catecholamines, neither process alone is sufficient to cause relaxation. However, at very high levels cAMP appears to be capable of producing relaxation in the absence of other inhibitory processes. Consequently, forskolin, although it lacks the ability to activate the cAMP-independent process, is capable of causing relaxation once sufficiently high levels of cAMP have been attained, i.e., much higher than those accompanying similar degrees of relaxation caused by isoproterenol. If the cAMP-independent process is not capable of causing relaxation by itself, as has been suggested,[58] then some degree of cAMP elevation should always be required for β-adrenoceptor-induced relaxation. However, in many of the reports discussed above for catecholamine-induced relaxation of rat and rabbit myometrium (see for example refs. 48, 49, 51, and 53), marked relaxation could be demonstrated with no measurable increase in tissue cAMP levels. In the absence of any direct evidence for compartmentalization of cAMP, the possibility should therefore be considered that low doses of catecholamines can relax myometrial preparations by a cAMP-independent process alone. With higher doses of these agents, the large increases in cAMP levels would presumably contribute to the relaxant response.

In a very recent report by Tanfin and Harbon,[30] the coupling of adrenergic and prostaglandin receptors to other components of the adenylate cyclase complex has been studied in uteri obtained from pregnant rats at different stages of gestation. When compared with the responses in estrogen-dominated myometrium, there was a marked attenuation at midgestation (day 12) of the adenylate cyclase activation seen in response to the receptor-mediated activators, isoproterenol and PGE_2, as well as to forskolin and cholera toxin. A progressive restoration of adenylate cyclase responsiveness, with full responsiveness returning by day 22, was seen with isoproterenol, forskolin, and cholera toxin but not with PGE_2. In uteri from term pregnant rats, PGE_2 markedly inhibited the ability of isoproterenol to increase the production of cAMP. Possible mechanisms responsible for these interactions were investigated in an elegant series of experiments in which the contributions of the guanine nucleotide-dependent regulatory proteins, G_s and G_i, to the responses were evaluated. Evidence was provided that the attenuated responses to isoproterenol and cholera toxin seen during midgestation were caused by an increase in the amount of G_i relative to G_s. Treatment of the tissues with pertussis toxin (an inhibitor of G_i) restored the responses to isoproterenol and cholera toxin but not to PGE_2. The ability of PGE_2 to inhibit activation of adenylate cyclase by isoproterenol in uteri from rats in the later stages of pregnancy was also prevented by pertussis toxin. It was suggested that the response to PGE_2 was converted from a G_s-mediated stimulation of adenylate cyclase in the estrogen-dominated myometrium to a G_i-mediated inhibition of cAMP production in the final stages of gestation. An inhibition of cAMP production by prostaglandins at the end of gestation might be important physiologically, since it should favor the increase in uterine contractility required for delivery. Thus, this report not only documents the presence of guanine-nucleotide-dependent regulatory proteins in myometrial tissues but provides some important information on possible roles for these proteins in regulating the responses to receptor-mediated agonists under differing hormonal conditions.

It should be noted, however, that some of the results presented by Tanfin and Harbon[30] are difficult to reconcile with previous observations in the literature. For example, an early study using intact segments of rat uterus showed that the responses to norepinephrine were predominantly relaxant (β adrenergic) in progesterone-treated uteri but predominantly excitatory (α adrenergic) in uteri under the influence of estrogen alone.[19] Similarly, in more recent studies where the effects of catecholamines on longitudinal and circular muscle layers were differentiated,[20,21] mainly inhibitory responses to norepinephrine were seen in longitudinal muscles obtained from rat uteri at midgestation (progesterone-dominated), whereas contractile responses were more prominent in longitudinal muscles obtained from rats at term pregnancy (estrogen-dominated). These results do not appear to be consistent with Tanfin and Harbon's observation that the cAMP-elevating effects of β agonists are attenuated in longitudinal smooth muscle cells from progesterone-dominated rat myometrium (i.e., day 12 of gestation) but are enhanced under conditions of estrogen dominance (i.e., estradiol-injected or term pregnant animals).[30] If it is assumed that cAMP mediates the uterine relaxant effects of the catecholamines, exactly the opposite results would have been expected. The relaxant effects of the catecholamines should be more prominent in estrogen-dominated tissues where the cAMP-elevating effects of the drugs are enhanced. However, as noted above, the excitatory or contractile effects are most prominent under these conditions. This implies that factors other than changes in adenylate cyclase activity or responsiveness must be important in the modulation of uterine responses to catecholamines by ovarian hormones.

A similar conclusion can be inferred from a recent study by Ohia and Boyle.[59] These authors presented evidence that the variations in response of the rat uterus to the β-adrenoceptor agonist salbutamol that occurred during the natural estrous cycle were not caused by differences in the response of adenylate cyclase to salbutamol or by an effect of endogenous prostaglandins on cAMP production. Thus, there are still many unanswered questions in this area, and further experiments will be necessary to understand fully the nature of the interactions between cAMP and the ovarian hormones in the control of uterine motility.

2. Role of Cyclic GMP in Control of Uterine Motility

In addition to cAMP, most mammalian tissues including uterine smooth muscle contain another cyclic nucleotide, guanosine 3',5'-cyclic monophosphate (cGMP). In analogy with the cAMP system, the enzymes responsible for the synthesis and degradation of cGMP are also present in these tissues. However, in contrast to the situation with adenylate cyclase, guanylate cyclase does not appear to be coupled to drug receptors via G proteins. Although cGMP has not been as widely studied in uterine smooth muscle as cAMP, a number of reports have been published since 1973 attempting to describe a role for cGMP in control of uterine motility. Cyclic GMP was initially believed to be a mediator of smooth muscle contraction but was later suggested to be involved in smooth muscle relaxation. A brief review of some of the studies that led to these divergent hypotheses is presented in the following sections.

2.1. Cyclic GMP as a Possible Mediator of Uterine Contraction

During the early 1970s, a series of reports appeared in the literature suggesting that cGMP might be responsible for the initiation of contractions by a variety of smooth muscle stimulants in various types of smooth muscles.[60-67] For example, in estrogen-primed rat uteri, cGMP levels were reported to be elevated by 150% to 500% after exposure to oxytocin, serotonin, methacholine, and prostaglandin $F_{2\alpha}$.[60] Marked increases in cGMP levels were also observed in guinea pig ileum in response to acetylcholine and bethanechol,[61] in rat vas deferens in response to carbachol and acetylcholine,[62,63] in human umbilical artery in response to bradykinin, serotonin, acetylcholine, and potassium ions,[64] in canine veins in response to prostaglandin $F_{2\alpha}$,[65,66] and in guinea pig trachea in response to acetylcholine, carbachol, and histamine.[67] Although simultaneous tension measurements were not made in most of these studies, all of the agents that were found to increase cGMP levels in the various smooth muscles were also known to be capable of producing contractions in the respective muscles. Therefore, these results tended to support the hypothesis, originally suggested by Goldberg and his co-workers,[60] that increases in tissue levels of cGMP were responsible for the promotion of smooth muscle contractions. This suggestion was part of the much broader Yin–Yang hypothesis described by Dr. Goldberg, in which it was suggested that the two cyclic nucleotides, cAMP and cGMP, exerted opposing effects in the control of a variety of biological processes. As discussed in Section 1.3, cAMP was assumed to be a mediator of uterine relaxation, and this, together with the observation that several uterine stimulants could increase cGMP levels in rat uterus, was a key factor in the formulation of the Yin–Yang hypothesis.

Other subsequent studies, however, failed to support this hypothesis with respect to the role of cGMP as a mediator of smooth muscle contraction. For example, attempts by several laboratories to confirm the ability of stimulant drugs to increase cGMP levels in rat uteri were unsuccessful.[68-70] No increases in cGMP levels were detected in strips of rat myometrium contracted by any of the uterine stimulants tested, including carbachol, angiotensin, prostaglandin $F_{2\alpha}$, and A23187. In contrast to these results in the rat uterus, marked elevations of cGMP were observed during carbachol-induced contractions of guinea pig myometrium[68,70] and taenia coli.[68] However, the increases in cGMP in these tissues did not occur until after the onset of the contractions.[68] For example, in the guinea pig myometrium, cGMP levels did not begin to increase until 15 sec after the onset of carbachol-induced contractions, at which time the muscles had already developed close to maximal tension.[68] These results, together with previous reports indicating that the cGMP elevations accompanying contractions of other types of smooth muscles were calcium dependent,[63,64] suggested that the cGMP increases that occurred during contractions of guinea pig myometrium were not responsible for initiation of the contractions but were a result of the increased cytoplasmic calcium concentrations that are known to accompany the contractions. The elevation of cGMP during drug-induced contractions in guinea pig uterus but not in rat uterus is therefore consistent with a later report demonstrating the existence of a calcium-dependent mechanism for elevation of cGMP in the former but not in the latter.[70] Dissociations between cGMP content and contraction were also reported in other types of smooth muscle such as the canine femoral artery.[71] In this tissue, no changes in cGMP levels were seen at any time during contractions induced by maximally effective concentrations of phenylephrine. Thus, results from several different tissues failed to support the proposed role for cGMP as a mediator of smooth muscle contraction.

Perhaps the strongest evidence against a role for cGMP as a mediator of smooth muscle contraction was provided by early reports that elevation of cGMP in response to drugs such as papaverine and nitroglycerin was accompanied by smooth muscle relaxation rather than contraction.[71,72] In the initial experiments on rat myometrial strips, cGMP levels were decreased rather than increased during potassium-induced contractures.[72] Addition of the vasodilator nitroglycerin increased cGMP levels by more than 80% and relaxed the muscles by about 50%. In phenylephrine-contracted canine femoral arteries, 100 μM carbachol increased cGMP levels by more than 100% but had no effect on tension.[71] High concentrations of papaverine and nitroglycerin completely relaxed the arteries and increased cGMP levels by 75% and 1540%, respectively. The 16-fold elevation of cGMP caused by nitroglycerin in this experiment was the largest increase in cGMP seen in any tissue up to that point and argued strongly against a role for cGMP as a mediator of smooth muscle contraction, since the muscles were relaxed by the nitroglycerin.

In summary, cGMP levels are increased during agonist-induced contractions in some but not all smooth muscles. In those muscles in which cGMP levels are increased during contractions, the increases in cGMP are calcium dependent and occur after the onset of the contractions. This indicates that the increases in cGMP are not responsible for the contractions but probably occur as a result of the increases in cytoplasmic calcium concentration that accompany the contractions.

2.2. Cyclic GMP as a Possible Mediator of Uterine Relaxation

As noted above, one of the strongest pieces of evidence against a role for cGMP as a mediator of smooth muscle contraction was the observation that nitroglycerin and papaverine could produce marked increases in cGMP levels in uterine and vascular smooth muscles and that these cGMP elevations were accompanied by relaxation of the muscles rather than contraction.[71,72] The ability of nitroglycerin to increase smooth muscle cGMP levels has since been confirmed in other types of smooth muscle.[73,74] In addition, it was found that a variety of other smooth muscle relaxants could increase cGMP levels in these preparations, and it was suggested that not only was cGMP not a mediator of smooth muscle contraction, but it might in fact be a mediator of the smooth muscle relaxant effects of certain drugs.[73,74] Since that time a large body of evidence has appeared in the literature supporting this hypothesis.

As a corollary to this hypothesis, it is suggested that the calcium-dependent increases in cGMP that occur during contractions of several types of smooth muscle (see previous section) may be part of a negative feedback mechanism functioning to lower cytoplasmic calcium to limit the contractions and/or prevent damage to the cell caused by excess calcium.

The evidence for a general role for cGMP in smooth muscle relaxation has been thoroughly reviewed elsewhere[75,76] and is not covered in any detail here. This evidence appears to be most convincing for vascular smooth muscle, and many of the criteria usually used to determine whether a cyclic nucleotide is involved in a particular response have been satisfied for cGMP as a second messenger for relaxation in vascular tissues. Several types of agents have been suggested to exert their vascular relaxant effects via elevation of cGMP. One group of drugs, collectively referred to as the nitrovasodilators[76] and including nitroglycerin, sodium nitroprusside, sodium nitrite, and others, appears to act through a common intermediate (nitric oxide) that activates soluble guanylate cyclase and increases tissue levels of cGMP. Another group of agents, which includes acetylcholine, bradykinin, ATP, and others, appears to act by releasing a smooth muscle relaxing factor from vascular endothelial cells. This factor in turn activates soluble guanylate cyclase in the smooth muscle cells, thus increasing the levels of cGMP in the muscle cells. The precise mechanism by which cGMP might bring about smooth muscle relaxation is not clear, although several possibilities have been suggested. References for the above points can be found in the reviews cited earlier.[75,76] As described in these reviews, excellent temporal and dose–response relationships have been reported for the abilities of these drugs to increase cGMP levels and to relax vascular preparations. Furthermore, procedures that can prevent the activation of guanylate cyclase by these agents can also block the vascular relaxant effects of the drugs. Selective inhibitors of cGMP phosphodiesterase have been reported to relax vascular preparations and to potentiate the relaxant effects of guanylate cyclase activators. Finally, the 8-bromo derivative of cGMP has been shown to relax a variety of smooth muscle preparations. These and other lines of evidence all tend to support a role for cGMP in smooth muscle relaxation. The volume and diversity of evidence in favor of this hypothesis is quite substantial, and it seems likely that cGMP plays a role in mediating the relaxant effects of a variety of drugs in vascular smooth muscle. However, the evidence for such a role in uterine smooth muscle is not as convincing. The remainder of this section focuses on the evidence for and against a role for cGMP specifically as a mediator of uterine relaxation.

A possible role for cGMP in uterine relaxation is supported by several lines of evidence: (1) nitrovasodilators such as nitroglycerin, nitroprusside, and hydroxylamine have been shown to increase cGMP levels in rat uterine preparations[72,77,78]; (2) phosphodiesterase inhibitors such as theophylline[77,79] and 1-methyl-3-isobutylxanthine,[77] which are known to relax many smooth muscle preparations, can increase cGMP levels in rat uterus; and (3) exogenous administration of 8-bromo cGMP can relax a variety of smooth muscle preparations including rat myometrial strips.[78,80]

These observations all provide circumstantial evidence that cGMP may be involved in uterine relaxation. However, none of these studies noted good temporal or dose–response relationships between the ability of the guanylate cyclase activators or phosphodiesterase inhibitors to elevate cGMP levels and their ability to relax the uterus. One of the most effective compounds at increasing uterine cGMP levels in these studies was sodium nitroprusside, but this compound was unable to relax spontaneous or KCl-induced contractions in rat myometrial strips at any concentration tested.[78] For example, 5 mM nitroprusside increased cGMP levels almost sixfold in isolated strips of rat myometrium but had no effect on spontaneous contractions in these preparations.[78] Nitroglycerin, on the other hand, completely inhibited spontaneous contractions in the uterine muscle at a concentration (0.5 mM) that had no significant effect on tissue levels of cGMP. Similar dissociations were also observed with two other nitrovasodilators, hydroxylamine and sodium azide.[78] For example, 1 mM hydroxylamine increased cGMP levels approximately 3.3-fold but had no effect on spontaneous contractions. A higher concentration of the drug (5 mM) completely inhibited the contractions and elevated cGMP levels about 3.9-fold. A similarly high concentration of sodium azide (5mM) completely relaxed the uterine strips but increased cGMP levels by less than 1.4-fold. All of these agents are believed to relax other smooth muscles by a common mechanism involving release or production of nitric oxide, which in turn activates guanylate cyclase in the muscle cells.[75] All of the drugs should therefore increase cGMP levels in the same intracellular pool(s) and should produce the same amount of relaxation for a given increase in cGMP. However, when cGMP elevation was plotted against the degree of uterine relaxation produced by various concentrations of these agents, no correlation was found between the two parameters (Fig. 2). The fact that nitroprusside was incapable of relaxing the muscles at any concentration, even though it produced the largest elevations of cGMP, indicates that the uterine relaxant effects of high concentrations of nitroglycerin, hydroxylamine, or sodium azide must be mediated via cGMP-independent mechanisms.

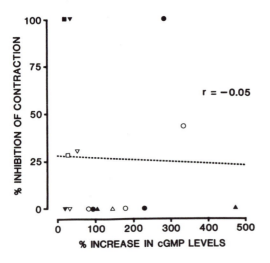

Figure 2. Lack of correlation between cGMP elevation and relaxation of rat myometrium by various concentrations of sodium nitroprusside (▲, △), nitroglycerin (■, □), sodium azide (▼, ▽), and hydroxylamine (●, ○). Filled symbols represent experiments on spontaneously contracting muscles, and open symbols represent experiments on KCl-contracted muscles (mean values of four to nine experiments). The calculated correlation coefficient (r) is −0.05. Modified from Diamond.[78]

The ability of a cyclic nucleotide or its analogues to mimic the effects of a drug or hormone is usually taken as strong evidence that the cyclic nucleotide plays a role as a mediator of the effects of the drug. As noted above, one of the pieces of evidence supporting a role for cGMP in uterine relaxation was the observation that 8-bromo cGMP can relax isolated strips of rat myometrium.[78,80] However, this ability of 8-bromo cGMP to relax uterine smooth muscle is difficult to reconcile with the observation that marked elevation of uterine cGMP by nitroprusside is not accompanied by relaxation. It is possible that, in the rat uterus, the cGMP increases caused by nitroprusside and the other nitrovasodilators occur in a compartment or pool that is not involved in the regulation of contractile activity. Alternatively, it is possible that cGMP does not play a role in the regulation of contraction in this tissue and that high concentrations of 8-bromo cGMP can relax the uterus by some mechanism independent of cGMP. In the absence of direct evidence to the contrary, this seems to be the simplest explanation for the available data. It is obvious that exogenous administration of 8-bromo cGMP does not necessarily mimic the effects of an increase in tissue levels of the parent cyclic nucleotide, thus indicating that caution must be used in interpreting the results of studies with this compound. Previous reports have also questioned the validity of using other cyclic nucleotide analogues such as dibutyryl cAMP in studies on the role of the cyclic nucleotides in smooth muscle.[81]

In summary, when all of the available evidence is considered, it appears that cGMP plays a minor role, if any, in the control of uterine motility.

3. Use of β-Adrenoceptor Agonists in Premature Labor

Preterm birth is the largest single contributor to infant morbidity and mortality in North America.[82] Since many premature deliveries are preceded by one or more episodes of preterm labor, a great deal of attention has been focused on attempts to suppress labor and prevent preterm delivery. However, the benefits of aggressive treatment of established preterm labor (i.e., with uterine relaxants or tocolytics) are not universally accepted.[83–85] Aggressive tocolytic therapy may expose both the mother and the fetus to increased and possibly unnecessary risks. One problem is that the diagnosis of preterm labor is very difficult. No satisfactory screening test or marker is available to establish firmly the diagnosis of impending labor. If the diagnosis is based solely on the basis of uterine contractions, many patients may be treated with potentially toxic drugs for a condition that may not exist. Even when significant delays in the onset of labor can be attained, this may not necessarily be associated with a reduction in perinatal mortality.[84,86] In spite of the availability of several effective tocolytic agents and more aggressive management approaches in recent years, the overall incidence of preterm delivery has remained constant, and prematurity still accounts for the vast majority of neonatal morbidity and mortality. Thus, the value of tocolytic therapy in premature labor remains controversial. A detailed discussion of this area is beyond the scope of this chapter, and the reader is referred to recent reviews[83-85] for further discussion and references.

In spite of the potential problems alluded to above, tocolytic therapy is generally accepted as standard care in the treatment of preterm labor in North America.[85,87] Several tocolytic agents have been used in this condition, including magnesium sulfate, indomethacin, and nifedipine in addition to the β-sympathomimetic drugs. As discussed in Section 1.1.1, the adrenoceptors involved in mediating the uterine relaxant effects of sympathomimetic amines appear to be of the β_2 subtype. Therefore, in order to minimize the potential side effects of these agents, β agonists with some selectivity for β_2-adrenoceptors have been most widely used. These include terbutaline, salbutamol, and ritodrine, among others. At the present time ritodrine is the only drug approved by the FDA for the treatment of preterm labor, and it is considered by many institutions to be the tocolytic drug of choice for the treatment of this condition.[86,87] The efficacy of β sympathomimetics in suppressing premature labor appears to be well established. In a recent "metaanalysis" of the data from 16 well-controlled trials of these agents in preterm labor,[86] the β agonists had an unequivocal effect in delaying the onset of delivery, and this was reflected in a reduction in the frequency of preterm birth and low birth weight. However, as noted above, no beneficial effect of this treatment on perinatal mortality or severe neonatal respiratory disorders could be demonstrated.[86] Ritodrine appears to be less effective as a tocolytic agent in patients with premature rupture of the membranes, and its usefulness in these patients has been questioned.[88]

In comparative trials, several of the other drugs available for tocolysis have been shown to be as effective as ritodrine in the suppression of preterm labor.[85,89] Therefore, the incidence and severity of side effects may be the deciding factor in choosing which drug or combination of drugs to use. The β sympathomimetics, including ritodrine, have been reported to have a number of relatively common and potentially dangerous side effects, including pulmonary edema, myocardial ischemia, cardiac arrhythmias, hypotension, and hyperglycemia.[90] It was therefore recommended that these drugs be used with caution in patients with diabetes, angina, or other cardiac problems.[90]

It should also be noted that β agonists fail to suppress labor in some patients, and in others their labor-inhibiting effect is transient.[91] The latter phenomenon may be related to the mode of administration of the drugs.[92] Concern for these potential problems has led some investigators to suggest that other agents should replace the β agonists as the first line of tocolytic therapy. For example, in a randomized comparison of ritodrine and magnesium sulfate, the efficacy and overall number of side effects were similar for the two agents, but the side effects were judged to be less serious in the patients treated with magnesium.[89] On the basis of this and other considerations, it was suggested by these authors that magnesium sulfate should be the agent of first choice in tocolytic therapy, with ritodrine as its pharmacological backup.[89] Thus, although the β agonists are still considered by many investigators to be the drugs of choice for the inhibition of preterm labor, this view is not universally accepted.

4. Summary and Conclusions

4.1. β-Adrenoceptors and cAMP in Uterine Relaxation

It is well established that stimulation of β-adrenoceptors promotes relaxation of the uterine smooth muscle, whereas stimulation of α-adrenoceptors results in uterine contraction. The relative preponderance of α- and β-adrenoceptors in the uterus can be altered by changes in concentrations of estrogen and progesterone. In most species, at least in longitudinal smooth muscle, the relaxant effects of adrenergic agonists are most prominent in progesterone-dominated uteri, and the contractile effects are enhanced in estrogen-dominated preparations. The underlying mechanisms by which the ovarian hormones can influence uterine responses to catecholamines have not been completely elucidated.

Coupling of β-adrenoceptors to adenylate cyclase via a stimulatory guanine nucleotide-dependent regulatory protein has been demonstrated in rat myometrium, and β-adrenoceptor agonists have been shown to increase cAMP levels in uterine preparations from a variety of species. Levels of the cyclic nucleotide can also be increased by nonadrenergic agents such as forskolin, a direct activator of adenylate cyclase, and papaverine, an inhibitor of cAMP breakdown. All of these agents are capable of relaxing myometrial preparations, and it has been suggested that elevation of cAMP is directly responsible for the uterine relaxant effects of these drugs. Although there is a large body of evidence in favor of this hypothesis, particularly for β-adrenoceptor agonists, there is also a substantial amount of data in the literature that indicates that cAMP elevation is not the sole mechanism responsible for the uterine relaxation caused by these agents. Most of the available evidence is consistent with the existence of two mechanisms for β-adrenoceptor-induced relaxation of the uterus, one of which is mediated via elevations in tissue levels of cAMP and one of which is independent of changes in cAMP levels. The precise underlying mechanisms by which either of these pathways could bring about uterine relaxation have not been determined. Finally, although other interpretations are possible, it seems likely that the cAMP-independent mechanism is of primary importance and may be sufficient to account for the relaxant effects of low concentrations of β agonists. The marked cAMP elevations produced by higher concentrations of the β agonists would presumably contribute to the relaxant effects of these agents.

4.2. Role of cGMP in Control of Uterine Motility

Current thinking regarding the putative role of cGMP in the control of smooth muscle motility has undergone a complete reversal since the mid-1970s. Cyclic GMP was originally assumed to be a mediator of smooth muscle contraction. This was based on early reports that increases in cGMP levels accompanied contractions induced by a variety of smooth muscle stimulants. However, subsequent reports provided evidence that the increases in cGMP were a result of, rather than the cause of, the contractions, and this led to a reevaluation of the role of cGMP in smooth muscle. Part of the evidence against a role for this cyclic nucleotide as a mediator of contraction was the observation that elevation of cGMP in uterine and vascular smooth muscles by drugs such as nitroglycerin and papaverine was accompanied by relaxation of the muscles rather than contraction. Several laboratories subsequently reported good correlations between the cGMP elevation and the relaxation caused by a variety of drugs, and it was suggested that cGMP might be responsible for the smooth-muscle-relaxing effects of these drugs. Since then, a considerable body of evidence has accumulated in the literature in support of this hypothesis, and it is now widely accepted that cGMP is a mediator of relaxation in some types of smooth muscle. However, the evidence for such a role in uterine smooth muscle is not as convincing. In fact, most of the available evidence indicates that cGMP does not function as a mediator of contraction or relaxation in this tissue. Thus, in contrast to the apparent situation in vascular smooth muscle, cGMP does not appear to play an important role in the control of uterine motility.

4.3. Use of β-Adrenoceptor Agonists in Premature Labor

For the present, ritodrine remains the standard tocolytic therapy in many institutions. As noted above, it has been clearly demonstrated to be effective in suppressing preterm labor, at least for short periods of time (i.e., 24–48 hr). Because of potentially troublesome side effects and other problems associated with β agonists, the use of ritodrine as the initial tocolytic agent in the treatment of preterm labor has been supplanted in some institutions by the use of magnesium sulfate. However, it is clear that no universally effective tocolytic strategy is currently available for controlling human premature labor.

Even if a significant delay in the onset of labor can be attained, it may not be accompanied by an improvement in neonatal morbidity or mortality, and the value of aggressive tocolytic therapy in premature labor has been questioned by some authors. It may be, as recently suggested by King *et al.*,[86] that the main benefit to be derived from tocolytic therapy may be to provide sufficient short-term delay of delivery for other effective measures to be implemented. This might allow time, for example, for the administration of corticosteroids to enhance fetal lung maturation or for the transfer of the patient to a hospital with facilities for obstetric or neonatal intensive care.

References

1. Ahlquist, R. P., 1948, A study of the adrenotropic receptors, *Am. J. Physiol.* **153**:583–600.
2. Ahlquist, R. P., 1962, The adrenotropic receptor–detector, *Arch. Int. Pharmacodyn.* **139**:38–41.
3. Miller, M. D., and Marshall, J. M., 1965, Uterine response to nerve stimulation: Relation to hormonal status and catecholamines, *Am. J. Physiol.* **209**:859–863.
4. Davidson, W. J., and Ikoku, C., 1966, The adrenergic receptors in the guinea pig uterus. *Can J. Physiol. Pharmacol.* **44**:491–493.

5. Tsai, T. H., and Fleming, W. W., 1964, The adrenotropic receptors of the cat uterus, *J. Pharmacol. Exp. Ther.* **143:**268–272.

6. Diamond, J., and Brody, T. M., 1966, Effect of catecholamines on smooth muscle motility and phosphorylase activity, *J. Pharmacol. Exp. Ther.* **152:**202–211.

7. Wansbrough, H., Nakanishi, H., and Wood, C., 1967, Effect of epinephrine on human uterine activity *in vitro* and *in vivo, Obstet. Gynecol.* **30:**779–789.

8. Lands, A. M., Arnold, A., McAuliff, J. P., Luduena, F. P., and Brown, T. G., 1967, Differentiation of receptor systems activated by sympathomimetic amines, *Nature* **214:**597–598.

9. Langer, S. Z., 1974, Presynaptic regulation of catecholamine release, *Biochem. Pharmacol.* **23:**1793–1800.

10. Bottari, S. P., Vokaer, A., Kaivez, E., Lescrainier, J. P., and Vauquelin, G., 1985, Regulation of alpha- and beta-adrenergic receptor subclasses by gonadal steroids in human myometrium, *Acta Physiol. Hung.* **65:**335–346.

11. O'Donnell, S. R., Persson, C. G. A., and Wanstall, J. C., 1978, An *in vitro* comparison of β-adrenoceptor stimulants on potassium-depolarized uterine preparations from guinea pigs, *Br. J. Pharmacol.* **62:**227–233.

12. Johansson, S. R. M., Andersson, R. G. G., and Wikberg, J. E. S., 1980, Comparison of β₁- and β₂-receptor stimulation in oestrogen or progesterone dominated rat uterus, *Acta Pharmacol. Toxicol.* **47:**252–258.

13. Lefkowitz, R. J., Stadel, J. M., and Caron, M. G., 1983, Adenylate cyclase-coupled beta-adrenergic receptors: Structure and mechanisms of activation and desensitization, *Annu. Rev. Biochem.* **52:**159–186.

14. Hoffman, B. B., Lavin, T. N., Lefkowitz, R. J., and Ruffolo, R. R., 1981, Alpha adrenergic receptor subtypes in rabbit uterus: Mediation of myometrial contraction and regulation by estrogens, *J. Pharmacol. Exp. Ther.* **219:**290–295.

15. Ariens, E. J., and Simonis, A. M., 1983, Physiological and pharmacological aspects of adrenergic receptor classification, *Biochem. Pharmacol.* **32:**1539–1545.

16. Bulbring, E., and Tomita, T., 1987, Catecholamine action on smooth muscle, *Pharmacol. Rev.* **39:**49–96.

17. Marshall, J. M., 1970, Adrenergic innervation of the female reproductive tract: Anatomy, physiology and pharmacology, *Ergeb. Physiol.* **62:**6–67.

18. Graham, J. P. D., and Gurd, M. R., 1960, Effects of adrenaline on the isolated uterus of the cat, *J. Physiol. (Lond.)* **152:**243–249.

19. Diamond, J., and Brody, T. M., 1966, Hormonal alteration of the response of the rat uterus to catecholamines, *Life Sci.* **5:**2187–2193.

20. Chow, E. H. M., and Marshall, J. M., 1981, Effects of catecholamines on circular and longitudinal muscle of the rat, *Eur. J. Pharmacol.* **76:**157–165.

21. Kishikawa, T., 1981, Alterations in the properties of the rat myometrium during gestation and post partum, *Jpn. J. Physiol.* **31:**515–536.

22. Roberts, J. M., Insel, P. A., and Goldfien, A., 1981, Regulation of myometrial adrenoceptors and adrenergic response by sex steroids, *Mol. Pharmacol.* **20:**52–58.

23. Roberts, J. M., Insel, P. A., Goldfien, R. D., and Goldfien, A., 1977, α-Adrenoceptors but not β-adrenoceptors increase in rabbit uterus with estrogen, *Nature* **270:**624–625.

24. Williams, L. T., and Lefkowitz, R. J., 1977, Regulation of rabbit myometrial alpha-adrenergic receptors by estrogen and progesterone, *J. Clin. Invest.* **60:**815–818.

25. Rodbell, M., 1980, The role of hormone receptors and GTP-regulatory proteins in membrane transduction, *Nature* **284:**17–22.

26. Codina, J., Hildebrandt, J., Sunyer, T., Sekura, R. D., Manclark, C. R., Iyengar, R., and Birnbaumer, L., 1984, Mechanisms in the vectorial receptor–adenylate cyclase signal transduction, *Adv. Cyclic Nucleotide Protein Phosphoryl. Res.* **17:**111–125.

27. Gilman, A. G., 1987, G proteins: Transducers of receptor-generated signals, *Annu. Rev. Biochem.* **56:**615–649.

28. Levitzki, A., 1988, From epinephrine to cyclic AMP, *Science* **241**:800–806.
29. Krall, J. F., Leshon, S. C., Frolich, M., Jamgotchian, N., and Korenman, S. G., 1982, Adenylate cyclase activation. Characterization of guanyl nucleotide requirements by direct radioligand-binding methods, *J. Biol. Chem.* **257**:10582–10586.
30. Tanfin, Z., and Harbon, S., 1987, Heterologous regulations of cAMP responses in pregnant rat myometrium. Evolution from a stimulatory to an inhibitory prostaglandin E_2 and prostacyclin effect, *Mol. Pharmacol.* **32**:249–257.
31. Dobbs, J. W., and Robison, G. A., 1968, Functional biochemistry of β-receptors in the uterus, *Fed. Proc.* **27**:352.
32. Triner, L., Vulliemoz, Y., Verosky, M., and Nahas, G. G., 1970, The effect of catecholamines on adenyl cyclase activity in rat uterus, *Life Sci.* **9**:707–712.
33. Triner, L., Nahas, G. G., Vulliemoz, Y., Overweg, N. I. A., Verosky, M., Habif, D. V., and Ngai, S. H., 1971, Cyclic AMP and smooth muscle function, *Ann. N.Y. Acad. Sci.* **185**:458–476.
34. Marshall, J. M., and Kroeger, E. A., 1973, Adrenergic influences on uterine smooth muscle, *Phil. Trans. R. Soc. Lond. [Biol.]* **265**:135–148.
35. Mitznegg, P., Heim, F., and Meythaler, B., 1970, Influence of endogenous and exogenous cyclic 3′,5′-AMP on contractile responses induced by oxytocin and calcium in isolated rat uterus, *Life Sci.* **9**:121–128.
36. Mitznegg, P., Hach, B., and Heim, F., 1971, The influence of cyclic 3′,5′-AMP on contractile responses induced by vasopressin in isolated rat uterus, *Life Sci.* **10**:169–174.
37. Polacek, I., and Daniel, E. E., 1971, Effect of α- and β-adrenergic stimulation on the uterine motility and adenosine 3′,5′-monophosphate level, *Can. J. Physiol. Pharmacol.* **49**:988–998.
38. Hardman, J. G., 1981, Cyclic nucleotides and smooth muscle contraction: Some conceptual and experimental considerations, in: *Smooth Muscle: An Assessment of Current Knowledge* (E. Bulbring, A. F. Brading, A. W. Jones, and T. Tomita, eds.), Edward Arnold, London, pp. 249–262.
39. Hardman, J. G., 1984, Cyclic nucleotides and regulation of vascular smooth muscle, *J. Cardiovasc. Pharmacol.* **6**:S639–S645.
40. Krall, J. F., Swensen, J. L., and Korenman, S. G., 1976, Hormonal control of uterine contraction. Characterization of cyclic AMP-dependent membrane properties in the myometrium, *Biochim. Biophys. Acta* **448**:578–588.
41. Nishikori, K., and Maeno, H., 1979, Close relationship between adenosine 3′:5′-monophosphate-dependent endogenous phosphorylation of a specific protein and stimulation of calcium uptake in rat uterine microsomes, *J. Biol. Chem.* **254**:6099–6106.
42. Scheid, C. R., Honeyman, T. W., and Fay, F. S., 1979, Mechanism of β-adrenergic relaxation of smooth muscle, *Nature* **277**:32–36.
43. Adelstein, R. S., and Hathaway, D. R., 1979, Role of calcium and cyclic adenosine 3′:5′ monophosphate in regulating smooth muscle contraction, *Am. J. Cardiol.* **44**:783–787.
44. Diamond, J., 1978, Role of cyclic nucleotides in control of smooth muscle contraction, *Adv. Cyclic Nucl. Res.* **9**:327–340.
45. Polacek, I., Bolan, J., and Daniel, E. E., 1971, Accumulation of adenosine 3′,5′-monophosphate and relaxation in the rat uterus *in vitro, Can. J. Pharmacol.* **49**:999–1004.
46. Diamond, J., and Holmes, T. G., 1975, Effects of potassium chloride and smooth muscle relaxants on tension and cyclic nucleotide levels in rat myometrium, *Can. J. Physiol. Pharmacol.* **53**:1099–1107.
47. Verma, S. C., and McNeill, J. H., 1976, Isoproterenol-induced relaxation, phosphorylase activation and cyclic adenosine monophosphate levels in the polarized and depolarized rat uterus, *J. Pharmacol. Exp. Ther.* **198**:539–547.
48. Izumi, H., and Kishikawa, T., 1982, Effects of ritodrine, a $β_2$-adrenoceptor agonist, on smooth muscle cells of the myometrium of pregnant rats, *Br. J. Pharmacol.* **76**:463–471.
49. Smith, D. D., and Marshall, J. M., 1986, Forskolin effects on longitudinal myometrial strips from the pregnant rat: Relationship with membrane potential and cyclic AMP, *Eur. J. Pharmacol.* **122**:29–35.

50. Diamond, J., and Marshall, J. M., 1969, Smooth muscle relaxants: Dissociation between resting membrane potential and resting tension in rat myometrium, *J. Pharmacol. Exp. Ther.* **168**:13–20.

51. Nesheim, B.-I., Osnes, J.-B., and Oye, I., 1975, Role of cyclic adenosine 3′,5′-monophosphate in the isoprenaline-induced relaxation of the oestrogen dominated rabbit uterus, *Br. J. Pharmacol.* **53**:403–407.

52. Nesheim, B.-I., and Sigurdson, S. G., 1978, Effects of isoprenaline and dibutyryl-cAMP on the electrical and mechanical activity of the rabbit myometrium, *Acta Pharmacol. Toxicol.* **42**:371–376.

53. Marshall, J. M., and Fain, J. N., 1985, Effects of forskolin and isoproterenol on cyclic AMP and tension in the myometrium, *Eur. J. Pharmacol.* **107**:25–34.

54. Harbon, S., and Clauser, H., 1971, Cyclic adenosine 3′,5′ monophosphate levels in rat myometrium under the influence of epinephrine, prostaglandins and oxytocin. Correlations with uterus motility, *Biochem. Biophys. Res. Commun.* **44**:1496–1503.

55. Vesin, M.-F., and Harbon, S., 1974, The effects of epinephrine, prostaglandins, and their antagonists on adenosine cyclic 3′,5′ monophosphate concentrations and motility of the rat uterus, *Mol. Pharmacol.* **10**:457–473.

56. Harbon, S., DoKhac, L., and Vesin, M.-F., 1976, Cyclic AMP binding to intracellular receptor proteins in rat myometrium. Effect of epinephrine and prostaglandin E_1, *Mol. Cell. Endocrinol.* **6**:17–34.

57. Harbon, S., Vesin, M.-F., DoKhac, L., and Leiber, D., 1978, Cyclic nucleotides in the regulation of rat uterus contractility, in: *Molecular Biology and Pharmacology of Cyclic Nucleotides* (G. Folco and R. Paoletti, eds.), Elsevier North-Holland, Amsterdam, pp. 279–296.

58. DoKhac, L., Mokhtari, A., and Harbon, S., 1986, A re-evaluated role for cyclic AMP in uterine relaxation. Differential effect of isoproterenol and forskolin, *J. Pharmacol. Exp. Ther.* **239**:236–242.

59. Ohia, S. E., and Boyle, F. C., 1988, Role of cyclic AMP in rat uterine inhibitory response to salbutamol during the natural oestrous cycle, *Arch. Int. Pharmacodyn.* **293**:245–256.

60. Goldberg, N. D., Haddox, M. K., Dunham, E., Lopez, C., and Hadden, J. W., 1974, The Yin Yang hypothesis of biological control: Opposing influences of cyclic GMP and cyclic AMP in the regulation of cell proliferation and other biological processes, in: *The Cold Spring Harbor Symposium on the Regulation of Proliferation in Animal Cells* (B. Clarkson and R. Baserga, eds.), Cold Spring Harbor Laboratory, New York, pp. 609–625.

61. Lee, T. P., Kuo, J. F., and Greengard, P., 1972, Role of muscarinic cholinergic receptors in regulation of guanosine 3′ : 5′-cyclic monophosphate content in mammalian brain, heart muscle and intestinal smooth muscle, *Proc. Natl. Acad. Sci. U.S.A.* **69**:3287–3291.

62. Schultz, G., Hardman, J. G., Schultz, K., Davis, J. W., and Sutherland, E. W., 1973, A new enzymatic assay for guanosine 3′ : 5′-cyclic monophosphate and its application to the ductus deferens of the rat, *Proc. Natl. Acad. Sci. U.S.A.* **70**:1721–1725.

63. Schultz, G., Hardman, J. G., Schultz, K., Davis, J. W., and Sutherland, E. W., 1973, The importance of calcium ions for the regulation of guanosine 3′ : 5′-cyclic monophosphate levels, *Proc. Natl. Acad. Sci. U.S.A.* **70**:3889–3893.

64. Clyman, R. I., Sandler, J. A., Manganiello, V. C., and Vaughan, M., 1975, Guanosine 3′,5′-monophosphate and adenosine 3′,5′-monophosphate content of human umbilical artery. Possible role in perinatal arterial patency and closure, *J. Clin. Invest.* **55**:1020–1025.

65. Dunham, E. W., Haddox, M. K., and Goldberg, N. D., 1974, Alteration of vein cyclic 3′ : 5′ nucleotide concentrations during changes in contractility, *Proc. Natl. Acad. Sci. U.S.A.* **71**:815–819.

66. Kadowitz, P. J., Joiner, P. D., Hyman, A. L., and George, W. J., 1975, Influence of prostaglandins E_1 and $F_{2\alpha}$ on pulmonary vascular resistance, isolated lobar vessels and cyclic nucleotide levels, *J. Pharmacol. Exp. Ther.* **192**:677–687.

67. Murad, F., and Kimura, H., 1974, Cyclic nucleotide levels in incubations of guinea pig trachea, *Biochim. Biophys. Acta* **343**:275–286.

68. Diamond, J., and Hartle, D. K., 1976, Cyclic nucleotide levels during carbachol-induced smooth muscle contractions, *J. Cyclic Nucleotide Res.* **2:**179–188.

69. Angles d'Auriac, G., and Worcel, M., 1976, Cellular levels of cAMP and cGMP in rat uterine smooth muscle. Effects of angiotensin, carbachol and various metabolic conditions, in: *Smooth Muscle Pharmacology and Physiology* (M. Worcel and G. Vassort, eds.), INSERM, Paris, pp. 101–111.

70. Leiber, D., and Harbon, S., 1982, The relationship between the carbachol stimulatory effect on cyclic GMP content and activation by fatty acid hydroperoxides of a soluble guanylate cyclase in the guinea pig myometrium, *Mol. Pharmacol.* **12:**654–663.

71. Diamond, J., and Blisard, K. S., 1976, Effects of stimulant and relaxant drugs on tension and cyclic nucleotide levels in canine femoral artery, *Mol. Pharmacol.* **12:**688–692.

72. Diamond, J., and Holmes, T. G., 1975, Effects of potassium chloride and smooth muscle relaxants on tension and cyclic nucleotide levels in rat myometrium, *Can. J. Physiol. Pharmacol.* **53:**1099–1107.

73. Schultz, K. D., Schultz, K., and Schultz, G., 1977, Sodium nitroprusside and other smooth muscle relaxants increase cyclic GMP levels in rat ductus deferens, *Nature* **265:**750–751.

74. Katsuki, S., Arnold, W. P., and Murad, F., 1977, Effects of sodium nitroprusside, nitroglycerin and sodium azide on levels of cyclic nucleotides and mechanical activity of various tissues, *J. Cyclic Nucleotide Res.* **3:**239–247.

75. Ignarro, L. J., and Kadowitz, P. J., 1985, The pharmacological and physiological role of cyclic GMP in vascular smooth muscle relaxation, *Annu. Rev. Pharmacol.* **25:**171–191.

76. Waldman, S. A., and Murad, F., 1987, Cyclic GMP synthesis and function, *Pharmacol. Rev.* **39:**163–196.

77. Leiber, D., Vesin, M.-F., and Harbon, S., 1978, Regulation of guanosine 3′,5′-cyclic monophosphate levels and contractility in rat myometrium, *FEBS Lett.* **86:**183–187.

78. Diamond, J., 1983, Lack of correlation between cyclic GMP elevation and relaxation of nonvascular smooth muscle by nitroglycerin, nitroprusside, hydroxylamine and sodium azide, *J. Pharmacol. Exp. Ther.* **225:**422–426.

79. Diamond, J., and Hartle, D. K., 1974, Cyclic nucleotide levels during spontaneous uterine contractions, *Can. J. Physiol. Pharmacol.* **52:**763–767.

80. Schultz, K. D., Bohme, E., Kreye, V. A. W., and Schultz, G., 1979, Relaxation of hormonally stimulated smooth muscular tissues by the 8-bromo derivative of cyclic GMP, *Naunyn Schmiedebergs Arch. Pharmacol.* **306:**1–9.

81. Bulbring, E., and Hardman, J. G., 1976, Effects on smooth muscles of nucleotides and the dibutyryl analogues of cyclic nucleotides, in: *Smooth Muscle Pharmacology and Physiology* (M. Worcel and G. Vassort, eds.), INSERM, Paris, pp. 125–131.

82. McCormick, M. C., 1985, The contribution of low birth weight to infant mortality and morbidity, *N. Engl. J. Med.* **312:**82–90.

83. Lumley, J., 1988, The prevention of preterm birth: Unresolved problems and work in progress, *Aust. Paediatr. J.* **24:**101–111.

84. King, J. F., Keirse, M. J. N. C., Grant, A., and Chalmers, I., 1985, Tocolysis—the case for and against, in: *Preterm Labour and Its Consequences* (R. W. Beard and F. Sharp, eds.), Royal College of Obstetricians and Gynaecologists, London, pp. 199–208.

85. Eggleston, M. K., 1986, Management of preterm labor and delivery, *Clin. Obstet. Gynecol.* **29:**230–239.

86. King, J. F., Grant, A., Kierse, M. J. N. C., and Chalmers, I., 1988, Beta-mimetics in preterm labor: An overview of the randomized controlled trials, *Br. J. Obstet. Gynaecol.* **95:**211–222.

87. Gonik, B., and Creasy, R. K., 1986, Preterm labor: Its diagnosis and management, *Am. J. Obstet. Gynecol.* **154:**3–8.

88. Garite, T. J., Keegan, K. A., Freeman, R. K., and Nageotte, M. P., 1987, A randomized trial of ritodrine tocolysis versus expectant management in patients with premature rupture of membranes at 25 to 30 weeks of gestation, *Am. J. Obstet. Gynecol.* **157:**388–393.

89. Hollander, D. I., Nagey, D. A., and Pupkin, M. J., 1987, Magnesium sulfate and ritodrine hydrochloride: A randomized comparison, *Am. J. Obstet. Gynecol.* **156:**631–637.
90. Benedetti, T. J., 1983, Maternal complications of parenteral β-sympathomimetic therapy for premature labor, *Am. J. Obstet. Gynecol.* **145:**1–6.
91. Caritis, S. N., Chiao, J. P., Moore, J. J., and Ward, S. M., 1987, Myometrial desensitization after ritodrine infusion, *Am. J. Physiol.* **253:**E410–E417.
92. Casper, R. F., and Lye, S. J., 1986, Myometrial desensitization to continuous but not to intermittent β-adrenergic agonist infusion in the sheep, *Am. J. Obstet. Gynecol.* **154:**301–305.

9

Physiological Roles of Gap Junctional Communication in Reproduction

Robert C. Burghardt and William H. Fletcher

1. Introduction

The uterus exhibits a unique structural and physiological plasticity when responding to ovarian cues with orderly cycles of tissue differentiation, glandular secretion, and regional turnover of terminally differentiated cells. These cyclic changes are programmed to be developmentally synchronous with blastocyst development, such that arrival of a blastocyst in the uterus, followed by interaction of blastocyst trophectoderm with uterine epithelial cells, results in the progressive phases of implantation.

Beginning with implantation and continuing throughout pregnancy, the plasticity of the uterus is even more dramatic. Interaction of the blastocyst with uterine luminal epithelium initiates an orderly morphological and physiological alteration of the endometrium (decidualization) involving extensive reorganization of endometrial cell populations. Subsequently, fetal growth is accommodated by growth of the uterus including considerable increase in the smooth muscle content of longitudinal and circular myometrial layers, which are maintained in a quiescent state until the onset of labor.

The role of estrogen and progesterone, as proximal initiators of these structural and functional properties in the uterus, has long been known.[1,2] Integration of cyclic uterine function and pregnancy-associated events involves these steroid hormones plus additional hormonal, humoral, and neural controls, many of which are addressed in this volume. In addition, cell–cell communication pathways mediated

Robert C. Burghardt • Department of Veterinary Anatomy, College of Veterinary Medicine, Texas A&M University, College Station, Texas 77843.   *William H. Fletcher* • Department of Anatomy, School of Medicine, Loma Linda University and Jerry L. Pettis Memorial Veterans Medical Center, Loma Linda, California 92357.

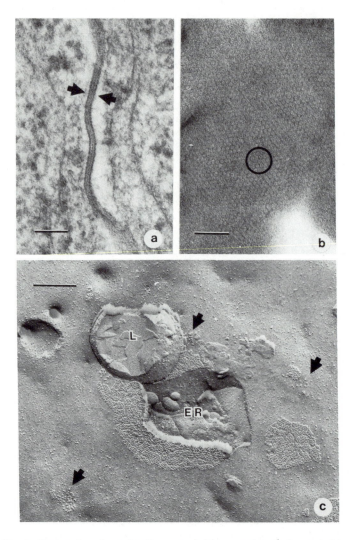

Figure 1. Structural properties of gap junctions revealed by several methods. **a:** A thin section cut perpendicular to the plasma membrane in an ovarian granulosa cell reveals close apposition of two plasma membranes at a gap junction (arrowheads) separated by a 2-nm gap. Line scale, 50 nm. **b:** Gap junction membrane purified from rat liver and viewed *en face* following negative staining procedures reveals aggregates of connexons. In favorable views (circle), an electron-dense core containing uranyl acetate stain is visible at the center of several connexons and is thought to be the hydrophilic channel. Line scale, 50 nm. **c:** Several gap junctions are visible in an *en face* preparation of granulosa cell membrane (P-face) obtained using the freeze fracture technique, which splits the membrane bilayer and exposes integral membrane proteins or aggregates of protein. The varied sizes of gap junctional aggregates are thought to represent stages in gap junction development. The larger, more dispersed particles in the vicinity of small gap junctions (arrowheads) are termed formation plaques and are thought to be connexons frozen in the process of aggregation into a gap junctional plaque. The large gap junctions observed in ovarian granulosa cells are often associated with cytoplasmic lipid droplets (L) and profiles of endoplasmic reticulum (ER). Line scale, 100 nm. **d:** Schematic diagram of gap junction in which two adjacent plasma membranes are linked by paired protein cylinders (connexons) that contain hydrophilic channels. Each connexon is comprised of six identical protein subunits called connexins. Adapted from Darnell *et al.*[184]

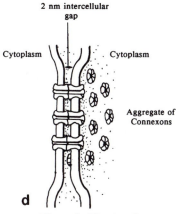

2 nm intercellular gap

Cytoplasm

Cytoplasm

Aggregate of Connexons

d

Figure 1. (*Continued*)

by gap junctions may constitute a segment of the regulatory spectrum involved in the integration of uterine function.

Gap junctions are thought to provide communicating cells with a common pool of small ions and metabolites that may have regulatory and/or informational properties when shared.[3,4] These cell–cell contacts are composed of aggregates of well-insulated hydrophilic channels that penetrate the plasma membrane of adjacent cells. Each channel in the aggregate is formed by alignment along a common axis of two identical protein cylinders, called connexons, with one connexon contributed by each of the communicating cell pair[5] (Fig. 1).

It is the purpose of this review to examine the experimental evidence for the possible transmission of regulatory signals between coupled cells and to identify uterine tissues in which the potential for signal exchange is developed. In the uterus, gap junctions are particularly dynamic structures, for they are inducible in certain regions of the uterus and subject to modulation and turnover throughout the organ.

2. Functional Implications of Junctional Communication

Direct intercellular contact by means of gap junctions appears to be a nearly ubiquitous property of cells in metazoan tissues.[6–11] With the exceptions of skeletal myocytes and many neurons, differentiated cells of epithelial, fibroblastic, or contractile nature are able to express the gap junction phenotype. In so doing, adjacent cells of histologically simple to complex tissues are capable of communicating directly with neighbors without using extracellular pathways.[12,13] Experimentally, this process is observed as the bidirectional exchange of ions (ionic or *electrotonic* coupling) or fluorescent dye molecules up to about 1 kDa maximum (dye transfer), which have been microinjected into one cell and passively diffuse through junctions to recipient cells.[14]

Because ionic coupling was elucidated in smooth muscle[15] and neurons,[16] its physiological significance seemed evident: that is, it could furnish a means for electrogenic cells to utilize each other as ionic sinks and sources.[12,13,17] However, when it was found that fluorescent dyes with no apparent bioactivity could also be exchanged between nonexcitable epithelial cells, the concept of intercellular communication took on a potentially broader significance.[6-11,18]

Initially, it was hypothesized that the ability to exchange small molecules could provide a mechanism by which growth-regulating factors could pass among cells united by gap junctions.[19] Support for this came from the finding that cells joined by the junctions could cooperate metabolically by passing nucleotides that had been metabolized in one cell to neighbor cells that were genetically defective and unable to use the nucleotide precursor (for review, see Pitts and Finbow[10]).

Collectively, the results from ionic coupling, fluorescent dye transfer, and metabolic cooperation experiments have indicated that these capabilities coincide with the presence of gap junctions[12,20] and are lost when the junctions are genetically deficient,[21] depleted by tumor promoters[22] or carcinogens,[23,24] or functionally blocked by antibodies directed against the junctional protein[25] or changes in the intracellular ionic environment.[26,27] Although a great deal of important information has been gleaned from such studies, the results have been mainly a documentation of phenomenology. The functional relevance of intercellular communication, especially of the specific mechanisms that it subserves, has been difficult to define.

Early experiments by Goshima[28] and Sheridan (reviewed by Sheridan and Atkinson[11]) indicated that second messengers like cyclic 3',5'-adenosine monophosphate (cAMP) would be good candidates for cell–cell communication molecules. Loewenstein and co-workers have long championed similar concepts and have formulated theories that provide a reasonable explanation for how the intercellular communication of signal molecules might influence cellular growth, differentiation, neoplastic transformation, and metabolic processes.[9]

If this diversity of events is modulated by gap junctional communication, it is reasonable to assume that a variety of regulatory molecules may be able to transfer among cells via junctional channels. In order to examine this possibility it was necessary first to devise procedures that would allow a direct means of visualizing intercellular communication and its physiological consequences in individual cells. In most instances where the biological role of cell communication has been examined, the results have suggested that cAMP is most likely able to act as the signal molecule.[20]

All evidence to date indicates that actions requiring cAMP mediation are actually carried out by the cAMP-dependant protein kinases.[29,30] Thus, if communication is involved in actions of cAMP, the protein kinases should be activated in cells that receive communicated signals. The cAMP-dependent protein kinases are unique among all known regulatory enzymes because in order to become active they must dissociate according to the simple reaction: $R_2C_2 + 4\,cAMP = R_2 \cdot cAMP_4 + 2C$.[29] The regulatory subunit (R) is the cAMP-binding protein, and the free catalytic subunits (C) are able to phosphorylate substrate proteins whose function is thereby altered, allowing them to carry out responses appropriate to the initial stimulus.[29,30] Because of this unique mechanism of activation, it was possible to develop direct cytochemical approaches for localizing the active C and R subunits in individual cells.[31–34] Free C subunits are localized using the fluoresceinated protein kinase inhibitor protein (F:PKI), which binds tightly ($K_i \sim 0.5$ nM) and exclusively[31] to the free C subunit but not to C in the holoenzyme form.[29–31] Fluoresceinated C (F:C) subunit is used to complex ($K_a \sim 0.1$ nM) with free R, thereby defining its intracellular sites.[33,34] Although it is not possible to use both F:PKI and F:C in the same preparation, by using them separately in identical preparations the intracellular kinetics of C and R subunits can be followed and compared. With these approaches, a number of observations have been made relevant to the function of contact-dependent cell–cell communication.

Initially,[35] primary preovulatory-phase ovarian granulosa cells were cocultured with the clonal ACTH-sensitive adrenocortical tumor cell line Y-1. Cultures were stimulated either with 8Br-cAMP or with hormone specific for one of the cell types. Granulosa and Y-1 cells both responded to 8Br-cAMP by dissociating holoenzyme with similar temporal kinetics and in about the same amounts, two- to fourfold above that of unstimulated controls. On exposure to follicle-stimulating hormone (FSH), which binds only to the granulosa cells, they dissociated protein kinase, and the amounts of free C subunit and its intracellular distribution were comparable to those observed after 8Br-cAMP stimulation. Y-1 cells in contact with responding granulosa cells also dissociated enzyme, whereas adrenal cells not in such contact never responded to hormone. Similar but reciprocal results were obtained in cocultures exposed to ACTH. That is, granulosa cells dissociated protein kinase if, and only if, they contacted responding Y-1 cells. Thus, each cell type was able to respond normally to its specific hormonal stimulus and at the same time to communicate a signal to its heterotypic partner, which otherwise was unable to respond. Importantly, in cocultures treated with FSH, the Y-1 cells produced corticosteroids, which granulosa cells cannot secrete, indicating that the signal they received from granulosa cell partners was bioactive and induced a response in Y-1 cells as though they had been stimulated with ACTH. Gap junctions were frequently seen at points of contact between Y-1 and granulosa cells, making it likely they were the sites where this bidirectional signal transfer occurred.

In a series of somewhat more physiological studies, it was found that only about one-third of the granulosa cells harvested from preovulatory follicles of cycling sows expressed receptors for luteinizing hormone (LH) as determined by immunochemistry[36] and more recently by flow cytometry.[37] These cultures were exposed to LH, receptor content defined as before, and protein kinase activation followed with the F:PKI probe. In these preparations, cells that bound hormone dissociated protein kinase, as did contacting neighbor cells that lacked receptors. However, cells that neither had receptors nor contacted receptor-bearing partners never dissociated protein kinase.

These observations demonstrate two important points. The first is that although ostensibly homogeneous, not all granulosa cells express the same receptor phenotype; however, that deficiency does not impair their ability to mount an apparently normal biological response to hormones. This conclusion may be broadly applicable to hormone-responsive cells that unite via gap junctions to neighbors. Clearly, uterine epithelial and smooth muscle cells fall into this category.

The second conclusion to be drawn is that gap junctions do mediate a defined mechanism: they allow the cell–cell transmission of signals that can activate the cAMP-dependent protein kinases.

The literature is replete with evidence that these enzymes are important to the regulation of all the processes indicated above[29,30] where gap junctional communication is thought to have biological significance. It is reasonable, therefore, to suggest that the mechanism of protein kinase regulation by signals transmitted through the junctions is the means by which these membrane specializations can influence a broad spectrum of cellular mechanisms.

There are numerous questions that this postulate brings forth. What is the signal that regulates the protein kinase, or is there more than one signal? Is regulation of the cAMP-dependent protein kinases the only mechanism subserved by gap junctions? Is cell–cell communication involved in the response to all hormones whose action is mediated by cAMP? Currently, there is no firm answer to any of these questions, but some new observations suggest that intercellular communication may be influential in other mechanisms.[38]

3. Structural and Functional Studies of Uterine Gap Junctions

As noted previously, the differentiated state and responsiveness of each of the tissue-level compartments of the mammalian uterus are profoundly affected by estrogen and progesterone. The organization of these tissue-level compartments in a cross section of the uterine wall is illustrated in Fig. 2. The wall of the uterus is composed of structurally and functionally distinct cellular layers. The innermost layer, the endometrium, lines the lumen of the organ and consists of two tissue compartments including a simple columnar luminal epithelium resting on the endometrial stroma. Invaginations of the luminal epithelium give rise to the endometrial glands, which may penetrate deep into the stroma. The myometrium and associated vasculature varies in organization between species but in rodents is generally organized into two layers including inner circular and outer longitudinal bundles; it generally comprises the greater part of the thickness of the uterine wall. In humans, muscle layers are much less clearly demarcated, with three or four layers present. A serous membrane covers most of the uterus.

A number of studies have been directed toward analysis of the intercellular junctional complement within uterine tissues during the estrous (and menstrual) cycle and during pregnancy, corresponding to periods in which circulating estrogen-to-progesterone ratios vary widely. It is probably not surprising that the gap junctional contacts in the various tissue compartments of the uterus appear to be influenced by local levels of estrogen and progesterone. What is unexpected is that several tissue compartments of the uterus normally lack gap junctional contacts, but these tissues are programmed to develop gap junctions in the hormonal milieu associated with certain stages of pregnancy. There is evidence that the ratio of nuclear receptor-bound progesterone to estrogen determines tissue responsiveness and the appearance of gap junctions in uterine cells. In the following sections, the presence of gap junctions and the factors regulating their appearance in endometrial, myometrial, and serosal layers of the uterus are described.

3.1. Uterine Endometrium

The luminal epithelium and stroma are the two main tissues of the endometrium that respond to fluctuating levels of ovarian hormones and to the physical stimuli provided by a blastocyst. Both tissues exhibit unique control mechanisms regulating the presence and amount of gap junctional membrane.

3.1.1. Uterine Luminal Epithelium

Uterine epithelium oscillates between proliferative and secretory activity in response to cyclic variation in the secretion of ovarian steroids. The uterus is primed by a period of estrogen dominance near ovulation, followed by a shift to progesterone dominance and permissive estrogen action to produce an endometrial epithelium and underlying stroma with structural and physiological properties necessary for nidation.[39] During early pregnancy, sensitivity of the endometrium to the blastocyst stimulus is present for a limited period of time. In ovariectomized mice and rats pretreated with progesterone for at least 48 hr, a very small dose of estrogen subsequently controls the timing and duration of sensitivity to a deciduogenic stimulus, which peaks about 24 hr after estrogen administration and lasts for about the same time period.[40] Intercellular junctions of the uterine epithelium have been studied in rodent and human tissues, but despite dynamic trophic responses to changing steroid hormone levels, gap junctions are normally small and few in number, and changes in the gap junctional content were initially considered to be very modest.[41] Until recently, the only suggestion that gap junctions might be influenced by the ovarian steroids was the observation that, following an appropriate hormone priming regimen, apical projections from uterine luminal epithelial cells in the rat often formed gap-junction-like contacts with other projections from the same cell, which were referred to as "reflexive" gap junctions.[42] These unusual cell surface contacts appear identical to gap junctions in both thin section and freeze-fracture electron microscopy and are termed reflexive gap junctions to identify such junctions formed by the plasma membrane of the same cell.[43] No functional significance is known for reflexive gap junctions. They may simply be an anomalous junction formed when two surface projections of the same cell containing junctional precursors became positioned at a proximity sufficient to permit the cell surface events leading to channel formation between adjacent membranes. Reflex-

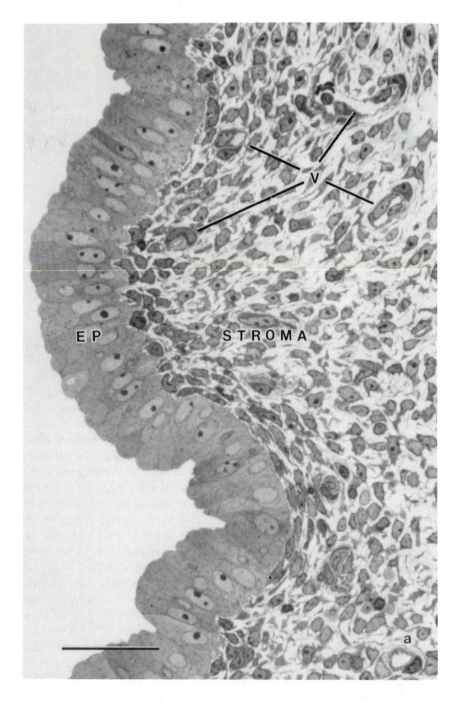

Figure 2. A light micrograph of rat uterine wall in cross section. **a:** Uterine endometrium of a hypophysectomized juvenile rat that was treated daily for five consecutive days with 50 μg of estradiol benzoate. The columnar luminal epithelium (EP) reflects the estrogen dominance. Stromal cells are spindle shaped and appear to be freely suspended in an extracellular matrix. Note the rich vascular supply

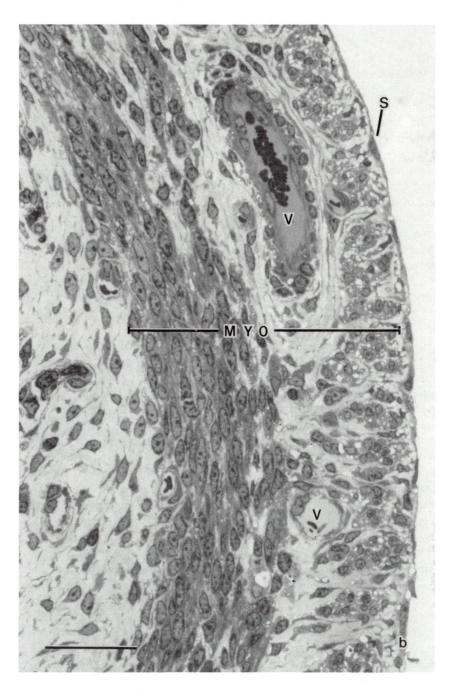

of stromal tissue (V). **b:** Outer wall of uterus exhibiting innermost components of uterine stroma. An inner circular and outer longitudinal layer of smooth muscle bundles separated by a vascular plexus (V) comprises the myometrium (MYO). An outer squamous epithelial cell layer, the serosa (S), covers the uterine wall. Line scale, 50 μm.

ive gap junctions have also been detected on human ovarian and uterine decidual cells.[44,45] However, in rat uterine epithelial cells, reflexive gap junctions have only been seen in the apical membrane after treatment with both progesterone and estrogen.[42]

The apparent limited responsiveness of uterine luminal epithelial cell gap junctions in cycling and pseudopregnant animals was surprising, since tight junctions between these cells exhibit significant modulation by steroid hormones. Tight junctions form a continuous belt around the apical region of the uterine epithelial cell and are thought to play a role as an intercellular barrier, regulating the fluid content of the lumen and, thereby, the environment of the blastocyst. Changes in both organization and numbers of junctional strands occur in tight junctions as a function of steroid hormone treatment in experimental animals,[46] during pseudopregnancy,[41] and in early pregnancy.[47] Similar kinds of structural alterations occur in tight junctions of human tissues during the menstrual cycle.[48] Proliferation of the tight junctional belt requires both progesterone and estrogen in ovariectomized animals[46] and is consistent with the increase in tight junctional complexity occurring in rodents during the preimplantation phase of pregnancy and pseudopregnancy.[41,47]

In a comprehensive series of studies recently completed by Winterhager *et al.*,[49] a dramatic amplification of uterine luminal epithelial cell gap junctions has been detected in response to epithelial cell interaction with a blastocyst. Notably, initiation of this response is precisely regulated, accounting for the fact that it escaped earlier detection. Rabbit uterine epithelial cells become junctionally coupled at the time of implantation as demonstrated by several techniques including freeze-fracture microscopy, immunodetection of the gap junction protein, and analysis of cellular dye transfer. The presence of the blastocyst, however, is a necessary condition for the induction of gap junctions, since the epithelium remains uncoupled in nonpregnant, pseudopregnant, and blastocyst-free uterine horn of tubal-ligated, unilaterally pregnant animals. Further, junctional coupling is initiated locally in the vicinity of the blastocyst and spreads outward to the blastocyst-free segments of the uterus.

In this context, the previous observation of uterine luminal epithelial reflexive gap junctions in progesterone- and estrogen-primed animals discussed earlier[42] may be significant. Although reflexive gap junctions probably do not confer any unique function within the uterine luminal epithelial cell, they may reflect the capability of uterine epithelial cells to exchange signals with a blastocyst or with other epithelial cells during the critical stage of embryo recognition. That is, the ability of uterine epithelial cells to form reflexive gap junctions could be related to the presence of gap junction precursors sequestered in the apical epithelial cell membrane in preparation for interaction with the blastocyst.

It is probably relevant that embryonic and uterine epithelial cells develop a variety of intercellular junctions during early implantation in rodents. Punctate desmosomes have been detected at the implantation site of the rabbit, bat, and ferret.[50] Desmosomes, which are intercellular junctions that function in cell adhesion and not intercellular communication, have also been observed between trophoblast and luminal epithelial cells in the mouse,[51] but as suggested by Tachi and Tachi,[52] it is likely that areas identified in electron micrographs in the former study as desmosomes also included gap-junction-like contacts. Tachi and Tachi[52] have also demonstrated that rat blastocysts could successfully undergo the stages of ovum implantation in the mouse uterus from the attachment phase to early trophoblast invasion of the endometrium. Further, during the attachment stage, gap junctions as well as adhering junctions including desmosomes and intermediate junctions developed between trophoblast and uterine cells of the host. However, with increasing motility of mural trophoblast, such junctions are ultimately disrupted, and uterine cells are consequently displaced from the immediate vicinity of the embryo. Therefore, although a variety of intercellular junctions may form during the early attachment of the blastocyst in rodents, these junctions are interrupted during subsequent stages of implantation.

The significance of these observations is not yet clear, but it is interesting that there is a temporal correlation between the transient and regional junctional development in luminal epithelial cells and uterine sensitivity for blastocyst attachment and early implantation. In addition to supporting the metabolism of the blastocyst by controlling the intrauterine environment, luminal epithelial cells are thought to transmit information from the blastocyst to the underlying stroma that may initiate decidualization.[53] Because the luminal epithelium appears intact during the initiation of the decidual response, epithelial transduction of a blastocyst message is proposed as the basis for uterine sensitivity to deciduogenic stimuli.[54]

The regional gap junctional response to embryo recognition recently demonstrated by Winterhager *et al.*[49] might also be associated with the mechanism for message transmission through luminal epithelial cells to underlying tissues. Precedent for this kind of signal transfer occurs between vascular endothelium and the underlying smooth muscle.[55]

3.1.2. Endometrial Stroma

During histogenesis of endometrial stroma, undifferentiated mesenchyme cells differentiate into fibroblasts, which are capable of further differentiation into decidual cells. Characteristically, stromal cells are spindle shaped (see Fig. 2) but can vary considerably with the hormonal state of the animal. They are normally free cells, not making physical contacts with neighboring cells.[56] Implantation of the blastocyst during normal pregnancy elicits an increase in endometrial vascular permeability and a highly ordered wave of mitosis and stromal cell differentiation known as decidualization, which spreads outward from the normal implantation site.[40] This maternal response to pregnancy, referred to as the deciduoma, can be mimicked by a variety of artificial physical stimuli if the stimulus is introduced into the uterine lumen of an appropriately sensitized pseudopregnant animal.[57]

Ultrastructural aspects of the antimesometrial endometrium, where implantation normally occurs in the rat uterus, have also been well documented during deciduoma formation.[58,59] Similar to the naturally decidualized uterus,[50] the anti-mesometrial deciduoma is characterized by the differentiation of large, tightly packed polyploid cells that exhibit morphological indications of active protein synthesis.[59] A striking feature of the decidual reaction in the antimesometrial region is the sharp reduction of intercellular spaces and corresponding increase of cellular surface area associated with the induction of numerous large gap junctions.[58,60] In addition, many circular or "annular" profiles of gap junction membrane are also frequently observed within decidual cells examined by thin-section electron microscopy. Although it is not possible to verify that all annular profiles of gap junctional membrane in thin sections have lost their continuity with the cell surface, other methods such as serial sectioning, lanthanum impregnation, and the appearance of annular gap junctions in single disaggregated cells support the fact that at least some of these structures are removed from the cell surface (for review, see Larsen and Risinger[8]). Annular gap junctions result from internalization of surface junctions and are frequently encountered in cells stimulated to increase the area of gap junctional membrane. The elaborate gap junctional contacts that are induced between stromal cells during deciduoma formation are illustrated in Fig. 3.

Because of the proximity of the decidua to the developing conceptus, the dramatic and orderly proliferation, cellular differentiation, and degeneration of stromal cells within a specific life span are probably related to the accommodation of coordinate changes taking place in the developing embryo during implantation and placentation. However, the basic functions of decidual cells remain poorly understood. Reduction in intercellular space and formation of extensive intercellular junctions may afford maternal protection against invasion. Other proposed functions include nutrition of the embryo,[61] immunologic protection of the embryo and/or mother,[62] and production of a decidual cell luteotrophin.[63] In rats, the decidual luteotrophin possesses several biochemical and physiological characteristics of prolactin, but it differs structurally and immunologically.[64] A human decidual prolactin, which is identical to pituitary prolactin, has been identified.[65,66] Release of this decidual prolactin appears to be stimulated by a placental protein[67] and inhibited by a decidual protein, both of which have no effect on pituitary prolactin.[68]

Without a better understanding of the function of decidual cells, the significance of the induction and dramatic proliferation of gap junctions between decidual cells is unknown. Based on functional studies in other systems described previously, we can only speculate that this transient gap junctional response may be involved in the coordination of the uterine epithelium and underlying stromal cell population as a preliminary to the implantation reaction. Since the synthesis of the decidual luteotrophin appears highly controlled, metabolic cooperation may also support the regulation of decidual luteotrophin secretion.

3.1.3. Hormonal Control of Endometrial Gap Junctions

As noted above, progesterone and estrogen priming alone are insufficient to elicit decidualization and the associated induction of luminal epithelial and decidual

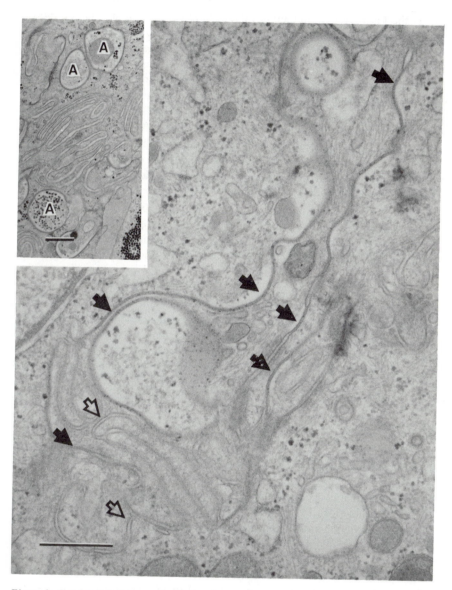

Figure 3. Gap junctions between decidual cells of the rat uterus, day 9 of pseudopregnancy. Note the close apposition of plasma membranes (open arrowheads) and the extensive gap junctional contacts that form between decidual cells (dark arrowheads) in the antimesometrial region of the deciduoma. Gap junctional contacts do not form between the stromal cells in the absence of a decidualizing stimulus. Inset: Annular gap junctions (A) are prominent in these cells indicating internalization of surface junctions. Line scales, 500 nm. (Micrographs courtesy of Dr. Ruth G. Kleinfeld.)

cell gap junctions. However, sensitivity of the uterus to a stimulus, like that provided by the blastocyst, to initiate decidualization requires an appropriate ratio of progesterone to estrogen to permit the subsequent transduction of the blastocyst signal from uterine luminal epithelium to underlying stroma. The possible biochemical basis for mechanisms of message transduction has recently been reviewed by Moulton and Koenig.[54] These investigators have provided evidence for the transmethylation of epithelial cell phospholipids in progestin- and estrogen-primed pseudopregnant rats and propose that phospholipid methylation may be involved in the primary response of epithelial cells to the blastocyst.[69] According to a model developed by Hirata and Axelrod,[70] the methylation of membrane phospholipids increases membrane fluidity, thereby facilitating coupling of the hormone receptor with adenylate cyclase. This could also result in arachidonic acid release from membrane phospholipids.[54]

Considerable evidence implicates the role of uterine prostaglandins in mediating changes in endometrial vascular permeability and subsequent decidualization in response to a deciduogenic stimulus.[71],[72] Such stimuli in sensitive uteri initiate rapid increases in uterine levels of cAMP that can be inhibited by indomethacin, an inhibitor of prostaglandin synthesis.[73]

The presence and amount of gap junction membrane in a variety of cell types can be regulated by cAMP and cAMP-mediated hormone action including prostaglandin E_1.[74-79] The induction of gap junctions and development of membrane permeability in a gap-junction-deficient cancer cell type (C1-1D) by administration of exogenous cAMP or by treatments elevating endogenous cAMP levels is significant, since the response, like the induction of endometrial epithelial and stromal cell gap junctions, also arises from a zero base.[78] Western blot analysis (electrophoretic transfer of proteins from polyacrylamide gels to nitrocellulose, followed by detection of specific proteins with an antibody-labeling step) of extracts from both induced and noninduced cultures, however, indicate that the gap junction peptide is present in roughly equal amounts, suggesting that, at least in this tissue, posttranslational factors are involved.[80] The report of the spontaneous assembly of gap junctions in cultured rat prostatic epithelial cells (a steroid-hormone-responsive cell type) following treatments that involve cytoskeletal perturbations suggests that induction of gap junctions in certain cell types may result from the assembly of a preexisting pool of junction precursors.[81] These investigators proposed that gap junction assembly resulted from a convergent migration of preexisting junctional precursors whose location in the membrane is released by modifications of cytoskeletal elements.

The involvement of the adenylate cyclase system as the principal modulator of the cytoskeleton as well as the association of cytoskeleton with the plasma membrane[82] may be a link between cAMP stimulation and the increase in gap junction size and membrane permeability in this system. A hypothesis that might account for the induction and modulation of gap junctions in uterine endometrium is that progesterone and/or estrogen may stimulate the synthesis and possibly the insertion of connexons into the plasma membrane. The assembly of connexons into junctional plaques could subsequently proceed following elevation of intracellular levels of cAMP resulting from the action of locally produced uterus-activating agents such as prostaglandins produced as a consequence of embryo recognition. Analysis of this hypothesis is certainly accessible to experimental investigation because of the highly regulated hormonal priming regimen and the time course of events associated with nidation.

3.2. Uterine Myometrium

The structure and physiology of uterine smooth muscle has been the subject of considerable research because of the importance of this tissue in the outcome of the birth process. Prematurity and birth injury constitute major hazards of the process, with prematurity associated with over 60% of all infant deaths.[83] Therefore, considerable effort continues to be expended on analysis of the factors that maintain pregnancy and initiate labor.

Although often considered to be a rather homogeneous tissue, uterine myometrium in animals with bicornuate uteri is actually composed of muscle bundles arranged in at least two layers oriented at right angles to each other and separated by a medial vascular plexus (see Fig. 2). The physiological properties of the two myometrial layers differ throughout pregnancy but become more similar near term.[84,85] In the simplex uterus of primates, the arrangement of muscle bundles is more complex and consists of three or four muscle layers, one of which is richly supplied with blood vessels.[56] During pregnancy, smooth muscle cells increase in number and size under the influence of both steroid hormones and physical stretch of the uterine wall to accommodate the growth of the fetus. On the basis of animal studies, primarily in sheep and rats, contractility of uterine smooth muscle is thought to be controlled by a balance of humoral factors such that those dictating myometrial quiescence predominate during pregnancy but are replaced by stimulatory factors shortly before the onset of labor.[86] As discussed in detail elsewhere in this volume, current understanding of smooth muscle contraction, as well as the therapies designed to block premature labor, stem from the relationships that link actin and myosin interaction, inositol-trisphosphate-induced calcium release, myosin light chain phosphorylation, and the involvement of cAMP in the regulation of myosin light chain kinase and of the cytoplasmic calcium levels.[87,88]

In recent years, the importance of smooth muscle conductivity has also been factored into our knowledge of uterine activity at term, largely because of the contributions of Garfield and co-workers. These investigators first observed the sudden appearance of gap junctions conjoining myometrial cells in rats just prior to and during parturition.[89] Subsequent demonstrations of an increase in number and size of gap junctions correlated with term and preterm labor in a variety of species[90–92] including humans[93] and the observation that prepartum myometrium undergoes a transition from asynchronous electrical activity to coordinated electrical activity just prior to parturition[94] have provided a biophysical basis for understanding the behavior of the myometrium during parturition. Because gap junctions provide excitable cells (neurons, cardiac and smooth muscle) with channels for the exchange of ions between coupled cells, the sudden development of myometrial gap junctions preceding the onset of labor permits the propagation of action potentials to coordinate waves of contraction necessary to expel the contents of gestational uterus.[92,94]

Although it appears that myometrial gap junctions are normally absent in immature, cycling, and early pregnant rats,[95] and gap junctions between myometrial cells are encountered in very low frequency in humans and a number of other species, in all species examined to date parturition is preceded by an increase in both the number and size of these intercellular contacts. However, within 24 hr after parturition, gap junctions between myometrial cells decrease (or disappear) to preterm levels.[96] Garfield *et al.*[97] proposed early that gap junction formation is a necessary step in the activation of myometrium and that knowledge of the conditions that regulate myometrial gap junctions would improve therapeutic approaches to terminating pregnancy and/or treating premature labor.

3.2.1. Myometrial Gap Junction Structure

When present, gap junctions between myometrial cells are typical of the cell–cell channels linking smooth muscle cells in other tissues and are most frequently seen connecting small cytoplasmic processes, although blunt lateral projections between cells are occasionally encountered (Fig. 4). Annular gap junctions have been observed *in vitro* within myometrial cells obtained from pregnant rats[98] and *in vivo* in hypophysectomized rats receiving high doses of exogenous estrogens.[99,100] The significance of these structures in relation to myometrial cells is discussed below.

3.2.2. Biochemical Analysis of Myometrial Gap Junctions

Biochemical characterization of gap junctions from pregnant uterus has only recently been initiated. Experimental approaches to study rat myometrial cell gap junctions have included purification and biochemical analysis of gap junctional membrane from pregnant myometrium, the development of antibody probes to recognize the myometrial gap junction protein, and functional analysis of mRNA populations obtained from myometrial cells actively expressing the gap junction protein. In studies directed toward analysis of the major gap junctional protein in rat uterus, homology in morphological properties and in two-dimensional tryptic peptide maps has been found in gap junction fractions isolated from rat heart, liver, and pregnant uterus.[101] Further, a polyclonal antibody prepared against a synthetic peptide corresponding to an amino-terminal sequence of the liver gap junction protein recognized the major junctional proteins in all three tissues.

Recombinant DNA technology has been applied to analysis of cell–cell channels. In these studies, a mRNA fraction isolated from pregnant rat myometrium has been delivered by either liposomes[102,103] or microinjection[104] into cells lacking detectable gap junctions. These procedures have provided circumstantial evidence for the expression of this foreign mRNA containing a putative message for the myometrial junction protein. The development of electrical coupling[102,103] in a channel-deficient cell line (C1-1D) was attributed to the development of cell–cell channels. In a mRNA expression system developed by Werner *et al.*,[104] paired *Xenopus* oocytes became electrically coupled within 4 hr after microinjection of a rat myometrial mRNA and exhibited size-limited dye transfer with reversible uncoupling in the presence of carbon dioxide. (Carbon dioxide causes cytoplasmic acidification, and gap junctions can be uncoupled by elevation of intracellular H^+.[27,105])

The molecular cloning of cDNAs (single-stranded DNA copies of purified mRNA, synthesized using reverse transcriptase) for gap junction proteins from rat and human liver[106,107] and rat heart[108] has recently been reported. These studies have indicated the existence of at least two distinct mRNAs encoding for connexons of different predicted molecular mass, designated "connexin32" and "connexin43" to distinguish the junctional peptide products of 32,000 and 43,000 Da, respectively. Northern blot analysis (a gel transfer hybridization technique in which electrophoretically fractionated RNA molecules are transferred or blotted on to a solid substrate such as nitrocellulose and then analyzed by hybridization with a radiolabeled DNA probe) has indicated that the cDNA originally isolated from liver encoding for connexin32 recognizes mRNAs in stomach, brain, and kidney, whereas the cDNA isolated from rat heart encoding for connexin43 recognizes mRNAs in ovary, term pregnant uterus, lens epithelium, and kidney.[108] Thus, it appears that myometrial gap junctions are related to the connexin43 family based on hybridization of the cDNA probe prepared from rat heart with RNA purified from term pregnant rat uterus.

Using the *Xenopus* oocyte expression system described above, Dahl *et al.*[109] have demonstrated the ability of rat liver cDNA encoding for the liver gap junction protein to induce cell–cell channels in paired oocytes. Induction of the channels was blocked by simultaneous injection of total liver mRNA and antisense RNA (RNA transcribed from the liver gap junction cDNA). A comparable analysis of total rat myometrial cell polyadenylated mRNA (3'-terminal sequences of adenylate residues characteristic of eukaryotic mRNA that are added posttranscriptionally prior to reaching the cytoplasm) with the cDNA encoding for connexin43 in this expression system would be very informative. Given the availability of powerful antibody and cDNA probes constructed to analyze gap junctions, our understanding of the molecular control and functional significance of myometrial gap junctions is within reach.

3.2.3. Control of Myometrial Gap Junction Formation

At present, our understanding of the involvement and regulation of myometrial gap junctions during parturition in humans is incomplete, but, as noted previously, animal studies have contributed much to our awareness of the importance of cell–

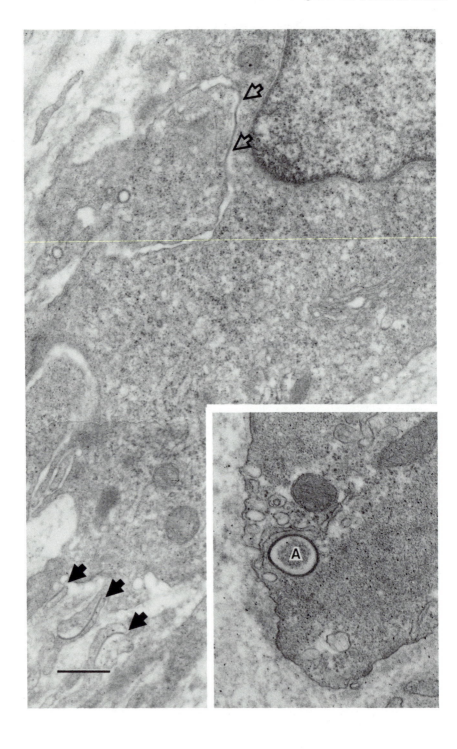

cell coupling and its regulation in uterine smooth muscle during normal and preterm labor. Uterine contractility is influenced by steroid hormones, neurohypophyseal peptides, and locally produced humoral agents such as the prostaglandins and a variety of biogenic amines. Other elements including nervous activity and purely physical factors, such as stretch of the uterine wall, play a role in the control of myometrial activity.[110] Each of these factors has also been shown to influence myometrial gap junctions as well, although experimental evidence points to the hormonal and humoral agents, including steroids and prostaglandins, as the major factors regulating cell–cell coupling in myometrium.

3.2.3a. Steroid Hormones. The myometrium-stimulating effect of estrogen and the inactivating effect of progesterone have been known for more than 50 years.[111,112] The effects of progesterone opposing the excitability caused by estrogen is the basis for the "progesterone block" of uterine contractility during pregnancy described by Csapo.[86] Many of the myometrial responses of pregnancy and labor have therefore been related to the ratio of estrogen to progesterone, which in many species has been shown to increase just prior to parturition. In rats and sheep, declining circulating levels of progesterone and increasing estrogen shift the regulatory balance that maintains the quiescent uterus of midpregnancy.[113,114] The phase of uterine excitability at the end of pregnancy begins when the concentration of progesterone falls below a critical level and that of estrogens exceeds threshold.

Although the primacy of estrogen in the formation of myometrial gap junctions is now well established, results of early studies in rats were equivocal. Bergman[115] first reported the detection of myometrial gap junctions subsequent to administration of diethylstilbestrol (DES) or estradiol in ovariectomized rats. Several subsequent reports[95,116] questioned the role of gap-junction-mediated electrotonic coupling in uterine smooth muscle and the ability of estrogen to induce these junctions *in vivo,* since gap junctions could not be detected in either mature pregnant rats "near" term or in immature, ovariectomized rats injected with DES. The apparent discrepancy in these studies was partially resolved subsequent to the observation that the induction of myometrial gap junctions in pregnant rats occurs immediately prior to parturition.[89] Further, the induction of myometrial gap junctions in immature[117] or hypophysectomized[99] rats required much higher doses of estrogen than expected despite the low levels of progesterone in the animal models. In addition, pregnant rats subjected to ovariectomy on day 16 of pregnancy and treated with estradiol developed myometrial gap junctions and delivered their fetuses prematurely on day 18.[118,119]

←——————————————————————————————————

Figure 4. Gap junctions between myometrial cells of rat uterus. Animals were given five consecutive daily injections of 50 μg estradiol benzoate 60 days post-hypophysectomy. Gap junctions are absent in animals not receiving exogenous estrogen administration but develop between small cytoplasmic processes (dark arrowheads) and between lateral projections of myometrial cells (open arrowheads) following administration of estrogens. Inset: Annular gap junctions (A) are a feature of myometrial cells receiving higher doses of exogenous estrogens. Line scale, 500 nm. (Micrograph reprinted from Burghardt *et al.*[121] with permission of the editor.)

Several studies have been directed toward analysis of the involvement of estrogens in the formation of myometrial gap junctions by examining the relationships of estrogen-receptor-binding ligands on the induction of junctional responses in myometrial cells from hypophysectomized rats.[100,120,121] These studies have provided the following observations regarding the action of estrogens on myometrial cell gap junctions: (1) There is a dose-dependent increase in the number of both cell surface and annular (probably internalized) gap junctions following daily administration of estrogens for five consecutive days. Annular gap junctions, however, are induced only with higher doses of estrogen. (2) The time course required for the induction of myometrial gap junctions is based on the amount of estrogen administered; i.e., a single injection of 500 μg estradiol benzoate, dissolved in 200 μl sesame oil and administered intraperitoneally, results in the induction of gap junctions within 24 hr, whereas with lesser amounts, three consecutive daily injections of 50 μg estradiol or five consecutive daily injections of 5 μg estradiol were required to induce gap junctions. (3) Weak or "impeded" estrogens such as estrone and estriol are able to stimulate gap junction formation, but at much higher doses than estradiol. (4) Simultaneous administration of estradiol and impeded estrogens produced a junctional response that was intermediate between estradiol and estriol or estrone. (5) Antagonistic triphenylethylene antiestrogens administered along with estradiol are able to completely block the induction of myometrial gap junctions. (6) One triphenylethylene derivative known to be a potent estrogen agonist (zuclomiphene citrate) was able to stimulate the formation of myometrial gap junctions.

Taken together, these observations have indicated that graded junctional responses of myometrial cells to both potent and impeded estrogens correlate well with the ability of the hormone to occupy activated nuclear receptors for a sufficient duration to elicit full estrogenic responses.[122] The antagonistic effects of estriol, estrone, and certain triphenylethylene antiestrogens on estradiol-induced gap junction formation in myometrium are probably related to competitive interactions of these ligands at the receptor level and may reflect antagonism of estradiol by reducing the number and/or duration of estrogen-associated estrogen–receptor complexes that are bound to nuclear acceptor sites.[123] It appears, therefore, that the regulatory control of myometrial gap junctions is exerted at the level of mRNA as suggested also by the *Xenopus* oocyte gap junction expression system described above. In this system, myometrial cell mRNA was prepared from estrogen-primed rats and used to induce new cell-to-cell channels in the oocyte membrane that were functionally competent in paired oocytes.[104] As noted by Dahl *et al.*,[103] it is only assumed that new gap junction formation in myometrium reflects the *de novo* synthesis of channel proteins; however, an alternative hypothesis that would be equally exciting is that estrogen may induce the synthesis of mRNA coding for a protein that regulates gap junction expression.

What is not clear yet from the above studies is why induction of myometrial gap junctions requires chronic high levels of estrogen stimulation[100,119,121] whereas other estrogen-dependent uterine responses require much less hormone. There is recent experimental evidence that certain estrogenic responses previously considered interrelated can be stimulated independently, depending on the chemical nature of the ligand–receptor complex and its interaction at genomic sites.[124] It is also likely that other uterus-activating agents play an important role in the appearance of gap junctions. As discussed below, several reports by Garfield and co-workers[125–127] provide evidence for the role of prostanoid products in modulating the appearance of myometrial gap junctions. In addition, physical factors, such as stretch or contraction of the uterine wall, appear to cause an increase in the number of myometrial gap junctions during parturition or in an appropriately estrogen-primed animal.[128,129] The action of these physical factors to increase the number of gap junctions may be mediated via their effects on prostaglandin synthesis.[130]

The role of progesterone on myometrial gap junctions, consistent with other actions of the hormone on the pregnant uterus, is to oppose the stimulatory action of estrogen.[89,100,118,121,125,127] Evidence from studies in both rats and sheep provides excellent correlation between the timing of gap junction formation and the increasing estrogen and decreasing progesterone production.[131,132] In pregnant rats, administration of progesterone prior to normal term can prevent birth.[97] In immature hypophysectomized rats in which myometrial gap junctions can be induced following daily injections of estrogen, increasing ratios of progesterone to junction-inducing doses of estrogen can literally titrate out the gap junctional response.[121]

The mechanisms by which estrogen–progesterone ratios are regulated toward the end of gestation vary among species, but current thinking has largely been influenced by elegant studies conducted in sheep,[133] where the action of a third steroid hormone, cortisol, is involved. In sheep the fetoplacental unit is steroidogenically active, and it appears that the fetus itself controls the timing of parturition.[134] Maturation of the fetal adrenal gland plays an important role in initiating the process through secretion of cortisol, which causes the subsequent induction of placental enzymes responsible for steroid metabolism. The fetus thereby controls the timing of birth through the effect of fetal cortisol on maternal estrogen–progesterone ratios, which in turn regulate prostaglandin secretion and the myometrial oxytocin receptor population[135,136] as well as the induction of gap junctions between uterine smooth muscle cells.[90]

However, the extent to which estrogens and progesterone control myometrial activity during parturition in humans remains controversial, since maternal plasma progesterone levels do not fall just prior to parturition, nor is there a sudden increase in fetal cortisol. In recent years it has been recognized that attempts to correlate myometrial contractility with serum estrogen–progesterone ratios suffer from the limitation that serum steroids do not necessarily reflect a similar balance of steroid receptors in target tissues. Saito *et al.*[137] have demonstrated that, even in rats, where both gap junction formation and the onset of labor correlates well with serum estrogen–progesterone ratios,[131] the relationship between gap junctions and nuclear estrogen and progesterone receptors is far clearer than could be forecast by measurement of serum steroids alone. Therefore, the fact that lower levels of myometrial progesterone receptors are detected in women at term than in nonpregnant myometrium[138] is relevant in this context. As suggested by Casey and MacDonald,[139] the possibility still exists that there is some form of progesterone withdrawal in women at term that is not necessarily reflected in plasma progesterone levels.

The action of the progesterone antagonist RU 486, which causes the termination of pregnancy in women,[140] indicates the importance of this steroid in the maintenance of pregnancy. RU 486 acts as an antihormone by binding with high affinity to the progesterone receptor but is unable to provoke receptor activation and progesterone-directed biological responses.[141] Notably, RU 486 is effective in producing delivery in all species studied. This is interpreted to result from the competitive interaction of RU 486 at the receptor level to effectively reduce the number of activated nuclear progesterone receptors, thereby producing the equivalent of a progesterone withdrawal. This same compound, when administered during the last trimester in rats, is capable of stimulating both the preterm development of myometrial gap junctions and initiation of labor and delivery.[140]

3.2.3b. Prostaglandins. Considerable evidence has accumulated over the past 20 years to indicate that contractility of the pregnant uterus and the synthesis of uterine prostaglandins are functionally related and that prostaglandin synthesis is the cause rather than the result of uterine contractility.[142] The central role of prostaglandins in the onset of normal as well as preterm labor is widely accepted.[139] The regulation of prostaglandin secretion appears to be primarily linked to the increased production of estrogen toward the end of gestation and, in species where progesterone withdrawal occurs, indirectly through progesterone-modulated estrogen production.[143–145] There is evidence that the mammalian uterus is also capable of metabolizing arachidonic acid via both the lipoxygenase and cyclooxygenase pathways and is responsive to products of both pathways.[146]

The specific involvement of prostanoid products on the formation of myometrial gap junctions is not clear, although there is experimental evidence suggesting influence of these compounds on the modulation of gap junctions in rat myometrium. Indomethacin, an inhibitor of prostaglandin synthesis, has been reported to inhibit formation of gap junctions in myometrial tissues *in vitro*.[98,125] An inhibitor of thromboxane synthesis and an agonistic prostacyclin analogue (carbacyclin) similarly inhibit gap junction formation in this *in vitro* system.[125] However, in contrast to the *in vitro* data,[98,125] both indomethacin and meclofenamate potentiate the estrogen-induced amplification of myometrial gap junctions *in vivo*.[126,127] Further studies are needed to establish whether products of the cyclooxygenase and lipoxygenase pathways modulate the expression of gap junctions during parturition or exert their effects primarily through modification of uterine contractile machinery.

3.2.4. Gap Junction Degradation in Myometrial Cells

As rapidly as gap junctions form just prior to labor, they are removed from the myometrial cell surface following parturition.[89,96,97] Both the mechanism controlling the disappearance of gap junctions within 24 hr post-partum and the fate of previously established gap junctional plaques are uncertain. The fact that ovariectomy on day 21 of pregnancy in rats did not prevent loss of gap junctions has been interpreted to indicate that ovarian hormones are not required for degradation of the junctions after parturition.[96] However, given the role of steroid hormones in the formation of gap junctions, it is possible that the postpartum decrease in gap junctions is related to the decline in activated nuclear estrogen receptors. This mechanism has been challenged since exogenous estrogen administration was shown to be insufficient to sustain the presence of these junctions when administered on the day before, during, or the day after parturition.[127] However, when high estrogen conditions are sustained through exogenous estrogen administration, Wathes and Porter[128] have shown that mechanical distention of the uterine wall can maintain the number of gap junctions observed during parturition for as long as 70 hr after delivery. These observations lend support to the role of estrogen as a proximal initiator of myometrial gap junction synthesis and to the involvement of other factors in modulating their expression.

Just as estrogen appears to be the proximal regulatory agent of gap junction formation through the transcription of a gap junction message, cessation of transcription with declining estrogen levels should down-regulate gap junctions regardless of the presence or absence of other agents that modulate myometrial gap junctions. The major gap junction protein in mammalian cells is synthesized and assembled into channels with times estimated at 3 to 4 hr[147,148] and turns over with a half-life estimated at 5 hr.[149,150] If the half-life of myometrial gap junctions is comparable to that reported in other cell types, the absence of new synthesis could account for their rapid disappearance from the cell surface.

The mechanism by which gap junctions are degraded once new synthesis is terminated also remains to be determined. Possible mechanisms include uncoupling followed by dispersal of gap junction components in the membrane[151] or internalization of surface junctions to form cytoplasmic annular profiles and subsequent digestion by a lysosomal process.[152] The latter mechanism has been suggested by Garfield[153] to account for the rapid loss of junctional membrane following parturition.

Our studies do not favor this mechanism of postpartum bulk removal of gap junctional membrane. Although the fate of interiorized gap junctions may be degradation through lysosomal activity, we believe that the appearance of annual junctions in myometrial cells reflects hyperstimulation of gap junction synthesis and that annular gap junctions are a mechanism by which excessive junctional membrane is removed in bulk during estrogen stimulation. At lower doses of exogenous estrogen administration, annular gap junctions are not induced, although gap junctions conjoining myometrial cells are numerous.[121] Stimulation of gap junction turnover by internalization can be induced in other cell types such as ovarian granulosa, thecal, and interstitial cells and in uterine serosal cells (see below) by exogenous hormone administration.[154–156]

Evaluation of myometrial cells following cessation of estrogen stimulation in hypophysectomized rats receiving doses insufficient to induce annular gap junctions reveals a progressive loss of junctional contacts over several days, yet we have been unable to detect any annular gap junctions in these tissues. In contrast, in animals induced to form annular gap junctions by high doses of estrogens, profiles of annular junctions are visible within myometrial cells for several days subsequent to estrogen withdrawal (R. C. Burghardt, unpublished results), suggesting that if annular gap junctions are involved in removal of cell surface gap junctional contacts, their destruction in animals treated with lower doses of estrogens would be possible to detect. There are other cell types in which disappearance of gap junctions is not linked to internalization.[157–159] Disassembly of junctional contacts and dispersion in the membrane prior to turnover is a difficult process to detect and may account for the loss of the gap junctions from the surface of myometrial cells in the postpartum period.

3.2.5. Regulation of Gap Junctional Conductance

Until fairly recently, an appreciation of the role that gap junction channels might play in the regulation of physiological processes was limited by the belief that conductance of the channel was unmodifiable beyond the ability of gap junction channels to form and disappear.[160] However, there is now considerable experimental evidence that gap junction channels can also be gated (i.e., fluctuate between an open and a closed state) by a variety of mechanisms.[161] Cytoplasmic conductance in most cell types is rapidly and reversibly reduced by subtle changes in intracellular pH and Ca^{2+}.[3,26,27,162] The actions of pH and Ca^{2+} are presumably on the same gating structure, although the sensitivity of most cells to pH is much greater than to Ca^{2+} because of higher H^+ binding affinity.[161]

In a number of cultured mammalian cell types, cAMP and agents that increase intracellular levels of cAMP have been found to increase coupling and/or the incidence of gap junctions.[76–79,163,164] The time course for these changes is much slower than for H^+- and Ca^{2+}-induced permeability changes and is generally dependent on protein synthesis.[76–79] However, cAMP and cAMP-mediated hormone action appear also to be involved in short-term modulation of conductance between mammalian cardiac cells[165] and between rat hepatocytes.[166] The level at which cAMP exerts a rapid effect on the modulation of junctional permeability is not yet established but has led to a search for a direct action of cAMP-dependent protein kinase on gap junctional communication. Saez et al.[166] have demonstrated that cAMP-dependent protein kinase catalyzes phosphorylation of the 27,000-Da protein of isolated liver junctions and suggest that this phosphorylation of the gap junction channel may modulate the conductance of liver gap junctions. In cultured mouse hepatocytes, cAMP has been shown to stimulate both the biosynthesis and the phosphorylation of the gap junction protein.[167] Therefore, gap junctions may not only provide a means of regulating cAMP-dependent protein kinase through junctionally mediated signal transfer as detailed in Section 2 but may also be an important substrate for the enzyme in a cellular response amplification process.

In contrast to the effects of increased junctional permeability that appear to be mediated by cAMP-dependent protein kinase, an opposite effect on junctional permeability has been reported subsequent to treatment of a number of mammalian cells with phorbol ester tumor promoters or diacylglycerol.[22,168-170] Phorbol esters and diacylglycerol are direct activators of protein kinase C and, like cAMP-dependent protein kinase, play an important role in signal transduction. Protein kinase C has also been shown to phosphorylate the rat liver junctional protein *in vitro,* and both kinases are virtually specific for serine residues.[167,171] The stoichiometry of phosphorylation, however, is tenfold greater for protein kinase C than for cAMP-dependent protein kinase.[171]

Additional circumstantial evidence for the role of protein phosphorylation in the modulation of gap junctional permeability has been provided by demonstrations of decreased junctional permeance without change in gap junctional area in cells expressing the protein products (pp60$^{v\text{-}src}$ and pp60$^{c\text{-}src}$) of the viral and cellular *src* genes.[172-174] Although the mechanisms underlying alterations in junctional permeability are unknown, a possibility under consideration is that the *src* gene product, which is a tyrosine-specific protein kinase, may act directly on the junctional protein. As yet, there is no evidence that tyrosine residues in the junctional protein are phosphorylated. Since pp60$^{v\text{-}src}$ also promotes phosphorylation of membrane phosphoinositide precursors of diacylglycerol and inositol trisphosphate,[175] the action of this protein kinase may modify the signal transduction route that utilizes diacylglycerol and inositol trisphosphate as intracellular messengers.[176] It has therefore been proposed that intercellular communication is under dual physiological control, with the cAMP signal route effecting increased junctional permeability and the diacylglycerol route reducing permeability.[170] Although phosphorylation of gap junction protein from intact cells has been demonstrated,[176] further analysis is needed to confirm whether gap junction phosphorylation results in a conformational change in the junctional protein, thereby altering channel permeability. Protein kinase modulation of junctional permeability may also be mediated by indirect mechanisms such as phosphorylation of gap-junction-associated proteins or by alteration in intracellular pH, Ca^{2+}, and/or calmodulin (for review, see Atkinson and Sheridan[177]).

3.2.6. Regulation of Gap Junctional Conductance in Myometrial Cells

From the foregoing discussion of uterine gap junctions it is clear that the potential for cell–cell communication within uterine tissues is regulated to a large extent by the formation and removal of gap junction channels. Analysis of functional modifications in junctional coupling in uterine myometrial cells has only recently been initiated, but experimental evidence has been provided for physiological control of cell–cell communication independent of gap junction formation and turnover. In a cell–cell diffusion assay to monitor the transfer of a radiolabeled metabolite, [^{3}H]2-deoxyglucose, a tenfold increase in the apparent diffusion coefficient was observed in myometrial cells obtained from parturient compared to nonparturient rats.[178] This same intercellular communication assay system was subsequently utilized to demonstrate a reduction in the permeability of gap junctions from parturient myometrium subsequent to elevation of intracellular Ca^{2+} or H^+.[179] In contrast to other cells, elevation of cAMP produced by treatment with dibutyryl cAMP or 8-bromo cAMP, by stimulation of adenylate cyclase with forskolin, or by inhibiting phosphodiesterase activity with theophylline reduced the apparent diffusion coefficient of [³H]2-deoxyglucose.[180] Similarly, several receptor-binding agonists that elevate intracellular cAMP in myometrium also reduced [³H]2-deoxyglucose diffusion.

These observations are significant in that elevation of intracellular cAMP in myometrial cells has an effect on junctional permeability that is opposite that of other mammalian cell types in which cAMP modulates the permeability of the channel.[165,166] However, the most widely used drugs in the treatment of preterm labor are the β-adrenergic agonists.[181] The inhibitory effects on myometrial contractility that result from administration of β_2-adrenergic agonists and other receptor-biding ligands such as relaxin are mediated by elevation of intracellular cAMP.[182,183] Therefore, in addition to the action of cAMP in reducing myometrial contractility through the regulation of myosin light chain kinase activity, cAMP may also exert a suppressive effect on myometrial electrical conductivity through down-regulation of myometrial gap junction permeability.

3.3. Uterine Serosal Epithelium

The uterus, like other organs resting in the peritoneal cavity, is covered by a squamous layer of mesothelial cells resting on loose connective tissue. Consistent with the intrinsic growth capacity of the uterus, this serous membrane is equipped to accommodate the cyclic expansion of the uterus and pregnancy-associated distention of the uterine wall. Junctional contacts connecting serosal cells would therefore be expected to exhibit structural alterations when adaptive modifications of the uterus to the products of conception occur.

Uterine serosal cells exhibit junctional complexes typical of other epithelial cells, which include zonula occludens or tight junctions as well as gap junctional contacts. The gap junction composition of this cell layer in rats has been noted to exhibit dynamic properties based on analysis of the effects of hypophysectomy and replacement therapy with exogenous estrogens.[100,121] Unlike uterine myometrial and endometrial stromal cells, serosal cells retain small but detectable gap junctions in the absence of hormones of the pituitary–ovarian axis. However, a dose-dependent increase in the number and size of gap junctions, as well as the induction of cytoplasmic annular gap junctions, results from administration of exogenous estrogens to hypophysectomized rats *in vivo* (Fig. 5). The induction of annular gap junctions following administration of estrogen requires considerably less exogenous hormone than is required to induce cell surface or internalized annular gap junctions between myometrial cells, and this observation was used to predict reliably the dose of potent or impeded estrogen necessary to induce the formation of gap junctions

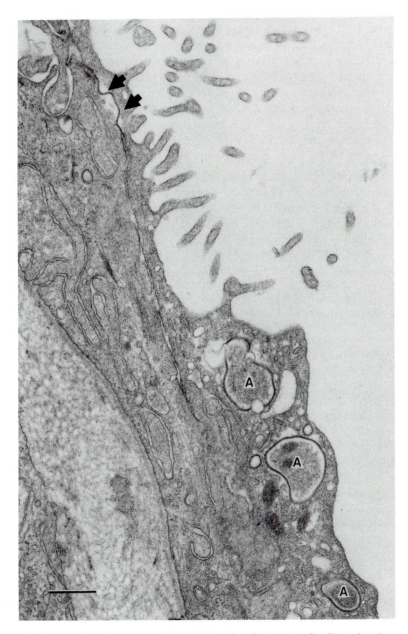

Figure 5. Gap junctions between serosal epithelial cells of rat uterus. Small gap junction contacts (arrowhead) are present between serosal cells in the absence of hormones from the pituitary–ovarian axis. Administration of exogenous estrogen causes increases in the number and size of gap junctions as well as induction of cytoplasmic annular gap junctions in a dose-dependent fashion. Line scale, 500 nm. (Micrograph reprinted from Burghardt *et al.*[100] with permission of editor.)

between myometrial cells.[121] The lower doses of exogenous estrogen needed to detect junctional responses in serosa are more consistent with those necessary to stimulate other endpoints of estrogen action in uterine tissues. In contrast to myometrial cells, however, simultaneous administration of estrogen with triphenylethylene antiestrogens did not result in significant antagonism of gap junction growth and turnover. Therefore, estrogen-stimulated modulation of serosal cell gap junctions was interpreted as resulting from an indirect action of estrogen on junctional membrane in this cell type.[100] One possibility is that estrogens may exert their effects on serosal gap junctions, as in myometrium, by the action of estrogen-induced distention of the uterine wall and/or prostaglandin synthesis.[128]

The significance of serosal cell gap junction modulation in response to estrogen stimulation has not yet been established. We believe that the dramatic amplification of gap junctional membrane in serosal cells may be one of the adaptive alterations of the uterine wall during pregnancy needed to accommodate distention while maintaining sufficient junctional membrane to facilitate effective metabolic coupling. Since the uterus can more than triple in size in a matter of days during early pregnancy, the induction of annular junctions may reflect overproduction of junctional membrane in estrogen-treated animals without the pregnancy-associated distention of the uterine wall. Overstimulation of surface junctions in hormone-treated animals could therefore be compensated for by internalization.

4. Conclusions

Several tissue compartments in the mammalian uterus have been described in which the potential for direct intercellular signal exchange mediated by gap junctions is developed. Although species-specific variability exists, it appears that cell–cell communication in the uterus is highly regulated. Gap junctions in uterine epithelium, stroma, and myometrium are normally either poorly developed or absent in cycling animals but are programmed to undergo ordered spatial and temporal patterns of induction, increase in number and size, and turnover in response to pregnancy. These patterns of gap junction development are influenced by the balance of activated nuclear estrogen and progesterone receptors, with the specific ratio of activated progesterone and estrogen receptors needed to cause modulation of junctional membrane being a function of the tissue type. Uterine luminal epithelium and underlying stroma are programmed to develop gap junctions in response to embryo recognition and implantation during a period of progesterone dominance and permissive estrogen action. Toward the end of pregnancy, when estrogen dominance is restored, gap junctions between myometrial cells form to permit propagation of action potentials and thereby coordinate myometrial contractile activity at birth.

The significance of gap junction expression in myometrial cells seems reasonably obvious with regard to electrical coupling. Less clear is the significance of gap junction formation in uterine cell types such as epithelium and stroma, which are not electrically excitable. Studies with cultured cells have provided evidence that gap junctions can transmit signals between coupled cells that can regulate the activity of cAMP-dependent protein kinases. The involvement of gap junctions and the physiological consequences resulting from the development of the potential for signal transfer in uterine endometrium during implantation and decidualization remain poorly understood, yet the highly orderly tissue responses initiated by embryo recognition and the regulation of decidual luteotrophin secretion may also be integrated by the cell-to-cell transmission of signals that can activate the cAMP-dependent protein kinases. Given the receptor population of myometrial cells, the induction of gap junctions could also provide a means by which myometrial cells could use each other as reservoirs for molecules to regulate the activity of cAMP-dependent protein kinases, a function that may be of equal importance to electrical coupling in myometrial cells.

Although recognition of the importance of electrical coupling in myometrial cells has led to more extensive analysis of gap junctions in this tissue, we continue to seek a better understanding of the physiological regulation of their appearance at term. Recent studies suggesting that the appearance and permeability of myometrial gap junctions can be experimentally manipulated provide a basis for further investigation of gap junctions as a possible target site for clinical intervention in the onset or delay of labor.

ACKNOWLEDGMENTS. We wish to thank Dr. Ruth G. Kleinfeld for providing micrographs of uterine decidual cells and Dr. Elke Winterhager for providing a preprint of her work. Supported by Basic Research Grant No. 1-1052 from the March of Dimes Birth Defects Foundation and HD 14781 from NIH (to R.C.B.) and by NIH grants HD 13704, HD 21318, AM 21019, AM 07310, and CA 23743 (to W.H.F.).

References

1. Loeb, L., 1908, The production of deciduomata and the relation between the ovaries and the formation of the decidua, *J.A.M.A.* **50:**1897–1901.
2. Reynolds, S. R. M., 1949, *Physiology of the Uterus,* 2nd, ed., Harper & Brothers, New York.
3. Loewenstein, W. R., 1981, Junctional intercellular communication: The cell-to-cell membrane channel, *Physiol. Rev.* **61:**829–913.
4. Hertzberg, E. L., Lawrence, T. S., and Gilula, N. B., 1981, Gap junctional communication, *Annu. Rev. Physiol.* **43:**479–491.
5. Peracchia, C., 1980, Structural correlates of gap junction permeation, *Int. Rev. Cytol.* **66:**81–146.
6. Bennett, M. V. L., and Goodenough, D. A., 1978, Gap junctions, electronic coupling and intercellular communication, *Neurosci. Res. Prog. Bull.* **16:**373–486.
7. Hertzberg, E. L., Lawrence, T. S., and Gilula, N. B., 1981, Gap junctional communication, *Annu. Rev. Physiol.* **43:**479–491.
8. Larsen, W. J., and Risinger, M. A., 1985, The dynamic life histories of intercellular membrane junctions, in: *Modern Cell Biology,* Volume 4 (B. Satir, ed.), Alan R. Liss, New York, pp. 151–216.

9. Loewenstein, W. R., 1968, Junctional intercellular communication and the control of growth, *Biochim. Biophys. Acta* **560**:1–65.
10. Pitts, J. D., and Finbow, M. E., 1986, The gap junction. *J. Cell. Sci. (Suppl.)* **4**:239–266.
11. Sheridan, J. D., and Atkinson, M. M., 1985, Physiological roles of permeable junctions: Some possibilities, *Annu. Rev. Physiol.* **47**:337–353.
12. Bennett, M. V. L., and Trinkaus, J. P., 1970, Electrical coupling between embryonic cells by way of extracellular space and specialized junctions, *J. Cell Biol.* **44**:592–606.
13. Gilula, N. B., Reeves, O. R., and Steinbach, A., 1972, Metabolic coupling, ionic coupling and cell contacts, *Nature* **235**:262–265.
14. Simpson, J., Rose, B., and Loewenstein, W. R., 1977, Size limit of molecules permeating the junctional membrane channels, *Science* **195**:294–296.
15. Dewey, M. M., and Barr, L., 1964, Intercellular connection between smooth muscle cells: The nexus, *Science* **137**:670–672.
16. Furshpan, E. J., and Potter, D. D., 1962, Intracellular and extracellular responses of several regions of the Mauthner cell of the goldfish, *J. Neurophysiol.* **25**:732–771.
17. Trachtenberg, M. C., and Pollen, D. A., 1970, Neuroglia: Biophysical properties and physiological function, *Science* **167**:1248–1252.
18. Fletcher, W. H., Byus, C. V., and Walsh, D. A., 1987, Receptor-mediated action without receptor occupancy: A function for cell–cell communication in ovarian function, in: *Regulation of Ovarian and Testicular Function* (V. B. Mahesh, D. S. Dhindsa, E. Anderson, and S. P. Kalra, eds.), Plenum Press, New York, pp. 299–323.
19. Loewenstein, W. R., 1968, Communication through cell junctions. Implications in growth control and differentiation, *Dev. Biol.* **19**:(Suppl. 2):151–183.
20. Lawrence, T. S., Beers, W. H., and Gilula, N. B., 1978, Transmission of hormonal stimulation by cell-to-cell communication, *Nature* **272**:501–506.
21. Azarnia, R., and Larsen, W. J., 1976, Intercellular communication and cancer, in: *Intercellular Communication* (W. C. DeMello, ed.), Plenum Press, New York, pp. 145–172.
22. Yotti, L. P., Chang, C. C., and Trosko, J. E., 1979, Elimination of metabolic cooperation in Chinese hamster cells by a tumor promoter, *Science* **206**:1089–1091.
23. Entomoto, T., and Yamasaki, H., 1984, Lack of intercellular communication between chemically transformed and surrounding nontransformed BALBc/3T3 cells, *Cancer Res.* **44**:5200–5203.
24. Janssen-Timmen, U., Traub, O., Dermietzel, R., Rabes, H. M., and Willecke, K., 1986, Reduced number of gap junctions in rat hepatocarcinomas detected by monoclonal antibody, *Carcinogenesis* **7**:1475–1482.
25. Hertzberg, E. L., Spray, D. C., and Bennett, M. V. L., 1985, Reduction of gap junctional conductance by microinjection of antibodies against the 27-kDa liver gap junction polypeptide, *Proc. Natl. Acad. Sci. U.S.A.* **82**:2412–2416.
26. Peracchia, C., and Peracchia, L. L., 1980, Gap junction dynamics: Reversible effects of divalent cations, *J. Cell Biol.* **87**:708–718.
27. Peracchia, C., and Peracchia, L. L., 1980, Gap junction dynamics: Reversible effects of hydrogen ions, *J. Cell Biol.* **87**:719–727.
28. Goshima, K., 1969, Synchronized beating of and electrotonic transmission between myocardial cells mediated by heterotypic strain cells in monolayer culture, *Exp. Cell Res.* **65**:161–169.
29. Walsh, D. A., and Cooper, R. H., 1979, The physiological regulation and function of cAMP-dependent protein kinases, in: *Biochemical Actions of Hormones*, Volume 6 (G. Litwack, ed.), Academic Press, New York, pp. 1–75.
30. Beebe, S. J., and Corbin, J. D., 1986, Cyclic nucleotide-dependent protein kinases, in: *The Enzymes*, Volume 17 (P. D. Boyer, ed.) Academic Press, New York, pp. 43–111.
31. Fletcher, W. H., and Byus, C. V., 1982, Direct cytochemical localization of catalytic subunits dissociated from cAMP-dependent protein kinase in Reuber H-35 hepatoma cells. I. Development and validation of fluorescinated inhibitor, *J. Cell Biol.* **93**:719–726.
32. Byus, C. V., and Fletcher, W. H., 1982, Direct cytochemical localization of catalytic subunits

dissociated from cAMP-dependent protein kinase in Reuber H-35 hepatoma cells. II. Temporal and spatial kinetics, *J. Cell Biol.* **93**:727–734.

33. Fletcher, W. H., Van Patten, S. M., Cheng, H.-C., and Walsh, D. A., 1986, Cytochemical identification of the regulatory subunit of the cAMP protein kinase by use of fluorescently labeled catalytic subunit, *J. Biol. Chem.* **261**:5504–5513.

34. Van Patten, S. M., Fletcher, W. H., and Walsh, D. A., 1986, The inhibitor protein of the cAMP-dependent protein kinase–catalytic subunit interaction, *J. Biol. Chem.* **261**:5514–5523.

35. Murray, S. A., and Fletcher, W. H., 1984, Hormone-induced intercellular signal transfer dissociates cyclic AMP-dependent protein kinase, *J. Cell Biol.* **98**:1710.

36. Fletcher, W. H., and Greenan, J. R. T., 1985, Receptor mediated action without receptor occupancy, *Endocrinology.* **116**:1660–1662.

37. Ware, C. F., Coffman, F., Green, L. M., and Fletcher, W. H., 1988, Hierarchy of molecular mechanisms controlling the sensitivity of tumor cells to cytolysis by lumphotoxin and tumor necrosis factor, in: *Proceedings of International Conference on Tumor Necrosis Factor/Cachepin, Lymphotoxin and Related Cytotoxins* (B. Bonavida, G. Gifford, H. Kirchner, and N. Old, eds.), F. Karger, Basel, pp. 26–31.

38. Fletcher, W. H., Shiu, W. W., Ishida, T. A., Haviland, D. L., and Ware, C. F., 1987, Resistance to the cytolytic action of lymphotoxin and tumor necrosis factor coincides with the presence of gap junctions uniting target cells, *J. Immunol.* **139**:956–962.

39. Anderson, W. A., Kang, Y.-H., and DeSombre, E., 1975, Estrogen and antagonist-induced changes in endometrial topography of immature and cycling rats, *J. Cell Biol.* **64**:692–703.

40. Psychoyos, A., 1973, Endocrine control of egg-implantation, in: *Handbook of Physiology, Endocrinology,* Volume 2, part 2 (R. O. Greep and E. B. Astwood, eds.), American Physiological Society, Washington, DC, pp. 187–215.

41. Winterhager, E., and Kuhnel, W., 1982, Alterations in intercellular junctions of the uterine epithelium during preimplantation phase in the rabbit, *Cell Tissue Res.* **224**:517–526.

42. Murphy, C. R., Swift, J. G., Mukherjee, T. M., and Rogers, A. W., 1982, Reflexive gap junctions on uterine luminal epithelial cells, *Acta Anat.* **112**:92–96.

43. Herr, J. C., 1976, Reflexive gap junctions. Gap junctions between processes arising from the same ovarian decidual cell, *J. Cell Biol.* **69**:495–501.

44. Herr, J. C., and Heidger, P. M., 1978, A freeze-fracture study of exocytosis and reflexive gap junctions in human ovarian decidual cells, *Am. J. Anat.* **152**:29–44.

45. Lawn, A. M., Wilson, E. W., and Finn, C. A., 1971, The ultrastructure of human decidual and predecidual cells, *J. Reprod. Fertil.* **26**:85–90.

46. Murphy, C. R., Swift, J. G., Mukherjee, T. M., and Rogers, A. W., 1981, Effects of ovarian hormones on cell membranes in the rat uterus. II. Freeze-fracture studies on tight junctions of the lateral plasma membrane of the luminal epithelium, *Cell Biophysics* **3**:57–69.

47. Murphy, C. R., Swift, J. G., Mukherjee, T. M., and Rogers, A. W., 1982, The structure of tight junctions between uterine luminal epithelial cells at different stages of pregnancy in the rat, *Cell Tissue Res.* **223**:281–286.

48. Murphy, C. R., Swift, J. G., Need, J. A., Mukherjee, T. M., and Rogers, A. W., 1982, A freeze-frature electron microscopic study of tight junctions of epithelial cells in the human uterus, *Anat. Embryol.* **163**:367–370.

49. Winterhager, E., Brummer, F., Dermietzel, R., Hulser, D. F., and Denker, H.-W., 1988, Gap junction formation in rabbit uterine epithelium in response to embryo recognition, *Dev. Biol.* **126**:203–211.

50. Enders, A. E., and Schlafke, S., 1967, A morphological analysis of the early implantation in the rat, *J. Anat.* **120**:185–225.

51. Potts, D. M., 1969, The ultrastructure of egg implantation, *Adv. Reprod. Physiol.* **4**:241–267.

52. Tachi, S., and Tachi, C., 1979, Ultrastructural studies on maternal–embryonic cell interaction during experimentally induced implantation of rat blastocysts to the endometrium of the mouse, *Dev. Biol.* **68**:203–223.

53. Lejeune, B., Van Hoeck, J., and Leroy, F., 1981, Transmitter role of the luminal uterine epithelium in the induction of decidualization in rats, *J. Reprod. Fertil.* **61:**235–240.

54. Moulton, B. C., and Koenig, B. B., 1986, Biochemical responses of the luminal epithelium and uterine sensitization, *Ann. N.Y. Acad. Sci.* **476:**95–109.

55. Furchgott, R. F., 1983, Role of endothelium in responses of vascular smooth muscle, *Circ. Res.* **53:**557–573.

56. Finn, C. A., and Porter, D. G., 1975, *The Uterus*, Elek Science, London.

57. DeFeo, V. J., 1967, Decidualization, in: *Cellular Biology of the Uterus* (R. M. Wynn, ed.), Meredith Publishing, New York, pp. 191–290.

58. Kleinfeld, R. G., Morrow, H. A., and DeFeo, V. J., 1976, Intercellular junctions between decidual cells in the growing deciduoma of the pseudopregnant rat uterus, *Biol. Reprod.* **15:**593–603.

59. O'Shea, J. D., Kleinfeld, R. G., and Morrow, H. A., 1983, Ultrastructure of decidualization in the pseudopregnant rat, *Am. J. Anat.* **166:**271–298.

60. Finn, C. A., and Lawn, A. M., 1967, Specialized junctions, between decidual cells in the uterus of the pregnant mouse, *J. Ultrastruct. Res.* **20:**321–327.

61. Finn, C. A., 1971, The biology of decidual cells, *Adv. Reprod. Physiol.* **5:**1–26.

62. Bernhard, O., and Rachman, F., 1980, Immunological aspects of the decidual reaction, *Prog. Reprod. Biol.* **7:**135–142.

63. Gibori, G., Rothchild, I., Pepe, G. J., Morishige, W. K., and Lam, P., 1974, Luteotrophic action of decidual tissue in the rat, *Endocrinology* **95:**1113–1118.

64. Jayatilak, P. G., Glaser, L. A., Basuray, R., Kelly, P. A., and Gibori, G., 1985, Identification and partial characterization of a prolactin-like hormone produced by rat decidual tissue, *Proc. Natl. Acad. Sci. U.S.A.* **82:**217–221.

65. Golander, A., Hurley, T., Barrett, J., Hize, A., and Handwerger, S., 1978, Prolactin synthesis by human chorion-decidual tissue: A possible source of amniotic fluid prolactin, *Science* **202:**311–312.

66. Tomita, K., McCoshen, J. A., Friesen, H. G., and Tyson, J. E., 1982, Quantitative comparison between biological and immunological activities of prolactin derived from human fetal and maternal sources, *J. Clin. Endocrinol. Metab.* **55:**269–271.

67. Handwerger, S., Barry, S., Markoff, E., Barrett, J., and Conn, P. M., 1983, Stimulation of the synthesis and release of decidual prolactin by a placental polypeptide, *Endocrinology* **112:**1370–1374.

68. Markoff, E., Howell, S., and Handwerger, S., 1983, Inhibition of decidual prolactin by a decidual peptide, *J. Clin. Endocrinol. Metab.* **56:**962–968.

69. Moulton, B. C., and Koenig, B. B., 1986, Hormonal control of phospholipid methylation in uterine luminal epithelial cells during sensitivity to deciduogenic stimuli, *Endocrinology* **118:**244–249.

70. Hirata, F., and Axelrod, J., 1980, Phospholipid methylation and biological signal transmission, *Science* **209:**1082–1090.

71. Kennedy, T. G., 1983, Embryonic signals and the initiation of blastocyst implantation, *Aust. J. Biol. Sci.* **36:**521–543.

72. Kennedy, T. G., 1985, Evidence for the involvement of prostaglandins throughout the decidual cell reaction in the rat, *Biol. Reprod.* **33:**140–146.

73. Kennedy, T. G., 1983, Prostaglandin E_2, adenosine 3′:5′-cyclic monophosphate and changes in endometrial vascular permeability in rat uteri sensitized for the decidual cell reaction, *Biol. Reprod.* **29:**1069–1076.

74. Johnson, R., Hammer, M., Sheridan, J., and Reval, J. P., 1974, Gap junction formation between reaggregating Novikoff hepatoma cells, *Proc. Natl. Acad. Sci. U.S.A.* **71:**4536–4543.

75. Flagg-Newton, J. L., 1980, The permeability of the cell-to-cell membrane channel and its regulation in mammalian cell junction, *In Vitro* **16:**1043–1048.

76. Flagg-Newton, J. L., and Loewenstein, W. R., 1981, Cell junction and cyclic AMP: II. Modulations of junctional membrane permeability, dependent on serum and cell density, *J. Membr. Biol.* **63:**121–131.

77. Flagg-Newton, J. L., Dahl, G., and Loewenstein, W. R., 1981, Cell junction and cyclic AMP: I. Upregulation of junctional membrane permeability and junctional membrane particles by administration of cyclic nucleotide or phosphodiesterase inhibitor, *J. Membr. Biol.* **63**:105–121.

78. Azarnia, R., Dahl, G., and Loewenstein, W. R., 1981, Cell junction and cyclic AMP: III. Promotion of junctional membrane permeability and junctional membrane particles in a junction deficient cell type, *J. Membr. Biol.* **63**:133–146.

79. Radu, A., Dahl, G., and Loewenstein, W. R., 1982, Hormonal regulation of cell junction permeability: Upregulation by catecholamine and prostaglandin E_1, *J. Membr. Biol.* **70**:239–251.

80. Hertzberg, E. L., 1985, Antibody probes in the study of gap junctional communication, *Annu. Rev. Physiol.* **47**:305–318.

81. Tadvalkar, G., and Pinto da Silva, P., 1983, *In vitro,* rapid assembly of gap junctions is induced by cytoskeletal disrupters, *J. Cell Biol.* **96**:1279–1287.

82. Zor, U., 1983, Role of cytoskeletal organization in the regulation of adenylate cyclase–cyclic adenosine monophosphate by hormones, *Endocrinol. Rev.* **4**:1–21.

83. Child Health and Human Development, 1981, *An Evaluation and Assessment of the State of the Sciences,* NIH Publication No. (PHS) 82-2304, National Institutes of Health, Bethesda.

84. Bengtsson, B., Chow, E. M. H., and Marshall, J. M., 1984, Calcium dependency of pregnant rat myometrium: Comparison of circular and longitudinal muscle, *Biol. Reprod.* **30**:869–878.

85. Leroy, M.-J., Ferre, F., Filliatreau, F. G., Cabrol, D., and Breuiller, M., 1986, Cyclic AMP metabolism in the inner and outer layers of human myometrium near term, *Acta Physiol. Hung.* **67**:83–94.

86. Csapo, A. I., 1956, Progesterone "block," *Am. J. Anat.* **98**:273–291.

87. Huszar, G., and Roberts, J. M., 1981, Biochemistry and pharmacology of the myometrium and labor: Regulation at the cellular and molecular levels, *Am. J. Obstet. Gynecol.* **142**:225–237.

88. Carsten, M. E., and Miller, J. D., 1987, A new look at uterine muscle contraction, *Am. J. Obstet. Gynecol.* **157**:1303–1315.

89. Garfield, R. E., Sims, S. M., and Daniel, E. E., 1977, Gap Junctions: Their presence and necessity in myometrium during parturition, *Science* **198**:958–960.

90. Garfield, R. E., Rabideau, S., Challis, J. R. G., and Daniel, E. E., 1979, Hormonal control of gap junctions in sheep myometrium, *Biol. Reprod.* **21**:999–1007.

91. Garfield, R. E., Daniel, E. E., Dukes, M., and Fitzgerald, J. D., 1982, Changes in gap junctions in myometrium of guinea pig at parturition and abortion, *Can. J. Physiol. Pharmacol.* **60**:335–341.

92. Demianczuk, N., Towell, M. E., and Garfield, R. E., 1984, Myometrial electrophysiologic activity and gap junctions in the pregnant rabbit, *Am. J. Obstet. Gynecol.* **149**:485–491.

93. Garfield, R. E., and Hayashi, R. H., 1981, Appearance of gap junctions in the myometrium of women during labor, *Am. J. Obstet. Gynecol.* **140**:254–260.

94. Krishnamurti, C. R., Kitts, D. D., Kitts, W. D., and Tompkins, J. G., 1982, Myoelectrical changes in the uterus of the sheep around parturition, *J. Reprod. Fertil.* **64**:59–67.

95. Garfield, R. E., and Daniel, E. E., 1974, The structural basis of electrical coupling (cell-to-cell contacts) in rat myometrium, *Gynecol. Invest.* **5**:284–300.

96. Berezin, I., Daniel, E. E., and Garfield, R. E., 1982, Ovarian hormones are not necessary for postpartum regression of gap junctions, *Can. J. Physiol. Pharmacol.* **60**:1567–1572.

97. Garfield, R. E., Sims, S. M., Kannan, M. S., and Daniel, E. E., 1978, Possible role of gap junctions in activation of myometrium during parturition, *Am. J. Physiol.* **235**:C168–C179.

98. Garfield, R. E., Merrett, D., and Grover, A. K., 1980, Gap junction formation and regulation in myometrium, *Am. J. Physiol.* **239**:C217–C228.

99. Merk, F. B., Kwan, P. W. L., and Leav, I., 1980, Gap junctions in the myometrium of hypophysectomized estrogen-treated rats, *Cell Biol. Int. Rep.* **4**:287–294.

100. Burghardt, R. C., Matheson, R. L., and Gaddy, D., 1984, Gap junction modulation in rat uterus. I. Effects of estrogen on myometrial and serosal cells, *Biol. Reprod.* **30**:239–248.

101. Zervos, A. S., Hope, J., and Evans, W. H., 1985, Preparation of a gap junction fraction from uteri

of pregnant rats: The 28-kD polypeptides of uterus, liver, and heart gap junctions are homologous, *J. Cell Biol.* **101**:1363–1370.

102. Dahl, G., Azarina, R., and Werner, R., 1980, *De novo* construction of cell-to-cell channels, *In Vitro* **16**:1068–1075.

103. Dahl, G., Werner, R., and Azarnia, R. 1983, Studies on the biogenesis of cell–cell channels, *Methods Enzymol.* **98**:537–545.

104. Werner, R., Miller, T., Azarnia, R., and Dahl, G., 1985, Translation and functional expression of cell–cell channel mRNA in *Xenopus* oocytes, *J. Membr. Biol.* **87**:253–268.

105. Turin, L., and Warner, A. E., 1980, Intracellular pH in early *Xenopus* embryos: Its effect on current flow between blastomeres, *J. Physiol. (Lond.)* **300**:489–504.

106. Paul, D. L., 1986, Molecular cloning of cDNA for rat liver gap junction protein, *J. Cell Biol.* **103**:123–134.

107. Kumar, N. M., and Gilula, N. B., 1986, Cloning and characterization of human and rat liver cDNAs coding for a gap junction protein, *J. Cell Biol.* **103**:767–776.

108. Byer, E. C., Paul, D. L., and Goodenough, D. A., 1987, Connexin43: A protein from rat heart homologous to a gap junction protein from liver, *J. Cell Biol.* **105**:2621–2629.

109. Dahl, G., Miller, T., Paul, D., Voellmy, R., and Werner, R.. 1987, Expression of functional cell–cell channels from cloned rat liver gap junction complementary DNA, *Science* **236**:1290–1293.

110. Fuchs, A. R., 1978, Hormonal control of myometrial function during pregnancy and parturition, *Acta Endocrinol (Kbh.) [Suppl.]* **221**:1–70.

111. Frank, R. T., Bonham, C., and Gustavson, R. G., 1925, A new method of assaying the potency of female sex hormone based upon its effect on spontaneous contraction of the uterus of the white rat, *Am. J. Physiol.* **74**:395–399.

112. Reynolds, S. R. M., and Allen, W. M., 1932, The effect of progestin-containing extracts of corpora lutea on uterine motility in the unanesthetized rabbit, with observations on pseudopregnancy, *Am. J. Physiol.* **102**:39–55.

113. Bedford, C. A., Challis, J. R. G., Harrison, F. A., and Heup, R. B., 1972, The role of estrogen and progesterone in the onset of parturition in various species, *J. Reprod. Fertil. [Suppl.]* **16**:1–23.

114. Csapo, A. I., 1975, The "see-saw" theory of the regulatory mechanism of pregnancy, *Am. J. Obstet. Gynecol.* **121**:578–581.

115. Bergman, R. A., 1968, Uterine smooth muscle fibers in castrate and estrogen-treated rats, *J. Cell Biol.* **36**:639–648.

116. Daniel, E. E., Daniel, V. P., Duchon, G., Garfield, R. E., Nichols, M., Malhotra, S. K., and Oki, M., 1976, Is the nexus necessary for cell-to-cell coupling of smooth muscle? *J. Membr. Biol.* **28**:207–239.

117. Dahl, G., and Berger, W., 1978, Nexus formation in the myometrium during parturition and induced by estrogen, *Cell Biol. Int. Rep.* **2**:381–387.

118. Garfield, R. E., Puri, C. P., and Csapo, A. I., 1982, Endocrine, structural, and functional changes in the uterus during premature labor, *Am. J. Obstet. Gynecol.* **142**:21–27.

119. MacKenzie, L. W., and Garfield, R. E., 1985, Effects of 17β-estradiol on myometrial gap junctions and pregnancy in the rat, *Can. J. Physiol. Pharmacol.* **64**:462–466.

120. Burghardt, R. C., Mitchell, P. A., and Kurten, R. C., 1984, Gap junction modulation in rat uterus. II. Effects of antiestrogens on myometrial and serosal cells, *Biol. Reprod.* **30**:249–255.

121. Burghardt, R. C., Gaddy-Kurten, D., Burghardt, R. L., Kurten, R. C., and Mitchell, P. A., 1987, Gap junction modulation in rat uterus. III. Structure–activity relationships of estrogen receptor-binding ligands on myometrial and serosal cells, *Biol. Reprod.* **36**:741–751.

122. Markaverich, B. M., and Clark, J. H., 1979, Two binding sites for estradiol in rat uterine nuclei: Relationship to uterotropic response, *Endocrinology* **105**:1458–1462.

123. Clark, J. H., and Markaverich, B. M., 1984, The agonistic and antagonistic actions of estriol, *J. Steroid Biochem.* **20**:1005–1013.

124. Korach, K. S., Fox-Davies, C. Quarmby, V. E., and Swaisgood, M. H., 1985, Diethylstilbestrol metabolites and analogs. Biochemical probes for differential stimulation of uterine estrogen responses, *J. Biol. Chem.* **260**:15420–15426.

125. Garfield, R. E., Kannan, M. S., and Daniel, E. E., 1980, Gap junction formation in myometrium: Control by estrogens, progesterone, and prostaglandins, *Am. J. Physiol.* **238**:C81–C89.
126. MacKenzie, L. W., Puri, C. P., and Garfield, R. E., 1983, Effect of estradiol-17β and prostaglandins on rat myometrial gap junctions, *Prostaglandins* **26**:925–941.
127. MacKenzie, L. W., and Garfield, R. E., 1985, Hormonal control of gap junctions in the myometrium, *Am. J. Physiol.* **248**:C296–C308.
128. Wathes, D. C., and Porter, D. G., 1982, Effect of uterine distention and oestrogen treatment on gap junction formation in the myometrium of the rat, *J. Reprod. Fertil.* **65**:497–505.
129. Ikeda, M., Shibata, Y., and Yamamoto, T., 1987, Rapid formation of myometrial gap junctions during parturition in the unilaterally implanted rat uterus, *Cell Tissue Res.* **248**:297–303.
130. Kloeck, F. K., and Jung, H., 1973, *In vitro* release of prostaglandins from the human myometrium under the influence of stretching, *Am. J. Obstet. Gynecol.* **115**:1066–1069.
131. Puri, C. P., and Garfield, R. E., 1982, Changes in hormone levels and gap junctions in the rat uterus during pregnancy and parturition, *Biol. Reprod.* **27**:967–975.
132. Verhoeff, A., Garfield, R. E., Ramondt, J., and Wallenburg, H. C. S., 1985, Electrical and mechanical uterine activity and gap junctions in peripartal sheep, *Am. J. Obstet. Gynecol.* **153**:447–454.
133. Liggins, G. C., Fairclough, R. J., Grieves, S. A., Kendall, J. Z., and Knox, B. S., 1973, The mechanism of parturition in the ewe, *Recent Prog. Horm. Res.* **29**:111–137.
134. Kitts, D. D., Anderson, G. B., Bon Durant, R. H., Kindahl, H., and Stabenfeldt, G. H., 1985, Studies on the endocrinology of parturition: Steroidogenesis in coexisting genetically dissimilar ovine fetuses, concomitant with the temporal patterns of maternal C_{18} and C_{19} steroids and prostaglandin $F_{2\alpha}$ release, *Biol. Reprod.* **33**:67–78.
135. Kendall, J. Z., Challis, J. R. G., Hart, I. C., Jones, C. T., Mitchell, M. D., Ritchie, J. W. K., Robinson, J. S., and Thorburn, G. D., 1977, Steroid and prostaglandin concentrations in the plasma of pregnant ewes during infusion of adrenocorticotrophin or dexamethasone to intact or hypophysectomized foetuses, *J. Endocrinol.* **75**:59–71.
136. Alexandrova, M., and Soloff, M. S., 1980, Oxytocin receptors and parturition. I. Control of oxytocin receptor concentration in the rat myometrium at term, *Endocrinology* **106**:730–735.
137. Saito, Y., Sakamoto, H., MacLusky, N. J., and Naftolin, F., 1985, Gap junctions and myometrial steroid hormone receptors in pregnant and postpartum rats: A possible cellular basis for the progesterone withdrawal hypothesis, *Am. J. Obstet. Gynecol.* **151**:805–812.
138. Giannopoulos, G., and Tulchinsky, D., 1979, Cytoplasmic and nuclear progestin receptors in human myometrium during the menstrual cycle and in pregnancy at term, *J. Clin. Endocrinol. Metab.* **49**:100–106.
139. Casey, M. L., and MacDonald, P. C., 1984, Endocrinology of preterm birth, *Clin. Obstet. Gynecol.* **27**:562–571.
140. Garfield, R. E., and Baulieu, E. E., 1987, The antiprogesterone steroid RU 486: A short pharmacological and clinical review, with emphasis on the interruption of pregnancy, *Baillieres Clin. Endocrinol. Metab.* **1**:207–221.
141. Baulieu, E. E., and Ulmann, A., 1986, Antiprogesterone activity of RU 486 and its contragestive and other applications, *Hum. Reprod.* **1**:107–110.
142. Williams, K. I., 1984, Prostaglandin synthesis and uterine contractility, in: *Uterine Contractility* (S. Bottari, J. P. Thomas, A. Vokaer, and R. Vokaer, eds.), Masson, New York, pp. 282–288.
143. Ham, E. A., Cirillo, V. J., Zanetti, M. E., and Kuehl, F. A., Jr., 1975, Estrogen directed synthesis of specific prostaglandins in uterus, *Proc. Natl. Acad. Sci. U.S.A.* **72**:1420–1424.
144. Thorburn, G. D., and Challis, J. R. G., 1979, Endocrine control of parturition, *Physiol. Rev.* **59**:863–918.
145. Taylor, M. J., Webb, R., Mitchell, M. D., and Robinson, J. S., 1982, Effect of progesterone withdrawal in sheep during late pregnancy, *J. Endocrinol.* **92**:85–93.
146. Hahn, D. W., McGuire, J. L., Carraher, R. P., and Demers, L. M., 1985, Influence of ovarian steroids on prostaglandin- and leukotriene-induced uterine contractions, *Am. J. Obstet. Gynecol.* **153**:87–91.

147. Revel, J. P., Yancey, S. B., Meyer, D. J., and Nicholson, B., 1980, Cell junctions and intercellular communication, *In Vitro* **16**:1010–1017.
148. Dahl, G., Asarnia, R., and Werner, R., 1981, Induction of cell–cell channel formation by mRNA, *Nature* **303**:435–439.
149. Fallon, R. F., and Goodenough, D. A., 1981, Five hour half-life of mouse liver gap junction protein, *J. Cell Biol.* **90**:521–526.
150. Traub, O., Druge, P. M., and Willecke, K., 1983, Degradation and resynthesis of gap junction protein in plasma membranes of regenerating liver after partial hepatectomy or cholestasis, *Proc. Natl. Acad. Sci. U.S.A.* **80**:755–759.
151. Greipp, E. B., and Revel, J. P., 1977, Gap junctions in development, in: *Intercellular Communication* (W. C. DeMello, ed.), Plenum Press, New York, pp. 1–32.
152. Larsen, W. J., and Tung, H. N., 1978, Origin and fate of cytoplasmic gap junctional vesicles in rabbit granulosa cells, *Tissue Cell* **10**:585–598.
153. Garfield, R. E., 1985, Cell-to-cell communication in smooth muscle, in: *Calcium and Contractility: Smooth Muscle* (A. K. Grover and E. E. Daniel, eds.), Humana Press, Clifton Manor, NJ, pp. 143–173.
154. Burghardt, R. C., and Anderson, E., 1979, Hormonal modulation of ovarian interstitial cells with particular reference to gap junctions, *J. Cell Biol.* **81**:104–114.
155. Burghardt, R. C., and Anderson, E., 1981, Hormonal modulation of gap junctions in rat ovarian follicles, *Cell Tissue Res.* **214**:181–193.
156. Burghardt, R. C., and Matheson, R. L., 1982, Gap junction amplification in rat ovarian granulosa cells. I. A direct response to follicle stimulating hormone, *Dev. Biol.* **94**:206–215.
157. Yee, S. B., and Revel, J. P., 1978, Loss and reappearance of gap junctions in regenerating livers, *J. Cell Biol.* **78**:554–564.
158. Lane, N. J., and Swales, L. S., 1980, Dispersal of junctional particles, not internalization, during the *in vivo* disappearance of gap junctions, *Cell* **19**:579–586.
159. Lee, W. M., Cran, D. G., and Lane, N. J., 1982, Carbon dioxide induced disassembly of gap junctional plaques, *J. Cell Sci.* **57**:215–228.
160. Spray, D. C., Harris, A. L., and Bennett, M. V. L., 1982, Comparison of pH and calcium dependence of gap junctional conductance, in: *Intracellular pH: Its Measurement, Regulation, and Utilization in Cellular Functions*, Alan R. Liss, New York, pp. 445–461.
161. Spray, D. C., White, R. L., Mazet, F., and Bennett, M. V. L., 1985, Regulation of gap junctional conductance, *Am. J. Physiol.* **248**:H753–H764.
162. Spray, D. C., Stern, J. H., Harris, A. L., and Bennett, M. V. L., 1982, Gap junctional conductance: Comparison of sensitivities to H and Ca ions, *Proc. Natl. Acad. Sci. U.S.A.* **79**:441–445.
163. in't Veld, P., Schuit, F., and Pipeleers, D., 1985, Gap junctions between pancreatic B-cells are modulated by cyclic AMP, *Eur. J. Cell Biol.* **36**:269–276.
164. Azarnia, R., and Russell, T. R., 1985, Cyclic AMP effects on cell-to-cell junctional membrane permeability during adipocyte differentiation of 3T3-L1 fibroblasts, *J. Cell Biol.* **100**:265–269.
165. De Mello, W. C., 1983, The role of cAMP and Ca on the modulation of junctional conductance: An integrated hypothesis, *Cell Biol. Int. Rep.* **7**:1033–1040.
166. Saez, J. C., Spray, D. C., Nairn, A. C., Hertzberg, E., Greengard, P., and Bennett, M. V. L., 1986, cAMP increases junctional conductance and stimulates phosphorylation of the 27-kDa principal gap junction polypeptide, *Proc. Natl. Acad. Sci. U.S.A.* **83**:2473–2477.
167. Traub, O., Look, J., Paul, D., and Willecke, K., 1987, Cyclic adenosine monophosphate stimulates biosynthesis and phosphorylation of the 26-kDa gap junction protein in cultured mouse hepatocytes, *Eur. J. Cell Biol.* **43**:48–54.
168. Enomoto, T., Sasaki, Y., Shiba, Y., Kanno, Y., and Yamasaki, H., 1981, Tumor promotors cause a rapid and reversible inhibition of the formation and maintenance of electrical cell coupling in culture, *Proc. Natl. Acad. Sci. U.S.A.* **78**:5628–5632.
169. Kanno, Y., 1985, Modulation of cell communication and carcinogenesis, *Jpn. J. Physiol.* **35**:693–707.

170. Yada, T., Rose, B., and Loewenstein, W. R., 1985, Diacylglycerol downregulates junctional membrane permeability. TMB-8 blocks this effect, *J. Membr. Biol.* **88:**217–232.

171. Takeda, A., Hashimoto, E., Yamamura, H., and Shimazu, T., 1987, Phosphorylation of liver gap junction protein by protein kinase C, *FEBS Lett.* **210:**169–172.

172. Atkinson, M. M., Menko, A. S., Johnson, R. G., Sheppard, J. R., and Sheridan, J. D., 1981, Rapid and reversible reduction of junctional permeability in cells infected with a temperature-sensitive mutant of avian sarcoma virus, *J. Cell Biol.* **91:**573–578.

173. Azarnia, R., and Loewenstein, W. R., 1984, Intercellular communication and the control of growth: X. Alteration of junctional permeability by the *src* gene. A study with temperature-sensitive mutant Rous sarcoma virus, *J. Membr. Biol.* **82:**191–205.

174. Azarnia, R., and Loewenstein, W. R., 1984, Intercellular communication and the control of growth: XI. Alteration of junctional permeability by the src gene in a revertant cell with normal cytoskeleton, *J. Membr. Biol.* **82:**207–212.

175. Sugimoto, Y., Whitman, M., Cantley, L. C., and Erikson, R. L., 1984, Evidence that Rous sarcoma virus transforming gene product phosphorylates phosphotidylinositol and diacylglycerol, *Proc. Natl. Acad. Sci. U.S.A.* **81:**2117–2121.

176. Berridge, M. J., 1987, Inositol trisphosphate and diacylglycerol: Two interacting second messengers, *Annu. Rev. Biochem.* **56:**159–193.

177. Atkinson, M. M., and Sheridan, J. D., 1985, Reduced junctional permeability in cells transformed by different viral oncogenes, in: *Gap Junctions* (M. V. L. Bennett and D. C. Spray, eds.), Cold Spring Harbor Press, Cold Spring Harbor, NY, pp. 205–213.

178. Cole, W. C., Garfield, R. E., and Kirkaldy, J. S., 1985, Gap junctions and direct intercellular communication between rat uterine smooth muscle cells, *Am. J. Physiol.* **18:**C20–C31.

179. Cole, W. C., and Garfield, R. E., 1985, Alterations in coupling in uterine smooth muscle, in: *Gap Junctions* (M. V. L. Bennett and D. C. Spray, eds.), Cold Spring Harbor Press, Cold Spring Harbor, NY, pp. 215–230.

180. Cole, W. C., and Garfield, R. E., 1986, Evidence for the physiological regulation of myometrial gap junction permeability, *Am. J. Physiol.* **251:**C411–C420.

181. Huszar, G., 1986, Cellular regulation of myometrial contractility and essentials of tocolytic therapy, in: *The Physiology and Biochemistry of the Uterus in Pregnancy and Labor* (G. Huszar, ed.), CRC press, Boca Raton, FL, pp. 107–126.

182. Nishikori, K., Weisbrodt, N. W., Sherwood, O. D., and Sanborn, B. M., 1983, Effects of relaxin on rat uterine myosin light chain kinase activity and myosin light chain phosphorylation, *J. Biol. Chem.* **258:**2468–2474.

183. Scheid, C. R., Honeyman, T. W., and Fay, F. S., 1979, Mechanism of β-adrenergic relaxation of smooth muscle, *Nature* **277:**32–36.

184. Darnell, J. E., Lodish, H. F., and Baltimore, D., 1986, *Molecular Cell Biology*, W. H. Freeman, New York.

10

Molecular Mechanisms of Steroid Hormone Action in the Uterus

Elwood V. Jensen

1. Introduction

As in other organs of the female reproductive tract, the growth and function of uterine tissues are controlled by endocrine factors, primarily the steroid sex hormones secreted by the ovary (and, during pregnancy, the placenta). The initial development of the uterus during puberty and its maintenance in a mature state are principally dependent on the estrogenic hormone estradiol, produced by the ripening ovarian follicle during the follicular phase of the menstrual or estrous cycle. The further proliferation of the uterine endometrium in preparation for the implantation of a fertilized ovum is brought about during the luteal phase by the action of progesterone, originating in the corpus luteum derived from the ruptured follicle after ovulation.

Because of the striking growth response elicited by minute amounts of estrogens in the uterus of the immature rodent, this organ has provided a useful system, first for the assay of hormonal activity in the original isolation and identification of estrogenic hormones and later for studying the biochemical mechanism of their action. For the most part, the basic concepts of steroid hormone interaction with receptors in hormone-responsive tissues were developed for estrogens in rat uterus, with subsequent demonstration that other classes of steroid hormones, including vitamin D, as well as retinoids and thyroid hormone, follow similar interaction patterns in their respective target cells.[1-8] In more recent investigations of the detailed molecular biology of hormone action, much information has been first obtained with other steroid hormones, mainly glucocorticoids, and then extended to the female sex hormones.

This chapter summarizes current understanding of the action of estrogenic and

Elwood V. Jensen • Ben May Institute, University of Chicago, Chicago, Illinois 60637.

progestational hormones in target tissues such as uterus. Studies of other types of hormones are included when the findings have had particular relevance to estrogen and progestin action.

2. Historical Perspective

During the two decades that followed the discovery and isolation of the female sex hormones, research in this area was mainly devoted to: determining chemical structures; devising procedures for the practical synthesis of these hormones and of analogues with pharmacological utility; correlating structure with biological activity; and elucidating the pathways of their biosynthesis and metabolism. By the middle 1950s it was well established what these hormones are, how they arise, and what they do in the body, but the mechanism by which they carry out their hormonal actions remained a mystery. It was generally considered that they must somehow influence the action of enzymes, probably by participating in processes related to their own metabolism. During the late 1950s and early 1960s, studies initiated principally by Mueller with the rat uterus[9,10] and by O'Malley and Korenman with the chick oviduct[11] led to the realization that an early action of sex hormones in reproductive tissues is to enhance the synthesis of mRNAs for certain cellular proteins, indicating an effect of these agents on transcriptive processes.

During this same period, insight complementing that derived from studying what the hormone does to the tissue was obtained from an opposite approach, namely, by asking what target cells do with the hormone. The synthesis of estrogenic hormones labeled with carrier-free tritium[12,13] provided a means for determining the fate of a physiological dose of administered steroid in hormone-responsive cells as compared to other tissues. This led to the discovery of steroid hormone receptors, as discussed in section 3.1.

3. Biochemical Studies

3.1. Receptor Proteins in Target Cells

3.1.1. Estrogen Binding in Vivo and in Vitro

A major advance in understanding steroid hormone action came with the demonstration that target cells contain characteristic hormone-binding components, now recognized as receptors, and that it is a steroid–receptor complex rather than the hormone itself that exerts a regulatory effect in the cell nucleus. The presence of such receptors in female reproductive tissues was first evident from their striking ability to take up and retain tritiated hexestrol[12] or estradiol[13] after the administration of physiological amounts of these substances to immature animals. It was later shown that most mammalian tissues contain small amounts of this binding protein, often called estrophilin, and it is the magnitude of their receptor content that is the unique characteristic of hormone-dependent tissues.[14] Estradiol was found to combine reversibly with the receptor and to initiate growth of the immature rat uterus without itself undergoing chemical change.[15] This behavior suggested that estradiol acts by influencing the properties of the receptor protein rather than participating in reactions of steroid metabolism as had once been assumed. *In vitro* exposure of excised uterine tissue[16-18] or of uterine homogenates[19] to dilute solutions of tritiated estradiol at physiological temperature gives an interaction of hormone with receptor similar to that observed *in vivo*. This includes binding to chromatin[20] and formation of the same estradiol–receptor complexes[21] described below (Section 3.1.2). This interaction, as well as the integrity of the complexes, depends on the presence of sulfhydryl groups.[22]

The specific uptake and retention of estradiol by reproductive tissues are inhibited, both *in vivo*[23-25] and *in vitro*,[21,26] by a class of estrogen antagonists such as nafoxidine, Parke-Davis CI-628, and tamoxifen, which are themselves very weak estrogens but inhibit the uterotropic action of estradiol. Correlation was observed between the reduction in hormone uptake and the inhibition of uterine growth when different amounts of nafoxidine were given together with estradiol to the immature rat.[25,26] This provided the first evidence that the estrogen-binding substances in reproductive tissues are true receptors in that their association with steroid actually is involved in hormonal action. In contrast, puromycin and actinomycin D, substances that also prevent uterine growth response to estrogen,[27,28] do not inhibit estradiol uptake and retention. Hence, association of the steroid with the receptor is an early step in the uterotropic process, initiating a sequence of biochemical events that can be blocked at later stages by these inhibitors of RNA and protein synthesis.[25]

3.1.2. Estrogen–Receptor Complexes

After the administration of physiological amounts of tritiated estradiol to immature rats, two forms of the estrogen receptor were identified in uterine tissue. As first shown by differential centrifugation of homogenates[18,29] and confirmed by autoradiography,[21,30] the great majority of the hormone is present in the nucleus. It is bound to chromatin,[31] from which it can be extracted with 300 or 400 mM KCl as an estradiol–receptor complex that sediments at about 5 S in salt-containing sucrose gradients.[21,32] A smaller amount (20–30%) appears in the high-speed supernatant or cytosol fraction bound to a protein sedimenting at 8–10 S in sucrose gradients of low ionic strength[33] but dissociated into a 4 S steroid-binding subunit in the presence of salt.[34,35] In uterine homogenates from rats that have not been exposed to estrogen, essentially all of the receptor appears in the cytosol, where it reacts with estradiol in the cold to form the 8 S/4 S steroid–receptor complex.[19,21,36]

The relationship between the two forms of the receptor protein has been the subject of much investigation, for it proved to be of fundamental importance in understanding what the hormone actually does. On the basis of a variety of experimental evidence, it was shown[19,37] that the estradiol–receptor complex bound in target cell nuclei is not produced directly but arises in a two-step process via the initial formation of the cytosol complex. This was soon supported by observations that exposure of uterine cytosol to estradiol at temperatures near physiological causes gradual conversion of the 4 S complex to the 5 S form[38–40] and that this change occurs more rapidly in the presence of DNA.[41] Moreover, administration of estradiol *in vivo* induces temporary depletion of cytosol receptor,[14,19,42] consistent with its utilization to produce nuclear complex.

The 5 S "transformed" receptor, like that extracted from the uterine nuclei of estrogen-treated animals but unlike the 4 S cytosol complex, has the ability to bind to isolated nuclei,[43] chromatin,[44] and DNA[45] as well as to phosphocellulose.[46] The importance of nuclear binding in biological action was demonstrated in experiments showing that full uterotropic response requires the continued presence of estrogen–receptor complex in the nucleus for a period of several hours.[47,48] Moreover, it was found that enhancement of RNA polymerase activity in isolated uterine nuclei, previously observed on incubation with estradiol–cytosol mixtures,[49] is elicited by treatment of the nuclei with the 5 S estradiol–receptor complex but not with the 4 S form.[50,51] This stimulation is specific for hormone-dependent tissues and tumors. Only the transformed complex was found to increase the binding of actinomycin D by isolated rat uterine[52,53] and breast cancer[53] nuclei. This effect is also produced by estradiol *in vivo*[54] and indicates enhanced transcriptional activity.[55]

From these and other observations it became apparent that it is the transformed receptor protein that serves as a regulator of gene expression in target cells and that an important role of the steroid hormone is to convert the native form of this protein to this biochemically functional modification.[56] Thus, receptor transformation represents a key step in estrogen action.

Though the principal biological effects of estrogens, as well as of other classes of steroid hormones, appear to result from the regulation of transcription in target genes, it should be pointed out that there also may be other types of action that do not involve the genome (for review see Liao and Hiipakka[7]). In particular, certain very rapid effects of estrogens in the uterus, such as hyperemia and water imbibition,[57] are not inhibited by actinomycin D. They may be related to histamine release,[58,59] possibly involving uterine eosinophils, which have been shown to concentrate estradiol.[60,61] That estrogenic hormones may influence the properties of biological membranes or other cellular constituents by a process independent of their genomic action is quite possible.

3.1.3. Progestin–Receptor Interaction

Early attempts to demonstrate specific binding of labeled progesterone by female reproductive tissues proved unsuccessful until it was shown that the uptake of progesterone by the mouse vagina is dependent on prestimulation by estrogen.[62] It has since been established that in most target cells a major gene product of estrogen action is the progestin receptor protein. In uteri of ovariectomized guinea pigs,[63] rabbits,[64] and hamsters[65] prestimulated with estrogen, several research groups reported a selective uptake of administered progesterone, two to seven times greater than that of the nonstimulated controls. In the guinea pig and rabbit, the great majority of the labeled steroid bound in the uterus is unchanged progesterone; in the hamster it is somewhat less but still considerably more than the few percent found in nontarget tissues. Progesterone binding was also observed in uteri of pregnant rats[66] and in uteri taken from intact rats and rabbits exposed to tritiated progesterone *in vitro.*[67]

Because of the interaction pattern already established for the estrogens, studies of progestin receptors moved rapidly to the use of broken cell systems and the search for receptor proteins in the various fractions of target tissue homogenates. Specific progesterone–receptor complexes were observed in the cytosol fraction of uterine homogenates from the guinea pig,[68–72] rat,[66,67,73–77] mouse,[75] rabbit,[64,67,74,78,79] dog,[80] and human,[64,81–84] either after exposure of tissue to labeled progesterone *in vivo* or *in vitro* or on addition of hormone directly to cytosol. In studies with ovariectomized animals, it was observed that pretreatment with estrogen not only increases the level of progesterone binding in uterine cytosol but also changes the nature of the receptor molecule. In cytosols from ovariectomized animals, the progesterone–receptor complex was found to sediment at 3.5 to 5 S. In contrast, in uterine cytosol from estrogen-treated castrate animals, the complex was reported to sediment at 6.7 to 8 S in gradients of low ionic strength and to be dissociated to a 3.5–5 S entity in the presence of salt. Because part of the binding in the 4–5 S region, unlike that at 7–8 S, was found to be insensitive to sulfhydryl-blocking reagents,[68] it is not certain whether some of the progestin binding in uterine cytosols from unstimulated animals may reflect interaction with substances other than receptor.

After administration of tritiated progesterone to estrogen-primed ovariectomized rabbits, more than half of the uterine radioactivity was found in the nuclear fraction of homogenates, from which it could be extracted by 500 mM KCl as a complex sedimenting at about 4 S.[78] In similarly treated guinea pigs, predominantly nuclear localization of bound steroid was indicated by autoradiography,[85] and in rat[86] and dog[87] uterus, the presence of nuclear complex was demonstrated by exchange of tritiated progesterone for nuclear-bound endogenous hormone.

The earliest and most comprehensive investigations of progesterone interaction with target cells were carried out with the chick oviduct, an organ that resembles the mammalian uterus in that its growth and development are dependent on estrogenic hormone. In estrogen-prestimulated chicks, administration of progesterone induces the synthesis of specific egg-white proteins, in particular avidin.[11] Although chick oviduct does not seem to concentrate administered progesterone above the blood level, specific binding, which is increased 20-fold by estrogen treatment, could be demonstrated in both the cytosol and nuclear fractions,[11,88] extractable from the latter by 300 mM KCl. Hormone–receptor complexes were identified both in cytosol[89] and in nuclear extract.[90] The cytosol complex, obtained either from progesterone injection *in vivo* or by direct addition of hormone to oviduct cytosol, was found to sediment as a single 3.8 S peak in salt-containing sucrose gradients but as a mixture of 5 S and 8 S entities in low-salt gradients. The nuclear complex, resulting either from hormone administration *in vivo* or from incubating oviduct tissue with progesterone at 37°C (but not 0°C) *in vitro,* sediments in salt-containing gradients at about 4 S. Thus it is indistinguishable by this criterion from the cytosol complex. It is shown that in both complexes, the steroid is unchanged progesterone and its binding is destroyed by sulfhydryl-blocking reagents.[90–92] It is evident that pre-stimulation with estrogen induces the formation of a progestin receptor that, like estrogen receptor, is a sulfhydryl-containing protein, sedimenting at about 8 S in low-salt gradients and dissociated into a 4 S binding unit at higher salt concentrations.

The relationship between the cytosol and nuclear hormone–receptor complexes of target cells was elucidated by experiments carried out with *in vitro* systems.[90,93] In oviduct tissue exposed to progesterone in the cold, the binding is primarily cytosolic, but on warming of the tissue to 37°C, cytosolic receptor is gradually depleted as nuclear complex increases. Salt-extractable nuclear complex is readily formed by incubating oviduct nuclei at 25°C with a combination of progesterone and oviduct cytosol but not with either alone. Similarly, incubation of cytosol from guinea pig[45] or rabbit[94] uterus or hen oviduct[95] with progestational hormone at 15–25°C transforms the receptor complex as indicated by its ability to bind to DNA, phosphocellulose, or ATP-sepharose, respectively. It is clear that the steroid–receptor complex, bound in target cell nuclei of progestin-treated animals, is derived from the initial formation of the cytosol complex, followed by the temperature-dependent, hormone-induced conversion of the receptor protein to an activated form that has acquired affinity for chromatin.

Thus, the estrogenic and progestational hormones appear to interact with their respective target cells by generally similar mechanisms, but with two interesting differences. First, estrogen receptors are unique among the family of steroid hormone receptors in that the nuclear-binding modification can be distinguished from the native or cytosol form by a difference in sedimentation rate. As discussed later, the transformed estrogen receptor appears to dimerize spontaneously, whereas transformed receptors for other steroid hormones may dimerize only on interacting with the response elements of target genes. This behavior of estrogen receptors was fortunate, for the difference in sedimentation rates between the nuclear and cytosol receptors was the initial clue that led to the recognition of the phenomenon of receptor transformation.[38,56]

In turn, progestin receptors are unique in that, at least in chick oviduct[96,97] and human breast (T47D cells)[98–100] or endometrial[101] cancer, they come in two sizes, A and B. Although reported values for the molecular weights of the two proteins vary, the B form of the chick receptor is usually described as 108 kDa,[102] and the A form as 79 kDa,[103] whereas the sum of the amino acids, indicated by the structure of the cloned cDNA, corresponds to 86 and 72 kDa.[104] Human B and A receptors are described as 109–119 and 89–93 kDa, respectively[99] and as 114 and 94 kDa when newly synthesized in T47D breast cancer cells.[100] The A form appears to be a truncated modification of the B that is immunochemically similar[105–107] but lacks a moiety at the amino-terminal region. Though the A form may be an artifact of proteolytic cleavage, there is evidence that it may arise from the alternate initiation of translation of receptor mRNA.[100,107-109] The respective roles of these two receptor forms in the interaction with target genes have been the subject of much stimulating speculation,[3,110] but at present the relative contributions of A and B receptors in transcription enhancement remain unclear. In contrast to chick oviduct and human cancer, two forms of the progestin receptor are not found in rabbit uterus, where only the full-length (110 kDa) receptor is seen when strict precautions are taken to avoid proteolysis.[111-113]

3.2. Receptor Transformation

In the first reports of the hormone-induced conversion of the native estrogen receptor to its nuclear-binding modification,[38-40] this phenomenon was referred to as receptor "transformation." However, with the demonstration of its occurrence with other classes of steroid hormones[45] and an appreciation of its biological significance, many investigators began to call it "activation." More recently there has been a trend back to the use of transformation, reserving the term activation for the conversion of the receptor protein from a nonbinding to a hormone-binding form.[114] Because receptor transformation was originally discovered with estrogens, most of the earlier studies were carried out with these hormones. More recently, investigations with glucocorticoid and progestin receptors have provided information that has greatly enhanced understanding of the transformation process in general.[115]

Hormone-induced transformation of the native receptor protein to a nuclear-binding form has been shown to be a general phenomenon for all classes of steroid hormones[2-5,116-119] Only in the case of estrogen receptor is this change accompanied by an increase in the sedimentation rate of the transformed as compared to the native complex. Early studies of the transformation process established that the 5 S entity is a dimeric form of a modified 4 S binding unit.[120-124] This observation has been substantiated by more recent studies[125] and, as discussed below, is consistent with the interaction of transformed estrogen receptor with a palindromic (i.e., displaying mirror image symmetry) sequence of amino acids believed to comprise the attachment site in the hormone response elements of target genes.[126] Even though transformed progestin receptors do not appear to dimerize *in vitro*, there is evidence to suggest that they, as well as glucocorticoid receptors, interact in a dimeric form with the hormone response elements of target genes.[127] In the intact cell, a dimer appears to be formed from the hormone-induced reaction of a cytoplasmic monomer with another monomer already in the nucleus.[128] Because the steroid-binding units of native and transformed progestin receptors can not be differentiated by sedimentation in salt-containing sucrose gradients, their transformation is often said to involve a change from 8 S to 4 S in low-salt gradients, whereas that with estrogen receptor is described as a change from 4 S to 5 S as seen in high-salt gradients.

Though much has been learned about the transformation reaction with the various classes of steroid hormone receptors,[129,130] the exact chemical details of this phenomenon are not completely understood. At physiological pH and ionic strength, transformation of the estrogen receptor requires the presence of hormone and temperatures near physiological.[38-41] The process is accelerated by dilution, higher salt concentrations or pH, and by the presence of DNA, and it is inhibited by EDTA. It is also inhibited by aprotinin and diisopropyl fluorophosphate.[131] Conversion to the nuclear-binding form is effected in the absence of steroid by precipitation with ammonium sulfate[132] and by dialysis or gel filtration[133]; transformation by salt precipitation but not by dialysis is accompanied by an increase in sedimentation rate from 4 S to 5 S. The transformed receptor binds estradiol more tightly than does the native receptor.[134] Transformation of the progestin receptor in oviduct[93] or uterine[94] cytosol or of purified receptor[135] also requires warming with hormone. However, the complex can be transformed in the cold by dilution or exposure to salt[94] and in the absence of steroid by precipitation with ammonium sulfate[93] or by treatment with salt,[136] ribonuclease,[135] heparin,[136] or ATP.[137] Glucocorticoid–receptor complexes likewise are transformed in the cold by dilution, dialysis, or gel filtration.[138-142]

The transformation of all classes of steroid hormone receptors is inhibited by molybdate and other transition metal oxyanions,[115,143] which provides a useful means for stabilizing the native form of the receptor. This agent also retards inactivation of crude receptor preparations *in vitro,* an action believed to result from its inhibition of endogenous phosphatases.[144] This interpretation has been questioned for progestin receptor.[145] The use of molybdate as a stabilizing agent, discovered during its investigation as a putative phosphatase inhibitor,[144-147] has greatly facilitated receptor purification as well as studies of structure and function. As first shown with glucocorticoid receptor,[148] pyridoxal 5'-phosphate blocks nuclear binding of estrogen[149] and progestin[150] receptors; although this may be an effect on transformation itself,[151] more likely it involves reaction with a DNA-binding site exposed by transformation.[147,148] Sulfhydryl groups, required for steroid binding by all classes of receptors,[2] appear to participate in the transformation of progestin[92,152] and glucocorticoid receptors.[114,153] It has been suggested that phosphorylation or dephosphorylation may be involved in the transformation process, but as yet no definite role has been established.[115,154] With estrogen receptors, the acquisition of nuclear binding ability and dimerization to form 5 S complex appear to be separate processes that proceed at different rates.[155,156] Presumably only the first of these takes place *in vitro* with progestin and glucocorticoid receptors.

In early studies of the kinetics of receptor transformation, this reaction was found to proceed more rapidly as the concentration was reduced. This is contrary to what would be expected for a dimerization process and suggested that a rate-limiting step may be the initial dissociation from a macromolecule.[123] This phenomenon was elucidated by the finding[157-160] that the 8–10 S forms of untransformed receptors for at least four classes of steroid hormones consist of a steroid-binding subunit associated with another protein of molecular weight 90,000, which is present as a dimer[161,162] and is lost on transformation.[163] In the case of glucocorticoids, this component appears to be required for initial hormone binding by the receptor moiety.[164] This substance was soon identified[165,166] as a heat shock protein (hsp90), which itself does not bind hormone but which appears to associate with the receptor to obscure the domain(s) in the transformed receptor responsible for binding to DNA or for formation of dimers. Interaction with the steroid hormone is considered to displace the heat-shock protein, making the DNA-binding domain of the receptor available for interaction with target genes.[115,167−170]

In addition to the 90-kDa heat shock protein, cytosols with untransformed receptors for estrogens[133] and glucocorticoids[138−142] have been shown to contain a low-molecular-weight substance that stabilizes the native form of the receptor. This can be removed by dialysis or gel filtration. In the case of estrogens, dialysis or gel filtration transforms the receptor in the absence of hormone,[133] though with glucocorticoids, transformation enhancement by dilution, dialysis, or gel filtration has been demonstrated only with the hormone−receptor complex. In addition to this small inhibitory factor, which for glucocorticoids is heat stable and found in a variety of animal tissues,[142] cytosols with glucocorticoid receptor also contain a macromolecular substance that impedes nuclear binding of the transformed complex,[171−176] an effect that can be counteracted by the addition of ATP.[177] Hepatoma cells appear to contain higher levels of this factor than do normal liver cells.[173]

Although the precise details of the transformation phenomenon remain to be elucidated, the concept of dissociation from heat-shock protein, probably coupled with removal of a low-molecular-weight inhibitory factor, serves to explain many of the heretofore puzzling observations. These include enhancement of transformation by dilution, dialysis, or gel filtration, its occurrence in the absence of steroid on salt precipitation, and the curious kinetics observed with the estrogen receptor, where an apparently second-order dimerization reaction proceeds faster on dilution. However, it does not directly address the fact that 400 mM KCl dissociates the 8 S native estrogen-receptor complex to a 4 S steroid-binding unit that is not nucleophilic until it undergoes further change[56] or that the 5 S transformed complex, extracted by salt from uterine nuclei of estrogen-treated animals, aggregates to an 8−10 S form when the ionic strength is reduced.[35] Nor does it explain activation of progestin receptor by treatment with ribonuclease[135] or ATP.[137] Nevertheless, these findings have greatly enhanced understanding of the transformation process. Still to be elucidated is how its association with hormone effects these changes in the receptor, although it has been proposed that transformation involves a self-proteolytic action of the receptor that is stimulated by interaction with the steroid.[131]

The relationship between the various complexes of the estrogen receptor, present in either cytosol or nuclear extract from target cells, is shown schematically in Fig. 1. As discussed in Section 3.5, both the steroid-binding unit and the heat shock protein appear to be phosphorylated. In Fig. 1, the estrogen receptor is shown hypothetically to contain RNA. Although this has not been demonstrated for estrogen receptors, untransformed progestin[135] and glucocorticoid[178] receptors have been found to contain ribonucleic acid.

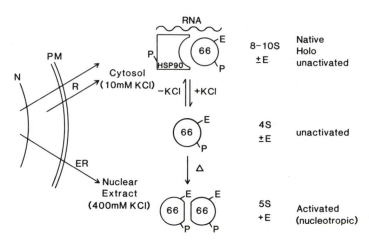

Figure 1. Schematic representation of various forms of the estrogen receptor isolated from target tissue. N, nuclear border; PM, plasma membrane; E, estrogen; P, phosphorylation; △, warming to physiological temperature. Molecular mass of the binding unit, as indicated by the structure of cloned cDNA for human receptor, is 66 kDa,[179] in agreement with values reported for purified rat uterine receptor.[180,181] Reproduced with permission from Greene *et al.*[179]

3.3. Intracellular Localization

Because the untransformed form of the estrogen receptor, like receptors for progestins, androgens, and glucocorticoids, usually appears in the cytosol fraction of homogenates prepared in a variety of media, it was originally assumed that before exposure to hormone the native receptor is located in the cytoplasm. Thus, the hormone-induced conversion of the native receptor to the transformed state was considered to be accompanied by its migration to a nuclear binding site.[19,37] This was thought to occur before transformation, since DNA was found to accelerate the transformation process,[41] and, after estrogen treatment *in vivo,* 4 S and 5 S complexes could both be found in rat uterine nuclei, with the 5 S form progressively increasing.[182] Although this scheme appeared consistent with most experimental findings, several observations indicated that much of the native receptor may be in the nucleus before exposure to hormone.

As reviewed recently,[183] unfilled and presumably untransformed estrogen receptors have been found in the nuclear fraction of certain tissue homogenates, including pig,[184] rat,[185] and human[186,187] uterus, human mammary[188–190] and rat pituitary[191] cancer, rat brain and pituitary,[192] chick[193,194] and toad[194] liver, and fish testis.[195] However, the high level of unfilled receptors observed in nuclei of MCF-7 human breast cancer cells was shown to result from cytoplasmic contamination.[196] Untransformed progestin receptor was found in nuclei from hen oviduct by extraction with molybdate to stabilize the native form.[197] Unoccupied receptors in target cell nuclei have also been observed for 5-androstene-3β,17β-diol,[198] aldosterone,[199] calcitriol,[200,201] and the ecdysteroid ponasterone A.[202]

The foregoing observations, taken with the evidence suggesting that receptor transformation may take place preferentially within the nuclear compartment,[41,182] raised a question as to whether the native estrogen receptor might be distributed throughout the target cell. In this case its fixation in the nucleus after hormone treatment would depend on its conversion to the transformed state that binds to chromatin, changing the equilibrium between nuclear and extranuclear receptor.[119,203−205] The possibility of a nuclear–cytoplasmic equilibrium of estrogen receptor distribution was suggested by autoradiographic studies of rat uteri exposed to tritiated estradiol in the cold.[205,206] In these experiments, somewhat more radioactivity (75–85%) was found to be associated with the nucleus than the 20–30% reported in earlier studies.[19] Autoradiography of rat uterus[207] or chick oviduct[208,209] exposed to labeled progestins in the cold also showed nearly all the radioactivity to be in the nucleus.

Additional evidence for predominantly nuclear localization of untransformed receptors came from studies with calcitriol (1α,25-dihydroxycholecalciferol), the biologically active modification of vitamin D_3. Unlike receptors for most classes of steroid hormones, nearly all the native calcitriol receptor is found in the nuclear fraction after homogenization of target tissue in hypotonic medium.[200,201] In buffers of physiological ionic strength, most of the untransformed calcitriol receptor appears in the cytosol,[210] in contrast to the transformed complex, which requires salt concentrations of 300 mM or higher for its solubilization.[211] It appears that untransformed receptors for the steroid and thyroid hormones present a spectrum of affinities for the nuclear compartment in target cells[212]; receptors for gonadal and adrenal hormones are loosely bound, those for ecdysteroids and vitamin D are intermediate, and, as had previously been known,[213,214] those for thyroid hormone are associated tightly with chromatin.

As discussed in Section 4, the availability of antibodies, and especially monoclonal antibodies, to steroid hormone receptors permitted their detection by means other than ligand binding, and it became possible to study intracellular localization by the techniques of immunocytochemistry. Early reports indicated substantial immunostaining or immunofluorescence of estrogen receptors in the extranuclear region in rat pituitary cells[215] and in frozen[216] or embedded[217] sections of human breast cancers. In contrast, most recent immunocytochemical studies have found only nuclear staining in frozen sections of human breast cancers,[218−226] human,[219,227] rabbit,[219] monkey,[228] and rat[229] uterus, rat neurons[229,230] and anterior pituitary,[229] and human ovarian cancer,[231] as well as in embedded specimens of human breast cancer.[232−234] However, two recent investigations found that, in addition to predominantly nuclear staining, embedded or frozen sections of estrogen-receptor-rich human breast cancers show a significant amount of specific cytoplasmic staining as well.[235,236] Immunocytochemical studies have detected progestin receptor only in the nuclei of chick oviduct,[208,209,237] rabbit,[238] guinea pig,[238] and human[101,239] uterus, rabbit pituitary gland,[238] and human breast,[239–241] ovarian,[239] and endometrial[239] cancer. Immunocytochemistry with electron microscopy showed only nuclear localization of estrogen receptor in human uterus,[242] although with rabbit uterus, an immunogold technique detected a small but definite amount of extranuclear progestin receptor in addition to nuclear receptor, apparently associated with ribosomes.[243]

Nuclear localization of estrogen receptor in GH$_3$ rat pituitary tumor cells was also demonstrated by cell enucleation, either with[244] or without[245] cytochalasin B, a fungal metabolite that disrupts microfilaments and permits nuclear removal while maintaining the integrity of the remaining cytoplast.[246] Most of the estrogen-binding activity was recovered in the nucleoplasts, with only 5–10% found in the cytoplast fraction. Similar results were obtained for progestin and glucocorticoid receptors in GH$_3$ cells[245] Cytochalasin enucleation of human endometrical carcinoma (HEC-50) cells showed 86% of the estrogen receptor to be in the nucloplast fraction.[247]

The foregoing observations, involving a variety of experimental approaches, leave little doubt that native receptors for estrogens and progestins, as well as for other steroid hormones, reside predominantly within the nuclear compartment in target cells. This is consistent with the finding that untransformed estrogen receptor has a weak affinity for DNA *in vitro*.[248] Furthermore, native progestin receptor contains a nuclear-binding signal sequence, which when deleted results in the receptor being cytoplasmic.[128] In most cases these receptors are so loosely bound that cell disruption in almost any homogenizing medium results in their extraction into the cytosol. What is not certain is whether the native receptor is confined exclusively to the nucleus or whether, as was suggested earlier,[203–207,212] a portion exists in the cytoplasm in equilibrium with a nuclear pool. Because in most experiments essentially no extranuclear immunostaining is observed in target tissues, and only a small amount of binding capacity is found in the cytoplast fraction of enucleated cells, there has been an erroneous assumption that it has been definitely established that untransformed receptor resides exclusively within the nucleus. Actually, the immunocytochemical report usually cited[219] states carefully that the estrogen receptor "may reside primarily in target cell nuclei."

Unfortunately, none of the experimental techniques that demonstrate the preponderance of nuclear localization is capable of quantitatively excluding the presence of small amounts of extranuclear receptor, which may be sufficient to function in an uptake–translocation system. For estrogens in the uterus, such an amount would be a minor fraction of the total receptor. Not only does the receptor content far exceed that used by a physiological dose of hormone,[19,37] but if it were continually replenished from a nuclear pool, only a small quantity of extranuclear receptor need be present at any time. This may be difficult to detect in immunocytochemical and autoradiographic experiments, which underestimate extranuclear receptor because they indicate concentrations rather than amounts, and in most target cells, especially in myometrium, the cytoplasmic volume greatly exceeds the nuclear volume. For example, in a roughly spherical cell with a diameter twice that of its nucleus, if half of the receptor were in the cytoplasm, the immunostaining observed would be 88% nuclear, and a 20% content of cytoplasmic receptor would appear as less than 4% extranuclear staining.

Just as there were findings not compatible with a totally cytoplasmic localization of untransformed receptor, there also are experimental observations not consistent with an exclusively nuclear localization. The autoradiographic studies that provided evidence that most of the native estrogen receptor of rat uterus is in the nucleus also indicate that 15–25% is extranuclear,[205,206] and quantitative autoradiography shows that 2 hr after the injection of labeled estradiol *in vivo,* 14% of the radioactivity in rat uterine endometrium is in the cytoplasm.[30] Immunocytochemical studies with the electron microscope[243] have detected extranuclear progestin receptor not seen with the light microscope.

Although it has been reported[219,227] that processing and fixing of tissue for immunocytochemical studies do not remove substantial amounts of receptor from frozen sections, these experiments do not rule out the possibility of some extraction, perhaps preferentially from the extranuclear region. This is suggested by the ability of hormone treatment to increase the intensity of nuclear staining. Though no increase in staining was seen in ovariectomized rabbit uterus or MCF-7 breast cancer cells after exposure to estradiol *in vivo* or *in vitro,* respectively,[219] incubation of spayed monkey uterus with estradiol prior to immunocytochemistry was found to give much stronger nuclear staining than that observed with untreated tissue.[228] Uterus and pituitary from ovariectomized or immature rats exhibit only weak nuclear staining, and brain shows none, but after injection of estradiol, all three tissues were found to give substantial staining of the nucleus.[229] Moreover, MCF-7 human breast cancer cells grown in medium with charcoal-stripped calf serum and no phenol red (which induces receptor transformation in cultured cells[249,250]) show little or no nuclear immunostaining, in contrast to those exposed to estradiol (or phenol red) in which there is nuclear staining.[249] This is not because of estrogen-induced synthesis of new receptor, since similar results are obtained in the presence of cycloheximide.

These findings imply that there can be estrogen receptor in target cells that is not seen immunocytochemically until it is converted to the more tightly bound form that either resists loss during experimental manipulation or is concentrated in the nucleus to detectable levels from a dilute state in the cytoplasm. With fluorescent polyclonal antibody, it was found that MCF-7 breast cancer cells and human endometrial and endometrial cancer cells, grown in the absence of estrogen, show cytoplasmic immunofluorescence, which shifts to predominantly nuclear fluorescence after incubation of the cells with estradiol at 37°C.[251] Administration of progestin was found to increase the nuclear receptor seen with the electron microscope in rat pituitary[252] and to cause rearrangement of the nuclear receptor distribution from a random pattern in the condensed chromatin to sites in the border between dense chromatin and nucleoplasm.[243] However, with the light microscope, chick oviduct nuclei were found to immunostain with the same intensity for progestin receptor whether or not there had been exposure to hormone.[209,237]

In contrast to the enucleation experiments with GH_3 rat pituitary[244,245] and

HEC-50 human endometrial[247] cancer cells, in which most of the binding capacity for estrogens, progestins, and glucocorticoids (86–95%) was found in the nucleoplast fraction, enucleation of bovine and porcine kidney cells showed only 30–40% of the calcitriol receptors to be in the nucleoplast fraction.[253] Only in cells first exposed to calcitriol was the majority of the receptor in the nucleoplasts. The reason for these conflicting results is not clear. The cells used in all these experiments were grown in culture with fetal calf serum, known to contain estrogens, progestins, and glucocorticoids (but not calcitriol), all of which could convert receptors to the nuclear-binding state. Nevertheless, transformation appears not to be a major factor, since nearly all the receptor of the GH_3 cells and two-thirds of that in the HEC-50 cells could be extracted from the nucleoplasts on homogenization in low-salt buffers. It is noteworthy that for glucocorticoid receptors the localization observed by enucleation is in marked disagreement with that found in the immunocytochemical studies described below.

One of the most disturbing aspects of the assumption that untransformed receptors for estrogens and progestins reside exclusively within the nucleus is the fact that this is not true for glucocorticoids. Several immunocytochemical studies have established clearly that native receptors for these hormones are found both in the nucleus and in the cytoplasm of target cells, in some cases with the cytoplasmic receptor predominating. Early experiments with fluorescent polyclonal antibodies to glucocorticoid receptor showed extranuclear antigen in rat thymocytes,[254] in rat liver and hepatoma,[255] and in human lymphocyte blast and HeLa cells,[254] with the fluorescence shifting to the nucleus after exposure of the cells to dexamethasone at 37°C *in vitro*. With polyclonal antibodies and immunoperoxidase staining, antigen was detected in the cytoplasm of some cells and in the nuclei of others in frozen sections of rat adrenal medulla.[256] It is seen both in the cytoplasm and in the nuclei of embedded specimens of rat liver[257,258] and pituitary[257]; with the liver it was shown that nuclear staining is markedly decreased after adrenalectomy and restored by the subsequent administration of glucocorticoids.

In more recent studies using monoclonal antibodies, staining was found to be mostly nuclear in nerve and glial cells from intact rats but predominantly or exclusively cytoplasmic in those from adrenalectomized animals, with nuclear staining restored by the administration of corticosterone.[230,259] In another study with monoclonal antibodies, using either frozen or embedded sections of various rat tissues, antigen was found in both nucleus and cytoplasm, with nuclear staining higher in liver, brain, and kidney, and cytoplasmic staining predominating in heart, lung, kidney tubules, thymus, and spleen.[260] By use of monoclonal antibodies and either immunofluorescence or immunoperoxidase techniques, glucocorticoid receptor was found in both the nucleus and the cytoplasm of normal rat hepatocytes grown in culture as well as in rat hepatoma (H-4-II-E) and human uterine carcinoma (NHIK 3025) cells.[261] In the hepatoma cells, incubation with dexamethasone at 37°C, but not at 4°C, was found to cause a marked increase in nuclear staining. In immunocytochemical studies with both the light microscope and the electron microscope, glucocorticoid receptor was found predominantly in the cytoplasm of human lymphoid cells and cell lines; incubation at 37°C with dexamethasone causes a shift of the receptor to the nucleus in a steroid-sensitive but not a steroid-resistant cell line.[262]

The foregoing observations provide strong evidence that in target cells for glucocorticoids a substantial portion of the untransformed receptor resides in the extranuclear region until association with the hormone converts it to the nuclear-binding form. Recent *in vitro* mutagenesis experiments have confirmed that the native glucocorticoid receptor is predominantly cytoplasmic.[263] In view of the similarities between the biochemical mode of action of glucocorticoids and that of the estrogenic and progestational hormones (i.e., hormone-induced conversion of the native receptor to a nucleophilic entity that interacts with a hormone-response element in target genes) and the analogies in receptor structure discussed later, it would be unlikely for glucocorticoids to follow an intracellular pathway different from that of other steroid hormones.

With such uncertainties in the quantitative assignment of native estrogen and progestin receptors to the nuclear compartment, and the fact that much of the glucocorticoid receptor is extranuclear, it would seem prudent to regard the exclusive nuclear localization of untransformed estrogen and progestin receptors as still unresolved. Hence, one must keep open the possibility that there may be some cytoplasmic receptor in equilibrium with a nuclear pool. Considering the ease with which transformed receptors for gonadal and adrenal hormones move in and out of isolated nuclei, even in the cold, it seems probable that these proteins would be diffusible to some extent under physiological conditions in the intact cell, as is indicated for progestin receptor by recent deletion mutation experiments.[128]

It should be mentioned that there are entities in both plasma and microsomal membranes of target cells that strongly bind estrogenic and progestational hormones (for reviews see refs. 264 and 265). Recent studies with fluorescent estradiol conjugate have demonstrated what appear to be high-affinity, low-capacity binding sites in plasma membranes of estrogen-sensitive but not of insensitive human breast cancer cells.[266] Progestin action at the plasma membrane of frog oocytes is indicated by the ability of polymer-bound hormone to induce maturation[267,268] and by photoaffinity labeling of a progestin-binding substance in the membrane.[269] The relationship of these phenomena to the receptor pathways already discussed is not clear, although certain of the microsomal substances may represent intermediate stages of receptor synthesis.[121] The estrophilic substances in plasma membranes of rat hepatocytes[270] and in pig uterine microsomes[121,271] show properties similar to those of cytosol and nuclear receptor proteins. However, in immunocytochemical studies with the electron microscope, monoclonal antibodies to estrogen or progestin receptors failed to detect antigen in the plasma membranes of human[242] or rabbit[243] uterine cells.

3.4. Current Status of the "Two-Step" Mechanism

With the recognition that much of the native receptor for steroid hormones is within the nucleus, it was obvious[183,249,272–276] that some revision was needed in the original concept that the untransformed or "cytosol" receptor is an extranuclear entity until interaction with the hormone converts it to the nuclear-bound form. However, the rush to proclaim that the two-step mechanism is no longer valid and that one must search for a new model for steroid hormone action[204,206,209,277−279] reflects a misunderstanding of what is meant by "two-step mechanism." This term was originally proposed[19] to indicate that the 5 S estradiol–receptor complex, bound in target cell nuclei after hormone administration *in vivo*,[21] is not generated directly but is derived from an initially formed complex of estradiol with the receptor protein identified in the cytosol.[33] When it was found that the 4 S binding unit of the cytosol receptor actually becomes the 5 S nuclear receptor under the influence of estradiol[38−41] and that only this transformed complex has the ability to bind to chromatin and influence the transcriptional activity of target cell nuclei,[50−52] the two-step concept took on additional meaning. In this refinement, it came to denote the hormone-induced conversion of a native receptor protein to its biochemically functional form that can then interact with target genes.[56] Translocation is involved only in that the transformed receptor complex must assume its genomic binding site from wherever it was originally located.

If the native receptor is entirely within the nucleus, the translocation to the target gene will be only intranuclear, but if it is partly in the cytoplasm, intracellular translocation will take place as well. These are important features, but they are details rather than differences in basic mechanism.[8] In spite of repeated emphasis on hormone-induced receptor transformation followed by genomic binding as the key steps in steroid hormone action,[40,43,56] some investigators seem to regard translocation from cytoplasm to nucleus as the essence of the two-step mechanism, which it is not. If it is to be claimed that the two-step model is invalid, on the basis of the belief that, except for glucocorticoids, native receptors for steroid hormones are entirely confined to the nucleus, it should be made clear that it is the intracellular translocation feature and not the two-step transformation mechanism itself that is being abandoned.

Whether the translocation accompanying transformation of estrogen and progestin receptors is intracellular or only intranuclear has no bearing on the validity of the two-step mechanism. If it should turn out that there is a cytoplasmic–nuclear equilibrium of receptor distribution, as there appears to be with glucocorticoid receptors, the only refinement needed in the original concept is that the amount of cytoplasmic receptor is less than originally considered, but it is continually replenished from a nuclear pool (Fig. 2A). On the other hand, if it can be established that native receptors for estrogens and progestins are indeed confined to the nucleus, there will be need for elucidation of how the steroid makes its way so rapidly from the blood transport proteins to the cell nucleus *in vivo*[18]; perhaps the low-affinity, high-capacity "type II" binding substances[280] may be involved. But once the hormone is in the nucleus, the two-step mechanism, involving receptor transformation followed by genomic binding, would function as previously conceived (Fig. 2B).

On the basis of considerations here and elsewhere,[8,247,249,274−276] it appears that the proposed invalidation of the two-step mechanism is ill-founded and tends to present a distorted picture, especially to investigators new to the field. Neither of the

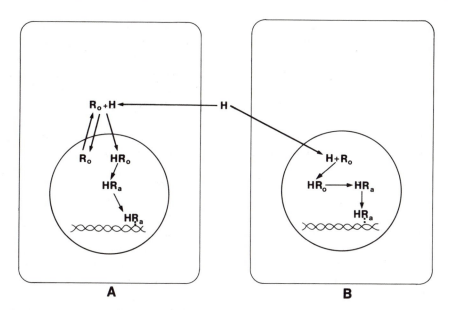

Figure 2. Schematic representation of hormone–receptor interaction in target cell. H, hormone; R_o, untransformed (native) receptor; R_a, transformed (activated) receptor. **A:** Small amount of extranuclear receptor in equilibrium with nuclear pool of loosely bound native receptor. **B:** Native receptor confined to the nucleus.

pathways shown in Fig. 2 represents more than a modification of the original concept. As has been aptly stated,[274] "if steroid receptors are indeed weakly held non-histone nuclear proteins, . . . the change [for the 'two-step' hypothesis] may be more cosmetic than substantive."

3.5. Receptor Phosphorylation

As has been reviewed recently,[154] the steroid-binding units of estrogen, progestin, glucocorticoid, and vitamin D receptors, as well as the associated heat shock protein, are phosphoproteins. The first suggestion that receptors might be phosphorylated came from observations that the ability of cultured thymus cells to bind glucocorticoids varies in proportion to their ATP levels. Cells grown in the absence of oxygen or glucose contain little ATP, and such cells,[281] as well as cytosol fractions of their homogenates,[282] show depressed binding of cortisol, which can be restored by growing the cells aerobically. Inhibition of ATP synthesis by 2,4-dinitrophenol likewise reduces hormone binding.[283] These findings, as well as studies with mouse fibroblasts,[284] suggested that the glucocorticoid-binding substance in the cytosol is continually being generated by an energy-requiring process not involving protein synthesis and led to the concept of receptor cycling. An inert form of the receptor is produced from transformed receptor in the nucleus; it then undergoes an energy-dependent reactivation and is recycled back into the receptor system.[284–286] Such inactive or "null" receptors were later detected in the nuclei of ATP-depleted mouse thymoma cells.[287] Similar inhibition of dihydrotestosterone binding in prostatic tissue *in vitro* by 2,4-dinitrophenol, cyanide, or azide[288] and enhancement of this binding in prostatic cytosol by ATP or GTP[289] suggested that phosphorylation may be involved in androgen receptor function with the possibility of a recycling process.[290]

Evidence suggesting a role of receptor phosphorylation in steroid hormone binding came from several observations. The loss in hormone-binding capacity for glucocorticoids,[144,146] estrogens,[291,292] and progestins,[145,147] that occurs on warming homogenates or extracts of target cells is prevented or retarded by phosphatase inhibitors such as molybdate, fluoride, glucose 1-phosphate, or 4-nitrophenyl phosphate. Receptor inactivation occurs on treatment of glucocorticoid receptor with a purified alkaline phosphatase preparation.[293] It was also found that glucocorticoid-binding ability in mouse fibroblast[294] or rat thymocyte[295] cytosols that had been depleted by warming can be restored by the action of ATP. Similarly, addition of ATP to preincubated cytosol–nuclei mixtures from mouse or calf uterus[296] or of ATP, GTP, or cGMP to human endometrial cells, homogenates, or cytosol[297,298] was found to increase the estrogen-binding ability of the receptor.

The foregoing indications for receptor phosphorylation, though suggestive, are still indirect. Early attempts to demonstrate directly that receptors are phosphorylated in target cells were limited by the difficulty in separating receptor from the variety of phosphorylated proteins present in larger quantities and by the presence of endogenous phosphatase and kinase activities. By using purified receptor and an exogenous cAMP-dependent protein kinase, it was first demonstrated that both the A and B forms of the progestin receptor from hen oviduct can undergo phosphorylation with labeled ATP *in vitro*.[299] Soon thereafter, purified estrogen receptor from calf uterus, after inactivation with endogenous phosphatase, was shown to be phosphorylated by labeled ATP and an endogenous calcium-stimulated kinase.[300,301] Purified progestin receptor from chick oviduct was likewise phosphorylated by an endogenous magnesium-dependent kinase[302,303] or an exogenous cAMP-dependent kinase.[137]

Direct evidence for receptor phosphorylation in target cells was first provided by the incorporation of labeled phosphate into progestin receptor on incubation of chick oviduct mince with [^{32}P]orthophosphate.[304] Subsequently, various investigators have reported phosphorylation of progestin receptor by incubating minces of chick oviduct[158,305–309] or rabbit uterus[310] and cultured chick oviduct[311,312] or T47D human breast cancer[313–316] cells with labeled orthophosphate. A and B forms of the receptor are phosphorylated to approximately the same degree[308,309] although the B receptors have been reported to have a unique phosphorylation site in addition to those in common with the A form.[306,315] The nonbinding 90-kDa component of untransformed progestin receptor also undergoes phosphorylation in the same systems.[158,304–306]

Phosphorylation of estrogen receptors was similarly effected in rat uterine tissue[317,318] and mouse Leydig tumor cells,[319] and that of glucocorticoid receptors in mouse fibroblasts, rat hepatocytes, mouse thymoma and pituitary cells, and human breast epithelial cells (for references see ref. 154). As with glucocortcoid receptors, phosphorylation of progestin receptors in target tissues or in whole cells is entirely on serine.[303,305,313,315] This is also the case with the associated heat-shock protein.[305] With purified oviduct receptor *in vitro,* phosphorylation can be effected on either serine[137] or tyrosine[320] by using the appropriate exogenous kinase. In contrast, estrogen receptor is reported to phosphorylate on tyrosine in calf uterine extracts using endogenous kinase[301,321] and in rat uterine tissue,[317,318] although recent studies have found estrogen receptors in calf uterus and MC7F-7 human breast cancer cells to be phosphorylated on serine.[322]

Phosphorylation of progestin receptors in whole cells or tissues is enhanced by the presence of the hormone.[307-310,313-316] However, no hormonal stimulation of receptor phosphorylation was observed with cultured chick oviduct cells.[311,312] Phosphorylation *in vitro* of a partially purified preparation of calf uterine estrogen receptor, using an endogenous kinase, was likewise found to be increased in the presence of estrogen.[321] In the case of progestin receptors of rabbit uterus[310] and T47D cancer cells,[314-316] there appear to be two phosphorylation processes: one is hormone independent and occurs with the untransformed receptor; the second is progestin stimulated and takes place with the transformed receptor, probably after nuclear binding. The latter may play a role in the action on gene expression.[316]

It is tempting to consider that the initial phosphorylation of untransformed receptors is involved in their ability to bind hormone. This is supported by the aforementioned observations of receptor inactivation by what appears to be endogenous phosphatase action and restoration of hormone-binding ability by ATP. It is also consistent with a recent finding using synthetic estrogen receptor.[323] When receptor is obtained by expressing cloned cDNA in a rabbit reticulocyte lysate system, it possesses only a few percent of the hormone-binding capacity of the original MCF-7 cell receptor or of recombinant receptor expressed in eukaryotic cells.[324] It can be endowed with full binding activity by phosphorylation with ATP and purified calf uterine kinase.[325] However, expression of cDNA for progestin receptor of chick oviduct in a bacterial system gives recombinant receptor protein that is fully active in hormone binding without need for any posttranslational modification.[326] Similar results were obtained with recombinant glucocorticoid receptor produced in a reticulocyte lysate system.[327] Moreover, studies of the synthesis of progestin receptor in T47D cells have shown that both the A and B receptors are first produced in a nonphosphorylated form that is fully active, both in binding hormone and in being converted to the transformed state which can then undergo secondary phosphorylation.[100] Thus, a general importance of primary phosphorylation in steroid binding by the native receptor is not entirely clear.

Likewise, a precise role for phosphorylation in receptor transformation and in nuclear binding is not definitely established. In contrast to transformed progestin receptor, which undergoes additional phosphorylation in rabbit uterus and T47D cells, there appears to be dephosphorylation of the transformed estrogen receptor in the nuclei of mouse[328] and calf[322] uterus and MCF-7 cancer cells.[322] With glucocorticoids, it has been reported that there is no net change in phosphorylation of the receptor as it undergoes transformation in target cells,[329] nor of the 90-kDa heat-shock protein as it associates with the native receptor or is separated during transformation *in vitro*.[330] The native and transformed glucocorticoid receptors appear to be phosphorylated to the same extent.[331] However, it has recently been found that, in intact mouse thymoma cells, treatment with hormone increases the average number of phosphates on both the untransformed and salt-extractable transformed receptor from three to five, but about 10% of the nuclear-bound receptor is not salt extractable, and this has only three phosphate groups.[332] Thus, no common pattern has emerged. In spite of the many observations suggesting that phosphorylation and/or dephosphorylation are somehow involved in the nuclear interaction of steroid hormone receptors, and the fact that receptor transformation is blocked by phosphatase inhibitors (see Section 3.2), a precise biochemical role for phosphorylation in these processes is still elusive.

4. Immunochemical Studies

4.1. Antibodies to Estrogen Receptor

The original recognition of steroid hormone receptors and most of the earlier information concerning their properties and function in target cells depended on the use of a radioactive steroid as a marker for the receptor protein to which it binds. Despite the wealth of information obtained using the ligand-binding approach, it has definite limitations, and there has been a need for an alternative means of receptor detection to complement the use of steroid-binding techniques. For many years, the attempted preparation of specific antibodies to steroid hormone receptors was remarkably unsuccessful. In target cells the receptors are present in minute amounts, and in crude tissue extracts they are rather labile; before the advent of molybdate stabilization, they readily lost their ability to bind the steroid needed as a marker during purification and concentration to provide an effective immunogen. Moreover, antibodies to steroid hormone receptors generally form nonprecipitating immune complexes, so that most conventional techniques of immunochemistry were ineffective in their detection. Fortunately, most antibody preparations that have been described do not interfere with the binding of hormone to its receptor, so the labeled steroid can be used as a marker for the receptor protein in the soluble immune complex.

Although there had been earlier suggestions of antisera to estrogen receptor, it was not until 1977 that definitive antibodies were obtained by immunizing rabbits[333] and later a goat[334] with partially purified transformed receptor from calf uterus. The increase in sedimentation rate of the labeled estradiol–receptor complex was used as the principal criterion of its association with one or in some instances two molecules of immunoglobulin. These polyclonal antibodies against calf uterus were cross-reactive with estrogen receptor from a number of mammalian species as well as from chick oviduct but not with receptors for other steroid hormones. Cross-reacting polyclonal antibodies were soon reported for estrogen receptor from calf uterus,[335] human myometrium,[336] and human breast cancer.[337]

The first application of the hybridoma procedure,[338] using transformed estrogen receptor of calf uterus, gave monoclonal antibodies that recognized only calf receptor.[339] Similar experiments with cytosol receptor from MCF-7 human breast cancer cells furnished cross-reacting monoclonal antibodies,[340] as did subsequent studies with untransformed calf uterine receptor.[341] Of 13 different monoclonal antibodies to the estrogen receptor of MCF-7 cells,[218] eight were found to react with receptor from every source tested, including hen oviduct; four reacted only with mammalian receptor (rat, calf, monkey, human), and one was specific for receptor from primates. This indicates that some but not all regions of the receptor molecule are conserved over a variety of species. However, these antibodies do not cross-react with receptors for other steroid hormones. In the presence of antiestrogens such as tamoxifen or its derivatives, an additional epitope is exposed in the human receptor molecule that is recognized uniquely by one of the monoclonal antibodies.[342] All polyclonal and monoclonal antibodies recognized both native and transformed receptor, although the species-specific monoclonal antibodies to calf uterine receptor reacted much more strongly with transformed receptor,[339,343] the form used as immunogen. Binding of one antibody molecule to each untransformed complex and two to each transformed complex observed in some instances is consistent with the dimeric structure of the transformed receptor.[341]

Certain of the monoclonal antibodies have been of considerable value. They have been used to: purify estrogen receptors by immunoaffinity methods; assay receptors in human cancers[218,220−226]; measure occupied receptors in cell nuclei as an alternative to exchange techniques[344]; initially characterize different domains in the receptor molecule[218]; localize untransformed receptor in the nucleus, as previously described; clone the cDNA of the human estrogen receptor as discussed in Section 5.

4.2. Antibodies to Progestin Receptor

As methods were developed for receptor stabilization and purification, polyclonal antibodies were prepared to progestin receptor from rabbit[345] and guinea pig[346] uterus and chick oviduct[347] and to purified A[105] and B[105,348,349] forms of the receptor from chick oviduct. These antibodies react with progestin receptors from the other species tested, with the exception of the antibody to rabbit uterine receptor, which recognizes receptor from other mammalian but not from avian species.[345] One preparation, raised against the purified B form of receptor from chick oviduct, reacted only weakly with mammalian receptor; it also gave a weak cross-reaction with estrogen receptor from chick oviduct.[348] All the other antibodies proved specific for progestin receptor, and whether raised against the A or the B form of the avian receptor, they recognize both the A and B proteins.

The first monoclonal antibody obtained with the avian progestin receptor[350] turned out to be to the non-steroid-binding component of the untransformed receptor.[157] This antibody provided a valuable reagent for the recognition and characterization of this receptor component as a 90-kDa heat-shock protein. The antibody does not recognize the transformed avian progestin receptor or the untransformed human receptor. It cross-reacts with untransformed but not transformed receptors for estrogens, androgens, and glucocorticoids from avian tissues, indicating the similarity of the heat-shock protein in native receptors for different steroid hormones.[157,350] Subsequently, monoclonal antibodies were prepared directly to the purified 90-kDa protein from chick oviduct receptor, two of which recognize the intact native receptor and one of which was found to cross-react with avian glucocorticoid and androgen receptors.[351]

A large number of monoclonal antibodies have been prepared against purified progestin receptor from rabbit uterus,[352–354] hen or chick oviduct,[355,356] human endometrical cancer,[101] and T47D human breast cancer cells.[99,106] In two instances the purified B form was used as the immunogen,[106,355] and the first monoclonal antibodies to a steroid-binding unit of avian progestin receptor were obtained by an *in vitro* immunization procedure in which isolated mouse spleen cells were treated directly with the B protein.[355] As with polyclonal antibodies, monoclonals raised against rabbit receptor do not react with avian receptor[352,353]; in one instance, they recognized several other mammalian progestin receptors, but in another they reacted with receptor only from human breast cancer but not from rat or guinea pig uterus.[353] In contrast to antibodies to rabbit receptor that do not recognize avian receptor, those raised against avian receptor cross-react with that from mammalian tissues, although one monoclonal was obtained that reacted with rabbit but not human receptor.[356] Of the many monoclonal antibodies raised against human progestin receptor, some were found specific for human receptor, whereas others recognize that from rabbit and, in one instance, from chick as well.[99,106] Most monoclonal preparations react with both the A and B forms of the avian or human receptor, but some were obtained that recognize only the B protein.[99,101,106,356] In no instance did any antibody recognize the A form but not the B, supporting the view that the two proteins are similar except for an additional moiety in the B form.

In addition to their use in characterizing different components of the receptor complex, antibodies to progestin receptor have been valuable in establishing the nuclear localization of the native receptor, in the immunoassay of progestin receptors in human cancers, and for the cloning of the cDNA for progestin receptor and determining its detailed structure, as described in the next section.

5. Molecular Biological Studies

5.1. Receptor Cloning and Structure

With the availability of purified receptor proteins for steroid and related hormones and of specific antibodies that react with them, it became possible to employ techniques of molecular biology to clone and express the complementary DNA coding for receptor biosynthesis. From its nucleotide sequence, the amino acid sequence of the receptor protein was deduced (for summary see ref. 357). Expression cloning was first accomplished for the human glucocorticoid receptor,[327,358–361] soon followed by human estrogen receptor[324,362,363] and the estrogen receptors of chick oviduct[364,365] and frog liver.[366] In the same fashion, the cDNAs for chicken[107,367,368] and rabbit[369] progestin receptor were cloned and sequenced, as were those for receptors for mineralocorticoids, vitamin D, thyroid hormone, and retinoic acid (references in ref. 370) and, more recently, androgens.[371,372] It became evident that these receptors show striking homology in certain regions of the protein molecule and little homology in others (Fig. 3), leading to the recognition that they represent a family of gene regulatory agents.[370,373] Of considerable interest are the unexpected finding of a relationship of the glucocorticoid[374] and later the estrogen[324] receptor to the viral oncogene *erb*-A and the subsequent demonstration that the c-*erb*-A protein actually is the thyroid hormone receptor.[375,376]

With the aid of deletion mutants, it was possible to identify individual domains in the receptor molecule and to correlate each with an aspect of receptor function.[370,373,374,377–379] As shown in Figs. 3 and 4, each receptor appears to have a steroid-binding region (E) near the C-terminal end, specific for that class of hormone, a cysteine-rich DNA-binding region (C) that is highly conserved, a small "hinge" region (D) joining these two domains, which can be deleted without much effect on function, and a large variable region (A/B) near the N-terminal end that is not essential to function but that is required for maximum activity. Most of the antibodies that have been obtained with progestin[354] and glucocorticoid[380,381] receptors appear to be directed against the A/B region, in contrast to estrogen receptor where most monoclonals recognize the E region, even though they do not interfere with steroid binding.[218]

5.2. Interaction with Target Genes

When it was recognized that hormone-induced transformation of steroid–receptor complexes causes them to bind to chromatin in the nucleus, attempts were made to

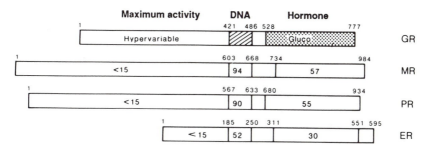

Figure 3. Comparison of domains in receptor molecules for glucocorticoids (GR), mineralocorticoids (MR), progestin, B form (PR), and estrogen (ER). The position of each domain boundary is given as the number of amino acids from the amino terminal (1). Numbers within the domains indicate the percentage homology with the corresponding domain in GR. Reproduced by permission from Evans.[370]

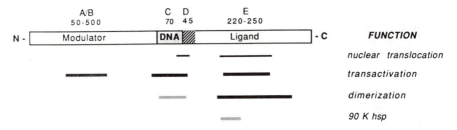

Figure 4. Functional organization showing hormone-binding and DNA-binding domains of a typical steroid hormone receptor. A/B, modulating region; C, DNA-binding region; D, "hinge" region; E, steroid-binding region. Horizontal bars represent putative areas participating in the various receptor functions. Transactivation denotes the enhancement by the receptor of transcriptional activity of the target genes. Reproduced with permission from Beato.[377]

identify the so-called "acceptor" site to which the transformed receptor becomes attached (for review see ref. 119). Although some evidence for selective nuclear binding has been reported, it has proved difficult to distinguish the selective interaction with target genes from the large amount of nonspecific binding to DNA in general. With the application of molecular biological approaches to receptor studies, and the identification of target genes, recombinant systems were constructed in which the effect of steroid–receptor complexes on gene expression could be investigated in greater detail. Although a complete description of the various gene systems that have been studied is beyond the scope of this chapter, some of the general patterns and concepts that have emerged may be mentioned (for reviews see refs. 8, 357, 370, 377, 382).

It is now evident that the effect of transformed hormone receptor complexes on gene expression does not involve an interaction at the site of transcription but rather with a "hormone response element" (HRE), an enhancer located in the promoter area in the 5'-flanking region of the target gene. Palindromic sequences of 13 to 15 base pairs appear to make up the actual receptor-binding site. The hormone response elements for glucocorticoids and for progestins appear to be similar in composition and to interact in some degree with both receptors, whereas those for estrogens and thyroid hormones, although related,[383] are somewhat different. Recognition of the response element appears to reside in the DNA-binding domain of the receptor (region C, Fig. 4), since the chimeric protein formed by replacing this region in the human estrogen receptor with the corresponding region of the human glucocorticoid receptor activates expression of a corticoid-inducible but not estrogen-inducible gene in the presence of estrogen.[384]

The precise nature of the interaction of steroid receptors with response elements is currently an active area of investigation, although many details are not yet clear. In intact cells or tissues the hormone is needed not only to transform the receptor but also to effect its binding to the enhancer element. However, certain mutant receptors, which lack the steroid-binding domain, have a constitutive ability to interact with DNA in the absence of hormone, and normal uncomplexed progesterone and glucocorticoid receptors are active in many *in vitro* systems. It has been observed that an effect of the steroid with these receptors is to increase the rates at which the receptor associates with and dissociates from DNA, thereby enhancing its efficiency in finding the proper location in the target gene.[385] Although they do not dimerize spontaneously on transformation, it appears that progestin and glucocorticoid receptors interact with their palindromic response elements in a dimeric form.[127,386]

There is a growing body of evidence that the interaction of the receptor with its response element is mediated through a pair of "zinc fingers" (Fig. 5) located in the highly conserved C region of all classes of receptors.[387−389] Such structures, involving zinc coordinated with appropriately located cysteine or histidine residues, have provided a general pattern for binding to DNA for a large number of regulatory proteins.[390] It has been shown that the binding of estrogen receptor to DNA-cellulose depends on the presence of receptor-bound metal that can be chelated by 1,10-phenanthrolene.[391] It has also been reported that estrogen receptor binds to the coding strand of DNA in the response element with a 60-fold higher affinity than it does to double-stranded DNA.[392] Finally, it appears that the enhancement of gene expression by interaction of the receptor at the response element may involve the participation of various transcription factors[393,394] including the chicken oviduct upstream promotor (COUP),[395] a protein that resembles the steroid hormone receptor family in its structure.[396]

6. Summary

In the three decades since the original discovery of steroid hormone receptors, much has been learned about the general nature of the processes by which these hormones exert their regulatory effects in target tissues, such as uterus. It seems clear that the intracellular receptor proteins are potential modulators of gene ex-

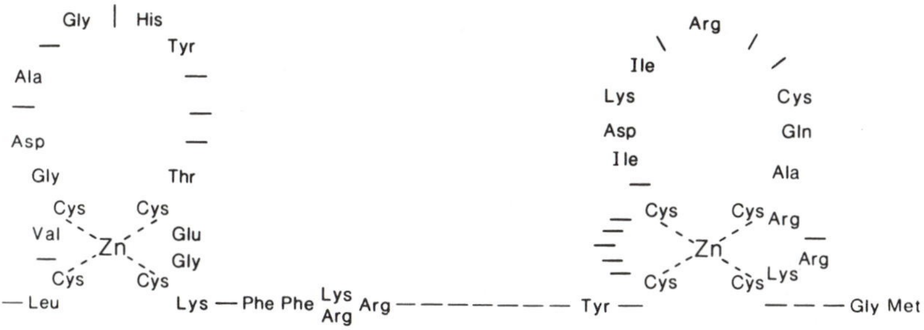

Figure 5. Hypothetical structure of the DNA-binding domain of hormone receptors with two putative zinc-binding fingers. Each zinc ion forms a tetrahedral coordination complex with Cys residues. Reproduced with permission from Evans.[370]

pression, needed for the effective operation of transcriptive processes in hormone-dependent cells, and that they are present in an inactive form until interaction with the steroid converts them to a functional state that can react with target genes. Receptor transformation appears to involve, at least in part, the removal of both macromolecular and micromolecular factors. For most classes of steroid hormones, much of the untransformed receptor is within the nuclear compartment, loosely held and readily extractable.

Methods have been developed for the stabilization, purification, and characterization of receptor proteins, and through cloning of their cDNAs, primary structures for essentially all these receptors are now known. This has led to an appreciation of structural similarities among the family of receptors for the different steroid hormones and to an identification of the regions in the protein molecule responsible for various aspects of their function. Monoclonal antibodies recognizing specific molecular domains are available for most receptors.

But despite the knowledge that has been accumulated, many key questions remain unsolved. How does association with a steroid hormone remove substances keeping the receptor in its nonfunctional state, and how does binding of the transformed complex to a hormone-response element in the promotor region enhance gene transcription? Once it has converted the receptor to the nuclear binding state, is there a further role for the steroid in modulating the transcription process? Still not entirely clear is the involvement of phosphorylation and/or dephosphorylation in hormone binding, receptor transformation, and transcriptional activation. Probably not as vital to basic understanding but still of importance in the overall picture is whether the native receptor for estrogens and progestins is entirely confined to the nucleus or whether there is an intracellular distribution equilibrium.

With the amount of study now being devoted to this field, and with the application of new experimental techniques, especially those of modern molecular biology, one can expect that soon there will be definitive answers to these and other pertinent questions, and that the precise mechanism of steroid hormone action will no longer be a mystery.

ACKNOWLEDGMENTS. Preparation of this article was begun while the author was a Scholar-in-Residence at the Fogarty International Center, National Institutes of Health, Bethesda, Maryland. Completion and publication of the review were supported by a grant (RDP-53A) from the American Cancer Society.

References

1. Raspé, G. (ed.), 1971, *Schering Workshop on Steroid Hormone Receptors Advances in the Biosciences,* Volume 7, Pergamon–Vieweg, Braunschweig.
2. Jensen, E. V., and DeSombre, E. R., 1972, Mechanism of action of the female sex hormones, *Annu. Rev. Biochem.* **41:**203–230.
3. O'Malley, B. W., and Means, A. R., 1974, Female steroid hormones and target cell nuclei, *Science* **183:**610–620.

4. Liao, S., 1975, Cellular receptors and mechanism of action of steroid hormones, *Int. Rev. Cytol.* **41**:87–172.

5. Gorski, J., and Gannon, F., 1976, Current models of steroid hormone action: A critique, *Annu. Rev. Physiol.* **38**:425–450.

6. Eriksson, H., and Gustafsson, J.-Å. (eds.), 1983, *Steroid Hormone Receptors: Structure and Function,* Elsevier/North Holland, Amsterdam.

7. Liao, S., and Hiipakka, R. A., 1984, Mechanism of action of steroid hormones at the subcellular level, in: *Biochemistry of Steroid Hormones,* 2nd ed. (H. L. J. Makin, ed.), Blackwell Scientific Publications, Oxford, pp. 633–680.

8. Ringold, G. M., 1985, Steroid hormone regulation of gene expression, *Annu. Rev. Pharmacol. Toxicol.* **25**:529–566.

9. Mueller, G. C., Herranen, A. M., and Jervell, K. J., 1958, Studies on the mechanism of action of estrogens, *Recent Prog. Horm. Res.* **14**:95–129.

10. Mueller, G. C., 1965, Role of RNA and protein synthesis in estrogen action, in: *Mechanisms of Hormone Action* (P. Karlson, ed.), Academic Press, New York, pp. 228–245.

11. O'Malley, B. W., McGuire, W. L., Kohler, P. O., and Korenman, S. G., 1969, Studies on the mechanism of steroid hormone regulation of synthesis of specific proteins, *Recent Prog. Horm. Res.* **25**:105–160.

12. Glascock, R. F., and Hoekstra, W. G., 1959, Selective accumulation of tritium-labelled hexoestrol by the reproductive organs of immature female goats and sheep, *Biochem. J.* **72**:673–682.

13. Jensen, E. V., and Jacobson, H. I., 1960, Fate of steroid estrogens in target tissues, in: *Biological Activities of Steroids in Relation to Cancer* (G. Pincus and E. P. Vollmer, eds.), Academic Press, New York, pp. 161–178.

14. Jensen, E. V., Numata, M., Smith, S., Suzuki, T., Brecher, P. I., and DeSombre, E. R., 1969, Estrogen–receptor interaction in target tissues, *Dev. Biol.* **3**(Suppl.):151–171.

15. Jensen, E. V., and Jacobson, H. I., 1962, Basic guides to the mechanism of estrogen action, *Recent Prog. Horm. Res.* **18**:387–414.

16. Stone, G. T., and Baggett, B., 1965, The *in vitro* uptake of tritiated estradiol and estrone by the uterus and vagina of the ovariectomized mouse, *Steroids* **5**:809–826.

17. Terenius, L., 1966, Specific uptake of oestrogens by the mouse uterus *in vitro, Acta Endocrinol. (Kbh.)* **53**:611–618.

18. Jensen, E. V., DeSombre, E. R., and Jungblut, P. W., 1967, Interaction of estrogens with receptor sites *in vivo* and *in vitro,* in: *Proceedings 2nd International Congress on Hormonal Steroids, Milan,* Excerpta Medica Foundation, Amsterdam, pp. 492–500.

19. Jensen, E. V., Suzuki, T., Kawashima, T., Stumpf, W. E., Jungblut, P. W., and DeSombre, E. R., 1968, A two-step mechanism for the interaction of estradiol with the rat uterus, *Proc. Natl. Acad. Sci. U.S.A.* **58**:632–638.

20. Maurer, H. R., and Chalkley, G. R., 1967, Some properties of a nuclear binding site of estradiol, *J. Mol. Biol.* **27**:431–441.

21. Jensen, E. V., DeSombre, E. R., Hurst, D. J., Kawashima, T., and Jungblut, P. W., 1967, Estrogen-receptor interactions in target tissues, *Arch. Anat. Microsc. Morphol. Exp.* **56**(Suppl.):547–569.

22. Jensen, E. V., Hurst, D. J., DeSombre, E. R., and Jungblut, P. W., 1967, Sulfhydryl groups and estradiol–receptor interaction, *Science* **158**:385–387.

23. Stone, G. M., 1964, The effect of estrogen antagonists on the uptake of tritiated estradiol by the uterus and vagina of the ovariectomized mouse, *J. Endocrinol.* **29**:127–136.

24. Roy, S., Mahesh, V. B., and Greenblatt, R. B., 1964, Inhibition of uptake of radioactive estradiol by the uterus and pituitary gland of immature rats, *Acta Endocrinol. (Kbh.)* **47**:669–675.

25. Jensen, E. V., 1965, Mechanism of estrogen action in relation to carcinogenesis, *Can. Cancer Conf.* **6**:143–165.

26. Jensen, E. V., Jacobson, H. I., Smith, S., Jungblut, P. W., and DeSombre, E. R., 1972, The use of estrogen antagonists in hormone receptor studies, *Gynecol. Invest.* **3**:108–122.

27. Mueller, G. C., Gorski, J., and Aizawa, Y., 1961, The role of protein synthesis in early estrogen action, *Proc. Natl. Acad. Sci. U.S.A.* **47:**164–169.

28. Ui, H., and Mueller, G. C., 1963, The role of RNA synthesis in early estrogen action, *Proc. Natl. Acad. Sci. U.S.A.* **50:**256–260.

29. Noteboom, W. D., and Gorski, J., 1965, Stereospecific binding of estrogens in the rat uterus, *Arch. Biochem. Biophys.* **11:**559–568.

30. Stumpf, W. E., 1968, Subcellular distribution of ^{3}H-estradiol in rat uterus by quantitative autoradiography. A comparison between ^{3}H-estradiol and ^{3}H-norethynodrel, *Endocrinology* **83:**777–782.

31. King, R. J. B., Gordon, J., Cowan, D. M., and Inman, D. R., 1966, The intranuclear localization of [6,7-^{3}H]-oestradiol-17β in dimethylbenzanthracene-induced rat mammary adenocarcinoma and other tissues, *J. Endocrinol.* **36:**139–150.

32. Puca, G. A., and Bresciani, F. 1968, Receptor molecule for oestrogens from rat uterus, *Nature* **218:**967–969.

33. Toft, D., and Gorski, J., 1966, A receptor molecule for estrogens: Isolation from the rat uterus and preliminary characterization, *Proc. Natl. Acad. Sci. U.S.A.* **55:**1574–1581.

34. Erdos, T., 1968, Properties of a uterine oestradiol receptor, *Biochem. Biophys. Res. Commun.* **37:**338–343.

35. Korenman, S. G., and Rao, B. R., 1968, Reversible disaggregation of the cytosol estrogen binding protein of uterine cytosol, *Proc. Natl. Acad. Sci. U.S.A.* **61:**1028–1033.

36. Toft, D., Shyamala, G., and Gorski, J., 1967, A receptor molecule for estrogens. Studies using a cell-free system, *Proc. Natl. Acad. Sci. U.S.A.* **57:**1740–1743.

37. Gorski, J., Toft, D., Shyamala, G., Smith, D., and Notides, A., 1968, Hormone receptors: Studies on the interaction of estrogen with the uterus, *Recent Prog. Horm. Res.* **24:**45–80.

38. Jensen, E. V., Numata, M., Brecher, P. I., and DeSombre, E. R., 1971, Hormone–receptor interaction as a guide to biochemical mechanism, *Biochem. Soc. Symp.* **32:**133–159.

39. Gschwendt, M., and Hamilton, T. H., 1972, The transformation of the cytoplasmic oestradiol–receptor complex into the nuclear complex in a uterine cell-free system, *Biochem. J.* **128:**611–616.

40. DeSombre, E. R., Mohla, S., and Jensen, E. V., 1975, Receptor transformation, the key to estrogen action, *J. Steroid Biochem.* **6:**469–473.

41. Yamamoto, K. R., and Alberts, B. M., 1972, *In vitro* conversion of estradiol–receptor protein to its nuclear form: Dependence on hormone and DNA, *Proc. Natl. Acad. Sci. U.S.A.* **69:**2105–2109.

42. Sarff, M., and Gorski, J., 1971, Control of estrogen binding protein concentration under basal conditions and after estrogen administration, *Biochemistry* **10:**2557–2563.

43. Jensen, E. V., Mohla, S., Gorell, T., Tanaka, S., and DeSombre, E. R., 1972, Estrophile to nucleophile in two easy steps, *J. Steroid Biochem.* **3:**445–458.

44. McGuire, W. L., Huff, K., and Chamness, G. C., 1972, Temperature-dependent binding of estrogen receptor to chromatin, *Biochemistry* **11:**4562–4565.

45. Milgrom, E., Atger, M., and Baulieu, E.-E., 1973, Acidophilic activation of steroid hormone receptors, *Biochemistry* **12:**5198–5205.

46. Atger, M., and Milgrom, E., 1976, Chromatographic separation on phosphocellulose of activated and nonactivated forms of steroid–receptor complex. Purification of the activated complex, *Biochemistry* **15:**4298–4304.

47. Gorski, J., and Raker, B., 1974, Estrogen action in the uterus: The requisite for sustained estrogen binding in the nucleus, *Gynecol. Oncol.* **2:**249–258.

48. Anderson, J. N., Peck, E. J., Jr., and Clark, J. H., 1975, Estrogen-induced uterine responses and growth: Relationship to receptor estrogen binding by uterine nuclei, *Endocrinology* **96:**160-167.

49. Raynaud-Jammet, C., and Baulieu, E.-E., 1969, Action de l'oestradiol *in vitro*: Augmentation de la biosynthèse d'acide ribonucléique dans les noyaux utérine, *C.R. Acad. Sci. [D] (Paris)* **268:**3211–3214.

50. Mohla, S., DeSombre, E. R., and Jensen, E. V., 1972, Tissue-specific stimulation of RNA

synthesis by transformed estradiol–receptor complex, *Biochem. Biophys. Res. Commun.* **46**:661-667.

51. Arbogast, L. Y., and DeSombre, E. R., 1975, Estrogen-dependent *in vitro* stimulation of RNA synthesis in hormone-dependent mammary tumors of the rat, *J. Natl. Cancer Inst.* **54**:483–485.

52. Leclercq, G., Hulin, N., and Heuson, J. C., 1973, Interaction of activated estradiol–receptor complex and chromatin in isolated uterine nuclei, *Eur. J. Cancer* **9**:681–685.

53. Verrijdt, A., Leclercq, G., Devleeschouwer, N., Danguy, A., 1985, Tritiated actinomycin-D staining method: A valuable tool to study oestrogen receptor-induced modifications of transcriptional activity in normal and neoplastic cells, *Arch. Int. Physiol. Biochim.* **93**:65–73.

54. Leroy, F., Preumont, A. M., Galand, P., and Brachet, J., 1972, Increased chromatin acid lability and actinomycin-D binding in endometrial cells under the action of sex steroids, *J. Endocrinol.* **52**:525–531.

55. Brachet, J., and Ficq, A., 1965, Binding sites of ^{14}C-actinomycin in amphibian ovocytes and an autoradiography technique for the detection of cytoplasmic DNA, *Exp. Cell Res.* **38**:153–159.

56. Jensen, E. V., and DeSombre, E. R., 1973, Estrogen–receptor interaction. Estrogenic hormones effect transformation of specific receptor proteins to a biochemically functional form, *Science* **182**:126-134.

57. Astwood, E. B., 1938, A six-hour assay for the quantitative determination of estrogen, *Endocrinology* **23**:25–31.

58. Spaziani, E., and Szego, C. M., 1959, Further evidence for mediation by histamine of estrogenic stimulation of the rat uterus, *Endocrinology* **64**:713–723.

59. Szego, C. M., 1965, Role of histamine in mediation of hormone action, *Fed. Proc.* **24**:1343–1352.

60. Tchernitchin, A., 1973, Fine structure of rat uterine eosinophils and the possible role of eosinophils in the mechanism of estrogen action, *J. Steroid Biochem.* **4**:277–282.

61. Tchernitchin, A., 1979, The role of eosinophil receptors in the non-genomic response to oestrogens in the uterus, *J. Steroid Biochem.* **11**:417–424.

62. Podratz, K. C., and Katzman, P. A., 1968, Effect of estradiol on uptake and retention of progesterone by the vagina of the ovariectomized mouse, *Fed. Proc.* **27**:497.

63. Falk, R. J., and Bardin, C. W., 1970, Uptake of tritiated progesterone by the uterus of the ovariectomized guinea pig, *Endocrinology* **86**:1059–1063.

64. Wiest, W. G., and Rao, B. R., 1971, Progesterone binding proteins in rabbit uterus and human endometrium, *Adv. Biosci.* **7**:251–266.

65. Leavitt, W. W., and Blaha, G. C., 1972, An estrogen-stimulated, progesterone-binding system in the hamster uterus and vagina, *Steroids* **19**:263–274.

66. Davies, I. J., and Ryan, K. J., 1972, The uptake of progesterone by the uterus of the pregnant rat *in vivo* and its relationship to cytoplasmic progesterone-binding protein, *Endocrinology* **90**:507–515.

67. McGuire, J. L., and DeDella, C., 1971, *In vitro* evidence for a progestogen receptor in the rat and rabbit uterus, *Endocrinology* **88**:1099–1103.

68. Milgrom, E., Atger, M., and Baulieu, E.-E., 1970, Progesterone in uterus and plasma. IV— Progesterone receptor(s) in guinea pig uterus cytosol, *Steroids* **16**:741–754.

69. Corvol, P., Falk, R., Freifeld, M., and Bardin, C. W., 1972, *In vitro* studies of progesterone binding proteins in guinea pig uterus, *Endocrinology* **90**:1464–1469.

70. Kontula, K., Jänne, O., Jänne, J., and Vihko, R., 1972, Partial purification and characterization of progesterone-binding protein from pregnant guinea pig uterus, *Biochem. Biophys. Res. Commun.* **47**:596–603.

71. Faber, L. E., Sandmann, M. L., and Stavely, H. E., 1972, Progesterone binding in uterine cytosols of the guinea pig, *J. Biol. Chem.* **247**:8000–8004.

72. Milgrom, E., Thi, L., Atger, M., and Baulieu, E.-E., 1973, Mechanisms regulating the concentration and the conformation of progesterone receptor(s) in the uterus, *J. Biol. Chem.* **248**:6366–6374.

73. Milgrom, E., and Baulieu, E.-E., 1970, Progesterone in uterus and plasma. I. Binding in rat uterus 105,000 *g* supernatant, *Endocrinology* **87**:276–287.
74. Reel, J. R., Van Dewark, S. D., Shih, Y., and Callantine, M. R., 1971, Macromolecular binding and metabolism of progesterone in the decidual and pseudopregnant rat and rabbit uterus, *Steroids* **18**:441–461.
75. Feil, P. D., Glasser, S. R., Toft, D. O., and O'Malley, B. W., 1972, Progesterone binding in the mouse and rat uterus, *Endocrinology* **91**:738–746.
76. Faber, L. E., Sandmann, M. L., and Stavely, H. E., 1972, Progesterone-binding proteins of the rat and rabbit uterus, *J. Biol. Chem.* **247**:5648–5649.
77. Saffran, J., Loeser, B. K., Haas, B. M., and Stavely, H. E., 1973, Binding of progesterone by rat uterus *in vitro. Biochem. Biophys. Res. Commun.* **53**:202–209.
78. Rao, B. R., Wiest, W. G., and Allen, W. M., 1973, Progesterone "receptor" in rabbit uterus. I. Characterization and estradiol-17β augmentation, *Endocrinology* **92**:1229–1240.
79. Davies, I. J., Challis, J. R. G., and Ryan, K. J., 1974, Progesterone receptors in the myometrium of pregnant rabbits, *Endocrinology* **95**:165–173.
80. Lessey, B. A., and Gorell, T. A., 1980, Analysis of the progesterone receptor in the beagle uterus and oviduct, *J. Steroid Biochem.* **13**:1173–1180.
81. Verma, U., and Laumas, K. R., 1973, *In vitro* binding of progesterone to receptors in the human endometrium and the myometrium, *Biochim. Biophys. Acta* **317**:403–419.
82. Kontula, K., Jänne, O., Luukkainen, T., and Vihko, R., 1973, Progesterone-binding protein in human myometrium. Ligand specificity and some physiocochemical characteristics, *Biochim. Biophys. Acta* **328**:145–153.
83. Rao, B. R., Wiest, W. G., and Allen, W. M., 1974, Progesterone "receptor" in human endometrium, *Endocrinology* **95**:1275–1281.
84. Jänne, O., Kontula, K., Luukkainen, T., and Vihko, R., 1975, Oestrogen-induced progesterone receptor in human uterus, *J. Steroid Biochem.* **6**:501–509.
85. Sar, M., and Stumpf, W. E., 1974, Cellular and subcellular localization of ^{3}H-progesterone or its metabolites in the oviduct, uterus, vagina and liver of the guinea pig, *Endocrinology* **94**:1116–1125.
86. Hsueh, A. J. W., Peck, E. J., Jr., and Clark, J. H., 1974, Receptor progesterone complex in the nuclear fraction of the rat uterus: Demonstration by ^{3}H-progesterone exchange, *Steroids* **24**:599–611.
87. Lessey, B. A., and Gorell, T. A., 1981, Nuclear progesterone receptors in the beagle uterus, *J. Steroid Biochem.* **14**:585–591.
88. O'Malley, B. W., Sherman, M. R., and Toft, D. O., 1970, Progesterone "receptors" in the cytoplasm and nucleus of chick oviduct target tissues, *Proc. Natl. Acad. Sci. U.S.A.* **67**:501–508.
89. Sherman, M. R., Corvol, P. L., and O'Malley, B. W., 1970, Progesterone-binding components of chick oviduct I. Preliminary characterization of cytoplasmic components, *J. Biol. Chem.* **245**:6085–6096.
90. O'Malley, B. W., Toft, D. O., and Sherman, M. R., 1971, Progesterone-binding components of chick oviduct II. Nuclear components, *J. Biol. Chem.* **246**:1117–1122.
91. Coty, W. A., 1980, Reversible dissociation of steroid hormone–receptor complexes by mercurial reagents, *J. Biol. Chem.* **255**:8035–8037.
92. Kalimi, M., and Banerji, A., 1981, Role of sulfhydryl modifying reagents in the binding and activation of chick oviduct progesterone-receptor complex, *J. Steroid Biochem.* **14**:593–597.
93. Buller, R. E., Toft, D. O., Schrader, W. T., and O'Malley, B. W., 1975, Progesterone-binding components of chick oviduct VIII. Receptor activation and hormone-dependent binding to purified nuclei, *J. Biol. Chem.* **250**:801–808.
94. Fleischmann, G., and Beato, M., 1979, Activation of the progesterone receptor of rabbit uterus, *Mol. Cell. Endocrinol.* **16**:181–197.
95. Toft, D. O., Lohmar, P., Miller, J., and Moudgil, V., 1976. The properties and functional significance of ATP binding to progesterone receptors, *J. Steroid Biochem.* **7**:1053–1059.

96. Schrader, W. T., and O'Malley, B. W., 1972, Progesterone-binding components of chick oviduct IV. Characterization of purified subunits, *J. Biol. Chem.* **247:**51–59.

97. Schrader, W. T., Toft, D. O., and O'Malley, B. W., 1972, Progesterone-binding protein of chick oviduct VI. Interaction of purified progesterone–receptor components with nuclear constituents, *J. Biol. Chem.* **247:**2401–2407.

98. Lessey, B. A., Alexander, P. S., and Horwitz, K. B., 1983, The subunit structure of human breast cancer progesterone receptors: Characterization by chromatography and photoaffinity labeling, *Endocrinology* **112:**1267–1274.

99. Greene, G. L., Harris, K., Bova, R., Kinders, R., Moore, B., and Nolan, C., 1988, Purification of T47D human progesterone receptor and immunochemical characterization with monoclonal antibodies, *Mol. Endocrinol.* **2:**714–726.

100. Sheridan, P. L., Francis, M. D., and Horwitz, K. B., 1989, Synthesis of human progesterone receptors in T47D cells. Nascent A- and B-receptors are active without a phosphorylation-dependent post-translational maturation step, *J. Biol. Chem.* **264:**7054–7058.

101. Clarke, C. L., Zaino, R. J., Feil, P. D., Miller, J. V., Steck, M. E., Ohlsson-Wilhelm, B. M., and Satyaswaroop, P. G., 1987, Monoclonal antibodies to human progesterone receptor: Characterization by biochemical and immunohistochemical techniques, *Endocrinology* **121:**1123–1132.

102. Kuhn, R. W., Schrader, W. T., Coty, W. A., Conn, P. A., and O'Malley, B. W., 1977, Progesterone-binding components of chick oviduct: Biochemical characterization of purified oviduct progesterone receptor B subunit, *J. Biol. Chem.* **252:**308–317.

103. Coty, W. A., Schrader, W. T., and O'Malley, B. W., 1979, Purification and characterization of the chick oviduct progesterone receptor A subunit, *J. Steroid Biochem.* **10:**1–12.

104. Conneely, O. M., Dobson, A. D. W., Tsai, M.-J., Beattie, W. G., Toft, D. O., Huckaby, C. S., Zarucki, T., Schrader, W. T., and O'Malley, B. W., 1987, Sequence and expression of a functional chicken progesterone receptor, *Mol. Endocrinol.* **1:**517–525.

105. Gronemeyer, H., Govindan, M. V., and Chambon, P., 1985, Immunological similarity between the chick oviduct progesterone receptor forms A and B, *J. Biol. Chem.* **260:**6916–6925.

106. Estes, P. A., Suba, E. J., Lawler-Heavner, J., Elashry-Stowers, D., Wei, L. L., Toft, D. O., Sullivan, W. P., Horwitz, K. B., and Edwards, D. P., 1987, Immunologic analysis of human breast cancer progesterone receptors. 1. Immunoaffinity purification of transformed receptors and production of monoclonal antibodies, *Biochemistry* **26:**6250–6262.

107. Gronemeyer, H., Turcotte, B., Quirin-Stricker, C., Bocquel, M. T., Meyer, M. E., Krozowski, Z., Jeltsch, J. M., Lerouge, T., Garnier, J. M., and Chambon, P., 1987, The chicken progesterone receptor: Sequence, expression and functional analysis, *EMBO J.* **6:**3985–3994.

108. Carson, M. A., Tsai, M.-J., Conneely, O. M., Maxwell, B. L., Clark, J. H., Dobson, A. D. W., Elbrecht, A., Toft, D. O., Schrader, W. T., and O'Malley, B. W., 1987, Structure–function properties of the chicken progesterone receptor A synthesized from complementary deoxyribonucleic acid, *Mol. Endocrinol.* **1:**791–801.

109. Conneely, O. M., Maxwell, B. L., Toft, D. O., Schrader, W. T., and O'Malley, B. W., 1987, The A and B forms of the chicken progesterone receptor arise by alternate initiation of translation of a unique mRNA, *Biochem. Biophys. Res. Commun.* **149:**493–501.

110. Buller, R. E., Schwartz, R. J., Schrader, W. T., and O'Malley, B. W., 1976, Progesterone-binding components of chick oviduct. *In vitro* effect of receptor subunits on gene transcription, *J. Biol. Chem.* **251:**5178–5186.

111. Loosfelt, H., Logeat, F., Vu Hai, M. T., and Milgrom, E., 1984, The rabbit progesterone receptor. Evidence for a single steroid-binding subunit and characterization of receptor mRNA, *J. Biol. Chem.* **259:**14196–14202.

112. Logeat, F., Pamphile, P., Loosfelt, H., Jolivet, A., Fournier, A., and Milgrom, E., 1985, One-step immunoaffinity purification of active progesterone receptor. Further evidence in favor of the existence of a single steroid-binding unit, *Biochemistry* **24:**1029–1035.

113. Lamb, D. J., Kima, P. E., and Bullock, D. W., 1986, Evidence for a single steroid-binding protein in the rabbit progesterone receptor, *Biochemistry* **25:**6319–6324.

114. Tienrungroj, W., Meshinchi, S., Sanchez, E. R., Pratt, S. E., Grippo, J. F., Holmgren, A., and Pratt, W. B., 1987, The role of sulfhydryl groups in permitting transformation and DNA binding of the glucocorticoid receptor, *J. Biol Chem.* **262**:6992–7000.

115. Pratt, W. B., 1987, Transformation of glucocorticoid and progesterone receptors to the DNA-binding state, *J. Cell. Biochem.* **35**:51–68.

116. Marver, D., Goodman, D., and Edelman, I. S., 1972, Relationships between renal cytoplasmic and nuclear aldosterone-receptors, *Kidney Int.* **1**:210–223.

117. Rousseau, G. G., Baxter, J. D., Higgins, S. J., and Tomkins, G. M., 1973, Steroid-induced nuclear binding of glucocorticoid receptors in intact hepatoma cells, *J. Mol. Biol.* **79**:539–554.

118. Munck, A., and Foley, R., 1979, Activation of steroid hormone–receptor complexes in intact target cells in physiological conditions, *Nature* **278**:752–754.

119. Jensen, E. V., 1979, Interaction of steroid hormones with the nucleus, *Pharmacol. Rev.* **30**:477–491.

120. Little, M., Szendro, P. I., and Jungblut, P. W., 1973, Hormone-mediated dimerization of microsomal estradiol receptor, *Hoppe-Seylers Z. Physiol. Chem.* **354**:1599–1610.

121. Little, M., Szendro, P., Teran, C., Hughes, A., and Jungblut, P. W., 1975, Biosynthesis and transformation of microsomal and cytosol estradiol receptors, *J. Steroid Biochem.* **6**:493–500.

122. Notides, A. C., and Nielsen, S., 1974, The molecular mechanism of the *in vitro* 4 S to 5 S transformation of the uterine estrogen receptor, *J. Biol. Chem.* **249**:1866–1873.

123. Notides, A. C., Hamilton, D. E., and Auer, H. E., 1975, A kinetic analysis of the estrogen receptor transformation, *J. Biol. Chem.* **250**:3945–3950.

124. Notides, A. C., Lerner, N., and Hamilton, D. E., 1981, Positive cooperativity of the estrogen receptor, *Proc. Natl. Acad. Sci. U.S.A.* **78**:4926–4930.

125. Miller, M. A., Mullick, A., Greene, G. L., and Katzenellenbogen, B. S., 1985, Characterization of the subunit nature of nuclear estrogen receptors by chemical cross-linking and dense amino acid labeling, *Endocrinology* **117**:515–522.

126. Kumar, V., and Chambon, P., 1988, The estrogen receptor binds tightly to its responsive element as a ligand-induced homodimer, *Cell* **55**:145–156.

127. Tsai, S. Y., Carlstedt-Duke, J., Weigel, N. L., Dahlman, K., Gustafsson, J.-Å., Tsai, M.-J., and O'Malley, B. W., 1988, Molecular interactions of steroid hormone receptor with its enhancer element: Evidence for receptor dimer formation, *Cell* **55**:361–369.

128. Guiochon-Mantel, A., Loosfelt, H., Lescop, P., Sar, S., Atger, M., Perrot-Applanat, M., and Milgrom, E., 1989, Mechanisms of nuclear localization of the progesterone receptor: Evidence for interaction between monomers, *Cell* **57**:1147–1154.

129. Milgrom, E., 1981, Activation of steroid–receptor complexes, in: *Biochemical Actions of Hormones,* Volume 8 (G. Litwack, ed.), Academic Press, New York, pp. 465–492.

130. Grody, W. W., Schrader, W. T., and O'Malley, B. W., 1982, Activation, transformation, and subunit structure of steroid hormone receptors, *Endocrine Rev.* **3**:141–163.

131. Puca, G. A., Abbondanza, C., Nigro, V., Armetta, I., Medici, N., and Molinari, A. M., 1986, Estradiol receptor has proteolytic activity that is responsible for its own transformation, *Proc. Natl. Acad. Sci. U.S.A.* **83**:5367–5371.

132. DeSombre, E. R., Mohla, S., and Jensen, E. V., 1972, Estrogen-independent activation of the receptor protein of calf uterine cytosol, *Biochem. Biophys. Res. Commun.* **48**:1601–1608.

133. Sato, B., Nishizawa, Y., Noma, K., Matsumoto, K., and Yamamura, Y., 1979, Estrogen-independent nuclear binding of receptor protein of rat uterine cytosol by removal of low molecular weight inhibitor, *Endocrinology* **104**:1474–1479.

134. Weichman, B. M., and Notides, A. C., 1977, Estradiol-binding kinetics of the activated and nonactivated estrogen receptor, *J. Biol. Chem.* **252**:8856–8862.

135. Thomas, T., and Kiang, D. T., 1986, Ribonuclease-induced transformation of progesterone receptor from rabbit uterus, *J. Steroid Biochem.* **24**:505–511.

136. Yang, C.-R., Mešter, J., Wolfson, A., Renoir, J.-M., and Baulieu, E.-E., 1982, Activation of chick oviduct progesterone receptor by heparin in the presence or absence of hormone, *Biochem. J.* **208**:399–406.

137. Singh, V. B., Eliezer, N., and Moudgil, V. K., 1986, Transformation and phosphorylation of purified molybdate-stabilized chicken oviduct progesterone receptor, *Biochim. Biophys. Acta* **888**:237–248.

138. Cake, M. H., and Goidl, J. A., Parchman, L. G., and Litwack, G., 1976, Involvement of a low molecular weight component(s) in the mechanism of action of the glucocorticoid receptor, *Biochem. Biophys. Res. Commun.* **71**:45–52.

139. Goidl, J. A., Cake, M. H., Dolan, K. P., Parchman, L. G., and Litwack, G., 1977, Activation of the rat liver glucocorticoid–receptor complex, *Biochemistry* **16**:2125–2130.

140. Bailly, A., Sallas, N., and Milgrom, E., 1977, A low molecular weight inhibitor of steroid receptor activation, *J. Biol. Chem.* **252**:858–863.

141. Sekula, B. C., Schmidt, T. J., and Litwack, G., 1981, Redefinition of modulator as an inhibitor of glucocorticoid receptor activation, *J. Steroid Biochem.* **14**:161–166.

142. Leach, K. L., Grippo, J. F., Housley, P. R., Dahmer, M. K., Salive, M. E., and Pratt, W. B., 1982, Characteristics of an endogenous glucocorticoid receptor stabilizing factor, *J. Biol. Chem.* **257**:381–388.

143. Dahmer, M. K., Housley, P. R., and Pratt, W. B., 1984, Effects of molybdate and endogenous inhibitors on steroid–receptor inactivation, transformation and translocation, *Annu. Rev. Physiol.* **46**:67–81.

144. Nielsen, C. J., Sando, J. J., Vogel, W. M., and Pratt, W. B., 1977, Glucocorticoid receptor inactivation under cell-free conditions, *J. Biol. Chem.* **252**:7568–7578.

145. Grody, W. W., Compton, J. G., Schrader, W. T., and O'Malley, B. W., 1980, Inactivation of chick oviduct progesterone receptors, *J. Steroid Biochem.* **12**:115–120.

146. Leach, K. L., Dahmer, M. K., Hammond, N. D., Sando, J. J., and Pratt, W. B. 1979, Molybdate inhibition of glucocorticoid receptor inactivation and transformation, *J. Biol. Chem.* **254**:11884–11890.

147. Toft, D., and Nishigori, H., 1979, Stabilization of the avian progesterone receptor by inhibitors, *J. Steroid Biochem.* **11**:413–416.

148. Cake, M. H., DiSorbo, D. M., and Litwack, G., 1978, Effect of pyridoxal phosphate on the DNA binding site of activated hepatic glucocorticoid receptor, *J. Biol. Chem.* **253**:4886–4891.

149. Müller, R. E., Traish, A., and Wotiz, H. H., 1980, Effects of pyridoxal 5'-phosphate on uterine estrogen receptor I. Inhibition of nuclear binding in cell-free system and intact uterus, *J. Biol. Chem.* **255**:4062–4067.

150. Nishigori, H., and Toft, D., 1979, Chemical modification of avian progesterone receptor by pyridoxal 5'-phosphate, *J. Biol. Chem.* **254**:9155–9161.

151. Traish, A., Müller, R. E., and Wotiz, H. H., 1980, Effects of pyridoxal 5'-phosphate on uterine estrogen receptor II. Inhibition of estrogen-receptor transformation, *J. Biol. Chem.* **255**:4068–4072.

152. Peleg, S., Schrader, W. T., and O'Malley, B. W., 1988, Sulfhydryl group content of chicken progesterone receptor: Effect of oxidation on DNA binding activity, *Biochemistry* **27**:358–367.

153. Kalimi, M., and Love, K., 1980, Role of chemical reagents in the activation of rat hepatic glucocorticoid–receptor complex, *J. Biol. Chem.* **255**:4687–4690.

154. Auricchio, F., 1989, Phosphorylation of steroid receptors, *J. Steroid Biochem.* **32**:613–622.

155. Bailly, A., LaFevre, B., Savouret, J.-F., and Milgrom, E., 1980, Activation and changes in sedimentation properties of steroid receptors, *J. Biol. Chem.* **255**:2729–2734.

156. Müller, R. E., Traish, A. M., and Wotiz, H. H., 1983, Estrogen receptor activation precedes transformation. Effects of ionic strength, temperature, and molybdate, *J. Biol. Chem.* **258**:9227–9236.

157. Joab, I., Radanyi, C., Renoir, M., Buchou, T., Catelli, M.-G., Binart, N., Mester, J., and Baulieu, E.-E., 1984, Common non-hormone binding component in non-transformed chick oviduct receptors of four steroid hormones, *Nature* **308**:850–853.

158. Dougherty, J. J., Puri, R. K., and Toft, D. O., 1984, Polypeptide components of two 8 S forms of chicken oviduct progesterone receptor, *J. Biol. Chem.* **259**:8004–8009.

159. Schuh, S., Yonemoto, W., Brugge, J., Bauer, V. J., Riehl, R. M., Sullivan, W. P., and Toft, D. O., 1985, A 90,000-dalton binding protein common to both steroid receptors and the Rous sarcoma virus transforming protein pp60$^{v\text{-}src}$, *J. Biol. Chem.* **260**:14292–14296.

160. Housley, P. R., Sanchez, E. R., Westphal, H. M., Beato, M., and Pratt, W. B., 1985, The molybdate-stabilized L-cell glucocorticoid receptor isolated by affinity chromatography or with a monoclonal antibody is associated with a 90–92-kDa nonsteroid-binding phosphoprotein, *J. Biol. Chem.* **260**:13810–13817.

161. Denis, M., Wikström, A.-C., and Gustafsson, J.-Å., 1987, The molybdate-stabilized nonactivated glucocorticoid receptor contains a dimer of M_r90,000 non-hormone-binding protein, *J. Biol. Chem.* **262**:11803–11806.

162. Radanyi, C., Renoir, J.-M., Sabbah, M., and Baulieu, E.-E., 1989, Chicken heat-shock protein of M_r = 90,000, free or released from progesterone receptor, is in a dimeric form, *J. Biol. Chem.* **264**:2568–2573.

163. Mendel, D. B., Bodwell, J. E., Gametchu, B., Harrison, R. W., and Munck, A., 1986, Molybdate-stabilized nonactivated glucocorticoid–receptor complexes contain a 90-kDa non-steroid-binding phosphoprotein that is lost on activation, *J. Biol. Chem.* **261**:3758–3763.

164. Bresnick, E. H., Dalman, F. C., Sanchez, E. R., and Pratt, W. B., 1989, Evidence that the 90-kDa heat shock protein is necessary for the steroid-binding conformation of the L cell glucocorticoid receptor, *J. Biol. Chem.* **264**:4992–4997.

165. Catelli, M. G., Binart, N., Jung-Testas, I., Renoir, J. M., Baulieu, E. E., Feramisco, J. R., and Welch, W. J., 1985, The common 90-kD protein component of non-transformed '8 S' steroid receptors is a heat shock protein, *EMBO J.* **4**:3131–3135.

166. Sanchez, E. R., Toft, D. O., Schlesinger, M. J., and Pratt, W. B., 1985, Evidence that the 90-kDa phosphoprotein associated with the untransformed L-cell glucocorticoid receptor is a murine heat shock protein, *J. Biol. Chem.* **260**:12398–12401.

167. Sanchez, E. R., Meshinchi, S., Tienrungroj, W., Schlesinger, M. J., Toft, D. O., and Pratt, W. B., 1987, Relationship of the 90-kDa murine heat shock protein to the untransformed and transformed states of the L cell glucocorticoid receptor, *J. Biol. Chem.* **262**:6986–6991.

168. Baulieu, E.-E., Binart, N., Cadepond, F., Catelli, M. G., Chambraud, B., Garnier, J., Gasc, J. M., Groyer-Schweizer, G., Oblin, M. E., Radanyi, C., Redeuilh, G., Renoir, J. M., and Sabbah, M., 1989, Do receptor-associated nuclear proteins explain earliest steps of steroid hormone function? in: *The Steroid/Thyroid Hormone Receptor Family and Gene Regulation* (J.-Å. Gustafsson, H. Eriksson, and J. Carlstedt-Duke, eds.), Birkhäuser Verlag, Basel, pp. 301–318.

169. Pratt, W. B., Sanchez, E. R., Bresnick, E. H., Meshinchi, S., Scherrer, L. C., Dalman, F. C., and Welsh, M. J., 1989, Interaction of the glucocorticoid receptor with the M_r90,000 heat shock protein: An evolving model of ligand-mediated receptor transformation and translocation, *Cancer Res.* **49**:2222s–2229s.

170. Denis, M., and Gustafsson, J.-Å., 1989, The $M_r \approx$ 90,000 heat shock protein: An important modulator of ligand and DNA-binding properties of the glucocorticoid receptor, *Cancer Res.* **49**:2275s–2281s.

171. Milgrom, E., and Atger, M., 1975, Receptor translocation inhibitor and apparent saturability of the nuclear acceptor, *J. Steroid Biochem.* **6**:487–492.

172. Simons, S. S., Jr., Martinez, H. M., Garcea, R. L., Baxter, J. D., and Tomkins, G. M., 1976, Interactions of glucocorticoid receptor–steroid complexes with acceptor sites, *J. Biol. Chem.* **251**:334–343.

173. Liu, S.-L. H., and Webb, T. E., 1977, Elevated concentration of a dexamethasone–receptor translocation inhibitor in Novikoff hepatoma cells, *Cancer Res.* **37**:1763–1767.

174. Atger, M., and Milgrom, E., 1978, Interaction of glucocorticoid–receptor complexes with rat liver nuclei, *Biochim. Biophys. Acta* **539**:41–53.

175. Taira, M., and Terayama, H., 1978, Comparison of corticoid receptor and other cytoplasmic factors among liver and hepatoma cell lines with different sensitivity to corticoid inhibition of cell growth, *Biochim. Biophys. Acta* **541**:45–58.

176. Isohashi, F., Terada, M., Tsukanaka, K., Nakanishi, Y., and Sakamoto, Y., 1980, A low-molecular-weight translocation modulator and its interaction with a macromolecular translocation inhibitor of the activated receptor–glucocorticoid complex, *J. Biochem.* **88:**775–781.

177. Horiuchi, M., Isohashi, F., Terada, M., Okamoto, K., and Sakamoto, Y., 1981, Interaction of ATP with a macromolecular translocation inhibitor of the nuclear binding of "activated" receptor–glucocorticoid complex, *Biochem. Biophys. Res. Commun.* **98:**88–94.

178. Unger, A. L., Uppaluri, R., Ahern, S., Colby, J. L., and Tymoczko, J. L., 1988, Isolation of ribonucleic acid from the unactivated rat liver glucocorticoid receptor, *Mol. Endocrinol.* **2:**952–958.

179. Greene, G. L., Gilna, P., and Kushner, P., 1989, Estrogen and progesterone receptor analysis and action in breast cancer, in: *Immunological Approaches to the Diagnosis and Therapy of Breast Cancer II* (R. L. Ceriani, ed.), Plenum Press, New York, pp. 119–129.

180. Katzenellenbogen, J. A., Carlson, K. E., Heiman, D. F., Robertson, D. W., Wei, L. L., and Katzenellenbogen, B. S., 1983, Efficient and highly selective covalent labeling of the estrogen receptor with [3H]tamoxifen aziridine, *J. Biol. Chem.* **258:**3487–3495.

181. Sakai, D., and Gorski, J., 1984, Reversible denaturation of the estrogen receptor and determination of polypeptide chain molecular weight, *Endocrinology* **115:**2379–2383.

182. Linkie, D. M., and Siiteri, P. K., 1978, A re-examination of the interaction of estradiol with target cell receptors, *J. Steroid Biochem.* **9:**1071–1078.

183. Walters, M. R., 1985, Steroid hormone receptors and the nucleus, *Endocrine Rev.* **6:**512–542.

184. Jungblut, P. W., Kallweit, E., Sierralta, W., Truitt, A. J., and Wagner, R. K., 1978, The occurrence of steroid-free, "activated" estrogen receptor in target cell nuclei, *Hoppe-Seylers Z. Physiol. Chem.* **359:**1259–1268.

185. Carlson, R. A., and Gorski, J., 1980, Characterization of a unique population of unfilled estrogen-binding sites associated with the nuclear fraction of immature rat uteri, *Endocrinology* **106:**1776–1785.

186. Levy, C., Mortel, R., Eychenne, B., Robel, P., and Baulieu, E.-E., 1980, Unoccupied nuclear oestradiol-receptor sites in normal human endometrium, *Biochem. J.* **185:**733–738.

187. Geier, A., Beery, R., Levran, D., Menczer, Y., and Lunenfeld, B., 1980, Unoccupied nuclear receptors for estrogen in human endometrial tissue, *J. Clin. Endocrinol. Metab.* **50:**541–545.

188. Zava, D. T., and McGuire, W. L., 1977, Estrogen receptor: Unoccupied sites in nuclei of a breast tumor cell line, *J. Biol. Chem.* **252:**3703–3708.

189. Panko, W. B., and MacLeod, R. M., 1978, Uncharged nuclear receptors for estrogen in breast cancers, *Cancer Res.* **38:**1948–1951.

190. Geier, A., Haimsohn, M., and Lunenfeld, B., 1982, Evidence for the origin of the unoccupied oestrogen receptor in nuclei of a breast-cancer cell line (MCF-7), *Biochem. J.* **202:**687–691.

191. Sonnenschein, C., Soto, A. M., Colofiore, J., and Farookhi, R., 1976, Estrogen target cells: Establishment of a cell line derived from the rat pituitary tumor MtT/F$_4$, *Exp. Cell Res.* **101:**15–22.

192. Clark, C. R., MacLusky, N. J., and Naftolin, F., 1982, Unfilled nuclear oestrogen receptors in the rat brain and pituitary gland, *J. Endocrinol.* **93:**327–338.

193. Mester, J., and Baulieu, E.-E., 1972, Nuclear estrogen receptor of chick liver, *Biochim. Biophys. Acta* **261:**236–244.

194. Ozon, R., and Belle, R., 1973, Recepteurs d'oestradiol-17β dans le foie de poule et de l'amhibien *Discoglossus pictus, Biochim. Biophys. Acta* **297:**155–163.

195. Callard, G. V., and Mak, P., 1985, Exclusive nuclear location of estrogen receptors in *Squalus* testis, *Proc. Natl. Acad. Sci. U.S.A.* **82:** 1336–1340.

196. Edwards, D. P., Martin, P. M., Horwitz, K. B., Chamness, G. C., and McGuire, W. L., 1980, Subcellular compartmentalization of estrogen receptors in human breast cancer cells, *Exp. Cell Res.* **127:**197–213.

197. Wolfson, A. J., 1984, Non-activated progesterone receptor extracted from nuclei of hen oviduct, *J. Steroid Biochem.* **21:**519–521.

198. Shao, T.-C., Castañeda, E., Rosenfield, R. L., and Liao, S., 1975, Selective retention and formation of a Δ⁵-androstenediol–receptor complex in cell nuclei of the rat vagina, *J. Biol. Chem.* **250:**3095–3100.

199. Pasqualini, J. R., Sumida, C., and Gelly, C., 1972, Mineralocorticosteroid receptors in the foetal compartment, *J. Steroid Biochem.* **3:**543–556.

200. Lawson, D. E. M., and Wilson, P. W., 1974, Intranuclear localization and receptor proteins for 1,25-dihydroxycholecalciferol in chick intestine, *Biochem. J.* **144:**573–583.

201. Kream, B. E., Reynolds, R. D., Knutson, J. C., Eisman, J. A., and DeLuca, H. F., 1976, Intestinal cytosol binders of 1,25-dihydroxyvitamin D_3 and 25-hydroxyvitamin D_3, *Arch. Biochem. Biophys.* **176:**779–787.

202. Yund, M. A., King, D. S., and Fristrom, J. W., 1978, Ecdysteroid receptors in imaginal discs of *Drosophila melanogaster*, *Proc. Natl. Acad. Sci. U.S.A.* **75:**6039–6043.

203. Williams, D., and Gorski, J., 1972, Kinetic and equilibrium analysis of estradiol in uterus: A model of binding-site distribution in uterine cells, *Proc. Natl. Acad. Sci. U.S.A.* **69:**3464–3468.

204. Sheridan, P. J., 1975, Is there an alternative to the cytoplasmic receptor model for the mechanism of action of steroids? *Life Sci.* **17:**497–502.

205. Sheridan, P. J., Buchanan, J. M., Anselmo, V. C., and Martin, P. M., 1979, Equilibrium: The intracellular distribution of steroid receptors, *Nature* **282:**579–582.

206. Martin, P. M., and Sheridan, P. J., 1982, Towards a new model for the mechanism of action of steroids, *J. Steroid Biochem.* **16:**215–229.

207. Sheridan, P. J., Buchanan, J. M., Anselmo, V. C., and Martin, P. M., 1981, Unbound progesterone receptors are in equilibrium between the nucleus and cytoplasm in cells of the rat uterus, *Endocrinology* **108:**1533–1537.

208. Gasc, J.-M., Ennis, B. W., Baulieu, E.-E., and Stumpf, W. E., 1983, Récepteur de la progestérone dans l'oviducte de poulet: Double révélation par immunohistochimie avec des anticorps antirécepteur et par autoradiographie à l'aide d'un progestagène tritié, *C.R. Acad. Sci. [D] (Paris)* **297:**477–482.

209. Ennis, B. W., Stumpf, W. E., Gasc, J.-M., and Baulieu, E.-E., 1986, Nuclear localization of progesterone receptor before and after exposure to progestin at low and high temperatures: Autoradiographic and immunohistochemical studies of chick oviduct, *Endocrinology* **119:**2066–2075.

210. Walters, M. R., Hunziker, W., Konami, D., and Norman, A. W., 1982, Factors affecting the distribution and stability of unoccupied 1,25-dihydroxyvitamin D_3 receptors, *J. Receptor Res.* **2:**331–346.

211. Walters, M. R., Hunziker, W., and Norman, A. W., 1980, Unoccupied 1,25-dihydroxyvitamin D_3 receptors: Nuclear/cytosol ratio depends on ionic strength, *J. Biol. Chem.* **255:**6799–6805.

212. Walters, M. R., Hunziker, W., and Norman, A. W., 1981, 1,25-Dihydroxyvitamin D_3 receptors: Intermediates between triiodothyronine and steroid hormone receptors, *Trends Biochem. Sci.* **6:**268–271.

213. Oppenheimer, J. H., Koerner, D., Schwartz. H. L., and Surks, M. I., 1972, Specific nuclear triiodothyronine binding sites in rat liver and kidney, *J. Clin. Endocrinol. Metab.* **35:**330–333.

214. DeGroot, L. J., and Strausser, J. L., 1974, Binding of T_3 in rat liver nuclei, *Endocrinology* **95:**74–83.

215. Morel, G., Dubois, P., Benassayag, C., Nunez, E., Radanyi, C., Redeuilh, G., Richard-Foy, H., and Baulieu, E.-E., 1981, Ultrastructural evidence of oestradiol receptor by immunochemistry, *Exp. Cell Res.* **132:**249–257.

216. Raam, S., Nemeth, E., Tamura, H., O'Briain, D. S., and Cohen, J. L., 1982, Immunohistochemical localization of estrogen receptors in human mammary carcinoma using antibodies to the receptor protein, *Eur. J. Cancer Clin. Oncol.* **18:**1–12.

217. Jensen, E. V., Greene, G. L., Closs, L. E., DeSombre, E. R., and Nadji, M., 1982, Receptors reconsidered: A 20-year perspective, *Recent Prog. Horm. Res.* **38:**1–40.

218. Greene, G. L., Sobel, N. B., King, W. J., and Jensen, E. V., 1984, Immunochemical studies of estrogen receptors, *J. Steroid Biochem.* **20:**51–56.

219. King, W. J., and Greene, G. L., 1984, Monoclonal antibodies localize oestrogen receptor in the nuclei of target cells, *Nature* 307:745–747.

220. King, W. J., DeSombre, E. R., Jensen, E. V., and Greene, G. L., 1985, Comparison of immunocytochemical and steroid-binding assays for estrogen receptor in human breast tumors, *Cancer Res.* 45:293–304.

221. Pertschuk, L. P., Eisenberg, K. B., Carter, A. C., and Feldman, J. G., 1985, Immunohistologic localization of estrogen receptors in breast cancer with monoclonal antibodies, *Cancer* 55:1513–1518.

222. McCarty, K. S., Jr., Szabo, E., Flowers, J. L., Cox, E. B., Leight, G. S., Miller, L., Konrath, J., Soper, J. T., Budwit, D. A., Creasman, W. T., Seigler, H. F., and McCarty, K. S., Sr., 1986, Use of a monoclonal anti-estrogen receptor antibody in the immunohistochemical evaluation of human tumors, *Cancer Res.* 46:4244s–4248s.

223. DeSombre, E. R., Thorpe, S. M., Rose, C., Blough, R. R., Andersen, K. W., Rasmussen, B. B., and King, W. J., 1986, Prognostic usefulness of estrogen receptor immunocytochemical assays for human breast cancer, *Cancer Res.* 46:4256s–4264s.

224. Charpin, C., Martin, P.-M., Jacquemier, J., Lavaut, M. N., Pourreau-Schneider, N., and Toga, M., 1986, Estrogen receptor immunocytochemical assay (ER-ICA): Computerized image analysis system, immunoelectron microscopy, and comparisons with estradiol binding assays in 115 breast carcinomas, *Cancer Res.* 46:4271s–4277s.

225. Jonat, W., Maass, H., and Stegner, H. E., 1986, Immunohistochemical measurement of estrogen receptors in breast cancer tissue samples, *Cancer Res.* 46:4296s–4298s.

226. Ozzello, L., De Rosa, C. M., Konrath, J. G., Yeager, J. L., and Miller, L. S., 1986, Detection of estrophilin in frozen sections of breast cancers using an estrogen receptor immunocytochemical assay, *Cancer Res.* 46:4303s–4307s.

227. Press, M. F., and Greene, G. L., 1984, An immunocytochemical method for demonstrating estrogen receptor in human uterus using monoclonal antibodies to human estrophilin, *Lab. Invest.* 50:480–486.

228. McClellan, M. C., West, N. B., Tacha, D. E., Greene, G. L., and Brenner, R. M., 1984, Immunocytochemical localization of estrogen receptors in the macaque reproductive tract with monoclonal antiestrophilins, *Endocrinology* 114:2002–2014.

229. Sar, M., and Parikh, I., 1986, Immunohistochemical localization of estrogen receptor in rat brain, pituitary and uterus with monoclonal antibodies, *J. Steroid Biochem.* 24:497–503.

230. Fuxe, K., Cintra, A., Agnati, L. F., Härfstrand, A., Wikström, A.-C., Okret, S., Zoli, M., Miller, L. S., Greene, G. L., and Gustafsson, J. Å., 1987, Studies on the cellular localization and distribution of glucocorticoid receptor and estrogen receptor immunoreactivity in the central nervous system of the rat and their relationship to the monoaminergic and peptidergic neurons of the brain, *J. Steroid Biochem.* 27:159–170.

231. Press, M. F., Holt, J. A., Herbst, A. L., and Greene, G. L., 1985, Immunocytochemical identification of estrogen receptor in ovarian carcinomas. Localization with monoclonal estrophilin antibodies compared with biochemical assays. *Lab. Invest.* 53:349–361.

232. Shimada, A., Kimura, S., Abe, K., Nagasaki, K., Adachi, I., Yamaguchi, K., Suzuki, M., Nakajima, T., and Miller, L. S., 1985, Immunocytochemical staining of estrogen receptor in paraffin sections of human breast cancer by use of monoclonal antibody: Comparison with that in frozen sections, *Proc. Natl. Acad. Sci. U.S.A.* 82:4803–4807.

233. Poulsen, H. S., Ozzello, L., King, W. J., and Greene, G. L., 1985, The use of monoclonal antibodies to estrogen receptors (ER) for immunoperoxidase detection of ER in paraffin sections of human breast cancer tissue, *J. Histochem. Cytochem.* 33:87–92.

234. De Rosa, C. M., Ozzello, L., Greene, G. L., and Habif, D. V., 1987, Immunostaining of estrogen receptor in paraffin sections of breast carcinomas using monoclonal antibody D75P3γ: Effects of fixation, *Am. J. Surg. Pathol.* 11:943–950.

235. Nadji, M., and Morales, A. R., 1986, *Immunoperoxidase Techniques: A Practical Approach to Tumor Diagnosis,* American Society of Clinical Pathologists Press, Chicago, pp. 76, 172, 173.

236. Marchetti, E., Querzoli, P., Moncharmont, B., Parikh, I., Bagni, A., Marzola, A., Fabris, G., and Nenci, I., 1987, Immunocytochemical demonstration of estrogen receptors by monoclonal antibodies in human breast cancer: Correlation with estrogen receptor assay by dextran-coated charcoal method, *Cancer Res.* **47**:2508–2513.

237. Gasc, J.-M., Renoir, J. M., Radanyi, C., Joab, I., Tuohimaa, P., and Baulieu, E.-E., 1984, Progesterone receptor in the chick oviduct: An immunohistochemical study with antibodies to distinct receptor components, *J. Cell Biol.* **99**:1193–1201.

238. Perrot-Applanat, M., Logeat, F., Groyer-Picard, M. T., and Milgrom, E., 1985, Immunocytochemical study of mammalian progesterone receptor using monoclonal antibodies, *Endocrinology* **116**:1473–1484.

239. Press, M. F., and Greene, G. L., 1988, Localization of progesterone receptor with monoclonal antibodies to the human progestin receptor, *Endocrinology* **122**:1165–1175.

240. Perrot-Applanat, M., Groyer-Picard, M.-T., Lorenzo, F., Jolivet, A., Vu Hai, M. T., Pallud, C., Spyratos, F., and Milgrom, E., 1987, Immunocytochemical study with monoclonal antibodies to progesterone receptor in human breast tumors, *Cancer Res.* **47**:2652–2661.

241. Pertschuk, L. P., Feldman, J. G., Eisenberg, K. B., Carter, A. C., Thelmo, W. L., Cruz, W. P., Thorpe, S. M., Christensen, J., Rasmussen, B. B., Rose, C., and Greene, G. L., 1988, Immunocytochemical detection of progesterone receptor in breast cancer with monoclonal antibody, *Cancer* **62**:342–349.

242. Press, M. F., Nousek-Goebl, N. A., and Greene, G. L., 1985, Immunoelectron microscopic localization of estrogen receptor with monoclonal estrophilin antibodies, *J. Histochem. Cytochem.* **33**:915–924.

243. Perrot-Applanat, M., Groyer-Picard, M.-T., Logeat, F., and Milgrom, E., 1986, Ultrastructural localization of progesterone receptor by an immunogold method: Effect of hormone administration, *J. Cell. Biol.* **102**:1191–1199.

244. Welshons, W. V., Lieberman, M. E., and Gorski, J., 1984, Nuclear localization of unoccupied oestrogen receptors, *Nature* **307**:747–749.

245. Welshons, W. V., Krummel, B. M., and Gorski, J., 1985, Nuclear localization of unoccupied receptors for glucocorticoids, estrogens and progesterone in GH$_3$ cells, *Endocrinology* **117**:2140–2147.

246. Carter, S. B., 1967, Effects of cytochalasins on mammalian cells, *Nature* **213**:261–264.

247. Gravanis, A., and Gurpide, E., 1986, Enucleation of human endometrial cells: Nucleo-cytoplasmic distribution of DNA polymerase α and estrogen receptor, *J. Steroid Biochem.* **24**:469–474.

248. Skafar, D. F., and Notides, A. C., 1985, Modulation of the estrogen receptor's affinity for estradiol, *J. Biol. Chem.* **260**:12208–12213.

249. Parikh, I., Rajendran, K. G., Su, J.-L., Lopez, T., and Sar, M., 1987, Are estrogen receptors cytoplasmic or nuclear? Some immunocytochemical and biochemical studies, *J. Steroid Biochem.* **27**:185–192.

250. Berthois, Y., Katzenellenbogen, J. A., and Katzenellenbogen, B. S., 1986, Phenol red in tissue culture media is a weak estrogen: Implications concerning the study of estrogen-responsive cells in culture, *Proc. Natl. Acad. Sci. U.S.A.* **83**:2496–2500.

251. Raam, S., Richardson, G. S., Bradley, F., MacLaughlin, D., Sun, L., Frankel, F., and Cohen, J. L., 1983, Translocation of cytoplasmic estrogen receptors to the nucleus: Immunohistochemical demonstration utilizing rabbit antibodies to estrogen receptors of mammary carcinomas, *Breast Cancer Res. Treat.* **3**:179–199.

252. Morel, G., Dubois, P., Gustafsson, J.-Å., Radojcic, M., Radanyi, C., Renoir, M., and Baulieu, E.-E., 1984, Ultrastructural evidence of progesterone receptor by immunochemistry, *Exp. Cell Res.* **155**:283–288.

253. Walters, S. N., Reinhardt, T. A., Dominick, M. A., Horst, R. L., and Littledike, E. T., 1986, Intracellular location of unoccupied 1,25-dihydroxyvitamin D receptors: A nuclear–cytoplasmic equilibrium, *Arch. Biochem. Biophys.* **246**:366–373.

254. Papamichail, M., Tsokos, G., Tsawdaroglou, N., and Sekeris, C. E., 1980, Immunocytochemical

demonstration of glucocorticoid receptors in different cell types and their translocation from the cytoplasm to the cell nucleus in the presence of dexamethasone, *Exp. Cell Res.* **125**:490–493.

255. Govindan, M. V., 1980, Immunofluorescence microscopy of the intracellular translocation of glucocorticoid–receptor complexes in rat hepatoma (HTC) cells, *Exp. Cell Res.* **127**:293–297.

256. Bernard, P. A., and Joh, T. H., 1984, Characterization and immunochemical demonstration of glucocorticoid receptor using antisera specific to transformed receptor, *Arch. Biochem. Biophys.* **229**:466–476.

257. Antakly, T., and Eisen, H. J., 1984, Immunocytochemical localization of glucocorticoid receptor in target cells, *Endocrinology* **115**:1984–1989.

258. Gasc, J.-M., and Baulieu, E.-E., 1987, From the structure of steroid receptors to their assessment by immunocytochemistry in target cells, *J. Steroid Biochem.* **27**:177–184.

259. Fuxe, K., Wikström, A.-C., Okret, S., Agnati, L. F., Härfstrand, A., Yu, Z.-Y., Granholm, L., Zoli, M., Vale, W., and Gustafsson, J.-Å., 1985, Mapping of glucocorticoid receptor immunoreactive neurons in the rat tel- and diencephalon using a monoclonal antibody against rat liver glucocorticoid receptor, *Endocrinology* **117**:1803–1812.

260. Teasdale, J., Lewis, F. A., Barrett, I. D., Abbott, A. C., Wharton, J., and Bird, C. C., 1986, Immunocytochemical application of monoclonal antibodies to rat liver glucocorticoid receptor, *J. Pathol.* **150**:227–237.

261. Wikström, A.-C., Bakke, O., Okret, S., Brönnegård, M., and Gustafsson, J.-Å., 1987, Intracellular localization of the glucocorticoid receptor: Evidence for cytoplasmic and nuclear localization, *Endocrinology* **120**:1232–1242.

262. Antakly, T., Thompson, E. B., and O'Donnell, D., 1989, Demonstration of intracellular localization and up-regulation of glucocorticoid receptor by *in situ* hybridization and immunocytochemistry, *Cancer Res.* **49**:2230s–2234s.

263. Picard, D., and Yamamoto, K. R., 1987, Two signals mediate hormone-dependent nuclear localization of the glucocorticoid receptor, *EMBO J.* **6**:3333–3340.

264. Haukkamaa, M., 1987, Membrane-associated steroid hormone receptors, in: *Steroid Hormone Receptors: Their Intracellular Localisation* (C. R. Clark, ed.), Ellis Horwood, Chichester, pp. 155–169.

265. Pietras, R. J., and Szego, C. M., 1979, Estrogen receptors in uterine plasma membrane. *J. Steroid Biochem.* **11**:1471–1483.

266. Berthois, Y., Pourreau-Schneider, N., Gandilhon, P., Mittre, H., Tubiana, N., and Martin, P. M., 1986, Estradiol membrane binding sites on human breast cancer cell lines. Use of a fluorescent estradiol conjugate to demonstrate plasma membrane binding systems, *J. Steroid. Biochem.* **25**:963–972.

267. Ishikawa, K., Hanaoka, Y., Kondo, Y., and Imai, K., 1977, Primary action of steroid hormone at the surface of amphibian oocyte in the induction of germinal vesicle breakdown, *Mol. Cell. Endocrinol.* **9**:91–100.

268. Godeau, J. F., Schorderet-Slatkine, S., Hubert, P., and Baulieu, E.-E., 1978, Induction of maturation in *Xenopus laevis* oocytes by a steroid linked to a polymer, *Proc. Natl. Acad. Sci. U.S.A.* **75**:2353–2357.

269. Sadler, S. E., and Maller, J. L., 1982, Identification of a steroid receptor on the surface of *Xenopus* oocytes by photoaffinity labeling, *J. Biol. Chem.* **257**:355–361.

270. Pietras, R. J., and Szego, C. M., 1980, Partial purification and characterization of oestrogen receptors in subfractions of hepatocyte plasma membranes, *Biochem. J.* **191**:743–760.

271. Little, M., Rosenfeld, G. C., and Jungblut, P. W., 1972, Cytoplasmic estradiol "receptors" associated with the "microsomal" fraction of pig uterus, *Hoppe-Seylers Z. Physiol. Chem.* **353**:231–242.

272. Walters, M. R., Hunziker, W., and Norman, A. W., 1981, A mathematical model describing the subcellular localization of non-membrane bound steroid, secosteroid and thyronine receptors, *J. Steroid Biochem.* **15**:491–495.

273. Clark, C. R., 1984, The cellular distribution of steroid hormone receptors: Have we got it right? *Trends Biochem. Sci.* **9**:207–208.
274. Schrader, W. T., 1984, New model for steroid hormone receptors? *Nature* **308**:17–18.
275. Jensen, E. V., 1984, Intracellular localization of estrogen receptors: Implications for interaction mechanism, *Lab. Invest.* **51**:487–488.
276. Gustafsson, J.-Å., Carlstedt-Duke, J., Poellinger, L., Okret, S., Wikström, A.-C., Brönnegård, M., Gillner, M., Dong, Y., Fuxe, K., Cintra, A., Härfstrand, A., and Agnati, L., 1987, Biochemistry, molecular biology, and physiology of the glucocorticoid receptor, *Endocrine Rev.* **8**:185–234.
277. Gorski, J., Welshons, W., and Sakai, D., 1984, Remodeling the estrogen receptor model, *Mol. Cell. Endocrinol.* **36**:11–15.
278. Gorski, J., Welshons, W. V., Sakai, D., Hansen, J., Walent, J., Kassis, J., Shull, J., Stack, G., and Campen, C., 1986, Evolution of a model of estrogen action, *Recent Prog. Horm. Res.* **42**:297–329.
279. Gorski, J., and Hansen, J. C., 1987, The "one and only" step model of estrogen action, *Steroids* **49**:461–475.
280. Clark, J. H., Markaverich, B., Upchurch, S., Eriksson, H., Hardin, J. W., and Peck, E. J., Jr., 1980, Heterogeneity of estrogen binding sites: Relationship to estrogen receptors and estrogen responses, *Recent Prog. Horm. Res.* **36**:89–134.
281. Munck, A., and Brinck-Johnsen, T., 1968, Specific and nonspecific physiocochemical interactions of glucocorticoids and related steroids with rat thymus cells *in vitro*, *J. Biol. Chem.* **243**:5556–5565.
282. Bell, P. A., and Munck, A., 1973, Steroid-binding properties and stabilization of cytoplasmic glucocorticoid receptors from rat thymus cells, *Biochem. J.* **136**:97–107.
283. Sloman, J. C., and Bell, P. A., 1976, The dependence of specific nuclear binding of glucocorticoids by rat thymus cells on cellular ATP levels, *Biochim. Biophys. Acta* **428**:403–413.
284. Ishii, D. N., Pratt, W. B., and Aronow, L., 1972, Steady-state level of the specific glucocorticoid binding component in mouse fibroblasts, *Biochemistry* **11**:3896–3904.
285. Munck, A., Wira, C., Young, D. A., Mosher, K. M., Hallahan, C., and Bell, P. A., 1972, Glucocorticoid–receptor complexes and the earliest steps in the action of glucocorticoids on thymus cells, *J. Steroid Biochem.* **3**:567–578.
286. Munck, A., and Holbrook, N. J., 1984, Glucocorticoid–receptor complexes in rat thymus cells. Rapid kinetic behavior and a cyclic model, *J. Biol. Chem.* **259**:820–831.
287. Mendel, D. B., Bodwell, J. E., and Munck, A., 1986, Glucocorticoid receptors lacking hormone-binding activity are bound in nuclei of ATP-depleted cells, *Nature* **324**:478–480.
288. Liao, S., and Fang, S., 1969, Receptor proteins for androgens and the mode of action of androgens on gene transcription in ventral prostate, *Vitam. Horm.* **27**:17–90.
289. Liao, S., Tymoczko, J. L., Castañeda, E., and Liang, T., 1975, Androgen receptors and androgen-dependent initiation of protein synthesis in the prostate, *Vitam. Horm.* **33**:297–317.
290. Liao, S., Rossini, G. P., Hiipakka, R. A., and Chen, C., 1980, Factors that can control the interaction of the androgen–receptor complex with the genomic structure in the rat prostate, in: *Perspectives in Steroid Receptor Research* (F. Bresciani, ed.), Raven Press, New York, pp. 99–112.
291. Auricchio, F., and Migliaccio, A., 1980, *In vitro* inactivation of oestrogen receptor by nuclei, *FEBS Lett.* **117**:224–226.
292. Auricchio, F., Migliaccio, A., and Rotondi, A., 1981, Inactivation of oestrogen receptor *in vitro* by nuclear dephosphorylation, *Biochem. J.* **194**:569–574.
293. Nielsen, C. J., Sando, J. J., and Pratt, W. B., 1977, Evidence that dephosphorylation inactivates glucocorticoid receptors, *Proc. Natl. Acad. Sci. U.S.A.* **74**:1398–1402.
294. Sando, J. J., La Forest, A. C., and Pratt, W. B., 1979, ATP-dependent activation of L cell glucocorticoid receptors to the steroid binding form, *J. Biol. Chem.* **254**:4772–4778.

295. Sando, J. J., Hammond, N. D., Stratford, C. A., and Pratt, W. B., 1979, Activation of thymocyte glucocorticoid receptors to the steroid-binding form. The roles of reducing agents, ATP, and heat-stable factors, *J. Biol. Chem.* **254**:4779–4789.

296. Auricchio, F., Migliaccio, A., Castoria, G., Lastoria, S., and Schiavone, E., 1981, ATP-dependent enzyme activating hormone binding of estradiol receptor, *Biochem. Biophys. Res. Commun.* **101**:1171–1178.

297. Fleming, H., Blumenthal, R., and Gurpide, E., 1982, Effects of cyclic nucleotides on estradiol binding in human endometrium, *Endocrinology* **111**:1671–1677.

298. Fleming, H., Blumenthal, R., and Gurpide, E., 1983, Rapid changes in specific estrogen binding elicited by cGMP or cAMP in cytosol from human endometrial cells, *Proc. Natl. Acad. Sci. U.S.A.* **80**:2486–2490.

299. Weigel, N. L., Tash, J. S., Means, A. R., Schrader, W. T., and O'Malley, B. W., 1981, Phosphorylation of hen progesterone receptor by cAMP dependent protein kinase, *Biochem. Biophys. Res. Commun.* **102**:513–519.

300. Migliaccio, A., Lastoria, S., Moncharmont, B., Rotondi, A., and Auricchio, F., 1982, Phosphorylation of calf uterus 17β-estradiol receptor by endogenous Ca^{2+}-stimulated kinase activating the hormone binding of the receptor, *Biochem. Biophys. Res. Commun.* **109**:1002–1010.

301. Migliaccio, A., Rotondi, A., and Auricchio, F., 1984, Calmodulin-stimulated phosphorylation of 17β-estradiol receptor on tyrosine, *Proc. Natl. Acad. Sci. U.S.A.* **81**:5921–5925.

302. Garcia, T., Tuohimaa, P., Mešter, J., Buchou, T., Renoir, J.-M., and Baulieu, E.-E., 1983, Protein kinase activity of purified components of the chicken oviduct progesterone receptor, *Biochem. Biophys. Res. Commun.* **113**:960–966.

303. Garcia, T., Buchou, T., Renoir, J.-M., Mešter, J., and Baulieu, E.-E., 1986, A protein kinase copurified with chick oviduct progesterone receptor, *Biochemistry* **25**:7937–7942.

304. Dougherty, J. J., Puri, R. K., and Toft, D. O., 1982, Phosphorylation *in vivo* of chicken oviduct progesterone receptor, *J. Biol. Chem.* **257**:14226–14230.

305. Puri, R. K., Dougherty, J. J., and Toft, D. O., 1984, The avian progesterone receptor: Isolation and characterization of phosphorylated forms, *J. Steroid Biochem.* **20**:23–29.

306. Puri, R. K., and Toft, D. O., 1986, Peptide mapping of the avian progesterone receptor, *J. Biol. Chem.* **261**:5651–5657.

307. Denner, L. A., Bingman, W. E. III, Greene, G. L., and Weigel, N. L., 1987, Phosphorylation of the chicken progesterone receptor, *J. Steroid Biochem.* **27**:235–243.

308. Sullivan, W. P., Madden, B. J., McCormick, D. J., and Toft, D. O., 1988, Hormone-dependent phosphorylation of the avian progesterone receptor, *J. Biol. Chem.* **263**:14717–14723.

309. Sullivan, W. P., Smith, D. F., Beito, T. G., Krco, C. J., and Toft, D. O., 1988, Hormone-dependent processing of the avian progesterone receptor, *J. Cell. Biochem.* **36**:103–119.

310. Logeat, F., Le Cunff, M., Pamphile, R., and Milgrom, E., 1985, The nuclear-bound form of the progesterone receptor is generated through a hormone-dependent phosphorylation, *Biochem. Biophys. Res. Commun.* **131**:421–427.

311. Garcia, T., Jung-Testas, I., and Baulieu, E.-E., 1986, Tightly bound nuclear progesterone receptor in not phosphorylated in primary chick oviduct cultures, *Proc. Natl. Acad. Sci. U.S.A.* **83**:7573–7577.

312. Garcia, T., Buchou, T., Jung-Testas, I., Renoir, J.-M., and Baulieu, E. E., 1987, Chick oviduct progesterone receptor phosphorylation: Characterization of a copurified kinase and phosphorylation in primary cultures, *J. Steroid Biochem.* **27**:227–234.

313. Rao, K. V. S., Peralta, W. D., Greene, G. L., and Fox, C. F., 1987, Cellular progesterone receptor phosphorylation in response to ligands activating protein kinases, *Biochem. Biophys. Res. Commun.* **146**:1357–1365.

314. Wei, L. L., Sheridan, P. L., Krett, N. L., Francis, M. D., Toft, D. O., Edwards, D. P., and Horwitz, K. B., 1987, Immunologic analysis of human breast cancer progesterone receptors. 2. Structure, phosphorylation, and processing, *Biochemistry* **26**:6262–6272.

315. Sheridan, P. L., Evans, R. M., and Horwitz, K. B., 1989, Phosphotryptic peptide analysis of human progesterone receptors. New phosphorylated sites formed in nuclei after hormone treatment, *J. Biol. Chem.* **264:**6520–6528.

316. Sheridan, P. L., Krett, N. L., Gordon, J. A., and Horwitz, K. B., 1988, Human progesterone receptor transformation and nuclear down-regulation are independent of phosphorylation, *Mol. Endocrinol.* **2:**1329–1342.

317. Migliaccio, A., Rotondi, A., and Auricchio, F., 1986, Estradiol receptor: Phosphorylation on tyrosine in uterus and interaction with anti-phosphotyrosine antibody, *EMBO J.* **5:**2867–2872.

318. Auricchio, F., Migliaccio, A., Castoria, G., Rotondi, A., Di Domenico, M., Pagano, M., and Nola, E., 1987, Phosphorylation on tyrosine of oestradiol-17β receptor in uterus and interaction of oestradiol-17β and glucocorticoid receptors with antiphosphotyrosine antibodies, *J. Steroid Biochem.* **27:**245–253.

319. Sato, B., Miyashita, Y., Maeda, Y., Noma, K., Kishimoto, S., and Matsumoto, K., 1987, Effects of estrogen and vanadate on the proliferation of newly established transformed mouse Leydig cell line *in vitro, Endocrinology* **120:**1112–1120.

320. Woo, D. D. L., Fay, S. P., Griest, R., Coty, W., Goldfine, I., and Fox, C. F., 1986, Differential phosphorylation of the progesterone receptor by insulin, epidermal growth factor, and platelet-derived growth factor receptor tyrosine protein kinases, *J. Biol. Chem.* **261:**460–467.

321. Auricchio, F., Migliaccio, A., Di Domenico, M., and Nola, E., 1987, Oestradiol stimulates tyrosine phosphorylation and hormone binding activity of its own receptor in a cell-free system, *EMBO J.* **6:**2923–2929.

322. Denton, R. R., and Notides, A. C., 1989, The nuclear form of the estrogen receptor is dephosphorylated, in: *Abstracts 71st Meeting,* The Endocrine Society, Seattle, p. 159.

323. Kumar, V., Green, S., Staub, A., and Chambon, P., 1986, Localisation of the oestradiol-binding and putative DNA-binding domains of the human oestrogen receptor, *EMBO J.* **9:**2231–2236.

324. Green, S., Walter, P., Kumar, V., Krust, A., Bornert, J.-M., Argos, P., and Chambon, P., 1986, Human oestrogen receptor cDNA: sequence, expression and homology to v-*erb*-A, *Nature* **320:**134–139.

325. Migliaccio, A., Di Domenico, M., Green, S., de Falco, A., Kajtaniak, E. L., Blasi, F., Chambon, P., and Auricchio, F., 1989, Phosphorylation on tyrosine of *in vitro* synthesized human estrogen receptor activates its hormone binding, *Mol. Endocrinol.* **3:**1061–1069.

326. Eul, J., Meyer, M. E., Tora, L., Bocquel, M. T., Quirin-Stricker, C., Chambon, P., and Gronemeyer, H., 1989, Expression of active hormone and DNA-binding domains of the chicken progesterone receptor in *E. coli, EMBO J.* **8:**83–90.

327. Hollenberg, S. M., Weinberger, C., Ong, E. S., Cerelli, G., Oro, A., Lebo, R., Thompson, E. B., Rosenfeld, M. G., and Evans, R. M., 1985, Primary structure and expression of a functional human glucocorticoid receptor cDNA, *Nature* **318:**635–641.

328. Auricchio, F., Migliaccio, A., Castoria, G., Lastoria, S., and Rotondi, A., 1982, Evidence that *in vivo* estradiol receptor translocated into nuclei is dephosphorylated and released into cytoplasm, *Biochem. Biophys. Res. Commun.* **106:**149–157.

329. Mendel, D. B., Bodwell, J. E., and Munck, A., 1987, Activation of cytosolic glucocorticoid-receptor complexes in intact WEHI-7 cells does not dephosphorylate the steroid-binding protein, *J. Biol. Chem.* **262:**5644–5648.

330. Orti, E., Mendel, D. B., and Munck, A., 1989, Phosphorylation of glucocorticoid receptor-associated and free forms of the ~90-kDa heat shock protein before and after receptor activation, *J. Biol. Chem.* **264:**231–237.

331. Tienrungroj, W., Sanchez, E. R., Housley, P. R., Harrison, R. W., and Pratt, W. B., 1987, Glucocorticoid receptor phosphorylation, transformation, and DNA binding, *J. Biol. Chem.* **262:**17342–17349.

332. Orti, E., Mendel, D. B., Smith, L. I., and Munck, A., 1989, Agonist-dependent phosphorylation and nuclear dephosphorylation of glucocorticoid receptors in intact cells, *J. Biol. Chem.* **264:**9728–9731.

333. Greene, G. L., Closs, L. E., Fleming, H., DeSombre, E. R., and Jensen, E. V., 1977, Antibodies to estrogen receptor: Immunochemical similarity of estrophilin from various mammalian species, *Proc. Natl. Acad. Sci. U.S.A.* **74**:3681–3685.
334. Greene, G. L., Closs, L. E., DeSombre, E. R., and Jensen, E. V., 1979, Antibodies to estrophilin: Comparison between rabbit and goat antisera, *J. Steroid Biochem.* **11**:333–341.
335. Radanyi, C., Redeuilh, G., Eigenmann, E., Lebeau, M.-C., Massol, N., Secco, C., Baulieu, E.-E., and Richard-Foy, H., 1979, Production et détection d'anticorps antirécepteur de l'oestradiol d'utérus de veau. Interaction avec le récepteur d'oviducte de poule, *C.R. Acad. Sci. [D] (Paris)* **288**:255–258.
336. Coffer, A. I., King, R. J. B., and Brockas, A. J., 1980, Antibodies to human myometrial oestrogen receptor, *Biochem. Int.* **1**:126–132.
337. Raam, S., Peters, L., Rafkind, I., Putnum, E., Longcope, C., and Cohen, J. L., 1981, Simple methods for production and characterization of rabbit antibodies to human breast tumor estrogen receptors, *Mol. Immunol.* **18**:143–156.
338. Köhler, G., and Milstein, C., 1975, Continuous cultures of fused cells secreting antibody of predefined specificity, *Nature* **256**:495–497.
339. Greene, G. L., Fitch, F. W., and Jensen, E. V., 1980, Monoclonal antibodies to estrophilin: Probes for the study of estrogen receptors, *Proc. Natl. Acad. Sci. U.S.A.* **77**:157–161.
340. Greene, G. L., Nolan, C., Engler, J. P., and Jensen, E. V., 1980, Monoclonal antibodies to human estrogen receptor, *Proc. Natl. Acad. Sci. U.S.A.* **77**:5115–5119.
341. Moncharmont, B., Su, J.-L., and Parikh, I., 1982, Monoclonal antibodies against estrogen receptor: Interaction with different molecular forms and functions of the receptor, *Biochemistry* **21**:6916–6921.
342. Martin, P. M., Berthois, Y., and Jensen, E. V., 1988, Binding of antiestrogens exposes an occult antigenic determinant in the human estrogen receptor, *Proc. Natl. Acad. Sci. U.S.A.* **85**:2533–2537.
343. Borgna, J.-L., Fauque, J., and Rochefort, H., 1984, A monoclonal antibody to the estrogen receptor discriminates between the nonactivated and activated estrogen- and anti-estrogen-receptor complexes, *Biochemistry* **23**:2162–2168.
344. Moncharmont, B., and Parikh, I., 1983, Binding of monoclonal antibodies to the nuclear estrogen receptor in intact nuclei, *Biochem. Biophys. Res. Commun.* **114**:107–112.
345. Logeat, F., Vu Hai, M. T., and Milgrom, E., 1981, Antibodies to rabbit progesterone receptor: Crossreaction with human receptor, *Proc. Natl. Acad. Sci. U.S.A.* **78**:1426–1430.
346. Feil, P. D., 1983, Characterization of guinea pig anti-progestin receptor antiserum, *Endocrinology* **112**:396–398.
347. Renoir, J.-M., Radanyi, C., Yang, C.-R., and Baulieu, E.-E., 1982, Antibodies against progesterone receptor from chick oviduct. Cross-reactivity with mammalian progesterone receptors, *Eur. J. Biochem.* **127**:81–86.
348. Tuohimaa, P., Renoir, J.-M., Radanyi, C., Mešter, J., Joab, I., Buchou, T., and Baulieu, E.-E., 1984, Antibodies against highly purified B-subunit of the chick oviduct progesterone receptor, *Biochem. Biophys. Res. Commun.* **119**:433–439.
349. Birnbaumer, M., Hinrichs-Rosello, M. V., Cook, R. G., Schrader, W. T., and O'Malley, B. W., 1987, Chemical and antigenic properties of pure 108,000 molecular weight chick progesterone receptor, *Mol. Endocrinol.* **1**:249–259.
350. Radanyi, C., Joab, I., Renoir, J.-M., Richard-Foy, H., and Baulieu, E.-E., 1983, Monoclonal antibody to chicken oviduct progesterone receptor. *Proc. Natl. Acad. Sci. U.S.A.* **80**:2854–2858.
351. Sullivan, W. P., Vroman, B. T., Bauer, V. J., Puri, R. K., Riehl, R. M., Pearson, G. R., and Toft, D. O., 1985, Isolation of steroid receptor binding protein from chicken oviduct and production of monoclonal antibodies, *Biochemistry* **24**:4214–4222.
352. Logeat, F., Vu Hai, M. T., Fournier, A., Legrain, P., Buttin, G., and Milgrom, E., 1983, Monoclonal antibodies to rabbit progesterone receptor: Crossreaction with other mammalian progesterone receptors, *Proc. Natl. Acad. Sci. U.S.A.* **80**:6456–6459.

353. Nakao, K., Myers, J. E., and Faber, L. E., 1985, Development of a monoclonal antibody to the rabbit 8.5 S uterine progestin receptor, *Can. J. Biochem. Cell Biol.* **63**:33–40.
354. Lorenzo, F., Jolivet, A., Loosfelt, H., Vu Hai, M. T., Brailly, S., Perrot-Applanat, M., and Milgrom, E., 1988, A rapid method of epitope mapping, *Eur. J. Biochem.* **176**:53–60.
355. Dicker, P. D., Tsai, S. Y., Weigel, N. L., Tsai, M.-J., Schrader, W. T., and O'Malley, B. W., 1984, Monoclonal antibody to the hen oviduct progesterone receptor produced following *in vitro* immunization, *J. Steroid Biochem.* **20**:43–50.
356. Sullivan, W. P., Beito, T. G., Proper, J., Krco, C. J., and Toft, D. O., 1986, Preparation of monoclonal antibodies to the avian progesterone receptor, *Endocrinology* **119**:1549–1557.
357. Stewart, H. J., Jones, D. S. C., Pascall, J. C., Popkin, R. M., and Flint, A. P. F., 1988, The contribution of recombinant DNA techniques to reproductive biology, *J. Reprod. Fert.* **83**:1–57.
358. Miesfeld, R., Okret, S., Wickström, A.-C., Wrange, Ö., Gustafsson, J.-Å., and Yamamoto, K. R., 1984, Characterization of a steroid hormone receptor gene and mRNA in wild-type and mutant cells, *Nature* **312**:779–781.
359. Weinberger, C., Hollenberg, S. M., Ong, E. S., Harmon, J. M., Brower, S. T., Cidlowski, J., Thompson, E. B., Rosenfeld, M. G., and Evans, R. M., 1985, Identification of human glucocorticoid receptor complementary DNA clones by epitope selection, *Science* **228**:740–742.
360. Govindan, M. V., Devic, M., Green, S., Gronemeyer, H., and Chambon, P., 1985, Cloning of the human glucocorticoid receptor cDNA, *Nucleic Acids Res.* **13**:8293–8304.
361. Miesfeld, R., Rusconi, S., Godowski, P. J., Maler, B. A., Okret, S., Wickström, A.-C., Gustafsson, J.-Å., and Yamamoto, K. R., 1986, Genetic complementation of a glucocorticoid receptor deficiency by expression of cloned receptor cDNA, *Cell* **46**:389–399.
362. Walter, P., Green, S., Greene, G., Krust, A., Bornert, J.-M., Jeltsch, J.-M., Staub, A., Jensen, E., Scrace, G., Waterfield, M., and Chambon, P., 1985, Cloning of the human estrogen receptor cDNA, *Proc. Natl. Acad. Sci. U.S.A.* **82**:7889–7893.
363. Greene, G. L., Gilna, P., Waterfield, M., Baker, A., Hort, Y., and Shine, J., 1986, Sequence and expression of human estrogen receptor complementary DNA, *Science* **231**:1150–1154.
364. Krust, A., Green, S., Argos, P., Kumar, V., Walter, P., Bornert, J.-M., and Chambon, P., 1986, The chicken oestrogen receptor sequence: Homology with v-*erb*A and the human oestrogen and glucocorticoid receptors. *EMBO J.* **5**:891–897.
365. Maxwell, B. L., McDonnell, D. P., Conneely, O. M., Schultz, T. Z., Greene, G. L., and O'Malley, B. W., 1987, Structural organization and regulation of the chicken estrogen receptor, *Mol. Endocrinol.* **1**:25–35.
366. Weiler, I. J., Lew, D., and Shapiro, D. J., 1987, The *Xenopus laevis* estrogen receptor: Sequence homology with human and avian receptors and identification of multiple estrogen receptor messenger ribonucleic acids, *Mol. Endocrinol.* **1**:355–362.
367. Jeltsch, J. M., Krozowski, Z., Quirin-Stricker, C., Gronemeyer, H., Simpson, R. J., Garnier, J. M., Krust, A., Jacob, F., and Chambon, P., 1986, Cloning of the chicken progesterone receptor, *Proc. Natl. Acad. Sci. U.S.A.* **83**:5424–5428.
368. Conneely, O. M., Sullivan, W. P., Toft, D. O., Birnbaumer, M., Cook, R. G., Maxwell, B. L., Zarucki-Schulz, T., Greene, G. L., Schrader, W. T., and O'Malley, B. W., 1986, Molecular cloning of the chicken progesterone receptor, *Science* **233**:767–770.
369. Loosfelt, H., Atger, M., Misrahi, M., Guiochon-Mantel, A., Meriel, C., Logeat, F., Benarous, R., and Milgrom, E., 1986, Cloning and sequence analysis of rabbit progesterone-receptor complementary DNA, *Proc. Natl. Acad. Sci. U.S.A.* **83**:9045–9049.
370. Evans, R. M., 1988, The steroid and thyroid hormone receptor superfamily, *Science* **240**:889–895.
371. Chang, C., Kokontis, J., and Liao, S., 1988, Molecular cloning of human and rat complementary DNA encoding androgen receptors, *Science* **240**:324–326.
372. Lubahn, D. B., Joseph, D. R., Sullivan, P. M., Willard, H. F., French, F. S., and Wilson, E. M., 1988, Cloning of human androgen receptor complementary DNA and localization to the X chromosome, *Science* **240**:327–330.
373. Green, S., and Chambon, P., 1986, A superfamily of potentially oncogenic hormone receptors, *Nature* **324**:615–617.

374. Weinberger, C., Hollenberg, S. M., Rosenfeld, M. G., and Evans, R. M., 1985, Domain structure of human glucocorticoid receptor and its relationship to the v-*erb*-A oncogene product, *Nature* 318:670–672.

375. Sap, J., Muñoz, A., Damm, K., Goldberg, Y., Ghysdael, J., Leutz, A., Beug, H., and Vennström, B., 1986, The c-*erb*-A protein is a high affinity receptor for thyroid hormone, *Nature* 324:635–640.

376. Weinberger, C., Thompson, C. C., Ong, E. S., Lebo, R., Gruol, D. J., and Evans, R. M., 1986, The c-*erb*-A gene encodes a thyroid hormone receptor, *Nature* 324:641–646.

377. Beato, M., 1989, Gene regulation by steroid hormones, *Cell* 56:335–344.

378. Kumar, V., Green, S., Stack, G., Berry, M., Jin, J.-R., and Chambon, P., 1987, Functional domains of the human estrogen receptor, *Cell* 51:941–951.

379. Dobson, A. D. W., Conneely, O. M., Beattie, W., Maxwell, B. L., Mak, P., Tsai, M.-J., Schrader, W. T., and O'Malley, B. W., 1989, Mutational analysis of the chicken progesterone receptor, *J. Biol. Chem.* 264:4207–4211.

380. Carlstedt-Duke, J., Okret, S., Wrange, Ö., and Gustafsson, J.-Å., 1982, Immunochemical analysis of the glucocorticoid receptor: Identification of a third domain separate from the steroid-binding and DNA-binding domains, *Proc. Natl. Acad. Sci. U.S.A.* 79:4260–4264.

381. Westphal, H. M., Moldenhauer, G., and Beato, M., 1982, Monoclonal antibodies to the rat liver glucocorticoid receptor, *EMBO J.* 1:1467–1471.

382. Berg, J. M., 1989, DNA binding specificity of steroid receptors, *Cell* 57:1065–1068.

383. Klock, G., Strähle, U., and Schütz, G., 1987, Oestrogen and glucocorticoid responsive elements are closely related but distinct, *Nature* 329:734–736.

384. Green, S., and Chambon, P., Oestradiol induction of a glucocorticoid-responsive gene by a chimaeric receptor, *Nature* 325:75–78.

385. Schauer, M., Chalepakis, G., Willmann, T., and Beato, M., 1989, Binding of hormone accelerates the kinetics of glucocorticoid and progesterone receptor binding to DNA, *Proc. Natl. Acad. Sci. U.S.A.* 86:1123–1127.

386. Tsai, S. Y., Tsai, M.-J., and O'Malley, B. W., 1989, Cooperative binding of steroid hormone receptors contributes to transcriptional synergism at target enhancer elements, *Cell* 57:443–448.

387. Evans, R. M., and Hollenberg, S. M., 1988, Zinc fingers: Gilt by association, *Cell* 52:1–3.

388. Freedman, L. P., Luisi, B. F., Korzun, Z. R., Basavappa, R., Sigler, P. B., and Yamamoto, K. R., 1988, The function and structure of the metal coordination sites within the glucocorticoid receptor DNA binding domain, *Nature* 334:543–546.

389. Danielsen, M., Hinck, L., and Ringold, G. M., 1989, Two amino acids within the knuckle of the first zinc finger specify DNA response element activation by the glucocorticoid receptor, *Cell* 57:1131–1138.

390. Klug, A., and Rhodes, D., 1987, 'Zinc fingers': A novel protein motif for nucleic acid recognition, *Trends Biochem. Sci.* 12:464–469.

391. Sabbah, M., Redeuilh, G., Secco, C., and Baulieu, E.-E., 1987, The binding activity of estrogen receptor to DNA and heat shock protein (M_r90,000) is dependent on receptor-bound metal, *J. Biol. Chem.* 262:8631–8635.

392. Lannigan, D. A., and Notides, A. C., 1989, Estrogen receptor selectively binds the "coding strand" of an estrogen responsive element, *Proc. Natl. Acad. Sci. U.S.A.* 86:863–867.

393. Schüle, R., Muller, M., Kaltschmidt, C., and Renkawitz, R., 1988, Many transcription factors interact synergystically with steroid receptors, *Science* 242:1418–1420.

394. Meyer, M.-E., Gronemeyer, H., Turcotte, B., Bocquel, M.-T., Tasset, D., and Chambon, P., 1989, Steroid hormone receptors compete for factors that mediate their enhancer function, *Cell* 57:433–442.

395. Wang, L.-H., Tsai, S. Y., Sagami, I., Tsai, M.-J., and O'Malley, B. W., 1987, Purification and characterization of chicken ovalbumin upstream promoter transcription factor from HeLa cells, *J. Biol. Chem.* 262:16080–16086.

396. Wang, L.-H., Tsai, S. Y., Cook, R. G., Beattie, W. G., Tsai, M.-J., and O'Malley, B. W., 1989, COUP transcription factor is member of the steroid receptor superfamily, *Nature* 340:163–166.

11

Oxytocin in the Initiation of Labor

Rosemary D. Leake

1. Synthesis and Secretion of Oxytocin

A single cell (magnocellular neuron) is responsible for synthesis, transport, and release of oxytocin. Oxytocin prohormone is synthesized in single neuronal cells of the supraoptic and paraventricular portions of the hypothalamus. The prohormone consists of the nonapeptide oxytocin, glycine-lysine-arginine, and a specific 93- to 95-amino-acid carrier protein, the neurophysin. The prohormone is synthesized in the rough endoplasmic reticulum of the neurosecretory neurons. It is then packaged into secretory granules for transport along the magnocellular neurons terminating in the posterior pituitary and median eminence. Collateral projections also extend to the brainstem and/or spinal cord. During axonal transport the prohormone is enzymatically converted to the nonapeptide hormone and its neurophysin. Ultimately, granules are stored in terminal bulbs in the posterior pituitary gland. Polarization of the neurosecretory neurons results in exocytic release of oxytocin and its neurophysin from the granules into the circulation.

Functionally, circulating oxytocin is uterotonic and affects milk ejection. Oxytocin is also found in the corpus luteum testis, and spinal cord; its function in these sites is unclear. The function of the oxytocin-specific neurophysin is also unclear. However, the oxytocin neurophysin, like the vasopressin-specific neurophysin, may be involved in the regulation of prolactin secretion.[1] This possibility is supported by the observation that prolactin immunoreactivity has been noted in the oxytocinergic portion of the magnocellular neurosecretory system.[2]

Rosemary D. Leake • Department of Pediatrics, School of Medicine, University of California Los Angeles, Harbor/UCLA Medical Center, Torrance, California 90509.

2. Oxytocin in Animal Parturition

Oxytocin has been known to be a powerful uterotonic since the description of its action by Sir Henry Dale in 1906.[3] This finding, coupled with the description by Ferguson[4] that cervical dilation in postpartum rabbits produces uterine contractions, suggested that maternal oxytocin secretion might initiate labor. Exogenous oxytocin infusion produces uterine contractions in women at term as well as in rats, pigs, mares, and guinea pigs when administered within a few hours of the onset of spontaneous labor.[5–8] Oxytocin infusions do not always produce human labor, however. Fetal oxytocin secretion also has been suggested to contribute to parturition, as the oxytocin content in the fetal pituitary decreases during birth in rats[5,9] and guinea pigs.[10]

Several studies have examined the possibility that the endogenous release of maternal or fetal oxytocin initiates labor. However, oxytocin levels measured by bioassay in plasma of maternal cows, horses, sheep, goats, and rabbits generally show no increase before or during the first stage of labor and increased levels only during fetal expulsion.[11,12] Circulating oxytocin levels measured by radioimmunoassay in the same species and in the pig also show increased maternal oxytocin levels only during fetal birth itself.[9–13]

Oxytocin secretion in ovine pregnancy, labor, and delivery has been studied extensively. Oxytocin plasma levels were measured serially in the perinatal period in the ovine fetus and maternal ewe utilizing chronically implanted vascular catheters.[18] Both maternal and fetal plasma oxytocin concentrations remain at basal levels throughout the third trimester including the 5 days prior to spontaneous labor. During the early stage of labor, oxytocin levels in maternal plasma do not differ from those of prelabor. At full cervical dilation there is a significant increase in maternal oxytocin compared to basal concentrations. Fetal plasma oxytocin levels increase only slightly during labor. Plasma oxytocin concentrations in newborn lambs are elevated for at least 15 min after delivery. Thus, the onset of ovine labor is not associated with increased maternal or fetal plasma oxytocin levels.

The increase in oxytocin plasma levels associated with fetal expulsion in many species may be secondary to vaginal/cervical distention or abdominal straining. As stated above, Ferguson[4] demonstrated that cervical distention in postpartum rabbits produces uterine contractions, and Summerlee[19] reported that abdominal straining in the laboring rat is associated with high-frequency firing of oxytocin neurons.

Interestingly, there appear to be times in the reproductive cycle of the sheep and goat when oxytocin secretion is suppressed in response to stimuli that would otherwise result in increased plasma oxytocin. Roberts and Share demonstrated a marked increase in oxytocin levels during uterine massage in lactating and cycling ewes but no response in the pregnant ewe.[20] Estrogen increased oxytocin release in cycling ewes, and progesterone inhibited oxytocin secretion.[21,22] Thus, differing responses during various times may result from changes in steroid production (during estrus, for example); the mechanism may well be related to steroid effects on the firing of the magnocellular neurons controlling oxytocin release.[23]

Mechanisms initiating labor in sheep appear to result from maturational changes in the maternal and fetal endocrine systems that do not involve oxytocin. There are hypothalamic and pituitary changes, producing adrenal stimulation and increased cortisol output. Placental enzymatic changes lead to a withdrawal of the progesterone block and to increased estrogen production. This series of events leads to prostaglandin release, stimulation of myometrial contractions, and the initiation of labor. As stated above, there is no increase in plasma oxytocin in either maternal or fetal ovine plasma until regular uterine contractions develop.

3. Maternal Oxytocin in Human Parturition

3.1. Plasma Oxytocin Levels

Circulating oxytocin levels were initially measured by bioassay. Hawkes[24] demonstrated significant oxytocin-like activity in cat pregnancy plasma. Using an early radioimmunoassay, Chard[25] was the first to detect oxytocin values in some samples of plasma of women in labor and in all cord blood samples examined. He postulated fetal secretion of oxytocin since umbilical arterial values were greater than umbilical venous levels.

There were additional early studies in human pregnancy reporting varying plasma levels of oxytocin, suggesting episodic secretion.[26] Later studies utilizing 1- to 15-min sampling protocols show no clear evidence of pulsatile secretion.[27,28] There is now general agreement that basal oxytocin levels do not vary among men, nonpregnant women, and pregnant women, generally approximating 1–5 μU/ml. Serial studies during human labor show an increase in plasma oxytocin only during the expulsive phase, to levels of 5–10 μU/ml.[28,29] Levels at this time approximate peak values in maternal plasma during infant suckling.[30] Neither manual nor mechanical breast stimulation increases plasma oxytocin levels during pregnancy, although both produce a vigorous response after delivery.[31,32] Moreover, mechanical breast stimulation evokes a significant increase in plasma oxytocin levels during the luteal but not the follicular phase of the menstrual cycle.[33] This suggests that similar to the sheep, oxytocin secretion in women is suppressed during certain phases of the reproductive cycle.

In addition to the nonapeptide form of oxytocin described above, there is recent evidence of a decapeptide form of circulating oxytocin. Studies utilizing high-pressure liquid chromatography demonstrate that this material is an extended form of oxytocin, oxytocin-glycine.[34] This novel peptide represents a portion of the oxytocin prohormone that in its entirety consists of an N-terminus oxytocin linked by glycine-lysine-arginine to a specific oxytocin neurophysin at the C-terminal end. Oxytocin-glycine appears to be a partially processed form of the oxytocin prohormone. Plasma levels of oxytocin-glycine have been reported to be elevated in estrogen-enriched states including human pregnancy,[35] in males given a single dose of estrogen,[36] and in human newborns.[37] The function of OT-Gly remains unclear.

It is not known, for example, if oxytocin-glycine acts as an oxytocic either directly or following possible degradation to oxytocin.

The identification of circulating OT-glycine in enriched-estrogen states clarified another series of observations. It had been known for some time that estrogen-stimulated neurophysins increase in women receiving oral contraceptives. In addition, in early reports estrogen administration appeared to produce an increase in oxytocin levels in parallel with estrogen-stimulated neurophysins.[38] More recent studies by the same authors report oxytocin levels to be similar to basal levels following estrogen administration. There are, however, significant increases in plasma levels of the extended form of oxytocin, oxytocin-glycine, under these conditions. It now appears that estrogen in the oral contraception preparation may release oxytocin-glycine rather than oxytocin. Early assays could not distinguish between these peptides.

3.2. Oxytocin Plasma Clearance Rate

Oxytocin clearance appears to occur largely in the kidney and to a lesser extent in the liver. Degradation by plasma enzymes does not appear to be an important pathway *in vivo,* even in the pregnant subject. Oxytocin plasma clearance rates are reported to be similar in men and in nonpregnant and pregnant women (21–27 ml/kg per min).[39] Oxytocin clearance rates in men and women in labor were also similar (approximately 17 ml/kg per min).[27] Moreover, the plasma half-life for oxytocin does not vary in pregnant and nonpregnant women, approximating 3 to 6 min.[40,41]

Studies of oxytocin metabolism in chronically catheterized ewes[42] showed basal oxytocin levels to be similar to that of the human. Oxytocin plasma clearance rate in the pregnant ewe and fetus is approximately 12 ml/kg per min. This value does not change when measured during oxytocin infusion rates of 6.4, 64, and 640 mU/kg per min.

3.3. Oxytocin in the Corpus Luteum

Wathef and Swann[43] first demonstrated that the corpus luteum of the cyclic ewe contains oxytocin. Concentrations of oxytocin in this organ decrease during early pregnancy[44]; this may facilitate luteal maintenance and may also play a role in reducing uterine contractility during pregnancy. Khan-Dawood[45] recently reported that oxytocin concentrations in the human corpus luteum are 100-fold less than that found in ovine corpus luteum. The role of luteal oxytocin and its possible secretion from this site remain unclear both in sheep and in women.

3.4. Cerebrospinal Fluid Oxytocin

Adult male rhesus monkeys exhibit a marked (three- to 12-fold) diurnal variation in CSF oxytocin levels.[46] Nonpregnant women and men also show this variation.[47] Elevated CSF levels appear to be linked to periods of light.[48] In both species there is a peak in cerebrospinal fluid oxytocin levels, for example, during the midday hours, with levels of cerebrospinal fluid oxytocin being higher than that of plasma. There is no diurnal rhythm for cerebrospinal fluid or plasma oxytocin in the cat, rat, or guinea pig.

There are no data supporting an effect of cerebrospinal fluid oxytocin in the initiation of labor. This possibility is unlikely, moreover, since it is known that there is an effective barrier for oxytocin and no evidence of exchange of oxytocin between blood and cerebrospinal fluid.[49]

3.5. Labor in Hypophysectomized Women

Hypophysectomy may result in diabetes insipidus; gestation in hypophysectomized women has been extensively reviewed.[50] Pregnancy for those with diabetes insipidus is generally uncomplicated, including the onset of labor, the delivery, and the ability to nurse. This is particularly interesting since oxytocin synthesis could be impaired if there are extensive lesions of the supraoptic and paraventricular nuclei. There is a report[51] describing a woman with a 23-year history of diabetes insipidus who delivered following the infusion of a quite minimal amount of oxytocin; presumably this represented increased uterine sensitivity. Thus, although exogenous oxytocin is clearly uterotonic, endogenous secretion of oxytocin appears to play a facilitative role in labor.

3.6. Labor Facilitation by Oxytocin

Although plasma levels of oxytocin do not appear to be elevated prior to the onset of labor, plasma oxytocin levels may not need to be increased markedly to play a facilitative role in labor. For example, interactions of prostaglandins and oxytocin are well known. Prostaglandins enhance myometrial sensitivity to oxytocin,[52] and prostaglandin inhibitors attenuate the effect of oxytocin on uterine contraction[53] (see Chapter 5). Increased biosynthesis of prostaglandins during the latter part of pregnancy and during labor may affect the uterine response to oxytocin.

Labor facilitation by oxytocin may occur by several other means as well. The possible role of increased numbers of oxytocin receptors prior to labor is reviewed in detail in Chapter 12. Moreover, in rat, sheep, guinea pig, and human pregnancy, myometrial gap junctions increase in frequency before, during, and after parturition. The increase in gap junction number immediately prior to the onset of parturition may lead to increased intracellular communication and increased frequency of uterine contractions in response to seemingly low levels of oxytocin. The physiological role of gap junctions and possible relationship to the onset of labor are outlined in Chapter 9 in this volume.

4. Special Considerations in Measuring Plasma Oxytocin

Blood-sampling schedules and sites could theoretically influence plasma oxytocin measurements. There appears to be no diurnal rhythm for plasma oxytocin. Sampling as frequently as every minute does not demonstrate marked variations; plasma oxytocin levels do not vary before, during, or following uterine contractions, for example.[28] The blood-sampling site (peripheral blood versus ovarian or central venous) does not significantly affect oxytocin levels.

Oxytocin levels measured by radioimmunoassay are known to be affected by peptidases in pregnancy plasma unless special techniques are employed. Cystine aminopeptidases (oxytocinase) capable of degrading oxytocin are secreted by trophoblasts into human maternal but not fetal plasma. The effect of these peptidases is largely inhibited by rapid acidification, icing, and the addition of an enzyme inhibitor such as phenanthrolene. However, no totally effective inhibitor of oxytocinase has been reported.

Although oxytocinase does not appear to have an important effect on oxytocin *in vivo,* the effect of oxytocinase on pregnancy plasma measurements can be significant. Pregnancy plasma enriched with synthetic oxytocin, for example, demonstrates an 85% reduction in oxytocin concentration during 1 hr of incubation at 37°C.[39] Plasma oxytocinase activity increases as human pregnancy advances[54] but does not appear to change immediately prior to labor or during various stages of labor.[55] There is no diurnal rhythm of oxytocinase activity.[56] Umbilical cord blood and plasma of men and nonpregnant women do not demonstrate degradative activity from oxytocinase. Moreover, the placentas of many animals including sheep, rabbits, and rats do not produce oxytocinase; basal levels of oxytocin are similar in these species to that of men, pregnant women, and nonpregnant women, suggesting that if oxytocinase activity has an effect on *in vivo* oxytocin metabolism, there are also concomitant effects on oxytocin release and/or metabolism.

The use of differing oxytocinase inhibitors could explain some of the marked variation in plasma oxytocin levels reported in human pregnancy plasma. Amico[27] initially reported that the addition of low concentrations of phenanthrolene to pregnancy plasma did not affect oxytocin measurements since values were the same as those with chilling alone. A more recent report[57] suggests that higher concentrations of phenanthrolene may produce a marked reduction in oxytocinase activity *in vitro.*

An extraction step is necessary to remove nonspecific factors interfering with the accuracy of the oxytocin radioimmunoassay and to improve assay sensitivity. Various techniques for extraction have been utilized, including use of acetone–ether, Fuller's earth, glass powder, and more recently octadecasilyl silica. These methods generally incorporate inhibition of oxytocinase in an acid medium, use of an absorbent for oxytocin, and recovery with an organic solvent, but there are slight differences in various laboratories.

Antibody specificity is also an important confounding factor in oxytocin measurements. Early reports of oxytocin plasma levels varied considerably, not only because of differences in oxytocinase inhibition and extraction steps but perhaps because some antisera measure both oxytocin and oxytocin-glycine. More recently, collaborative studies[58] conducted in several research laboratories (University of Pittsburgh School of Medicine; Cornell University Medical College; St. John's University, Jamaica, New York; Harbor–UCLA Medical Center; and the National Institutes for Medical Research in London, U.K) comparing bioassays and RIAs utilizing various antisera to measure tissue and plasma oxytocin levels reported a good correlation among these facilities. All report pregnancy plasma levels of oxytocin to be less than 3 μU/ml. Following extraction the sensitivity of the assay approximates 0.3 μU/ml per tube (0.7 μU/ml whole blood), and the intra- and interassay reproductibility 7% and 12%, respectively.

5. Placental Transfer of Oxytocin

Early studies suggested bidirectional placental transfer in the guinea pig.[59] Glatz,[42] utilizing chronic preparations of early third-trimester catheterized sheep, examined the possibility of transplacental passage of oxytocin by measuring paired fetal and maternal oxytocin levels. Oxytocin infusion to the maternal ewe produced no change in fetal oxytocin levels, indicating a lack of maternal-to-fetal transfer.

In human pregnancy, simultaneously obtained maternal and cord blood samples have been examined. Cord plasma oxytocin concentrations are similar to maternal levels at delivery in one report.[60] Studies by Chard,[25] Dawood,[61] and from our laboratory[62] measuring oxytocin levels simultaneously in umbilical arterial and umbilical venous plasma suggest that the fetus produces oxytocin, but there is no evidence that oxytocin crosses the placenta. The stimulus for fetal secretion of oxytocin is unclear.

6. Fetal Oxytocin in Labor

6.1. Fetal Oxytocin in Animal Parturition

Oxytocin content in the fetal pituitary decreases during birth in rats[5,9] and guinea pigs,[10] but it is not known whether this initiates birth in these species. The fetal sheep posterior pituitary appears to play no major role in the initiation of labor, although the fetal lamb (and neonates of nearly all species studied) shows transient elevations in plasma oxytocin levels at birth.

6.2. Fetal Oxytocin in Human Parturition

There appears to be increased fetal oxytocin secretion prior to or during human birth, since oxytocin levels in the umbilical artery exceed umbilical venous levels. Moreover, most authors report that cord blood oxytocin values are higher in deliveries following labor than after cesarian section without labor.[22] Following delivery there is a somewhat reduced level of oxytocin in peripheral venous blood of the newborn, but levels remain elevated over adult basal values throughout the first 4 days of life.[62] The stimulus for fetal and neonatal oxytocin secretion remains unclear but appears to continue beyond the immediate period following delivery.

7. Conclusion

Neither fetal nor maternal oxytocin appears to initiate human parturition, although increased end-organ responsiveness to oxytocin may well facilitate labor. Oxytocin increases appear to contribute to expulsion of the fetus. Certainly, postpartum maternal oxytocin secretion is associated with but not critical for lactation. Whether extended forms of oxytocin contribute to the initiation or completion of parturition is unknown.

References

1. Toubeau, G., Desclin, J., Parmentier, M., and Pasteels, J. L., 1979, Cellular localization of a prolactin-like antigen in the rat brain, *J. Endocrinol.* **83**:261–266.
2. Nagy, G., Mulchahey, J. J., Smyth, D. G., and Neill, J. D., 1988, The glycopeptide moiety of vasopressin-neurophysin precursor is neurohypophyseal prolactin releasing factor, *Biochem. Biophys. Res. Commun.* **151**:524–529.
3. Dale, H. H., 1906, On some physiologic actions of ergot, *J. Physiol (Lond.)* **34**:163–205.
4. Ferguson, J. K. W., 1941, A study of the motility of the intact uterus at term, *Surg. Gynecol. Obstet.* **73**:359–366.
5. Fuchs, A. R., and Poblete, V. F., Jr., 1970, Oxytocin and uterine function in pregnancy and parturition in rats, *Biol. Reprod.* **2**:387–400.
6. Muhrer, M. E., Shippen, O. T., and Lasley, J. F., 1955, Use of oxytocin for initiating parturition and reducing farrowing time in sows, *J. Anim. Sci.* **14**:1250–1254.
7. Hillman, R. B., 1975, Induction of parturition in mares, *J. Reprod. Fertil. (Suppl.)* **23**:641–644.
8. Bell, G. H., 1941, Behavior of pregnant uterus of guinea pig, *J. Physiol. (Lond.)* **100**:263–274.
9. Fuchs, A. R., and Saito, S., 1971, Pituitary oxytocin and vasopressin content of pregnant and parturient rats before, during and after parturition, *Endocrinology* **88**:574–578.
10. Burton, A. M., and Forsling, M., 1972, Hormone content of the neurohypophysis in foetal, newborn and adult guinea pigs, *J. Physiol. (Lond.)* **221**:6.
11. Folley, S. J., and Knaggs, G. S., 1965, Levels of oxytocin in the jugular vein blood of goats during parturition, *J. Endocrinol.* **33**:301–315.
12. Haldar, J., 1970, Independent release of oxytocin and vasopressin during parturition in the rabbit, *J. Physiol. (Lond.)* **206**:723–730.
13. Schams, D., Schmidt-Polex, B., and Krusc, V., 1979, Oxytocin determination by radioimmunoassay in cattle, *Acta Endocrinol. (Kbh.)* **92**:258–270.
14. Fuchs, A. R., and Dawood, M. Y., 1980, Oxytocin release and uterine contraction during parturition in rabbits, *Endocrinology* **107**:1117–1126.
15. Forsling, M. L., Taverne, M. A. M., Parbizi, N., Elsaesser, F., Smidt, D., and Ellendorf, F., 1979, Plasma oxytocin and steroid concentrations during late pregnancy, parturition and lactation in the miniature pig, *J. Endocrinol.* **82**:61–69.
16. Boer, K., Dogterom, J., and Pronker, H. F., 1980, Pituitary content of oxytocin, vasopressin and

alpha-melanocyte stimulating hormone in the fetus of the rat during labour, *J. Endocrinol.* **86**:221–229.

17. Allen, W. E., Chard, T., and Forsling, M. L., 1973, Peripheral plasma levels of oxytocin and vasopressin in the mare during parturition, *J. Endocrinol.* **57**:175–176.

18. Glatz, T. H., Weitzman, R. E., Eliot, R. J., Klein, A. H., Nathanielsz, P. W., and Fisher, D. A., 1981, Ovine maternal and fetal plasma oxytocin concentrations before and during parturition, *Endocrinology* **108**:1328–1332.

19. Summerlee, A. J. S., 1981, Extracellular recordings from oxytocin neurons during the expulsion phase of birth in unanesthetized rats, *J. Physiol. (Lond.)* **321**:1–9.

20. Roberts, J. S., and Share, L., 1969, Effects of progesterone and estrogen on blood levels of oxytocin during vaginal distention, *Endocrinology* **84**:1706–1710.

21. Roberts, J. S., 1973, Functional integrity of the oxytocin-releasing reflex in goats: Dependence on estrogen, *Endocrinology* **93**:1309–1312.

22. Roberts, J. S., 1975, Cyclical fluctuations in reflexive oxytocin release during the estrous cycle of the goat, *Biol. Reprod.* **13**:314–318.

23. Negoro, H., Visessuwan, S., and Holland, R. C., 1972, Reflex activation of paraventricular nucleus units during the reproductive cycle in ovariectomized rats treated with estrogen or progesterone, *J. Endocrinol.* **59**:559–567.

24. Hawkes, R. W., 1961, Oxytocin and unidentified oxytoxic substance in extracts of blood, in: *Oxytocin* (R. Caldeyro-Barcia and H. Eller, eds.), Pergamon Press, New York, pp. 428–432.

25. Chard, T., Hudson, C. N., Edwards, C. R. W., and Boyd, B. R. H., 1971, Release of oxytocin and vasopressin by the human fetus during labor, *Nature* **234**:352–354.

26. Kumaresan, P., Anandarangan, P. B., Dianzon, W., and Vasicka, E. A., 1974, Plasma oxytocin levels during human pregnancy and labor as determined by radioimmunoassay, *Am. J. Obstet. Gynecol.* **119**:215–233.

27. Amico, J. A., Seitchik, J., and Robinson, A. G., 1984, Studies of oxytocin in plasma of women during hypocontractile labor, *J. Clin. Endocrinol. Metab.* **58**:274–279.

28. Leake, R. D., Weitzman, R. E., Glatz, T. H., and Fisher, D. A., 1981, Plasma oxytocin concentrations in men, nonpregnant women before and during spontaneous labor, *J. Clin. Endocrinol. Metab.* **53**:730–733.

29. Goodfellow, C. F., Hull, M. G. R., Swaab, D. F., Dogteram, J., and Buijs, R. M., 1983, Oxytocin deficiency at delivery with epidural analgesia, *B. J. Obstet. Gynaecol.* **90**:214–219.

30. Weitzman, R. E., Leake, R. D., Rubin, R. T., and Fisher, D. A., 1980, The effect of nursing on neurohypophyseal hormones and prolactin in human subjects, *J. Clin. Endocrinol. Metab.* **51**:836–839.

31. Ross, M. G., Ervin, M. G., and Leake, R. D., 1986, Breast stimulation contraction stress test: Uterine contractions in the absence of oxytocin release, *Am. J. Perinatol.* **3**:35–37.

32. Leake, R. D., Fisher, D. A., Ross, M. G., and Buster, J. E., 1984, Oxytocin secretory response to breast stimulation in pregnant women, *Am. J. Obstet. Gynecol.* **148**:259–262.

33. Leake, R. D., Buster, J. E., and Fisher, D. A., 1984, The oxytocin secretory response to breast stimulation in women during the menstrual cycle, *Am. J. Obstet. Gynecol.* **148**:457–460.

34. Amico, J. A., and Hempel, I., 1990, A glycine extended form of oxytocin circulates in the plasma of individuals administered estrogen, *Neuroendocrinology* in press.

35. Amico, J. A., Ervin, M. G., Finn, F. N., Leake, R. D., Fisher, D. A., and Robinson, A. G., 1986, The plasma of pregnant women contains a novel oxytocin-vasotocin-like peptide. *Metabolism* **35**:596–601.

36. Amico, J. A., Ervin, M. G., Leake, R. D., Fisher, D. A., Finn, F. M., and Robinson, A. G., 1985, A novel oxytocin-like and vasotocin-like peptide in human plasma after administration of estrogen, *J. Clin. Endocrinol. Metab.* **60**:5–12.

37. Ervin, M. G., Amico, J. A., Leake, R. D., Ross, M. G., Robinson, A. G., and Fisher, D. A., 1988, Arginine vasotocin and a novel oxytocin–vasotocin like material in plasma of humans, *Biol. Neonate* **53**:17–22.

38. Amico, J. A., Seif, S. M., and Robinson, A. G., 1981, Oxytocin in human plasma: Correlation with neurophysin and stimulation with estrogen, *J. Clin. Endocrinol. Metab.* **52**:988–993.
39. Leake, R. D., Weitzman, R. E., and Fisher, D. A., 1980, Pharmacokinetics of oxytocin in human subjects, *Obstet. Gynecol.* **56**:701–703.
40. Fabian, M., Forsling, M. L., Jones, J. J., and Pryor, J. S., 1969, The clearance and antidiuretic potency of neurohypophysial hormones in man, and their plasma binding and stability, *J. Physiol. (Lond.)* **204**:653–655.
41. Ryden, G., and Sjoholm, I., 1971, The metabolism of oxytocin in pregnant and nonpregnant women, *Acta Obstet. Gynecol. Scand.* **50**:37–41.
42. Glatz, T. H., Weitzman, R. E., Nathanielsz, P. W., and Fisher, D. A., 1980, Metabolic clearance rate and transplacental passage of oxytocin in the pregnant ewe and fetus, *Endocrinology* **106**:1006–1011.
43. Walthes, D. C., and Swann, R. W., 1982, Is oxytocin an ovarian hormone? *Nature* **197**:225–227.
44. Sheldick, E. L., and Flint, A. P. F., 1983, Regression of the corpus lutea in sheep in response to loprestenol is not affected by loss of luteal oxytocin after hysterectomy, *J. Reprod. Fertil.* **68**:155–160.
45. Khan-Dawood, F. S., and Dawood, M. Y., 1983, Human ovaries contain immunoreactive oxytocin, *J. Clin. Endocrinol. Metab.* **57**:1129–1132.
46. Perlow, M. J., Reppert, S. M., Artman, H. G., Fisher, D. A., Seif, S. M., and Robinson, A. G., 1981, Oxytocin, vasopressin and estrogen-stimulated neurophysin: Daily patterns of concentration in cerebrospinal fluid, *Science* **216**:1416–1418.
47. Amico, J. A., Tenicela, R., Johnston, J., and Robinson, A. G., 1983, A time dependent peak of oxytocin exists in the cerebrospinal fluid but not in the plasma of humans, *J. Clin. Endocrinol. Metab.* **57**:947–951.
48. Artman, H. G., Repport, S. M., Perlow, M. J., Swaminathan, S., Odie, T. H., and Fisher, D. A., 1982, Characterization of the daily oxytocin rhythm in primate CSF, *J. Neurosci.* **2**:598–603.
49. Robinson, A. G., and Zimmerman, E. A., 1973, Cerebrospinal fluid and ependymal neurophysin, *J. Clin. Invest.* **52**:1260–1267.
50. Amico, J. A., 1985, Diabetes insipidus in pregnancy, in: *Frontiers in Hormone Research,* Volume 13 (P. Czernichow and A. G. Robinson, eds.), S. Karger, Basel, pp. 266–271.
51. Cobo, E., DeBernal, M., and Gaitan, E., 1972, Low oxytocin secretion in diabetes insipidus associated with normal labor, *Am. J. Obstet. Gynecol.* **114**:861–866.
52. Liggins, G. C., 1973, Hormone interactions in the mechanism of parturition, *Mem. Soc. Endocrinol.* **20**:119–123.
53. Dubin, N. H., Ghodaonkar, R. B., and King, T. M., 1979, Role of prostaglandin production in spontaneous and oxytocin induced uterine contractile activity in *in vitro* pregnant rat uteri, *Endocrinology* **105**:47–51.
54. Ances, I. G., 1972, Observations on the level of blood oxytocinase throughout the course of labor and pregnancy, *Am. J. Obstet. Gynecol.* **113**:291–295.
55. Titus, M. A., Reynolds, D. R., Glendening, M. B., and Page, E. W., 1960, Plasma aminopeptidase activity (oxytocinase) in pregnancy and labor, *Am. J. Obstet. Gynecol.* **113**:291–295.
56. Carter, E. R., Goodman, L. V., DeHaan, R. M., and Sobota, J. T., 1974, Serum oxytocinase levels, *Am. J. Obstet. Gynecol.* **80**:1124–1128.
57. Burd, J. M., Davison, J., Weightman, D. R., and Baylis, P. H., 1987, Evaluation of enzyme inhibitors of pregnancy associated oxytocinase: Application to the measurement of plasma immunoreactive oxytocin during human labor, *Acta Endocrinol. (Kbh.)* **114**:458–464.
58. Amico, J. A., Fuchs, A. R., Haldar, J., Leake, R. D., Robinson, I. C. A. F., and Robinson, A. G., 1985, Collaborative study of the radioimmunoassay of oxytocin, in: *Oxytocin: Clinical and Laboratory Studies* (J. A. Amico and A. G. Robinson, eds.), Excerpta Medica, Amsterdam, pp. 44–50.
59. Burton, G. M., Illingworth, D. V., and Challis, R. E., 1974, Placental transfer of oxytocin in the guinea pig and its release during parturition, *J. Endocrinol.* **60**:499–506.

60. Vasicka, A., Kumaresan, P., Han, G. S., and Kumaresan, M., 1978, Plasma oxytocin in initiation of labor, *Am. J. Obstet. Gynecol.* **130:**263–272.

61. Dawood, M. Y., Wang, C. F., Gupta, R. M., and Fuchs, F., 1977, Fetal contribution of oxytocin in human parturition, *Gynecol. Invest.* **8:**33–34.

62. Leake, R. D., Weitzman, R. E., and Fisher, D. A., 1981, Oxytocin concentrations during the neonatal period, *Biol. Neonate* **39:**127–131.

12

Oxytocin Receptors in the Uterus

Melvyn S. Soloff

1. Introduction

Oxytocin is a nonapeptide that is synthesized by magnocellular neurons of the hypothalamus and, as shown within the past 10 years, by ovarian luteal cells of certain species.[1-3] The peptide is similar in structure to arginine vasopressin, or antidiuretic hormone (Fig. 1), accounting for each having overlapping activities on the other's target cells. Thus, vasopressin will cause uterine contractions in experimental animals, but considerably higher doses are required to obtain the same response as with oxytocin.

Labor at term can be induced either by giving exogenous oxytocin or by stimulating endogenous oxytocin release.[4] Despite these findings, many investigators have felt that oxytocin is not a physiological initiator of labor because fluctuations in its concentration in the blood immediately before labor do not correspond to changes in uterine activity. It now seems clear, from data collected in several species, that oxytocin's activity during pregnancy and labor depends more on the sensitivity of myometrial cells than on the concentration of the hormone in the circulation.

Oxytocin also has been shown to stimulate the production of prostaglandin $F_{2\alpha}$ ($PGF_{2\alpha}$) by decidua.[15,16] In view of the strong evidence for an important role of prostaglandins in labor initiation, oxytocin may serve a dual function in initiating labor by acting on both decidua and myometrium. Labor may begin when the sensitivity of the myometrium and decidua to oxytocin increases sufficiently to respond to basal or slightly elevated levels of oxytocin in the blood.

Melvyn S. Soloff • Department of Biochemistry, Medical College of Ohio, Toledo, Ohio 43699.

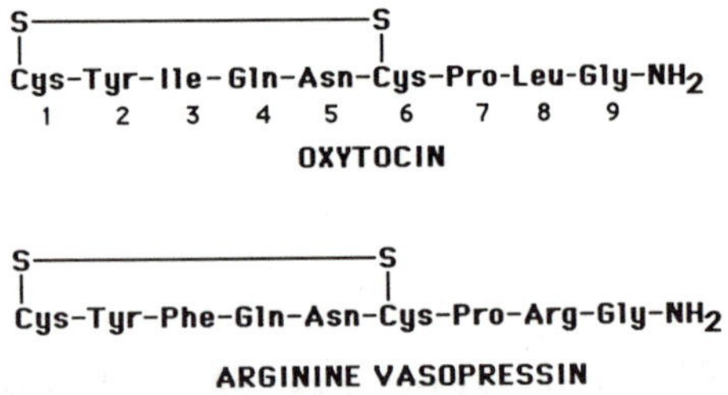

```
  S————————————————S
  |                |
Cys-Tyr-Ile-Gln-Asn-Cys-Pro-Leu-Gly-NH2
  1   2   3   4   5   6   7   8   9
              OXYTOCIN
```

```
  S————————————————S
  |                |
Cys-Tyr-Phe-Gln-Asn-Cys-Pro-Arg-Gly-NH2
          ARGININE VASOPRESSIN
```

Figure 1. Structures of oxytocin and arginine vasopressin. Lysine vasopressin, found in swine, differs from arginine vasopressin by the presence of a lysine residue in position 8.

2. The Relationship between Sensitivity and Oxytocin Receptors

The increase in myometrial sensitivity to oxytocin occurs concomitantly with an increase in the number of oxytocin receptors in the rat,[5-7] guinea pig,[8] human,[9,10] and rabbit.[11] In the rat studies, receptor measurements were made on the same uterine samples used to determine oxytocin sensitivity *in vivo*.[4] Premature labor in rats[12] and humans[13] was associated with premature rises in oxytocin receptor concentrations. With delayed parturition in rats[14] and humans,[13] there was a corresponding delay in the rise in oxytocin receptor concentrations in the myometrium. Decidual oxytocin receptors also may be important for regulating oxytocin-mediated $PGF_{2\alpha}$ release. Thus, the activity of the myometrium and decidua in response to oxytocin cannot be assumed to be a reflection of circulating levels of the peptide. Instead, the concentration of oxytocin receptors must be taken into account. Factors causing changes in oxytocin receptor concentrations, and hence sensitivity to oxytocin, are the subject of this chapter.

3. Criteria for Oxytocin Binding Sites Being Receptors

Binding sites for [³H]oxytocin have been demonstrated in plasma membrane fractions of myometrium and endometrium from several species. Several lines of evidence, reviewed elsewhere,[17,18] have indicated that the binding sites are receptor components. Half-maximal binding (apparent equilibrium dissociation constant, K_d) to uterine membranes occurred with low-nanomolar concentrations of the peptide, consistent with the dose of oxytocin eliciting half-maximal responses *in vitro*. Binding to myometrial sites was specific for a series of oxytocin analogues; the relative affinity of each analogue corresponded to its relative potency in eliciting uterine contractions. Binding also was target-tissue specific. Factors causing increases in oxytocin binding, such as estrogen treatment or the addition of specific metal ions, favored an increased myometrial response and/or sensitivity to oxytocin. Finally, occupancy of binding sites by oxytocin or analogues corresponded to effector functions, such as inhibition of Ca^{2+}, Mg^{2+}-ATPase activity of rat myometrial plasma membranes and $PGF_{2\alpha}$ release by sheep endometrium. These studies are discussed in more detail below.

4. Methods for Determining Oxytocin Receptor Concentrations

Quantitation of oxytocin binding sites has been determined with [³H]oxytocin, labeled in the tyrosyl ring at position 2, with specific activities about 10–40 Ci/mmol. Iodination of the tyrosyl residue results in the loss of oxytocic and binding activities.[19,20] Recently, however, the development of an iodinated antagonist of oxytocin[21] offers great promise in detecting, quantitating, and characterizing binding sites present either in low concentrations or in tissues that are available only in limited quantity.

[³H]Oxytocin binding sites were distributed with plasma membrane fractions after subcellular fractionation of rat myometrial homogenates.[22,23] Although extensive fractionation studies have not been carried out on myometrial samples from other species or from endometrium/decidua, it has been assumed that oxytocin receptors reside on the plasma membrane; and crude or enriched plasma membrane fractions have been used for oxytocin binding studies to give results that consistently have shown comparable binding constants.

The apparent equilibrium dissociation constant and the concentration of binding sites have been determined by Scatchard analysis.[24] In general, the presence of a single class of independent binding site with high affinity for oxytocin ($K_d = 1$–2 nM) has facilitated analysis of Scatchard plots as simple linear regressions. In some instances, the concentration of myometrial oxytocin receptors has been expressed as a relative measure, based on single-point determinations.[5] In order for relative concentrations to be valid, the affinity of receptor sites for oxytocin must remain constant.

5. Myometrial Oxytocin Receptor Regulation

5.1. Steroids

Up- and down-regulation of myometrial oxytocin receptors are pronounced during the estrous cycle and in pregnancy; the concentration of receptors varies in proportion to changes in sensitivity of the myometrium to oxytocin. Oxytocin receptor concentrations in ewe myometrial and endometrial plasma membranes were greatest at estrus and relatively low 4 or 5 days earlier or later.[25] Treatment of rats with estradiol-17β after ovariectomy on day 20 of pregnancy caused more than a fivefold increase in the concentration of myometrial oxytocin receptors 48 hr later.[7] Administration of progesterone with estradiol, however, prevented the estrogen-stimulated rise in oxytocin receptor concentrations.[7] Similar results have been obtained with uterine explants from immature rats.[26] Addition of physiological concentrations of estradiol to the medium resulted in about a fivefold increase in oxytocin receptor concentrations after 2 days of culture. The increase was blocked by the protein synthesis inhibitor cycloheximide. However, when cycloheximide was added 2 days after estrogen, oxytocin receptor levels remained elevated for at least another 2 days. These results indicated that once receptors were up-regulated, turnover was minimal. The addition of progesterone 2 days after estrogen, however, caused a sharp fall in oxytocin receptor concentrations.[26] The effects of progesterone presumably were not mediated by inhibition of estrogen action, which was already blocked by cycloheximide. The actions of estrogen and progesterone *in vitro* therefore mimic their *in vivo* effects, indicating that the steroids act directly on uterine cells. Steps leading to progesterone down-regulation of oxytocin receptors remain to be studied.

In the rat, the number of oxytocin binding sites per milligram of myometrial protein or per cell increased abruptly 24 hr or less before the onset of labor.[5,6] Receptor levels were greatest during labor and fell sharply after parturition, reaching base-line values by 1 or 2 days post-partum. These changes in receptor concentration were proportional to changes in plasma estrogen/progesterone concentration ratios.[5,6] Induction of premature labor in rats by a single subcutaneous dose of $PGF_{2\alpha}$ caused premature rises in both plasma estrogen/progesterone ratios and myometrial oxytocin receptor concentrations.[12] When the onset of labor was delayed by 1 or 2 days by pharmacological doses of luteinizing hormone-releasing hormone, increases in plasma estradiol/progesterone ratios and myometrial oxytocin receptor concentrations also were delayed.[14] Several lines of evidence therefore indicate that estrogens and progesterone, acting in opposing fashion, are involved in the regulation of oxytocin receptor concentrations in rat myometrium.

5.2. Metal Ions and Oxytocin Action

Specific metal ions might be important in the physiological regulation of oxytocin action at the receptor level. Calcium is required for the action of all myometrial stimulants, but Mg^{2+} is important specifically for the effects of oxytocin and analogues on the uterus. Removal of Mg^{2+} from the medium bathing isolated uterus reduced the stimulatory effects of oxytocin, and addition of Mg^{2+} in concentrations between 1 and 10 mM potentiated the effects of the peptide.[27–30] Normal concentrations of free extracellular Mg^{2+} are in the range of 0.75 to 1.25 mM. Magnesium appeared to increase the affinity of receptor sites for several oxytocin analogues because dose–response curves were parallel-shifted to the left in the presence of Mg^{2+}.[29,30] Other studies have indicated that Mg^{2+} increased the maximal response of the myometrium to oxytocin analogues,[31] suggesting that the intrinsic activities of the peptides were increased by metal ion. Bentley[29] suggested that Mg^{2+} might alter the conformation of oxytocin, producing a better fit with the receptor site. An alternative explanation is that metal ion might influence the binding of peptides to receptors by modifying the receptor site. These two postulated modes of potentiation of oxytocin action are not mutually exclusive. A third possibility is that metal ions form a bridge between oxytocin and its receptor.[32]

The binding of [³H]oxytocin by rat mammary gland membranes as a function of divalent metal ion concentration (Mg^{2+}, Mn^{2+}, or Co^{2+}) indicated that the receptor has two distinct regions for metal ion interaction.[33] The binding of metal ion to region A (availability) results in a receptor with the maximal number of binding sites but with low affinity for oxytocin. The binding of metal ion to region B (binding) ends in a receptor site of low availability but with potentially high affinity for oxytocin. Binding of metal ion to both sites results in a receptor of high affinity and full availability. This mechanism would explain why metal ions increase the response of target cells to oxytocin (increased number of receptors) and shift the dose–response curve to the left (increased affinity). Extensive studies have not been carried out with uterine myometrial receptors, but they have the same metal ion requirement for oxytocin binding as mammary gland membranes.[34]

5.3. Oxytocin Analogues and Metal Ions

The potentiating effect of metal ion on oxytocin analogues is generally inversely proportional to the potency of the peptide.[29,31,32,35−37] 7-Glycine oxytocin is relatively inactive, but its potency is increased about ten times when the Mg^{2+} concentration bathing uterine smooth muscle *in vitro* is increased from 0 to 0.5 mM.[38] The activity of a potent peptide like 4-threonine oxytocin, however, is actually less in the presence of 0.5 mM Mg^{2+}.[39] Soloff and Grzonka[40] found that there was a good correlation between the ability of oxytocin analogues to inhibit [³H]oxytocin binding by rat myometrial membranes and their uterotonic potencies. In these studies Mn^{2+} was used instead of Mg^{2+} because of its greater effect. An increase in Mn^{2+} concentration from 1 to 10 mM enhanced the affinity of uterine membranes for the analogues in inverse proportion to their affinities in 1 mM Mn^{2+} (Fig. 2). This was particularly well seen for the lower-affinity analogues. Based on the results with a series of oxytocin analogues modified in the 7 position, the metal ion effect does not appear to be related to its interaction with the peptide. The mechanisms of the metal ion effect are not entirely understood, but Mn^{2+} and other specific divalent cations might allow the conformation of myometrial receptors to adapt better to less-well-fitting ligands.

It is possible that metal ions join subunits of the receptor. The addition of 1 mM EDTA results in the rapid dissociation of [³H]oxytocin from mammary gland membranes[33] and myometrial membranes.[34] The effects of EDTA are reversible, and restoration of the metal ion concentration reestablished [³H]oxytocin binding activity. On the other hand, the addition of EDTA to solubilized oxytocin receptors results in an irreversible loss of about half of the binding sites on restoration of the free Mn^{2+} concentration (Fig. 3). Although there are several possible interpreta-

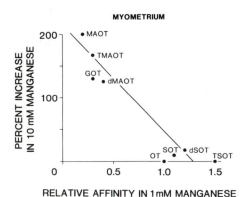

MYOMETRIUM

Figure 2. Effect of Mn^{2+} on potentiation of the binding of oxytocin analogues to myometrial membranes from estrogen-treated rats. The ability of each analogue to inhibit the binding of [^{3}H]oxytocin in the presence of 1 mM and 10 mM Mn^{2+} was determined, and the increase in inhibition in going from 1 mM to 10 mM Mn^{2+} is plotted against the inhibition in 1 mM Mn^{2+}. Inhibition with each analogue in the presence of 1 mM Mn^{2+} is expressed relative to that obtained with oxytocin (relative affinity of oxytocin = 1). The results show that increasing the metal ion concentration selectively increases the affinity of myometrial oxytocin receptors for lower-affinity analogues, in inverse proportion to their affinity in low concentrations of Mn^{2+}. Abbreviations used: OT, oxytocin; GOT, [7-glycine[oxytocin; SOT, [7-sarcosine]oxytocin; dSOT, [1-β-mercaptopropionic acid, 7-sarcosine]oxytocin; TSOT, [4-threonine, 7-sarcosine]oxytocin; MAOT, [7-N-methylalanine]oxytocin; TMAOT, [4-threonine, 7-N-methylalanine]oxytocin; dMAOT, [1-β-mercaptopropionic acid, 7-N-methylalanine]oxytocin. From Soloff and Grzonka.[40]

tions of these results, the high-affinity receptor site may involve the interaction of two or more subunits, which, when dissociated in intact membranes, can readily reassociate. Dissociation of the subunits in solution, however, results only in reassociation of about half of the corresponding subunits because they are no longer held in close proximity by an intact membrane.

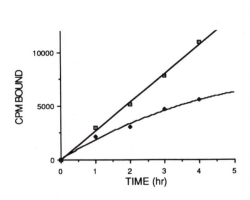

Figure 3. Partial reversibility of specific [^{3}H]oxytocin binding in CHAPSO {3-[(3-cholamidopropyl)dimethylammonio]-2-hydroxy-1-propane sulfonate}-solubilized extracts of regressed mammary gland of the rat after EDTA treatment. The extracts were incubated at 10°C for 30 min in the presence (♦) or absence (□) of 2 mM EDTA. Mn^{2+}, 12 mM, was then added, and specific oxytocin binding was determined. Binding reaches a plateau after about 4 hr under these conditions.[49] Each point is the mean of duplicates. The results show that binding capacity was reduced by almost half after incubation with EDTA. In contrast to these results, binding to intact mammary membranes is unaffected by treatment with EDTA.[33]

The physiological importance of metal ions as potential regulators of oxytocin action is not presently understood. Indeed, the means by which dramatic changes in oxytocin binding during the estrous cycle and at the end of pregnancy take place are not clear. It is possible that changes in uterine receptor concentrations are the result of synthesis and breakdown of receptor molecules. Alternatively, acute changes in oxytocin binding as a result of changes in the metal ion concentration *in vitro* suggest that changes in the conformation of existing receptor molecules may account for apparent changes in receptor concentration.

6. Distinct Vasopressin and Oxytocin Receptors in the Myometrium of Nonpregnant Women

Vasopressin is a more potent myometrial stimulant than oxytocin in nonpregnant women[41–43] and in women during the first trimester of pregnancy.[44] The high motility of the uterus during menstruation appears to be the result of vasopressin action because inhibition of the central release of vasopressin by water-induced diuresis reduced uterine motility.[45] Uterine contractions could be reestablished by intravenous administration of vasopressin but not oxytocin. Because of the greater sensitivity of the myometrium to oxytocin in the later stages of pregnancy,[46] the question arises as to whether there is a single receptor population with changing specificity for vasopressin and oxytocin during gestation. Alternatively, there could be separate vasopressin and oxytocin receptors, the relative proportions of which change during pregnancy. Consistent with its high uterotonic potency in nonpregnant uterus, vasopressin was as effective as oxytocin in inhibiting the binding of [^{3}H]oxytocin to fundus and corpus plasma membrane preparations.[47] Studies by Guillon *et al.*[48] suggested that there are separate receptors for vasopressin and oxytocin in the myometrium of nonpregnant women. Because the ligand specificity of myometrial vasopressin receptors was primarily of the V_1 subtype, which is characteristic of vascular smooth muscle, it is possible that vascular contamination of the membrane preparation contributed to the binding. High-affinity vasopressin binding sites had about six times the capacity for vasopressin as high-affinity oxytocin binding sites had for oxytocin. In addition, the order of potency of analogues inhibiting [^{3}H]oxytocin binding was different from that inhibiting [^{3}H]vasopressin binding. The greater sensitivity of the myometrium to vasopressin in the nonpregnant uterus may be the result of a greater number of vasopressin receptors. During pregnancy the shift in sensitivity of the myometrium to oxytocin may be related to the up-regulation of receptors for oxytocin but not for vasopressin. Until separate myometrial oxytocin and vasopressin receptors are isolated physically, however, we can only speculate on the mechanisms responsible for changing sensitivities of the myometrium to the two peptides during pregnancy.

7. Characterization of Oxytocin Binding Sites

7.1. Solubilization of Oxytocin Receptors

Of several detergents tested, only 3-[(3-cholamidopropyl)dimethylammo-nio]-2-hydroxy-1-propane sulfonate (CHAPSO) allowed solubilization of specific [^{3}H]oxytocin binding activity.[49] About 25% of the binding sites of membranes from either uterine myometrium from pregnant rats or involuted mammary gland from rats was solubilized.[49] Similar results were obtained from rabbit myometrium (S. Baldwin and M. S. Soloff, unpublished data). Repeated extraction of membranes did not result in any further solubilization of binding activity. The relatively low yield of extractable binding activity could be related to the association of receptors with relatively insoluble proteins. Edelman[50] proposed that lateral movement of intramembranous proteins results from their interaction with microfilaments and associated proteins. Mescher *et al.*[51] showed that membrane proteins that interact with the cytoskeleton will remain insoluble, even with high concentrations of de-tergent. Oxytocin receptors in smooth muscle or smooth-muscle-like target cells may be associated with the cytoskeleton and with portions of the membrane that anchor bundles of actin and myosin that are involved in cell contraction.[52]

The binding of [^{3}H]oxytocin in detergent extracts was shown to be similar or identical to that of oxytocin receptors on intact plasma membranes.[49] Several synthetic oxytocin analogues inhibited [^{3}H]oxytocin binding in the same rank order in both solubilized and intact membrane preparations. Both solubilized and intact membrane preparations also require Mn^{2+} for [^{3}H]oxytocin binding. The concentration of binding sites in solubilized extracts of uterine myometrium was substantially greater in uteri from rats in labor than from rats several days before labor, corroborating results with intact membranes.[5] These findings suggest that the marked rise in receptor concentration near the end of gestation is not caused by the unmasking of cryptic receptor sites, because solubilization would expose hidden binding sites. The affinity of the solubilized binding sites for oxytocin was comparable to that of intact myometrial membranes. The binding of [^{3}H]oxytocin to de-tergent-solubilized extracts from intestinal smooth muscle, which is not a target for oxytocin, was negligible.

Gel filtration of detergent-solubilized extracts of rat myometrium indicated the existence of multiple-sized binding sites.[49] Although the receptor may exist as a multimeric structure and interact with other membrane proteins, the absence of discrete forms on gel filtration made it difficult to draw any conclusions about the size of the receptor. The functional size of intact membrane receptors in both rat myometrium and mammary gland was estimated instead by radiation inactivation studies.[53] Exposure of frozen membranes for increasing lengths of time to ^{60}Co radiation resulted in the progressive inactivation of binding activity, determined after thawing of the membranes. From the rate of loss of binding activity, the functional sizes of oxytocin receptors in mammary gland and myometrial membranes were estimated to be 57,500 ± 3,000 (S.D.) and 58,800 ± 1,600 Da, respectively.[53] This knowledge will be useful in the purification and the characterization of the receptors and associated membrane components.

7.2. Information Gained from the Use of Oxytocin Analogues

Elimination or methylation of glycinamide from the carboxyl terminus of arginine vasopressin markedly reduces its vasopressor and antidiuretic activities.[54,55] On the other hand, removal of this residue along with other modifications of vasopressin antagonists did not diminish the antagonism of either the vasopressor or antidiuretic effects of vasopressin.[55] Thus, the carboxyl terminus of vasopressin is required for activity but not for binding to the receptor site.[55]

An oxytocin antagonist with an iodinated tyrosyl amide replacing the carboxy-terminal glycinamide had a pA_2* of 7.7 when tested on rat uterus *in vitro*, indicating an apparent K_d of about 20 nM.[21] The apparent K_d of the antagonist estimated from binding to uterine plasma membranes from estrogen-treated rats was 48 pM. Although the basis of the discrepancy in estimated K_d values from the bioassay and binding studies is not clear, the carboxy terminal of oxytocin, like that of vasopressin, does not appear to be required for interaction with receptor sites. As in vasopressin, the amino-terminal glycinamide of oxytocin is required for agonist activity.[56] Replacement of leucine with ornithine as the penultimate carboxy-terminal residue does not diminish antagonist activity.[21] The same substitution in oxytocin, however, results in a substantial loss in oxytocic activity.[56] These findings may be very helpful in the design of analogues for affinity chromatographic purification and cross-linking of oxytocin receptors.

8. Coupling between Oxytocin-Receptor Occupancy and Cell Contraction

8.1. Calcium

Oxytocin-induced contractions of rat myometrium are Ca^{2+} dependent, even after K^+-induced depolarization.[57] Contractions can be prevented by inhibition of Ca^{2+} gating with calcium channel blockers.[58,59] Oxytocin- but not acetylcholine-induced contractions of rat myometrium occurred when Ca^{2+} in the medium was replaced with Mg^{2+}.[59] In medium containing Ca^{2+}, contractions caused by acetylcholine but not by oxytocin were inhibited by the presence of Mn^{2+}. These results suggest that there is more than one type of Ca^{2+} channel and that the effects of oxytocin and acetylcholine are mediated by distinct channels.[59]

The relationship between oxytocin receptor occupancy and the gating of Ca^{2+}

---

*pA_2 is the negative log of the molar concentration of antagonist reducing the response of $2X$ units of agonist to that of $1X$, administered before antagonist. A greater pA_2 value indicates a more potent antagonist.

is not understood. If the occupancy of oxytocin receptors were tightly coupled to the opening of calcium channels, we might expect that modification of the channels would perturb the binding of oxytocin. For example, Cantor et al.[60] showed that dihydropyridine calcium channel blockers inhibited the binding of nonneuronal benzodiazepine ligands by membrane fractions from rat heart, kidney, and brain. Other types of calcium channel blockers, e.g., verapamil and diltiazem, were without effect. The effects of the dihydropyridine blockers were also specific for the ligand, as they did not affect the binding by neuronal site ligands. We were unable to demonstrate any effect of dihydropyridines on the binding of [^{3}H]oxytocin to myometrial plasma membranes from estrogen-treated rats (S. Baldwin and M. S. Soloff, unpublished data), indicating that if there is a relationship between oxytocin binding and the opening of Ca^{2+} channels, it is not of the same type as that shown by Cantor et al.

8.2. Calcium, Magnesium ATPase

The large difference in Ca^{2+} concentration between the outside and inside of cells is maintained in part by the unidirectional export of Ca^{2+} by a high-affinity Ca^{2+} extrusion pump.[61] The process, involving utilization of ATP via a Ca^{2+}, Mg^{2+}-ATPase has been demonstrated in a number of cell types (see ref. 62 for references), including stomach smooth muscle.[63] The Ca^{2+}, Mg^{2+}-ATPase from heart sarcolemma has been isolated by calmodulin affinity chromatography and shown to be about 140,000 mol. wt.[64] The purified ATPase was reconstituted in asolectin liposomes, which pumped Ca^{2+} with an approximate stoichiometry to ATP of 1.[65]

Åkerman and Wikström[66] demonstrated the presence of oxytocin-inhibited Ca^{2+}, Mg^{2+}-ATPase activity in myometrial plasma membrane fractions from estrogen-treated rabbits. Further characterization[67] of oxytocin inhibition of this activity in rat myometrial plasma membrane fractions showed that the concentration of oxytocin giving half-maximal inhibition, 1 nM, corresponded to the concentration occupying half of the receptor sites (Fig. 4). Several synthetic oxytocin analogues inhibited the ATPase in proportion to their ability to inhibit the binding of [^{3}H]oxytocin to myometrial receptor sites and to stimulate uterine contractions. Oxytocin-inhibited Ca^{2+}, Mg^{2+}-ATPase activity was specific for oxytocin target tissues, being present on fat cell membranes but absent on membranes from rat duodenum. Like oxytocin receptor concentrations, myometrial Ca^{2+}, Mg^{2+}-ATPase activity was stimulated by estrogen, and the effects of estrogen were opposed by progesterone treatment.[67] At the beginning of labor in the rat, when plasma estrogen/progesterone concentration ratios are the greatest,[5] oxytocin inhibition of Ca^{2+}, Mg^{2+}-ATPase activity was 10,000 times greater than on day 18.[68] This increase is no doubt the result of estrogen-induced increases in both oxytocin receptor concentration and Ca^{2+}, Mg^{2+}-ATPase activity. The Ca^{2+}, Mg^{2+}-ATPase activity in sarcolemma from human pregnant myometrium also was inhibited

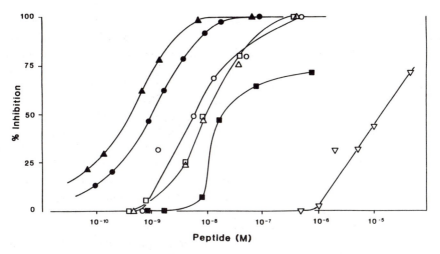

Figure 4. Inhibition of Ca^{2+},Mg^{2+}-ATPase activity of myometrial plasma membranes from estrogen-treated rats by oxytocin, oxytocin analogues, and bradykinin. Oxytocin (●), [1-(L-2-hydroxy-3-mercapto-propanoic acid), 4-threonine]oxytocin (▲), bradykinin (○), arginine vasopressin (□), lysine vasopressin (△), [1-(β-mercapto-β,β-cyclopentamethylenepropionic acid), 4-threonine]oxytocin (■), [2-O-meth-yltyrosine]oxytocin (▽). Inhibition was proportional to the rank order of uterotonic potencies of each analogue. From Soloff and Sweet.[67]

by oxytocin.[69] The ATPase was similar to the Ca^{2+}-extrusion ATPase of erythro-cyte membranes with respect to its high affinity for Ca^{2+}, inhibition by vanadate, and dependency on calmodulin.

Hormone-inhibited Ca^{2+}, Mg^{2+}-ATPase activities on plasma membrane frac-tions have also been demonstrated for insulin on rat fat cells,[70] calcitonin on rat liver,[71] and vasopressin on rat hepatocytes.[72] Regulation of Ca^{2+} efflux may be important in sustaining transient increases in intracellular Ca^{2+} concentrations, thereby prolonging the effects of agonists.

8.3. Phosphoinositol Metabolism

Several studies have suggested that the effects of oxytocin on the myometrium are mediated by phosphoinositol metabolism. Marc *et al.*[73] found that oxytocin caused the rapid formation of inositol trisphosphate by guinea pig myometrial strips. Similarly, oxytocin and vasopressin increased the production of inositol phosphates by human gestational myometrial samples.[74] The addition of inositol trisphosphate to the microsomal fraction of homogenates of myometrium from near-term pregnant cows caused the release of Ca^{2+}.[75] These results suggest that in myometrial cells, as in other cell types, the binding of inositol trisphosphate to receptors on the smooth endoplasmic reticulum releases Ca^{2+} into the cytosol.[76,77] Oxytocin also has been shown to stimulate the formation of phosphoinositol in slices of mammary tissue.[78,79] Arginine vasopressin, however, was more than twice as potent as oxytocin, raising the question of whether the stimulation of inositol phosphate formation by oxytocin in the mammary gland is the result of occupancy of vasopressin receptors rather than oxytocin receptors.[79] The presence of vasopressin receptor in mammary tissue was corroborated by binding studies with specific oxytocin and vasopressin analogues.[79] [³H]Vasopressin was bound with an apparent K_d of about 0.7 nM in contrast to a K_d of 2–5 nM for oxytocin. The concentration of binding sites was about 25% that of oxytocin binding sites.[79] Vasopressin stimulates inositol phosphate formation in a broad range of cell types,[76,77,80–83] including cells derived from a mammary tumor.[84,85] Vasopressin also has been shown to stimulate the formation of inositol phosphates in vascular cells,[86–88] which may be present in significant numbers in uterine tissue samples. The potency of oxytocin in stimulating inositol phosphate accumulation in the rat tail artery was about 0.1% that of vasopressin.[88] Whether the effects of oxytocin on uterine myometrial cells are mediated by polyphosphoinositide hydrolysis remains to be demonstrated by the use of a series of oxytocin and vasopressin analogues to demonstrate a correspondence between uterotonic potency and stimulation of phosphoinositol formation.

8.4. Lack of Evidence for Involvement of a G Protein

GTP-binding proteins are involved in coupling hormone receptors to adenylate cyclase and photoreceptors to cGMP phosphodiesterase[89] (see also Chapters 5 and 8). G proteins also appear to be involved in the regulation of phosphoinositol turnover by a phosphatidylinositol-specific phospholipase C. Guanine nucleotides reduce the affinity of G-protein-coupled receptor sites for agonist, as shown with several cell types and agonists.[90–94] The nonhydrolyzable GTP analogues guanosine-5'-O-3-thiotriphosphate (GTPγS) and guanylyl imidodiphosphate [Gpp(NH)p], in concentrations between 1 and 250 μM, had no effect on the binding of [³H]oxytocin to plasma membranes from the rat myometrium (M. S. Soloff, unpublished data). In contrast, 250 μM GTPγS reduced the concentration of oxytocin binding sites on mammary gland membranes by about 24% (Fig. 5). It is not clear at the present time whether oxytocin receptors of the myometrium and mammary gland differ in their coupling to G proteins or whether there is a second population of receptor sites in the mammary gland that is coupled to G protein. Vasopressin receptors are present in mammary gland,[79] and it is possible that the reduction in mammary binding sites for [³H]oxytocin with GTPγS might be related to an effect on the vasopressin receptor population and not on oxytocin receptors. Guillon and co-workers[48] found that Gpp(NH)p reduced the binding of [³H]oxytocin and [³H]vasopressin to myometrial membrane fractions from nonpregnant women. Maximal inhibition of binding, however, was twice with [³H]vasopressin than [³H]oxytocin, suggesting that only vasopressin receptors are coupled to G proteins.

Figure 5. Effect of GTPγS, 250 μM (♦), on the binding of [³H]oxytocin to mammary gland membranes from the lactating rat. No GTPγS (□). Scatchard analyses were carried out with 500 μg of plasma membrane protein from mammary gland of lactating rat in 200 μl of assay bufffer, as outlined previously.[40] [³H]Oxytocin (1 nM) and increasing amounts of nonradioactive oxytocin were added to a series of tubes containing membranes to give a final incubation volume of 250 μl. The samples were incubated for 1 hr at 22°C, and bound oxytocin was separated from that unbound by filtration. Simple linear regressions of the ratio of bound to unbound oxytocin (*B/U*) versus the concentration of oxytocin bound indicated a single class of independent binding site, the concentration of which was reduced by about 24% by the presence of 250 μM GTPγS.

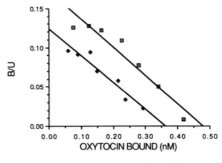

9. Endometrium

9.1. Endometrial Oxytocin Receptors

The specific binding of [³H]oxytocin was demonstrated in membrane fractions of the endometrium of ewes.[25] The concentration of binding sites increased at the end of the estrous cycle, corresponding to an increased sensitivity to oxytocin-induced release of $PGF_{2\alpha}$.[25] The number of endometrial oxytocin receptors may be regulated by estrogen and progesterone, much like the number of myometrial receptors in the rat[7,26,95] and rabbit.[96] Oxytocin binding sites also have been demonstrated in endometrium from nonpregnant human uterus[97] and decidua from term pregnant uteri.[13]

9.2. Effects of Oxytocin on Endometrial Prostaglandin Synthesis

Studies in several species have shown that oxytocin stimulates uterine prostaglandin synthesis when administered near estrus or after estrogen treatment.[98−102] Oxytocin also has been shown to stimulate $PGF_{2\alpha}$ synthesis by decidual tissue from the human[13,15] and uterus of the rat[103] at the end of pregnancy.

Like nonpregnant or early pregnant human myometrium, human endometrium was more responsive to vasopressin than to oxytocin. Perifusion of human endometrial samples with 0.1 μM vasopressin caused a greater amount of $PGF_{2\alpha}$ release than 10 μM oxytocin (Fig. 6). These findings support observations showing that changes in plasma vasopressin levels that occur during the menstrual cycle[104] might regulate endometrial $PGF_{2\alpha}$ production. Strömberg and co-workers[105] found that infusion of lysine vasopressin into women on the first or second day of menses

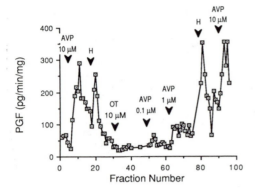

Figure 6. Arginine vasopressin (AVP) stimulation of $PGF_{2\alpha}$ release by human endometrium. The explant was perifused at a rate of 0.2 ml/min with AVP in the concentrations indicated, 10 μM histamine (H), or 10 μM oxytocin (OT). The amount of $PGF_{2\alpha}$ in each fraction (1.0 ml) of the perifusate was determined by radioimmunoassay.

resulted in a significant increase in uterine activity and plasma levels of prostaglandin metabolites. These workers have suggested that elevated prostaglandin levels associated with primary dysmenorrhea are brought about by enhanced vasopressin action.

The estrogen-induced onset of luteolysis in sheep (decline in circulating progesterone) was preceded by increases in endometrial oxytocin receptor concentration.[106] These observations led Hixon and Flint[106] to postulate that luteolysis, resulting from oxytocin-stimulated increases in $PGF_{2\alpha}$ release from endometrial cells, is dictated by increases in endometrial oxytocin receptor concentrations. The mechanism by which oxytocin stimulates uterine $PGF_{2\alpha}$ synthesis has been suggested to involve increased hydrolysis of phosphatidylinositol phosphates with concomitant production of diacylglycerol, which can be subsequently hydrolyzed to release arachidonic acid.[107] Estrogen treatment caused a marked increase in the ability of oxytocin to stimulate phosphoinositide synthesis and hydrolysis,[106] as might be expected from an estrogen-induced increase in oxytocin receptor concentration. However, there was a lag of more than 12 hr between receptor induction, as measured by [^{3}H]oxytocin binding, and increased phosphatidylinositol turnover. As was the case for oxytocin-inhibited Ca^{2+}, Mg^{2+}-ATPase activity in myometrium, estrogen appears to activate steps occurring after occupancy of oxytocin receptors in stimulating $PGF_{2\alpha}$ release from endometrial cells.

10. Concluding Remarks

The traditional approach of viewing the endocrine status of an organism as a reflection of circulating concentrations of hormones does not appear to be appropriate for oxytocin, at least with respect to the responsiveness of uterine tissues to the peptide. Hormone action involves setting into motion a chain of events that culminate in specific responses. Theoretically, any component of the chain can be an important point of regulation of hormone action. For some reason, yet unclear, changes in oxytocin receptor concentration appear to be a pivotal step. Oxytocin receptor concentrations in the uterus are up-regulated by estrogen and down-regulated by progesterone, resulting in changes in receptor concentrations during pregnancy and in the estrous/menstrual cycle. Certain divalent metal cations can modify oxytocin–receptor interactions in an acute manner, but the physiological significance of this type of regulation remains to be established.

Changes in the concentration of myometrial and endometrial/decidual oxytocin receptors appear to be coordinated, possibly accounting for both myometrial contractions and decidual $PGF_{2\alpha}$ release at the end of pregnancy. The endometrium may be the major target of oxytocin during the estrous cycle, as oxytocin-stimulated $PGF_{2\alpha}$ release may initiate luteolysis in some species while increases in myometrial contractility may not be as significant. During lactation, when blood levels of oxytocin are elevated, the relatively low concentrations of uterine oxytocin receptors may explain the unresponsiveness of both endometrium and myometrium to oxytocin. The changing sensitivity of the myometrium to vasopressin also suggests that regulation of vasopressin receptor levels may also occur. To understand neurohypophyseal hormone actions on the reproductive tract and mammary gland, therefore, it is important to study regulation of receptor levels as well as events mediating neurohypophysial hormone–receptor interaction resulting in biological responses of target cells.

ACKNOWLEDGMENTS. Supported by grant HD8406 from the National Institutes of Health.

References

1. Fields, M. J., Fields, P. A., Castro-Hernandez, A., and Larkin, L. H., 1980, Evidence for relaxin in corpora lutea of late pregnant cows, *Endocrinology* **107**:869–876.
2. Wathes, D. C., and Swann, R. W., 1982, Is oxytocin an ovarian hormone? *Nature* **297**:225–227.
3. Wathes, D. C., 1984, Possible actions of gonadal oxytocin and vasopressin, *J. Reprod. Fertil.* **71**:315–345.
4. Boer, K., Lincoln, D. W., and Swaab, D. F., 1975, Effects of electrical stimulation of the neurohypophysis on labour in the rat, *J. Endocrinol.* **65**:163–176.
5. Soloff, M. S., Alexandrova, M., and Fernström, M. J., 1979, Oxytocin receptors: Triggers for parturition and lactation? *Science* **204**:1313–1315.
6. Alexandrova, M., and Soloff, M. S., 1980, Oxytocin receptors and parturition. I. Control of oxytocin receptor concentration in the rat myometrium at term, *Endocrinology* **106**:730–735.
7. Fuchs, A.-R., Periyasamy, S., Alexandrova, M., and Soloff, M. S., 1983, Correlation between oxytocin receptor concentration and responsiveness to oxytocin in pregnant rat myometrium. Effect of ovarian steroids, *Endocrinology* **113**:742–749.
8. Alexandrova, M., and Soloff, M. S., 1980, Oxytocin receptors and parturition in the guinea pig, *Biol. Reprod.* **22**:1106–1111.
9. Sakamoto, H., Den, K., Yamamoto, K., Arai, T., Kawai, S., Oyama, Y., Yoshida, T., and Takagi, S., 1979, Study of oxytocin receptor in human myometrium using highly specific ³H-labeled oxytocin, *Endocrinol. Jpn.* **26**:515–522.

10. Fuchs, A.-R., Fuchs, F., Husslein, P., Soloff, M. S., and Fernström, M. J., 1982, Oxytocin receptors and human parturition. A dual role for oxytocin in the initiation of labor, *Science* **215**:1396–1398.

11. Riemer, R. K., Goldfien, A. C., Goldfien, A., and Roberts, J. M., 1986, Rabbit uterine oxytocin receptors and *in vitro* contractile response: Abrupt changes at term and the role of eicosanoids, *Endocrinology* **119**:669–709.

12. Alexandrova, M., and Soloff, M. S., 1980, Oxytocin receptors and parturition. III. Increases in estrogen receptor and oxytocin receptor concentrations in the rat myometrium during $PGF_{2\alpha}$-induced abortion, *Endocrinology* **106**:739–743.

13. Fuchs, A.-R., Fuchs, F., Husslein, P., and Soloff, M. S., 1984, Oxytocin receptors in the human uterus during pregnancy and parturition, *Am. J. Obstet. Gynecol.* **150**:734–741.

14. Bercu, B. B., Hyashi, A., Poth, M. Alexandrova, M., Soloff, M. S., and Donahoe, P. K., 1980, LHRH induced delay of parturition, *Endocrinology* **107**:504–508.

15. Fuchs, A.-R., Husslein, P., and Fuchs, F., 1981, Oxytocin and the initiation of human parturition. II. Stimulation of prostaglandin production in human decidua by oxytocin, *Am. J. Obstet. Gynecol.* **141**:694–697.

16. Wilson, T., Liggins, G. C., and Whittaker, D. J., 1988, Oxytocin stimulates the release of arachidonic acid and prostaglandin $F_{2\alpha}$ from human decidual cells, *Prostaglandins* **35**:771–780.

17. Soloff, M. S., 1979, Regulation of oxytocin action at the receptor level, *Life Sci.* **25**:1453–1460.

18. Soloff, M. S., 1985, Oxytocin receptors and mechanisms of oxytocin action, in: *Oxytocin: Clinical and Laboratory Studies* (J. A. Amico and A. G. Robinson, eds.), Elsevier Scientific, Amsterdam, pp. 259–276.

19. Morgat, J. L., Hung, L. T., Cardinaud, R., Fromageot, P., Bockaert, J., Imbert, M., and Morel, F., 1970, Peptidic hormone interactions at the molecular level—Preparation of highly labelled ^3H oxytocin, *J. Labelled Compd.* **6**:276–284.

20. Flouret, G., Terada, S., Yang, F., Nakagawa, S. H., Nakahara, T., and Hechter, O., 1977, Iodinated neurohypophyseal hormones as potential ligands for receptor binding and intermediates in synthesis of tritiated hormones, *Biochemistry* **16**:2119–2124.

21. Elands, J., Barberis, C., Jard, S., Tribollet, E., Dreifuss, J.-J., Bankowski, K., Manning, M., and Sawyer, W. H., 1987, ^{125}I-Labelled d$(CH_2)_5$[Tyr(Me)^2,Thr^4,Tyr^9]OVT: A selective oxytocin receptor ligand, *Eur. J. Pharmacol.* **147**:197–207.

22. Soloff, M. S., Schroeder, B. T., Chakraborty, J., and Pearlmutter, A. F., 1977, Characterization of oxytocin receptors in the uterus and mammary gland, *Fed. Proc.* **36**:1861–1866.

23. Matlib, M. A., Crankshaw, J., Garfield, R. E., Crankshaw, D. J., Kwan, C.-Y., Branda, L. A., and Daniel, E. E., 1979, Characterization of membrane fractions and isolation of purified plasma membranes from rat myometrium, *J. Biol. Chem.* **254**:1834–1840.

24. Scatchard, G., 1949, The attraction of proteins for small molecules and ions, *Ann. N.Y. Acad. Sci.* **51**:660–672.

25. Roberts, J. S., McCracken, J. A., Gavagan, J. E., and Soloff, M. S., 1976, Oxytocin-stimulated release of prostaglandin $F_{2\alpha}$ from ovine endometrium *in vitro*: Correlation with estrous cycle and oxytocin–receptor binding, *Endocrinology* **99**:1107–1114.

26. Soloff, M. S., Fernström, M. A., Periyasamy, S., Soloff, S., Baldwin, S., and Wieder, M., 1983, Regulation of oxytocin receptor concentration in rat uterine explants by estrogen and progesterone, *Can. J. Biochem. Cell Biol.* **61**:625–630.

27. van Dyke, H. B., and Hastings, A. B., 1928, The response of smooth muscle in different ionic environments, *Am. J. Physiol.* **83**:563–577.

28. Clegg, P. C., Hopkinson, P., and Pickles, V. R., 1963, Some effects of calcium and magnesium ions on guinea-pig uterine muscle, *J. Physiol. (Lond.)* **167**:1–17.

29. Bentley, P. J., 1965, The potentiating action of magnesium and manganese on the oxytocic effect of some oxytocin analogues, *J. Endocrinol.* **32**:215–222.

30. Krejci, I., and Polacek, I., 1968, Effect of magnesium on the action of oxytocin and a group of analogues on the uterus *in vitro*, Eur. J. Pharmacol. **2**:393–398.

31. Walter, R., Dubois, B. M., and Schwartz, I. L., 1968, Biological significance of the amino acid residue in position 3 of neurohypophyseal hormones and the effect of magnesium on their uterotonic action, *Endocrinology* **83**:979–983.

32. Schild, H. O., 1969, The effect of metals on the S-S polypeptide receptor in depolarized rat uterus, *Br. J. Pharmacol.* **36**:329–349.

33. Pearlmutter, A. F., and Soloff, M. S., 1979, Characterization of the metal ion requirement for oxytocin–receptor interaction in rat mammary gland membranes, *J. Biol. Chem.*, **254**:3899–3906.

34. Soloff, M. S., and Swartz, T. L., 1974, Characterization of a proposed oxytocin receptor in the uterus of the rat and sow, *J. Biol. Chem.* **249**:1376–1381.

35. Chan, W. Y., and Kelly, N., 1967, A pharmacological analysis on the significance of the chemical functional groups on oxytocin to its oxytocic activity and on the effect of magnesium on the *in vitro* and *in vivo* oxytocin activity of neurohypophysial hormones, *J. Pharmacol. Exp. Ther.* **156**:150–158.

36. Rudinger, J., 1968, Synthetic analogues of oxytocin: An approach to problems of hormone action, *Proc. R. Soc. Lond. [Biol.]* **170**:17–26.

37. Rudinger, J., Pliska, V., and Krejci, I., 1972, Oxytocin analogs in the analysis of some phases of hormone action, *Recent Prog. Horm. Res.* **28**:131–172.

38. Manning, M., Coy, E., and Sawyer, W. H., 1970, Solid phase synthesis of 4-threonine oxytocin. A more potent and specific oxytocic agent than oxytocin, *Biochemistry* **9**:3925–3930.

39. Sawyer, W. H., and Manning, M., 1971, 4-Threonine analogues of neurohypophysial hormones with selectively enhanced oxytocin-like activities, *J. Endocrinol.* **49**:151–165.

40. Soloff, M. S., and Grzonka, Z., 1986, Effect of manganese on relative affinities of receptor for oxytocin analogues. Binding studies with rat myometrial and mammary gland membranes, *J. Biol. Chem.* **254**:3899–3906.

41. Joelsson, I., Ingelman-Sundberg, A., and Sandberg, F., 1966, The *in vivo* effect of oxytocin and vasopressin on the nonpregnant human uterus, *J. Obstet. Gynaecol. Br. Commonw.* **73**:832–836.

42. Coutinho, E. M., and Lopes, A. C. V., 1968, Response of the nonpregnant uterus to vasopressin as an index of ovarian function, *Am. J. Obstet. Gynecol.* **102**:479–489.

43. Bengtsson, L. P., 1970, Effect of progesterone upon the *in vivo* response of the human myometrium to oxytocin and vasopressin, *Acta Obstet. Gynecol. Scand.* **49**(Suppl. 6):19–25.

44. Embrey, M. P., and Moir, J. C., 1967, A comparison of the oxytocic effects of synthetic vasopressin and oxytocin, *J. Obstet. Gynaecol. Br. Commonw.* **74**:648–652.

45. Cobo, E., Cifuentes, R., and de Villamizar, M., 1978, Inhibition of menstrual uterine motility during water diuresis. *J.Obstet. Gynecol.* **132**:313–320.

46. Caldeyro-Barcia, R., and Sereno, J., 1961, The response of the human uterus to oxytocin throughout pregnancy, in: *Oxytocin* (R. Caldeyro-Barcia and H. Heller, eds.), Pergamon Press, Oxford, pp. 177–202.

47. Fuchs, A.-R., Fuchs, F., and Soloff, M. S., 1985, Oxytocin receptors in nonpregnant human uterus, *J. Clin. Endorcinol. Metab.* **60**:37–41.

48. Guillon, G., Balestre, M. N., Roberts, J. M., and Bottari, S. P., 1987, Oxytocin and vasopressin: Distinct receptors in myometrium, *J. Clin. Endocrinol. Metab.* **64**:1129–1135.

49. Soloff, M. S., and Fernström, M. A., 1987, Solubilization and properties of oxytocin receptors in rat mammary gland membranes, *Endocrinology* **120**:2474–2482.

50. Edelman, G. M., 1976, Surface modulation in cell recognition and cell growth, *Science* **198**:218–226.

51. Mescher, M. R., Jose, M. J. L., and Balk, S. P., 1981, Actin-containing matrix associated with the plasma membrane of murine tumour and lymphoid cells, *Nature* **289**:139–144.

52. Small, J. V., 1974, Contractile units in vertebrate smooth muscle cells, *Nature* **249**:324–327.

53. Soloff, M. S., Beauregard, G., and Potier, M., 1988, Determination of the functional size of oxytocin receptors in plasma membranes from mammary gland and uterine myometrium of the rat by radiation inactivation, *Endocrinology* **122**:1769–1772.

54. Glass, J. D., and du Vigneaud, V., 1972, Synthesis and certain pharmacological properties of

lysine-vasopressinoic acid methylamide and lysine-vasopressinoic acid dimethylamide, *J. Med. Chem.* **15**:486–488.

55. Manning, M., Olma, A., Klis, W., Kolodziejczyk, A., Nawrocka, E., Misicka, A., Seto, J., and Sawyer, W. H., 1984, Carboxy terminus of vasopressin required for activity but not binding, *Nature* **308**:652–653.

56. Berde, B., and Boissonnas, R. A., 1968, Basic pharmacological properties of synthetic analogues and homologues of the neurohypophysial hormones, in: *Handbook of Experimental Pharmacology, Volume XXIII. Neurohypophysial Hormones and Similar Polypeptides* (B. Berde, ed.), Springer-Verlag, New York, pp. 802–870.

57. Marshall, J. M., 1974, Effects of neurohypophysial hormones on the myometrium, in: *The Pituitary Gland and its Neuroendocrine Control, Part 1, Handbook of Physiology, Section 7: Endocrinology, Volume IV* (R. O. Greep, E. B. Astwood, E. Knobil, W. H. Sawyer, and S. R. Geiger, eds.), American Physiological Society, Washington, DC, pp. 469–492.

58. Forman, A., Gandrup, P., Andersson, K. E., and Ulmsten, U., 1982, Effects of nifedipine on oxytocin- and prostaglandin $F_{2\alpha}$-induced activity in the postpartum uterus, *Am. J. Obstet. Gynecol.* **144**:665–670.

59. Sakai, K., Yamaguchi, T., Morita, S., and Uchida, M., 1983, Agonist-induced contraction of rat myometrium in Ca-free solution containing Mn, *Gen. Pharmacol.* **14**:391–400.

60. Cantor, E. H., Kenessey, A., Semenuk, G., and Spector, S., 1984, Interaction of calcium channel blockers with non-neuronal benzodiazepine binding sites, *Proc. Natl. Acad. Sci. U.S.A.* **81**:1549–1552.

61. Schatzmann, H. J., and Vincenzi, F. F., 1969, Calcium movements across the membrane of human red cells, *J. Physiol. (Lond.)* **201**:369–395.

62. Pershadsingh, H. A., and McDonald, J. M., 1980, A high affinity calcium-stimulated magnesium-dependent adenosine triphosphatase in rat adipocyte plasma membranes, *J. Biol. Chem.* **255**:4087–4093.

63. Wuytack, F., Raeymaekers, L., Verbist, J., De Smedt, H., and Casteels, R., 1984, Evidence for the presence in smooth muscle of two types of Ca^{2+}-transport ATPase, *Biochem. J.* **224**:445–451.

64. Caroni, P., and Carafoli, E., 1981, The Ca^{2+}-pumping ATPase of heart sarcolemma. Characterization, calmodulin dependence, and partial purification, *J. Biol. Chem.* **256**:3263–3270.

65. Caroni, P., Zurini, M., Clark, A., and Carafoli, E., 1983, Further characterization and reconstitution of the purified Ca^{2+}-pumping ATPase of heart sarcolemma, *J. Biol. Chem.* **258**:7305–7310.

66. Åkerman, K. E. O., and Wikström, M. K. F., 1979, $(Ca^{2+} + Mg^{2+})$-stimulated ATPase activity of rabbit myometrium plasma membrane is blocked by oxytocin, *FEBS Lett.* **97**:283–287.

67. Soloff, M. S., and Sweet, P., 1982, Oxytocin inhibition of $(Ca^{2+} + Mg^{2+})$ATPase activity in rat myometrial plasma membranes, *J. Biol. Chem.* **257**:10687–10693.

68. Huszar, G., 1986, Cellular regulation of myometrial contractility and essentials of tocolytic therapy, in: *The Physiology and Biochemistry of the Uterus in Pregnancy and Labor* (G. Huszar, ed.), CRC Press, Boca Raton, FL, pp. 107–126.

69. Popescu, L. M., Nutu, O., and Panoiu, C., 1985, Oxytocin contracts the human uterus at term by inhibiting the myometrial Ca^{2+}-extrusion pump, *Biosci. Rep.* **5**:21–28.

70. Pershadsingh, H. A., and McDonald, J. M., 1979, Direct addition of insulin inhibits a high affinity Ca^{2+}-ATPase in isolated adipocyte plasma membranes, *Nature* **281**:495–497.

71. Yamaguchi, M., 1979, Effect of calcitonin on Ca-ATPase activity of plasma membrane in liver of rats, *Endocrinol. Jpn.* **26**:605–609.

72. Lin, S.-H., Wallace, M. A., and Fain, J. N., 1983, Regulation of Ca^{2+}-Mg^{2+}-ATPase activity in hepatocyte plasma membranes by vasopressin and phenylephrine, *Endocrinology* **113**:2268–2275.

73. Marc, S., Lieber, D., and Harbon, S., 1986, Carbachol and oxytocin stimulate the generation of inositol phosphates in the guinea pig myometrium, *FEBS Lett.* **201**:9–14.

74. Schrey, M. P., Read, A. M., and Steer, P. J., 1986, Oxytocin and vasopressin stimulate inositol phosphate production in human gestational myometrium and decidual cells, *Biosci. Rep.* **6**:613–619.

75. Carsten, M. E., and Miller, J. D., 1985, Ca^{2+} release by inositol trisphosphate from Ca^{2+}-transporting microsomes derived from uterine sarcoplasmic reticulum, *Biochem. Biophys. Res. Commun.* **130**:1027–1031.

76. Streb, H., Bayerdorffer, E., Hasse, W., Irvine, R. F., and Schulz, I., 1984, Effect of inositol-1,4,5-trisphosphate on isolated subcellular fractions of rat pancreas, *J. Membr. Biol.* **81**:241–253.

77. Spät, A., Fabiato, A., and Rubin, R. P., 1986, Binding of inositol trisphosphate by a liver microsomal fraction, *Biochem. J.* **233**:929–932.

78. Zhao, X., and Gorewit, R. C., 1987, Inositol-phosphate response to oxytocin-stimulation in dispersed bovine mammary cells, *Neuropeptides* **10**:227–233.

79. Soloff, M. S., Fernström, M. A., and Fernström, M. J., 1989, Vasopressin and oxytocin receptors on plasma membranes from rat mammary gland. Demonstration of vasopressin receptors by stimulation of inositol phosphate formation, and oxytocin receptors by binding of a specific ^{125}I-labeled oxytocin antagonist, $d(CH_2)_5^1[Tyr(Me)^2,Thr^4,Tyr-NH_2^9]OVT$, *Biochem. Cell Biol.* **67**:152–162.

80. Tolbert, M. E. M., White, A. C., Aspry, K., Cutts, J., and Fain, J. N., 1980, Stimulation by vasopressin and α-catecholamines of phosphatidylinositol formation in isolated rat liver parenchymal cells, *J. Biol. Chem.* **255**:1938–1944.

81. Kirk, C. J., and Michell, R. H., 1981, Phosphatidylinositol metabolism in rat hepatocytes stimulated by vasopressin, *Biochem. J.* **194**:155–165.

82. Bhalla, T., Enyedi, P., Spät, A., and Antoni, F. A., 1985, Pressor-type vasopressin receptors in the adrenal cortex: Properties of binding, effects on phosphoinositide metabolism and aldosterone secretion, *Endocrinology* **117**:421–423.

83. Woodcock, E. A., McLeod, J. K., and Johnston, C. I., 1986, Vasopressin stimulates phosphatidylinositol turnover and aldosterone synthesis in rat adrenal glomerulosa cells: Comparison with angiotensin II, *Endocrinology* **118**:2432–2436.

84. Koreh, K., and Monaco, M. E., 1986, The relationship of hormone-sensitive and hormone-insensitive phosphatidylinositol to phosphatidylinositol 4,5-bisphosphate in the WRK-1 cell, *J. Biol. Chem.* **261**:88–91.

85. Kirk, C. J., Guillon, G., Balestre, M.-N., and Jard, S., 1986, Stimulation by vasopressin and other agonists, of inositol-lipid breakdown and inositol phosphate accumulation in WRK1 cells, *Biochem. J.* **240**:197–204.

86. Nabika, T., Velletri, P. A., Lovenberg, W., and Beaven, M. A., 1985, Increase in cytosolic calcium and phosphoinositide metabolism induced by angiotensin II and [Arg]vasopressin in vascular smooth muscle cells, *J. Biol. Chem.* **260**:4661–4670.

87. Aiyar, N., Nambi, P., Stassen, F. L., and Crooke, S. T., 1986, Vascular vasopressin receptors mediate phosphatidylinositol turnover and calcium efflux in an established smooth muscle cell line, *Life Sci.* **39**:37–45.

88. Fox, A. W., Friedman, P. A., and Abel, P. W., 1987, Vasopressin receptor mediated contraction and [^{3}H]inositol metabolisms in rat tail artery, *Eur. J. Pharmacol.* **135**:1–10.

89. Casey, P. J., and Gilman, A. G., 1988, G. protein involvement in receptor–effector coupling, *J. Biol. Chem.* **263**:2577–2580.

90. Fischer, J. B., and Schonbrunn, A., 1988, The bombesin receptor is coupled to a guanine nucleotide-binding protein which is insensitive to pertussis and cholera toxins, *J. Biol. Chem.* **263**:2808–2816.

91. Goodman, R. R., Cooper, M. J., Gavish, M., and Snyder, S. H., 1982, Guanine nucleotide and cation regulation of the binding of [^{3}H]cyclohexyladenosine and [^{3}H]diethylphenylxanthine to adenosine A_1 receptors in brain membranes, *Mol. Pharmacol.* **21**:329–335.

92. Koch, B. D., and Schonbrunn, A., 1984, The somatostatin receptor is directly coupled to adenylate cyclase in GH_4C_1 pituitary cell membranes, *Endocrinology* **114**:1784–1790.

93. Hinkle, P. M., and Kinsella, P. A., 1984, Regulation of thyrotropin-releasing hormone binding by monovalent cations and guanyl nucleotides, *J. Biol. Chem.* **259**:3445–3449.

94. U'Prichard, D. C., and Snyder, S. H., 1978, Guanyl nucleotide influences of ^{3}H-ligand binding to α-noradrenergic receptors in calf brain membranes, *J. Biol. Chem.* **253**:3444–3449.
95. Soloff, M. S., 1975, Uterine receptors for oxytocin: Effects of estrogens, *Biochem. Biophys. Res. Commun.* **65**:205–212.
96. Nissenson, R., Flouret, G., and Hechter, O., 1978, Opposing effects of estradiol and progesterone on oxytocin receptors in rabbit uterus, *Proc. Natl. Acad. Sci. U.S.A.* **75**:2044–2048.
97. Fuchs, A.-R., Fuchs, F., and Soloff, M. S., 1985, Oxytocin receptors in nonpregnant human uterus, *J. Clin. Endocrinol. Metab.* **60**:37–41.
98. Sharma, S. C., and Fitzpatrick, R. J., 1974, Effect of oestradiol-17β and oxytocin on prostaglandin F alpha release in the anoestrous ewe, *Prostaglandins* **6**:97–105.
99. Mitchell, M. D., Flint, A. P.F., and Turnbull, A. C., 1975, Stimulation by oxytocin of prostaglandin F levels in uterine venous effluent in pregnant and puerperal sheep, *Prostaglandins* **9**:47–56.
100. Newcomb, R., Booth, W. D., and Rowson, L. E. A., 1977, The effect of oxytocin treatment on the levels of prostaglandin F in the blood of heifers, *J. Reprod. Fertil.* **49**:17–24.
101. Milvae, R. A., and Hansel, W., 1980, Concurrent uterine venous and ovarian arterial prostaglandin F concentrations in heifers treated with oxytocin, *J. Reprod. Fertil.* **60**:7–15.
102. Oyedipe, E. O., Gustafsson, B., and Kindahl, H., 1984, Blood levels of progesterone and 15-keto-13,14-dihydroprostaglandin $F_{2\alpha}$ during the estrous cycle of oxytocin treated cows, *Theriogenology* **22**:329–339.
103. Chan, W. Y., 1977, Relationship between the uterotonic action of oxytocin and prostaglandins: Oxytocin action and release of PG-activity in isolated nonpregnant and pregnant rat uteri, *Biol. Reprod.* **17**:541–548.
104. Forsling, M. L., Åkerlund, M., and Strömberg, P., 1981, Variations in plasma concentrations of vasopressin during the menstrual cycle, *J. Endocrinology.* **89**:263–266.
105. Strömberg, P., Åkerlund, M., Forsling, M. L., and Kindahl, H., 1983, Involvement of prostaglandins in vasopressin stimulation of the human uterus, *Br. J. Obstet. Gynecol.* **90**:332–337.
106. Hixon, J. E., and Flint, A. P. F., 1988, Effects of a luteolytic dose of oestradiol benzoate on uterine oxytocin receptor concentrations, phosphoinositide turnover and prostaglandin $F_{2\alpha}$ secretion in sheep, *J. Reprod. Fertil.* **79**:457–467.
107. Flint, A. P. F., Leat, W. M. F., Sheldrick, E. L., and Stewart, H. J., 1986, Stimulation of phosphoinositide hydrolysis by oxytocin and the mechanism by which oxytocin controls prostaglandin synthesis in the ovine endometrium, *Biochem. J.* **237**:797–805.

13

Regulatory Peptides and Uterine Function

Bent Ottesen and Jan Fahrenkrug

1. Introduction

The existence of unidentified neurotransmitters participating in nonadrenergic, noncholinergic nervous control of uterine function has been known for several years.[1-3] It is now recognized that the nerve supply to the urogenital tract is more complex since, in addition to the classical cholinergic and adrenergic innervation, there are nerves that could be classified as noncholinergic and nonadrenergic. The demonstration of peptides in nervous structures of the urogenital system of several species including humans (Table I) suggests that these substances could be involved in the local control of noncholinergic, nonadrenergic physiological events. All peptides, except relaxin, were originally isolated from extragenital sources, mainly the brain or the gut, and the presence in the genital tract has been demonstrated by immunologic techniques (immunohistochemistry and/or radioimmunoassay of tissue extracts). Although chromatographic studies suggest that these immunoreactive peptides are identical with their extragenital counterparts, confirmation awaits isolation and sequencing.

In the present chapter we review the localization, biological action, physiological implications, and putative clinical application of these neuropeptides.

2. Peptide Biosynthesis

Regulatory peptides are diverse in their function and localization. However, peptides that are secreted from cells share a common property in that they all are

Bent Ottesen • Departments of Obstetrics and Gynecology, Hvidovre and Herlev Hospitals, University of Copenhagen, Copenhagen, Denmark.   ***Jan Fahrenkrug*** • Department of Clinical Chemistry, Bispebjerg Hospital, University of Copenhagen, Copenhagen, Denmark.

Table I. Peptide Immunoreactivities Present in the Female Genital Tract

Gut–brain peptides
    Substance P
    Vasoactive intestinal polypeptide (VIP)
    Peptide witn N-terminal histidine and C-terminal isoleucine amide
    Neuropeptide Y (NPY)
    Leucine-enkephalin
    Methionine-enkephalin
    Gastrin-releasing peptide (GRP, bombesin)
Hypothalamic releasing hormones
    Thyrotropin-releasing hormone (TRH)
    Somatostatin
Neurohypophyseal hormones
    Oxytocin
    Neurophysin
    Vasopressin
Others
    Calcitonin-gene-related peptide (CGRP)
    Renin
    Relaxin
    Galanin

initially synthesized as larger precursors (prepropeptides), which are processed proteolytically to form biologically active products. The proteolytic events include removal of the signal sequence, which is necessary for sequestration of the protein into the endoplasmic reticulum, as well as subsequent enzymatic cleavage by endopeptidases that usually occur at some dibasic residues and occasionally at monobasic sites. The C-terminally extended peptides are often processed further by carboxypeptidases that specifically removes basic amino acid residues. Specific regulatory peptides can also undergo a variety of other posttranslational modifications, which include disulfide bond formation, glycosylation, sulfation, phosphorylation, acetylation, and amidation. A precursor may contain more than one biologically active peptide in its sequence, which are then cosynthesized and coreleased. Often the remaining peptide sequences with unknown biological activity are secreted as well. Finally, flexibility in proteolytic processing could explain the tissue- and/or cell-specific processing of certain precursors.

3. Vasoactive Intestinal Peptide and Peptide with N-Terminal Histidine and C-Terminal Isoleucine Amide

Vasoactive intestinal peptide (VIP) is a 28-amino-acid peptide that was originally isolated from porcine intestine and recognized for its potent vasodilatory effects.[4,5] The amino acid sequence of VIP from pig, cow, man, and rat is identical, whereas guinea pig VIP differs in four amino acid residues (positions 5, 9, 12,

26).[6-8] Another 27-amino-acid peptide with structural resemblance to VIP, PHI (peptide with N-terminal histidine and C-terminal isoleucine amide) (Fig. 1) was later isolated from porcine intestine[9] and found to have a distribution similar to that of VIP.[10-13] In man and rat, VIP was, by gene technology, found to be derived from a 170-amino-acid precursor molecule that in its sequence contained PHI.[14,15] The human form of this peptide, PHM (C-terminal methionine amide), differs from the porcine PHI by two amino acids and from rat PHI by four amino acids (Fig. 1).

3.1. Localization

Both VIP and PHI/PHM are present in the female genital nerves, where they seem to be localized in neurons (Fig. 2) presumed to be mainly intrinsic or

	1	2	3	4	5	6	7	8	9	10	11	12	13	14	15
VIP	His	Ser	Asp	Ala	Val	Phe	Thr	Asp	Asn	Tyr	Thr	Arg	Leu	Arg	Lys-
PHM	**His**	Ala	**Asp**	Gly	**Val**	**Phe**	**Thr**	Ser	Asp	Phe	Ser	Lys	**Leu**	Leu	Gly-
PHI	**His**	Ala	**Asp**	Gly	**Val**	**Phe**	**Thr**	Ser	Asp	Phe	Ser	**Arg**	**Leu**	Leu	Gly-
Secretin	**His**	**Ser**	**Asp**	Gly	Thr	**Phe**	**Thr**	Ser	Glu	Leu	Ser	**Arg**	**Leu**	**Arg**	Asp-
Glucagon	**His**	**Ser**	Gln	Gly	Thr	**Phe**	**Thr**	Ser	Asp	**Tyr**	Ser	Lys	Tyr	Leu	Asp-
GIP	Tyr	Ala	Glu	Gly	Thr	**Phe**	Ile	Ser	Asp	**Tyr**	Ser	Ile	Ala	Met	Asp
GRF	Tyr	Ala	**Asp**	**Ala**	Ile	**Phe**	**Thr**	Asn	Ser	**Tyr**	Arg	Lys	Val	Leu	Gly-

	16	17	18	19	20	21	22	23	24	25	26	27	28	29
VIP	Gln	Met	Ala	Val	Lys	Lys	Tyr	Leu	Asn	Ser	Ile	Leu	Asn	NH_2
PHM	**Gln**	Leu	Ser	Ala	**Lys**	**Lys**	**Tyr**	**Leu**	Glu	**Ser**	Leu	Met	NH_2	
PHI	**Gln**	Leu	Ser	Ala	**Lys**	**Lys**	**Tyr**	**Leu**	Glu	**Ser**	Leu	Ile	NH_2	
Secretin	Ser	Ala	Arg	Leu	Gln	Arg	Leu	**Leu**	Gln	Gly	Leu	Val	NH_2	
Glucagon	Ser	Arg	Arg	Ala	Gln	Asp	Phe	Val	Gln	Trp	Leu	Met	Asn	Thr
GIP	Lys	Ile	His	Gln	Gln	Asp	Phe	Val	**Asn**	Trp	Leu	**Leu**	Ala	Gln-
GRF	**Gln**	Leu	Ser	Ala	Arg	**Lys**	Leu	**Leu**	Gln	Asp	**Ile**	Met	Ser	Arg-

	30	31	32	33	34	35	36	37	38	39	40	41	42	43	44
GIP	Lys	Gly	Lys	Lys	Asn	Asp	Trp	Lys	His	Asn	Ile	Thr	Gln		
GRF	Gln	Gln	Gly	Glu	Ser	Asn	Gln	Glu	Arg	Gly	Ala	Arg	Ala	Arg	Leu-NH_2

Figure 1. The amino acid sequences of human VIP, PHM, porcine PHI, secretin, human glucagon, gastric inhibitory polypeptide (GIP), and growth hormone-releasing factor (GRF). Residues identical to VIP are shown in boldface.

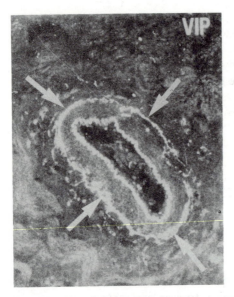

Figure 2. VIP- and PHI-immunoreactive nerve fibers (arrows) in an artery from the human vagina. The internal elastic membrane displays autofluorescence. Bar = 5 μm.

postganglionic parasympathetic. The localization of VIP immunoreactivity has been studied by immunohistochemistry, and the concentration of VIP immunoreactivity has been measured in tissue specimens from the female genital tract by radioimmunoassay.[16–28] Studies have been performed on the female genital organs from the pig, cat, rat, mouse, goat, rabbit, guinea pig, and human. Although the number of VIP-containing nerve fibers varies considerably among species, a common pattern of distribution seems to be present (Fig. 3). VIPergic nerve fibers are most abundant in the vagina, the cervix, and the clitoris, less numerous in the uterine body and tubes, and rare in the ovary. Locally, a rich nerve supply is observed around the natural sphincters, i.e., the internal and external cervical os and the isthmic part of the fallopian tube. Clusters of ganglion cells displaying VIP immunoreactivity are situated in the paracervical tissue of the uterovaginal junction[17,24] (Fig. 4).

Transection of the hypogastric nerve has no effect on the VIPergic nerve supply, whereas removal of the paracervical ganglia reduces the number of genital VIPergic nerve fibers except in the ovaries. Following transection of the uterine cervix, VIP-containing fibers disappear from the body of the uterus and the oviduct but persist in the ovaries. In the cervix below the level of deletion, however, the number of VIPergic nerves is unaffected.[17] These findings indicate that the major VIPergic nerve supply in the female genital tract is intrinsic, originating from the local ganglion. This hypothesis is supported by evidence from a recent study by Gu et al.[29] They demonstrated that the paracervical ganglion is the origin of the

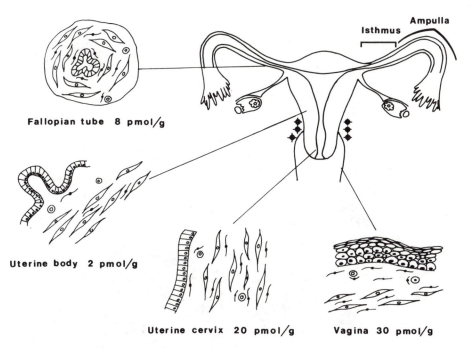

Fallopian tube 8 pmol/g

Uterine body 2 pmol/g

Uterine cervix 20 pmol/g

Vagina 30 pmol/g

Isthmus

Ampulla

Figure 3. The distribution of VIP immunoreactivity within the human female genital tract. Mean concentrations of VIP are given.

VIPergic nerves of the rat uterus, using a retrograde tracing technique combined with immunocytochemistry and denervation (Fig. 5). The VIPergic nerve fibers in the uterus seem to innervate epithelial cells, blood vessels, and smooth muscle cells (Figs. 2, 4, 6, 7).

At the ultrastructural level VIP immunoreactivity has been localized to large (100-nm) spherical dense-core (synaptic) vesicles in the nerve terminals.[30] The nerve endings are localized close to vascular and nonvascular smooth muscle and epithelial cells, indicating that these structures make up the target cells. These synaptic vesicles may be identical to the polypeptide type of nerves ("P terminals") previously described by Sporrong and associates in the cat uterus.[31]

The distribution and origin of VIP-containing nerves are very similar to those described for the adrenergic nerves supplying the female genital tract, and the numbers of VIPergic and adrenergic nerve fibers decrease toward term of pregnancy.[25,32] Cyclic fluctuations of VIP in the human fallopian tube have recently been reported.[33] The observed changes in VIP levels at the neuronal level are similar to those previously shown for norepinephrine.[34] In contrast to norepinephrine,[35,36] the total concentration of VIP in the uterus seems not to be influenced by treatment with sexual steroids.[37] In addition to the adrenergic and VIPergic innervation, acetyl-cholinesterase (AChE)-positive nerves (cholinergic) similarly innervate epithelial

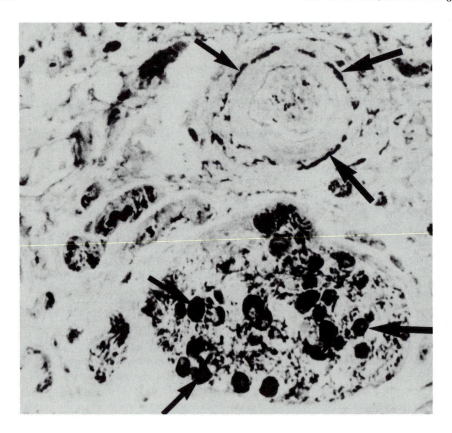

Figure 4. Section through paracervical tissue of cat. Ganglion containing VIP-immunoreactive cell bodies (lower part, arrows) as well as cross-sectioned artery surrounded by VIPergic fibers (upper part, arrows) (×200). (From Ottesen *et al.*[24] by courtesy of the editors.)

and smooth muscle cells in the human female genital tract.[38] The cholinergic innervation of vessels seems to be entirely restricted to the parametrial artery and present in only some species.

No studies are available concerning the coexistence of adrenergic/cholinergic transmitter and VIP/PHM within the same fiber of the female genital tract.

The distribution of PHI/PHM-immunoreactive nerves in the female genital tract is similar to the VIPergic distribution, although in some locations they seem to be less abundant.[12,39,40]

3.2. Biological Action

In the female genital tract VIP induces a dose-related increase in endometrial, myometrial, as well as total uterine blood flow,[41–48] and in the uterine vascular bed

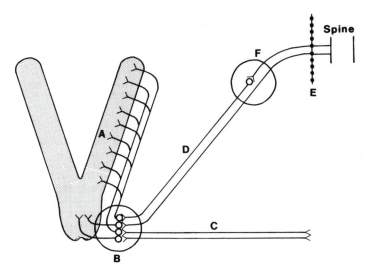

Figure 5. Schematic drawing illustrating the relationship between the uterine VIPergic neurons and hypogastric and pelvic nerves. VIPergic neurons (A) have their cell bodies in the paracervical ganglia (B). These are activated via nicotinic receptors by preganglionic fibers. They travel in the pelvic (C) and hypogastric nerves (D). The hypogastric fibers arise in either the lumbar sympathetic chain (E) or inferior mesenteric ganglion (F). Some fibers synapse in the inferior mesenteric ganglion, but the majority pass through. (From Ottesen[30] by courtesy of the editors.)

VIP seems to be more potent than other known vasodilators (Figs. 8, 9). The investigations on uterine blood flow reported have been performed on animals with intact ovaries or oophorectomized animals pretreated with estrogen and progesterone. The relationship between reproductive phase and the effect of VIP on blood flow still needs to be elucidated in detail. Clark *et al.*[46] have demonstrated that exogenously administered VIP in pregnant ewes has no effect on uterine blood flow. They concluded that either high local endogenous production of VIP prevents exogenous VIP from excerting its vascular effect or that VIP is a very weak uterine vasodilator in pregnant ewes.

The effect of PHI on myometrial blood flow has been studied in the female rabbit[49]; PHI displays a dose-dependent vasodilatory effect of the same magnitude as that of VIP (Fig. 10). Dose–response curves from the two peptides are superimposable with the same maximum and potency. When the two peptides are administered in equimolar doses, an additive effect is observed.

The effect of VIP on nonvascular smooth muscle has been studied in several tissues. A summary of the regions of the female genital tract in various species that have been investigated is given in Table II.[20,24,28,50,57] VIP has a dose-dependent inhibitory effect on mechanical activity of all smooth muscle preparations from the female genital tract. The inhibitory effect on electrical activity has been demonstrated on smooth muscle from the rabbit uterine body. The *in vivo* effect of VIP on

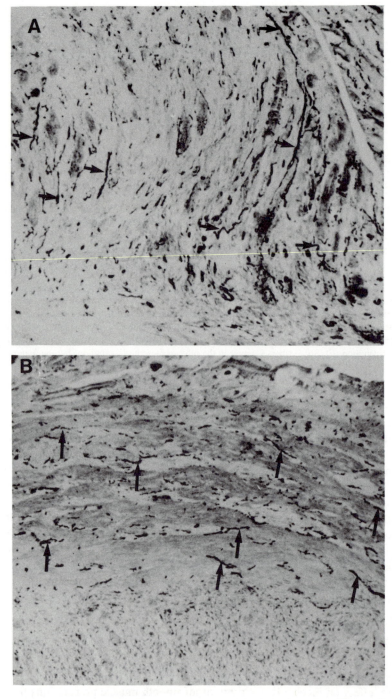

Figure 6. Cat uterus. Numerous VIPergic nerves (arrows) in the endometrium (**A**) as well as in the myometrium (**B**). PAP staining (×150). (Reproduced from Ottesen *et al.*[24] by courtesy of the editors.)

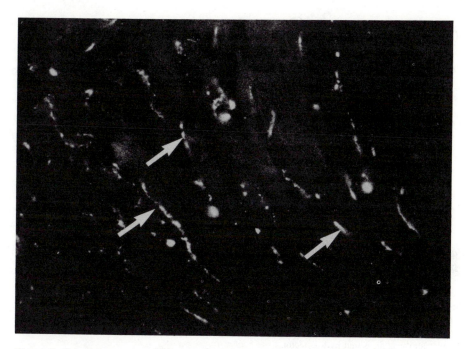

Figure 7. Cat uterine cervix. VIP-immunofluorescent nerves are numerous among bundles of smooth muscle (×300). (Reproduced from Ottesen *et al.*[24] by courtesy of the editors.)

intrauterine activity has been measured in human nonpregnant females.[58] An intrauterine microtip catheter was used to measure changes in pressure between the anterior and the posterior uterine walls. This pressure is primarily a result of the uterine mechanical activity but is also influenced by intraabdominal pressure. The results from these investigations confirm the relaxing effect of VIP.

The double- and single-sucrose-gap methods have been used to obtain information on the effect of VIP on action potential, changes in membrane resistance, and resting membrane potential in a small group of uterine smooth muscle cells from the rabbit and guinea pig.[54] The results from these studies suggest that the primary action of VIP is on the calcium balance of the myometrial smooth muscle cells, possibly to accelerate sequestration and/or extrusion of calcium from the cell. In some ways this is associated with the other effects of VIP observed in a sucrose-gap experiment, namely, inhibition of the generator potential, hyperpolarization, and a small increase in permeability of the membrane to potassium (Fig. 11).

The inhibitory action of VIP on nonvascular and vascular smooth muscle is not affected by α- and β-adrenoceptor blocking agents (phenoxybenzamine and propranolol), an anticholinergic antimuscarinic agent (atropine), or blockers of nerve transmission (tetrodotoxin, TTX). This indicates that VIP acts directly on the smooth muscle cell. The relaxant effect is not caused by specific antagonism at the

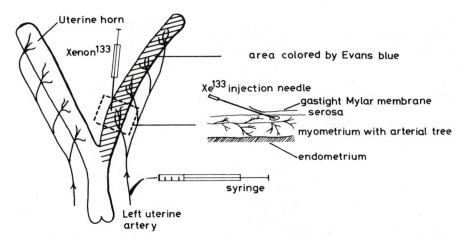

Figure 8. Myometrial blood flow (MBF) in nonpregnant rabbit measured by the washout of ^{133}Xe after local intramyometrial labeling. The abdomen was opened, and a branch of the uterine artery was exposed and ligated. A polyethylene catheter introduced into this branch was advanced retrogradely into the uterine artery, thereby allowing the hatched area to be exposed to an infused substance without occluding the blood supply to the uterine horn. Within this area, visualized by injection of Evans blue, an intramyometrial injection of ^{133}Xe was made, and the surface of the tissue was covered by a gas-tight mylar membrane. The tissue was then shielded with lead sheets, and the washout of xenon from the myometrium was measured by a detector.

oxytocin receptor, since VIP inhibits regular spontaneous contraction and contractions elicited by the addition of oxytocin, carbachol, prostaglandin $F_{2\alpha}$''' and substance p[44,53–54] (Fig. 12). Recently VIP binding sites have been identified in genital tissue by autoradiography[59] on tissue sections, and specific receptors have been reported in membrane preparations of the porcine uterine smooth muscle.[60]

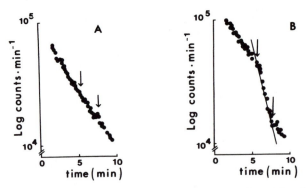

Figure 9. Typical curves showing clearance of ^{133}Xe from the myometrium of the rabbit. The effect of close intraarterial infusion of vehicle (**A**) [blood flow in A is 16 ml $\times$ min^{-1} $\times$ (100 g)$^{-1}$] and VIP, 50 pmol $\times$ min^{-1} $\times$ kg^{-1} [blood flow in **B** is 40 ml $\times$ min^{-1} $\times$ (100 g)$^{-1}$] for 2 min (between arrows).

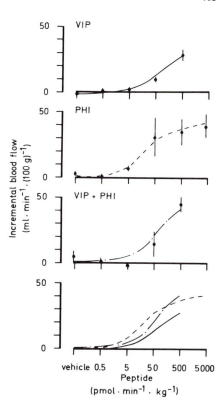

Figure 10. Relationship between dose of the peptide histidine isoleucine amide (PHI), vasoactive intestinal polypeptide (VIP), and PHI + VIP and incremental myometrial blood flow ($\triangle$MBF). Equimolar doses of PHI and VIP (0.25, 2.5, 25, and 250 pmol/min per kg body wt. given over 2 min) were used to examine possible interaction between the two peptides. Bottom panel is a combination of the three top panels. $\triangle$MBF = MBF2 − MBF1, where MBF2 is MBF during infusion of peptides or vehicle and MBF1 is MBF immediately before start of close intraarterial infusion. Data given as means ± S.E. of five to ten animals. (Reproduced from Bardrum *et al.*[49] by courtesy of the publishers.)

Table II. Summary of the Relaxant Effect of VIP on the Mechanical Activity of Nonvascular Smooth Muscle Specimens from the Nonpregnant Female Genital Tract

Species	Oviduct[b]	Uterine body[b]	Uterine cervix[b]
Woman	*	*	*
Rabbit[a]	*	**	*
Cat	n.t.	*	n.t.
Pig	n.t.	*	n.t.
Goat	n.t.	*	n.t.
Rat	n.t.	*	n.t.
Guinea pig[a]	n.t.	**	n.t.

[a] In these species the effects on both mechanical and myoelectrical activity have been studied.
[b] Symbols: *VIP inhibits mechanical activity; **VIP inhibits both myoelectric and mechanical activity; n.t., not tested.

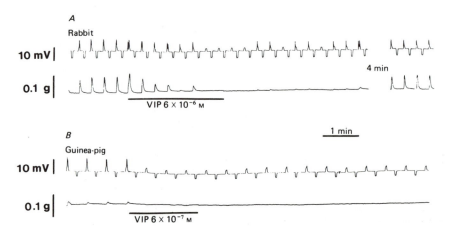

Figure 11. Effect of VIP on membrane potential and conductance of myometrial strips in a double sucrose gap. A: Rabbit; depolarizing impulses were 5×10^{-6} A, hyperpolarizing impulses were 5×10^{-6} A. B: Guinea pig; depolarizing impulses were 9×10^{-8} A, hyperpolarizing impulses were 5×10^{-7} A. (Reproduced from Bolton *et al.*[54] by courtesy of the publishers.)

The binding of VIP to its receptor activates adenylate cyclase.[59] These findings point toward the existence of a specific receptor for VIP in uterine smooth muscle.

The effect of PHI/PHM on myometrial smooth muscle strips from human and rabbit is identical to that of VIP[40,49] (Fig. 13). Recently, a larger C-terminally extended form of PHM designated PHV, which seems to occur in the genitalia, was shown to be more potent than VIP and PHM in relaxing isolated rat uterus.[61]

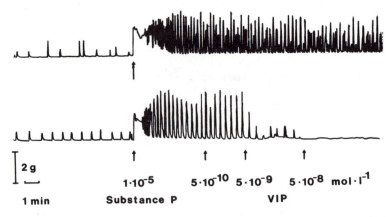

Figure 12. *In vitro* tension recordings on myometrial specimens from estradiol-pretreated rabbit. Upper trace shows the effect of substance P added to the organ bath (arrow) in a final concentration of 10^{-5} M. Lower trace shows the effect of VIP (5×10^{-10} to 5×10^{-8} M) on substance-P-evoked activity. (Reproduced from Ottesen *et al.*[44] by courtesy of the editors.)

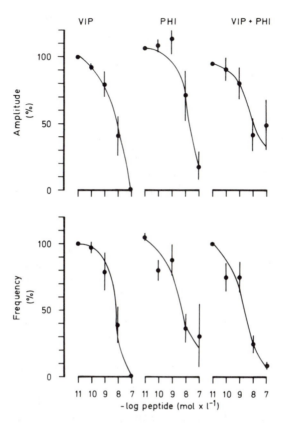

Figure 13. Inhibitory effect of peptide histidine isoleucine amide (PHI), vasoactive intestinal polypeptide (VIP), and PHI + VIP on spontaneous myometrial muscle activity *in vitro*. Effect is given as percentage of controls; i.e., no inhibition = 100%, and total relaxation equals = 0%. Changes in both amplitude (top) and frequency (bottom) are given. Figures are given as means and S.E. of six to 11 experiments. (Reproduced from Bardrum *et al.*[49] by courtesy of the editors.)

The VIP release during activation of the autonomic nerve supply to the uterus in the cat has been studied[62] (Fig. 14). The concentration of VIP was measured in the venous effluent from the uterus during electrical stimulation. There was a small spontaneous release of VIP during the basal condition as witnessed by a positive venoarterial concentration gradient. Efferent electrical stimulation of the parasympathetic nerves supplying the uterus, i.e., the pelvic nerves (S1–S3), caused a significant increase in the VIP release. Transmission of impulses from the pelvic nerves to the VIPergic neurons seems to involve cholinergic, nicotinic receptors, since the induced VIP release is completely abolished by hexamethonium. The pelvic-nerve-mediated VIP release does not conform to the classical (muscarinic) parasympathetic sequence of events, since the release of VIP during pelvic stimulation is resistant to atropine. Adrenergic fibers are not responsible for this atropine-

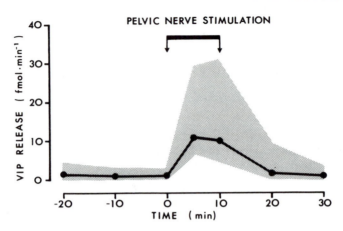

Figure 14. The effect of electrical pelvic nerve stimulation (10 Hz, 12 V, 2 msec) on VIP output from the uterus in anesthetized cats that had received atropine and α- and β-adrenoceptor blocking agents (phenoxybenzamine and propranolol) by local intraarterial infusion. Medians and total ranges are given ($N = 4$). (Reproduced from Fahrenkrug and Ottesen[62] by courtesy of the editors.)

resistant action, since neither α- nor β-adrenoceptor blocking agents inhibit the VIP release (Fig. 14). Efferent electrical stimulation of the sympathetic nerves supplying the uterus, i.e., the hypogastric nerve, also elicited a VIP release resistant to atropine and α- and β-adrenoceptor blocking agents, where hexamethonium was able to abolish the release. Here too, cholinergic, nicotinic receptors appear to be mediators. The VIP release induced by stimulation of the hypogastric nerve was accompanied by an increase in the uterovenous blood flow after muscarinic and adrenergic blockade, in accordance with the biological effect of exogenous VIP. It is therefore likely that VIP is a mediator of nonadrenergic, nonmuscarinic vasodilation in the uterus.

The release of VIP induced by electrical field nerve stimulation of superfused uterine smooth muscle from the cat has been studied *in vitro*.[63] Field stimulation in the presence of adrenergic and muscarinic blockade induces a relaxation of the smooth muscle preparation and a concomitant VIP release into the perfusate. These observations suggest that VIP is a mediator of nonadrenergic, noncholinergic, nonvascular smooth muscle relaxation within the female genital tract.[30,64]

3.3. Functional Significance

3.3.1. Ovum Transport

The ovum reaches the ampulla of the fallopian tube within a few hours after ovulation and remains there for 2–3 days before its rapid transit in the isthmus[65] (see Fig. 3). A sphincterlike function is therefore attributed to the isthmus of the oviduct. The existence of an unidentified neurotransmitter mediating non-adrenergic, noncholinergic oviductal smooth muscle relaxation has been known for some years.[1,2] The rich supply of VIPergic nerve fibers and the fact that VIP fulfills a number of the neurotransmitter criteria make it likely that VIP represents this hitherto unknown inhibitory neurotransmitter. Treatment of rabbits with estrogen and progesterone in order to imitate the postovulatory steroid concentration in women makes the uterine smooth muscle more sensitive to the relaxant effect of VIP,[37] thus favoring the opening of the isthmic sphincter. In favor of this hypothesis is the finding reported by Helm *et al.*[33] that in the isthmic part of the human fallopian tube, the highest concentration of VIP can be demonstrated in the luteal phase, thus facilitating the opening of the tubal sphincter. In the rabbit, however, the infusion of VIP over a period of 60–240 min in a dose of 75 pmol/min per kg relaxes the fallopian tube but has no effect on ovum transport.[66]

3.3.2. Uterine Motility

VIP may, as in the oviduct, participate in the local nervous control of myometrial motility. It inhibits myometrial activity in nonpregnant women, both *in vivo* and *in vitro,* but has no effect on preparations from the uterine body obtained at term.[25] Preliminary *in vitro* experiments have demonstrated that in the third month of pregnancy VIP is still able to inhibit myometrial activity (B. Ottesen, unpublished observation). Accordingly, the number of VIP receptors per milligram protein is significantly lower in the myometrium of rabbits at term compared both to early gestation and to nonpregnant animals.[37]

VIP could therefore have a progesteronelike effect[67] keeping the uterine smooth muscle relaxed during pregnancy until the time of delivery. This is supported by the antagonistic effect of VIP on myometrial activity induced by prostaglandin and oxytocin in rabbits.[50,52,53] Both prostaglandins and oxytocin seem to be involved in the initiation of labor.[68] A significant decrease in VIP content (pmol/g wet weight) within the uterine body toward full term has also been demonstrated in rabbits and rats.[32,37] In rats, pregnancy induces a marked, almost 50%, reduction in the total content of VIP in the uterine horns. The innervation normalizes within 25 days following delivery. No significant changes can be demonstrated in the VIP innervation of the cervical region in relation to pregnancy. In contrast to the body of the uterus, which must remain relaxed during gestation and contract at parturition, the cervix is closed throughout pregnancy and opens only at parturition. VIP may be involved in the relaxation of the cervix. Increased activity in the dense network of VIPergic nerves within the cervix may lead to relaxation of the cervical smooth muscle. This may explain the increased VIP concentration measured in peripheral venous blood of women during delivery.[25] This increase, however, may be secondary to other substances, which may cause release of VIP rather than increased activity of VIPergic nerves within the cervix.

3.3.3. Blood Flow

A noncholinergic vasodilation induced by nerve stimulation involving an unknown transmitter was reported by Bell in 1968.[3] Our present knowledge leads us to suggest that VIP may be this hitherto unknown transmitter. Although there is no doubt that the process of menstruation in primates is under hormonal control, suggestions have been put forward that nervous mechanism may be involved.[69] The vascular localization and the stimulatory effect of VIP on endometrial blood flow indicate a role for VIP in the regulation of menstruation. The menstrual bleeding is believed to be induced by contraction in the spiral arteries leading to ischemia of the epithelium. This phase is followed by a vasodilation of the arteries and subsequent shedding of the superficial layers of the endometrium. Although no menstruation-related changes in the VIP concentration of peripheral blood can be detected,[25] a local release of VIP may still occur.

4. Neuropeptide Y

Neuropeptide Y (NPY) is a 36-amino-acid peptide originally extracted from the brain and chemically characterized as having tyrosine both N- and C-terminally in the molecule (Fig. 15). Its amino acid sequence reveals strong similarities to other peptides of the pancreatic polypeptide family.[70,71]

4.1. Localization

Nerve fibers displaying NPY immunoreactivity have been demonstrated in all regions of the female genital tract in the guinea pig, cat, pig, man, rabbit, and rat.[71-77] The NPY fibers form a dense network in the uterine cervix, and they are quite numerous in the uterine body and the fallopian tubes. The majority of immunoreactive fibers are found in close association with nonvascular smooth muscle and

```
          1   2   3   4   5   6   7   8   9  10  11  12  13  14  15  16  17

NPY   Tyr-Pro-Ser-Lys-Pro-Asp-Asn-Pro-Gly-Glu-Asp-Ala-Pro-Ala-Glu-Asp-Met

APP   Gly-Pro-Ser-Gln-Pro-Thr-Tyr-Pro-Gly-Asp-Asp-Ala-Pro-Val-Glu-Asp-Leu


         18  19  20  21  22  23  24  25  26  27  28  29  30  31  32  33  34  35  36

NPY   Ala-Arg-Tyr-Tyr-Ser-Ala-Leu-Arg-His-Tyr-Ile-Asn-Leu-Ile-Thr-Arg-Gln-Arg-Tyr-NH2

APP   Ile-Arg-Phe-Tyr-Asn-Asp-Leu-Gln-Gln-Tyr-Leu-Asn-Val-Val-Thr-Arg-His-Arg-Tyr-NH2
```

Figure 15. The amino acid sequences of human neuropeptide Y (NPY) and avian pancreatic polypeptide (APP). Corresponding residues are shown in boldface.

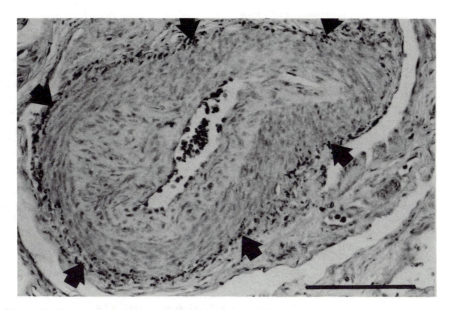

Figure 16. Cross section of human cervical blood vessel. NPY-immunoreactive nerve fibers (arrows) are located at the adventitia–median border (PAP technique). Bar = 100 μm.

around large and medium-sized arteries (Fig. 16). A few fibers occur in relation to epithelial cells. In the genital tract, NPY-like immunoreactivity seems to be present in norepinephrine-containing neurons, suggesting that NPY is a major peptide in the sympathetic nervous system.[73] A very dense occurrence of NPY-containing nerve cell bodies within the pelvic ganglia suggests that NPY nerve fibers originate from these local ganglia. This is supported by denervation experiments in the rat, which lead to increased amounts of NPY immunoreactivity within the ganglia, most likely because of the decreased axonal transport of the peptide.

4.2. Biological Action

In vitro experiments using the external longitudinal muscle of the isthmic part from the human fallopian tube and the cervical smooth muscle from estrogen-pretreated rats have shown that NPY inhibits dose-dependently the contractile response to transmural electrical stimulation, although the peptide does not affect the resting tension or the spontaneous activity.[75] In the isthmus preparation NPY inhibits tritiated norepinephrine release, suggesting a prejunctional inhibitory action on adrenergic transmission. In rabbits, however, NPY has a direct and dose-related stimulatory effect on nonvascular smooth muscle[76] (Fig. 17). Neuropeptide Y is also a potent vasoconstrictor, both *in vitro*[77] and on myometrial blood flow in nonpregnant rabbits[76] (Fig. 18).

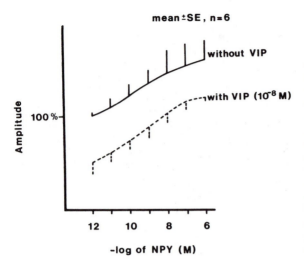

Figure 17. The effect of NPY and NPY + VIP on spontaneous myometrial muscle activity *in vitro*. Effect on the amplitude is given as percentage of a preceding control period, i.e., the amplitude when no peptides are present. The effect is expressed as mean and S.E. of experiments from six animals. A significant dose-related stimulatory effect of NPY is observed. Addition of VIP leads to a downward displacement of the curve. (Reproduced from Tenmuko *et al.*[76] by courtesy of the publishers.)

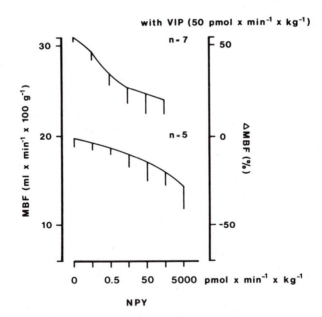

Figure 18. The effect of NPY and NPY + VIP on the myometrial blood flow. The relationship between NPY as well as NPY + VIP and incremental myometrial blood flow ($\triangle$MBF) has been established. $\triangle$MBF = MBF_2 − MBF_1, where MBF_2 is the myometrial blood flow related to the infusion of the peptide or vehicle, and MBF_1 is the myometrial blood flow immediately before the start of the infusion. Figures are given in ml × min^{-1} × 100 g^{-1} as mean and S.E. of five to seven experiments, respectively. NPY induces a significant dose-related decrease in the MBF. The addition of the vasodilator VIP (upper trace) increases MBF but does not prevent vasoconstriction. The maximum dose of NPY used does not fully overcome the vasodilating properties of VIP. (Modified from Tenmuko *et al.*[76] by courtesy of the publishers.)

4.3. Physiological Implications

Neuropeptide Y may participate in the local nervous control of the smooth muscle contraction and blood flow within the uterus.

5. Substance P

Substance P (SP) was discovered in 1931 by Euler and Gaddun,[79] thanks to its depressor effect. It was primarily demonstrated within the brain, but immunohistochemical studies have revealed that SP immunoreactivity is present in peripheral sensory and autonomic nerves, particularly in the primary sensory afferents.[80] The peptide consists of 11 amino acids and belongs to the family of tachykinins; it was the first peptide discovered to have a dual brain–gut distribution (Fig. 19).

5.1. Localization

Immunohistochemistry and immunochemistry have been used to study the localization and distribution of SP immunoreactivity within the female genital tract.[81–84] Several species have been studied, including man, rat, cat, and guinea pig (for recent reviews, see ref. 85). The SP-immunoreactive nerve fibers follow a common pattern in all species: a great number of fibers are observed within the vaginal wall, a moderate number within the uterus and oviduct, and hardly any in the ovary. Throughout the female genital tract, the SP-immunoreactive nerve fibers are localized in close relation to vascular and nonvascular smooth muscle as well as epithelial cells. In the fallopian tube the highest concentration of SP immunoreactivity has been demonstrated in the uterotubal junction, indicating a physiological role for SP in the nervous control of the isthmic sphincter.[82]

In the rat, no significant differences in the total content of SP immunoreactivity per organ can be detected during the estrous cycle, after ovariectomy, or during pregnancy.[81] Gu *et al.*[86] have demonstrated that hypogastric denervation significantly reduces uterine SP immunoreactivity, and a total pelvic denervation leads to complete disappearance of SP immunoreactivity in the rat uterus. Furthermore, it has been shown that the SP-immunoreactive nerve fibers in the uterus, cervix, and vagina are capsaicin sensitive. Capsaicin treatment leads to a selective and permanent destruction of primary afferent SP neurons and their terminals but spares SP neurons of the CNS. This suggests that SP immunoreactivity is associated with the primary afferent fibers that follow the pudendal, pelvic, and hypogastric nerves. The terminals are most likely located in sympathetic ganglia with the cell bodies in

Arg-Pro-Lys-Pro-Gln-Gln-Phe-Phe-Gly-Leu-Met-NH$_2$

Figure 19. The amino acid sequence of human substance P.

the dorsal root ganglia.[81,86] Consequently, capsaicin-sensitive afferents are able to release their transmitters at both their central and peripheral terminals, thereby mediating biological effects at both terminals.

5.2. Biological Action

The distribution of SP-immunoreactive nerve fibers suggests that SP may influence blood flow and smooth muscle activity. Substance P has been shown to be a vasodilator agent in the genital tract.[84,87] It also affects nonvascular smooth muscle of the reproductive tract by increasing both the frequency and amplitude of uterine smooth muscle contractions.[88] This effect can be antagonized by VIP[88] (Fig. 12). In addition, there is good evidence to suggest that SP is involved in sensory reflexes, especially those related to nociception. This and the distribution of SP-containing primary afferent nerves within the genital tract suggest that SP may participate in certain reproductive reflexes such as implantation and parturition.

5.3. Physiological Implications

The dense innervation of the isthmic sphincter of the fallopian tube by SP- and VIP-immunoreactive nerve fibers and the opposing actions of the two peptides suggest that these peptidergic nerve fibers participate in a dual local nervous control of the sphincter.

5.4. Labor

The presence of SP in the female genital tract and its pharmacological effects on smooth muscle raise the possibility that SP participates in the regulation of the rhythmic muscular activity and muscle tone of the uterus. It might be suggested that SP release within the uterus at term may in some way participate in the cascade of events leading to the onset of labor.

6. Enkephalins: Leu- and Met-Enkephalin

Leu- and Met-enkephalin are peptides consisting of five amino acids (Fig. 20). The enkephalins are neuropeptides with high affinity on opiate receptors and have been demonstrated immunohistochemically in neuronal elements of both the central and the peripheral nervous systems.

6.1. Localization

The distribution in the feline genitourinary tract has been described by Alm *et al.*[89] Enkephalin-immunoreactive nerve fibers occurs in the paracervical ganglia. These fibers seem to connect with the ganglia, and fine immunoreactive varicosities

Tyr-Gly-Gly-Phe-Leu

Tyr-Gly-Gly-Phe-Met

Figure 20. The amino acid sequences of human Leu-enkephalin and Met-enkephalin.

can be observed in close association with nonstaining cell bodies. In the female genital tract, enkephalin-immunoreactive nerve fibers are scarce, but they are regularly observed in the smooth muscle layer within the cervix.[89]

6.2. Biological Action

In the female genital tract, Leu-enkephalin stimulates myometrial blood following close intraarterial administration. The effect on smooth muscle *in vitro* is unknown.

7. Calcitonin-Gene-Related Peptide

Calcitonin-gene-related peptide (CGRP) consists of 37 amino acid residues. This peptide, which has been identified in central and peripheral neural tissue, is a product of mRNA formed by alternative tissue-specific RNA splicing and a transcription of the calcitonin gene.[90,91] Though originating from the same gene, CGRP bears no structural similarity to calcitonin (Fig. 21).

7.1. Localization

CGRP immunoreactivity is localized within nerve fibers throughout the female genital tract of the rat.[92] In general, the distribution of CGRP parallels that of

Ala-Cys-Asp-Thr-Ala-Thr-Cys-Val-Thr-His-Arg-Leu-Ala-Gly-Leu-

Leu-Ser-Arg-Ser-Gly-Gly-Val-Val-Lys-Asn-Asn-Phe-Val-Pro-Thr-

Asn-Val-Gly-Ser-Lys-Ala-Phe-NH$_2$

Cys-Gly-Asn-Leu-Ser-Thr-Cys-Met-Leu-Gly-Thr-Tyr-Thr-Gln-Asp-

Phe-Asn-Lys-Phe-His-Thr-Phe-Pro-Gln-Thr-Ala-Ile-Gly-Val-Gly-

Ala-Pro-NH$_2$

Figure 21. The amino acid sequences of human α-calcitonin-gene-related peptide (CGRP) and of human calcitonin.

substance P, and it is possible that the two peptides may coexist in some nerve fibers of the genital tract, although the number of CGRP-immunoreactive nerves markedly exceeds those of substance P. In the uterine cervix and the fallopian tube, immunoreactive nerve fibers are localized in the submucosa, particularly beneath the epithelium, but the fibers do not seem to penetrate the basal membrane of the epithelium. Fewer CGRP-immunoreactive nerve fibers are present in the uterus, where they are distributed mainly in the myometrium. CGRP-immunoreactive fibers are also seen adjacent to most blood vessels including both arteries and veins. The fact that capsaicin pretreatment results in a substantial depletion of CGRP content in the genital tract suggests that CGRP is present in afferent sensory nerve fibers.

7.2. Biological Action

The CGRP nerve fibers in the rat myometrium may be involved in the regulation of uterine mechanical activity, since CGRP has been shown to be a potent inhibitor of spontaneous smooth muscle activity in the rat uterus.[93]

8. Galanin

Galanin is a 29-amino-acid peptide recently isolated from porcine upper intestine[94] (Fig. 22). Galanin immunoreactivity is present in significant quantities in tissue from female human genitourinary tract and in the genitourinary tract of the female rat.[95] Most galanin-immunoreactive nerves are found within the smooth muscle, but the galanin fibers are also seen in nerve bundles and in close relation to arteries and veins. The location of galanin provides some clues to the possible target of the peptide, and experiments *in vitro* have demonstrated that galanin induces a slight contraction of smooth muscle preparations from the uterus.[93]

9. Oxytocin, Vasopressin, Renin, and Relaxin

Oxytocin, vasopressin, renin, and relaxin immunoreactivities have been localized in the genital tract but not within nervous structures (Fig. 23). Oxytocin, oxytocin neurophysin, and vasopressin immunoreactivity occur in the endometrium of the uterus and the fallopian tube.[96,97]

Renin immunoreactivity occurs in the apical region of endometrial cells and the perivascular cells of the myometrium.[98] Intramural renin may play a local role in

Gly-Trp-Thr-Leu-Asn-Ser-Ala-Gly-Tyr-Leu-Leu-Gly-Pro-His-Ala-

Ile-Asp-Asn-His-Arg-Ser-Phe-His-Asp-Lys-Tyr-Gly-Leu-Ala-NH$_2$

Figure 22. The amino acid sequence of human galanin.

Oxytocin   Cys-Tyr-Ile-Gln-Asn-Cys-Pro-Leu-Gly-NH$_2$

Arg-Vasopressin   Cys-Tyr-Phe-Gln-Asn-Cys-Pro-Arg-Gly-NH$_2$

Renin   Asp-Arg-Val-Tyr-Ile-His-Pro-Phe-His-Leu-
                                   Val-Ile-His

Relaxin (porcine)   <u>A chain</u>
                                  Arg-Met-Thr-Leu-Ser-Glu-Lys-Cys-Cys-Glu-Val-
                                  Gly-Cys-Ile-Arg-Lys-Asp-Ile-Ala-Arg-Leu-Cys
                                  <u>B chain</u>
                                  Ser-Thr-Asn-Asp-Phe-Ile-Lys-Ala-Cys-Gly-Arg-
                                  Glu-Leu-Val-Arg-Leu-Trp-Val-Glu-Ile-Cys-Gly-
                                  Ser-Val-Ser-Trp

Relaxin (rat)   <u>A chain</u>
                                  Ser-Gly-Ala-Leu-Leu-Ser-Glu-Gln-Cys-Cys-His-
                                  Ile-Gly-Cys-Thr-Arg-Arg-Ser-Ile-Ala-Lys-Leu-Cys
                                  <u>B chain</u>
                                  Arg-Val-Ser-Glu-Glu-Trp-Met-Asp-Gln-Val-Ile-Gln
                                  Val-Cys-Gly-Arg-Gly-Tyr-Ala-Arg-Ala-Trp-Ile-Glu
                                  Val-Cys-Gly-Ala-Ser-Val-Gly-Arg-Leu-Ala-Leu

Figure 23. The amino acid sequences of oxytocin, vasopressin, renin, and relaxin (human unless otherwise indicated).

the regulation of vascular tone and/or the tone of the uterine muscle itself, and endometrial renin could be involved in the bleeding arrest that follows abrasion. In pregnant mice, renin has been localized almost exclusively in the decidual endodermal epithelial linings of the yolk sack (uterine cavity) near the marginal sections of the placenta.[99] This localization of renin suggests a role in parturition or delivery of the placenta.

    Relaxin was originally isolated from the corpus luteum of pregnant sows.[100] Relaxin immunoreactivity occurs in both nongestational and gestational corpus luteum and in the secretory endometrium of nonpregnant women.[101,102] Relaxin

```
Ala-Gly-Cys-Lys-Asn-Phe-Phe-Trp-Lys-Thr-Phe-Thr-Ser-Cys
```

Figure 24. The amino acid sequence of human somatostatin.

has been clearly demonstrated to have an important role in parturition[103,104] in many species through an effect on the pelvic ligaments. Its importance in human parturition is unclear, although there is some support for a role in cervical ripening[105,106] and rupture of membranes.[100] The effect of relaxin on the human myometrium is inconsistent, and the problems may not be solved until human relaxin becomes available.

10. Somatostatin

Somatostatinlike immunoreactivity has been demonstrated in the chorionic villi and decidua of pregnant women (Fig. 24). The highest concentration is localized within the syncytiotrophoblast villi and in the stromal cells of the decidua.[107] The significance of these findings needs further elucidation.

11. Conclusion and Perspectives

By immunocytochemistry a number of the gut–brain peptides have been demonstrated in nerve fibers of the mammalian uterus. These peptides are localized to large vesicles and nerve terminals of afferent fibers or efferent nerves innervating blood vessels, nonvascular smooth muscle, lining epithelium, and glands. There is evidence that some neuropeptides (VIP, PHI, NPY) participate in the local noncholinergic, nonadrenergic nervous control of smooth muscle activity and blood flow and that other peptides (substance P, CGRP) seem to be sensory transmitters. It is likely that impaired function of the peptidergic nerves may be involved in pathophysiological conditions.

References

1. Murcott, C. A., and Carpenter, J. R., 1977, Inhibition in the rat oviduct mediated by an unidentified transmitter, *Br. J. Pharmacol.* **61**:498P.
2. Lindblom, B., Ljung, B., and Hamberger, L., 1979, Adrenergic and novel non-adrenergic neuronal mechanism in the control of smooth muscle activity in the human oviduct, *Acta Physiol. Scand.* **106**:215–220.
3. Bell, C., 1968, Dual vasoconstrictor and vasodilator innervation of the uterine arterial supply in the guinea pig, *Circ. Res.* **23**:279–289.
4. Mutt, V., and Said, S. I., 1974, Structure of the porcine vasoactive intestinal octacosapeptide. The amino-acid sequence. Use of kallikrein in its determination, *Eur. J. Biochem.* **42**:581–589.

5. Said, S. I., and Mutt, V., 1970, Polypeptide with broad biological activity: Isolation from small intestine, *Science* **169**:1217–1218.

6. Carlquist, M., Mutt, V., and Jörnvall, H., 1979, Isolation and characterization of bovine vasoactive intestinal peptide (VIP), *FEBS Lett.* **108**:457–460.

7. Carlquist, M., McDonald, T. J., Go, V. L. W., Bataille, D., Johansson, C., and Mutt, V., 1982, Isolation and amino acid composition of human vasoactive intestinal polypeptide (VIP), *Horm. Metab. Res.* **14**:28–29.

8. Du, B.-H., Eng, J., Hulmes, J. D., Chang, M., Pan, Y.-C. E., and Yalow, R. S., 1985, Guinea pig has a unique mammalian VIP, *Biochem. Biophys. Res. Commun.* **128**:1093–1098.

9. Tatemoto, K., and Mutt, V., 1981, Isolation and characterization of the intestinal peptide porcine PHI (PHI-27), a new member of the glucagon–secretin family, *Proc. Natl. Acad. Sci. U.S.A.* **78**:6603–6607.

10. Anand, P., Gibson, S. J., Yiangou, Y., Christofides, N. D., Polak, J. M., and Bloom, S. R., 1984, PHI-like immunoreactivity co-locates with the VIP-containing system in human lumbosacral spinal cord, *Neurosci. Lett.* **46**:191–196.

11. Bishop, A. E., Polak, J. M., Yiangou, Y., Christofides, N. D., and Bloom, S. R., 1984, The distribution of PHI and VIP in porcine gut and their co-localisation to a proportion of intrinsic ganglion cells, *Peptides* **5**:235–259.

12. Fahrenkrug, J., Bek, T., Lundberg, J. M., and Hökfelt, T., 1985, VIP and PHI in cat neurons: Colocalization but variable tissue content possibly due to differential processing, *Regul. Peptides* **12**:21–34.

13. Lundberg, J. M., Fahrenkrug, J., Hökfelt, T., Martling, C.-R., Larsson, O., Tatemoto, K., and Änggård, A., 1984, Co-existence of peptide HI (PHI) and VIP in nerves regulating blood flow and bronchial smooth muscle tone in various mammals including man, *Peptides* **5**:593–606.

14. Nishizawa, M., Hayakawa, Y., Yanaihara, N., and Okamoto, H., 1985, Nucleotide sequence divergence and functional constraint in VIP precursor mRNA evolution between human and rat, *FEBS Lett.* **183**:55–59.

15. Itoh, N., Obata, K.-I., Yanaihara, N., and Okamoto, H., 1983, Human preprovasoactive intestinal polypeptide contains a novel PHI-27 like peptide, PHM-27, *Nature* **304**:547–549.

16. Alm, P., Alumets, J., Håkanson, R., and Sundler, F., 1977, Peptidergic (vasoactive intestinal peptide) nerves in the genitourinary tract, *Neuroscience* **2**:751–754.

17. Alm, P., Alumets, J., Håkanson, R., Owman, C., Sjöberg, N.-O., Sundler, F., and Walles, B., 1980, Origin and distribution of VIP (vasoactive intestinal polypeptide)-nerves in the genitourinary tract, *Cell Tissue Res.* **205**:337–347.

18. Alm, P., Alumets, J., Håkanson, R., Helm, G., Owman, C., Sjöberg, N.-O., and Sundler, F., 1980, Vasoactive intestinal polypeptide nerves in the human female genital tract, *Am. J Obstet. Gynecol.* **136**:349–351.

19. Lundberg, J. M., Hökfelt, T., Änggård, A., Uvnäs-Wallensten, K., Brimijoin, S., Brodin, E., and Fahrenkrug, J., 1980, Peripheral peptide neurons: Distribution, axonal transport, and some aspects on possible function, in: *Neural Peptides and Neuronal Communication* (E. Costa and M. Trabucchi, eds.), Raven Press, New York, pp. 25–36.

20. Ström, C., Lundberg, J. M., Ahlman, H., Dahlström, A., Fahrenkrug, J., and Hökfelt, T., 1981, On the VIPergic innervation of the utero-tubal junction, *Acta Physiol. Scand* **111**:213–215.

21. Larsson, L.-I., Fahrenkrug, J., and Schaffalitzky de Muckadell, O. B., 1977, Vasoactive intestinal polypeptide occurs in nerves of the female genitourinary tract, *Science* **197**:1374–1375.

22. Humphrey, C., Munro, H., and Fischer, J. E., 1979, *In vitro* synthesis of vasoactive intestinal peptide: Discovery of a large precursor molecule, *Surg. Forum* **30**:355–357.

23. Ottesen, B., Ulrichsen, H., Fahrenkrug, J., Wagner, G., Jensen, N. E., Carter, A. M., Larsen, J.-J., and Stolberg, B., 1979, The effect of vasoactive intestinal polypeptide on myometrial electrical activity, contractility and blood flow, in: *Psychoneuroendocrinology in Reproduction* (L. Zichella and P. Pancheri, eds.), Elsevier/North-Holland Biomedical Press, Amsterdam, pp. 439–448.

24. Ottesen, B., Larsen, J.-J., Fahrenkrug, J., Stjernquist, M., and Sundler, F., 1981, Distribution and motor effect of VIP in female genital tract, *Am. J. Physiol.* **240**:E32–E36.
25. Ottesen, B., Ulrichsen, H., Fahrenkrug, J., Larsen, J.-J., Wagner, G., Schierup, L., and Søndergaard, F., 1982, Vasoactive intestinal polypeptide and the female genital tract: Relationship to reproductive phase and delivery, *Am. J. Obstet. Gynecol.* **143**:414–420.
26. Goodnough, J. E., O'Dorisio, T. M., Friedman, C. I., and Kim, M. H., 1979, Vasoactive intestinal polypeptide in tissues of the human female reproductive tract, *Am. J. Obstet. Gynecol.* **134**:579–580.
27. Lynch, E. M., Wharton, J., Bryant, M. G., Bloom, S. R., Polak, J. M., and Elder, M. G., 1980, The differential distribution of vasoactive intestinal polypeptide in the normal human female genital tract, *Histochemistry* **67**:169–177.
28. Helm, G., Ottesen, B., Fahrenkrug, J., Larsen, J.-J., Owman, C., Sjöberg, N.-O., Stolberg, B., Sundler, F., and Walles, B., 1981, Vasoactive intestinal polypeptide (VIP) in the human female reproductive tract: Distribution and motor effects, *Biol. Reprod.* **25**:227–234.
29. Gu, J., Polak, J. M., Su, H. C., Blank, M. A., Morrison, J. F. B., and Bloom, S. R., 1984, Demonstration of paracervical ganglion origin for the vasoactive intestinal peptide-containing nerves of the rat uterus using retrograde tracing techniques combined with immunocytochemistry and denervation procedures, *Neurosci. Lett.* **51**:377–382.
30. Ottesen, B., 1983, Vasoactive intestinal polypeptide as a neurotransmitter in the female genital tract, *Am. J. Obstet. Gynecol.* **147**:208–224.
31. Sporrong, B., Clase, L., Owman, C. H., and Sjöberg, N.-O., 1977, Electron microscopy of adrenergic, cholinergic, and "p-type" nerves in the myometrium and a special kind of synaptic contacts with the smooth muscle cells, *Am. J. Obstet. Gynecol.* **127**:811–817.
32. Stjernquist, M., Alm, P., Ekman, R., Owman, C., Sjöberg, N.-O., and Sundler, F., 1985, Levels of neural vasoactive intestinal polypeptide in rat uterus are markedly changed in association with pregnancy as shown by immunocytochemistry and radioimmunoassay, *Biol. Reprod.* **33**:157–163.
33. Helm, G., Ekman, R., and Owman, C., 1987, Cyclic fluctuation of vasoactive intestinal polypeptide measured radioimmunologically in various regions of the human fallopian tube, *Int. J. Fertil.* **32**:467–471.
34. Owman, C., and Sjöberg, N.-O., 1976, Influence of sex hormones on the amount of adrenergic transmitter in the rabbit oviduct, in: *Neuroendocrine Regulation of Fertility* (T. C. Anand Humar, ed.), S. Karger, Basel, pp. 260–267.
35. Owman, C., Sjöberg, N.-O., and Sjöstrand, N. O., 1974, Short adrenergic neurons, a peripheral neuroendocrine mechanism, in: *Amine Fluorescence Histochemistry* (M. Fujivara and C. Tamaka, eds.), Igaku Shoin, Tokyo, pp. 47–55.
36. Thorbert, G., Alm, P., and Rosengren, E., 1978, Cyclic and steroid-induced changes in adrenergic neurotransmitter level of guinea-pig uterus, *Arch. Obstet. Gynecol. Scand.* **57**:45–48.
37. Ottesen, B., Larsen, J.-J., Staun-Olsen, P., Gammeltoft, S., and Fahrenkrug, J., 1985, Influence of pregnancy and sex steroids on concentration, motor effect and receptor binding of VIP in the rabbit female genital tract, *Regul. Peptides* **11**:83–92.
38. Bell, C., 1972, Autonomic nervous control of reproduction: Circulatory and other factors, *Pharmacol. Rev.* **24**:657–735.
39. Blank, M. A., Gu, J., Allen, J. M., Huang, W. M., Yiangou, Y., Ch'ng, J., Lewis, G., Elder, M. G., Polak, J. M., and Bloom, S. R., 1986, The regional distribution of NPY-, PHM-, and VIP-containing nerves in the human female genital tract, *Int. J. Fertil.* **31**:218–222.
40. Palle, C., Ottesen, B., Jørgensen, J., and Fahrenkrug, J., 1989, Comparison of the occurrence and actions of peptide histidine methionin (PHM) and vasoactive intestinal polypeptide (VIP) in the human female reproductive tract, *Biol. Reprod.* **41**:1103–1111.
41. Ottesen, B., and Fahrenkrug, J., 1981, Effect of vasoactive intestinal polypeptide (VIP) upon myometrial blood flow in non-pregnant rabbit, *Acta Physiol. Scand.* **112**:195–201.

42. Carter, A. M., Einer-Jensen, N., Fahrenkrug, J., and Ottesen, B., 1981, Increased myometrial blood flow evoked by vasoactive intestinal polypeptide in the non-pregnant goat, *J. Physiol. (Lond.)* **310**:471–480.

43. Ottesen, B., and Einer-Jensen, N., 1984, Increased endometrial clearance of [85]Krypton evoked by VIP in rabbits, *Acta Physiol. Scand.* **121**:185–187.

44. Ottesen, B., Gram, B. R., and Fahrenkrug, J., 1983, Neuropeptides in the female genital tract: Effect on vascular and non-vascular smooth muscle, *Peptides* **4**:387–392.

45. Ottesen, B., Einer-Jensen, N., Carter, A. M., 1983, Vasoactive intestinal polypeptide and endometrial blood flow in goat, *Anim. Reprod. Sci.* **6**:217–222.

46. Clark, K. E., Mills, E. G., Stys, S. J., and Seeds, A. E., 1981, Effects of vasoactive polypeptides on the uterine vasculature, *Am. J. Obstet. Gynecol.* **139**:182–188.

47. Clark, K. E., Austin, J. E., and Stys, S. J., 1982, Effect of vasoactive intestinal polypeptide on uterine blood flow in pregnant ewes, *Am. J. Obstet. Gynecol.* **144**:497–502.

48. Hansen, V., Ottesen, B., Allen, J., Maigaard, S., and Forman, A., 1988, Effects of VIP and PHM in human intracervical arteries, *Acta Obstet. Gynecol. Scand.* **67**:699–701.

49. Bardrum, B., Ottesen, B., and Fahrenkrug, J., 1986, Peptides PHI and VIP: Comparison between vascular and non-vascular smooth muscle effect in rabbit uterus, *Am. J. Physiol.* **251**:E48–E51.

50. Ottesen, B., Ulrichsen, H., Wagner, G., and Fahrenkrug, J., 1979, Vasoactive intestinal polypeptide (VIP) inhibits oxytocin induced activity of the rabbit myometrium, *Acta Physiol. Scand.* **107**:285–287.

51. Ottesen, B., Fahrenkrug, J., Wagner, G., Ulrichsen, H., Einer-Jensen, N., Carter, A. M., Larsen, J.-J., and Stolberg, B., 1980, Effects of VIP in the female genital tract, *Endocrinol. Jpn. [Suppl.]* **1**:71–78.

52. Ottesen, B., Wagner, G., and Fahrenkrug, J., 1980, Vasoactive intestinal polypeptide (VIP) inhibits prostaglandin-$F_{2\alpha}$-induced activity of the rabbit myometrium, *Prostaglandins* **19**:427–435.

53. Ottesen, B., 1981, Vasoactive intestinal polypeptide (VIP): Effect on rabbit uterine smooth muscle *in vivo* and *in vitro*, *Acta Physiol. Scand.* **113**:193–199.

54. Bolton, T. B., Lang, R. J., and Ottesen, B., 1981, Mechanism of action of vasoactive intestinal polypeptide on myometrial smooth muscle of rabbit and guinea-pig, *J. Physiol. (Lond.)* **318**:41–56.

55. Helm, G., Ekman, R., Rydhström, H., Sjöberg, N.-O., and Walles, B., 1985, Changes in oviductal VIP content induced by sex steroids and inhibitory effect of VIP on spontaneous oviductal contractility, *Acta Physiol. Scand.* **125**:219–224.

56. Fredericks, C. M., and Ashton, S. H., 1982, Effect of vasoactive intestinal polypeptide (VIP) on the *in vitro* and *in vivo* motility of the rabbit reproductive tract, *Fertil. Steril.* **37**:845–850.

57. Stjernquist, M., and Owman, C., 1984, Vasoactive intestinal polypeptide (VIP) inhibits neurally evoked smooth muscle activity of rat uterine cervix *in vitro*, *Regul. Peptides* **8**:161–167.

58. Ottesen, B., Gerstenberg, T., Ulrichsen, H., Manthorpe, T., Fahrenkrug, J., and Wagner, G. 1983, Vasoactive intestinal polypeptide (VIP) increases vaginal blood flow and inhibits smooth muscle activity in women, *Eur. J. Clin. Invest.* **13**:321–324.

59. Inyama, C. O., Wharton, J., Davis, C. J., Jackson, R. H., Bloom, S. R., and Polak, J. M., 1987, Distribution of vasoactive intestinal polypeptide binding sites in guinea pig genital tissue, *Neurosci. Lett.* **81**:111–116.

60. Ottesen, B., Staun-Olsen, P., Gammeltoft, S., and Fahrenkrug, J., 1982, Receptors for vasoactive intestinal polypeptide on crude smooth muscle membranes from porcine uterus, *Endocrinology* **110**:2037–2043.

61. Yiangou, U., Di Marzo, V., Spokes, R. A., Pamico, M., Morris, H. R., and Bloom, S. R., 1987, Isolation, characterization and pharmacological actions of peptide histidine valine 42, a novel prepro-vasoactive intestinal peptide-derived peptide, *J. Biol. Chem.* **262**:14010–14013.

62. Fahrenkrug, J., and Ottesen, B., 1982, Nervous release of vasoactive intestinal polypeptide from the feline uterus: Pharmacological characteristics, *J. Physiol. (Lond.)* **331**:451–460.

63. Hansen, B. R., Ottesen, B., and Fahrenkrug, J., 1986, Neurotransmitter-role of VIP in non-adrenergic relaxation of feline myometrium, *Peptides* 7(suppl. 1):201–203.
64. Helm, G., Håkanson, R., Leandeer, S., Owman, C., Sjöberg, N.-O., and Sporrong, B., 1982, Neurogenic relaxation mediated by vasoactive intestinal polypeptide (VIP) in the isthmus of the human fallopian tube, *Regul. Peptides* 3:145–149.
65. Cheviakoff, S., Diaz, S., Carril, M., Patritti, N., Croxatto, H. D., Llados, C., Ortiz, M. E., and Croxatto, H. B., 1976, Ovum transport in women, in: *Ovum Transport and Fertility Regulation* (M. J. K. Harper, C. J. Pauerstein, C. E. Adams, E. M. Coutinho, H. B. Croxatto, and D. M. Paton, eds.), Scriptor, Copenhagen, pp. 416–424.
66. Fredericks, C. M., Lundquist, L. E., Mathur, R. S., Ashton, S. H., and Landgrebe, S. R., 1983, Effects of vasoactive intestinal polypeptide upon ovarian steroids, ovum transport and fertility in the rabbit, *Biol. Reprod.* 28:1052–1060.
67. Csapo, A., 1956, Progesterone "block," *Am. J. Anat.* 98:273–291.
68. Fuchs, F., 1977, Endocrinology of labor, in: *Endocrinology of Labor* (F. Fuchs and A. Klopper, eds.), Harper & Row, New York, pp. 327–349.
69. Bell, C., 1972, Autonomic nervous control of reproduction: Circulatory and other factors, *Pharmacol. Rev.* 24:657–735.
70. Tatemoto, K., Carlquist, M., and Mutt, V., 1982, Neuropeptide Y—a novel brain peptide with structural similarities to peptide YY and pancreatic polypeptide, *Nature* 296:659–660.
71. Tatemoto, K., 1982, Neuropeptide Y: Complete aminoacid sequence of the brain peptide, *Proc. Natl. Acad. Sci. U.S.A.* 79:5485–5489.
72. Polak, J. M., and Bloom, S. R., 1984, Regulatory peptides—the distribution of two newly discovered peptides: PHI and NPY, *Peptides* 5(Suppl. 1):79–89.
73. Lundberg, J. M., Terenius, L., Hökfelt, T., and Goldstein, M., 1983, High levels of neuropeptide Y in peripheral noradrenergic neurons in various mammals including man, *Neurosci. Lett.* 42:167–172.
74. Ekblad, E., Edvinsson, L., Wahlestedt, C., Uddmann, R., Håkanson, R., and Sundler, F., 1984, Neuropeptide Y co-exists and co-operates with noradrenaline in perivascular nerve fibers, *Regul. Peptides* 8:225–235.
75. Stjernquist, M., Emson, P., Owman, C., Sjöberg, N.-O., Sundler, F., and Tatemoto, K., 1983, Neuropeptide Y in the female reproductive tract of the rat. Distribution of nerve fibres and motor effects, *Neurosci. Lett.* 39:279–284.
76. Tenmuko, S., Ottesen, B., O'Hare, M. M. T., Sheikh, S., Bardrum, B., Hansen, B., Walker, B., Murphy, R. F., and Schwartz, T. W., 1988, Interaction of NPY and VIP in regulation of myometrial blood flow and mechanical activity, *Peptides* 9:269–275.
77. Jørgensen, J. J., Sheikh, S., Foreman, A., Nørgård, M., Schwartz, T. W., and Ottesen, B., 1988, Neuropeptide Y (NPY) in the human female genital tract: Localization, characterization and biological action, *Am. J. Physiol.* 257:E220–E227.
78. Samuelson, V. E., and Dalsgaard, C. J., 1985, Action and localization of neuropeptide Y in the human fallopian tube, *Neurosci. Lett.* 8:49–54.
79. Von Euler, U. S., and Gaddum, J. H., 1981, An unidentified depressor substance in certain tissue extracts, *J. Physiol. (Lond.)* 72:74–87.
80. Von Euler, U. S., 1981, The history of substance P, *Trends Neurosci.* 4:IV.
81. Traurig, H., Saria, A., and Lembeck, F., 1984, Substance P in primary afferent neurons of the female rat reproductive system, *Naunyn-Schmiedebergs Arch. Pharmacol.* 326:343–346.
82. Formann, A., Andersson, K.-E., Maigaard, S., and Ulmsten, U., 1985, Concentrations and contractile effects of substance P in the human ampullary–isthmic junction, *Acta Physiol. Scand.* 124:17–23.
83. Papka, R. E., Cotton, J. P., and Traurig, H. H., 1985, Comparative distribution of neuropeptide tyrosine-, vasoactive intestinal polypeptide-, substance P-immunoreactive, acetylcholinesterase-positive and noradrenergic nerves in the reproductive tract of the female rat, *Cell Tissue Res.* 242:475–490.

84. Alm, P., Alumets, J., Brodin, E., Håkanson, R., Nilsson, G., Sjöberg, N.-O., and Sundler, F., 1978, Peptidergic (substance P) nerves in the genito-urinary tract, *Neuroscience* **3:**419–425.

85. Skrabanek, P., and Powell, D., 1983, Substance P in obstetrics and gynecology, *Obstet. Gynecol.* **61:**641–646.

86. Gu, J., Huang, W., Islam, K., McGregor, G., Terenghi, G., Morrison, J., Bloom, S., and Polak, J., 1983, Substance-P containing nerves in the mammalian genitalia, in: *Substance P* (P. Skrabanek and D. Powell, eds.), Boole Press, Dublin, pp. 263–264.

87. Gram, B. R., and Ottesen, B., 1982, Increased myometrial blood flow evoked by substance P, *Pflugers Arch.* **395:**347–350.

88. Ottesen, B., Søndergaard, F., and Fahrenkrug, J., 1983, Neuropeptides in the regulation of female genital smooth muscle contractility, *Acta Obstet. Gynecol. Scand.* **62:**591–592.

89. Alm, P., Alumets, J., Håkanson, R., Owman, C., Sjöberg, N.-O., Stjernquist, M., and Sundler, F., 1981, Enkephalin-immunoreactive nerve fibers in the feline genito-urinary tract, *Histochemistry* **72:**351–355.

90. Amara, S. G., Jonas, V., Rosenfeld, M. G., Ong, E. S., and Evans, R. M., 1982, Alternative RNA processing in calcitonin gene expression generates RNA encoding different polypeptide products, *Nature* **298:**240–244.

91. Rosenfeld, M. G., Mermad, G.-J., Amara, S. G., Swanson, L. W., Sawchencko, P. E., Rivier, J., Vale, W. W., and Evans, R. M., 1983, Production of a novel neuropeptide encoded by the calcitonin gene via tissue specific RNA processing, *Nature* **304:**129–135.

92. Ghatei, M. A., Gu, J., Mulderry, P. K., Blank, M. A., Allen, J. M., Morrison, J. F. B., Polak, J. M., and Bloom, S. R., 1985, Calcitonin gene-related peptide (CGRP) in the female rat urogenital tract, *Peptides* **6:**809–815.

93. Bek, T., Ottesen, B., and Fahrenkrug, J., 1988, The effect of galanin, CGRP, and ANP on spontaneous smooth muscle activity of rat uterus, *Peptides* **9:**497–500.

94. Tatemoto, K., Rökaeus, Å., Jörnvall, H., McDonald, T. J., and Mutt, V., 1983, Galanin—a novel biologically active peptide from porcine intestine, *FEBS Lett.* **164:**124–128.

95. Bauer, F. E., Christofides, N. D., Haecker, G. W., Blank, M. A., Polak, J. M., and Bloom, S. R., 1986, Distribution of galanin immunoreactivity in the genitourinary tract of man and rat, *Peptides* **7:**5–10.

96. Schaeffer, J. M., Liu, J., Hsueh, A. J., and Yen, S. S. C., 1984, Presence of oxytocin and arginine vasopressin in human ovary, oviduct, and follicular fluid, *J. Clin. Endocrinol. Metab.* **59:**970–973.

97. Ciarochi, F. F., Robinson, A. G., Verbalis, J. G., Seif, S. M., and Zimmerman, E. A., 1985, Isolation and localization of neurophysin-like proteins in rat uterus, *Peptides* **6:**903–911.

98. Hackenthal, E., Metz, J., Poulsen, K., Rix, E., and Taugner, R., 1980, Renin in the uterus of non-pregnant mice. Immunocytochemical, ultrastructural and biochemical studies, *Histochemistry* **66:**229–238.

99. Rix, E., Hackenthal, E., Metz, J., Poulsen, K., and Taugner, R., 1980, Renin in the uterus of pregnant mice, immunocytochemical, ultrastructural and biochemical studies, *Histochemistry* **68:**253–263.

100. Weiss, G., 1984, Relaxin, *Annu. Rev. Physiol.* **46:**43–52.

101. Yki-Järvinen, H., Wahlström, T., and Seppälä, M., 1983, Immunohistochemical demonstration of relaxin in the genital tract of pregnant and nonpregnant women, *J. Clin. Endocrinol.* **57:**451–454.

102. Bryant-Greenwood, G., Ali, S., Mandel, M., and Greenwood, F., 1987, Ovarian and decidual relaxins in human pregnancy, *Adv. Exp. Med.* **219:**709–713.

103. Sanborn, B. M., Kuo, H. S., Weisbrodt, N. W., and Sherwood, O. D., 1980, The interaction of relaxin with the rat uterus. I. Effect on cyclic nucleotide levels and spontaneous contractile activity, *Endocrinology* **106:**1210–1215.

104. Anderson, L. L., 1987, Regulation of relaxin secretion and its role in pregnancy, *Adv. Exp. Med. Biol.* **219:**421–463.

105. MacLennan, A. H., Green, R. C., Grant, P., and Nicolson, R., 1986, Ripening of the human

cervix and induction of labor with intracervical purified porcine relaxin, *Obstet. Gynecol.* **68**:598–601.

106. Wiqvist, I., Norstrom, A., O'Byrne, E., and Wiqvist, N., 1984, Regulatory influence of relaxin on human cervical and uterine connective tissue, *Acta Endocrinol. (Kbh.)* **106**:127–132.

107. Kumasaka, T., Nishi, N., Yaoi, Y., Kido, Y., Saito, M., Okayasu, I., Shimizu, K., Hatakeyama, S., Sawano, S., and Kokusu, K., 1979, Demonstration of immunoreactive somatostatin-like substance in villi and decidua in early pregnancy, *Am. J. Obstet. Gynecol.* **134**:39–44.

14

Biosynthesis and Function of Eicosanoids in the Uterus

Gautam Chaudhuri

1. History of Prostaglandins

The credit for the discovery of prostaglandins belongs to the Swedish scientist U. S. von Euler. Other scientists had observed pharmacological effects with semen and prostatic extracts that can now, with hindsight, be attributed to the presence of prostaglandins. It was von Euler, however, who established beyond doubt that the active principle, which he named prostaglandin, belongs to a completely new group of naturally occurring substances.

In 1934, von Euler[1] observed that human semen and extracts of sheep vesicular glands lowered arterial blood pressure on intravenous injection and stimulated various isolated intestinal and uterine smooth muscle preparations. He showed that the active principle was a lipid-soluble acid, and this differed from all other known substances, e.g., histamine, acetylcholine, and adenylic compounds, with similar biological activity.[2] Goldblatt[3] independently described this active principle of seminal fluid. This active principle was named prostaglandin, as it was thought to be synthesized by the prostate gland. It was Eliasson[4] who, by fractionation of the ejaculates, made the important observation that human seminal prostaglandin is secreted mainly by the seminal vesicles and not by the prostate. Thus, the original assumption that gave rise to the name prostaglandin was proved incorrect. By that time, however, pure prostaglandins were being isolated, and the name had become firmly established in the literature; therefore, it was not changed.

Gautam Chaudhuri • Departments of Obstetrics–Gynecology and Pharmacology, School of Medicine, University of California Los Angeles, Los Angeles, California 90024.

2. Isolation and Structure of Prostaglandins

The impetus for further biological work resulted from the elegant isolation and chemical characterization achieved by Bergstrom and colleagues.[5,6] This group isolated two compounds that behaved differently on partition between ether and an aqueous phosphate buffer. The one more soluble in ether was called prostaglandin E (PGE); the other, more soluble in phosphate buffer (in Swedish, phosphate is spelled with *F*), was called prostaglandin F (PGF). Later, at an international convention, it was decided to name the new prostaglandins alphabetically, and since their initial discovery we have PGA, B, D, E, F, G, H, and I (see Fig. 1).

The basic 20-carbon skeleton of the prostaglandins has been named prostanoic acid. All the naturally occurring prostaglandins are hydroxylated in the 15 position and contain a 13,14-*trans* double bond. The degree of unsaturation of the side chain is indicated by the subscript numeral after the letter; thus, prostaglandin E_1 and F_1 have only the *trans* double bond, and PGE_2 and F_2 have, in addition, a *cis*-double bond in the 5,6 position (Fig. 1). Reduction of prostaglandin E yields two isomeric alcohols, Fα and Fβ.

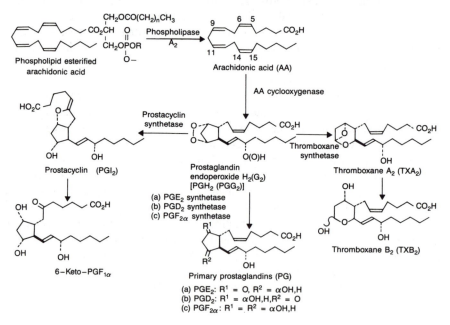

Figure 1. Structure of arachidonic acid and its subsequent conversion by the cyclooxygenase. R = inositol, choline, or other phospholipid component. Prostaglandin endoperoxide G_2 (PGG_2) has an -OOH group in position 15, whereas prostaglandin endoperoxide H_2 (PGH_2) is formed by the loss of one of the oxygens forming an -OH at position 15. Synthetases a, b, and c produce primary prostaglandins a, b, and c.

In 1964, Bergstrom and colleagues[7] and Van Dorp and colleagues[8] independently demonstrated that prostaglandins are biosynthesized from polyunsaturated fatty acids.

3. Arachidonic Acid Metabolism

Arachidonic acid is the precursor of the bisenoic (those with two unsaturated double bonds, e.g., PGE_2, $PGF_{2\alpha}$, PGI_2) prostaglandins and is the most common fatty acid present in cellular phospholipids. It can be obtained by desaturation and chain elongation of dietary linolenic acid or directly from the diet. Arachidonic acid is liberated from cellular phospholipids by the action of phospholipases, which can be activated by mechanical or chemical changes.[9] Corticosteroids inhibit prostaglandin synthesis by inhibiting phospholipase A_2.[10,11] This inhibition is mediated by the production of a specific macromolecule variously named as macrocortin,[10] lipocortin, or lipomodulin.[11] Biosynthesis and release of prostaglandins from tissues occur readily in response to a variety of physiological and pathological stimuli, and it would appear that any distortion of cell membrane is an adequate trigger mechanism.[12] Unlike many other biologically active substances, prostaglandins are formed immediately prior to release and are not stored in the body. The enzymes that synthesize prostaglandins are present in all organs studied so far except red blood cells.[13] Some tissues, such as seminal vesicles, kidneys, and lungs, have a greater capacity for prostaglandin synthesis than others.[13]

Once released from the membrane phospholipids, arachidonic acid is metabolized by cyclooxygenase or lipoxygenase (Fig. 1). The term "eicosanoids" is applied to all the 20-carbon derivatives, whereas "prostanoids" refers only to those with a ring like that in the prostanoic acid skeleton. The cyclooxygenase enzyme forms the prostaglandin endoperoxide PGG_2[14] (see Fig. 1), whereas the lipoxygenase forms the corresponding hydroperoxides.[15]

4. Cyclooxygenase Pathways of Arachidonic Acid Metabolism

Arachidonic acid and certain other polyunsaturated fatty acids[16] may be transformed into prostaglandins (PG) by the enzyme cyclooxygenase, which inserts two molecules of oxygen into arachidonate to yield a 15-hydroperoxy-9,11-endoperoxide with a substituted cyclopentane ring (PGG_2). A hydroperoxidase reduces PGG_2 to its 15-hydroxy analogue (PGH_2). It is thought that a single protein is responsible for both reactions and is called prostaglandin (endoperoxide) synthetase. The cyclic endoperoxides possess intrinsic biological activity as vasodilators and aggregators of platelets, but they are unstable and are normally rapidly converted into other products.[16]

The minimum molecular weight of the purified enzyme is about 72,000,[16] and it requires 1 mol of heme per mol of enzyme for maximal catalytic activity. Immunocytochemical studies indicate that cyclooxygenase is contained in endoplasmic reticulum and nuclear membrane but not plasma membrane or mitochondrial membrane of cultured fibroblasts. The cyclooxygenase-catalyzed fatty acid oxidation occurs slowly initially and later accelerates. Exogenous hydroperoxides eliminate this kinetic lag phase at concentrations (10^{-7} to 10^{-8} M) far below the K_m (10^{-5} M) of the peroxidase activity. It appears that there is a continuous requirement for the activator hydrogen peroxide, as termination of cyclooxygenase-catalyzed substrate oxidation occurs on addition of glutathione and glutathione peroxidase. The cyclooxygenase-catalyzed oxygen consumption declines to zero before complete consumption of fatty acid substrate, and a second burst of oxygen consumption occurs on addition of fresh enzyme. This process of self-deactivation of the cyclooxygenase appears to occur in intact cells as well as with purified enzyme preparations and may limit *in vivo* prostaglandin biosynthesis.[16] The hydroperoxidase activity is capable of catalyzing the reduction of other lipid hydroperoxides and hydrogen peroxide and also undergoes self-deactivation.[16] The nonsteroidal antiinflammatory agents inhibit the cyclooxygenase by preventing the abstraction of hydrogen from C-13 and thereby blocking peroxidation at C-11 or C-15. They do not directly block the hydroperoxidase activity[16] (see Fig. 2). The PGH_2 formed

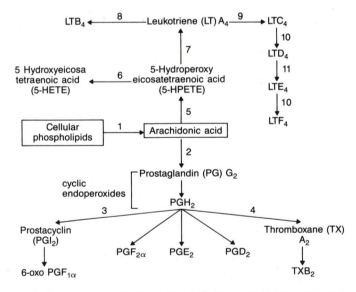

Figure 2. Steps in the conversion of cellular phospholipids to arachidonic acid and its subsequent conversion to other products and the enzymes involved. 1, phospholipase A_2 (inhibited by a phospholipase inhibitor whose synthesis is stimulated by corticosteroids); 2, cyclooxygenase (inhibited by nonsteroidal antiinflammatory drugs and BW-755C); 3, prostacyclin synthetase; 4, thromboxane synthetase; 5, 5-lipoxygenase (inhibited by BW-755C); 6, glutathione peroxidase; 7, dehydrogenase; 8, LTA_4 epoxide hydrolase; 9, glutathione transferase; 10, glutamyl transferase; 11, cysteinyl-glycine dipeptidase. From Chaudhuri,[16a] with permission.

isomerizes enzymically or nonenzymically to the stable substances PGE_2, $PGF_{2\alpha}$, and PGD_2. The prostaglandin endoperoxides are also transformed enzymically into two other unstable products: prostacyclin (PGI_2) and thromboxane A_2 (TXA_2).[16]

4.1. Prostacyclin

Different tissues give rise to different arachidonate metabolites. The main cyclooxygenase product of arachidonic acid in vascular tissue is prostacyclin[17] (Fig. 2), which is a potent vasodilator and inhibitor of platelet aggregation. It is chemically unstable with a half-life of 2–3 min, breaking down to 6-keto-$PGF_{1\alpha}$ (6-oxo-$PGF_{1\alpha}$). Endothelial cells are the most active producers of prostacyclin, and therefore the ability of large vessels to synthesize prostacyclin is greatest at the intimal surface and progressively decreases toward the adventitia. Many other tissues apart from blood vessels are now known to be capable of generating prostacyclin or its degradation product 6-keto-$PGF_{1\alpha}$. The organs include decidual tissue and myometrium of pregnant rat[18] and human chorion, amnion, and decidua.[19] Prostacyclin production has also been reported in the placenta,[20] but this has never been convincingly demonstrated.

4.1.1. Prostacyclin Synthetase

Immunofluorescence studies indicate that the enzyme is located in the plasma membrane and nuclear membrane of a wide variety of smooth muscle cells. The purified enzyme exhibits a minimum molecular weight of 50,000.[16] The optical behavior of the purified enzyme suggests that it may be a cytochrome P-450-type hemoprotein.[16] Prostacyclin synthetase is inactivated by a variety of lipid hydroperoxides, a process that is partially prevented by reducing compounds that are radical scavengers.[16]

4.1.2. Mechanism of PGI_2 Action

Prostacyclin elevates cAMP levels in platelets, vascular smooth muscle cells, and toad bladder epithelium.[16] Thus, activation of adenylate cyclase via a PGI_2 receptor may represent a general mechanism for the effects of PGI_2 on responsive cells. Prostacyclin combines with specific protein receptors that are present on membranes of PGI_2-responsive cells.[16]

4.2. Thromboxanes

In blood platelets, the principal metabolite of the cyclic endoperoxides PGG_2 and PGH_2 is thromboxane A_2 (TXA_2). This compound was so named because it was first demonstrated to form in the platelets (thrombocytes) and contains an oxetane ring in its structure. Thromboxane A_2 contracts vascular smooth muscle and induces platelet aggregation and serotonin release and has a very short half-life ($t_{1/2} = 30$

sec). The oxetane ring of TXA_2 spontaneously hydrolyzes to the less active hemi-acetal thromboxane B_2 (TXB_2). Synthesis of TXA_2 has been demonstrated in certain vascular tissues including rabbit pulmonary arteries, dog mesenteric arteries, bovine cerebral arteries, as well as human umbilical arteries and veins and saphenous veins.[16] Porcine, bovine, and human aortic endothelial cells in culture are also capable of generating TXA_2 from both exogenous and endogenously released substrate. Other cells and tissues in which TXA_2 synthesis has been detected include polymorphonuclear leukocytes (PMNs), macrophages, monocytes, fibroblasts, lungs, kidney, spleen, gastric mucosa, and inflammatory granuloma.[16]

The platelet thromboxane synthetase activity is associated with dense tubular membranes and catalyzes formation of TXA_2 from PGH_2. It has been suggested that this enzyme may also be a cytochrome P450-type hemoprotein.[16]

4.3. Prostaglandins D_2, E_2, and $F_{2\alpha}$

Enzymic (endoperoxide isomerase) isomerization of PGH_2 leads to the stable prostaglandins D_2 and E_2; nonenzymatic isomerization also occurs. The enzymatic reduction (endoperoxide reductase) leads to the stable prostaglandin $F_{2\alpha}$. These compounds are produced by vascular tissues, uterus, and many other cells and tissues. Their numerous and diverse effects include the relaxation or contraction of many smooth muscles such as vascular, bronchial, tracheal, uterine, and gastrointestinal smooth muscle. The relative production of these stable prostaglandins varies with the tissue and conditions under study.

These prostaglandins are metabolized at the 15 position by oxidation of the secondary alcohol group into a ketone; NAD^+ is required as cofactor. The 15-oxo-prostaglandins so formed, especially of the F series, may not lose much biological activity when compared with the parent compound. Subsequent reduction of the 13,14 double bond results in a much larger loss of biological activity. The 13,14-dihydro-15-oxo-prostaglandins formed are the main circulating metabolites of PGE_2 and $PGF_{2\alpha}$ and may exceed the plasma levels of the primary prostaglandins by some 20-fold. Useful information can be obtained concerning the release of $PGF_{2\alpha}$ by measuring the levels of its 13,14-dihydro-15-oxo-metabolite. The equivalent metabolite of PGE_2 is not so useful, as it readily dehydrates into the PGA_2 metabolite, making measurement difficult.[21,22]

5. Lipoxygenase Pathways of Arachidonic Acid Metabolism

In addition to the cyclooxygenase pathway, polyunsaturated fatty acids such as arachidonic acid can be oxidized via lipoxygenase enzymes to form corresponding hydroperoxides (Fig. 2). The first lipoxygenase enzyme was found in the platelets and lungs[23,24] and is responsible for the formation of lipid peroxides such as 12-hydroperoxyeicosatetraenoic acid (12-HPETE) and its degradation product 12-hydroxyeicosatetraenoic acid (12-HETE), which is a potent chemotactic agent for polymorphonuclear leucocytes (PMNs). Subsequently, a number of enzymes that peroxidize arachidonic acid in different positions have been discovered. The most important of these is a 5-lipoxygenase, which occurs in PMNs.[25] The action of this enzyme results in the formation of a complex group of compounds known collectively as leukotrienes[25] (Figs. 2 and 3).

The first step of the 5-lipoxygenase pathway is the formation of 5-HPETE (Fig. 3); this is converted either to monohydroxyeicosatetraenoic acid (5-HETE) or to a 5,6 epoxide known as leukotriene A_4 (LTA$_4$). Leukotriene A_4 may itself be transformed either to 5,12-dihydroxyeicosatetraenoic acid, known as leukotriene B_4 (LTB$_4$), or to leukotriene C_4 (LTC$_4$). Leukotriene C_4 is a glutathionyl derivative formed by the action of a glutathione-S-transferase. Leukotriene D_4 (LTD$_4$) is synthesized by the removal of glutamic acid from LTC$_4$. Leukotriene E_4 results from the subsequent cleavage of glycine by an aminopeptidase. The final step of these transformations produces LTF$_4$ by reincorporation of glutamic acid into the molecule to form a glutamyl, cysteinyl derivative.[26,27]

The 5-lipoxygenase enzyme is found in the cytosolic fraction and is a monomeric protein with a molecular weight of 68,000–70,000.[16] The subcellular localization of the leukotriene-forming enzymes differs from that of the microsomal enzymes responsible for prostaglandin biosynthesis. The enzymes required for LTA$_4$ and LTB$_4$ synthesis are cytosolic, whereas the enzymes involved in the peptidoleukotriene formation from LTA$_4$ are particulate.[16] Thus, LTA$_4$ has to reach the particulate enzyme in the granules or some other subcellular fraction to be converted into LTC$_4$.[16] Since the LTD$_4$-forming enzyme is in the plasma membrane, LTC$_4$ is most likely converted to LTD$_4$ as it passes through the plasma membrane to the outside of the cell.

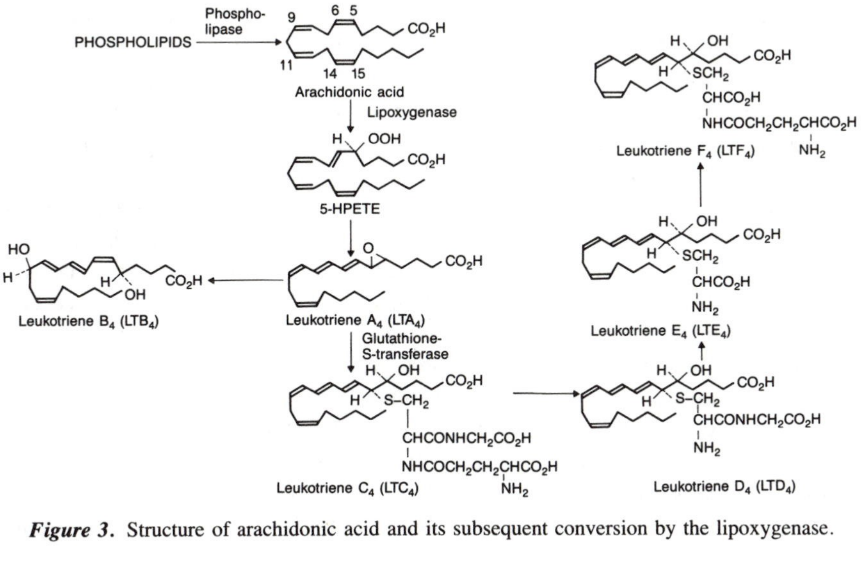

Figure 3. Structure of arachidonic acid and its subsequent conversion by the lipoxygenase.

Leukotriene B_4 is one of the most potent naturally occurring inducers of neutrophil chemotaxis both *in vitro* and *in vivo,* inducing leukocyte aggregation. It releases lysosomal enzymes and is released during the inflammatory process in animals and man.[28-30]

The bronchoconstriction mediator slow-reacting substance of anaphylaxis (SRS-A) discovered more than 40 years ago is now known to be a mixture of LTC_4, LTD_4, and LTE_4. There is now evidence to suggest that lipoxygenase activity is involved both in the generation of anaphylactic mediators and in the sensitization of smooth muscle to their contractile actions.[31,32]

6. Control of Eicosanoid Biosynthesis in the Uterus

The cyclooxygenase products, e.g., prostaglandins, are formed by both the myometrium and the endometrium and have important physiological roles. However, synthesis of lipoxygenase products by uterine tissues has not been studied extensively, and their physiological role is therefore less certain.

6.1. Control of Prostaglandin Biosynthesis in the Uterus

The control of prostaglandin biosynthesis in the uterus can be divided into factors controlling the biosynthesis of prostaglandins in the myometrium and factors contributing to prostaglandin biosynthesis in the endometrium.

6.1.1. Biosynthesis of Prostaglandins in the Myometrium

The myometrium in various species including humans contains the enzymes cyclooxygenase and prostacyclin synthetase. Immunohistochemical localization of the cyclooxygenase and PGI_2 synthetase clearly shows that both of these enzymes are located predominantly in the myometrial cell itself.[33] Prostacyclin seems to be the predominant prostaglandin formed in this tissue. This is in contrast to the endometrium, which predominantly forms the conventional prostaglandins (PGE and F series) and not prostacyclin.[34,35]

The factors controlling the concentration of enzymes in the myometrium have not been elucidated. However, the concentrations of the cyclooxygenase and prostacyclin synthetase enzymes in nonpregnant and pregnant uteri have been reported.[36] The pregnant myometrium contained significantly more cyclooxygenase per milligram microsomal protein than did the nonpregnant myometrium. The increase in the concentration of the cyclooxygenase enzyme appeared to be about threefold compared to the nonpregnant myometrium. Though there was an increase in the prostacyclin synthetase, it was much less than the increase in the cyclooxygenase and had a different subcellular distribution, i.e., in the nonmicrosomal fraction. The reason for this increase in PGI_2 synthetase in pregnancy is not known. This increase cannot at present be attributed entirely to the increased extent of myometrial vascularization during pregnancy. As has been previously mentioned, the PGI_2 synthetase was found to be located predominantly in the smooth muscle cells themselves and not in the myometrial blood vessels.[33] It was hypothesized that the higher relative amount of cyclooxygenase would lead to an increase in the synthesis of the contractile prostaglandins instead of prostacyclin. However, there was no evidence for an increase in cyclooxygenase over the last few weeks of pregnancy or with the onset of labor.[36]

It is interesting that major differences in the enzyme concentrations were found between different regions of a primigravida uterus removed at 34 weeks of gestation because of cervical carcinoma. In the myometrium, the cyclooxygenase concentrations were highest in close proximity to the placental bed. They decreased significantly from the inner to the outer layer.[37]

6.1.2. Prostaglandin Production by the Endometrium

A problem with many studies undertaken to evaluate prostaglandin concentrations in the endometrium arises because of the methodologies involved. Studies involving estimation of prostaglandin concentrations in endometrial samples obtained by biopsies represent the synthesizing capacity of the endometrium rather than basal prostaglandin concentrations, as it is well known that tissue trauma could lead to varying prostaglandin generation.[12] Different results are also obtained when endometrial explants that are composed predominantly of epithelial and stromal cells are used for studies[38] when compared to those in which the epithelial and stromal cells are isolated and therefore the possible interactions between these two cell types are removed.[39] Both cell types, i.e., epithelial and stromal components, have the capacity to synthesize prostaglandins.[39,40]

The ovarian steroids have been postulated to be the necessary stimulus for the production of $PGF_{2\alpha}$ by the primate endometrium.[41] Treatment of ovariectomized monkeys with estradiol resulted in an increase in $PGF_{2\alpha}$ levels in uterine flushings.[42] Treatment with both progesterone and estradiol was not significantly different from treatment with estradiol alone. In women, there was little increase in the levels of $PGF_{2\alpha}$ in the endometrium or uterine flushings during the proliferative phase when estradiol was being released from the ovary, though there was a significant increase in $PGF_{2\alpha}$ levels between early- and midproliferative-phase endometrium (Fig. 4), possibly reflecting increased estradiol secretion at this time.[41,41a]

A large rise in $PGF_{2\alpha}$ levels also occurred during the midluteal phase, when plasma progesterone and progesterone, levels are both high. Consequently, progesterone, acting on the estradiol-primed uterus, may be the physiological stimulus for $PGF_{2\alpha}$ production by the human uterus. Most likely in the human, the glandular epithelial cells rather than the stroma produce the majority of prostaglandins. Epithelial cells in cultures obtained from human proliferative-phase endometrium responded to estradiol by increasing $PGF_{2\alpha}$ output significantly in response to added arachidonic acid or calcium ionophore A-23187,[39] a phenomenon not observed with stromal cells. Since estradiol did not increase the uptake of [^{3}H]arachidonic acid

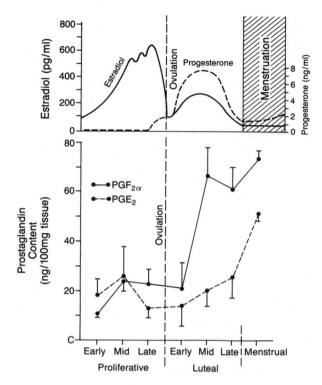

Figure 4. The top panel shows the serum estradiol and progesterone concentration throughout the menstrual cycle. The bottom panel shows the endometrial PGE_2 and $PGF_{2\alpha}$ concentration. The luteal-phase endometrium shows higher prostaglandin concentrations than the proliferative phase of the menstrual cycle. From refs. 41 and 41a, with permission.

into epithelial cells, the likely mechanism could be either an increase in the cyclooxygenase in these cells or inhibition of the metabolism of newly formed $PGF_{2\alpha}$ within the epithelial cells.[39] It thus appears that the stimulatory effects of estradiol on prostaglandin synthesis are restricted to the epithelial cellular component.

At present, there is general agreement that progesterone has an inhibitory action on prostaglandin synthesis *in vivo,* although it has been suggested that tissue exposed to progesterone *in vivo* has the potential to produce greater amounts of prostaglandins because of a "priming" process.[43] However, the results obtained from *in vitro* systems are less coherent. It has been shown that progesterone inhibited prostaglandin production in endometrial explants both with and without estradiol stimulation.[44] Recently, using endometrial fragments, progesterone has been shown to be considerably more effective in suppressing prostaglandin synthesis.[45,46] The results suggest that prostaglandin production in human secretory endometrium, e.g., during the luteal phase of the cycle, is likely to be very low despite the obvious ability of this tissue to synthesize prostaglandins in response to the trauma of obtaining the biopsy.[41] The exact site in the synthetic pathway of prostaglandins where progesterone exerts its inhibitory action is not known. Unlike glucocorticoids, which stimulate the production of the phospholipase A_2 inhibitory protein lipomodulin, progesterone decreases lipomodulin.[46] Recently, it has been shown that there is an increase in the synthesis of uteroglobulin in the secretory endometrium, and an inverse relationship exists between the concentration of uteroglobulin and prostaglandin $F_{2\alpha}$ in the endometrium.[47]

The adrenal steroids also affect prostaglandin output by human endometrium *in vitro*. Dehydroepiandrosterone sulfate (DHEA-S), unconjugated dehydroepiandrosterone, dehydroepiandrosterone acetate, and other C_{19} steroids enhanced prostaglandin $F_{2\alpha}$ output from human secretory endometrium cultured *in vitro*.[48] Dexamethasone, on the other hand, reduced prostaglandin $F_{2\alpha}$ output from human secretory endometrium in culture, and this was associated with an increase in lipomodulin.[46]

6.2. Synthesis of Lipoxygenase Products by the Uterus

Rat,[49] guinea pig,[40] rabbit,[51] and human uterine tissues[52] are all capable of producing lipoxygenase products. In the human, fetal cotyledon and fetal membrane mainly formed LTB_4, whereas maternal cotyledon and myometrium synthesized mainly 12-hydroxyeicosatetraenoic acid (12-HETE).[52] Leukotriene B_4 (LTB_4) is known to have immunosuppressive properties, and it has been speculated that its production predominantly by tissues of fetal origin is of importance in the immunologic adaptations of pregnancy.[52]

Some of the sulfidopeptidyl leukotrienes, in addition to their ability to contract the uterus,[53] also increase endometrial capillary permeability,[54] and it has been suggested that these substances may play a role in decidualization.[54−56]

7. Prostaglandin and Leukotriene Receptors

Binding sites for PGE_1 and PGE_2 are present in the hamster,[57] rat,[58] bovine,[59] monkey,[60] and human[61,62] uterus. These receptors are associated with the particulate matter of the cell, are protein in nature, have a K_d that has varied from 1.5^{62} to 3 nM,[59] and are situated in the myometrium. The binding affinity of these receptors for $PGF_{2\alpha}$ is 100 to 1000 times lower.[59,61−63]

Whether there are two distinct receptors for prostaglandin E and F remains debatable.[59,61−63] Though there are at least two receptor types, PGE binds better than $PGF_{2\alpha}$ to both sites. However, at the classic PGE site, $PGF_{2\alpha}$ must be present at approximately 200 times the PGE concentration for $PGF_{2\alpha}$ to remove PGE, whereas at the second site it takes only 20 times the PGE concentration for $PGF_{2\alpha}$ to remove PGE. Also PGE displaces $PGF_{2\alpha}$ by a factor of 10 at this second site, not the 200-fold difference seen at the classic site. The K_d is approximately 30–50 nM for $PGF_{2\alpha}$ at this site.

In nonpregnant human uteri,[62] the binding for PGE_2 was greatest in the fundus and corpus uteri and decreased toward the cervix.[59,61-63] Cervical binding was detectable only in the follicular phase, but the apparent affinity (K_d 11.4 nM) was significantly lower than in the rest of the uterus (K_d range 0.5–3 nM). In the pregnant uteri, high-affinity binding (K_d 1.5–2.6 nM) was present in all myometrial samples but not in the cervix (K_d 19.4 nM). Furthermore, the concentration of binding sites in the pregnant fundus–corpus area (range 222–270 fmol/mg protein) was lower than in the nonpregnant (725–843 fmol/mg protein) uterus. On the basis of their finding, the authors have speculated that the high-affinity binding sites may be responsible for the contracting effects of PGE, whereas the low-affinity sites in the cervix may be responsible for its remodeling, softening effect. Similar to observations with PGE_2 receptors, the concentration of $PGF_{2\alpha}$ binding sites in the corpus of pregnant human uteri (369 fmol/mg protein) is lower than that in nonpregnant (688 fmol/mg protein) uteri.[62] Cervical binding was present only in samples from pregnant uteri.[62] The decreased concentration of prostaglandin receptors in the pregnant myometrium may reflect the relatively greater increase in connective tissue in the uterus in pregnancy.[62] If the concentration of receptors for prostaglandins is compared to the DNA content or calculated per cell, there is little or no change between pregnant and nonpregnant uterus.[61]

The PGE_2 receptor concentration and distribution did not vary with the phase of the menstrual cycle, suggesting that physiological levels of sex steroids do not modulate the receptor, at least in women.[62] However, altered numbers of uterine PGE binding sites have been reported in women taking oral estrogen[63] and in hamsters during estrus.[64] It is interesting that there were no differences in the PGE_2 receptor concentration in uteri from women with severe dysmenorrhea (pain at the time of menstruation) when compared to women without dysmenorrhea.[62] However, a significant correlation between increased menstrual blood loss and an increase in myometrial PGE receptor concentration has been reported.[65] Some of the receptors may have come from blood vessels and or platelets, which would help explain these findings. Endometrium obtained from menorrhagic (increased menstrual blood loss) women synthesized more of the vasodilator PGEs than endometrium obtained from normal women.[66] This, together with an increase in PGE receptor concentration,[65] could account for the increased bleeding. The fenamate group of PG synthesis inhibitors, in addition to their action in inhibiting the enzyme cyclooxygenase, also block the action of PGE at the receptor level[67,68] and may therefore be more effective in reducing blood loss in unexplained menorrhagia. The myometrium and vascular smooth muscle also contain PGI_2 specific binding sites, whereas the endometrium and vascular endothelium contain very few or no binding sites.[69]

Leukotriene C_4 stimulates myometrial and vascular smooth muscle contraction,[70] and its receptors in the bovine[71] and human[69] uterus have been demonstrated in the luminal epithelial cells and stromal cells of the endometrium, myometrial smooth muscle, as well as arteriolar smooth muscles.[59,71] By contrast, the glandular epithelium and vascular endothelium contained few or no LTC_4 binding sites.

The number of LTC_4 binding sites in the luminal epithelial cells of the endometrium was as great as that in the lung tissue, which normally contains large amounts of these receptors.[69]

8. Prostaglandin Contractile Action on the Cellular Level

The contractile action of prostaglandins on the uterus appears to be based on calcium mobilization. The uterine contractile prostaglandins, PGE_2 and $PGF_{2\alpha}$, caused net calcium release from microsomal preparations enriched in sarcoplasmic reticulum.[72,73] In bovine term pregnant uterus, PGE_2 had a greater effect than $PGF_{2\alpha}$, consistent with pharmacological potency. Prostaglandin $F_{2\alpha}$ was more potent in preparations of pregnant than of nonpregnant bovine uterus.[74] Effects of PGE_2 were also observed in term pregnant human uterus obtained at cesarean hysterectomy.[75] The release of calcium from intracellular stores (sarcoplasmic reticulum) and the increase in intracellular free calcium is a first step in the events leading to uterine contraction. Additional evidence for the requirement of calcium release from intracellular stores in PGE_2-induced uterine contraction was obtained from experiments in calcium-free solution.[76]

9. Prostaglandin and Oxytocin Interaction

The first evidence of interaction between oxytocin and prostaglandins on the uterus was provided by Vane and Williams.[77] They demonstrated that indomethacin and meclofenamate, both of which are potent inhibitors of prostaglandin synthetase, antagonized the contractile effects of oxytocin on the isolated uterus from the pregnant rat. Physiological amounts of oxytocin infused via the arterial supply into the sheep uterus *in vivo* released $PGF_{2\alpha}$ within 3 min.[78] There was a simultaneous increase in uterine tone and amplitude of contraction. However, it appears that in this species the contractile effects of oxytocin are independent of prostaglandin output, as simultaneous administration of indomethacin blocks the release of $PGF_{2\alpha}$ without affecting the contractility of the uterus.[78] Oxytocin has a negligible effect on uterine $PGF_{2\alpha}$ release in anestrous sheep unless the sheep have been primed with estradiol.[79] Thus, at least in this species, oxytocin needs an estradiol-primed uterus before it stimulates $PGF_{2\alpha}$ synthesis and release.

Oxytocin enhances $PGF_{2\alpha}$ release from the sheep endometrium cultured *in vivo*. This enhancement reaches a maximum on the day of estrus. High-affinity binding sites for oxytocin are found in the sheep endometrium and myometrium. The binding sites in the endometrium are about twice the number found in the myometrium, but oxytocin enhances $PGF_{2\alpha}$ release only from the endometrium, suggesting that endometrium is a prime target for oxytocin.

Oxytocin at physiological concentration stimulates immediate release of free arachidonic acid (for further discussion see Chapter 5) from dispersed human decidual cells in a perfusion system.[81] This indicates that oxytocin activates phospholipases, thus enhancing prostaglandin synthesis. The effect of oxytocin on the release of [³H]arachidonic acid from decidual cells of women in labor was significantly greater than from women not in labor or with endometrial cells of nonpregnant women and correlates well with reported oxytocin receptor concentrations in these tissues.[81] This suggests a role for endogenous oxytocin in stimulating prostaglandin synthesis at the onset of parturition.

Although the interaction between oxytocin and prostaglandin $F_{2\alpha}$ is well accepted, the modulating role of prostaglandins on the oxytocin receptor is not clear. Binding of [³H]oxytocin to isolated myometrial plasma membranes was not affected by the presence of prostaglandin $F_{2\alpha}$ or PGE_2 in the incubation medium. Long-term treatment with $PGF_{2\alpha}$ or indomethacin had no effect on oxytocin receptor concentrations and dissociation constants of myometrial plasma membranes or on maximal contractility or K_m values of isolated uterine strips exposed to oxytocin. The authors concluded that the concerted effect of oxytocin and prostaglandins on myometrial contraction does not appear to involve modulation of the oxytocin receptor by prostaglandins.[82]

However, another study[83] reported a close temporal relationship between enhanced oxytocin responsiveness of pregnant rat uteri and increased prostaglandin synthesis but no significant differences in cyclooxygenase activities among the microsomes prepared from uteri of different gestational ages. Suppression of PG synthesis attenuated the oxytocin responsiveness and markedly reduced oxytocin binding sites from 242 to 70 fmol/mg protein. There was no change in binding affinity. The authors concluded that prostaglandins stimulate oxytocin receptor formation[83] (for further discussion see Chapter 5). The discrepancies in the results on the modulating role of prostaglandins on oxytocin receptors may be partly related to different physiological states of the animals utilized for the studies and to differences in the experimental protocols. Further well-controlled studies need to be done to clarify this discrepancy.

10. Role of Prostaglandins in Tubal Motility

Ovum pick-up and its subsequent transport through the ampulla take place within a few hours of follicular rupture.[84] Important factors for the transport of ova are thought to include ciliary activity as well as muscular contractions.[85] It appears that endogenous estrogens stimulate tubal contractility in humans, whereas progesterone is inhibitory.[86]

Information regarding the action of prostaglandins on the motility of the human fallopian tube and the role of prostaglandins, if any, on their spontaneous motility are of clinical importance. For example, could administration of prostaglandin analogues accelerate the passage of ova through the oviduct and exert a contraceptive action? If increased levels of prostaglandins are found in the peritoneal fluid, for example, in endometriosis, could these, by acceleration of ovum transport, be responsible for the associated infertility? If spontaneous motility of the fallopian tubes is mediated by endogenous prostaglandins, should aspirin and other non-steroidal antiinflammatory drugs then be contraindicated during a critical part of the menstrual cycle to prevent slowing down of the ovum transport leading to tubal pregnancy?

On the basis of studies carried out in rabbits, it was proposed that prostaglandins may play a role in oviductal motility. Prostaglandin $F_{2\alpha}$, when administered subcutaneously to rabbits soon after ovulation, accelerated ovum transport.[87] Later, Ellinger and Kirton[88] observed in the same species that administration of both PGE_1 and $PGF_{2\alpha}$ accelerated the transport of ova by 40 hr and appreciably reduced the implantation rate if administered 13 hr after ovulation. Furthermore, under estrogen dominance as is seen in rabbits, increased formation of $PGF_{2\alpha}$[89,90] and increased sensitivity of the musculature to $PGF_{2\alpha}$ stimulate tubal contractility.[91] It is interesting that administration of indomethacin to rabbits before and after ovulation does not affect ovum transport, thereby questioning the role of endogenous prostaglandins on the spontaneous oviductal motility in rabbits.[92,93]

However, one should exercise great caution in extrapolating data obtained from rabbits to humans. Ovum transport in women shows certain differences compared with the rabbit, which is the laboratory species most widely studied in this context. The duration of ovum transport in humans is in the range of 80 hr, and retention at the ampullary–isthmic junction is not so distinct as in the rabbit.[84] It is therefore not surprising that clinical trials to develop a contraceptive in humans using prostaglandin analogues to accelerate ovum transport have proved disappointing.[94]

Whether endogenous prostaglandins are responsible for the spontaneous motility of human fallopian tubes is controversial. We have shown[95] that the prostaglandins released *in vitro* from fallopian tube are a result and not a cause of spontaneous motility. Indomethacin was able to inhibit synthesis of the prostaglandin released *in vitro* without affecting spontaneous motility. Similarly, papaverine, a smooth muscle relaxant, was able to block both the spontaneous motility and the prostaglandin release. We therefore concluded that prostaglandin was released as a result of tissue distortion resulting from the spontaneous motility rather than being responsible for the spontaneous motility. On this basis, we concluded that it would be safe to use nonsteroidal antiinflammatory drugs during the menstrual cycle without increasing the risk of tubal pregnancy. However, another group was able to inhibit the contractions of the fallopian tube with indomethacin and suggested that prostaglandins may be involved in the spontaneous motility of the human fallopian tube.[96]

11. Role of Prostaglandins and Leukotrienes as Mediators of Decidualization

In mammals the endometrium undergoes certain changes to allow implantation of the fertilized ova, and these changes of the endometrium are termed decidualization. In the rat and the mouse, increased uterine stromal vascular permeability is considered one of the earliest prerequisite events for blastocyst implantation and subsequent decidualization.[97]

Prostaglandins, by virtue of their vasoactive nature, are considered to be one of the important mediators of this process.[98,99] Recently it has been shown that leukotrienes are also operative in the rat uterus during early pregnancy[100,101] and appear to interact with prostaglandins in the induction of decidualization in the rat.[100] The process of decidualization involves an interaction between luminal epithelium and adjacent stroma.[102] The receptors for PGE_2 have been found to be limited to the uterine stroma in the rat.[103] The role of prostaglandins and leukotrienes in implantation of the human embryo is not known.

12. Prostaglandins and the Intrauterine Contraceptive Device

The intrauterine contraceptive device (IUD) has been widely used for controlling fertility, and many studies have been undertaken to elucidate its mechanism of action. No unifying concept has been presented to explain its mechanism because of the wide range of biological actions and species variability.

In mice, after successful mating, the presence of a silk thread in one horn of the uterus not only prevents implantation in that horn but, in addition, causes a significant reduction in the number of implantations in the contralateral control horn.[104,105] It has been suggested that substances are produced in the lumen of the uterine horn in the presence of a foreign body and that these products pass through the communication that exists between the two horns in the upper part of the cervix in mice; this could adversely affect the implantations in the control horn as well.[104] When the uterine connection is eliminated by dividing the two horns above the cervix, the bilateral effect is abolished.[105]

In rats, there is no anatomic communication between the two uterine horns, and introduction of a silk thread in one horn prevents implantation on that side only.[106,107] However, in the presence of luminal continuity established by uterine anastomosis, the introduction of a silk thread prevents implantation in the contralateral control horn as well.[107,108]

In rabbits, the contraceptive action is less complete compared with that in rodents, with implantation occurring both above and below the device.[109,110]

In guinea pigs[111] and sheep,[112,113] the bilateral presence of the device results in a shortening of the normal length of the estrous cycle, thereby suggesting that the intrauterine contraceptive device may be responsible for releasing a luteolytic factor from the uterus in these species.

In rhesus monkeys and women, the intrauterine contraceptive device probably exerts its contraceptive action in the uterus, but the exact stage at which death or disappearance of the early embryo occurs is not known.[114] Chaudhuri[115] presented a unifying hypothesis that all the actions observed with the intrauterine contraceptive device, including the species variability, can be explained on the basis of prostaglandin release. He demonstrated such release in rats.[116] Since then, prostaglandin release has been demonstrated in virtually all the animal species studied as well as in women.[117,118] Indomethacin was able to reverse the contraceptive actions in rabbits[116] but not in rats,[116] suggesting that additional mechanisms may be involved in the rat. Prostaglandin release is important for decidualization,[98,99,119] which is necessary for the implantation process; thus, a marked diminution in synthesis of prostaglandins (as by indomethacin) would, by itself, lead to inhibition of the implantation. This may be the explanation for the inability of indomethacin to reverse the IUD-mediated contraceptive action in rats. It has been reported that the contraceptive action of the intrauterine contraceptive device is reversed by prostaglandin synthetase inhibitors not only in rabbits[117] but also in women.[120]

It is now accepted that most of the IUD-associated side effects, e.g., increased uterine motility,[121] increased bleeding,[122,124] and accelerated luteolysis seen in guinea pigs and sheep,[111,113] are related to prostaglandin release, as these effects can be reversed by simultaneous use of prostaglandin synthetase inhibitors.[125,126] The types of prostaglandins released by the intrauterine contraceptive device in earlier studies were demonstrated to be PGE and F.[116] At that time, the existence of the products of the lipoxygenase pathway leading to synthesis of leukotrienes were not known, and further studies need to be carried out to evaluate the possible role of these substances. Leukocyte infiltration is consistently present in association with the intrauterine contraceptive device, and the cyclooxygenase inhibitors were unable to prevent the influx of leukocytes in association with the intrauterine contraceptive device.[127,129,130] It is possible that the presence of the intrauterine contraceptive device stimulates arachidonic acid release, thus increasing the products of both the cyclooxygenase and the lipoxygenase pathways. Leukotriene B_4 is one of the most important chemotactic factors known,[131] and other leukotrienes also play a role in the inflammatory process alongside prostaglandin. This may be the explanation for the observed effects of inflammation and leukocyte infiltration in the endometrium of women containing the intrauterine contraceptive device, which have often been described as a "sterile" inflammatorylike response.

An important clinical application has been the use of the nonsteroidal antiinflammatory drugs for menorrhagia induced by the intrauterine contraceptive device.[122,124] Administration of nonsteroidal antiinflammatory drugs only at the time of menstruation avoids any potential for reversal of the contraceptive action. They may also be administered at the time of insertion of the intrauterine contraceptive device to prevent expulsion as well as associated dysmenorrhea.

13. Menorrhagia

There are reports that menorrhagia, in the absence of organic disease[132] as well as that induced by the intrauterine contraceptive device,[122,124] has been successfully treated with nonsteroidal antiinflammatory drugs. This reduction in bleeding from the endometrium is opposite to that encountered at other sites where, after administration of these drugs, there is usually an increased tendency to bleed. The reason for this paradoxical effect is not known.

14. Prostaglandins and Endometriosis

The role of prostaglandins in producing some symptoms of endometriosis is still a matter of controversy. The endometrium is known to produce prostaglandins, and it is therefore not surprising that ectopic endometrial tissue retains this intrinsic phenomenon. Moon and colleagues[133] have demonstrated that ectopic endometrium obtained from patients with endometriosis contains a higher concentration of prostaglandin F than normal endometrium. They also demonstrated that the ectopic endometrium in cultures produced higher amounts of PGF than normal endometrium. Similarly, production of 6-keto-PGF$_1$ and TXB$_2$ by endometriotic tissue has been reported.[134] Drake and colleagues[135] demonstrated the presence of high concentrations of TXB$_2$ and prostacyclin in the peritoneal fluid obtained from patients with endometriosis. By contrast, Rock and colleagues[136] were unable to demonstrate a significant difference in the peritoneal fluid concentrations of PGE$_2$, PGF$_{2\alpha}$, the PGF$_{2\alpha}$ metabolite, 15-keto-13,14-dihydroprostaglandin F$_2$, and TXB$_2$ when compared with peritoneal fluid obtained from disease-free women. These differences in results by different investigators are not readily explainable.

Macrophages are reported to be present in higher numbers in peritoneal fluid of patients with endometriosis.[137] Also, the peritoneal fluid that is collected at the time of laparoscopy may be contaminated by blood trickling into the cul-de-sac from the laparoscopic insertion site. Leukocytes[138] and macrophages[139,140] produce prostaglandins, and leukocytes and platelets produce TXA$_2$,[16] and these cells are high-capacity synthetic sites. It is therefore possible that a significant amount of prostaglandins could be formed during the handling of the sample, and the addition of prostaglandin synthesis inhibitors to the sampling tube after collecting the fluid may not be sufficient to overcome the rapid artifactual production. It is interesting that in some of the studies, no mention is made of indomethacin or a cyclooxygenase inhibitor being added to the sample tube at the time of collection of the sample. Therefore, the prostaglandins being measured could well have been synthesized after collection.

Administration of cyclooxygenase inhibitors does not uniformly reduce the symptoms of endometriosis,[134,141] nor has it been reported to increase the pregnancy rate in patients with endometriosis. The lipoxygenase products of arachidonic acid metabolism would not be inhibited by these nonsteroidal antiinflammatory drugs. Leukotriene B$_4$ is one of the most potent chemotactic factors known.[28,30] In recent years, a great deal of interest has centered around the role of macrophages in infertility. Macrophages have been found to be increased in the peritoneal fluid of patients with endometriosis.[142] It has been speculated that phagocytosis of sperm by the macrophages may be an important factor responsible for infertility in these cases.[142] However, it has recently been demonstrated that there is no increase in LTB_4 in the peritoneal fluid obtained from women with endometriosis when compared to those obtained from disease-free women.[143]

15. Cervical Ripening

The transformation of the cervix, which is normally a long, firm, closed fibrous tube, to a soft compliant structure allowing for effacement and dilation is termed cervical ripening. Danforth[144,145] has over the years stressed the importance of collagen in the ripening of the cervix, as collagen is more abundant than smooth muscle in the cervix. It is now thought that for the ripening process there is an alteration in the spatial arrangements of collagen fibers resulting from changes in the composition of the ground substance (in particular glycosaminoglycans and proteoglycans) causing a reduction in the tensile strength of the tissue. These changes, which all appear to be under the control of fibroblasts, have been elegantly reviewed.[146]

Prostaglandins stimulate the synthesis of proteoglycans in cultured fibroblasts[147] as well as in the human uterine cervix *in vitro*.[148] Prostaglandins also increase the concentration of sulfated glycosaminoglycans.[149] There is controversy as to whether prostaglandins increase the protease or collagenase activity of the cervix.[150,151] It has been suggested that changes in the content of various glycosaminoglycans and the resulting physiochemical interactions with collagen may play a predominant role in the process of prostaglandin-induced cervical ripening.[151] Further studies need to be done to clarify more precisely the mechanism of prostaglandin-induced cervical ripening.

16. Conclusion

It is fair to conclude that although we now know a lot more about factors that influence the synthesis of prostaglandins in the uterus, we are still ignorant about factors that control its synthesis in certain key physiological events as in premature labor and parturition.

The physiological role of products of arachidonic acid metabolism by the lipoxygenase pathway as it relates to reproduction has still not been clearly elucidated, and it is hoped that significant progress will be made in this area.

Other metabolic pathways for arachidonic acid continue to be described.[152,153] Very little information is available on the physiological role of products of these pathways, e.g., epoxyeicosatrienoic acid (EET). Future work should address the interactions of the various products produced by different pathways and the importance of such interactions in reproductive physiology. Thus, one could hope to look for exciting work related to eicosanoids in the future.

References

1. Von Euler, U. S., 1934, Zur Kenntnis der pharmakologischen Wirkungen von nativsekreten and extrakten mannlicher accessorischer Geschlechtsdrusen, *Arch. Exp. Pathol. Pharmakol.* **175**:78–84.
2. Von Euler, U. S., 1936, On the specific vasodilating and plain muscle stimulating substances from accessory genital glands in man and certain animals (prostaglandin and vesiglandin), *J. Physiol. (Lond.)* **88**:213–234.
3. Goldblatt, M. W., 1933, A depressor substance in seminal fluid, *J. Soc. Chem. Ind.* **52**:1056–1057.
4. Eliasson, R., 1959, Studies on prostaglandin. Occurrence, formation and biological actions, *Acta Physiol. Scand [Suppl.]* **158**:1–73.
5. Bergstrom, S., and Sjovall, J., 1960, The isolation of prostaglandin F from sheep prostate glands, *Acta Chem. Scand.* **14**:1693–1700.
6. Bergstrom, S., and Sjovall, J., 1960, The isolation of prostaglandin E from sheep prostate glands, *Acta Chem. Scand.* **14**:1701–1705.
7. Bergstrom, S., Danielsson, J., and Samuelsson, B., 1964, The enzymatic formation of prostaglandin E_2 from arachidonic acid, *Biochem. Biophys. Acta* **90**:207–210.
8. van Dorp, D. A., Beerthuis, R. K., Nugteren, D. H., and Vonkeman, H., 1964, The biosynthesis of prostaglandins, *Biochim. Biophys. Acta* **90**:204–206.
9. Flower, R. J., and Blackwell, G. J., 1976, The importance of phospholipase A_2 in prostaglandin biosynthesis, *Biochem. Pharmacol.* **25**:285–291.
10. Blackwell, C. J., Carnuccio, R., DiRosa, M., Flower, R. J., Parente, L., and Persico, P., 1980, Macrocortin: A polypeptide causing the anti-phospholipase effect of glucocorticoids, *Nature* **287**:147–149.
11. DiRosa, M., Flower, R. J., Hirata, F., Parente, L., and Russo-Marie, F., 1984, Nomenclature announcement: Anti-phospholipase proteins, *Prostaglandins* **28**:441–442.
12. Piper, P. J., and Vane, J. R., 1971, The release of prostaglandins from lung and other tissues, *Ann. N.Y. Acad. Sci.* **180**:363–385.
13. Christ, E. J., and Van Dorp, D. A., 1972, Comparative aspects of prostaglandin biosynthesis, *Biochim Biophys. Acta* **270**:537–545.
14. Hamberg, M., Svensson, J., and Samuelsson, B., 1974, A new concept concerning the mode of action and release of prostaglandins, *Proc. Natl. Acad. Sci. U.S.A.* **71**:3824–3828.
15. Moncada, S., and Higgs, E. A., 1988, Metabolism of arachidonic acid, *Ann. N.Y. Acad. Sci.* **522**:454–463.
16. Needleman, P., Turk, J., Jakschik, B. A., Morrison, A. R., and Lefkowitch, J. B., 1986, Arachidonic acid metabolism, *Annu. Rev. Biochem.* **55**:69–102.
16a. Chaudhuri, G., 1985, Physiologic aspects of prostaglandins and leukotrienes, in: *Seminars in Reproductive Endocrinology*, Volume III (R. Subir, ed.), Thieme Medical Publishers, New York, pp. 219–230.
17. Tuvemo, T., Strandberg, K., Hamberg, M., and Samuelsson, B., 1981, Maintenance of the tone of the human umbilical artery by prostaglandin and thromboxane formation, *Adv. Prostaglandin Thromboxane Res.* **1**:425–428.
18. Williams, K. I., Dembinska-Kiec, A., Zmuda, A., and Gryglewski, R. J., 1978, Prostacyclin formation by myometrial and decidual fractions of the pregnant rat uterus, *Prostaglandins* **15**:343–350.
19. Mitchell, M. D., Bibby, J. G., Hicks, B. R., and Turnbull, A. C., 1978, Possible role of prostacyclin in human parturition, *Prostaglandins* **16**:931–937.
20. Myatt, L., and Elder, M. G., 1977, Inhibition of platelet aggregation by a placental substance with prostacyclin-like activity, *Nature* **268**:159–160.
21. Lands, W. E. M., 1979, The biosynthesis and metabolism of prostaglandins, *Annu. Rev. Physiol.* **41**:633–652.

22. Samuelsson, B., Granstrom, E., Green, K., Hamberg, M., and Hammerstrom, S., 1975, Prostaglandins, *Annu. Rev. Biochem.* **44**:669–695.
23. Hamberg, M. J., Svensson, J., Wakabayashi, T., and Samuelsson, B., 1974, Isolation and structure of two prostaglandin endoperoxides that cause platelet aggregation, *Proc. Natl. Acad. Sci. U.S.A.* **71**:345–349.
24. Nugteren, D. H., Arachidonate lipoxygenase in blood platelets, *Biochim. Biophys. Acta* **380**:299–307.
25. Borgeat, P., Hamberg, M., and Samuelsson, B., 1976, Transformation of arachidonic acid and homo-β-linolenic acid by rabbit polymorphonuclear leukocytes, *J. Biol. Chem.* **251**:7816–7820.
26. Samuelsson, B., 1983, Leukotrienes: Mediators of immediate hypersensitivity reactions and inflammation, *Science* **220**:568–575.
27. Hammarstrom, S., 1983, Leukotrienes, *Annu. Rev. Biochem.* **52**:355–377.
28. Bray, M. A., 1983, The pharmacology and pathophysiology of leukotriene B_4, *Br. Med. Bull.* **39**:249–254.
29. Lewis, R. A., and Austen, K. F., 1984, The biologically active leukotrienes, *J. Clin. Invest.* **73**:889–897.
30. Higgs, G. A., and Moncada, S., 1985, Leukotrienes in disease: Implications for drug development, *Drugs* **30**:1–5.
31. Adcock, J. J., and Garland, L. G., 1980, A possible role for lipoxygenase products as regulators of airway smooth muscle activity, *Br. J. Pharmacol.* **69**:167–169.
32. Patterson, R., Pruzansky, J. J., and Harris, K. E., 1981, An agent that releases basophil and mast cell histamine but blocks cyclo-oxygenase and lipoxygenase metabolism of arachidonic acid inhibits immunoglobulin E-mediated asthma in rhesus monkeys, *J. Allergy Clin. Immunol.* **67**:444–449.
33. Moonen, P., Klok, G., and Kierse, M. J. N. C., 1985, Immunohistochemical localization of prostaglandin endoperoxide synthase and prostacyclin synthase in pregnant human myometrium, *Eur. J. Obstet. Gynecol. Reprod. Biol.* **19**:151–158.
34. Abel, M. H., and Kelly, R. W., 1979, Differential production of prostaglandins within the human uterus, *Prostaglandins* **18**:821–828.
35. Christensen, N. J., and Green, K., 1983, Bioconversion of arachidonic acid in human pregnant reproductive tissues, *Biochem. Med.* **30**:162–180.
36. Moonen, P., Klok, G., and Keirse, M. J. N. C., 1984, Increase in concentrations of prostaglandin endoperoxide synthase and prostacyclin synthase in human myometrium in late pregnancy, *Prostaglandins* **28**:309–321.
37. Keirse, M. J. N. C., Moonen, P., and Klock, G., 1985, The influence of uterine anatomy on the concentrations of prostaglandin endoperoxide and prostacyclin synthases during human pregnancy, *Eur. J. Obstet. Gynecol. Reprod. Biol.* **19**:327–331.
38. Schatz, F., and Gurpide, E., 1983, Effects of estradiol on $PGF_{2\alpha}$ levels in primary monolayer cultures of epithelial cells from human proliferative endometrium, *Endocrinology* **113**:1274–1279.
39. Schatz, F., Markiewicz, L., and Gurpide, E., 1987, Differential effects of estradiol, arachidonic acid and A-23187 on prostaglandin $F_{2\alpha}$ output by epithelial and stromal cells of human endometrium, *Endocrinology* **120**:1465–1471.
40. Lumsden, M. A., Brown, A., and Baird, D. T., 1984, Prostaglandin production from homogenates of separated glandular epithelium and stroma from human endometrium, *Prostaglandins* **28**:485–496.
41. Downie, J., Poyser, N. I., and Wunderlich, M., 1974, Levels of prostaglandins in human endometrium during the normal menstrual cycle, *J. Physiol. (Lond.)* **236**:465–472.
41a. Guyton, A. C., 1986, *Medical Physiology*, 7th ed., W. B. Saunders, Philadelphia.
42. Demers, I. M., Yoshuga, K., and Greep, R. O., 1974, Prostaglandin F in monkey uterine fluid during the menstrual cycle and following steroid treatment, *Prostaglandins* **5**:513–520.
43. Horton, E. W., and Poyser, N. I., 1976, Uterine luteolytic hormone: A physiological role for prostaglandin $F_{2\alpha}$, *Physiol. Rev.* **56**:595–651.

44. Abel, M. H., and Baird, D. T., 1980, The effect of 17β estradiol and progesterone on prostaglandin production by human endometrium maintained in organ culture, *Endocrinology* **106**:1599–1606.

45. Schatz, F., Markiewicz, L., Barg, P., and Gurpide, E., 1985, *In vitro* effects of ovarian steroids on prostaglandin output by human endometrium and endometrial epithelial cells, *J. Clin. Endocrinol. Metab.* **61**:361–367.

46. Gurpide, E., Markiewicz, I., Schatz, F., and Hirata, F., 1986, Lipocortin output by human endometrium *in vitro*, *J. Clin. Endocrinol. Metab.* **63**:162–166.

47. Kikukawa, T., Cowan, B. D., Tejada, R., and Mukherjee, A. B., 1988, Partial characterization of a uteroglobulin-like protein in the human uterus and its temporal relationship to prostaglandin levels in this organ, *J. Clin. Endocrinol. Metab.* **67**:315–321.

48. Markiewicz, L., and Gurpide, E., 1988, C-19 adrenal steroids enhance prostaglandin $F_{2\alpha}$ output by human endometrium *in vitro*, *Am. J. Obstet. Gynecol.* **159**:500–504.

49. Malathy, P. V., Cheng, H. C., and Dey, S. K., 1986, Production of leukotrienes and prostaglandins in the rat uterus during periimplantation period, *Prostaglandins* **32**:605–614.

50. Carraher, R., Hahn, D., Ritchie, D., and McGuire, J., 1983, Involvement of lipoxygenase products in myometrial contractions, *Prostaglandins* **26**:23–32.

51. Pakrasi, P. L., Becka, R., and Dey, S. K., 1985, Cyclo-oxygenase and lipoxygenase pathways in the preimplantation rabbit uterus and blastocysts, *Prostaglandins* **29**:481–495.

52. Mitchell, M. D., and Grzyboski, C. F., 1987, Arachidonic acid metabolism by lipoxygenase pathways in intrauterine tissues of women at term pregnancy, *Prostaglandins Leukotrienes Med.* **28**:303–312.

53. Chegini, N., and Rao, C. V., 1988, Quantitative light microscopic autoradiographic study on [^{3}H]leukotriene C_4 binding to non-pregnant bovine uterus tissue, *Endocrinology* **122**:1732–1736.

54. Tawfik, O. W., Sagrillo, C., Johnson, D. C., and Dey, S. K., 1987, Decidualization in the rat: Role of leukotrienes and prostaglandins, *Prostaglandins Leukotrienes Med.* **29**:221–227.

55. Tawfik, O. W., and Dey, S. K., 1988, Further evidence for role of leukotrienes as mediators of decidualization in the rat, *Prostaglandins* **35**:379–402.

56. Pakrasi, P. L., and Dey, S. K., 1986, Evidence for an inverse relationship between cyclo-oxygenase and lipoxygenase pathway in the pregnant rabbit endometrium, *Prostaglandins Leukotrienes Med.* **18**:347–352.

57. Kimball, F. A., Kirton, K. T., and Wyngarden, I. J., 1975, Prostaglandin specific binding in hamster myometrial low speed supernatant, *Prostaglandins* **9**:413–429.

58. Johnson, M., Jessup, R., and Ramwell, P. W., 1974, Correlation of prostaglandin E_1 receptor binding with evoked uterine contraction: Modification by disulfide reduction, *Prostaglandins* **6**:433–449.

59. Carsten, M., and Miller, J., 1981, Prostaglandin E_2 receptor in the myometrium. Distribution in subcellular fractions, *Arch. Biochem. Biophys.* **212**:700–704.

60. Kimball, F. A., Kirton, K. T., and Wyngarden, I. J., 1975, Prostaglandin E_1 specific binding in rhesus myometrium, *Prostaglandins* **10**:853–864.

61. Giannopoulos, G., Jackson, K., Kredentser, J., and Tulchinsky, D., 1985, Prostaglandin E and $F_{2\alpha}$ receptors in human myometrium during the menstrual cycle and in pregnancy and labor, *Am. J. Obstet. Gynecol.* **153**:904–910.

62. Adelantado, J. M., Lopez Bernal, A., and Turnbull, A. C., 1988, Topographical distribution of prostaglandin E receptors in human myometrium, *Br. J. Obstet. Gynaecol.* **95**:348–353.

63. Bauknecht, T., Krahe, B., Rechenbach, U., Zahradnik, H. P., and Breckwoldt, M., 1981, Distribution of prostaglandin E_2 and prostaglandin $F_{2\alpha}$ receptors in human myometrium, *Acta Endocrinol. (Kbh.)* **98**:446–450.

64. Wakeling, A. E., Kirton, K. E., and Wyngarden, L. J., 1973, Prostaglandin receptors in the hamster uterus during the estrous cycle, *Prostaglandins* **4**:1–9.

65. Adelantado, J. M., Rees, M. C. P., Bernal, A. L., and Turnbull, A. C., 1988, Increased uterine prostaglandin E receptors in menorrhagic women, *Br. J. Obstet. Gynaecol.* **95**:162–165.

66. Smith, S. K., Abel, M. H., Kelly, R. W., and Baird, D. T., 1981, Prostaglandin synthesis in the

endometrium of women with ovular dysfunctional uterine bleeding, *Br. J. Obstet. Gynaecol.* **88:**434–442.

67. Sanger, G. J., and Bennett, A., 1979, Fenamates may antagonize the actions of prostaglandin endoperoxides in human myometrium, *Br. J. Clin. Pharmacol.* **8:**479–482.

68. McClean, J. R., and Gluckman, M. I., 1983, On the mechanism of the pharmacologic activity of meclofenamate sodium, *Arzneimittelforschung* **33:**627–632.

69. Chegini, N., and Rao, C. V., 1988, The presence of leukotriene C_4 and prostacyclin binding sites in non-pregnant human uterine tissue, *J. Clin. Endocrinol. Metab.* **66:**76–87.

70. Weichman, B. M., and Tucker, S. S., 1982, Contractions of guinea pig uterus by synthetic leukotrienes, *Prostaglandins* **24:**245–253.

71. Chegini, N., and Rao, C. V., 1988, Quantitative light microscopic autoradiographic study on [^{3}H]leukotriene C_4 binding to non-pregnant bovine uterine tissue, *Endocrinology* **122:**1732–1736.

72. Carsten, M. E., 1974, Hormonal regulation of myometrial calcium transport, *Gynecol. Invest.* **5:**269–275.

73. Carsten, M. E., and Miller, J. D., 1977, Effects of prostaglandins and oxytocin on calcium release from a uterine microsomal fraction, *J. Biol. Chem.* **252:**1576–1581.

74. Carsten, M. E., 1974, Prostaglandins and oxytocin: Their effects on uterine smooth muscle, *Prostaglandins* **5:**33–40.

75. Carsten, M. E., 1973, Prostaglandins and cellular calcium transport in the pregnant human uterus, *Am. J. Obstet. Gynecol.* **117:**824–832.

76. Villar, A., D'Ocon, D., and Anselmi, E., 1985, Calcium requirement of uterine contraction induced by PGE_1: Importance of intracellular calcium stores, *Prostaglandins* **30:**491–496.

77. Vane, J. R., and Williams, K. I., 1973, The contribution of prostaglandin production to contractions of the isolated uterus of the rat, *Br. J. Pharmacol.* **48:**629–639.

78. Roberts, J. S., Barcikowski, B., Wilson, I., Skarnes, R. C., and McCracken, J. A., 1975, Hormonal and related factors affecting the release of prostaglandin $F_{2\alpha}$ from the uterus, *J. Steroid Biochem.* **6:**1091–1097.

79. Sharma, S. C., and Fitzpatrick, R. J., 1974, Effect of estradiol 17-β and oxytocin treatment on prostaglandin F release in the anoestrous ewe, *Prostaglandins* **6:**97–105.

80. Roberts, J. S., McCracken, J. A., Gavagan, J. E., and Soloff, M. S., 1976, Oxytocin stimulated release of prostaglandin $F_{2\alpha}$ from ovine endometrium *in vitro:* Correlation with estrous cycle and oxytocin receptor binding, *Endocrinology* **99:**1107–1114.

81. Wilson, T., Liggins, C. C., and Whittaker, D. J., 1988, Oxytocin stimulates the release of arachidonic acid and prostaglandin $F_{2\alpha}$ from human decidual cells, *Prostaglandins* **35:**771–780.

82. Engstrom, T., Atke, A., and Vilhardt, H., 1988, Oxytocin receptors and contractile responses of the myometrium after long term infusion of prostaglandin $F_{2\alpha}$, indomethacin, oxytocin and an oxytocin antagonists in rats, *Regul. Peptides* **20:**65–72.

83. Chan, W. Y., 1987, Enhanced prostaglandin synthesis in the parturient rat uterus and its effects on myometrial oxytocin receptor concentrations, *Prostaglandins* **34:**888–902.

84. Croxatto, H. B., Ortiz, M. E., Diaz, S., Hess, R., Balmaceda, J., and Croxatto, H. D., 1978, Studies on the duration of egg transport by the human oviduct, *Am. J. Obstet. Gynecol.* **132:**629–634.

85. Hafez, E. S. E., 1979, Function of the fallopian tube in human reproduction, *Clin. Obstet. Gynecol.* **22:**61–79.

86. Lindblom, B., Hamberger, L., and Ljung, B., 1980, Contractile patterns of isolated oviductal smooth muscle under different hormonal conditions, *Fertil. Steril.* **33:**283–287.

87. Chang, M. D., and Hunt, D. M., 1972, Effect of prostaglandin $E_{2\alpha}$ on the early pregnancy of rabbits, *Nature* **236:**120–121.

88. Ellinger, J. V., and Kirton, K. T., 1971, Ovum transport in rabbits injected with prostaglandin E_1 and $F_{2\alpha}$, *Biol. Reprod.* **11:**93–96.

89. Saksena, S. K., and Harper, M. J. K., 1975, Relationship between concentration of prostaglandin F (PGF) in the oviduct and egg transport in rabbits, *Biol. Reprod.* **13:**68–76.

90. Spilman, C. H., 1974, Oviduct response to prostaglandins: Influence of estradiol and progesterone, *Prostaglandins* **7**:465–472.

91. Coutinho, E. M., and Maia, H. S., 1971, The contractile response of the human uterus, fallopian tubes and ovary to prostaglandins *in vivo*, *Fertil. Steril.* **22**:539–543.

92. El-Banna, A. A., Sacher, B., and Schilling, E., 1976, Effect of indomethacin on egg transport and pregnancy in the rabbit, *J. Reprod. Fert.* **46**:375–378.

93. Hodgson, B. J., 1976, Effects of indomethacin and ICI 46,474 administered during ovum transport on fertility in rabbits, *Biol. Reprod.* **14**:451–457.

94. Croxatto, H. B., Ortiz, M. E., Guiloff, E., Ibarra, A., Salvatierra, A. M., Croxatto, H. D., and Spilman, C. H., 1978, Effect of 15(*S*)-15-methyl prostaglandin $F_{2\alpha}$ on human oviductal motility and ovum transport, *Fertil. Steril.* **30**:408–414.

95. Elder, M. G., Myatt, L., and Chaudhuri, G., 1977, The role of prostaglandins in the spontaneous motility of the fallopian tube, *Fertil. Steril.* **28**:86–90.

96. Lindblom, B., Wilhelmsson, M., Wikland, M., Hamberger, L., and Wiquist, N., 1983, Prostaglandins and oviductal function, *Acta Obstet. Gynecol. Scand. [Suppl.]* **113**:43–46.

97. Psychoyos, A., 1973, Endocrine control of egg implantation, in: *Handbook of Physiology,* Section 7, Vol. II (R. O. Greep, ed.), American Physiological Society, Washington, DC, pp. 187–215.

98. Kennedy, T. G., and Lukash, L. A., 1982, Induction of decidualization in the rats by the intraluminal infusion of prostaglandins, *Biol. Reprod.* **27**:253–260.

99. Kennedy, T. G., 1985, Evidence for the involvement of prostaglandins throughout the decidual cell reaction in the rat, *Biol. Reprod.* **33**:140–146.

100. Tawfik, O. W., Sagrillo, C., Johnson, D. C., and Dey, S. K., 1987, Decidualization in the rat: Role of leukotrienes and prostaglandins, *Prost. Leuk. Med.* **29**:221–227.

101. Tawfik, O. W., and Dey, S. K., 1988, Further evidence for role of leukotrienes as mediators of decidualization in the rat, *Prostaglandins* **35**:379–386.

102. Lejeune, B., VanHoeck, J., and Leroy, F., 1981, Transmitter role of luminal uterine epithelium in the induction of decidualization in rats, *J. Reprod. Fertil.* **61**:235–240.

103. Kennedy, T. G., Martel, D., and Psychoyos, A., 1983, Endometrial prostaglandins E_2 binding during the estrous cycle and its hormonal control in ovariectomized rats, *Biol. Reprod.* **29**:565–571.

104. Doyle, L. L., and Margolis, A. J., 1966, Effect of an IUFB on reproduction in mice, *J. Reprod. Fertil.* **11**:27–32.

105. Marston, J. H., and Kelly, W. A., 1969, The time and site of contraceptive action of an intrauterine device in the mouse, *J. Endocrinol.* **43**:83–93.

106. Doyle, L. L., and Margolis, A. J., 1964, Intrauterine foreign body: 1. Effect on reproductive processes in the rat, *Fertil. Steril.* **15**:597–606.

107. Batta, S. K., and Chaudhury, R. R., 1968, Antifertility effect of an intrauterine silk thread in rats with a connection between the two uterine horns, *J. Reprod. Fertil.* **16**:371–379.

108. Marston, J. H., and Kelly, W. A., 1969, The effect of uterine anastomosis on the action of an intrauterine device in the rat, *J. Endocrinol.* **43**:95–103.

109. Adams, C. E., and Eckstein, P., 1965, Effect of intrauterine silk threads on location and survival of conceptuses in the rabbit, *J. Reprod. Fertil.* **9**:351–354.

110. Adams, C. E., and Eckstein, P., 1965, Effects of IUFBs on pregnancy in the rabbit, *Fertil. Steril.* **16**:508–521.

111. Donovan, B. T., and Traczyk, W., 1962, The effect of uterine distention on the oestrus cycle of the guinea pig, *J. Physiol. (Lond.)* **161**:227–236.

112. Moore, W. W., and Nalbandov, A. V., 1953, Neurogenic effects of uterine distention on the oestrus cycle of the ewe, *Endocrinology* **53**:1–11.

113. Nalbandov, A. V., Moore, W. W., and Norton, W. W., 1955, Further studies on the neurogenic control of the oestrus cycle by uterine distention, *Endocrinology* **56**:225–231.

114. Eckstein, P., 1970, Mechanisms of action of intrauterine contraceptive devices in women and other mammals, *Br. Med. Bull.* **26**:52–59.

115. Chaudhuri, G., 1971, Intrauterine device: Possible role of prostaglandins, *Lancet* **1**:480.

116. Chaudhuri, G., 1973, Release of prostaglandins by the IUCD, *Prostaglandins* **3**:773–784.
117. Saksena, S. K., and Harper, M. J. K., 1974, Prostaglandin-mediated action of intrauterine devices: F prostaglandins in the uterine horn of pregnant rabbits with unilateral intrauterine devices, *Fertil. Steril.* **25**:121–126.
118. Hillier, K., and Kasonde, J. M., 1976, Prostaglandin E and F concentrations in human endometrium after insertion of intrauterine contraceptive device, *Lancet* **1**:15–16.
119. Chaudhuri, G., 1976, Mechanisms of IUD-induced suppression of deciduoma—the role of prostaglandins, *Adv. Prostaglandin Thromboxane Res.* **2**:893–894.
120. Buhler, M., and Papiernik, E., 1983, Successive pregnancies in women fitted with intrauterine devices who take anti-inflammatory drugs, *Lancet* **1**:483.
121. Chaudhuri, G., 1977, The role of prostaglandins on the increase in motility of the rat uterine horn containing a silk suture, *Int. J. Fertil.* **22**:44–47.
122. Damarawy, H., and Toppazada, M., 1976, Control of bleeding due to IUDs by a prostaglandin biosynthesis inhibitor, *IRCS Med. Sci.* **4**:5–10.
123. Guillebaud, J., Anderson, A. B. M., and Turnbull, A. C., 1978, Reduction by mefenamic acid of increased menstrual blood loss associated with intrauterine contraception, *Br. J. Obstet. Gynaecol.* **85**:53–62.
124. Roy, S., and Shaw, S. T., Jr., 1981, Role of prostaglandins in IUD-associated uterine bleeding-effect of a prostaglandin synthetase inhibitor (ibuprofen), *Obstet. Gynecol.* **58**:101–106.
125. Donovan, B. T., 1975, Indomethacin, ketoprofen and corpus luteum regression in the guinea pig, *Br. J. Pharmacol.* **53**:225–227.
126. Spilman, C. H., and Duby, R. T., 1972, Prostaglandin-mediated luteolytic effect of an intrauterine device in sheep, *Prostaglandins* **4**:57–64.
127. Parr, E. L., Schaedler, R. W., and Hirsch, J. G., 1967, The relationship of polymorphonuclear leucocytes to infertility in uteri containing large foreign bodies, *J. Exp. Med.* **126**:523–537.
128. Sagiroglu, N., and Sagiroglu, E., 1970, Biologic mode of the Lippes loop in intrauterine contraception, *Am. J. Obstet. Gynecol.* **106**:506–515.
129. Myatt, L., Chaudhuri, G., Gordon, D., and Elder, M. S., 1977, Prostaglandin production by leucocytes attached to intrauterine devices, *Contraception* **15**:589–599.
130. Greenwald, G. S., 1965, Interruption of pregnancy in the rat by the uterine suture, *J. Reprod. Fertil.* **9**:9–17.
131. Ford Hutchinson, A. W., Bray, M. A., and Dorg, M. V., 1980, Leukotriene B$_4$, a potent chemokinetic and aggregating substance released from polymorphonuclear leucocytes, *Nature* **286**:264–265.
132. Anderson, A. B. M., Guillebaud, J., Haynes, P. J., and Turnbull, A. C., 1976, Reduction of menstrual blood loss by prostaglandin synthetase inhibitors, *Lancet* **1**:774–776.
133. Moon, Y. S., Leung, P. C. S., Yuen, B. H., and Gomel, V., 1981, Prostaglandin F in human endometriotic tissue, *Am. J. Obstet. Gynecol.* **141**:344–345.
134. Ylikorkala, O., and Viinikka, L., 1983, Prostaglandins and endometriosis, *Acta Obstet. Gynecol. Scand. [Suppl.]* **113**:105–107.
135. Drake, T. S., O'Brien, W. F., Ramwell, P. W., and Metz, S. A., 1981, Peritoneal fluid thromboxane B$_2$ and 6-keto-prostaglandin F$_{2\alpha}$ in endometriosis, *Am. J. Obstet. Gynecol.* **140**:401–404.
136. Rock, J. A., Dubin, N. H., Ghodgaonkar, R. B., Bergquist, C. A., Erozan, Y. S., and Kimball, A. W., 1982, Cul-de-sac fluid in women with endometriosis. Fluid volume and prostanoid concentration during the proliferative phase of the cycle—days 8–12, *Fertil. Steril.* **37**:747–750.
137. Halme, J., Becker, S., Hammond, M. G., Raj, M. H., and Raj, S., 1983, Increased activation of pelvic macrophages in infertile women with mild endometriosis, *Am. J. Obstet. Gynecol.* **145**:333–337.
138. Higgs, G. A., and Youlten, L. J. F., 1972, Prostaglandin production by rabbit peritoneal polymorphonuclear leucocytes *in vitro*, *Br. J. Pharmacol.* **44**:330–332.
139. Humes, J. L., Bonney, R. J., Pelus, L., Dahlgren, M. E., Sadowski, S. J., Kuhl, F. A., and Davis, R., 1977, Macrophages synthesize and release prostaglandins in response to inflammatory stimuli, *Nature* **269**:149–151.

140. Dy, M., Astoin, M., Rigaud, M., and Hamberger, J., 1980, Prostaglandin (PG) release in the mixed lymphocyte culture; effect of presensitization by a skin allograft: Nature of the PG-producing cell, *Eur. J. Immunol.* **10:**121–126.

141. Kauppila, A., Puolakka, J., and Ylikorkala, O., 1979, Prostaglandin biosynthesis inhibitors and endometriosis, *Prostaglandins* **18:**655–661.

142. Haney, A. D., Muscato, J. J., and Weinberg, J. B., 1981, Peritoneal fluid cell populations in infertility patients, *Fertil. Steril.* **35:**696–698.

143. Rapkin, A., and Bhattacharya, P., 1989, Peritoneal fluid eicosanoids in chronic pelvic pain, *Prostaglandins* **38:** 447–452.

144. Danforth, D. N., and Buckingham, J. C., 1964, Connective tissue changes and their relation to pregnancy, *Obstet. Gynecol. Surv.* **19:**715–732.

145. Danforth, D. N., Veis, A., Breen, M., Weinstein, H. G., Buckingham, J., and Manalo, P., 1974, The effect of pregnancy and labour on the human cervix: Changes in collagen, glycoproteins and glycosaminoglycans, *Am. J. Obstet. Gynecol.* **120:**641–651.

146. Liggins, G. C., 1978, Ripening of the cervix, in: *Seminars in Perinatology*, Volume II, No. 3 (T. K. Oliver and T. H. Kirschbaum, eds.), Grune & Stratton, New York, pp. 261–271.

147. Murota, S., Abe, M., and Otsuka, K., 1977, Stimulatory effects of prostaglandins on the production of hexosamine containing substance by cultured fibroblasts, *Prostaglandins* **14:**983–991.

148. Norstrom, A., 1982, Influence of prostaglandin E_2 on the biosynthesis of connective tissue constituents in the pregnant human cervix, *Prostaglandins* **23:**361–367.

149. Uldbjerg, N., Ekman, G., Malmstrom, A., Ulmsten, U., and Wingerup, L., 1983, Biochemical changes in human cervical connective tissue after local application of PGE_2, *Gynecol. Obstet. Invest.* **15:**291–299.

150. Goshowaki, H., Ito, A., and Mori, Y., 1988, Effects of prostaglandins on the production of collagenase by rabbit uterine cervical fibroblasts, *Prostaglandins* **36:**107–114.

151. Rath, W., Adelmann-Grill, B. C., Pieper, U., and Kuhn, W., 1987, The role of collagenases and proteases in prostaglandin induced cervical ripening, *Prostaglandins* **34:**119–127.

152. Capdevila, J., Chacos, N., Werringloer, J., Prough, R. A., and Estabrook, R. W., 1981, Liver microsomal cytochrome P450 and the oxidative metabolism of arachidonic acid, *Proc. Natl. Acad. Sci. U.S.A.* **78:**5362–5366.

153. Capdevila, J., Chacos, N., Falck, J. R., Manna, S., Negro-Vilar, A., and Ojeda, S., 1983, Novel hypothalamic arachidonate products stimulate somatostatin release from the median eminence, *Endocrinology* **113:**421–423.

15

Pharmacological Application of Prostaglandins, Their Analogues, and Their Inhibitors in Obstetrics

Robert H. Hayashi

1. Introduction

The central role of eicosanoids, particularly prostaglandin (PG) E_2 and $PGF_{2\alpha}$, in human parturition has been well documented in other chapters of this book. The following conclusions can be drawn: (1) prostaglandins reliably cause cervical ripening and uterine contractions, which result in pregnancy termination at any gestational age; the uterine sensitivity increases as term gestation is approached; (2) increasing concentration of arachidonic acid, precursor of prostaglandins, occurs in amniotic fluid of women in labor; (3) intraamniotic injection of arachidonic acid stimulates labor; and (4) inhibitors of prostaglandin synthesis can diminish or prevent uterine contractions.[1,2]

The availability of the products of the cyclooxygenase pathway of arachidonic acid metabolism in commercial quantities has encouraged the pharmacological application of these products in pregnancy termination. Inhibitors of this pathway have been useful in the abatement of premature uterine contractions.

Three generations of synthetic prostaglandins have been produced thus far. The first generation is represented by the classic compounds PGE_2 and $PGF_{2\alpha}$ (Dinoprost or Prestin). About 90% of these prostaglandins are metabolized in one pass through the pulmonary circulation.[3] This rapid clearance results in about 3% of an intravenous dose remaining in the systemic circulation after 90 sec.[4] Continuous and high-dose systemic administration is required to achieve a therapeutic effect on the uterus.

Robert H. Hayashi • Department of Obstetrics and Gynecology, University of Michigan Medical School, Ann Arbor, Michigan 48109.

Early pregnancy termination requires high circulating levels of prostaglandins because of the low sensitivity of the uterus at that time. The result is a greater than 50% incidence of significant side effects. These include nausea, vomiting, diarrhea, uterine contraction-induced pain, and fever. Prostaglandins E and F stimulate longitudinal muscle contraction in the gastrointestinal tract and relax the sphincter muscles, thus shortening the gut transit time.[6] Prostaglandins stimulate intestinal mucosal cells to secrete fluid and electrolytes, resulting in nausea and emesis, which may eventually result in fluid and electrolyte imbalance. The prostaglandin F series can provoke increased airway resistance in normal patients and intense bronchospasm in asthmatic patients.[7] The fever is a result of an effect of prostaglandin on the central nervous system thermoregulatory center. However, prostaglandins rarely cause life-threatening maternal adverse effects. There are only two reported maternal deaths, both from cardiovascular collapse, that are directly attributable to prostaglandin $F_{2\alpha}$ administration.[5]

It has been demonstrated that reproductive tissue, i.e., the uterus, decidua, cervix, and fetal membranes, all contain precursors of prostaglandins. Physiologically, prostaglandins appear to be locally generated in the target organ, producing minimal systemic effects. Thus, prostaglandin was administered locally, that is, intravaginally, intracervically, or intra- and extraamniotically. The dosage of prostaglandin required for a target organ effect decreased dramatically, as did the side effects. Currently, only PGE_2 is used clinically and is administered in a gel or suppository vehicle into the vagina or cervix. Its primary use is for cervical ripening prior to induction of term labor. These classic compounds have been replaced by newer formulations with fewer side effects when used for early pregnancy termination.

The second generation of synthetic prostaglandins was introduced in the early 1970s. These have increased potency and longer duration of action because of less pulmonary degradation.[8] Only one analogue, 15-methyl-$PGF_{2\alpha}$ [$PGF_{2\alpha}$ methyl ester, (15S)-15 methyl-$PGF_{2\alpha}$ tromethamine) prestin 15/M carbprost], is being used currently, and that is for the management of atonic postpartum hemorrhage.

The third and current generation of synthetic prostaglandins are PGE analogues. They offer high efficacy, reduced rates of side effects, and improved stability. Their greatest clinical potential is for induction of abortion. One is used systemically, 16-phenoxy-PGE_2 ester (Sulprostone, developed by Pfizer-Schering). Two others, developed for use vaginally, are 16,16-dimethyl-*trans*-Δ^2-PGE_1 methyl ester (ONO-802, developed by ONO Pharmaceuticals) and 9-deoxo-16,16-dimethyl-9-methylene-PGE_2 (developed by Upjohn).

2. Termination of Pregnancy

Pregnancy termination with prostaglandins was first reported by Karim and Filshie in 1970 when abortion was successfully induced in 13 of 15 patients with intravenous $PGF_{2\alpha}$.[9] About half the patients had significant gastrointestinal side effects. Subsequent reports of pregnancy termination by intravenous and oral routes had similar incidences of unpleasant side effects.

2.1. Mechanism

The mechanism of prostaglandin-induced abortions is still controversial. It appears that the action of stimulating myometrial contractions is not singularly the cause of the abortion, since oxytocin, which can evoke myometrial contractions in early gestation, is not an efficient abortifacient. Other effects produced by prostaglandins must play additive roles. For instance, induction of endogenous prostaglandin production plays a complementary role in pregnancy termination by prostaglandins. Women with intrauterine fetal demise undergoing pregnancy termination by the use of intramuscular injections of 15-methyl-$PGF_{2\alpha}$ were divided into two groups matched for age, gravity, and duration of pregnancy.[10] One group received indomethacin suppositories to inhibit endogenous prostaglandin production before and during the induction. These women had a longer induction-to-termination interval, required more repeated injections, and had a higher number of failed cases in comparison to the group who received the prostaglandin analogue only.

Csapo observed that progesterone levels fell rapidly in pregnant patients receiving prostaglandin for termination.[11] He proposed that fetoplacental unit function was altered, resulting in decreased estrogen and progesterone production, thereby lowering the threshold for the uterotonic actions of prostaglandins. Schulman's group concluded that in women at term with an unripe cervix, prostaglandin increases myometrial oxytocin sensitivity. He compared intravaginal suppositories of 1 to 2 mg of prostaglandin E_2 12 hr before oxytocin induction to a constant dose of oxytocin for 12 hr before increasing oxytocin and performing amniotomy for induction.[12] Less oxytocin was required to effect delivery in the prostaglandin group. Norstrom incubated cervical tissue in a solution of tritiated proline and glucosamine. He demonstrated an increased incorporation of these substances and new ground substance formation in tissues obtained from first-trimester pregnant women who had 0.5 mg PGE_2 in gel placed intracervically prior to termination.[13] This suggested cervical collagen remodeling and may be the most important noncontractile effect promoting pregnancy termination.

2.2. Early Pregnancy Termination

The use of vaginal prostaglandins for very early pregnancy termination (<8 weeks) has been reported with variable results. Studies using 25 to 50 mg intravaginal $PGF_{2\alpha}$,[14] another using a second-generation $PGF_{2\alpha}$, 4 mg intravaginal $PGF_{2\alpha}$ 15-methyl ester,[15] and a third using two vaginal suppositories of prostaglandin $F_{2\alpha}$ totaling 3 mg[16] all were effective in terminating early pregnancy. However, over half experienced significant side effects of nausea and vomiting. About half the patients required a curettage to complete the abortion. The use of third-generation analogues of PGE in early pregnancy termination resulted in more encouraging results. In 240 patients using a PGE_2 derivative vaginally[17] and in 45 patients using a PGE_1 derivative[18] administered by the intrauterine route, more than 93% of abortions were complete, with side effects in fewer than 20%. Nevertheless, prostaglandins have not been considered more useful than surgical dilation and evacuation in terminating early pregnancy because of their variable efficacy and significant side effects at the dosages required for efficacy.

2.3. Midtrimester Pregnancy Termination

During the late first and early second trimesters of pregnancy, the termination method most utilized is the intraamniotic instillation of prostaglandins. After 14 weeks' gestation, the cost, morbidity, and mortality of dilation and evacuation are high enough to balance the increased side effects of prostaglandins.[19] In Denmark, intraamniotic instillation of 40 mg $PGF_{2\alpha}$ has become the predominant method of second-trimester abortions.[20] A large multicenter experience found this method to be 80% to 90% effective, and there was only a 2% incidence of side effects. However, about half of the patients required removal of the placenta manually or by instrumentation, and 25% required 48 hr to abort.[21] This particular study compared 15-methyl-$PGF_{2\alpha}$ with the parent compound and found them equivalent, whereas in a multicenter study in India, a similar comparison found $PGF_{2\alpha}$ 15-methyl ester more effective although it produced more side effects.[22] The induction-to-abortion interval averaged 19 hr in both groups. Although this method is effective and well tolerated, its prolonged induction-to-abortion interval is not desirable and has prompted simultaneous use of 100 to 200 mU/min of intravenous oxytocin[23] or 40 to 80 g of intraamniotic urea[24] to reduce this interval. There is an increased risk of uterine rupture associated with augmenting $PGF_{2\alpha}$ in this fashion, especially with oxytocin.[25] Use of the second- and third-generation prostaglandin analogues in midtrimester abortions given by various routes of administration such as extraamniotic, intramuscular,[26] and vaginal[27] has been efficacious. However, there were side effects including a 1% to 6% incidence of cervical laceration, especially when drug was administered by the intramuscular route. The higher dosage of the intramuscular route perhaps resulted in more forceful uterine contractions, making the cervix more liable for trauma.

2.4. Pregnancy Termination in Abnormal Pregnancy

In abnormal pregnancies where amniotic fluid is either scant, bloody, or absent, such as in the case of a missed abortion, fetal death, or hydatidiform mole, the preferred route of administration of prostaglandin is vaginal. Under these circumstances, the practice of giving 20 mg of PGE_2 (suppository) every 3 to 6 hr until labor is established has been popularized.[28] The mean induction time for patients with intrauterine fetal demise was 10.9 hr. Premedication with antidiarrheal, antiemetic, and antipyretic drugs can ameliorate the side effects. Uterine rupture has been reported with use of 20 mg PGE_2 vaginal suppository to terminate a pregnancy complicated by fetal death in the third trimester.[29,30] In this circumstance, a smaller dosage of PGE_2 as a vaginal suppository (3–5 mg) or intracervical gel (1 mg) is advised, since the sensitivity to prostaglandin of the term myometrium is much greater than earlier in pregnancy.[31] 9-Deoxo-16,16-dimethyl-9-methylene-PGE_2, 15-methyl $PGF_{2\alpha}$,[32] and ONO-802[33] were all very effective with mild side effects. Some differences in side effects were found between compounds. The use of doses until abortion may increase side effects without improving efficacy (three doses were as effective as five).[33]

2.5. Current Use of Prostaglandins

Current practice in the use of prostaglandins for pregnancy termination in America is dictated by two issues. The first is the incidence and severity of side effects. A higher dosage is required for efficacy in the first than in the second trimester; hence, a higher incidence of more severe side effects occurs. The second issue is limited approval by the Food and Drug Administration of the use of prostaglandins and their analogues.[34] Prostaglandin $F_{2\alpha}$ 15-methyl ester is approved for intramuscular use as a primary abortifacient between gestational weeks 13 and 20 and as an adjunctive agent up to 26 weeks in cases of failed abortion by other methods or with rupture of membranes in a previable pregnancy. Prostaglandin $F_{2\alpha}$ given intraamniotically is approved for use between gestational weeks 16 and 20 for abortion. Intravaginal PGE_2 suppository is approved for abortion between gestational weeks 12 and 20 and for completion of missed abortion or termination of intrauterine fetal demise until week 28. It is also approved as an adjunct in the management of hydatidiform mole.

2.6. Cervical Priming

Surgical termination of pregnancy either by dilatation and curettage in the first trimester or dilatation and evacuation in the second trimester may be the procedure of choice (up to 24 weeks) because of the incidence of side effects or prolonged induction time of prostaglandin use. Mechanical dilatation of the cervix is an important disadvantage. This can result in sequelae including immediate cervical lacerations (one to two per 100 abortions)[35,36] and subsequent cervical stenosis, incompetence, miscarriage, or preterm labor.[37,38] A large, well-documented prospective study of 31,917 women was performed in the Kaiser medical plan. This revealed a definite increased risk of second-trimester fetal loss in nulliparous women who had previous instrumental cervical dilatation for abortion.[39] A prospective controlled study was done to determine the outcome of subsequent pregnancies in a cohort of patients who underwent a second-trimester prostaglandin-induced abortion. In contrast to the previous study, there was no increased incidence of preterm labor or cervical problems as described for mechanical dilatation.[40]

Preparation of the cervix prior to surgical induction has become an important adjunct. Preoperative intravaginal administration of PGE_2 has been shown to soften the cervix. One milligram of PGE_2 gel was successful in allowing an 8-mm Hegar dilator to pass intracervically in 85% versus 0% for the placebo group in a study of primigravida abortions.[41] In a Norwegian clinical trial, 290 primigravid patients were randomized to a 0.5-mg intracervical PGE_2 group or a placebo group administered the day before a surgical abortion was performed.[42] There was significantly increased cervical dilatation (8.0 mm versus 5.7 mm) in the PGE_2-treated group, but almost 20% were admitted to the hospital for abortion, bleeding, or painful contractions. These were considered treatment failures. Prostaglandin E_2 can produce preabortion cervical priming but falls short of an ideal agent because of side effects. If the dose of prostaglandin is lowered or the treatment interval is shortened to lessen side effects, therapeutic benefit may be lost.[43]

Prostaglandin analogues have the most promising results regarding cervical priming thus far. In an early clinical trial, 354 primigravid patients were treated with a vaginal suppository containing 1 mg of 15-methyl-$PGF_{2\alpha}$ 12 hr prior to vacuum aspiration for first-trimester abortion. It was reported, however, that 80% of the women were hospitalized for abdominal pain.[44] The mean cervical dilation was 8.5 mm at time of surgery. In a later study on 80 patients 3 hr prior to surgical termination, side effects were reduced to 10% by decreasing the dose to 0.5 mg.[45] Two studies were performed using 16,16-dimethyl-trans-Δ^2-prostaglandin E_1 methyl ester (ONO-802) in double-blind, placebo-controlled, randomly assigned trials, with 1 mg inserted 3 hr or more before first-trimester surgical abortion with a spring gauge used to evaluate the force required to dilate the cervix in each of the 54 women in each group, it was demonstrated that significantly less force was required in the treated group.[46] Only 47% of this group had mild pain and cramping with slight bleeding. Only one patient required pain medication. With a different vehicle, ONO-802 was administered to 121 women undergoing first-trimester induced abortion at 0 and 3 hr after treatment.[47] No significant change was demonstrated in the nulliparous group, and a significant, but of marginal clinical value, decrease of cervical resistance was found in the parous group. There were no side effects in the treated group. These authors note that other studies, in which the cervical changes are significant demonstrate systemic absorption of the prostaglandin as reflected in the incidence of side effects, i.e., cramping and discomfort. They conclude that the local action of the ONO-802 suppository on softening the cervix is minimal. Thus, it appears that the vehicle carrying the prostaglandin is important in allowing the prostaglandin to be absorbed. The absorption carries the liability of an incidence of side effects in order to be effective.

Sulprostone, a PGE_2 analogue, was evaluated using different routes of administration (intracervical, intramuscular, or subcutaneous) at different dosages in 170 primigravidas. The drug was administered at 14 to 16 hr before a late first-trimester abortion. This study demonstrated that the intracervical administration of 25 to 100 μg of sulprostone resulted in no side effects, and the highest dosages achieved a cervical dilation of 7 mm or more in only 40% of patients, whereas in patients receiving 350 μg by intramuscular injections with a 15% incidence of side effects, 90% achieved a cervical dilation of 7 mm or more.[48] In a prospective randomized dose-finding study in 60 women, sulprostone gel was administered intracervically at doses of 25, 50, and 100 µg in 60 women at 6–8 hr before first-trimester abortion.[49] Objective cervical softening was evaluated using a special tonometer before gel application and just before surgery. The efficiency of cervical priming correlated with the rate of vaginal bleeding and with the frequency of contraction-related lower abdominal pain. The findings of the above studies support the concept of dose-related effects and side effects in the use of synthetic prostaglandins (third generation) to prime the cervix before surgical termination of pregnancy.

3. Term Cervical Ripening and Induction of Labor

3.1. Physiological Background

During gestation, the pregnant cervix must maintain a firm consistency in order to keep the pregnancy intact. At the end of pregnancy it undergoes a process called cervical ripening, which allows the passage of the products of conception. This process involves a breakdown of the collagen matrix of the cervix and a change in the glycosoaminoglycans and water content of the ground substance.[50] The clinical picture is a cervix that softens, demonstrates increased compliance, and dilates in the face of coordinated myometrial contractions. Cervical ripening prior to the induction of labor is imperative. Attempts to induce term cervical ripening have been made with mechanical cervical dilatation,[51] estrogens,[52] relaxin,[53] and prostaglandin.[54]

3.2. Prostaglandin E_2 for Cervical Ripening

Embrey in 1969 was first to observe cervical changes after systemic administration of prostaglandins for labor induction.[55] Although prostaglandins of both the F and E series have been applied, the E series has been favored since it is felt to be physiological. This is based on the observation that incubation of term ovine cervical tissues with prostaglandin precursors results in augmented production of prostaglandin E and prostacyclin but not of PGF.[56] Also, prostaglandin E_2 is more potent than prostaglandin $F_{2\alpha}$ at reducing stiffness and increasing elasticity as evaluated by the stretch modulus of human cervical tissue.[57] Application of prostaglandin E_2 as a cervical ripening agent does mimic spontaneous priming in both clinical (cervical dilatation rate) and, as demonstrated in cervical biopsies, morphological and biochemical terms.[58,59] Prostaglandin E_2 administered intravenously for cervical ripening was associated with significant side effects. Therefore, the oral route, 1 mg PGE_2 given hourly for three doses, was evaluated for tolerance and efficacy. Weiss *et al.*[60] and Golbus and Creasy[54] demonstrated significant cervical ripening with minimal side effects. The latter study was placebo controlled and double blinded. However, Friedman and Sachtleben, using a shorter preinduction period than used by Golbus and Creasy (<12 hr), could not show priming in the term cervix.[61]

3.3. Prostaglandin E_2 Preparations for Local Application

Clinical investigators found an increasing degree of reliability and efficacy as the route of administration approached the cervix itself. Prostaglandin E_2 carried in tablets, lipid-based suppositories, embedded in pessaries, and mixed in gel vehicles has been applied intravaginally in dosages from 1 to 5 mg with high degree of efficacy.[62-65] When applied extraamniotically or intracervically as a viscous gel, only one-tenth the dosage is required for high-efficacy cervical priming.[66,67] The development of a viscous gel as a vehicle for prostaglandin E_2 has been through several stages. Initially, Calder, Embrey, and Tait used hydroxymethylcellulose (Tylose) as the vehicle.[66] The Swedish group initially used hydroxypropylmethylcellulose (Hypromellose) but felt that the distribution of prostaglandin was not as uniform as desired and that long-term stability of the prostaglandin was not reliable when using a cellulose vehicle.[67] They mixed the prostaglandin E_2 solution with a special cross-linked starch polymer, lyophilized it, and dispersed it into syringes at a dose of 0.5 mg of prostaglandin E_2 (Perstorp AB, Sweden).[68] The powder could be stored for over 1 year without decreased activity and is reconstituted to a firm sticky viscous gel with 2 ml of sterile saline 15 sec prior to use. The Swedish studies published using this product report the best efficacy rates for cervical ripening.[69-71] A more recent study in the United States showed that only three of 47 patients failed induction, requiring cesarean section, in contrast to 11 of 47 in the control group. A significantly shorter active phase and second stage of labor were observed in the treated patients. Only two patients went into labor in the 12 hr before induction with oxytocin. No fetal heart rate abnormalities or uterine hyperstimulation occurred during the priming.[69]

Another commercially prepared viscous gel formulation is prostaglandin E_2 dissolved in triacetin gel base with colloid silicone dioxide added (Prepidil, Upjohn). Prepidil is a prepackaged unidose syringe; when stored at 4°C it is stable for up to 1 year as shown by bioavailability studies on monkeys. A prepackaged unidose water-soluble vaginal film of prostaglandin E_2 has also been evaluated.[72] A review of the more recent publications dealing with local cervical ripening agents shows that Prepidil is used because of its practicality, stability, and the difficulty of manufacturing the other products. In fact, the Swedish group that introduced the starch formulation found the triacetin-based gel easier to use and more practical.[73] There are six recently published prospective, randomized, controlled studies demonstrating the high efficacy of intracervical application of 0.5 mg of Prepidil to ripen the term cervix.[74-79] Four studies used placebo control.[74,77-79] Of interest in these studies is that a third to over one-half of the treated subjects went into labor during the preinduction period. Because of one case of uterine hyperactivity and five cases of severe fetal heart rate decelerations in the treatment group ($n = 30$),[78] it is recommended to monitor fetal heart rate for 1 to 2 hr following instillation of the PGE_2 gel.

A recent study measured the prostaglandin E metabolite in the plasma from patients at term who received Prepidil gel or placebo for cervical ripening. Endogenous prostaglandin production contributed to the increased metabolite. The levels were variable but corresponded to the clinical response.[80] The authors postulated that the variable response to treatment could include the possibility of partial placement of gel into the endocervical canal or its partial extrusion and that the physiological status of these subjects affected absorption. Factors that could affect absorption include vaginal pH and the amount and concentration of mucus in the cervical canal. Also, the manipulation of the cervix required to gain access to the endocervical canal may increase local endogenous production of prostaglandin E_2. The deposition of Prepidil intravaginally would obviate some of the aforementioned concerns. There is only one recently published report of intravaginal Prepidil as a cervical ripening agent.[81] It was a randomized prospective placebo-controlled dose-finding study using 1-mg, 2-mg, and 3-mg dosages of Prepidil. Increasing doses were effective in ripening the cervix, and 3-mg doses induced labor in 50%. Of 60 treated patients, two had fetal distress caused by uterine hypertonus. Seven had the gel removed because of uterine hypertonus. The relatively high incidence of side effects in this study suggests the possibility of increased absorption from intravaginal triacetin gel. Another randomized placebo-controlled study was recently reported from Canada using 2.5 ml of commercially prepared methylcellulose gel. The study resulted in 33% induction of labor in the preinduction period, with 3.8% incidence of hyperstimulation.[82] The study supported high efficacy for cervical ripening.

Since these preparations are not FDA approved, physicians who wish to use prostaglandin for cervical ripening or labor induction must make arrangements with the manufacturer to begin a clinical trial. Alternatively, PGE_2 suppositories may be reformulated. There are six recently published clinical trials using reformulation of PGE_2 suppositories. It is interesting to examine the different methods used for preparation, storage stability, and techniques of application and compare their results.

A group in Oregon prepared the prostaglandin gel by thawing 20-mg vaginal suppositories and homogeneously mixing them with a previously prepared methylhydroxyethylcellulose (Tylose) gel. Aliquots were stored at 4°C for 30 days or at −20° for 90 days.[83] They used a single application of 5 mg of PGE_2 intravaginally to demonstrate efficacy for cervical ripening in 30 patients in a prospective, placebo-controlled fashion. They then studied 150 consecutive patients and had a 46% incidence of spontaneous labor. The authors make no mention of excessive uterine activity.

A group in Pennsylvania dissolved a 20-mg vaginal PGE_2 suppository in 100 ml of hydroxymethylcellulose (K-Y jelly) heated to 85°C and stored the gel in 2-mg aliquots at 4°C for a maximum of 4 hr.[84] The authors claim to have performed radioimmunoassays to confirm stability and reproducible dosage. Using a randomized placebo-controlled design of 2 mg PGE_2 gel placed intravaginally in a single application, the study was unable to demonstrate efficacy for cervical ripening in 44 subjects. The authors admit that preparation methods might have altered biological activity of the PGE_2.

A group in Tampa prepared prostaglandin E_2 gel by grinding a 20-mg suppository and mixing it with hydroxyethylcellulose (K-Y jelly) gel to a volume of 20 ml.[85] The gel was prepared individually for each subject and inserted intravaginally in either 3-mg or 5-mg dosage within 15 min of preparation. Using a randomized placebo-controlled design in 60 high-risk patients, the study demonstrated significant efficacy for cervical ripening for both dosages. Uterine contractions were stimulated in 87.5% of the patients, and 37.5% went into active labor during the preinduction period.

A group from Denver prepared prostaglandin E_2 gel by dissolving a 20-mg suppository in warm ($\leq 37°C$) Tylose solution and stored 3 mg/10 ml gel aliquots at $-4°C$.[86] The gel was thawed at room temperature and injected via catheter into a size-70 contraceptive diaphragm (Ortho All-flex) previously positioned over the cervix. If cervical score was not significantly altered at 6 hr, gel placement was repeated. High efficacy was demonstrated in 240 patients; in 29% labor was induced with gel alone, and three patients (1.4%) required gel removal because of uterine hypertonus.

A group from California prepared 3-mg suppositories by slicing a 20-mg vaginal suppository into sixths (3-mm-wide slices).[87] The authors state that homogeneous distribution of PGE_2 throughout the original 20-mg suppository was confirmed by performing prostaglandin assays on representative cut sections (but no description of such assays exists in the text). The authors, using a randomized placebo-controlled design, demonstrated efficacy for cervical ripening in 63 patients using a single intravaginal 3-mg PGE_2 suppository. Forty-two percent were induced during the preinduction period. Eight percent of patients experienced uterine hyperstimulation. Two studies found that repeated application of prostaglandin E_2 gel did not significantly increase cervical change beyond that effected by the first application.[88,89]

This literature review reveals that prostaglandins are efficacious as cervical ripening agents. Although efficacy has been established, the large number of variables, only some of which have been controlled, makes it impossible to define the best vehicle or dose. However, as many as 50% of patients treated will experience labor within several hours. Some may regard the distinction between the use of prostaglandin for cervical ripening and for labor induction as academic. Inadequate monitoring during the period of cervical ripening and the possibility that labor is converted to hypertonus might make the academic point clinically relevant. As an *ideal* cervical ripening agent, prostaglandin fails, since it causes uterine contractions and a high incidence of induction of labor.

3.4. Present Status of PGE₂ Use for Induction of Labor

Labor induction by repetitive application of 3-mg vaginal PGE_2 suppositories every 6 hr up to three times was tried in two studies.[88,89] A 50–75% success rate was obtained. Goodlin *et al.* studied patients with an unripe cervix at first application and had a 31% labor induction rate.[90] No greater incidence of maternal or fetal side effects was noted than in women undergoing spontaneous or oxytocin-induced labor. In patients with a ripe cervix at first application, a 65% labor induction rate was found.[91] Multiparous patients had a 82% rate of successful induction, whereas primiparous patients had a 29% rate. The dose of subsequently used oxytocin, if required, was found to be less than usually needed.

Oral, vaginal, and intracervical PGE_2 are routinely used to induce labor in Europe and Great Britain. Currently, oral PGE_2 is favored. A recent review of the literature comparing oral PGE_2 and intravenous oxytocin has demonstrated no clear differences.[92] The reasons that the prostaglandins have not achieved greater acceptance or commercial availability in the United States are complex. The current drug regulatory process and medicolegal climate have played a role. Case reports of maternal and fetal disasters using prostaglandin to induce labor are present in the literature. These include uterine rupture[93] and fetal[94] and neonatal[95] demise. The incidence of these disasters must be quite rare with thoughtful use, proper dosages, and avoidance of concomitant oxytocin usage, but the incidence of uterine hyperactivity and/or hypertonus is disconcerting. A recent report has noted that 0.25 mg of terbutaline given subcutaneously was effective in decreasing the uterine contraction intensity following 5 mg of PGE_2 intravaginally.[96]

In summary, the use of a single application of PGE_2 placed intravaginally appears to have reasonable efficacy as a cervical ripening agent. However, with the lack of commercial preparation of a stable vehicle, issues of bioactivity of reconstituted PGE_2 remain important. The use of prostaglandin to induce labor has no significant advantage over the already accepted role of incremental oxytocin infusions.

4. Postpartum Hemorrhage

Obstetric hemorrhage caused by postpartum uterine atony is responsible for about 25% of maternal deaths annually. The use of prostaglandins has added an important adjunct in its management.

4.1. Use of $PGF_{2\alpha}$

When conventional ecbolic agents have failed to correct uterine atony, prostaglandin use has been quite successful in avoiding the need for major surgical intervention. The pharmacological treatment of uterine hemorrhage with prostaglandin was first suggested by Bygdeman *et al.* in 1965.[97] However, they were unable to demonstrate that intravenous injection of prostaglandin E was more efficacious than methyl ergonovine maleate in decreasing postpartum blood loss following normal deliveries.[98] Use of prostaglandins to treat postpartum hemorrhage related to uterine atony was first shown to be effective by Takagi *et al.* in Japan in 1976.[99] They demonstrated that $PGF_{2\alpha}$ administered by the intramyometrial route was more effective than intramuscular or continuous intravenous routes. The mean injection-to-uterine-contraction time for the intramuscular route was 35 min, that for the intravenous route was 10 min, and that for the intramyometrial route was 4 min. Prostaglandin $F_{2\alpha}$ was more effective than ergot alkaloids. They also demonstrated in dose–response experiments that as little as 0.25 mg of $PGF_{2\alpha}$ was effective, but they recommended a 1-mg dose to be given intramyometrially for high efficacy. Several case reports in America confirmed the efficacy and ease of the transabdominal intramyometrial injection of 1 mg of $PGF_{2\alpha}$ for control of postpartum hemorrhage. Jacobs and Arias report a successful outcome in three cases using 1 mg of intramyometrial $PGF_{2\alpha}$ injection and a response time of within 2 min.[100] Andrinopoulos and Mendenhall reported successful management in a patient with delayed postpartum hemorrhage using a transabdominal intramyometrial injection of $PGF_{2\alpha}$ after failure of injection into the cervix.[101] The response time was less than 3 min. Experimentation with myometrial injections of $PGF_{2\alpha}$ in nonhuman primates with follow-up histological evaluation failed to reveal any long-lasting local tissue effects of the injection.[102]

4.2. Use of Analogues of $PGF_{2\alpha}$

An analogue of prostaglandin $F_{2\alpha}$, (15-S)-15-methylprostaglandin $F_{2\alpha}$ tromethamine (Prostin 15/M), has been shown to be a more potent (tenfold) uterotonic agent with a much longer duration of action than the parent compound.[103] In 1977, Corson and Bolognese reported that Prostin 15/M was effective in treating postpartum hemorrhage when $PGF_{2\alpha}$ was not.[104] A second report confirmed the efficacy.[105] Both studies suggested that systemic administration (intramuscular) might provide better uterine distribution of the drug, since the analogues are protected from a one-pass pulmonary metabolism. A low dosage (250 µg) could be used to decrease unwanted side effects. In 1981, two larger clinical trials were reported. Toppozada *et al.* from Egypt reported management of severe postpartum hemorrhage related to uterine atony using 15-methyl-$PGF_{2\alpha}$ given intramuscularly.[106] In 15 of 16 patients hemorrhage was controlled by medication. Similar results were reported by Hayashi *et al.* in America in 18 of 20 patients.[107] Interestingly, the three patients who failed to respond in the two studies all had chorioamnionitis. In a subsequent report, Hayashi *et al.* achieved successful control in 44 of 51 patients (86%) using the same regimen.[108] Four of the seven failures had chorioamnionitis. Thirty (68%) patients required one intramuscular injection, nine (20%) had two, and three (7%) had multiple injections. Two additional patients, following an intramuscular injection, responded to a single intramyometrial injection.[108] The mean response time for the intramuscular injection route was 45 min. Mild transient side effects occurred in fewer than 10% of subjects. Butino and Garite also reported successful control of postpartum hemorrhage in 22 of 26 (84.6%) patients.[109] Two failures had placenta accreta. Mild side effects were reported in half of the patients. In the larger study by Hayashi *et al.*, five of six patients responded dramatically to intramyometrial injection.[108] This experience was better than one reported in California, where two of five patients treated with 15-methyl-$PGF_{2\alpha}$ by the intramyometrial route required surgery.[110] Recently, Thiery and Parewijck reported a large study using the intramyometrial route with success in 48 of 50 patients.[111]

They reported no side effects. Thus, the 15-methyl-PGF$_{2\alpha}$ given as a 250-μg intramuscular or intramyometrial injection is successful in 85% or more cases in controlling postpartum hemorrhage associated with uterine atony. The Food and Drug Administration has approved the use of Prostin 15/M (15-methyl-PGF$_{2\alpha}$) for management of postpartum atonic uterine hemorrhage.

4.3. Use of PGE$_2$

Vaginal prostaglandin E$_2$ suppositories (20 mg) have been used successfully to control immediate[112] and delayed[113] postpartum hemorrhage. In a recent case report intravenous infusion of PGE$_2$ at 20 μg/min, followed by maintenance at 10 μg/min for 12 hr, was used to control postpartum hemorrhage.[114]

5. Inhibition of Preterm Labor with Prostaglandin Synthetase Inhibitors

The concept of using inhibitors of prostaglandin synthetase as tocolytic agents in the management of preterm labor assumed that prostaglandins played a major role in the onset of preterm labor. In 1972, there were two publications suggesting this.[115,116] Both demonstrated significant prolongation of pregnancy in rats exposed to inhibitors of prostaglandin synthesis. Several years later the same result, delayed parturition, was obtained in primates using indomethacin.[117] In a large retrospective study of pregnant patients taking salicylates chronically for arthritis, there was a high incidence of gestation lasting beyond 42 weeks.[118] The notion that a similar delay could occur earlier in gestation was supported. A study demonstrated a delay of midtrimester abortion induced by hypertonic saline in patients exposed to inhibitors of prostaglandin synthetase.[119]

5.1. Efficacy of Indomethacin

Zuckerman et al. in 1974 were the first to report the use of indomethacin for inhibition of human preterm labor.[120] They were successful in stopping uterine contractions in 40 of 50 patients with clinical evidence of premature labor, using a 100-mg indomethacin rectal suppository followed by 25 mg orally every 6 hr up to 24 hr after contractions ceased. A year later, Wiqvist et al. reported reduction of uterine activity as measured by external tocometry by use of indomethacin in six patients in premature labor.[121] Neither study reported adverse fetal effects attributable to indomethacin; however, there were no neonatal follow-up data. In later years, two other large noncontrolled experience studies were reported using indomethacin as a tocolytic agent. In 1984, Zuckerman et al. reported that indomethacin completely inhibited premature labor in 83% of 297 women in preterm labor,[122] and Dudley and Hardie reported 41% were preterm deliveries in their series of 167 patients.[123] These uncontrolled studies and other smaller studies, however, do not satisfactorily establish efficacy of indomethacin for this indication. To date, there are only two prospective randomized placebo-controlled clinical trials. Niebyl *et al.* administered indomethacin (150 mg daily for 1 to 2 days) orally for the treatment of premature labor in a prospective randomized, double-blind fashion.[124] Treatment success was observed in 14 of 15 cases in the indomethacin group (93.3%), which was significantly better than six of 15 cases in the placebo group (40%). The second study by Zuckerman *et al.* was of a similar design with 18 cases in each group.[125] Treatment consisted of an indomethacin rectal suppository of 100 mg followed by 25 mg orally four times daily for 24 hr. In 15 of 18 indomethacin-treated patients (83.2%), preterm labor was arrested, compared with four of 18 in the placebo group (22.2%). No cases of neonatal pulmonary hypertension were noted. Although the numbers are small in these two placebo-controlled studies, both groups concluded that indomethacin was significantly more effective than a placebo in inhibiting premature labor.

5.2. Side Effects of Indomethacin

In fetal lambs, prostaglandins of the E series produced pulmonary vasodilatation and a decrease in pulmonary vascular resistance. Prostacyclin has a similar effect.[126] Administration of prostaglandin synthetase inhibitors has produced constriction and even closure of the fetal ductus arteriosus in rats, rabbits,[127] and lambs.[128] The presumed consequence of prolonged constriction of the fetal ductus arteriosus is pulmonary hypertension in the newborn. The first report of a possible relationship of indomethacin used to treat premature labor and pulmonary hypertension of the human newborn was by Manchester *et al.* in 1976.[129] Since then, there have been a number of reports suggesting this relationship. It is important to note that the syndrome of progressive pulmonary hypertension of the newborn or the syndrome of "persistent fetal circulation" may have several other causes including antenatal or intrapartum hypoxia, acidosis, hypoglycemia, maternal diabetes, hypocalcemia, perinatal sepsis, hypovolemia, or hyperviscosity.[130] Placental transfer and plasma clearance data on indomethacin in fetuses are lacking, but a serum half-life of 14.7 hr in the newborn compared to 2.2 hr in the adult has been measured by Traeger *et al.*[131] One disquieting fact is that in human newborns, a uniquely small dose of indomethacin (0.1–0.3 mg/kg) usually produces closure of the ductus arteriosus in cases with ductal patency[132]; however, five to six times that dose is required to cause constriction or closure of the fetal lamb ductus arteriosus.[128]

Recently Gamissans and Balasch have done an in-depth literature analysis of this very important issue.[133] They noted that in each of the nine fetuses from six anecdotal case reports of neonatal pulmonary hypertension syndrome in which the mother was given prostaglandin synthetase inhibitors (PGSI) in pregnancy, there were other confounding factors in the etiology of this entity, e.g., fetal hypoxia. Of 14 case reports, the incidence of neonatal pulmonary hypertension syndrome in uncontrolled studies using PGSI in preterm labor was 18 in 801 (2.24%) cases. There were three neonatal deaths (0.36%). These authors also examined 8 controlled studies using PGSI in preterm labor and found no difference in incidence of neonatal pulmonary hypertension (0.23% and 0.25%) or deaths from neonatal pulmonary hypertension (0%) in the treatment or control groups. The total number of cases was 434 in the test group versus 406 in the control group. Two recent reports focus on the perinatal outcome of infants whose mothers received short courses of indomethacin for premature labor. Dudley and Hardie in 167 patients had an overall perinatal mortality of 1.7% and no cases of premature closure of the ductus arteriosus or persistent fetal circulation.[123] Niebyl and Witter extended their earlier study to include 46 infants exposed to indomethacin *in utero* and compared perinatal outcome to two control groups.[134] One control group consisted of the next consecutive patient treated with another tocolytic agent, and the other control group was formed by the next consecutive patient matched by gestational age who did not receive any tocolytic agent. There were no differences in neonatal complications and no cases of premature closure of the ductus arteriosus or persistent fetal circulation or pulmonary hypertension. In the prospective randomized placebo-controlled study by Zuckerman *et al.* using indomethacin cited earlier, neonatal follow-up revealed no cases of pulmonary hypertension.[125] Despite all available evidence to the contrary, the use of indomethacin or other prostaglandin synthetase inhibitors for preterm labor is still proscribed in North America. This author will use it in situations where other tocolytics are contraindicated or fail, and only for short-term therapy. There is a need for a large-scale randomized prospective comparison to β-mimetic therapy.

6. Conclusions

In conclusion, the use of prostaglandins and their analogues as well as their inhibitors in clinical obstetrics has not only brought about significant advances in the care of women but has added to the knowledge of the physiological roles of prostaglandin in reproductive medicine. Prostaglandin analogues have become highly efficacious as abortifacients for midtrimester pregnancy termination and the control of severe postpartum hemorrhage from uterine atony. Their use as agents to ripen the term cervix prior to labor induction and to induce labor has been popularized outside the United States. Finally, use of the prostaglandin synthetase inhibitor to abate premature labor requires more study regarding its safety for the fetus.

References

1. Bleasdale, J. E., and Johnston, J. M., 1984, Prostaglandins and human parturition: Regulation of arachadonic acid mobilization, *Perinatal. Med.* **5**:151–191.
2. MacDonald, P. C., Porter, J. C., Schwarz, B. E., and Johnston, J. M., 1978, Initiation of parturition in the human female, *Semin. Perinatol.* **2**:273–286.
3. Ferreira, S. H., and Vane, J. R., 1967, Prostaglandins: Their disappearance from and release into the circulation, *Nature* **216**:868–873.

4. Hamberg, M., and Samuelsson, B., 1971, On the metabolism of prostaglandin E_1 and E_2 in man, *J. Biol. Chem.* **246:**6713–6721.

5. Cates, W., and Jordon, H. V. F., 1979, Sudden collapse and death of women obtaining abortions induced with prostaglandin F_2 (alpha), *Am. J. Obstet. Gynecol.* **133:**398–400.

6. Wilson, D. C., 1984, Prostaglandins, their actions on the gastrointestinal tract, *Arch. Intern. Med.* **133:**112–118.

7. Parker, C. W., and Snider, D. E., 1973, Prostaglandins and asthma, *Ann. Intern. Med.* **78:**963–965.

8. Takaki, N., Tredway, D., Toomer, P., Murray, W., and Daane, T., 1976, Therapeutic abortion of early human gestation with intramuscular 15-methyl prostaglandin F_2 (alpha), *Contraception* **13:**319–332.

9. Karim, S. M. M., and Filshie, G. M., 1970, Therapeutic abortion using prostaglandin F_2 alpha, *Lancet* **1:**157–159.

10. Souka, A. R., Karsoon, O., Shams, A., and Toppozada, M., 1983, Role of endogenous prostaglandins in pregnancy termination by 15-methyl PGF_2 alpha, *Prostaglandins* **25:**711–714.

11. Csapo, A. I., 1972, On the mechanism of the abortifacient action of prostaglandin F_2 alpha, in: *The Prostaglandin* (E. M. Southern, ed.), Futura, Mt. Kisco, NY, pp. 337–365.

12. Jagni, N., Schulman, H., Fleischer, A., Mitchell, J., and Blattner, P., 1984, Role of prostaglandin-induced cervical changes in labor induction, *Obstet. Gynecol.* **63:**225–229.

13. Norstrom, A., 1983, Effects of endocervical administration of prostaglandin E_2 on cervical dilatation and connective tissue biosynthesis in the first trimester of pregnancy, *Adv. Prostaglandin Thromboxane Leukotriene Res.* **12:**461–464.

14. Sato, T., Ami, K., and Matsumoto, S., 1973, The induction of abortion and menstruation by the intravaginal administration of prostaglandin F_2 alpha, *Am. J. Obstet. Gynecol.* **116:**287–289.

15. Foster, H. W., Smith, M., McGruder, C. E., Richard, F., and McIntyre, J., 1985, Post-conception menses induction using prostaglandin vaginal suppositories, *Obstet. Gynecol.* **65:**682–685.

16. Borten, M., and Friedman, E. A., 1984, Postconceptional induction of menses with double prostaglandin F_2 alpha impact, *Am. J. Obstet. Gynecol.* **150:**1006–1007.

17. Karim, S. M. M., Rao, B., Ratnam, S. S., Prasad, R. N., Wong, Y. M., and Ilancheran, A., 1977, Termination of early pregnancy (menstrual induction) with 16-phenoxy tetranor PGE_2 methylsulfonylamide, *Contraception* **16:**377–381.

18. Takagi, S., Sakato, H., and Yoshida, T., 1977, Termination of very early pregnancy by ONO-802 (16,16-dimethyl-*trans* PGE_1 methyl ester), *Prostaglandins* **14:**791–798.

19. Cates, W., Schulz, K. F., and Grimes, D. A., 1977, The effect of delay and choice of method on the risk of abortion morbidity, *Fam. Plan. Perspect.* **9:**266.

20. Lange, A. P., 1983, Prostaglandins as abortifacients in Denmark, *Acta Obstet. Gynecol. Scand. [Suppl.]* **113:**117–124.

21. Who Task Force on the Use of Prostaglandins for the Regulation of Fertility, 1977, Prostaglandins and abortions. III. Comparison of single intraamniotic injection of 15-methyl prostaglandin F_2 alpha for termination of second trimester pregnancy: An international multicenter study, *Am. J. Obstet. Gynecol.* **129:**601–606.

22. Tejuda, S., Choudhury, S. D., and Manchanda, P. K., 1978, Use of intra- and extraamniotic prostaglandins for the termination of pregnancies. Report of multicentric trials in India, *Contraception* **18:**641–653.

23. Kochenour, N., Engel, T., Henry, G., and Droegmueller, W., 1972, Midtrimester abortion produced by intra-amniotic prostaglandin F_2 alpha augmented with intravenous oxytocin, *Am. J. Obstet. Gynecol.* **114:**516–519.

24. King, T. M., Atienza, M. F., Burkman, R. T., Burnett, L., and Bell, W. R., 1974, The synergistic activity of intra-amniotic prostaglandin F_2 alpha and urea in the midtrimester elective abortion, *Am. J. Obstet. Gynecol.* **120:**704–718.

25. Hayashi, R. H., Rothwell, R. O., and Weinberg, P. C., 1975, Uterine rupture complicating midtrimester abortion in a young woman of low parity, *Int. J. Gynecol. Obstet.* **13:**229–232.

26. Who Task Force on the Use of Prostaglandins for the Regulation of Fertility, 1977, Prostaglandins and abortion. I. Intramuscular administration of 15-methyl prostaglandin F_2 alpha for induction of abortion in weeks 10 to 20 of pregnancy, *Am. J. Obstet. Gynecol.* **129:**593–596.

27. Cameron, I. T., Michie, A. F., and Baird, D. T., Prostaglandin-induced pregnancy termination: Further studies using Gemeprost vaginal pessaries in the early second trimester, *Prostaglandins* **34:**111–117.

28. Southern, E. M., and Gutknecht, G. D., 1976, Management of intrauterine fetal demise and missed abortion using prostaglandin E_2 suppositories, *Obstet. Gynecol.* **47:**602–606.

29. Valenzuela, G., Hayashi, R. H., Lackritz, R., and Soriero, O. M., 1980, Uterine rupture at term with vaginal prostaglandin E_2, *Am. J. Obstet. Gynecol.* **138:**1223–1226.

30. Sawyer, M. M., Lipschitz, J., Anderson, G. D., and Dilts, P. V., Jr., 1980, Third trimester uterine rupture associated with vaginal prostaglandin E_2, *Am. J. Obstet. Gynecol.* **140:**710–711.

31. Eckman, G., Uldbjers, N., Wingerup, L., and Ulmsten, U., 1983, Intracervical instillation of PGE_2-gel in patients with missed abortion or intrauterine fetal death, *Arch. Gynecol.* **233:**241–245.

32. Christensen, N. J., and Bygdeman, M., 1983, The use of prostaglandins for termination of abnormal pregnancy, *Acta Obstet. Gynecol. Scand. [Suppl.]* **113:**153–157.

33. Garcia, N., Dargenio, R., Panetta, V., Moneta, E., Tancredi, G., and Giannitelli, A., 1987, A prostaglandin analogue (ONO-802) in treatment of missed abortion, intrauterine fetal death and hydatidiform mole: A dose-finding trial, *Eur. J. Obstet. Gynecol. Reprod. Biol.* **25:**15–22.

34. Jacobs, M. M., 1986, Clinical obstetric use of arachidonic acid metabolites and potential adverse effects, *Semin. Perinatol.* **10:**299–315.

35. Nemec, D. K., Pendergast, T. J., and Trumbower, W., 1978, Medical abortion complications: An epidemiologic study at a mid-Missouri clinic, *Obstet. Gynecol.* **51:**433–434.

36. Schulz, K. F., Grimes, D. A., and Cates, W., Jr., 1983, Measures to prevent cervical injury during suction curettage abortion, *Lancet* **1:**1182–1185.

37. Cates, W., 1977, Late effects of abortion: Hypothesis or knowledge, *J. Reprod. Med.* **22:**207–212.

38. Pantelakis, S. N., Papadimitriou, D. C., and Doxidia, S. A., 1973, Influence of induced and spontaneous abortion on subsequent pregnancy, *Am. J. Obstet. Gynecol.* **116:**799–805.

39. Harlap, S., Shiono, P. H., Ramcharau, S., Berendes, H., and Pellegrin, F., 1979, A prospective study of spontaneous fetal losses after induced abortions, *N. Engl. J. Med.* **301:**677–681.

40. McKenzie, I. Z., and Hillier, K., 1977, Prostaglandin-induced abortion and outcome of subsequent pregnancies: A prospective controlled study, *Br. Med. J.* **2:**1114–1117.

41. Heinzl, S., and Andor, J., 1981, Preoperative administration of prostaglandins to avoid dilatation induced damage in first trimester pregnancy termination, *Gynecol. Obstet. Invest.* **12:**29–36.

42. Iversen, T., and Skjeldstad, F. E., 1985, Intracervical administration of prostaglandin E_2 prior to vacuum aspiration: A prospective double-blind randomized study, *Int. J. Gynecol. Obstet.* **23:**95–99.

43. McKenzie, I. Z., and Fry, A., 1981, Prostaglandin E_2 pessaries to facilitate first trimester aspiration termination, *Br. J. Obstet. Gynecol.* **88:**1033–1037.

44. Nilsson, S., Johnell, H., Langhoff-Roos, J., Nyberg, R., and Green, K., 1983, Vaginal administration of 15-methyl-PGF_2 alpha-methyl ester prior to vacuum aspiration. Peri- and postoperative complications, *Acta Obstet. Gynecol. Scand.* **62:**599–602.

45. Arias, F., 1984, Efficacy and safety of low-dose 15-methyl prostaglandin F_2 alpha for cervical ripening in the first trimester of pregnancy, *Am. J. Obstet. Gynecol.* **149:**100–101.

46. Chen, J. K., and Elder, M. G., 1983, Preoperative cervical dilatation by vaginal pessaries containing prostaglandin E_1 analogue, *Obstet. Gynecol.* **62:**339–342.

47. Hulka, J. F., and Chepko, M., 1987, Vaginal prostaglandin E_1 analogue (ONO-802) to soften the cervix in first trimester abortion, *Obstet. Gynecol.* **69:**57–60.

48. Jerve, F., and Fylling, P., 1983, Sulprostone for preoperative cervical dilatation in primigravidas scheduled for late trimester termination of pregnancy, *Arch. Gynecol.* **233:**199–203.

49. Rath, W., Dennemark, N., and Godicke, H. D., 1985, Preoperative cervical priming by intracervical application of a new sulprostone gel contraception, *Contraception* **31**:207–216.
50. Danforth, D. N., Bockingham, J. D., and Roddick, J. W., 1960, Connective tissue changes incident to cervical effacement, *Am. J. Obstet. Gynecol.* **80**:939–945.
51. Gower, R., Toraya, J., and Miller, J. M., Jr., 1982, Laminaria for preinduction cervical ripening, *Obstet. Gynecol.* **60**:617–619.
52. Gordon, A., and Calder, A., 1977, Estradiol applied locally to ripen the unfavorable cervix, *Lancet* **2**:1319–1321.
53. Evans, M., Dugan, M. B., Moad, A., Evans, W. J., Bryant-Greenwood, G. D., and Greenwood, F. C., 1983, Ripening of the human cervix with porcine ovarian relaxin, *Am. J. Obstet. Gynecol.* **147**:410–414.
54. Golbus, M., and Creasy, R., 1977, Uterine priming with oral prostaglandin E_2 prior to elective induction with oxytocin, *Prostaglandins* **14**:577–581.
55. Embrey, M. P., 1969, The effect of prostaglandins on the human pregnant uterus, *J. Obstet. Gynaecol. Br. Commonw.* **76**:783–789.
56. Ellwood, D. A., Anderson, A. B. M., and Mitchell, M. D., 1981, Prostanoids, collagenase and cervical softening in the sheep, in: *The Cervix in Pregnancy and Labor* (D. A. Ellwood and A. B. M. Anderson, eds.), Churchill Livingston, New York, pp. 57–73.
57. Spatling, L., Neuman, M. R., Huch, R., and Huch, A., 1985, Influence of different prostaglandin applications on cervical rheology, *Int. J. Gynecol. Obstet.* **23**:369–376.
58. Uldbjerg, N., Ekman, G., Malmstrom, A., Sporrong, B., Ulmsten, U., and Wingerup, L., 1981, Biochemical and morphological changes of human cervix after local application of prostaglandin E_2 in pregnancy, *Lancet* **1**:267–268.
59. Ekman, G., Malmstrom, A., Uldbjerg, N., and Ulmsten, U., 1986, Cervical collagen: An important regulator of cervical function in term labor, *Obstet. Gynecol.* **67**:633–636.
60. Weiss, R. R., Tejeni, N., Israeli, J., Evans, M. J., Bhakthauathsalan, A., and Mann, L. I., 1975, Priming of the uterine cervix with oral prostaglandin E_2 in term multigravida, *Obstet. Gynecol.* **46**:181–184.
61. Friedman, E. A., and Sachtleben, M. R., 1975, Preinduction priming with oral prostaglandin E_2, *Am. J. Obstet. Gynecol.* **121**:521–523.
62. Gordon-Weight, A. P., and Elder, M. G., 1979, Prostaglandin E_2 tablets used intravaginally for the induction of labour, *Br. J. Obstet. Gynaecol.* **86**:32–36.
63. Liggens, G., 1979, Controlled trial of induction of labor by vaginal suppositories containing prostaglandin E_2, *Prostaglandins* **18**:167–172.
64. Shepard, J., Pearce, J., and Sims, C., 1979, Induction of labour using prostaglandin E_2 pessaries, *Br. Med. J.* **2**:108–110.
65. MacKenzie, I. Z., and Embrey, M. P., 1978, The influence of preinduction vaginal prostaglandin E_2 gel upon subsequent labour, *Br. J. Obstet. Gynaecol.* **85**:657–661.
66. Calder, A. A., Embrey, M. P., and Tait, T., 1977, Ripening of the cervix with extraamniotic prostaglandin E_2 in viscous gel before an induction of labour, *Br. J. Obstet. Gynaecol.* **84**:264–268.
67. Wingerup, L., Andersson, K.-E., and Ulmsten, U., 1978, Ripening of the uterine cervix and induction of labour at term with prostaglandin E_2 in viscous gel, *Acta Obstet. Gynecol. Scand.* **57**:403–406.
68. Harris, A. S., Kirstein-Pederson, A., Stenberg, P., Ulmsten, U., and Wingerup, L., 1980, The preparation, characterization and stability of a new prostaglandin E_2 gel for local administration, *J. Pharm. Sci.* **69**:1271–1273.
69. Ekman, G., Forman, A., Marsal, K., and Ulmsten, U., 1983, Intravaginal versus intracervical application of prostaglandin E_2 in viscous gel for cervical priming and induction of labor at term in patients with an unfavorable cervical state, *Am. J. Obstet. Gynecol.* **147**:657–661.
70. Wingerup, L., Ekman, G., and Ulmsten, U., 1983, Local application of prostaglandin E_2 in gel, A new technique to ripen the cervix during pregnancy, *Acta Obstet. Gynecol. Scand. [Suppl.]* **113**:131–136.

71. Ferguson, J. E., Ueland, F. R., Stevenson, D. K., and Ueland, K., 1988, Oxytocin induced labor characteristics and uterine activity after preinduction cervical priming with prostaglandin E₂ intracervical gel, *Obstet. Gynecol.* **72**:739–745.

72. Viegas, O. A., Singh, K., Adaikan, P. G., Karim, S. M., and Ratnam, S. S., 1986, Preinduction cervical priming in high risk pregnancy—experience with a new sustained release PGE₂ vaginal film, *Prostaglandin Leukotriene Med.* **21**:61–68.

73. Floberg, J., Allen, J., Belfrage, P., Bygdeman, M., and Ulmsten, U., 1983, Experience with an industrially manufactured gel PGE₂ for cervical priming, *Arch. Gynecol.* **233**:225–229.

74. Wiqvist, I., Norstrom, A., and Wiqvist, N., 1986, Induction of labor by intracervical PGE₂ in viscous gel, *Acta Obstet. Gynecol. Scand.* **65**:485–492.

75. Kiebach, D. G., Zahradnik, H. P., Quaas, L., Kroner-Fehmel, E. E., and Lippert, T. H., 1986, Clinical evaluation of endocervical prostaglandin E₂-triacetin gel for preinduction cervical softening in pregnant women at term, *Prostaglandins* **32**:81–85.

76. Noah, M. L., DeCoster, J. M., Fraser, T. J., and Orr, J. D., 1987, Preinduction cervical softening with endocervical PGE₂ gel: A multicenter trial, *Acta Obstet. Gynecol. Scand.* **66**:3–7.

77. Nimrod, C., Currie, J., Yee, J., Dodd, G., and Persaud, D., 1984, Cervical ripening and labor induction with intracervical triacetin base prostaglandin E₂ gel: A placebo-controlled study, *Obstet. Gynecol.* **64**:476–479.

78. Laube, D. W., Zlatnick, F. J., and Pitkin, R. M., 1986, Preinduction cervical ripening with prostaglandin E₂ intracervical gel, *Obstet. Gynecol.* **68**:54–57.

79. Yonekura, M. L., Songster, G., and Smith-Wallace, T., 1985, Preinduction cervical priming with PGE₂ intracervical gel, *Am. J. Perinatol.* **2**:305–310.

80. Kimball, F. A., Ruppel, P. L., Noah, M. L., Decoster, J. M., delaFuente, P., Castillo, J. M., and Hernandez, J. M., 1986, The effect of endocervical PGE₂-gel (Prepidil) gel on plasma levels of 13,14-dihydro-15-keto-PGE₂ (PGEM) in women at term, *Prostaglandins* **32**:527–537.

81. Graves, G. R., Baskett, T. F., Gray, J. H., and Luther, E. R., 1985, The effect of vaginal administration of various doses of prostaglandin E₂ gel on cervical ripening and induction of labor, *Am. J. Obstet. Gynecol.* **151**:178–181.

82. Bernstein, P., Leyland, N., Gurland, P., and Gare, D., 1987, Cervical ripening and labor induction with prostaglandin E₂ gel: A placebo-controlled study, *Am. J. Obstet. Gynecol.* **156**:336–340.

83. Prins, R. P., Bolton, R. N., Mark, C., Nielson, D. R., and Watson, P., 1983, Cervical ripening with intravaginal prostaglandin E₂ gel, *Obstet. Gynecol.* **61**:459–462.

84. Lorenz, R. P., Botti, J. J., Chez, R. A., and Bennett, N., 1984, Variations of biologic activity of low-dose prostaglandin E₂ on cervical ripening, *Obstet. Gynecol.* **64**:123–127.

85. Williams, J. K., Wilkerson, W. G., O'Brien, W. F., and Knuppel, R. A., 1985, Use of prostaglandin E₂ topical cervical gel in high-risk patients: A critical analysis, *Obstet. Gynecol.* **66**:769–773.

86. Turner, J. E., Shannon-Burke, M., Porreco, R. P., and Weiss, M. A., 1987, Prostaglandin E₂ in Tylose gel for cervical ripening before induction of labor, *J. Reprod. Med.* **32**:815–821.

87. Buchanan, B., Macer, J., and Yonekura, M. L., 1984, Cervical ripening with prostaglandin F₂ vaginal suppositories, *Obstet. Gynecol.* **63**:659–663.

88. Mainprize, T., Nimrod, C., Dodd, G., and Persaud, D., 1987, Clinical utility of multiple-dose administration of prostaglandin E₂ gel, *Am. J. Obstet. Gynecol.* **156**:341–343.

89. Prins, R. P., Nielson, D. R., Bolton, R. N., Mark, C. III, and Watson, P., 1986, Preinduction cervical ripening with sequential use of prostaglandin E₂ gel, *Am. J. Obstet. Gynecol.* **154**:1275–1279.

90. Goodlin, R. C., Reschl, L., and Clewell, W. H., 1986, Absence of maternal side effects from prostaglandins use for cervical ripening, *J. Reprod. Med.* **31**:1095–1097.

91. Macer, J., Buchanan, D., and Yonekura, M. L., 1984, Induction of labor with prostaglandin E₂ vaginal suppositories, *Obstet. Gynecol.* **63**:664–668.

92. Lange, A. P., Westergaard, J. G., Secher, N. J., and Pedersen, G. T., 1983, Labor induction with prostaglandins, *Acta Obstet. Gynecol. Scand. [Suppl.]* **113**:177–185.

93. Larsen, J. V., Brits, C. I., and Whittal, D., 1984, Uterine hyperstimulation and rupture after induction of labour with prostaglandin E₂: Case reports, *S. Afr. Med. J.* **65**:615–616.

94. Quinn, M. A., and Murphy, M. A., 1981, Fetal death following extraamniotic prostaglandin gel, *Br. J. Obstet. Gynaecol.* **88**:650–651.
95. Simmons, K., and Savage, W., 1984, Neonatal death associated with induction of labour with intravaginal prostaglandin E_2: Case report, *Br. J. Obstet. Gynaecol.* **91**:598–599.
96. Chamberlin, R. O., Cohen, G. R., and Knuppel, R. A., 1987, Uterine hyperstimulation resulting from intravaginal prostaglandin E_2: A case report, *J. Reprod. Med.* **32**:233–235.
97. Bygdeman, M., Kwon, S. U., and Mukherjee, T., 1965, Effect of intravenous infusion of prostaglandin E_1 and E_2 on motility of the pregnant human uterus, *Am. J. Obstet. Gynecol.* **102**:317–326.
98. Roth-Brandel, U., Bydegman, M., and Wiquist, N., 1970, A comparative study on the influence of prostaglandin E_1, oxytocin and ergometrine on the pregnant human uterus, *Acta Obstet. Gynecol. Scand. [Suppl.]* **5**:19–25.
99. Takagi, S., Yoshida, T., Togo, Y., Tochigi, H., Abe, M., Sakata, H., Fujii, T. K., Takahashi, H., and Tochigi, B., 1976, The effects of intramyometrial injection of prostaglandin F_2 alpha on severe post-partum hemorrhage, *Prostaglandin* **12**:565–579.
100. Jacobs, M. M., and Arias, F., 1980, Intramyometrial prostaglandin F_2 alpha in the treatment of severe postpartum hemorrhage, *Obstet. Gynecol.* **55**:665–666.
101. Andrinopoulis, G. C., and Mendenhall, H. W., 1983, Prostaglandin F_2 in the management of delayed postpartum hemorrhage, *Am. J. Obstet. Gynecol.* **146**:217–218.
102. Wentz, A. C., and King, T. M., 1972, Intramyometrial prostaglandin F_2 alpha, *Am. J. Obstet. Gynecol.* **114**:112–114.
103. Kirton, K. T., and Forbes, A. D., 1972, Activity of 15(s)15-methyl prostaglandin F_2 and F_2 alpha as stimulants of uterine contractility, *Prostaglandin* **1**:319–326.
104. Corson, S. L., and Bolognese, R. L., 1977, Postpartum uterine atony treated with prostaglandins, *Am. J. Obstet. Gynecol.* **129**:918–919.
105. Zhradnik, H. P., Steiner, H., Hillemans, H. G., Brekwoldt, M., and Ardelt, W., 1977, Prostaglandin F_2 alpha and 15-methyl prostaglandin F_2 alpha treatment of massive uterine bleeding, *Geburtsh. Frauenheilkd.* **37**:493.
106. Toppozada, M., El-Bossaty, M., El-Rahman, H. A., Shans, E. L., and Din, A. H., 1981, Control of intractible atonic postpartum hemorrhage by 15-methyl prostaglandin F_2 alpha, *Obstet. Gynecol.* **58**:327–330.
107. Hayashi, R. H., Castillo, M. S., and Noah, M. I., 1981, Management of severe postpartum hemorrhage due to uterine atony using an analogue of prostaglandin F_2 alpha, *Obstet. Gynecol.* **58**:426–429.
108. Hayashi, R. H., Castillo, M. S., and Noah, M. L., 1984, Management of severe postpartum hemorrhage with a prostaglandin F_2 alpha analogue, *Obstet. Gynecol.* **63**:806–808.
109. Buttino, L., Jr., and Garite, T. J., 1986, The use of 15 methyl F_2 alpha prostaglandin (Prostin 15M) for the control of postpartum hemorrhage, *Am. J. Perinatol.* **3**:241–243.
110. Bruce, S. L., Paul, R. H., and VanDorsten, J. B., 1983, Control of postpartum uterine atony by intramyometrial prostaglandin, *Obstet. Gynecol. [Suppl.]* **59**:47–50.
111. Thiery, M., and Parewijck, W., 1985, Local administration of (15S)-15 methyl PGF_2 alpha for management of hypotonic postpartum hemorrhage, *Z. Geburtshilfe. Perinatal.* **189**:179–180.
112. Hertz, R. H., Sokol, R. J., and Dierker, L. J., 1980, Treatment of postpartum uterine atony with prostaglandin F_2 vaginal suppositories, *Obstet. Gynecol.* **56**:129–130.
113. Goldstein, A. J., Kent, D. R., and David, A., 1983, Prostaglandin E_2 vaginal suppositories in the treatment of intractable late-onset postpartum hemorrhage. A case report, *J. Reprod. Med.* **28**:425–426.
114. Hensen, G., Gough, J. D., and Gillmer, M. D. G., 1983, Control of persistent primary postpartum hemorrhage due to uterine atony with intravenous prostaglandin E_2. Case report, *Br. J. Obstet. Gynecol.* **90**:280–282.
115. Aiken, J. W., 1972, Aspirin and indomethacin prolong parturition in rats, *Nature* **240**:21.
116. Chester, R., Dukos, M., Slater, S. R., and Walpole, A. L., 1972, Delay of parturition in the rat by anti-inflammatory agents which inhibit the biosynthesis of prostaglandins, *Nature* **240**:37.

117. Novey, M. J., Cook, M. J., and Manaugh, L., 1974, Indomethacin block of normal onset of parturition in primates, *Am. J. Obstet. Gynecol.* **118:**412–416.
118. Collins, E., and Turner, G., 1973, Salicylates and pregnancy, *Lancet* **2:**1494.
119. Waltman, R., Tricomi, V., and Palav, A., 1973, Aspirin and indomothin: Effect on instillation–abortion time of midtrimester hypertonic saline induced abortion, *Prostaglandin* **3:**47–58.
120. Zuckerman, H., Reiss, U., and Reubenstein, I., 1974, Inhibition of human premature labor by indomethacin, *Obstet. Gynecol.* **44:**787–792.
121. Wiqvist, N., Lundstrom, V., and Green, K., 1975, Premature labor and indomethacin, *Prostaglandin* **10:**515–526.
122. Zuckerman, H., Shalev, E., Gilad, G., and Katzuni, E., 1984, Further study of the inhibition of premature labor by indomethacin. Part I, *J. Perinatol. Med.* **12:**19–23.
123. Dudley, D. K. L., and Hardie, M. J., 1985, Fetal and neonatal effects of indomethacin used as a tocolytic agent, *Am. J. Obstet. Gynecol.* **151:**181–184.
124. Niebyl, J. R., Blake, D. A., White, R. D., Kumor, K. M., Dubin, N. H., Courtland-Robinson, J., and Egner, P. G., 1980, The inhibition of premature labor with indomethacin, *Am. J. Obstet. Gynecol.* **136:**1014–1019.
125. Zuckerman, H., Shalar, E., Gilad, G., and Katzuni, E., 1984, Further study of the inhibition of premature labor by indomethacin. Part II. Double-blind study, *J. Perinatal. Med.* **12:**25–29.
126. Cassin, S., 1980, Role of prostaglandins and thromboxanes in the control of the pulmonary circulation in the fetus and newborn, *Semin. Perinatol.* **4:**101–107.
127. Sharpe, G. L., Larsson, K. S., and Thalme, B., 1975, Studies on closure of the ductus arteriosis XII: *In utero* effect of indomethacin and sodium salicylate in rats and rabbits, *Prostaglandins* **9:**585–596.
128. Coceani, F., Olley, P. M., and Bodach, E., 1976, A possible reulation of muscle tone in the ductus arteriosus, *Adv. Prostaglandin Thromboxane Res.* **1:**417–419.
129. Manchester, D., Margolis, H. S., and Sheldon, R. E., 1976, Possible association between maternal indomethacin therapy and primary pulmonary hypertension of the newborn, *Am. J. Obstet. Gynecol.* **126:**467–469.
130. Siassi, B., Goldberg, J. J., Emmanouilides, G. C., Higashino, S. M., and Louis, E., 1971, Persistent pulmonary vascular obstruction in the newborn, *J. Pediatr.* **82:**1103–1106.
131. Traeger, A., Noschel, H., and Zaumsiel, J., 1973, The pharmacokinetics of indomethacin in pregnant and parturient women and their newborn infants, *Zentralbl. Gynakol.* **95:**635–641.
132. Heyman, M. A., Rudolph, A. M., and Silverman, N. H., 1976, Closure of the ductus arteriosus in premature infants by inhibition of prostaglandin synthesis, *N. Engl. J. Med.* **295:**530–533.
133. Gamissans, O., and Balasch, J., 1984, Prostaglandin synthetase inhibitors in the treatment of preterm birth, in: *Preterm Birth: Causes, Prevention and Management* (F. Fuchs and P. G. Stubblefield, eds.), Macmillan, New York, pp. 223–248.
134. Niebyl, J. R., and Witter, F. R., 1986, Neonatal outcome after indomethacin treatment for preterm labor, *Am. J. Obstet. Gynecol.* **155:**747–749.

16

Fetal Tissues and Autacoid Biosynthesis in Relation to the Initiation of Parturition and Implantation

Marlane J. Angle and John M. Johnston

1. Introduction

The concept that the fetus may regulate the initiation or course of parturition was first expounded over 100 years ago by Speigelberg (cited by Thornburn[1]). The communication between the fetus and the maternal compartment, particularly in humans, may well be mediated via the amniotic fluid to the amnion tissue. Amnion, in turn, may transduce a signal or signals via the chorion laeve and decidua vera tissues to the myometrium, which ultimately results in the rhythmic contraction of this tissue as well as softening of the cervix. As will be discussed, amnion tissue is ideally suited to be the recipient of such a signal from the fetus via the amniotic fluid. In addition, the amplification of such information may also be a function of this tissue. Some of the characteristic features of amnion tissue that may be involved in its biological function are:

1. This tissue contains an abundance of lipid droplets (the importance of lipids in parturition is discussed subsequently) as visualized by oil red O and hematoxylin and eosin staining.[2]
2. When these cells are examined by electron microscopy, one can see that these cuboidal epithelial cells contain an abundance of microvilli, intracellular junctions, and cytoplasmic filaments. The presence of the numer-

Marlane J. Angle and John M. Johnston • Departments of Biochemistry and Obstetrics–Gynecology and The Cecil H. and Ida Green Center for Reproductive Biology Sciences, University of Texas Southwestern Medical Center, Dallas, Texas 75235.

ous microvilli on the fetal side of the amnion cell renders this cell ideally suited for the absorptive process as found in other absorptive cells such as the small intestine and the proximal convoluted renal tubule of the kidneys. In addition, an increased density of receptors per cell may occur as a consequence of numerous microvilli and may be involved in the transduction of a signal(s).

3. Evidence to support the view that a communication system does exist between amniotic fluid and decidual tissue and vice versa has been provided by the fact that prostaglandin E_2 can be translocated *in vitro* from the fetal compartment to the decidual side and to the maternal compartment.[3] We have recently added [³H]PGE$_2$ to the fetal side of an amnio-chorion-decidua preparation and observed that the radioactivity that reaches the maternal side is completely metabolized by the prostaglandin dehydrogenase known to be present in chorion laeve and decidua tissue(s). Translocation from the decidua to the amniotic fluid has been shown to occur with prolactin.[4]

4. Amnion tissue has a significantly higher specific activity of prostaglandin synthetase than chorion laeve and decidua vera tissues. Furthermore, this activity is increased in amnion tissue obtained at term and in labor compared to the activity present at term but not in labor. No significant change in the specific activity of prostaglandin synthetase was found in chorion laeve and decidua vera tissues. In contrast, amnion tissue has a limited capacity to inactivate prostanoids because of the absence or very low activity of 15-hydroxyprostaglandin dehydrogenase.[5] The other intrauterine tissues, such as chorion laeve and decidua vera, have significant activities of prostaglandin dehydrogenase.

5. Phospholipase A_2 and C are present in fetal membranes (Fig. 1). Their specific activities increase markedly during gestation in amnion tissue.[6] No change in the specific activities of these enzymes was found in chorion laeve or decidua vera tissues during gestation. This is illustrated in Fig. 2. When amnion cells were placed in primary culture, the activities of these two enzymes returned to levels similar to those found in early amnion tissue (13–17 weeks).[7] The nature of the stimulus that regulates the activities in amnion tissue is not known.

We have focused on the importance of amnion tissue in relation to human

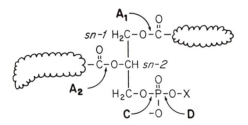

Figure 1. The structure of a generalized glycerophospholipid. The various alcohols that may be found in phosphomonoester linkage are depicted as X. The sites of hydrolysis catalyzed by the various phospholipases are indicated by the letters. (From Bleasdale and Johnston,[2] with permission.)

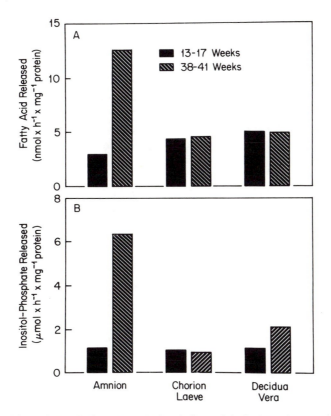

Figure 2. Activities of phospholipase A_2 and phospholipase C in fetal membranes and decidua vera obtained either early or late in gestation. The specific activities of (**A**) phospholipase A_2 and (**B**) phospholipase C were measured in amnion, chorion laeve, and uterine decidua vera tissues obtained either at 13–17 weeks or 38–41 weeks of gestation. Phospholipase A_2 was assayed employing the 750-g supernatant fraction of each tissue. Phospholipase C activity was assayed using the 105,000-g supernatant fraction of each tissue.

parturition. It should be emphasized, however, that the interrelationship of multiple tissues is no doubt of fundamental importance in the final phase of the initiation of parturition for rhythmic contraction of the pregnant uterus and cervical softening. In the subsequent sections of this review we emphasize the role of fetal membranes in initiation of parturition. The precise role of the fetus in the onset of labor is currently unknown, but several hormones and autacoids have been demonstrated to be involved in regulating parturition (see Chapter 17).[8-12] These hormones include oxytocin, progesterone, estrogens, cortisol, and catecholamines. In addition to these hormonal regulators, two families of compounds, arachidonic acid and its metabolites and platelet-activating factor (PAF), have recently been implicated in the initiation of labor.[2,13] This review considers the possible sites of origin of these two autacoids and their subsequent interactions with fetal and maternal tissues in view of their potential roles in the regulation of the onset of labor.

2. Autacoids in Amniotic Fluid

2.1. Prostaglandins

An involvement of lipids in the initiation of parturition was first suggested by Luukkainen and Csapo,[14] who reported that the intravenous infusion of a lipid emulsion into pregnant rabbits resulted in increased responsiveness of the rabbit uteri to oxytocin. Lanman and co-workers[15-17] later reported that the active component of these emulsions was phosphatidylcholine containing an essential fatty acid such as linoleic acid or arachidonic acid. It was demonstrated subsequently that parturition in the rabbit is induced by the intravenous administration of arachidonic acid.[18] Likewise, intrauterine injection of arachidonic acid induces premature oviposition in the quail,[19,20] and instillation of an arachidonate–albumin complex into the amniotic sac results in termination of human pregnancy.[21]

Karim and co-workers reported in 1966[22] that prostaglandins were present in amniotic fluid and that prostaglandin-rich extracts of amniotic fluid obtained during labor increased the intensity of spontaneous contractions of the uterus.[23] Extracts of amniotic fluid obtained during labor were found to contain prostaglandins of the E and F series in concentrations much greater than those found in extracts of amniotic fluid that were obtained prior to the onset of labor.[23] Subsequently, it was demonstrated[24-26] that the intravenous infusion of prostaglandin E_2 or prostaglandin $F_{2\alpha}$ induced labor in women at or near term. These two prostaglandins, when administered in increased amounts, cause the induction of abortion in women during the first or second trimester.[27,28]

The use of inhibitors of prostaglandin biosynthesis has yielded information about the role of prostaglandins in parturition. For instance, it was demonstrated that the chronic ingestion of aspirin, an inhibitor of cyclooxygenase (the enzyme that catalyzes the initial reaction in the conversion of arachidonic acid to various prostanoids), resulted in prolonged gestation in women.[29,30] Furthermore, the administration of either aspirin or indomethacin to pregnant rats[31] and indomethacin to monkeys[32] resulted in delayed parturition. Likewise, it has been reported that the administration of indomethacin to women several hours after induction of midtrimester abortion (by extraamniotic administration of hypertonic saline) resulted in a significantly increased delay between induction and eventual abortion.[33]

The concentrations of prostaglandin metabolites in amniotic fluid increase during labor. The approximate increases in concentrations of prostaglandins and their metabolites in amniotic fluid during labor were found to be as follows: prostaglandin E_2 tenfold,[23,34,35] prostaglandin $F_{2\alpha}$ 12-fold,[36,37] 6-ketoprostaglandin $F_{1\alpha}$ (a stable metabolite of prostacyclin) twofold,[38] and thromboxane B_2 (a stable metabolite of thromboxane A_2) fourfold.[39] The concentrations of various prostaglandins and their metabolites in the plasma of pregnant women have also been determined throughout pregnancy. A small but significant increase in the concentration of 13,14-dihydro-15-ketoprostaglandin $F_{2\alpha}$ in plasma was found during the final month of pregnancy.[40] Recently, Mitchell *et al.*[41] have reported that the concentration of a stable metabolite of prostaglandin E_2 (11-deoxy-13,14-dihydro-15-keto-11,16-cycloprostaglandin E_2) was significantly greater in the plasma of women in labor than in the plasma of the same women prior to the onset of labor. Hamberg[42] measured the daily urinary excretion of the major metabolites of prostaglandin $F_{2\alpha}$ in women before pregnancy, throughout pregnancy, and in the postpartum period. He reported that the daily excretion rate of prostaglandin $F_{2\alpha}$ metabolites increased during pregnancy and that the excretion rate doubled just before the onset of labor. Based on the reported daily excretion rates of the major metabolites of prostaglandins during human parturition and the arachidonic acid mobilized into amniotic fluid during parturition, Okita *et al.*[43] computed that the decrease in arachidonic acid content of fetal membranes during labor could account for both the increased prostaglandin production and increased concentration of arachidonic acid in amniotic fluid during labor.

2.2. Platelet-Activating Factor

Platelet-activating factor (PAF), a unique glycerophospholipid mediator of a variety of physiological events, is produced by a number of cell types including white cells of the granulocyte series, macrophages, endothelial cells, and mast cells.[44] Platelet-activating factor is structurally related to phosphatidylcholine, differing from the parent compound by the presence of a fatty alcohol in an ether linkage at the *sn*-1 position of the glycerol backbone and an acetyl group at the *sn*-2 position. It has been suggested to be involved in several diverse pathophysiological conditions including asthma, allergic responses, inflammation, myocardial infarctions, systemic hypertension, and gastric ulceration.[45] In addition to these pathophysiological conditions, we and others have suggested that PAF is involved in the initiation and maintenance of human labor, fetal lung maturation, and physiological events associated with implantation.

The first suggestion that PAF may be involved in parturition was based on the observation that PAF appeared in human amniotic fluid in association with labor.[46] Amniotic fluid was obtained at term from women before or after the start of labor; PAF activity was quantified by measuring PAF-induced aggregation of horse platelets (see Fig. 3). Platelet aggregation occurred in a dose-dependent manner to PAF purified from amniotic fluid obtained from women at term and in active labor and was indistinguishable from that obtained with authentic PAF. Recently, Nishihira *et al.*,[47] using gas chromatography and mass spectrometry to analyze amniotic fluid samples from ten women in labor and ten women not in labor, confirmed and extended these observations: PAF was present in all ten samples obtained from women in labor but was absent from ten samples from women not in labor. On the basis of these and previous findings,[46] it is concluded that PAF appears in amniotic fluid in association with labor.

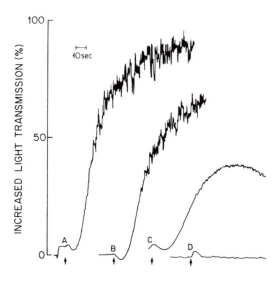

Figure 3. Platelet aggregation in response to amniotic fluid PAF at various concentrations. The phospholipids in the lipid extract of amniotic fluid obtained from women in labor were separated by thin-layer chromatography. Thin-layer areas corresponding to PAF were eluted, and the extracts were tested for platelet-aggregating activity. A, B, and C represent lipid extracts from 10, 5, and 2 ml of amniotic fluid, respectively, and D a lipid extract from 50 ml of amniotic fluid obtained from a woman at term but not in labor. (From Billah and Johnston,[46] with permission.)

3. Tissue Origins of Autacoids Found in Amniotic Fluid in Association with Parturition

3.1. Fetal Membranes

Possible sources of the prostaglandins found in association with parturition are the fetal membranes and decidua vera. It is well accepted that various injuries to the fetal membranes and decidua (e.g., premature rupture, infections, and exposure to hypertonic solutions) lead to the onset of parturition.

As discussed, amnion, chorion laeve, and decidua vera all contain prostaglandin synthetase activity, but whereas prostaglandin E_2 is by far the major product in amnion and chorion laeve, decidua vera produces prostaglandin E_2 and prostaglandin $F_{2\alpha}$ in approximately equal amounts.[5,48–51] The specific activity of prostaglandin synthetase in amnion tissue was greater than that of chorion laeve or decidua vera tissues.[5] In addition, human fetal membranes and decidua vera contain lipoxygenase activity,[52] and small amounts of hydroxy- and hydroperoxyeicosanoic acids and leukotrienes are produced in these tissues.

In an attempt to define the sources of the arachidonic acid and prostanoids found in amniotic fluid during early labor, the fatty acid composition of fetal membranes[53] was determined. It was observed that there was a small but significant decrease in the total arachidonic acid content (expressed as a percentage of the total fatty acid) of fetal membranes obtained from women in labor compared to those from women not in labor. In amnion tissue obtained before labor, arachidonic acid constitutes 14.2% of the total fatty acids in this tissue. During early labor this decreased to approximately 10.3% (i.e., a 27% decrease in the arachidonic acid content). Significant changes in the arachidonic acid content of (diacyl)phosphatidylethanolamine and phosphatidylinositol were observed during early labor.[43] Before labor, arachidonic acid constituted 17.2% of all fatty acids esterified in (diacyl)phosphatidylethanolamine. After early labor this value had decreased to approximately 9.9%. Thus, during early labor there was a 42% decrease in the arachidonic acid content of (diacyl)phosphatidylethanolamine of amnion. Similarly, the arachidonic acid content of phosphatidylinositol declined from 22.7% of total fatty acids esterified in this glycerophospholipid before labor to 14.8% during early labor.[43] Therefore, there was an approximate decrease of 35% in the arachidonic acid content of phosphatidylinositol of amnion tissue. However, it should be noted that the total amount of phospholipid or individual phospholipid did not change, but other saturated fatty acids were substituted for arachidonate at the *sn*-2 position of phosphatidylethanolamine and phosphatidylinositol during early labor (Fig. 1). In chorion laeve, smaller but statistically significant decreases in the arachidonic acid content of (diacyl)phosphatidylethanolamine (20%) and phosphatidylinositol (16%) were observed. We have computed that almost an equal amount of arachidonic acid is released from phosphatidylethanolamine and phosphatidylinositol during early labor.[43] The heterogeneity of decidua vera tissue and its contamination with various blood elements prevented a meaningful analysis of this tissue. It was originally suggested that the lysoglycerophospholipids increase in fetal membranes during early labor.[11] Subsequent investigations by Okita *et al.*,[43] however, revealed that the increased lyso formation was caused by the nonenzymatic breakdown of the plasmalogen ethanolamine because of the acid conditions employed for extraction.[43]

On the basis of these experiments, it was concluded that there is a selective release of arachidonic acid from (diacyl)phosphatidylethanolamine and phosphatidylinositol in fetal membranes during early labor.

Platelet-activating factor found in amniotic fluid may arise from several possible sources, one of which is the fetal membranes. When the concentration of PAF was measured in amnion obtained at term after the onset of labor, it was found to be more than twofold greater than PAF found in amnion obtained before labor.[54] In addition, Ca^{2+} ionophore stimulation of amnion tissue in culture increased the amount of PAF tenfold.[54] These observations provided the first evidence that PAF could be synthesized in amnion tissue. The biosynthesis of PAF was then systematically examined in fetal membranes and decidua vera tissues early and later in gestation. The PAF is synthesized by the remodeling pathway (Fig. 4) from alkylacylglycerophosphocholine (alkylacyl-GPC) after hydrolysis of the fatty acid at the *sn*-2 position by phospholipase A_2, followed by addition of an acetate group as catalyzed by acetyltransferase to form alkylacetyl-GPC or PAF. The PAF can also be inactivated by an acetylhydrolase to form the biologically inactive lysoPAF, which in turn can be reacylated by an acyltransferase or a transacylase. As will be discussed, the latter enzyme has a specificity for an arachidonoyl substrate. We demonstrated that PAF, lysoPAF, and alkylacyl-*sn*-glycero-3-phosphocholine (GPC) are present in human amnion tissue. Although PAF was synthesized in

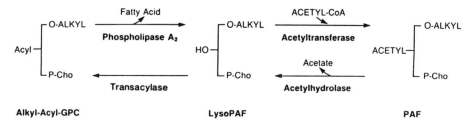

Figure 4. The remodeling pathway of PAF metabolism.

amnion, the release of PAF from amnion disks in culture could not be detected.[54] Thus, another source for the PAF found in amniotic fluid was sought.

3.2. Other Fetal Tissues as Sources of PAF and Prostaglandins

Amniotic fluid at term is enriched with surfactant, a lipoprotein complex that is synthesized by type II pneumonocytes of the fetal lung and transported to amniotic fluid in the form of lamellar bodies as a result of fetal breathing movements. It has recently been reported that surfactant will stimulate PGE_2 production in human amnion.[54a] When the distribution of PAF between the lamellar body fraction and a supernatant fraction isolated from amniotic fluid obtained from women in labor was determined, approximately 44% of the PAF in amniotic fluid was associated with the lamellar-body-enriched fraction.[46] PAF precursors, lysoPAF and alkylacyl-GPC, were also associated with the lamellar body fraction of amniotic fluid. In addition, Nishihira et al.[47] analyzed amniotic fluid by gas chromatography and mass spectrometry and demonstrated that the alkyl group of PAF in amniotic fluid was exclusively the octadecyl species. Based on this finding, the authors suggested that amniotic-fluid-associated PAF was not of renal origin, since the kidney primarily synthesizes the hexadecyl species.[55] They proposed instead that PAF may be derived primarily from fetal lung as opposed to other fetal tissues and fetal or maternal membranes.

The observation that a significant amount of PAF was localized in the lamellar-body-enriched fraction of amniotic fluid, together with the failure to observe the secretion of PAF from amnion tissue and the suggestion by Nishihira et al.[47] that the PAF in amniotic fluid was not of renal origin, prompted us to examine the fetal lung as a possible source of PAF present in the amniotic fluid. A pulmonary origin of PAF in amniotic fluid was also attractive because PAF has been shown to stimulate glycogen breakdown.[56–58] It is well established that glycogen can serve as the precursor to both the glycerol and fatty acids of the major glycerophospholipid present in the surfactant produced by fetal lungs.[59–61] In addition, PAF synthesized and secreted by fetal lung, the last major organ system to develop, may also be involved in initiation of parturition by regulation of prostaglandin production in fetal membranes, since we have previously found that PAF is a potent stimulus for PGE_2 formation in amnion tissue disks (Fig. 5).[62]

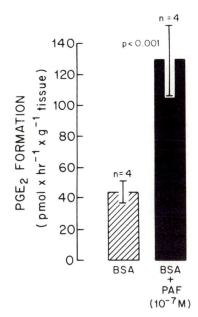

Figure 5. PAF-induced efflux of prostaglandin E_2 (PGE_2) from disks of amnion tissue. Tissue disks were maintained for 2 hr in pseudoamniotic fluid that contained fatty-acid-free bovine serum albumin (0.18%) or albumin plus PAF (10^{-7} M). The PGE_2 was quantitated by radioimmunoassay. (From Billah *et al.*,[62] with permission from Butterworth Publishers.)

Changes in the concentration of PAF and lipid precursors of PAF in lung, kidney, and liver of fetal rabbits during gestation were determined[63] in order to assess whether the lung was a potential source of PAF found in amniotic fluid. The concentration of PAF per milligram of protein in fetal kidney decreased significantly between the 21st and 24th day of gestation and was further decreased by day 31, one day prior to parturition. In fetal liver, the amount of PAF per milligram of protein did not change significantly between the 21st and 31st days of gestation. Although the concentration of PAF per milligram of protein in fetal rabbit lung was significantly lower than that in fetal kidney on day 21 of gestation, the amount of pulmonary PAF tripled between day 21 and 31 of gestation to reach amounts that were severalfold greater than those in fetal kidney. We observed that the lysoPAF and alkylacyl-GPC, both precursors of PAF, decreased by 60% between day 21 and 31 of gestation in fetal rabbit lung. This lends further support to the theory that PAF increases in this tissue during the later stages of gestation.

Arachidonic acid and prostaglandins present in amniotic fluid may also originate from the fetal lung, kidney, skin, and other tissues in addition to fetal membranes. Fetal bovine and rabbit lungs have been shown to be active in converting arachidonic acid to PGE_2 both before and after birth. Maximal production of PGE_2 occurred at 28 days of gestation in the rabbit.[64] The fetal sheep also appears to have a "critical prostaglandin period" in which PGE_2 and $PGF_{2\alpha}$ biosynthesis and catabolism change dramatically in lung and other tissues with increasing age of the fetus.[65] In addition, human as well as fetal and neonatal rabbit lungs have been shown to produce prostacyclins.[66] The kidney may also be a potential source for the prostaglandins found in the amniotic fluid via fetal urine. The presence in fetal urine of prostaglandins and a number of factors that stimulate prostaglandin synthesis in fetal membranes has been the subject of much discussion. The identification and characterization of the biological function of such factors in relation to the events of parturition await future investigations.

4. Enzymatic Synthesis and Degradation of Autacoids during Parturition

4.1. Biosynthesis of Prostaglandins in Fetal Membranes

Prostaglandins are not known to be stored in tissues; therefore, increased formation of prostaglandins from these tissues must reflect increased biosynthesis, decreased degradation, or a combination of both. As discussed previously, amnion has a very low capacity for prostaglandin degradation.[5] Thus, mechanisms that regulate prostaglandin formation probably regulate prostaglandin concentration in this tissue. The obligate precursor of prostaglandins of the 2 series is unesterified arachidonic acid.[67,68] In most tissues, however, the intracellular concentration of unesterified arachidonic acid is low, and much of the arachidonic acid is esterified in the form of glycerophospholipids. Arachidonic acid is esterified primarily in the *sn*-2 position of glycerophospholipids. Lands and Samuelsson[69] and Vonkeman and van Dorp[70] were the first to demonstrate that unesterified arachidonic acid was required for prostaglandin formation and to suggest that the enzymatic release of arachidonic acid from esterified forms may be the rate-limiting step in prostaglandin biosynthesis.

Release of arachidonic acid from the glycerophospholipid backbone can be accomplished in a single step by the action of phospholipase A_2. The activity of the various phospholipases is illustrated in Fig. 1. Phospholipase A_2 activity was shown to be present in various intrauterine tissues.[71–73] It was found that human amniochorion contained a phospholipase A_2 that was distributed largely in the microsomal and cytosolic fractions. Most of the amniochorion phospholipase A_2 activity was Ca^{2+} dependent, and the pH optimum was found to be approximately 8,[74] suggesting that this enzyme was not lysosomal in origin.

The substrate specificity of phospholipase A_2 of fetal membranes was investigated using phosphatidylcholine and phosphatidylethanolamine as substrates (each with palmitic acid at the *sn*-1 position and with various radiolabeled fatty acids at the *sn*-2 position).[74] The rate of hydrolysis of 1-palmitoyl-2-arachidonoyl-*sn*-glycerophosphoethanolamine was approximately four times that observed with the corresponding molecular species of phosphatidylcholine as substrate. Furthermore, phosphatidylethanolamine containing arachidonic acid in the *sn*-2 position was a better substrate than phosphatidylethanolamine containing oleic acid at the *sn*-2 position. When a mixture of these two molecular species of phosphatidylethanolamine was used as substrate in a mixed micelle, the phospholipase A_2 in amniochorion still exhibited preference for phosphatidylethanolamine molecules that contained arachidonic acid.[74] Thus, the presence in fetal membranes of a Ca^{2+}-

dependent phospholipase A_2 with substrate preference for phosphatidylethanolamine containing arachidonate can account for the release of arachidonic acid from phosphatidylethanolamine that occurs during early labor.[43] As previously discussed, the release of arachidonic acid in fetal membranes from phosphatidylethanolamine was approximately 50% of the arachidonic acid lost during early labor.

The remaining 50% of the arachidonate was released from phosphatidylinositol. The phospholipase A_2 of amnion tissue did not catalyze the hydrolysis of phosphatidylinositol.[2] Therefore, alternative mechanisms for the release of arachidonic acid from phosphatidylinositol have to be considered (Fig. 1). Consequently, we determined the activity of phospholipase C in these tissues.[75] An assay procedure was developed in which the release of [³H]inositol phosphates from phosphatidyl-[³H]inositol was quantified[75] in order to measure phospholipase C activity. When the [³H]substrate was incubated with various subcellular fractions of amnion and chorion laeve tissues, water-soluble ³H-labeled products (inositol-1,2-cyclic phosphate and inositol 1-phosphate) and diacylglycerol were formed as a result of phospholipase C action.[75] The pH optimum of this phosphatidylinositol-specific phospholipase C was 6.5–7.5, and the enzyme required Ca^{2+} for activity. The formation of monoacylglycerol was also detected. This observation was the first evidence that amnion tissue contained a diacylglycerol lipase. Initially, there was a 1 : 1 molar stoichiometry between the amount of radiolabeled water-soluble products released and the amount of diacylglycerol formed. In amnion, chorion laeve, and decidua vera, most activity (>90%) was associated with the cytosolic fraction.

The action of phospholipase C alone would not release unesterified arachidonic acid from phosphatidylinositol but rather would produce diacylglycerol rich in arachidonic acid (since phosphatidylinositol in amnion tissue is itself enriched with arachidonic acid).[43] In fact, the diacylglycerol content of amnion tissue increased two- to threefold during early labor[76] and had high arachidonic acid composition almost identical to that of phosphatidylinositol.[76] Therefore, the further metabolism of the diacylglycerol was examined in amnion tissue. Two metabolic fates of diacylglycerol were identified: (1) hydrolysis by diacylglycerol lipase to a 2-monoacylglycerol with subsequent release of arachidonic acid as catalyzed by a monoacylglycerol lipase and (2) phosphorylation by diacylglycerol kinase to form phosphatidic acid (Fig. 6).

The preferred substrate of the diacylglycerol lipase is a diacylglycerol containing arachidonic acid in the *sn*-2 position. The monoacylglycerol produced does not accumulate because the monoacylglycerol lipase activity was also detected in amnion, chorion laeve, and decidua vera tissues[77] and has a specific activity approximately seven times greater than that of diacylglycerol lipase. The monoacylglycerol lipase in fetal membranes and decidua vera preferentially hydrolyzed the 2-arachidonoyl glycerol species. On the basis of the subcellular distribution of the activities of diacylglycerol lipase and monoacylglycerol lipase as well as other characteristics of these lipases, it was concluded that these enzymatic activities were attributable to two distinct enzymes.[77]

Alternatively, diacylglycerol produced in amnion may be phosphorylated to

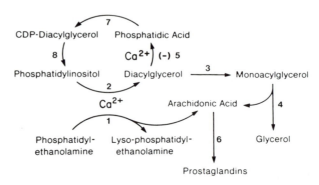

Figure 6. Enzymatic pathways for the mobilization of arachidonic acid in amnion. Reactions are catalyzed by (1) phospholipase A_2, (2) phosphatidylinositol-specific phospholipase C, (3) diacylglycerol lipase, (4) monoacylglycerol lipase, (5) diacylglycerol kinase, (6) prostaglandin synthase complex, (7) CTP : phosphatidate cytidylyltransferase, and (8) CDP-diacylglycerol : inositol 3-phosphatidyltransferase. (From Bleasdale and Johnston,[2] with permission.)

produce phosphatidic acid in a reaction catalyzed by diacylglycerol kinase. Diacylglycerol kinase activity was demonstrated in amnion, chorion laeve, and decidua vera tissues.[78] The enzyme is inhibited by Ca^{2+} and in the absence of Ca^{2+} had a higher affinity for diacylglycerol (apparent $K_m = 0.6$ mM) than did diacylglycerol lipase. In the absence of calcium and in the presence of low concentrations of diacylglycerols, diacylglycerol substrates are primarily utilized via diacylglycerol kinase and recycled into phosphatidylinositol. However, at higher concentrations of diacylglycerols (1 mM) and in the presence of Ca^{2+}, the diacylglycerol lipase pathway predominates, thus leading to the generation of free arachidonic acid. Though mobilization of arachidonate from phosphatidylinositol requires three enzymes (phosphatidylinositol-specific phospholipase C, diacylglycerol lipase, and monoacylglycerol lipase), similar quantities of arachidonic acid are released from phosphatidylinositol as from diacylphosphatidylethanolamine, which requires only the action of one enzyme (phospholipase A_2). Collectively, all four enzymes account for the selective mobilization of arachidonic acid from phosphatidylethanolamine and phosphatidylinositol during early labor as is depicted in Fig. 6.

4.2. Biosynthesis of PAF in Amnion Tissue

Platelet-activating factor can be synthesized by two pathways, a remodeling pathway (see Fig. 3) or the *de novo* pathway (see Fig. 7). The first step in the remodeling pathway for PAF biosynthesis is the removal of a long-chain fatty acid from membrane-associated alkylacyl-GPC by the action of phospholipase A_2. The product, lysoPAF, is then acetylated by PAF acetyltransferase, forming PAF. The degradation and inactivation of PAF occurs primarily by the removal of the *sn*-2 acetate group by PAF acetylhydrolase, thus regenerating lysoPAF. This product can then be reacylated by at least two mechanisms.[79,80] The transacylase pathway prefer-

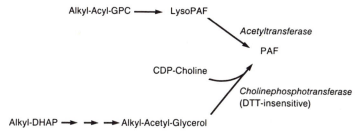

Figure 7. The remodeling and *de novo* pathways for PAF synthesis. The enzymes of the remodeling pathway (top of figure) are phospholipase A$_2$ (alkylacyl-GPC to lysoPAF) and lysoPAF : acetyl-CoA acetyltransferase (lysoPAF to PAF). The *de novo* pathway (bottom of figure) for PAF biosynthesis utilizes dihydroxyacetone phosphate (DHAP) and acyl-CoA as the starting substrates. The ether linkage is formed by an exchange reaction with the long-chain fatty alcohol, which is acetylated by acetyl-CoA and then dephosphorylated to the alkylacetylglycerol. This diacylglycerol analogue is utilized for PAF by the transfer of phosphaticylcholine from CDP-choline as catalyzed by a dithiothreitol-insensitive choline phosphotransferase.

entially utilizes the arachidonoyl substrate, leading to an alkylacyl-GPC that is enriched with the arachidonoyl species. In the second pathway of PAF synthesis, PAF is formed *de novo* from dihydroxyacetone phosphate (DHAP) in an extensive series of enzymatic conversions as elucidated and described by Snyder and co-workers.[81]

We have investigated the enzymes involved in PAF metabolism in amnion tissue and their regulation. A phospholipase A$_2$ activity was demonstrated in amnion tissue that cleaves alkylacyl (long-chain) GPC.[54] The enzyme activity is not altered by Ca^{2+} and is distinctly different from the phospholipase A$_2$ that we have previously characterized in this tissue.[82] Amnion tissue also contains PAF : acetyl-CoA acetyltransferase activity, the final enzyme in the remodeling pathway for PAF biosynthesis. This enzyme was activated by low Ca^{2+} concentrations and was associated with the microsomal fraction of amnion tissue. PAF acetylhydrolase, the enzyme that degrades PAF, was also present in amnion tissue and was associated with the cytosolic fraction. Acetylhydrolase activity was also demonstrated in amniotic fluid and was not affected by Ca^{2+}. Acetyltransferase and acetylhydrolase activities in fetal membranes and decidua were similar and were unchanged with gestational age when samples obtained from early and late gestation were compared. In addition, activity did not change in samples from women not in labor as compared to activity in samples from women late in labor.

4.3. Biosynthesis of PAF in Fetal Lung

The activities of several of the enzymes involved in PAF biosynthesis (see Fig. 7) have been investigated in various fetal tissues, including fetal lung. A calcium-independent phospholipase A$_2$ with a substrate specificity for alkyl-containing glycerophospholipids has been found in fetal rabbit lungs.[83] Its activity in the day-31 fetal rabbit lung is at least twofold higher than the adult phospholipase A_2 activity. Like activity in amnion, acetyltransferase in fetal lung was microsomal and activated by low concentrations of Ca^{2+}.[63] The specific activity of this enzyme increased in fetal rabbit lung tissue between day 21 and 24 of gestation and remained elevated through day 31. Acetyltransferase activity was also elevated in human fetal lung tissues when placed in organ culture.[84] Acetyltransferase activity in human and rabbit fetal lung was severalfold higher than activity in fetal kidney or liver[63] and increased activity correlated with an increase in the amount of pulmonary PAF. The specific activities in whole lung tissue, type II pneumonocytes, and alveolar macrophages were also measured.[63] The activity of acetyltransferase was greater in subcellular fractions prepared from type II pneumonocytes than in corresponding fractions prepared from either whole lung tissue or alveolar macrophages. Thus, type II pneumonocytes, the site of lung surfactant biosynthesis, have a higher specific activity of acetyltransferase than whole lung tissue or alveolar macrophages. In contrast, the activity of acetylhydrolase, the primary enzyme involved in degradation of PAF, did not change between day 21 and day 31 of gestation in the rabbit or in explants of human fetal lung tissue maintained for 6 days in culture.[63,84]

At least one enzyme in the *de novo* pathway has also been demonstrated to be elevated in the fetus prior to parturition. Hoffman *et al.*[85] have reported an increase in the activity of dithiothreitol-insensitive choline phosphotransferase, the final enzyme in the *de novo* pathway for PAF biosynthesis (Fig. 7) in fetal rabbit lung. Thus, it can be concluded that the fetal lung has the capacity for the increased synthesis of PAF by both the remodeling and *de novo* pathways, thereby providing a mechanism for the biosynthesis of PAF at a critical time in development. Based on these observations it is suggested that increased secretion of PAF in association with lamellar bodies secreted by type II pneumonocytes occurs during the latter stages of gestation. In addition, PAF also may facilitate secretion of surfactant by type II pneumonocytes in the fetal lung.[13,86] During the early stages of gestation, when low concentrations of PAF are secreted in association with surfactant, PAF is rapidly converted into biologically inactive lysoPAF by acetylhydrolase known to be present in amniotic fluid.[46] During the latter stages of fetal lung development, when surfactant and PAF biosynthesis is rapidly increasing, the rate of secretion of PAF by the type II pneumonocytes exceeds the capacity of the acetylhydrolase to inactivate PAF. This mechanism may account for both the presence of PAF in the amniotic fluid obtained from women at term and in labor as well as our observation that significant amounts of lysoPAF were always present in amniotic fluid at all stages of gestation.[46]

4.4. Maternal Plasma PAF Acetylhydrolase Activity

In the previous sections, we have described the biochemical mechanisms involved in the release of arachidonic acid that is utilized for regulation of prostaglandin synthesis as well as the mechanisms involved in the biosynthesis of PAF in various fetal tissues during parturition. We suggest that the production of both of these autacoids is increased and integrated during labor.

As discussed, it was suggested that PAF acetylhydrolase served a preventive role in amniotic fluid by allowing PAF accumulation only when the capacity of PAF secretion by the fetal lung exceeded the capacity of the acetylhydrolase. In consideration of the role of acetylhydrolase in inactivating PAF, we recently investigated the activity of this enzyme in the maternal plasma of the rabbit throughout gestation. The activity of this enzyme in plasma was described several years ago,[87] and more recently its activity has been reported to be elevated in spontaneously hypertensive rats[88] and in white males who were suffering from hypertension.[89] Both an intracellular and plasma form of the enzyme have been reported.[88] Farr *et al.*[90,91] were the first to describe this enzyme in human plasma (acid-labile factor, ALF). The characteristic features were acid lability, Ca^{2+} independence, inactivation by heating at 65°C for 30 min, and pronase and trypsin sensitivity but papain resistance. These investigators also reported that the activity of this enzyme was associated with a lipoprotein.[90] Stafforini *et al.*[92,93] have purified PAF acetylhydrolase from human plasma. They reported that 30% of the activity was associated with HDL and 70% with the LDL fractions and suggested that activity could be transferred from one lipoprotein to another.

In consideration that the activity of PAF acetylhydrolase in maternal plasma may regulate the amount of PAF that reaches the myometrium, we have investigated the activity of this enzyme in plasma from five rabbits throughout pregnancy and the postnatal period (see Fig. 8).[94] The specific activity of this enzyme in the five nonpregnant rabbits before mating was approximately 131 nmol $\times$ min^{-1} $\times$ ml^{-1} plasma. The activity of acetylhydrolase in rabbits with uncomplicated pregnancies (rabbits 1–3) increased slightly during the first 13 days and then decreased to a minimum between 23 and 29 days (approximately 10 nmol $\times$ min^{-1} $\times$ ml^{-1} plasma). However, within 24–48 hr following delivery, specific activity increased to prepregnancy values. In contrast, no change in activity was found in rabbit 4, which proved not to be pregnant. Rabbit 5 delivered two dead offspring at term. The size of these dead fetuses approximated those found at 28 days of gestation. The maternal plasma hydrolase activity in this rabbit decreased to some extent after 15 days of gestation, then stabilized. In contrast to the human,[92] PAF acetylhydrolase in the rabbit was associated primarily with the HDL fraction of plasma.

On the basis of these data we speculate that during early pregnancy, when PAF acetylhydrolase is high, maternal plasma hydrolase may inactivate any PAF of fetal origin. However, during the later stages of gestation, when fetal tissues such as the lung increase synthesis of PAF, the increased PAF in amniotic fluid may stimulate its own biosynthesis in fetal membranes. A similar amplification action of PAF on its biosynthesis has recently been reported[95] in neutrophils, possibly because of an increase in acetyltransferase activity. This increased production of PAF coincides with a decrease in the capacity of maternal plasma to inactivate PAF. Consequently, at this time, PAF concentrations exceed the capacity of acetylhydrolase to inactivate PAF, and concentrations may become sufficiently elevated to initiate myometrial

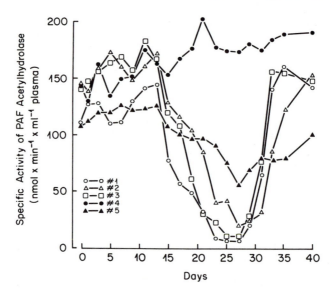

Figure 8. Specific activities of plasma PAF acetylhydrolase in rabbits before (day 0) and during pregnancy (days 1–31) and lactation (day 32–40). Rabbits 1–3 were uncomplicated pregnancies. Rabbit 4 showed no signs of pregnancy during study period and subsequently was examined by x ray on day 34; no fetuses were found. The fifth rabbit delivered two dead offspring on day 34 (weight similar to 28-day fetuses).[94]

contractions or to interact synergistically with other mediators such as prostaglandins to enhance myometrial contractility. Since, in many species, the fetal amnion and chorion are avascular membranes, we envision the highly vascularized decidua vera as playing a critical role in regulating the amount of PAF that reaches the maternal myometrium.

5. Regulation of Autacoid Production during Parturition

5.1. Calcium

Prostanoids and PAF have both been shown to initiate myometrial contractions[47,96] and thus may be critical regulators of parturition. We have suggested that Ca^{2+} plays a central role in the regulation of autacoid preoduction. Recently Challis and colleagues extended these observations by demonstrating that the formation of prostaglandins E_2 and $F_{2\alpha}$ by collagenase-dispersed amnion cells is dependent in part on the presence of extracellular Ca^{2+}.[97] Prostaglandin release into the medium was greater from amnion cells obtained after spontaneous labor than it was from cells prepared from amnion tissue obtained following elective cesarean section prior to the onset of labor. The release of prostaglandins was decreased in both cell populations when Ca^{2+} was omitted from the incubation medium or when a Ca^{2+} channel blocker was included in the incubation. In addition to the role of Ca^{2+} in the activation of phospholipase A_2 and C, Ca^{2+} also prevents the recycling of diacylglycerols back to phosphatidylinositol, further directing the release of arachidonic acid for prostanoid formation (see Fig. 6). Furthermore, acetyltransferase, a critical enzyme in PAF biosynthesis, is maximally activated at low intracellular calcium concentrations (1 μM). Thus, intracellular calcium may simultaneously regulate formation of both families of autacoids at two or more critical enzymatic steps.

In addition, prostanoid biosynthesis and PAF metabolism are closely related, since it was demonstrated that PAF will stimulate prostaglandin release from amnion tissue disks[62] as it does in other tissues.[13,54] The effect of PGE_2 production was similar to that elicited by addition of the Ca^{2+} ionophore A23187. Furthermore, in some cell types, e.g., neutrophils and macrophages, both arachidonic acid and alkyllysoglycerophosphocholine, the PAF precursor, may arise from a common precursor.[98]

5.2. Positive and Negative Effectors

Saeed *et al.*[99] reported the presence of an inhibitor of prostaglandin synthesis in amniotic fluid. The molecular mass of the inhibitor was 50–60 kDa, and the substance was recovered in increased amounts prior to the onset of labor and in decreased amounts at term during labor. Recently, Liggins and co-workers[100] have reported inhibitors of prostaglandin biosynthesis in human endometrial cells that were present in amniotic fluid of term, nonlaboring patients. These inhibitors appeared to be proteinaceous in nature. It was suggested that phospholipase A_2 activity was altered by their actions. The origin and chemical nature of this compound have yet to be established. Strickland *et al.*[101] reported the presence in human fetal urine of a factor that stimulates prostaglandin formation when added to bovine seminal vesicle microsomes. Furthermore, the stimulatory effect of fetal urine on prostaglandin production was greater for urine obtained from fetuses delivered at term after labor than for urine of fetuses delivered by cesarean section at term before the onset of labor. The stimulatory activity of fetal urine was not significantly altered by boiling, was dialyzable, and was substantially lost by treatment of the urine samples with activated charcoal. It was concluded that the stimulatory factor may be a steroid or other small lipid molecule, but the identification of this factor awaits further investigation. Casey *et al.*[102] have described another substance present in human fetal urine that causes a tenfold to 600-fold increase in prostaglandin E_2 production by amnion cells in culture. The factor appears to be related to the epidermal growth factor family.

Inhibitors of PAF action have also been reported. At least two inhibitors of PAF-induced platelet aggregation have been isolated from the nonpregnant rat uterus,[103] and an additional inhibitor has been reported to exist in liver.[104] The uterine inhibitors have been reported to be mixtures of various molecular species of acyllysoglycerophosphocholine, alkyllysoglycerophosphocholine, and sphingomyelin.[103] The relationship between PAF inhibitors and the role of PAF in parturition await future investigation.

5.3. Cyclic AMP

Another factor that may regulate both prostaglandin and PAF formation during parturition is cAMP. Incubation of amnion tissue disks with the β agonist isoproterenol (10^{-5} M) initiates a large but transitory increase in amnion cAMP accompanied by a sustained stimulation of the release of arachidonic acid and prostaglandin E_2 from disks.[105] The elevation in cAMP levels and the amount of prostaglandin released were dependent on the dose of isoproterenol used.

6. Functional Significance of Autacoid Production during Parturition

6.1. Myometrial Contraction

Maintenance of a quiescent uterus until the appropriate time, followed by initiation of myometrial reactivity, is required for successful gestation. Both prostaglandins and PAF have been shown to initiate myometrial contractility (for further discussion of the effects of prostaglandin on myometrial contractions see Chapter 14). Nishihara and colleagues[47] have shown that PAF (either an authentic sample or that purified from amniotic fluid obtained from women in labor) could initiate and propagate contractions of the rat uterus (see Fig. 9). Similar results have been reported for guinea pig[106] and human[107] myometrial strips.

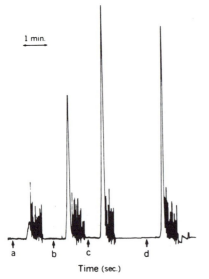

Figure 9. The PAF-induced contraction of rat uterine muscle. A uterine muscle strip was exposed to (a) 500 pM, (b) 750 pM, (c) 1000 pM authentic PAF, or (d) PAF isolated from amniotic fluid (equivalent to 30 ml) collected during parturition from women having a normal pregnancy. (Data are reproduced from Nishihira *et al.*,[47] with permission.)

6.2. Lung Maturation

Several authors have speculated that prostaglandins may be involved in surfactant biosynthesis, as there is a correlation between the onset of PGE_2 biosynthesis and surfactant biosynthesis in the rabbit.[64,108] In addition, this increase in PGE_2 and surfactant synthesis also correlates with an increase in lung cAMP and has been shown to stimulate adenylate cyclase in rat and monkey lungs.[109,110]

In addition, a role for PAF has been proposed in the glycogenolysis that occurs in conjunction with surfactant biosynthesis.[13] Shukla *et al.*[58] observed that PAF at concentrations as low as 2×10^{-10} M rapidly increased glycogenolysis and glucose release by perfused rat liver. The effect of PAF on glycogenolysis in liver tissue was only demonstrable in the perfused liver system[56,58] and not in isolated hepatocytes.[57] The observed glycogenolysis in liver may be pertinent to regulation of surfactant production by fetal lungs. Before the onset of surfactant production, fetal type II pneumonocytes contain numerous glycogen granules.[111–114] Subsequently, the glycogen granules disappear from fetal type II pneumonocytes and are replaced by lamellar bodies.[115] There is considerable evidence that glycogen provides much of the carbon and ATP necessary to support the synthesis of surfactant lipids.[59–61] In consideration of the established effect of PAF on glycogenolysis in rat liver and the precursor–product relationship between fetal glycogen and surfactant lipids during fetal lung maturation, the observed increase in PAF concentration in fetal lung tissue was compared to changes in the glycogen content of rabbit fetal lung tissue during gestation.[84] The amount of glycogen per milligram of fetal rabbit lung protein decreased from 159 μg on day 21 of gestation to 34 μg on day 31. The most dramatic change in glycogen content occurred after day 24 of gestation. The amount of PAF in fetal rabbit lung increases just prior to the observed decrease of glycogen. Thus, the increase in PAF and surfactant synthesis and the decrease in glycogen content are temporally related. In contrast, the glycogen content of fetal liver increased dramatically (9.5 μg/mg of protein on day 21 to 239 μg/mg of protein on day 31).

Similar results have been observed when explants of human fetal lung have been placed in culture for 7 days.[63] During the first 24 hr in culture, the glycogen content of the explants decreased rapidly to approximately one-half that found in the initial culture period. The rapid loss of glycogen that occurred during the first 24 hr is accompanied by an increase in PAF content.[63] Thus, as in fetal rabbit lung tissue, an inverse relationship exists between the amount of PAF in human fetal lung and the glycogen content. In addition, direct evidence for an effect of PAF on glycogen breakdown in fetal lung tissue has been obtained by the intraperitoneal injection of PAF into fetal rabbits at various stages of gestation. Injection of PAF resulted in a decrease in lung glycogen and an increase in lung and serum lactate concentrations.[116]

Glycerophospholipid biosynthesis in fetal type II pneumonocytes may depend on endogenous glycogen because of the low blood supply to the fetal lung.[117] Thus, an increased rate of glycogenolysis may be expected to precede the initiation of surfactant lipid synthesis. Consistent with this view is the observed temporal relationship between an increase in lysoPAF : acetyl-CoA acetyltransferase activity,[63,84] an increase in the amount of PAF, a decrease in lysoPAF, and an increase in the synthesis of disaturated phosphatidylcholine and other surfactant lipids.[2,116] These observations are consistent with a mechanism in which PAF stimulates glycogenolysis in fetal lung as it does in adult rat liver.

Thus, PAF and prostanoid formation may initiate a cascade of events. As the fetal lung matures it synthesizes increasing amounts of PAF and prostaglandins, which in turn enhance fetal lung development including glycogen breakdown and surfactant biosynthesis. As a result of the increased biosynthesis and secretion of PAF along with surfactant, the PAF concentration in amniotic fluid may exceed the capacity for hydrolysis by PAF acetylhydrolase in amniotic fluid. Accumulating PAF may in turn act directly on the fetal membranes, resulting in an increase in arachidonate mobilization and prostaglandin biosynthesis as well as a further stimulation of PAF biosynthesis in these tissues. Therefore, prostaglandins and PAF may both act directly on myometrium to initiate contractility and thereby facilitate parturition. For a further description of these events see Fig. 10.

6.3. Preterm Labor

We have also investigated the role of PAF in complicated pregnancies.[117a] As discussed, we found that PAF was present in the amniotic fluid in normal pregnancies at term and in labor (mean 200 pM) and was absent or in low concentrations in those samples obtained from women not in labor. We have recently determined the PAF concentration in amniotic fluid obtained from a group of patients with preterm labor or premature rupture of membranes. The mean values were 800 pM. With the limited sample number ($N = 8$), it would appear that increased PAF concentrations are found in amniotic fluid samples obtained from pregnancies that may be destined to early delivery.

7. Early Pregnancy

In addition to a role for PAF in parturition, PAF has also been proposed to participate in events associated with maternal recognition of pregnancy and implantation. Mild systemic thrombocytopenia is one of the earliest measurable maternal responses to fertilization,[118] and the presence of a viable embryo has been shown to cause a reduction in peripheral blood platelet counts in mice[118] and women.[119] Embryonic production of a factor that induces thrombocytopenia has been demonstrated by culturing 8- to 16-cell mouse embryos for 24 hr followed by injecting the culture media into splenectomized, nonpregnant mice and monitoring platelet disappearance. Preliminary characterization of this embryo-derived factor has suggested it to be PAF.[120] Production of PAF during the early days of pregnancy appears to be

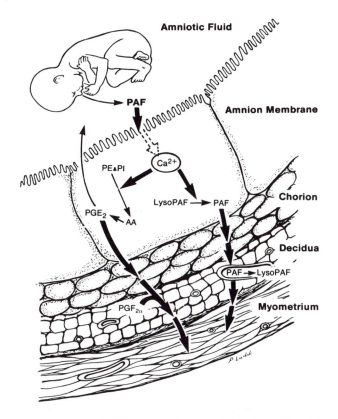

Figure 10. Proposed interrelations of PAF synthesized and secreted by the fetal lung into amniotic fluid. The PAF in amniotic fluid stimulates prostaglandin and PAF biosynthesis in amnion tissue (an avascular fetal membrane in the human), possibly by a Ca^{2+}-mediated event. Production of PGE_2 and PAF may be further magnified by the stimulation of PGE_2 and PAF in chorion laeve and PGE_2, $PGF_{2\alpha}$, and PAF in decidua vera tissues. Tissue PAF may be inactivated by PAF acetylhydrolase during early and midgestation because of its high specific activity in maternal plasma. However, late in gestation PAF escapes inactivation in the vascular decidua because of the decrease in maternal plasma acetylhydrolase activity.[13]

critical as women, monitored during *in vitro* fertilization and embryo transfer studies, whose embryos did not produce PAF failed to become pregnant, whereas approximately 50% of the women whose embryos did produce PAF were subsequently identified as pregnant.[121]

Further studies using a highly specific and sensitive platelet serotonin release assay have shown that mouse embryos in culture produce significant quantities of platelet-activating factor at a time in culture that corresponds to development of the morula. However, by the time the zygote had become a blastula, no PAF production could be detected in culture (M. J. Angle, E. W. Byrd, and J. M. Johnston, unpublished results). This drop in PAF production in culture corresponds to the observed return to normal of systemic platelet counts seen around the time of implantation in mice.[118]

In addition to embryonic production of PAF, the endometrium of the pre-implantative uterus also produces PAF.[122] PAF has been found in the nonpregnant rat uterus[123] and rabbit uterus.[124] Its activity in the uterus changes during early pregnancy; uterine PAF concentration in the rabbit is low (<2 pmol/g tissue) for the first 2 days of pregnancy, increases to approximately 30 pmol/g tissue by day 4, where it remains until the time of implantation, then drops precipitously to values similar to those found in the nonpregnant uterus. Thus, the drop in uterine PAF production also corresponds with the implantation-associated return of systemic platelet counts to normal values. Further support for the role of PAF in maintenance of early pregnancy can be illustrated by the PAF antagonist studies of Spinks and O'Neill[125] and Acker et al.[126] Both groups demonstrated that administration of PAF antagonists could inhibit implantation in the rat or mouse. Consequently, it is reasonable to speculate that production of PAF by both the embryo and uterus is a necessary prerequisite for successful implantation. The precise function(s) of early pregnancy-associated PAF in the maintenance of pregnancy is yet to be established. However PAF has been shown to induce the production of early pregnancy factor (EPF) in mice.[127] EPF is a protein found in normal maternal serum only during pregnancy. The presence of EPF in serum at any other time is pathological and is associated with tumors of germ cell origin.[128]

Prostaglandins as well as PAF appear to be associated with implantation in several species. Indomethacin treatment blocked implantation in mice[129,130] and rats,[131,132] and subcutaneous injections of prostaglandins reversed this block.[129-131] In rabbits, indomethacin treatment did not completely prevent implantation but reduced the number of implantation sites.[133-135] This drug also inhibited implantation-associated increases in vascular permeability at uterine sites adjacent to the embryo. Furthermore, prostaglandin synthesis by blastocysts[136-140] as well as by uterine endometrium[137,138,140,141] has been demonstrated in several species. Animal studies appear to implicate the endometrium as the site of action of prostaglandins.[135,142] Consistent with this idea is the demonstration of PGE binding sites within the rat[143,144] and human[145] endometrium as well as endometrial epithelial cell uptake of prostaglandins in the rabbit.[146] Specifically, prostaglandins, particularly of the E series, appear to play a critical role in the regulation of increased uterine vascular permeability, a necessary prerequisite for successful implantation.[131,147-149]

8. Conclusion

Both PAF and prostaglandins are important lipid mediators of a variety of physiological responses. In addition, they appear to play critical roles during two vital stages of pregnancy, i.e., maternal recognition of pregnancy and parturition. Furthermore, biochemical evidence suggests that biosynthesis of both mediators involved in parturition is intimately associated. Thus, further elucidation of the regulation of PAF and eicosanoid biosynthesis and their interactions during all stages of pregnancy in fetal membranes and other tissues may provide us with additional information to enhance the percentage of successful pregnancies following *in vitro* fertilization and embryo transfer, decrease the incidence of premature labor, and further understand the relationship between lung maturation, the last major organ system to develop, and the initiation of parturition.

ACKNOWLEDGMENTS. The second author (J.M.J.) gratefully acknowledges the research contributions of his colleagues, Drs. T. Okazaki, J. R. Okita, G. C. Di Renzo, N. Sagawa, M. M. Billah, M. M. Anceschi, C. Ban, D. R. Hoffman, and N. Maki while in my laboratory. We also gratefully acknowledge the editorial assistance of Ms. Dolly Tutton. Investigations performed in the authors' laboratories were supported by U.S. Public Health Service grants HD13912 and HD11149 and the Robert A. Welch Foundation, Houston, Texas. Dr. Angle is a recipient of a Chilton Fellowship.

References

1. Thorburn, G. D., 1979, Physiology and control of parturition: Reflections on the past ideas for the future, *Anim. Reprod.* **2**:1–27.
2. Bleasdale, J. E., and Johnston, J. M., 1984, Prostaglandins and human parturition: Regulation of arachidonic acid mobilization, in: *Reviews in Perinatal Medicine,* Volume 5, Alan R. Liss, New York, pp. 151–191.
3. Nakla, S., Skinner, K., Mitchell, B. F., and Challis, J. R. G., 1986, Changes in prostaglandin transfer across human fetal membranes obtained after spontaneous labor, *Am. J. Obstet. Gynecol.* **155**:1337–1341.
4. Golander, A., Hurley, T., Barrett, J., Hizi, A., and Handwerger, S., 1978, Prolactin synthesis by human chorion–decidual tissue: A possible source of prolactin in the amniotic fluid, *Science* **202**:311–313.
5. Okazaki, T., Casey, M. L., Okita, J. R., MacDonald, P. C., and Johnston, J. M., 1981, Initiation of human parturition: XII. Biosynthesis and metabolism of prostaglandins in human fetal membranes and uterine decidua, *Am. J. Obstet. Gynecol.* **139**:373–381.
6. Okazaki, T., Sagawa, N., Bleasdale, J. E., Okita, J. R., MacDonald, P. C., and Johnston, J. M., 1981, Initiation of human parturition: XIII. Phospholipase C, phospholipase A_2, and diacylglycerol lipase activities in fetal membranes and decidua vera tissues from early and late gestation, *Biol. Reprod.* **25**:103–109.
7. Okita, J. R., Sagawa, N., Casey, M. L., and Snyder, J. M., 1983, A comparison of human amnion tissue and amnion cells in primary culture by morphological and biochemical criteria, *In Vitro* **19**:117–126.
8. Casey, M. L., Winkel, C. A., Porter, J. C., and MacDonald, P. C., 1983, Endocrine regulation of the initiation and maintenance of parturition in women, *Clin. Perinatol.* **10**:709–721.
9. Challis, J. R. G., 1984, Characteristics of parturition, in: *Maternal and Fetal Medicine* (R. K. Creasy and R. Resnik, eds.), W. B. Saunders, Philadelphia, pp. 401–414.
10. Liggins, G. C., 1981, Endocrinology of parturition, in: *Fetal Endocrinology* (M. J. Novy and J. A. Resko, eds.), Academic Press, New York, pp. 211–237.
11. MacDonald, P. C., Porter, J. C., Schwarz, B. E., and Johnston, J. M., 1978, Initiation of parturition in the human female, *Sem. Perinatol.* **2**:273–286.

12. Thorburn, G. D., and Challis, J. R. G., 1979, Endocrine control of parturition, *Physiol. Rev.* **59**:863–918.

13. Johnston, J. M., Bleasdale, J. E., and Hoffman, D. R., 1987, Functions of PAF in reproduction and development: Involvement of PAF in fetal lung maturation and parturition, in: *Platelet Activating Factor and Related Lipid Mediators* (F. Snyder, ed.), Plenum Press, New York, pp. 375–402.

14. Luukkainen, T. U., and Csapo, A. I., 1963, Induction of premature labor in the rabbit after pretreatment with phospholipids, *Fertil. Steril.* **14**:65–72.

15. Lanman, J. T., Herod, L., and Thau, R., 1972, Premature induction of labor with dilinoleyl lecithin in rabbits, *Pediatr. Res.* **6**:701–704.

16. Lanman, J. T., Herod, L., and Thau, R., 1974, Phospholipids and fatty acids in relation to the premature induction of labor in rabbits, *Pediatr. Res.* **8**:1–4.

17. Ogawa, Y., Herod, L., and Lanman, J. T., 1970, Phospholipids and the onset of labor in rabbits, *Gynecol. Invest.* **1**:240–248.

18. Nathanielsz, P. W., Abel, M., and Smith, G. W., 1973, Hormonal factors in parturition in the rabbit, in: *Foetal and Neonatal Physiology* (K. S. Comline, K. W. Cross, and G. S. Dawes, eds.), Cambridge University Press, Cambridge, pp. 594–601.

19. Hertelendy, F., 1972, Prostaglandin-induced premature oviposition in the coturnix quail, *Prostaglandins* **2**:269–279.

20. Hertelendy, F., Yeh, M., and Bielier, H. V., 1974, Induction of oviposition in the domestic hen by prostaglandins, *Gen. Comp. Endocrinol.* **22**:529–531.

21. MacDonald, P. C., Schultz, F. M., Duenhoelter, J. H., Gant, N. F., Jimenez, J. M., Pritchard, J. A., Porter, J. C., and Johnston, J. M., 1974, Initiation of human parturition. I. Mechanism of action of arachidonic acid, *Obstet. Gynecol.* **44**:629–636.

22. Karim, S. M. M., 1966, Identification of prostaglandins in human amniotic fluid, *J. Obstet. Gynaecol. Br. Commonw.* **73**:903–908.

23. Karim, S. M. M., and Devlin, J., 1967, Prostaglandin content of amniotic fluid during pregnancy and labour, *J. Obstet. Gynaecol. Br. Commonw.* **74**:230–234.

24. Bygdeman, M., Kwon, S. W., Mukherjee, T., and Miqvist, N., 1968, Effect of intravenous infusion of prostaglandin E_1 and E_2 on motility of the pregnant human uterus, *Am. J. Obstet. Gynecol.* **102**:317–326.

25. Karim, S. M. M., Trussell, R. R., Patel, R. C., and Hillier, K., 1968, Response of pregnant human uterus to prostaglandin-$F_{2\alpha}$-induction of labour, *Br. Med. J.* **4**:621–623.

26. Karim, S. M. M., Hillier, K., Trussell, R. R., Patel, R. C., and Tamusange, S., 1970, Induction of labour with prostaglandin E_2, *J. Obstet. Gynaecol. Br. Commonw.* **77**:200–210.

27. Karim, S. M. M., and Filshie, G. M., 1972, The use of prostaglandin E_2 for therapeutic abortion, *J. Obstet. Gynaecol. Br. Commonw.* **79**:1–13.

28. Karim, S. M. M., and Filshie, G. M., 1970, Therapeutic abortion using prostaglandin $F_{2\alpha}$, *Lancet* **1**:157–159.

29. Collins, E., and Turner, G., 1975, Maternal effects of regular salicylate ingestion in pregnancy, *Lancet* **2**:335–338.

30. Lewis, R. B., and Schulman, J. D., 1973, Influence of acetylsalicylic acid, an inhibitor of prostaglandin synthesis, on the duration of human gestation and labour, *Lancet* **2**:1159–1161.

31. Aiken, J. W., 1972, Aspirin and indomethacin prolong parturition in rats: Evidence that prostaglandins contribute to expulsion of foetus, *Nature* **240**:21–25.

32. Novy, M. K., Cook, M. J., and Manaugh, L., 1974, Indomethacin block of normal onset of parturition in primates, *Am. J. Obstet. Gynecol.* **118**:412–416.

33. Waltman, R., Tricomi, V., and Palav, A. B., 1972, Mid-trimester hypertonic saline-induced abortion: Effect of indomethacin on induction/abortion time, *Am. J. Obstet. Gynecol.* **114**:829–831.

34. Dray, F., and Frydman, R., 1976, Primary prostaglandins in amniotic fluid in pregnancy and spontaneous labor, *Am. J. Obstet. Gynecol.* **126**:13–19.

35. Keirse, M. J. N. C., and Turnbull, A. C., 1973, E prostaglandins in amniotic fluid during late pregnancy and labour, *J. Obstet. Gynaecol. Br. Commonw.* **80**:970–973.

36. Kierse, M. J. N. C., Mitchell, M. D., and Turnbull, A. C., 1977, Changes in prostaglandin F and 13,14-dihydro-15-keto-prostaglandin F concentrations in amniotic fluid at the onset of and during labour, *Br. J. Obstet. Gynaecol.* **84**:743–746.

37. Satoh, K., Yasumizu, T., Fukuoka, H., Kinoshita, K., Kaneko, Y., Tsuchiya, M., and Sakamoto, S., 1979, Prostaglandin $F_{2\alpha}$ metabolite levels in plasma, amniotic fluid, and urine during pregnancy and labor, *Am. J. Obstet. Gynecol.* **133**:886–890.

38. Mitchell, M. D., Keirse, M. J. N. C., Brunt, J. D., Anderson, A. B. M., and Turnbull, A. C., 1979, Concentrations of the prostacyclin metabolite, 6-keto-prostaglandin $F_{1\alpha}$, in amniotic fluid during late pregnancy and labour, *Br. J. Obstet. Gynaecol.* **86**:350–353.

39. Mitchell, M. D., Keirse, M. J. N. C., Anderson, A. B. M., and Turnbull, A. C., 1978, Thromboxane B_2 in amniotic fluid before and during labour, *Br. J. Obstet. Gynaecol.* **85**:442–445.

40. Green, K., Bygdeman, M., Toppozada, M., and Wiqvist, N., 1974, The role of prostaglandin $F_{2\alpha}$ in human parturition, *Am. J. Obstet. Gynecol.* **120**:25–31.

41. Mitchell, M. D., Ebenhack, K., Kraemer, D. L., Cox, K., Cutrer, S., and Strickland, D. M., 1982, A sensitive radioimmunoassay for 11-deoxy-13,14-dihydro-15-keto-11,16-cyclo-prostaglandin E_2: Application as an index of prostaglandin E_2 biosynthesis during human pregnancy and parturition, *Prostaglandins Leukotrienes Med.* **9**:549–557.

42. Hamberg, M., 1974, Quantitative studies on prostaglandin synthesis in man. III. Excretion of the major urinary metabolite of prostaglandins $F_{1\alpha}$ and $F_{2\alpha}$ during pregnancy, *Life Sci.* **14**:247–252.

43. Okita, J. R., MacDonald, P. C., and Johnston, J. M., 1982, Mobilization of arachidonic acid from specific glycerophospholipids of human fetal membranes during early labor, *J. Biol. Chem.* **257**:14029–14034.

44. McManus, L. M., 1986, Pathobiology of platelet-activating factors, *Pathol. Immunopathol. Res.* **5**:104–117.

45. Braquet, P., Tougui, L., Shen, T. Y., and Vargaftig, B. B., 1987, Perspectives in platelet-activating factor research, *Pharmacol. Rev.* **39**:97–146.

46. Billah, M. M., and Johnston, J. M., 1983, Identification of phospholipid platelet-activating factor (1-O-alkyl-2-acetyl-*sn*-glycero-3-phosphocholine) in human amniotic fluid and urine, *Biochem. Biophys. Res. Commun.* **113**:51–58.

47. Nishihira, J., Ishibashi, T., Mai, Y., and Muramatsu, T., 1984, Mass spectrometric evidence for the presence of platelet-activating factor (1-O-alkyl-2-acetyl-*sn*-glycero-3-phosphocholine) in human amniotic fluid during labor, *Lipids* **19**:907–910.

48. Keirse, M. J. N. C., and Turnbull, A. C., 1976, The fetal membranes as a possible source of amniotic fluid prostaglandins, *Br. J. Obstet. Gynaecol.* **83**:146–151.

49. Kinoshita, K., Satoh, K., and Sakamoto, S., 1977, Biosynthesis of prostaglandin in human decidua, amnion, chorion and villi, *Endocrinol. Jpn.* **24**:343–350.

50. Mitchell, M. D., Bibby, J., Hicks, B. R., and Turnbull, A. C., 1978, Specific production of prostaglandin E by human amnion *in vitro*, *Prostaglandins* **15**:377–382.

51. Willman, E. A., and Collins, W. P., 1978, The metabolism of prostaglandin E_2 by tissues from the human uterus and foeto-placental unit, *Acta Endocrinol.* (Kbh.) **87**:632–642.

52. Saeed, S. A., and Mitchell, M. D., 1982, Formation of arachidonate lipoxygenase metabolites by human fetal membranes, uterine decidua vera and placenta, *Prostaglandins Leukotrienes Med.* **8**:635–640.

53. Schwarz, B. E., Schultz, F. M., MacDonald, P. C., and Johnston, J. M., 1975, Initiation of human parturition: III. Fetal membrane content of prostaglandin E_2 and $F_{2\alpha}$ precursor, *Obstet. Gynecol.* **46**:564–568.

54. Ban, C., Billah, M. M., Truong, C. T., and Johnston, J. M., 1986, Metabolism of platelet-activating factor (1-O-alkyl-2-acetyl-*sn*-glycero-3-phosphocholine) in human fetal membranes and decidua vera, *Arch. Biochem. Biophys.* **246**:9–18.

54a. Lopez-Bernal, A., Newman, G. E., Phizackerly, P. J., and Turnbull, A. C., 1988, Surfactant stimulates prostaglandin E production in human amnion, *Br. J. Obstet. Gynecol.* **95**:1013–1017.

55. Smith, K. A., Prewitt, R. L., Byers, L. W., and Muirhead, E. E., 1981, Analogs of phosphatidylcholine: α-Adrenergic antagonists from the renal medulla, *Hypertension* **3**:460–470.

56. Buxton, D. B., Shukla, S. D., Hanahan, D. J., and Olson, M. S., 1984, Stimulation of hepatic glycogenolysis by acetylglyceryl ether phosphorylcholine, *J. Biol. Chem.* **259:**1468–1471.

57. Fisher, R. A., Shukla, S. D., Debuysere, M. S., Hanahan, D. J., and Olson, M. S., 1984, The effect of acetylglyceryl ether phosphorylcholine on glycogenolysis and phosphatidylinositol 4,5-bisphosphate metabolism, *J. Biol. Chem.* **249:**8685–8688.

58. Shukla, S. D., Buxton, D. B., Olson, M. S., and Hanahan, D. J., 1983, Acetylglyceryl ether phosphorylcholine. A potent activator of hepatic phosphoinositide metabolism and glycogenolysis, *J. Biol. Chem.* **258:**10212–10214.

59. Bourbon, J. R., Rieutort, M., Angle, M. J., and Farrell, P. M., 1982, Utilization of glycogen for phospholipid synthesis in fetal rat lung, *Biochim. Biophys. Acta* **712:**382–389.

60. Farrell, P. M., and Bourbon, J. R., 1986, Fetal lung surfactant lipid synthesis from glycogen during organ culture, *Biochim. Biophys. Acta* **878:**159–167.

61. Maniscalco, W. M., Wilson, C. M., Gross, I., Cobran, L. S. A., Rooney, S. A., and Warshaw, J. B., 1978, Development of glycogen and phospholipid metabolism in fetal and newborn rat lung, *Biochim. Biophys. Acta* **530:**333–346.

62. Billah, M. M., Di Renzo, G. C., Ban, C., Troung, C. T., Hoffman, D. R., Anceschi, M. M., Bleasdale, J. E., and Johnston, J. M., 1985, Platelet-activating factor metabolism in human amnion and the responses of this tissue to extracellular platelet-activating factor, *Prostaglandins* **30:**841–850.

63. Hoffman, D. R., Truong, T. C., and Johnston, J. M., 1986, Metabolism and function of platelet-activating factor in rabbit fetal lung development, *Biochim. Biophys. Acta* **879:**88–95.

64. Powell, W. S., and Solomon, S., 1978, Biosynthesis of prostaglandins and thromboxane B_2 in fetal lung homogenate, *Prostaglandins* **15:**351–364.

65. Pace-Asciak, C. R., 1977, Prostaglandin biosynthesis and catabolism in the developing fetal sheep lung, *Prostaglandins* **13:**649–660.

66. Strickland, D. N., and Mitchell, M. D., 1983, Biosynthesis of prostaglandins by tissues of the human fetus, *Prostaglandins* **26:**983–989.

67. Bergstrom, S., Danielsson, H., and Samuelsson, B., 1964, The enzymatic formation of prostaglandin E_2 from arachidonic acid. Prostaglandins and related factors, *Biochim. Biophys. Acta* **90:**207–210.

68. Blank, M. L., Snyder, F., Byers, L. W., Brooks, B., and Muirhead, E. E., 1979, Antihypertensive activity of an alkyl ether analog of phosphatidylcholine, *Biochem. Biophys. Res. Commun.* **90:**1194–1200.

69. Lands, W. E. M., and Samuelsson, B., 1968, Phospholipid precursors of prostaglandins, *Biochim. Biophys. Acta* **164:**426–529.

70. Vonkeman, H., and van Dorp, D. A., 1968, The action of prostaglandin synthetase on 2-arachidonyl-lecithin, *Biochim. Biophys. Acta* **164:**430–432.

71. Gustavi, B., A delayed menstruation? *Lancet* **2:**1149–1150.

72. Schultz, F. M., Schwarz, B. E., MacDonald, P. C., and Johnston, J. M., 1975, Initiation of human parturition: II. Identification of phospholipase A_2 in fetal chorio–amnion and uterine decidua, *Am. J. Obstet. Gynecol.* **123:**650–653.

73. Grieves, S. A., and Liggins, G. C., 1976, Phospholipase A activity in human and ovine uterine tissues, *Prostaglandins* **12:**229–241.

74. Okazaki, T., Okita, J. R., MacDonald, P. C., and Johnston, J. M., 1978, Initiation of human parturition: X. Substrate specificity of phospholipase A_2 in human fetal membranes, *Am. J. Obstet. Gynecol.* **130:**432–438.

75. Di Renzo, G. C., Johnston, J. M., Okazaki, T., Okita, J. R., MacDonald, P. C., and Bleasdale, J. E., 1981, Phosphatidylinositol-specific phospholipase C in fetal membranes and uterine decidua, *J. Clin. Invest.* **67:**847–867.

76. Okita, J. R., MacDonald, P. C., and Johnston, J. M., 1982, Initiation of human parturition: XIV. Increase in the diacylglycerol content of amnion during parturition, *Am. J. Obstet. Gynecol.* **142:**432–435.

77. Okazaki, T., Sagawa, N., Okita, J. R., Bleasdale, J. E., MacDonald, P. C., and Johnston, J. M.,

1981, Diacylglycerol metabolism and arachidonic acid release in human fetal membranes and decidua vera, *J. Biol. Chem.* **256:**7316–7321.

78. Sagawa, N., Okazaki, T., MacDonald, P. C., and Johnston, J. M., 1982, Regulation of diacylglycerol metabolism and arachidonic acid release in human amniotic tissue, *J. Biol. Chem.* **257:**8158–8162.

79. Kramer, R. M., Patton, G. M., Pritzker, C. R., and Deykin, D., 1984, Metabolism of platelet-activating factor in human platelets. Transacylase-mediated synthesis of 1-O-alkyl-2-arachidonoyl-*sn*-glycero-3-phosphocholine, *J. Biol. Chem.* **259:**13316–13320.

80. Sugiura, T., and Waku, K., 1985, CoA-independent transfer of arachidonic acid from 1,2-diacyl-*sn*-glycero-3-phosphocholine to 1-O-alkyl-*sn*-glycero-3-phosphocholine (lyso platelet activating factor) by macrophage microsomes, *Biochem. Biophys. Res. Commun.* **127:**384–390.

81. Lee, T.-C., and Snyder, F., 1985, Function, mechanism and regulation of platelet-activating factor and related ether lipids, in: *Phospholipids and Cellular Regulation,* Volume II (J. F. Kuo, ed.), CRC Press, Boca Raton, FL, pp. 1–39.

82. Okazaki, T., Okita, J. R., MacDonald, P. C., and Johnston, J. M., 1978, Initiation of human parturition: X. Substrate specificity of phospholipase A_2 in human fetal membranes, *Am. J. Obstet. Gynecol.* **130:**432–438.

83. Angle, M. J., Paltauf, F., and Johnston, J. M., 1988, Selective hydrolysis of ether-containing glycerophospholipids by phospholipase in rabbit lung, *Biochim. Biophys. Acta* **962:**234–240.

84. Hoffman, D. R., Truong, T. C., and Johnston, J. M., 1986, The role of platelet-activating factor in human fetal lung maturation, *Am. J. Obstet. Gynecol.* **155:**70–75.

85. Hoffman, D. R., Bateman, M., and Johnston, J. M., 1988, Synthesis of platelet activating factor by cholinephosphotransferase in developing fetal rabbit lung, *Lipids* **23:**96–100.

86. Kumar, R., King, R. J., and Hanahan, D. J., 1985, Occurrence of glyceryl ethers in the phosphatidylcholine fraction of surfactant from dog lungs, *Biochim. Biophys. Acta* **836:**19–26.

87. Pinckard, R. N., Farr, R. S., and Hanahan, D. J., 1979, Physiochemical and functional identity of rabbit platelet-activating factor (PAF) released *in vitro* during IgE anaphylaxis with PAF released *in vitro* from IgE sensitized basophils, *J. Immunol.* **123:**1847–1857.

88. Blank, M. L., Hall, M. N., Cress, E. A., and Snyder, F., 1983, Inactivation of 1-alkyl-2-acetyl-*sn*-glycero-3-phosphocholine by a plasma acetylhydrolase: Higher activities in hypertensive rats, *Biochem. Biophys. Res. Commun.* **113:**666–671.

89. Crook, J. E., Mroczkowski, P. J., Cress, E. E., Blank, M. L., and Snyder, F., 1986, Serum platelet-activating factor acetylhydrolase activity in white and black essential hypertensive patients, *Circulation* **74:**329.

90. Farr, R. S., Cox, C. P., Wardlow, M. J., and Jorgensen, R., 1980, Preliminary studies of an acid-labile factor (ALF) in human sera that inactivates platelet-activating factor (PAF), *Clin. Immunol. Immunopathol.* **15:**318–330.

91. Farr, R. S., Wardlow, M. L., Cox, C. P., Meng, K. E., and Greene, D. E., 1983, Human serum acid-labile factor is an acylhydrolase that inactivates platelet-activating factor, *Fed. Proc.* **42:**3120–3122.

92. Stafforini, D. M., Prescott, S. M., and McIntyre, T. M., 1987, Human plasma platelet-activating factor acetylhydrolase. Purification and properties, *J. Biol. Chem.* **262:**4223–4230.

93. Stafforini, D. M., McIntyre, T. M., Carter, M. E., and Prescott, S. M., 1987, Human plasma platelet-activating factor acetylhydrolase. Association with lipoprotein particles and role in the degradation of platelet-activating factor, *J. Biol. Chem.* **262:**4215–4222.

94. Maki, N., Hoffman, D. R., and Johnston, J. M., 1988, Platelet-activating factor acetylhydrolase activity in maternal, fetal and newborn rabbit plasma during pregnancy and lactation, *Proc. Natl. Acad. Sci. U.S.A.* **85:**728–732.

95. Doebber, T. W., and Wu, M. S., 1987, Platelet-activating factor (PAF) stimulates the PAF-synthesizing enzyme acetyl-CoA : 1-alkyl-*sn*-glycero-3-phosphocholine O^2-acetyltransferase and PAF synthesis in neutrophils, *Proc. Natl. Acad. Sci. U.S.A.* **84:**7557–7561.

96. Carsten, M. E., and Miller, J. D., 1983, Regulation of myometrial contractions, in: *Initiation of Parturition: Prevention of Prematurity, Report of the Fourth Ross Conference on Obstetric*

Research (P. C. MacDonald and J. Porter, eds.), Ross Laboratories, Columbus, OH, pp. 166–171.

97. Olson, D. M., Opavsky, M. A., and Challis, J. R. G., 1983, Prostaglandin synthesis by human amnion is dependent upon extracellular calcium, Can. J. Physiol. Pharmacol. **61**:1089–1092.

98. Wykle, R. L., 1987, Interrelationships in the metabolism of platelet-activating factor and arachidonate in neutrophils, in: Platelet-Activating Factor and Related Lipid Mediators (F. Snyder, ed.), Plenum Press, New York, pp. 273–280.

99. Saeed, S. A., Strickland, D. M., Young, D. C., Dang, A., and Mitchell, M. D., 1982, Inhibition of prostaglandin synthesis by human amniotic fluid: Acute reduction in inhibitory activity of amniotic fluid obtained during labor, J. Clin. Endocrinol. Metab. **55**:801–803.

100. Wilson, T., Liggins, G. C., Aimer, G. P., and Skinner, S. J. M., 1985, Partial purification and characterization of two compounds from amniotic fluid which inhibit phospholipase activity in human endometrial cells, Biochem. Biophys. Res. Commun. **131**:22–29.

101. Strickland, D. M., Saeed, S. A., Casey, M. L., and Mitchell, M. D., 1983, Stimulation of prostaglandin biosynthesis by urine of the human fetus may serve as a trigger for parturition, Science **220**:521–522.

102. Casey, M. L., MacDonald, P. C., and Mitchell, M. D., 1983, Stimulation of prostaglandin E_2 production in amnion cells in culture by a substance(s) in human fetal and adult urine, Biochem. Biophys. Res. Commun. **114**:1056–1063.

103. Nakayama, R., Katsuhiko, Y., and Saito, K., 1987, Existence of endogenous inhibitors of platelet-activating factor (PAF) with PAF in rat uterus, J. Biol. Chem. **262**:13174–13179.

104. Miwa, M., Hill, C., Kumar, R., Sugatani, J., Olson, M. S., and Hanahan, D. J., 1987, Occurrence of an endogenous inhibitor of platelet-activating factor in rat liver, J. Biol. Chem. **262**:527–530.

105. Di Renzo, G. C., Anceschi, M. M., and Bleasdale, J. E., 1984, Beta-adrenergic stimulation of prostaglandin production by human amnion tissue, Prostaglandins **27**:37–49.

106. Montrucchio, G., Alloatti, G., Tetta, C., Roffinello, C., Emanuelli, G., and Camussi, G., 1986, In vitro contractile effect of platelet-activating factor on guinea pig myometrium, Prostaglandins **32**:539–554.

107. Tetta, G., Montrucchio, G., Alloatti, G., Roffinello, C., Emanuelli, G., Benedetto, C., Camussi, G., and Massobrio, M., 1986, Platelet-activating factor contracts human myometrium in vitro, Proc. Soc. Exp. Biol. Med. **183**:376–381.

108. Gluck, L., Sribney, M., and Kulovich, M. V., 1967, The biochemical development of surface activity in mammalian lung. II. The biosynthesis of phospholipids in the lung of the developing rabbit fetus and newborn, Pediatr. Res. **1**:247–265.

109. Barrett, C. T., Sevanian, A., Lavin, N., and Kaplan, S. A., 1976, Role of adenosine $3',5'$-monophosphate in maturation of fetal lungs, Pediatr. Res. **10**:621–625.

110. White, G. J., 1974, Stimulation of rat and monkey lung adenyl cyclase by various prostaglandins, Fed. Proc. **33**:590.

111. Brandstrup, N., and Kretchmer, N., 1965, The metabolism of glycogen in the lungs of the fetal rabbit, Dev. Biol. **11**:202–216.

112. Kikkawa, Y., 1975, Morphology and morphologic development of the lung, in: Pulmonary Physiology of the Fetus, Newborn, and Child (E. Scarpelli, ed.), Lea & Febiger, Philadelphia, pp. 35–60.

113. Shelley, H. J., 1961, Glycogen reserves and their changes at birth and in anoxia, Br. Med. Bull. **17**:137–143.

114. Williams, M. C., and Mason, R. J., 1977, Development of the type II cell in the fetal rat lung, Am. Rev. Respir. Dis. **115**:37–47.

115. Snyder, J. M., Mendelson, C. R., and Johnston, J. M., 1985, The morphology of lung development in the human fetus, in: Pulmonary Development. Transition from Intrauterine to Extrauterine Life (G. H. Nelson, ed.), Marcel Dekker, New York, pp. 19–46.

116. Hoffman, D. R., White, R. G., Angle, M. J., Maki, N., and Johnston, J. M., 1988, Platelet-activating factor induces glycogen degradation in fetal rabbit lung in utero, J. Biol. Chem. **263**:9316–9319.

117. Rudolph, A. M., Itskouitz, J., Iwamoto, H., Reuss, M. L., and Heymann, M. A., 1981, Fetal cardiovascular responses to stress, *Semin. Perinatol.* **5:**109–120.

117a. Hoffman, D. R., Romero, R., and J. M. Johnston, 1990, Detection of platelet activating factor in amnionic fluid of complicated pregnancies, *Am. J. Obstet. Gynecol.* (in press).

118. O'Neill, C., 1985, Examination of the causes of early pregnancy-associated thrombocytopenia in mice, *J. Reprod. Fertil.* **73:**567–577.

119. O'Neill, C., Pike, I. L., Porter, R. N., Gidley-Baird, A. A., Sinosich, M., and Saunders, D., 1985, Maternal recognition of pregnancy prior to implantation: Methods for monitoring embryonic viability *in vitro* and *in vivo, Ann. N.Y. Acad. Sci.* **442:**429–439.

120. O'Neill, C., 1985, Partial characterization of the embryo-derived platelet-activating factor in mice, *J. Reprod. Fertil.* **75:**375–380.

121. O'Neill, C., Gidley-Baird, A. A., Pike, I. L., and Saunders, D. M., 1987, Use of a bioassay for embryo derived platelet-activating factor as a means of assessing quality and pregnancy potential of human embryos, *Fertil. Steril.* **47:**969–975.

122. Angle, M. J., Jones, M. A., McManus, L. M., Pinckard, R. N., and Harper, M. J. K., 1988, Platelet-activating factor in the rabbit uterus during early pregnancy, *J. Reprod. Fertil.* **83:**711–722.

123. Yasuda, K., Satouchi, K., and Saito, K., 1986, Platelet-activating factor in normal rat uterus, *Biochem. Biophys. Res. Commun.* **138:**1231–1236.

124. Angle, M. J., Jones, M. A., Pinckard, R. N., McManus, L. M., and Harper, M. J. K., 1985, Platelet-activating factor (PAF) in the rabbit uterus during early pregnancy, *Biol. Reprod.* **32(Suppl. 1):**143.

125. Spinks, N. R., and O'Neill, C., 1987, Embryo-derived platelet-activating factor is essential for establishment of pregnancy in the mouse, *Lancet* **1:**106–107.

126. Acker, G., Hecquet, F., Etienne, A., Braquet, P., and Mencia-Huerta, J. M., 1988, Role des mediateirs lipidiques de l'inflammation dans l'ovoimplantation chez la ratte, in: *Immunologie de la Grossesse* (G. de Chaouat, ed.), INSERM, Elsevier, Amsterdam (in press).

127. Orozco, C., Perkins, T., and Clarke, F. M., 1986, Platelet-activating factor induces the expression of early pregnancy factor activity in female mice, *J. Reprod. Fertil.* **78:**549–555.

128. Morton, H., Rolfe, B., and Cavanaugh, A., 1982, Early pregnancy factor: Biology and clinical significance, in: *Pregnancy Proteins* (J. G. Gondzenskos, B. Tersnei, M. Seffala, eds.), Academic Press, Sydney, pp. 391–405.

129. Lau, I. F., Saksena, S. K., and Chang, M. C., 1973, Progesterone blockade by indomethacin, an inhibitor of prostaglandin synthesis: Its reversal by prostaglandins and progesterone in mice, *Prostaglandins* **4:**795–803.

130. Saksena, S. K., Lau, I. F., and Chang, M. C., 1976, Relationship between oestrogen, prostaglandin $F_{2\alpha}$ and histamine in delayed implantation in the mouse, *Acta Endocrinol.* (Kbh.) **81:**801–807.

131. Kennedy, T. G., 1977, Evidence for a role of prostaglandins in the initiation of blastocyst implantation in the rat, *Biol. Reprod.* **16:**286–291.

132. Phillips, C. A., and Poyser, N. L., 1981, Studies on the involvement of prostaglandins in implantation in the rat, *J. Reprod. Fertil.* **62:**73–81.

133. El-Banna, A. A., 1980, The degenerative effect in rabbit implantation sites by indomethacin. I. Timing of indomethacin action, possible effect in uterine proteins and the effect of replacement doses of $PGF_{2\alpha}$, *Prostaglandins* **20:**587–599.

134. Hoffman, L. H., 1978, Antifertility effects of indomethacin during early pregnancy in the rabbit, *Biol. Reprod.* **18:**148–153.

135. Snabes, M. C., and Harper, M. J. K., 1984, Sites of action of indomethacin on implantation in the rabbit, *J. Reprod. Fertil.* **71:**559–565.

136. Pakrasi, P. L., and Dey, S. K., 1982, Blastocyst is the source of prostaglandins in the implantation site in the rabbit, *Prostaglandins* **24:**73–77.

137. Harper, M. J. K., Norris, C. J., and Rajkumar, D., 1983, Prostaglandin release by zygotes and endometria of pregnant rabbits, *Biol. Reprod.* **28:**350–362.

138. Lewis, G. S., Thatcher, W. W., Bazer, F. W., and Curl, J. S., 1982, Metabolism of arachidonic acid *in vitro* by bovine blastocysts and endometrium, *Biol. Reprod.* **27**:431–439.
139. Chepenik, K. P., and Smith, J. B., 1980, Synthesis of prostaglandins by mouse embryos, *IRCS Med. Sci.* **8**:783–784.
140. Hyland, J. H., Manns, J. G., and Humphrey, W. D., 1982, Prostaglandin production by ovine embryos and endometrium *in vitro, J. Reprod. Fertil.* **65**:299–304.
141. Tsang, B. K., and Ooi, T. C., 1982, Prostaglandin secretion by human endometrium *in vitro, Am. J. Obstet. Gynecol.* **142**:626–633.
142. Kennedy, T. G., 1983, Embryonic signals and the initiation of blastocyst implantation, *Aust. J. Biol. Sci.* **36**:531–543.
143. Kennedy, T. G., Martel, D., and Psychoyos, A., 1983, Endometrial prostaglandin E$_2$ binding: Characterization in rats sensitized for the decidual cell reaction and changes during pseudopregnancy, *Biol. Reprod.* **29**:556–564.
144. Kennedy, T. G., Martel, D., and Psychoyos, 1983, Endometrial prostaglandin E$_2$ binding during the estrous cycle and its hormonal control in ovariectomized rats, *Biol. Reprod.* **29**:565–571.
145. Hofmann, G. E., Rao, C. V., DeLeon, F. D., Toledo, A. A., and Santilippo, J. S., 1985, Human endometrial prostaglandin E$_2$ binding sites and their profiles during the menstrual cycle in pathological stages, *Am. J. Obstet. Gynecol.* **151**:369–375.
146. Jones, M. A., and Harper, M. J. K., 1983, Prostaglandin accumulation by isolated uterine endometrial epithelial cells from six-day pregnant rabbits, *Biol. Reprod.* **29**:1201–1209.
147. Psychoyos, A., 1973, Endocrine control of egg implantation, in: *Handbook of Physiology,* Volume II, Part 2, (R. O. Greep, E. B. Astwood, and S. R. Geiger, eds.), American Physiological Society, Bethesda, MD, pp. 187–215.
148. Evans, C. A., and Kennedy, T. G., 1978, The importance of prostaglandin synthesis for the initiation of blastocyst implantation in the hamster, *J. Reprod. Fertil.* **54**:255–261.
149. Hoffman, L. H., DiPietro, D. L., and McKenna, T. J., 1978, Effects of indomethacin on uterine capillary permeability and blastocyst development in rabbits, *Prostaglandins* **15**:823–828.

17

Endocrinology of Pregnancy and Parturition

M. Linette Casey and Paul C. MacDonald

1. Overview

Much progress has been recorded in the study of the endocrinology of pregnancy and parturition in mammalian species in the past two decades. Interestingly, the most completely defined endocrine physiology of pregnancy among mammalian species is that of the human. And, the endocrine changes that occur during pregnancy in women are among the most remarkable known in mammalian physiology or pathophysiology. Consider, for example, the following: In pregnant women at or near term, 75 to 125 mg of estrogen, 250 to 600 mg of progesterone, 1 to 2 mg of aldosterone, and 3 to 12 mg of deoxycorticosterone are produced each day. There also are phenomenal increases in the levels of plasma renin, angiotensinogen, and angiotensin II during pregnancy; and enormous amounts of protein hormones such as chorionic gonadotropin, placental lactogen, chorionic ACTH, chorionic thyrotropin, and others are produced during pregnancy.[1] But more than that, recent evidence is supportive of a role for growth factors, e.g., epidermal growth factor, transforming growth factors, and immunohormonal agents, e.g., interleukin-1 and colony-stimulating factors, in the events that are associated with labor.[2] All investigators agree that maternal and fetal adaptations to this distinctive hormonal milieu are profound. And many if not all investigators also propose a role for one or a number of hormones, steroid and protein, in the coordinated set of events that facilitate the maintenance of pregnancy and lead to the timely onset of labor and successful parturition.

There is little doubt that great advances have been made in a definition of the

M. Linette Casey and Paul C. MacDonald • The Cecil H. and Ida Green Center for Reproductive Biology Sciences and Departments of Biochemistry and Obstetrics–Gynecology, University of Texas Southwestern Medical Center, Dallas, Texas 75235.

physiological mechanisms involved in the maintenance of pregnancy and the regulation of parturition in sheep, whereas greater progress may have been made in defining the biomolecular events of labor in women. Unfortunately, however, we still are unable to solve several major riddles that are posed by apparent differences in the processes involved in parturition in these two species; seemingly, an explanation for these conundrums would provide us with insights into a biochemical definition of the physiological events of pregnancy and labor in both (perhaps all) species. It is not totally clear why it is so difficult to come to a complete and satisfactory definition of the sequence of events involved in the successful progression of pregnancy that culminates in the timely initiation of parturition in women. No doubt it is because our understanding of this momentous process in the human is fragmentary, but equally likely, it may be that our interpretation of the biomolecular events of pregnancy and parturition in other species is less precise than we believe.

2. Hypotheses

Three major hypotheses for the initiation of labor have been proposed and considered in some detail by a number of investigators; these are (1) the "progesterone withdrawal" hypothesis, (2) the oxytocin theory, and (3) the "fetal–maternal organ communication system" proposition. There are many diverse combinations of selected features of one or more of these three theories in the speculations of most investigators. Commonly, for example, it is presumed that pregnancy is maintained until the mature fetus, in some manner, delivers some sort of signal that initiates parturition; alternatively, and more likely, the fetus retreats from the support of pregnancy maintenance. And, in most if not all theories of parturition currently being considered, a role for prostaglandins (PGs)—be it a primary or a secondary role—is proposed.

In a number of mammalian species, there is incontrovertible evidence in favor of the proposition that the fetus is in control of its destiny with respect to the timely onset of parturition at term. And teleologically, it is satisfying to believe that the fetus, after maturation of key organs and systems, is able to inaugurate a chain of events that culminate in labor. In the sheep, goat, and cow, for example, the maturation of the fetal adrenal appears to be of signal importance in the chain of events that initiates labor. Cortisol, secreted in increased amounts by the adrenals of ovine and bovine fetuses, acts (directly or indirectly) to bring about changes in placenta that eventuate in alterations in the relative rates of progesterone and estrogen secretion.[3,4] Progesterone withdrawal and a surge in estrogen herald the onset of labor; these endocrine changes appear to effect the accelerated formation of prostaglandins in uterine tissues. In the goat, accelerated production of prostaglandin in the uterus gives rise to luteolysis and thence progesterone withdrawal and the onset of labor.[5] In the horse, the fetal gonad (ovary or testis) serves as an extraplacental, adrenal-independent source of the C_{19} steroid dehydroisoandrosterone sulfate (also called dehydroepiandrosterone sulfate), which serves as the precursor for placental estrogen biosynthesis.[6] Removal of the fetal gonads in the horse does not change the time of onset of labor but leads to dysfunctional and prolonged labor. Thus, it is clear that in a number of species the successful initiation of labor is brought about by an interaction between fetal endocrine tissues and placenta. Indeed, there are many investigators who take the view that ultimately, when the biochemical and physiological processes involved in parturition in all species are understood, we will recognize that there are astounding similarities in the molecular events of labor in all species.

3. Endocrinology of Pregnancy and Parturition in Sheep

Perhaps the most complete evidence for any of the theorems set forward for the initiation of parturition is that for "progesterone withdrawal," especially in the sheep, in which this dramatic endocrine change precedes the onset of labor. The series of endocrine events that are believed to be involved in the initiation of parturition in the sheep are summarized briefly as follows. At a critical time in fetal organ maturation, there is increased responsiveness of the fetal adrenal to ACTH, which is secreted by the fetal pituitary. Thereafter, there is an increase in the rate of secretion of cortisol from the sheep fetal adrenal. Fetal cortisol acts in some manner to bring about in trophoblasts an increase in steroid 17α-hydroxylase and 17,20-desmolase activities,[7] which are catalyzed by a single enzyme (steroid 17,20-desmolase is also known as steroid 17,20-lyase). Increases in these activities cause decreased progesterone secretion and increased estrogen secretion (Fig. 1). Progesterone withdrawal, together with adequate or increased estrogen secretion, seems to provoke (by mechanisms that are not yet defined) increased prostaglandin formation, the formation of gap junctions in myometrium, sensitization of the uterus to uterotonic agents, cervical ripening, and the commencement of labor. Thus, in sheep, it is probable that the mature sheep fetus provides the initial signal for the commencement of parturition. Specifically, there is a fetal-initiated retreat from the maintenance of pregnancy that is mediated by way of progesterone withdrawal.

An understanding of the mechanisms that serve to regulate steroid enzyme synthesis and activity in the fetal adrenals and placenta and an evaluation of the interdependence of these two tissues in the control of estrogen synthesis and progesterone secretion by the placenta are believed to be of paramount importance in the definition of the endocrinology of parturition in the sheep. Recently, the biomolecular basis for the alterations in the rates of steroid secretion by the sheep placenta that occur just prior to term have been defined. It appears that these modifications are essential for the onset of labor in this species.

In most mammalian species other than the primate, there is a reasonably well-defined relationship between decreased progesterone secretion and increased estrogen formation before the commencement of labor. In general, a close functional relationship exists between fetal endocrine glands (adrenals or gonads) and the placenta in the regulation of estrogen formation. For example, it is firmly established

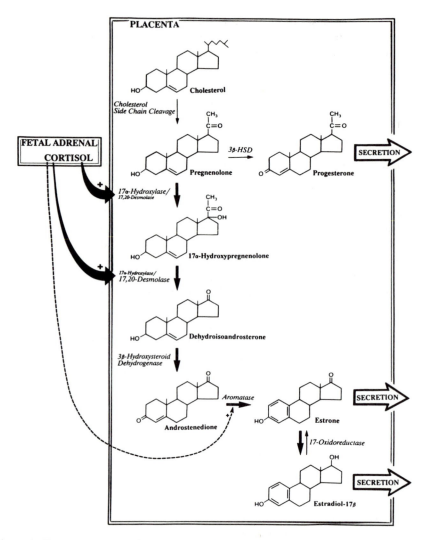

Figure 1. The endocrinology of pregnancy in sheep. Prior to term, cholesterol is metabolized in placenta, principally to progesterone. Near term, cortisol, produced by the fetal adrenal, acts to facilitate the redirection of pregnenolone to estrogens. Thus, at term, increased fetal cortisol production serves to decrease secretion of progesterone and to increase estrogen secretion.

that after adrenalectomy of the sheep fetus, the pregnant ewe fails to go into labor.[8] Furthermore, administration of dexamethasone, a potent glucocorticosteroid, to the fetal lamb results in the onset of labor within 2 days.[8] These findings lead to the conclusion that, in the sheep, an increase in the rate of secretion of cortisol by the fetal adrenal serves as the signal for the initiation of parturition. Dexamethasone, administered to the fetal lamb, also causes, within 48 hr, a 200-fold increase in placental steroid 17α-hydroxylase activity and a similar increase in the content of the cytochrome P-450$_{17\alpha}$ component of this enzyme complex.[9]* On the other hand, dexamethasone treatment is associated with only a slight increase in the activity of steroid aromatase activity in the sheep placenta. Accordingly, France *et al.*[9] suggest that cortisol, produced in increased amounts by the sheep fetal adrenal late in pregnancy, leads to expression of cytochrome P-450$_{17\alpha}$ in placenta. Progesterone withdrawal in the pregnant ewe at term is effected by the diversion of placental metabolism of pregnenolone away from progesterone and into 17α-hydroxypregnenolone and thence dehydroisoandrosterone, both of these steroid conversions catalyzed by P-450$_{17\alpha}$ (Fig. 1). Thus, it is proposed that in the sheep, the surge in estrogen synthesis occurs as a result of a greatly enhanced rate of C$_{19}$ steroid synthesis in the placenta. At the molecular level, the increased expression of cytochrome P-450$_{17\alpha}$ in sheep placentomes is clearly linked to the progression of labor. Therefore, it is speculated that there is molecular regulation of expression of the P-450$_{17\alpha}$ gene, possibly at a 5'-flanking site, which may correspond to a glucocorticosteroid receptor binding region.[9]

Increasing rates of cortisol secretion in the fetal sheep commencing about day 120 of gestation are correlated with the development of increased ACTH responsiveness and adenylate cyclase activity in the fetal sheep adrenal.[10] Interestingly, there is little change in the specific content of P-450$_{scc}$ and P-450$_{11\beta}$ enzyme protein (as assessed by immunoblotting) in adrenals from sheep fetuses of gestational ages 85, 127, or 140 days.[9] On the other hand, as gestation progresses, there is a dramatic increase in the levels of P-450$_{17\alpha}$.[9] These findings are interpreted to mean that the expression of P-450$_{17\alpha}$ in the fetal sheep adrenal is dependent on ACTH-mediated, cAMP-dependent regulatory processes, whereas the expression of other steroidogenic enzymes, namely, P-450$_{scc}$ and P-450$_{11\beta}$, is much less dependent on such regulatory mechanisms.

Other investigators have suggested that fetal cortisol induction of placental microsomal steroid 17α-hydroxylase/17,20-lyase activity would cause increased conversion of progesterone to estrone[7] and thereby a decline in progesterone secretion and an increase in estrogen secretion that precedes the onset of labor. To investigate this issue, France and colleagues[9] evaluated 17α-hydroxylase/17,20-lyase activity and enzyme content in placenta of pregnant ewes late in gestation; to do so, dexamethasone was infused into the chronically instrumented fetal sheep until active labor was established. The infusion of dexamethasone was associated

*The cellular content of specific enzymes (proteins) is assessed by use of antibodies specific for the protein of interest. For example, the content of the specific cytochrome P-450$_{17\alpha}$, steroid 17α-hydroxylase, in a specific tissue is evaluated by electrophoretic separation, transfer to nitrocellulose paper, and immunoblotting. Immunoblotting, also referred to as "western" blotting, is a technique whereby an antibody to the protein of interest is used, together with a radiolabeled or "enzyme-labeled" second antibody to IgG (or the immunoglobulin class of the first antibody). Thereby, the amount of protein of interest can be evaluated either qualitatively or quantitatively (by densitometry of radiographic or colorimetric data).

with a 200-fold increase in progesterone 17α-hydroxylase activity in placental microsomes. By immunoblotting placental homogenates with specific polyclonal antibodies to cytochrome P-450$_{17\alpha}$ and the microsomal NADPH : P-450 reductase, they found that the activity was associated with specific synthesis of the P-450$_{17\alpha}$ component. The level of P-450$_{scc}$ in placenta did not change appreciably after dexamethasone treatment. France *et al.*[11] found, as have others, that placental microsomal steroid aromatase activity increased 1.5- to twofold after dexamethasone treatment.

Finally, the results of studies conducted to evaluate the substrate specificity of the steroid 17α-hydroxylase/17,20-lyase enzyme in sheep placenta after the infusion of dexamethasone into the fetus are supportive of the conclusion that the glucocorticosteroid-induced enzyme catalyzes the conversion of pregnenolone to dehydroisoandrosterone preferentially compared with the conversion of progesterone to androstenedione. Specifically, pregnenolone and 17-hydroxypregnenolone are better substrates for 17α-hydroxylase/17,20-lyase than are progesterone and 17-hydroxyprogesterone. This pathway serves, therefore, as the principal pathway of synthesis of C_{19} steroids and aromatizable substrates in sheep placenta. Thus, progesterone withdrawal may occur because of increased metabolism of pregnenolone to C_{19} steroids at the expense of progesterone formation. In addition, increased conversion of progesterone to 17α-hydroxyprogesterone contributes to the withdrawal of progesterone. In turn, increased estrogen formation occurs as a result of the provision of substrate to the placental aromatase enzyme as well as a relatively small increase in aromatase enzyme content *per se.*

4. Endocrinology of Pregnancy and Parturition in Women

4.1. Introduction

As stated, much progress has been made toward a definition of the sequence of physiological changes that transpire, largely by way of the fetal adrenal–placental interactions, before the onset of parturition in sheep. There are, however, important differences between the events that occur in the human and those that occur in the sheep in fetal adrenal–placental function and the relationship of adrenal–placental steroidogenesis to the initiation of parturition in these two species. These crucial differences in key events in sheep parturition and those of the human seem to be pivotal in the physiological scheme in each species. Among the most important of these are that (1) there is no demonstrable withdrawal of progesterone prior to the initiation of labor in women, (2) cortisol infused into the fetus or the mother does not cause labor in the human, and (3) steroid 17α-hydroxylase activity is not demonstrable in human placental tissue. Yet, it is quite likely that there are striking similarities between the biomolecular processes involved in parturition in all species.

4.2. The Human Fetal Adrenal

In pursuit of this tenet, it is fascinating to consider the potential role of the human fetal adrenal in a variety of very important pregnancy phenomena. The attraction of investigators to the study of this gland is easily understood. The human fetal adrenal cortex at term is, relative to body size, 25 times larger than that of the adult. And the steroidogenic capacity of the fetal adrenal is equivalent to its enormous size. Indeed, the fetal adrenal gland is the principal source of C_{19} steroids used by the placenta in the formation of the large amounts of estrogen that are characteristic of human pregnancy.[12] Soon after birth, there is a marked involution of the gland, which involves primarily the loss of the unique fetal zone, which comprises 85% of the mass of this organ just prior to birth. These attributes alone were sufficient to evoke an in-depth consideration of the role of the fetal adrenal in the onset of parturition.

The secretions of the human fetal adrenal cortex are believed to be important in the maintenance of pregnancy, the timely maturation of fetal organs, and the preparation for successful parturition. In the past 10 years, considerable progress has been made in defining the biomolecular basis for the regulation of steroidogenesis in this unique endocrine gland. Implicitly, the role of the definitive or neocortical zone of the adrenal is to produce optimal amounts of glucocorticosteroids, which may be instrumental in maturational processes important for successful extrauterine life. Today, it also is clear that secretions of the fetal adrenal are important in the final maturational processes of other organ systems in the human, for example, in the development of low-density lipoprotein (LDL) receptors in liver[13] and in the biosynthesis of surfactant in the mature fetal lung.[14] Diminished adrenal production of or placental utilization of dehydroisoandrosterone sulfate (e.g., in anencephaly and placental sulfatase deficiency) commonly results in a lack of cervical ripening or delayed labor or both. It is reasonably easy to envision, therefore, that the human fetal adrenal is important in the maintenance of pregnancy, the maturation of fetal organs, and in the development of systems important for the commencement of parturition.

4.3. Estrogen

The endocrine physiology of human pregnancy is characterized by a state of hyperestrogenism of almost unimaginable magnitude. As stated in Section 1, it can be computed that the amount of bioactive estrogen (17β-estradiol and estriol) produced in most near-term pregnant women is more than 100 mg/day.[1] In the human, estrogen is produced in the placenta by utilization of C_{19} steroids that are derived from the fetal adrenal gland.[12,15] In pregnancies in which there is fetal anencephaly, placental sulfatase deficiency, or fetal death, there is, relative to normal pregnancy, severe hypoestrogenism. This state of hypoestrogenism results from decreased availability of precursor for placental estrogen biosynthesis. In the case of anencephaly and fetal death, the production of dehydroisoandrosterone sulfate by the fetal adrenal is decreased markedly, and in the case of placental sulfatase deficiency, the hydrolysis of dehydroisoandrosterone sulfate to the nonconjugated steroid is precluded. Yet, even in these pregnancies, more than 10 mg of bioactive estrogens may be produced daily, an amount ten times that secreted by the ovaries of young women just prior to the midcycle LH surge.

Thus, on the one hand, it is difficult to envision cellular "hypoestrogenism" in pregnant women except by the production of an antiestrogen(s), e.g., progesterone. But on the other hand, there is direct evidence for relative biological hypoestrogenism in women who are pregnant with an anencephalic fetus or a dead fetus. Namely, we have demonstrated that estrogen, administered to women pregnant with a dead fetus. Namely, we have demonstrated that estrogen, administered to women pregnant with a dead fetus or an anencephalic fetus, causes a significant increase in extraadrenal steroid 21-hydroxylase activity.[16] Moreover, as reviewed in Chapters 9, 10, 12, and 14, it is known that estrogen acts *in vivo* and *in vitro* to promote a number of events that are temporally related to the onset of labor in women and in other mammalian species; these include the stimulation of prostaglandin production in decidua/endometrium, the formation of gap junctions, and the synthesis of oxytocin receptors.[17,18] It is likely, therefore, that there is some form of dynamic interplay between estrogen and progesterone—at least, estrogen and progesterone action—even when the concentrations of both agents are enormous. Be that as it may, it is clear that certain physiological–biochemical–morphological changes in the uterus that are reminiscent of those that occur with progesterone withdrawal in some species also come to pass in the uterus of pregnant women just before or at the time of the onset of labor.

4.4. Progesterone

In the past, however, we have been consumed and perhaps obsessed by the seemingly insurmountable impasse created by the absence of "progesterone withdrawal" prior to the onset of labor in women. Because of this, we have sought to identify alternatives that would permit us to merge or else to separate forever the endocrinology and physiology of the sheep and human models of parturition.

4.4.1. Progesterone in the Maintenance of Pregnancy

Progesterone is believed to be essential for the maintenance of pregnancy, hence the name *pro-gestation steroid*. Generally, this attribute of progesterone is not one that is well defined from a biomolecular perspective. And, we submit, it is our ignorance of the precise mechanism by which progesterone is active in the maintenance of pregnancy that has created many misconceptions concerning the initiation of parturition in all species. It seems likely that one function of progesterone in the maintenance of pregnancy is through an action that facilitates the inhibition of decidual/endometrial prostaglandin formation or release or both.[19-21] But even this presumed action of progesterone is unsettling because in certain physiological and pharmacological circumstances, progesterone appears to bring about an increase in the capacity for prostaglandin secretion/release from endometrium/decidua.[22-24] Thus, it may be that other (fetally derived) agents act to promote decidual maturation and maintain pregnancy by suppressing $PGF_{2\alpha}$ formation, but such processes likely are progesterone dependent. Progesterone withdrawal does cause labor, but this may be true because the primary pregnancy maintenance system directed by the fetus, i.e., the prevention of decidual prostaglandin production, is progesterone dependent. Recently, it has been particularly surprising—indeed, shocking—to learn that normal conception, pregnancy, and parturition can occur in women with abetalipoproteinemia in whom the plasma levels of progesterone are extremely low, but this in fact is the case.[25]

4.4.2. Regulation of Uterotonin Production by Progesterone

There is considerable evidence that pregnancy is a physiological state in which uterine (decidual) prostaglandin formation is rigidly inhibited; likely, this iron-handed regulation is effected by a paracrine system [through some fundamental joint venture with contiguous fetal membranes (and amniotic fluid)] and an endocrine system (the action of progesterone).[17] And it is most likely that these two systems are interdependent. We have proposed that the timely interruption of this coordinated state of intense inhibition of prostaglandin formation is synchronous with parturition. Such interruption can be inaugurated by the withdrawal of endocrine (progesterone) support or else by the withdrawal of paracrine support provided by the conceptus (by way of the fetal membranes).[17]

4.4.3. Placental Contributions to the Maintenance of Pregnancy

Compared with all other species, the endocrinology of the human trophoblast is truly exceptional and astounding. Phenomenal quantities of steroid and protein hormones are produced in the human placenta. Nonetheless, during the last two decades, many investigators have set aside any direct elementary role for the placenta in the penultimate events of human parturition, and seemingly for good reasons. The placenta is extremely rich in oxytocinase and prostaglandin dehydrogenase; communication between fetal and maternal tissues (decidua vera and myometrium) seems less direct if transmitted by way of placenta; most placental products do not enter the amniotic fluid directly; and a number of additional lines of evidence can be cited in support of the view that the placenta is not pivotal in the final events of human parturition. On the other hand, we recognize that some of the well-established physiological events of parturition, as defined in the sheep and in other species, do develop after fundamental modifications in placenta. Regrettably, no such placental happening has been identified in the human.

4.4.4. Contributions of the Fetal Adrenal to the Steroid Hormonal Milieu of Pregnancy and Parturition

In fact, in the human fetus, there is no clear-cut increase in the secretion of cortisol prior to the onset of labor. In addition, there is an entity, i.e., congenital adrenal hyperplasia, in the human in which cortisol secretion is precluded. This disorder is characterized by deficiencies in steroidogenic enzyme activities that lead to decreased cortisol secretion by the adrenal; in pregnancies in which the fetus is afflicted with this disorder, labor apparently occurs on time.[26] Notwithstanding these differences between the endocrine physiology of human and sheep pregnancy, we have been captivated by the probable importance of the enormous human fetal adrenal glands in parturition. There is great attraction to the likelihood that the human fetal adrenal is involved in maturational processes important for the timely onset of labor. Indeed, it was almost 25 years ago that the concept of a fetal–maternal organ communication system was strengthened by our discovery that the human fetal adrenal supplies the C_{19} steroid precursors used for placental estrogen biosynthesis.[15] In the human, the fetal adrenal cortex produces incredible quantities of dehydroisoandrosterone sulfate by way of the consumption of large quantities of fetal plasma low-density lipoprotein (LDL). The human placenta utilizes prodigious quantities of maternal plasma LDL for the production of progesterone.[27] Indeed, there seem to be many redundancies in the sex hormone endocrine physiology of human pregnancy. Possibly for this reason, it has been disappointingly impossible to relate all of these important endocrine findings of human pregnancy into one single, unifying scheme similar to that of the sheep, in which progesterone withdrawal is preeminent.

4.4.5. Events in Parturition and Labor Common among Species

To continue to assess the expectation that ultimately we will recognize that the biomolecular events involved in the initiation of parturition are remarkably similar in all species, one could focus on those events that seem unambiguously to be the same in all species. Rhythmic contractions of the uterus bring about the delivery of the fetus and other products of conception in all mammalian species. And during labor at least, if not before, there is increased prostaglandin production in all mammalian (and avian) species tested.[28,29] So far as is presently known, the three events related temporally to the onset of labor—(1) cervical ripening, (2) the development of oxytocin receptors in myometrium,[30] and (3) the formation of gap junctions between myometrial cells[31]—occur in all species. If we could accept the proposition, for example, that the formation of gap junctions is requisite for coordinated contractions of the smooth muscle of the uterus irrespective of the uterotropic agent(s) ultimately involved, it would follow that a precise definition of the biochemical regulation of gap junction formation is elementary to an understanding of the chain of events that cause labor. Similarly, it can be argued that successful parturition is achieved only after cervical ripening. The same logic may hold for the development of oxytocin receptors, but for quite different reasons. Teleologically, the development of oxytocin receptors may be consequential for reasons independent of the initiation of parturition; the most compelling reason could be the institution of a means for ensuring contraction of the uterus after delivery of the fetus. In regard to cervical ripening and the formation of gap junctions between myometrial cells, it seems reasonably clear that estrogen treatment (at least in very large doses) promotes and progesterone inhibits these processes.[32,33] It seems possible that prostaglandins (produced in response to endocrine manipulations) are the agents involved most directly in cervical ripening[3] and in the formation of gap junctions[33]; seemingly, PGE_2 and $PGF_{2\alpha}$ promote cervical ripening, and $PGF_{2\alpha}$ facilitates and PGI_2 prevents gap junction formation.[34]

After progesterone withdrawal and a surge in estrogen secretion in the pregnant ewe, cervical ripening commences, myometrial oxytocin receptors appear, and gap junctions develop between myometrial cells.[32] Thus, in some manner, progesterone withdrawal is associated (at least temporally) with preparation and sensitization of the myometrium to the action of uterotonic agents; and this modification in uterine preparedness for labor is associated temporally with discrete morphological and biochemical changes in myometrium. It also is well established that there is sensitization of the human uterus to uterotonins late in gestation; refractoriness to oxytocin in early pregnancy (and even until the last few weeks or even days of pregnancy) is well recognized. It also is known that gap junctions and oxytocin receptors appear in the myometrium during (before) labor in women irrespective of whether the onset of labor occurs on time or prematurely.[35,36] Generally, the same is true of cervical ripening. Therefore, it seems that whatever biochemical phenomena come to bear to prepare the uterus for labor, the same phenomena are operative in all mammalian species. But once again, we are faced with the dilemma of the absence of "progesterone withdrawal" in the case of the human. We submit that an understanding of this important point of divergence in the endocrinology of parturition in the sheep and human is fundamental to a final resolution of the precise regulation of the labor process in all species.

4.4.6. Progesterone Withdrawal in Women

Even though there is no decrease in the level of progesterone in maternal or fetal blood in the human before the onset of labor, the progesterone withdrawal hypothesis is not easily abandoned; in fact, there is evidence that myometrial contractions will commence in women if progesterone withdrawal is effected. The classic example of this, of course, is ovulatory (progesterone-withdrawal) menses. Generally, dysmenorrhea (presumably caused principally by the discomfort of uterine contractions) occurs primarily in women with progesterone-withdrawal-induced menses. In fact, Gustavii[37] referred to parturition as "delayed menstruation." He postulated that destabilization of lysosomes in decidua in response to progesterone withdrawal would cause activation of phospholipase A_2, the liberation of arachidonic acid, the formation of prostaglandins, and thence the onset of labor.

Another example of progesterone-withdrawal-induced uterine contractions in women may be that of the occurrence of spontaneous labor in some instances after fetal demise. After death of the human fetus, it is common for progesterone levels to remain quite elevated despite a rapid and precipitous decline in the concentrations of estrogens. This, no doubt, is explained by a dependence of the human placenta on C_{19} steroids from the fetal adrenal for estrogen formation but a dependence of the placenta on maternal plasma LDL-cholesterol for placental progesterone formation.[27] Thus, after fetal death but with continuing placental function, progesterone levels remain high while estrogen levels are low. And after fetal death, oxytocin induction of labor commonly is difficult. After unpredictable times following fetal death, we have observed that the levels of plasma progesterone may begin to decline, and this decline commonly heralds the spontaneous onset of labor—sometimes many weeks after death of the fetus.

Yet another example of progesterone withdrawal in women is the removal of the corpus luteum in early (before 8 weeks) human pregnancy, which is followed by spontaneous abortion.[38] Abortion also follows the inhibition of progesterone action,[39] and treatment of women in early pregnancy with inhibitors of progesterone synthesis causes increased sensitivity of myometrium to oxytocin.[40] It seems reasonably clear, therefore, that in the human, as in the sheep, labor could ensue if progesterone withdrawal were effected. All of these lines of evidence could be cited in support of the possibility that some form of progesterone deprivation is operative in the initiation of spontaneous parturition in the human. We doubt the validity of this proposition, however, for reasons to be presented in detail.

Thus, it is apparent that progesterone withdrawal or deprivation as an important event in the biomolecular processes that lead to the onset of parturition could not be set aside easily. But what could be involved in such a process? If progesterone levels in maternal and fetal blood in human pregnancy do not decline prior to (or even during) labor, a number of possible alternatives have been considered.

4.4.6a. Progesterone Binding. The sequestration of progesterone in plasma in such a manner as to limit or preclude its entry into cells. Possibilities include progesterone-binding protein(s) as in the guinea pig, uteroglobin, enzymes, or lipoproteins, e.g., LDL, in which lipophilic progesterone is sequestered. These possibilities have not been excluded completely, but the fact that the metabolic clearance rate of progesterone from plasma is not altered significantly during human pregnancy mitigates against the probability of progesterone sequestration in plasma of pregnant women.[41] A progesterone-binding protein, seemingly unique, was found in association with fetal membranes obtained from near-term but not from early-gestation pregnancies.[42] After careful investigation, we concluded, as have others,[43] that this progesterone-binding protein likely was cortisol-binding globulin (CGB) and that the unique characteristics of its steroid-binding profile in our studies were probably related to the presence of denatured albumin in the preparations. As yet, however, we have no satisfactory explanation for the apparent association of this CBG-like protein with amnion in late pregnancy.

4.4.6b. Progesterone Receptors. A timely decrease in the number of progesterone receptors in key tissues. There is no convincing evidence in favor of this proposition.

4.4.6c. Antiprogestins. The major metabolites of progesterone in pregnant women, 5α-dihydroprogesterone and 20α-dihydroprogesterone, are not characterized as antiprogestins.

4.4.6d. Metabolism of Progesterone in Tissue Sites of Action. The possibility that, near term, there is increased metabolism of progesterone in or near key tissues of progesterone action has been considered by a number of investigators. In most tissues, however, the reverse obtains. Namely, there is a decline in the specific activities of most progesterone-metabolizing enzymes as pregnancy approaches term.[44]

A strong argument that has been presented against the hypothesis that there is progesterone deprivation of some sort as an important event in the initiation of labor in women is that progesterone treatment in the human will neither inhibit the onset of labor nor act to arrest preterm labor.[45] Nonetheless, the importance of progesterone deprivation in so many other mammalian species seems so clear. And in these other species, progesterone withdrawal occurs irrespective of whether the principal site of progesterone production is the corpus luteum or the placenta or both.

But alas, we must conclude, however reluctantly, that there is no progesterone withdrawal to herald the onset of labor in women. There is no sequestration of progesterone in blood; there is no unusual metabolism of progesterone in tissues; there is no important extraplacental site of progesterone production. In short, we conclude and accept the probability that progesterone withdrawal or deprivation does not occur in women until after delivery of the placenta. But perhaps, teleologically, this is when it should occur, because the most consistent sequel of progesterone withdrawal (in terms of a temporal relationship to a physiological event) in all mammalian species examined is the initiation of lactogenesis.[46,47] There is a very close relationship in time between the onset of progesterone withdrawal and the onset of lactogenesis, and this is true in species in which progesterone withdrawal seems to occur before parturition and in species in which progesterone withdrawal occurs after delivery of the placenta. In fact, it can be argued that the single best biological endpoint of progesterone withdrawal in all species is the initiation of lactogenesis. In women, this occurs after delivery—not before. So we feel obliged, finally, to set aside progesterone withdrawal or deprivation as a fundamental prerequisite for the initiation of labor in women.

It is possible, however, that the initiation of parturition in the human could come about by "overcoming" or "subverting" or "abrogating" the action of progesterone. The importance of these distinctions resides not only in the acceptance of the fact that progesterone withdrawal does not occur in women before labor commences but also in recognizing that whatever the process is that permits an increase in prostaglandin formation, it must be operative on the background of continued progesterone action. There is ample evidence that such is the case in women. We surmise, therefore, that we must search for mechanisms whereby increased prostaglandin formation in the human is permitted in progesterone-conditioned tissues. The changes that transpire in the uterine cervix, the formation of gap junctions between myometrial cells, the increase in the number of oxytocin receptors, as well as increases in the rates of formation of PGE_2 and $PGF_{2\alpha}$ during (before) labor can be taken as evidence that the action of progesterone has been subverted. Yet another explanation is that the salutary effect of progesterone in the maintenance of pregnancy is mediated indirectly. In particular, we suggest that the inhibition of decidual prostaglandin formation throughout most of pregnancy is imposed by a fetal paracrine system that is dependent on progesterone action.

5. Endocrinology of the Fetal–Maternal Organ Communication System

For several years, we[48] and others[18] have emphasized the potential importance of a fetal–maternal organ communication system in the biomolecular events that culminate in the timely onset of labor in women. According to these models, it is envisioned that the human fetus is in control of its own destiny with respect to the onset of labor at term. Over the years, most investigators have been so concerned with maternal contributions to fetal growth, development, protection, and well-being that we may have been blinded to the breadth of the role of the fetus in the direction of pregnancy responses in maternal tissues. We accept, of course, the importance of fetal (placental) tissue production of human chorionic gonadotropin (hCG) and human placental lactogen (hPL) as well as the sex steroid hormones estrogen and progesterone. But considerably less attention has been given to a potentially essential role of the fetus (or fetal membranes) at the interface among the amniotic fluid, fetal membranes, and decidua vera (or equivalent spaces in other species) in the maintenance of pregnancy. Fortunately, however, largely through the pioneering experiments of Bazer and colleagues[49] and others, this field is rapidly developing. They have emphasized that the mother must first recognize and accept pregnancy. These summary descriptions involve many fundamental concepts including but not limited to implantation, prevention of endometrial formation of luteolysin, production of a luteotropin, immunologic acceptance, and inhibition of prostaglandin secretion/release (at least into the uterine vein). But more than that, the decidua must thereafter be regulated—and this, we suggest, is fundamental to the maintenance of pregnancy (i.e., the prevention of labor). We suggest that the fetus is the primary moving force in this process, i.e., the maintenance of pregnancy. And we suggest further that the prevention of labor, i.e., the prevention of uterotonin formation/action, is accomplished in large measure by the fetal-directed growth, maturation, and function of the decidua.

According to this concept, the role of progesterone/estrogen is placed in a new light. It is envisioned that the fetal-directed maintenance of decidual function is progesterone dependent. The importance of this distinction is simply that parturition in all mammalian species may occur in a timely manner when the fetal-directed maintenance of pregnancy is relinquished. If this process were progesterone dependent, it follows that progesterone withdrawal would cause labor, but progesterone withdrawal is not necessarily essential for the commencement of parturition; indeed, it may not be optimal.

Thus, a fundamental role for the conceptus in decidual function and in the maintenance of pregnancy may be by way of inhibition of decidual uterotonin formation. It is likely that this important process is instituted and perpetuated by the interdependent actions of more than one system, i.e., a paracrine system (fetus by way of amnion and chorion laeve) that is dependent on the endocrine system—fetal, maternal, and placental. We suggest that components of this paracrine system include protein hormones, growth factors, and immunohormones derived from the fetus, fetal membranes, and placentas, whereas the endocrine system is comprised principally of estrogen and progesterone of placental origin.

References

1. Casey, M. L., MacDonald, P. C., and Simpson, E. R., 1985, Endocrinologic changes in pregnancy, in: *Williams Textbook of Endocrinology* (D. W. Foster and J. D. Wilson, eds.), W. B. Saunders, Philadelphia, pp. 422–437.
2. Casey, M. L., and MacDonald, P. C., 1988, Biomolecular processes in the initiation of parturition: Decidual activation, *Clin. Obstet. Gynecol.* **31:**533–552.
3. Liggins, G. C., Fairclough, R. J., Grieves, S. A., Forster, C. S., and Knox, B. S., 1977, Parturition in the sheep, in: *The Fetus and Birth,* Ciba Foundation Symposium (J. Knight and M. O'Connor, eds.), Elsevier/North-Holland, Amsterdam, pp. 5–30.
4. Hoffmann, B., Wagner, W. C., Rattenberger, E., and Schmidt, J., 1977, Endocrine relationships during late gestation and parturition in the cow, in: *The Fetus and Birth,* Ciba Foundation Symposium (J. Knight and M. O'Connor, eds.), Elsevier/North-Holland, Amsterdam, pp. 107–125.
5. Currie, W. B., and Thorburn, G. D., 1977, The fetal role in timing the initiation of parturition in the goat, in: *The Fetus and Birth,* Ciba Foundation Symposium (J. Knight and M. O'Connor, eds.), Elsevier/North-Holland, Amsterdam, pp. 49–72.
6. Allen, W. R., and Pashen, R. L., 1981, The role of prostaglandins during parturition in the mare, *Acta Vet. Scand. [Suppl].* **77:**279–298.
7. Anderson, A. B. M., Flint, A. P. F., and Turnbull, A. C., 1975, Mechanism of action of glucocorticoids in induction of ovine parturition: Effect on placental steroid metabolism, *J. Endocrinol.* **66:**61–70.
8. Liggins, G. C., Fairclough, R. J., Grieves, S. A., Kendall, J. Z., and Knox, B. S., 1973, The mechanism of initiation of parturition in the ewe, *Recent Prog. Horm. Res.* **29:**111–150.
9. France, J. T., Magness, R. R., Murry, B. A., Rosenfeld, C. R., and Mason, J. I., 1988, The regulation of ovine placental steroid 17α-hydroxylase and aromatase by glucocorticoid, *Mol. Endocrinol.* **2:**193–199.
10. Wintour, E. M., 1984, Developmental aspects of the hypothalamic–pituitary–adrenal axis, *J. Dev. Physiol.* **6:**291–299.
11. France, J. T., Mason, J. I., Magness, R. R., Murry, B. A., and Rosenfeld, C. R., 1980, Ovine placental aromatase: Studies of activity levels, kinetic characteristics and effects of aromatase inhibitors, *J. Steroid Biochem.* **28:**155–160.

12. Siiteri, P. K., and MacDonald, P. C., 1966, Placental estrogen biosynthesis during human pregnancy, *J. Clin. Endocrinol. Metab.* **26**:751–761.
13. Carr, B. R., and Simpson, E. R., 1984, Cholesterol synthesis by human fetal hepatocytes: effects of hormones, *J. Clin. Endocrinol. Metab.* **58**:111–116.
14. Ballard, P. L., 1986, *Monographs in Endocrinology,* Volume 28, Springer-Verlag, New York.
15. Siiteri, P. K., and MacDonald, P. C., 1963, The utilization of circulating dehydroisoandrosterone sulfate for estrogen synthesis during human pregnancy, *Steroids* **2**:713–730.
16. MacDonald, P. C., Cutrer, S., MacDonald, S. C., Casey, M. L., and Parker, C. R., 1982, Regulation of extraadrenal steroid 21-hydroxylase activity: Increased conversion of plasma progesterone to deoxycorticosterone during estrogen treatment of women pregnant with a dead fetus, *J. Clin. Invest.* **69**:469–478.
17. Casey, M. L., and MacDonald, P. C., 1988, The role of a fetal–maternal paracrine system in the maintenance of pregnancy and the initiation of parturition, in: *Fetal and Neonatal Development* (C. T. Jones, ed.), Perinatology Press, Ithaca, NY, pp. 521–532.
18. Challis, J. R. G., and Olson, D. M., 1988, Parturition, in: *The Physiology of Reproduction* (E. Knobil and J. D. Neill, eds.), Raven Press, New York, pp. 2177–2216.
19. Liggins, G. C., Campos, G. A., Roberts, C. M., and Skinner, S. J., 1980, Production rates of prostaglandin F, 6-keto-PGF$_{1\alpha}$, and thromboxane B$_2$, by perifused human endometrium, *Prostaglandins* **19**:461–477.
20. Abel, M. H., and Baird, D. T., 1980, The effect of 17β-estradiol and progesterone on prostaglandin production by human endometrium maintained in organ culture, *Endocrinology* **106**:1599–1606.
21. Schatz, F., Markiewicz, L., and Gurpide, E., 1986, Hormonal effects of PGF$_{2\alpha}$ output by cultures of epithelial and stromal cells in human endometrium, *J. Steroid Biochem.* **24**:297–301.
22. Downie, J., Poyser, N. L., and Wunderlich, M., 1974, Levels of prostaglandins in human endometrium during the menstrual cycle, *J. Physiol. (Lond.)* **236**:465–472.
23. Singh, E. J., Baccarini, I. M., and Zuspan, F. P., 1975, Levels of prostaglandins F$_{2\alpha}$ and E$_2$ in human endometrium during the menstrual cycle, *Am. J. Obstet. Gynecol.* **121**:1003–1006.
24. Maathuis, J. B., and Kelly, R. W., 1978, Concentrations of prostaglandins F$_{2\alpha}$ and E$_2$ in the endometrium throughout the human menstrual cycle, after the administration of clomiphen or an oestrogen–progestogen pill and in early pregnancy, *J. Endocrinol.* **77**:361–371.
25. Parker, C. R., Jr., Illingworth, D. R., Bissonnette, J., and Carr, B. R., 1986, Endocrine changes during pregnancy in a patient with homozygous familial hypobetalipoproteinemia, *N. Engl. J. Med.* **314**:557–560.
26. Price, H. V., Cone, B. A., and Keogh, M., 1971, Length of gestation in congenital adrenal hyperplasia, *J. Obstet. Gynaecol. Br. Commonw.* **78**:430–434.
27. Carr, B. R., and Simpson, E. R., 1981, Lipoprotein utilization and cholesterol synthesis by the human fetal adrenal gland, *Endocrine Rev.* **2**:306–326.
28. Novy, M. J., and Liggins, G. C., 1980, Role of prostaglandins, prostacyclin, and thromboxanes in the physiologic control of the uterus and in parturition, *Semin. Perinatol.* **4**:45–66.
29. Hertelendy, F., 1983, Regulation of oviposition, in: *Report of the Fourth Ross Conference on Obstetric Research* (J. C. Porter and P. C. MacDonald, eds.), Ross Laboratories Publications, Columbus, OH, pp. 43–51.
30. Reimer, R. K., Goldfien, A. C., Goldfien, A., and Roberts, J. M., 1986, Rabbit uterine oxytocin receptors and *in vitro* contractile response: Abrupt changes at term and the role of eicosanoids, *Endocrinology* **119**:699–709.
31. Garfield, R. E., and Hayashi, R. H., 1981, Appearance of gap junctions in the myometrium of women during labor, *Am. J. Obstet. Gynecol.* **140**:254–260.
32. Liggins, G. C., 1978, Ripening of the cervix, *Semin. Perinatol.* **2**:261–271.
33. Garfield, R. E., Kannan, M. S., and Daniel, E. E., 1980, Gap junction formation in myometrium: Control by estrogens, progesterone, and prostaglandins, *Am. Physiol. Soc.* **238**:C81–C89.
34. Garfield, R. E., Merrett, D., and Grover, A. K., 1980, Gap junction formation and regulation in myometrium, *Am. J. Physiol.* **239**:C217–C228.

35. Fuchs, A. R., Fuchs, F., Husslein, P., Soloff, M. S., and Fernstrom, M. J., 1982, Oxytocin receptors and human parturition: A dual role for oxytocin in the initiation of labor, *Science* **215:**1396–1398.
36. Garfield, R. E., 1984, Control of myometrial function in preterm versus term labor, *Clin. Obstet. Gynecol.* **27:**572–599.
37. Gustavii, B., 1972, Labour: A delayed menstruation? *Lancet* **2:**1149–1150.
38. Csapo, A. I., Pulkkinen, M. O., and Wiest, W. G., 1973, Effects of luteectomy and progesterone replacement therapy in early pregnant patients, *Am. J. Obstet. Gynecol.* **115:**759–765.
39. Kovacs, L., Sas, M., Resch, B. A., Ugocsai, G., Swahn, M. L., Bygdeman, M., and Rowe, P. G., 1984, Termination of very early pregnancy by UR-486—an antiprogestational compound, *Contraception* **29:**299–310.
40. Webster, M. A., Pattison, N. S., Phipps, S. L., and Gillmer, M. D. G., 1985, Myometrial activity in first trimester human pregnancy after epostane therapy. Effect of intravenous oxytocin, *Br. J. Obstet. Gynaecol.* **92:**957–962.
41. Lin, T. J., Billiar, R. B., and Little, B., 1972, Metabolic clearance of progesterone in the menstrual cycle, *J. Clin. Endocrinol. Metab.* **35:**879–886.
42. Schwarz, B. E., Milewich, L., Johnston, J. M., Porter, J. C., and MacDonald, P. C., 1976, Initiation of human parturition: V. Progesterone binding substance in fetal membranes, *Obstet. Gynecol.* **48:**685–689.
43. Kossmann, J. C., Bard, H., and Gibb, W., 1982, Characterization of specific steroid binding in human amnion at term, *Biol. Reprod.* **27:**320–326.
44. Milewich, L., Gant, N. F., Schwarz, B. E., Chen, G. T., and MacDonald, P. C., 1977, Initiation of human parturition: VIII. Metabolism of progesterone by fetal membranes of early and late human gestation, *Obstet. Gynecol.* **50:**45–48.
45. Fuchs, F., and Stakemann, G., 1960, Treatment of threatened premature labor with large doses of progesterone, *Am. J. Obstet. Gynecol.* **79:**173–176.
46. Kulski, J. K., Smith, M., and Hartmann, P. E., 1977, Perinatal concentrations of progesterone, lactose and α-lactalbumin in the mammary secretion of women, *J. Endocrinol.* **74:**509–510.
47. Convey, E. M., 1973, Serum hormone concentrations in ruminants during mammary growth, lactogenesis, and lactation: A review, *J. Dairy Sci.* **57:**905–917.
48. Casey, M. L., and MacDonald, P. C., 1986, Initiation of labor in women, in: *The Physiology and Biochemistry of the Uterus in Pregnancy and Labor* (G. Huszar, ed.), CRC Press, Boca Raton, FL, pp. 155–161.
49. Bazer, F. W., Vallet, J. L., Roberts, R. M., Sharp, D. C., and Thatcher, W. W., 1986, Role of conceptus secretory products in establishment of pregnancy, *J. Reprod. Fertil.* **76:**841–850.

18

Circulation in the Pregnant Uterus

Charles R. Brinkman III

1. Vascular Anatomy

1.1. Nonpregnant

The primary blood supply to the uterus is the uterine artery, a major branch of the hypogastric or internal iliac artery. In the fetus this same hypogastric artery gives rise to the umbilical artery, which delivers nearly 50% of the cardiac output to the placenta. This fact is mentioned in order to stress the remarkable capability of this particular portion of the vascular system to accommodate a tremendous increase in blood flow during pregnancy. The uterine arteries descend for a short distance (about 4 cm) after dividing from the hypogastric (internal iliac) artery. In the base of the broad ligament, the uterine artery progresses medially, crossing anterior to the ureter about 2 cm lateral to its division into the cervicovaginal artery and the ascending uterine artery at the level of the supravaginal portion of the cervix. The ascending branch runs a tortuous course along the lateral margin of the uterus, providing a major branch to the upper portion of the cervix and eight to 24 arcuate branches to the uterus[1] (Fig. 1). The tortuosity of the uterine artery as it traverses the broad ligament and as it ascends the uterus is directly related to the woman's parity. The arteries of parous women assume a much more convoluted and tortuous course.

The arcuate arteries are usually paired, passing to both the anterior and posterior, traversing the uterus between the outer third and the inner two-thirds of the myometrium to anastomose with its counterpart from the opposite side. The origins of these branches are nonsymmetric and of varying size and supply larger or smaller areas of the uterine wall.[2] Radial arteries branch from the arcuates to supply the inner two-thirds of the uterus. They terminate in the spiral arteries, which supply the endometrium, decidua, and placenta during pregnancy (Fig. 2).

Charles R. Brinkman III • Department of Obstetrics and Gynecology, School of Medicine, University of California Los Angeles, Harbor/UCLA Medical Center, Torrance, California 90509.

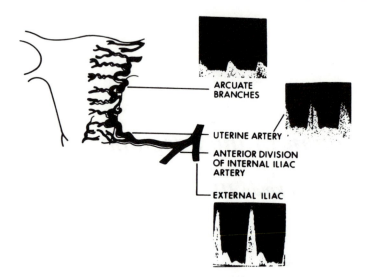

Figure 1. The left side of this figure illustrates the vascular anatomy of the uterus. Note the tortuous nature of the ascending portion of the uterine artery. The right-hand portion illustrates the velocity wave patterns of three different pelvic arteries. Note that the arcuate vessels have proportionally the highest diastolic frequency. (From Schulman *et al.*,[18] courtesy of Dr. Harold Schulman and *The American Journal of Obstetrics and Gynecology.*)

The ascending branch of the uterine artery terminates as it reaches the fallopian tube, trifurcating into fundal, tubal, and ovarian branches. The ovarian branch anastomoses with the terminal branch of the ovarian artery. The tubal branch passes through the mesosalpinx and supplies the fallopian tube. The fundal branch supplies the upper portion of the uterus.

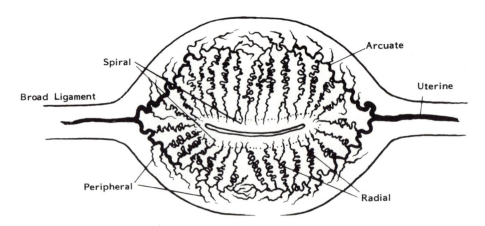

Figure 2. Schematic drawing of intrinsic arteries of the uterus. From Burchell *et al.*,[2a] with permission.

The ovarian artery, a direct branch of the aorta, traverses the broad ligament, entering from the infundibulopelvic ligament. Numerous branches enter the ovary from the ovarian hilum, and the main portion of the artery proceeds medially to anastomose with the uterine branch. The rich anastomosis between the right and left ascending branches of the uterine arteries and the right and left branches of the ovarian arteries allows this well-perfused organ to tolerate sacrifice of one or more of these vessels without apparent effect on uterine function.

1.2. Pregnant

Pregnancy results in remarkable and unique changes in the uterine vasculature, not the least of which is a tenfold increase in blood flow, which we more fully discuss in Section 2.1. Probably the most remarkable change in pregnancy is the histological transformation of the spiral arteries, as described by the work of Brosens, Robertson, and Dixon.[3-5] This physiological change in the placental bed normally extends from the decidua into the inner myometrium. The first stage of this phenomenon occurs in the first trimester with endovascular trophoblasts invading the maternal spiral arteries. These vessels are converted into distensible funnel-shaped vessels opening into the intervillous space. In the second trimester, the myometrial segments of the spiral artery undergo changes in which the internal elastic lamella and the inner media are replaced by a thick layer of convoluted fibrinoid material in which cytotrophoblasts are embedded. This disappearance of the musculoelastic layers of these vessels has important implications when one comes to discuss the response of the uteroplacental circulation to neural and humoral stimulation.

The diameter of the uterine artery varies considerably from woman to woman and from side to side. The combined diameter of the left and right uterine arteries in the reproductive age group is 2–5 mm in the nonpregnant state. Although everyone agrees there is enlargement during pregnancy, no specific figures have been discovered by this reviewer. Data derived from the study on nonpregnant and pregnant sheep demonstrate a doubling of the diameters of the uterine artery and its major branches.[6]

1.3. Experimental Animals

Because of the difficulty in studying the physiology and pathophysiology of the human uterine and uteroplacental circulation, much of our understanding has been derived from animal studies. The sheep uterine and uteroplacental hemodynamics has probably been most extensively studied. The sheep has a bicornuate uterus in which a fetus may be located in one or both horns. Commercially bred experimental animals have about a 30% occurrence of twins. The sheep has a common internal iliac artery arising at the terminal aorta. Figure 3 illustrates the aorta actually trifurcating into the left and right external iliac and the common internal iliac. The common internal iliac is usually less than 1 cm in length before it divides into what

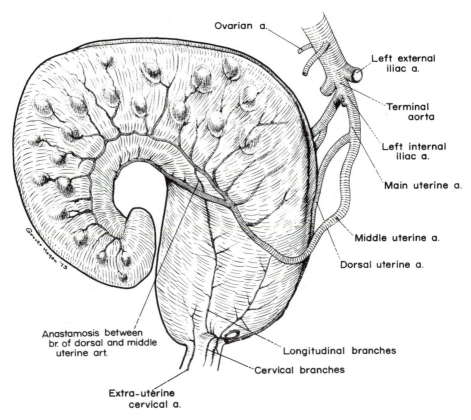

Figure 3. Vascular anatomy of the sheep uterine circulation. Note that main uterine artery divides into the middle and dorsal uterine arteries, which are comparable to the ascending and decending or cervical branches, respectively, in the human. (From Fuller *et al.*,[6] courtesy of Dr. Ellen Fuller and the *American Journal of Physiology*.)

is termed either the left and right internal iliac or main uterine artery, which has a single posterior branch called the sacral artery. The main uterine artery divides into the larger middle uterine artery and smaller dorsal uterine artery. The middle uterine artery is comparable to the ascending uterine in the human and supplies the major portion of the uterine horn. The dorsal uterine artery supplies the lower uterine corpus and the cervix. Fuller, to whom we owe most of the descriptive anatomy of ovine uterine vasculature, demonstrated a nearly doubling of the diameter of the middle and dorsal uterine artery in pregnancy.[6] The ovarian artery also significantly increases in diameter during pregnancy. Minor anastomoses with arcuate branches of the middle uterine artery have been demonstrated. For a more detailed discussion of the comparative anatomy of uterine circulation, the reader is referred to the excellent review by Reynolds.[7]

2. Methods for Study of Uterine Circulation and Blood Flow in the Human

In this section, I review the more common methods that have been used to study the physiology and pathophysiology of human uterine blood flow. Immediately it will become apparent to the reader that many of these methods would no longer be acceptable to Institutional Review Boards. The difficulty in studying uterine circulation accounts for the paucity of detailed and generally accepted data.

The methods to be discussed can be considered under two general categories: those that determine or estimate total uterine blood flow and those that estimate only maternal placental blood flow. In the case of total uterine blood flow, either total inflow or outflow from all sources is determined without distinction to its distribution to placenta, myometrium, endometrium, or cervix. Those techniques that estimate maternal placental blood flow choose to ignore other uterine distribution sites.

An important consideration for any student of uterine circulatory physiology is the units in which the value is expressed. This of course is quite relevant in pregnancy, when the weight of the uterus and its contents is continuously changing, as is the total weight of the organism, as a function of duration of gestation. Because the clinically applicable methods are relatively noninvasive, much of the data are reported per unit volume or as relative changes in some arbitrary units.

2.1. Total Uterine Blood Flow

Assali was able to measure total uterine blood flow using electromagnetic flow transducers placed around a single uterine artery in a few patients undergoing elective termination of pregnancy between 10 and 28 weeks.[8] He assumed that the blood flow would be equal in both uterine arteries and that the contribution of the ovarian arteries would be negligible. These studies stand alone as the only direct data to be accumulated in the human and will probably never be repeated.

The most frequently used method has been some type of indicator dilution technique. In the 1950s a nitrous oxide method that originally had been developed to study cerebral blood flow was applied to the uteroplacental circulation.[9,10] This method also necessitated relatively invasive techniques with arterial and venous outflow blood samples needed. The technique requires what is called a "steady state," which, under the circumstances of a laparotomy, in presumably the supine position, and under general or conduction anesthesia, may be difficult to achieve and probably accounts for the widely variable results. Radioiodinated serum albumin (Risa-131) also has been used as an indicator dilution technique with continuous monitoring of scintillation counts over placental and nonplacental sites.[11] Indicator clearance techniques were first adapted to the study of uterine perfusion in the nonpregnant state using hydrogen gas in the late 1960s.[12]

2.2. Placental Blood Flow

With the availability of isotopes such as indium-113m and xenon-133 and sophisticated counting devices, studies were extended to pregnancy with greater control over variables.[13,14] These isotopes may be administered either intravenously, in which case the subject must hold her breath during the first pass through the lung, or locally into the tissue to be studied, which for the placenta means the intervillous space. There is, of course, some clinical risk associated with direct intervillous space injection. Analysis of available data would indicate that intervillous space perfusion is nonuniform, and therefore the reliability of this method is questionable.

Another radiolabeled clearance technique for quantitating uteroplacental blood flow is via the conversion rate of androstenedione to estradiol.[15] This method requires the intravenous infusion of [4-^{14}C]androstenedione and depends on the unique placental aromatization to [^{14}C]estradiol. Quantitation of this conversion over a specific time course combined with the volume distribution and excretion rate allows calculation of the rate of precursor (androstenedione) being supplied to the placenta.

2.3. Other Methods

Thermistor probes have been placed in the cervix and between the decidua and fetal membranes to estimate the relative changes in local tissue blood flow.[16] This technique depends on heating a thermistor above body temperature and measuring either heat loss or current necessary to maintain a stable temperature. The rate of heat loss is proportional to tissue blood flow and can be qualitated but not quantitated.

Modern technology has in recent years developed sophisticated techniques of Doppler velocimetry. Initially utilized to study cardiac, cerebral, and peripheral vascular disease states, they are now receiving wide attention for the study of uterine and fetal umbilical hemodynamics.[17,18] In essence, either pulsed or continuous-wave Doppler signals are directed at the vessel under study in order to determine the velocity of blood within that vessel during systole and diastole. Ratios relating systolic and diastolic velocities can be calculated; these ratios reflect the resistance to blood flow in the vascular beds being supplied by the vessel under study. The right side of Fig. 1 illustrates the Doppler frequency detected from three different vessels supplying or adjacent to the uterus. Note the relatively high diastolic component when related to systolic peak in the arcuate branches as compared to the external iliac vessel. The higher the diastolic signal, the greater is the blood flow during diastole, and the lower the vascular resistance.

In addition to the systolic/diastolic ratios, volume flow can be calculated if the angle at which the sound waves strike the vessel and the cross-sectional area of the vessel are known. This latter requirement presents a significant problem in a pulsatile arterial system. Problems notwithstanding, Doppler velocimetry holds great promise for the study of uteroplacental hemodynamics throughout normal and pathological pregnancy.

3. Physiology of Uterine Circulation

From the preceding section it is obvious that many different methods have been used to study uteroplacental hemodynamics with widely different applications and thus varying results. In this section, I discuss general physiological principals and behavior relying on consensus opinion rather than presenting exhaustive arguments that may serve to cloud the underlying issues. Methods of study are discussed where they are deemed important.

3.1. Neural Control

The uterus is richly supplied from the sacral plexus with sympathetic and parasympathetic fibers. Much of our information regarding the activity of the autonomic nervous system in the human comes from histological studies that demonstrate a reduction in immunofluorescent transmitter substance and neural elements in pregnancy. Although the majority of our physiological information comes from animal experiments, there are some limited data in the human that tend to confirm experimental animal results.

Several authors have demonstrated reduced stainable transmitter substance and adrenergic neural elements in tissue sections from pregnant uteri as compared to nonpregnant.[19,20] Autonomic agonists such as epinephrine, norepinephrine, and isoproterenol introduced directly into the uterine circulation have much less effect on uterine blood flow in the pregnant than in the nonpregnant animal.[21,22] Direct administration into the uterine circulation allows observation of the response of local uterine receptors unclouded by systemic cardiovascular effects that may alter cardiac output and resistance in other vascular beds. It is essential that one attempt to separate these local responses to a given neural transmitter from the systemic effects that may be altering distribution of blood flow and supersede the local uterine response. Indeed, it appears from *in vitro* studies that vessels from several different vascular beds have reduced sensitivity to autonomic agonists during pregnancy.[23]

Most investigators have been unable to demonstrate a uterine vasodilatory effect from local β-adrenergic stimulation in the normal pregnant animal.[24,25] One must keep in mind that the uterine circulation in normal pregnancy is probably near maximally dilated and therefore has limited margin for further dilatory response.

There is a paucity of studies in the human related to the response of the uterine circulation to various vasoactive substances in the nonpregnant uterus. The thermistor method of determining changes in uterine blood flow demonstrated an increased flow with the systemic administration of terbutaline, a selective β_2 agonist that has been demonstrated to have some systemic β_1 effects.[16] Increased blood flow was also demonstrated in a 16-week pregnancy using the same agent.[16] Reduced blood flow was produced in the nonpregnant uterus using intravenous vasopressin. These observations must be interpreted in light of our previous warning regarding systemic responses to vasoactive substances. Epinephrine mixed with local anesthesia and administered into the epidural space did not cause any significant change[26] in intervillous blood flow as determined by the ^{133}Xe technique (see Chapter 19). It is important to recognize that this study introduced variables that may have tended to maintain or increase placental flow.

3.2. Autoregulation

The question of autoregulation of the uteroplacental circulation has for many years been a hotly debated subject. Autoregulation is the ability of an organ or tissue to alter its vascular resistance under the stimulus of increased or decreased perfusion in order to maintain a constant or stable blood flow. This phenomenon is best represented by the kidney, which when subjected to decreased perfusion will vasodilate, lowering vascular resistance, and thereby return blood flow toward "normal." Teleologically, autoregulation of uteroplacental circulation is an attractive hypothesis. On the other hand, there is very little evidence for it being an active mechanism in experimental animals and virtually no evidence or data in the human. With a variety of measures used to increase or decrease uterine perfusion, including pharmacological, pathophysiological (such as renal hypertension), and mechanical methods, there have been no consistent data to demonstrate that the uterus has the ability to alter its own vascular resistance.[27-29]

The very limited data available in the human are also interpreted as not demonstrating autoregulation of the placental components of blood flow. Two studies using the ^{133}Xe clearance method have demonstrated a maintenance of the myometrial component of uterine blood flow under the stimulus or stress of hypertension and the supine position.[30,31] On the other hand, intervillous perfusion is reportedly reduced.[32,33] The newer Doppler methods for monitoring uteroplacental vascular resistance offer an opportunity to follow this parameter longitudinally and noninvasively over a variety of physiological and pharmacological stresses. Early reports indicate that patients with hypertension have reduced diastolic blood velocity, which is interpreted to mean that the uteroplacental vascular bed is increasing its vascular resistance and reducing blood flow. Because the blood flow is reduced rather than maintained, this cannot be called autoregulation. All in all, autoregulation does not appear to be an active mechanism for maintaining uteroplacental perfusion in the animal species studied or with the limited data available in the human.

3.3. Labor and Delivery

With a good working knowledge of the uterine vascular anatomy and an understanding of labor mechanics, one can appreciate the effects of uterine contractions on uteroplacental circulation. Because the arcuate and radial arteries course within the myometrium, they and their accompanying veins are immediately and directly affected by changes in uterine tone and uterine contractions. As the intramyometrial pressure increases, compression of arcuate and radial veins will occur; complete occlusion or compression follows as the contraction reaches its maximum or exceeds intraluminal vascular pressure. The contraction does squeeze out some of the blood already within the uterine venous bed. These capacitance vessels with their thin walls, large volume, and low pressure would be quite sensitive to external forces. On the arterial side, which has thicker walls and higher transmural pressure, it would take a more forceful compression or contraction to influence blood flow. During maximal uterine contractions the intramyometrial pressure may very well exceed the intraarterial pressure of these vessels, resulting in compression and reduced flow.

Studies done in pregnant sheep using the electromagnetic flow technique have demonstrated a fall in uterine blood flow that is roughly proportional to the magnitude of the contraction, whether spontaneous or induced by oxytocin.[34,35] Since in these studies perfusion pressure did not change during uterine contractions, it was concluded that the fall in uterine blood flow was secondary to increased vascular resistance within the uterus. There were no observed changes in central hemodynamics or other regional blood flows. Cineradioangiographic visualization in the pregnant rhesus monkey demonstrated that blood flow through the uterus and intervillous space is greatly reduced or abolished when the uterus contracts.[36] It also appeared that the flow through individual spiral arteries varied independent of uterine contractions. The important variables were thought to be intrauterine pressure, frequency and duration of contractions, and the contour of the contraction waveform as measured by intrauterine catheter. Arteriography carried out at term in women with malformed fetuses essentially confirmed the observations in the rhesus.[37] It was also observed in the human that the degree of blood flow reduction during uterine contractions varied within different areas of the uterine wall.

Borotanek and Hendricks studied 16 term pregnant women in either spontaneous or oxytocin-induced labor with a thermistor implanted in the cervix.[38] They also monitored intrauterine pressure and femoral arterial and venous pressure. An initial decrease in uterine blood flow was noted prior to any sequence of contractions; the etiology was unknown, but this may represent a subtle undetected increase in myometrial tone. They also observed that when intrauterine pressure reaches 30 mm Hg, the blood flow begins to decrease further, reaching its lowest point at the peak of intrauterine pressure or shortly thereafter. Blood flow then began to return toward precontraction levels as the contraction abated. Recovery was incomplete with hypertonic or hyperactive contractions. Supine or lateral positioning did not appear to affect the response to uterine contractions. They noted an increase in femoral arterial and venous pressure with contractions.

Uterine artery blood velocity has been studied with the pulsed Doppler technique in laboring patients.[39] Velocity was measured at two locations on the uterine artery and in a vessel that was probably an arcuate artery. The lower portion of the uterine artery had a 20–40% decreased mean velocity during contractions, whereas the upper portion had a 50–60% decrease. The artery traversing the myometrium had 100% reduction in velocity during contractions. In similar studies arcuate artery blood velocity was correlated with intrauterine pressure. A near linear relationship up to a 60% reduction in velocity was observed as intrauterine pressure increased from 0 to 50 mm Hg.[40] Velocity was also noted to decrease by an average of 35% during prelabor contractions. It is not stated whether different locations were studied, but from what we have learned from the angiographic techniques different intramyometrial arteries are probably being affected to varying degrees.

Expulsion of the fetus reduces total uterine cavity volume by over 50% and markedly increases myometrial tone, which in turn must significantly reduce uterine vascular capacitance. These changes would result in increased vascular resistance, reduced blood flow, and reduced venous volume. With placental separation and passage shortly following delivery in the human, the hemodynamic changes are even further exaggerated. We do know that acute volume shifts are evident in women undergoing central hemodynamic monitoring at the time of delivery. In the sheep, which may not be an ideal model because the placenta does not separate and pass for 1 to 2 days after parturition, Assali reported that uterine blood flow did not decrease significantly until after passage of the placenta.[33]

The central hemodynamic changes post partum have been reasonably well studied in the 6 to 12 weeks following delivery. We know that the pregnancy-induced changes in cardiac output have essentially returned to the nonpregnant state by about 6 weeks. It is also well recognized that uterine involution is not complete until some time after 6 weeks and that the rate of involution is increased by lactation. Unfortunately, the circulatory and vascular involution of the human uterus has not been studied to any degree, and we therefore only assume the sequence of events. This is certainly an area that would be of great interest for future investigation.

3.4. Hormonal Effects

The effects of estrogen and progesterone on uterine circulation have intrigued investigators for many years. The question of what role they play in the circulatory and hemodynamic changes of pregnancy has been investigated primarily in research animals. Several investigators have demonstrated that when 17β-estradiol is administered to oophorectomized nonpregnant or pregnant sheep, there is a significant increase in uterine blood flow following a latent period of 20 to 30 min.[40–42] Constant infusions of estrogen or repeated bolus doses have demonstrated tachyphylaxis. The mechanism for this profound vasodilatation has been extensively studied. Although it is obvious that some vasodilator substance is either released or induced by the estrogen, the specific reaction is unknown. Such diverse mechanisms as the autonomic nervous system and histamine have been excluded by the use of receptor specific blocking agents and both H_1 and H_2 blockers.[42,43] In addition, it has been demonstrated that exogenous estrogens do not change cardiac output or alter other major regional blood flows. On the other hand, cycloheximide, an inhibitor of protein synthesis, will completely abolish the uterine circulatory response to estrogen.[44]

It is of interest that serum or urine estriol concentrations have been used to monitor fetal well-being in high-risk pregnancies during the third trimester. A fall in estriol concentration is interpreted as indicating fetal jeopardy, possibly related to either reduced uterine blood flow or fetal conversion of precursors. One may question whether the decreased estriol concentration is an effect of reduced uterine blood flow or whether it was the cause. However, the animal studies elucidated above strongly support the conclusion that estrogen concentration does not play a significant role in the maintenance of uterine blood flow during pregnancy. Studies in the pregnant baboon further confirm this conclusion by demonstrating that the estradiol production rate by the fetoplacental unit is dependent on uterine perfusion rather than being the factor that maintains perfusion.[45]

Progesterone also has been studied in oophorectomized sheep by several investigators.[46-48] It is clear that progesterone alone has no appreciable effect on uterine blood flow.[46] In contrast, when given with or before estrogen, there is a 25–30% attenuation of the estrogen response.

There are essentially no observations in humans to support these experimental animal data. It is safe to say that at the present time there is no evidence that the development or maintenance of uteroplacental perfusion during pregnancy is dependent on the major gestational hormones, estrogen and progesterone.

In recent years attention has turned to the prostaglandins as possible regulators of uterine blood flow during pregnancy. Several animal species have been demonstrated to have uterine production of large quantities of prostaglandins in both the pregnant and non pregnant state.[49-52] The hypothesis has evolved that the maintenance of vasodilation during pregnancy is dependent on a balance between vasodilator and vasoconstrictor prostaglandins. An inadequate uteroplacental production of vasodilator prostaglandins results in a variety of pregnancy-related pathology.[53,54] In addition to their direct effect on vascular smooth muscle, they may modify the response to neurotransmitters and other circulating vasoactive agents such as angiotensin. Because some of the prostaglandins have a direct effect on myometrial tone and contractility, they may also exert hemodynamic effects via this mechanism (see Section 3.3).

The experimental evidence for the effects of the various prostaglandins on uteroplacental blood flow is derived exclusively from animal studies. For these animal studies the radioactive microsphere method has been used to determine the distribution of blood flow to the various components of the uteroplacental vascular bed.[55] Prostacyclin (PGI_2), when delivered into the uterine circulation, has been demonstrated to dilate myometrial[57] but not placental vessels.[56] Prostaglandin (PG) E_2 is reported to be a weak uterine vasodilator in the nonpregnant animal but produces a dose-related reduction in uterine blood flow in pregnant animals related to an increase in uterine tonus.[57] Prostaglandin D_2 produced significant increases in uterine blood flow in both nonpregnant and pregnant sheep.[57]

Other authors have used a different approach to study the role of prostaglandins on uterine blood flow by administering agents that block prostaglandin production.[58,59] Although there is some difference of opinion, it appears that except for a transient vasoconstriction with the acute administration of indomethacin, uteroplacental vascular resistance does not significantly change despite reduced plasma prostaglandin levels.[60] One must keep in mind that prostaglandin synthesis inhibitors block production of both vasoconstrictor and dilator prostaglandins.

4. Pathophysiology

4.1. Growth Retardation

Uteroplacental blood flow is one of the most important rate-limiting factors for normal fetal growth and outcome. Many disease states have a direct or indirect effect on uterine blood flow and therefore have an impact on fetal well-being. Although there are obviously many factors that influence fetal growth and pregnancy outcome, decreased uteroplacental blood flow is recognized as being an important variable. Several investigators using a variety of techniques have demonstrated either reduced blood flow or increased uterine vascular resistance in women with growth-retarded infants when compared to normal.[61-63]

4.2. Effects of Position

It has been recognized for many years that the supine position has major impact on hemodynamics in the third trimester of pregnancy. Inferior vena caval compression to a degree that reduces cardiac output most assuredly has an adverse effect on uterine blood flow. Kauppila and colleagues studied 12 normotensive near-term pregnant women in both the supine and the 45° left tilt positions by the ^{133}Xe washout method.[31] They noted that there was a significant decrease in intervillous blood flow when the patient was moved into the supine position, although there was essentially no change in myometrial blood flow. Other investigators also have reported a 17% decrease in total uteroplacental blood flow in the supine position.[64]

Bed rest in the lateral position is a frequently used therapeutic modality in conditions such as hypertension and intrauterine growth retardation. It is of interest that this therapy frequently ameliorates of many of the findings associated with intrauterine growth retardation. This may be further evidence for the positive effect of lateral recumbency on uterine blood flow.

4.3. Hypertension

Both gestational hypertension and chronic hypertension may have an impact on uteroplacental blood flow and fetal growth. These conditions have a wide spectrum, from very mild elevations of blood pressure to very severe hypertension and proteinuria. The severity tends to be related to the degree of effect one sees on uteroplacental perfusion and fetal growth. The pathophysiology of severe gestational proteinuric hypertension is characterized by generalized vasospasm, reduced intravascular volume, and placental infarcts. Two specific lesions of the uteroplacental vascular bed have been described in hypertensive pregnant women.

The first pathognomonic change described was that of hyperplasia, fibrinoid necrosis, and accumulation of lipid-laden foam cells in the intima and vessel wall of the spiral arteries. Because of the lipid deposits the lesion has been called "acute artherosis."[65] The second lesion or change described is a failure of the second wave of endovascular trophoblast migration to occur into the myometrial segments of the spiral arteries.[4] This is usually detected from about the 16th week in normal pregnancies, whereas in pregnancies complicated by hypertension, it is present to lesser degrees and appears to be related to the severity of disease.

These pathological changes may very well account for the presence of placental infarcts and reduced uteroplacental perfusion reported in hypertensive diseases of pregnancy. There are limited human studies that report up to 60% reduction in intervillous blood flow in the presence of maternal hypertension.[66,67] Doppler velocimetry studies also confirm the presence of high vascular resistance and low diastolic flow in many hypertensive pregnant women.[68]

4.4. Smoking

Nicotine absorbed from cigarette smoking produces a systemic sympathomimetic effect characterized by elevated heart rate and blood pressure. The most obvious effect of cigarette smoking during pregnancy is the delivery of infants who are of lower birth weight than expected.[69,70] The question of how smoking causes this consistent observation remains unclear. Some authors have suggested that the effect is primarily related to reduced prenatal weight gain observed in smoking women.[71,72] Others hypothesize that reduced fetal growth is related to the direct fetal effects of nicotine or carbon monoxide. Acute studies in pregnant sheep failed to reveal any significant effect on uterine blood flow of either pharmacological doses of nicotine or cigarette smoking.[73] Other investigators using a chronically prepared sheep model to determine uterine blood flow demonstrated a significant decrease in uterine blood flow with a constant infusion of 0.025 mg/kg per min of nicotine.[74] Human intervillous blood flow has also been reported to decrease about 20% immediately after smoking a single cigarette.[75,76] Blood flow recovered to presmoking levels within 15 min. Interestingly, this reduced intervillous flow was not observed in a group of hypertensive women. The authors hypothesized that because intervillous blood flow is already reduced in the hypertensive patient secondary to increased sympathetic tone, the added effect of nicotine has little or no further constriction.[76]

4.5. Recreational Drugs

Use of illicit drugs has progressively increased in certain segments of our population. Some of these drugs are suspected to have a significant effect on uteroplacental perfusion. The most frequently abused substance, cocaine, appears to carry the highest perinatal risk and has the strongest evidence for decreasing uteroplacental blood flow. Cocaine stimulates the central nervous system from above downward, causing sympathetic stimulation, which results in increased heart rate and elevated blood pressure. Vasoconstriction is a prominent response to cocaine. Several clinical studies have reported an increased incidence of cerebral vascular accident and placental abruption associated with cocaine use.[77,78] Woods demonstrated a 52% increase in uterine vascular resistance in pregnant sheep at a relatively low dose of intravenous cocaine.[79] Preliminary data indicate that reduced diastolic velocity (increased vascular resistance) is present in the uterine arteries of women who use cocaine (C. Marshall, personal communication).

Although heroin has been shown to be associated with intrauterine growth retardation, there is no current evidence that this effect is related to a change in uteroplacental perfusion. Overall, the life style of patients who use illicit drugs is always a confounding variable.

4.6. Exercise

Measurement of uterine blood flow during and following exercise has been carried out in several different animal models.[80–82] All of these investigators have reported a decrease, which appears to be linearly related to the strenuousness of the exercise.[81] The presumed mechanism for this observation is a redistribution of cardiac output and an increase in endogenous catecholamines, both of which could result in reduced uteroplacental blood flow. Evidence for a significant effect in the human is more circumstantial. Steegers and his co-workers failed to demonstrate any significant change in uterine arterial flow velocity waveforms immediately following mild treadmill exercise during the last trimester.[83] These authors also did not observe any fetal heart rate changes, but others have reported both fetal bradycardia and tachycardia.[84] The last bit of indirect evidence that moderate to strenuous exercise may have some effect on uteroplacental perfusion is that women who exercise vigorously during pregnancy have infants who are lighter in weight than their sedentary controls.[85]

5. Conclusions

Uterine vascular anatomy undergoes remarkable compensatory changes in order to accommodate the nearly tenfold increase in blood flow during pregnancy. The mechanisms and control of these changes are poorly understood and a fruitful area for additional research. The quest to quantitate uteroplacental perfusion has extended over several decades and has included techniques and degrees of invasiveness that are not acceptable by current standards. The use of Doppler velocimetry promises to greatly expand our knowledge of the physiology and pathophysiology of uteroplacental hemodynamics.

It appears that in "normal" pregnancy the placental vascular bed is maximally dilated and not capable of a further reduction in vascular resistance. This concept is important as we consider the effect on uteroplacental perfusion of a wide variety of physiological and pathological stimuli and stress. I have attempted to review the most frequent and significant, but by no means have I included all of the known effectors of uteroplacental blood flow.

I believe that both the basic scientist and the clinician can, with a good working knowledge of uterine vascular anatomy and physiology, hypothesize the effect or response to a particular stress or condition. But only further research will let us elucidate the responses of this multifactorial system.

References

1. Fernstrom, I., 1955, Arteriography of the uterine artery, *Acta Radiol. [Suppl.]* **122**:7–128.
2. Itskovitz, Lindenbaum, S., and Brandes, M., 1980, Arterial anastomoses in the pregnant human uterus, *Obstet. Gynecol.* **55**:67–71.
2a. Burchell, R. C., Creed, F., Rasoulpour, M., and Whitcomb, M., 1978, Vascular anatomy of the human uterus and pregnancy wastage, *Br. J. Obstet. Gynecol.* **85**:698–706.
3. Brosens, I., Robertson, W. B., and Dixon, H. G., 1967, The physiological response of the vessels of the placental bed to normal pregnancy, *J. Pathol. Bacteriol.* **93**:569–579.
4. Brosens, I., Robertson, W. B., and Dixon, H. G., 1972, The role of the spiral arteries in the pathogenesis of preeclampsia, *Obstet. Gynecol. Annu.* **1**:177–191.
5. Brosens, I., Dixon, H. G., and Robertson, W. B., 1977, Fetal growth retardation and the arteries of the placental bed, *Br. J. Obstet. Gynecol.* **84**:656–663.
6. Fuller, E. O., Galletti, P. M., and Takeuchi, T., 1975, Major and collateral components of blood flow to pregnant sheep uterus, *Am. J. Physiol.* **229**:279–285.
7. Reynolds, S. R. M., 1963, Maternal blood flow in the uterus and placenta, in: *Handbook of Physiology, Circulation*, Volume II (W. F. Hamilton and P. Dow, eds.), American Physiological Society/Waverly Press, Baltimore, pp. 1585–1618.
8. Assali, N. S., Rauramo, L., and Peltonen, T., 1960, Measurement of uterine blood flow and uterine metabolism. VIII. Uterine and fetal blood flow and oxygen consumption in early human pregnancy, *Am. J. Obstet. Gynecol.* **79**:86–98.
9. Assali, N. S., 1953, Measurement of uterine blood flow and uterine metabolism I. Critical review of methods, *Am. J. Obstet. Gynecol.* **66**:3–10.
10. Metalfe, J., Rommey, S. L., Ramsey, L. H., Reid, D. E., and Burwell, S. C., 1955, Estimation of uterine blood flow in normal human pregnancy at term, *J. Clin. Invest.* **34**:1632–1638.
11. Scheffs, J., Vasicka, A., Li, C., Solomon, N., and Siler, W., 1971, Uterine blood flow during labor, *Obstet. Gynecol.* **38**:15–24.
12. Klingenberg, I., and Aukland, K., 1969, Measurement of human uterine cervical blood flow by local hydrogen gas clearance, *Acta Obstet. Gynecol. Scand.* **48**:455–469.
13. Lippert, T. H., Cloeren, S. E., Fridrich, R., and Hinselmann, M., 1974, Continuous recording of uteroplacental blood volume in the human, *Acta Obstet. Gynecol. Scand.* **52**:131–134.
14. Lippert, T. H., Fridrich, R., Hindelman, P., Kubli, F., and Ruettgers, H., 1973, Critical assessment of xenon washout technique for measurement of placental blood flow, *Gynecol. Invest.* **4**:14–23.
15. Edman, C. D., Toofanian, A., MacDonald, P. C., and Gant, N. F., 1981, Placental clearance rate of maternal plasma androstenedione through placental estriadiol formation. An indirect method of assessing uteroplacental blood flow, *Am. J. Obstet. Gynecol.* **191**:1029–1037.
16. Akerlund, M., Bengtsson, L. P., and Caster, A. M., 1975, A technique for monitoring endometrial or decidual blood flow with an intra-uterine thermistor probe, *Acta Obstet. Gynecol. Scand.* **54**:469–477.
17. Janbu, T., Koss, K. S., Nesheim, B. I., and Wesche, J., 1985, Blood velocities in the uterine artery in humans during labor, *Acta Physiol. Scand.* **124**:153–161.

18. Schulman, H., Fleischer, A., Farmakides, G., Bracero, L., Rochelson, B., and Grunfeld, L., 1986, Development of uterine artery compliance in pregnancy as detected by Doppler ultrasound, *Am. J. Obstet. Gynecol.* **155**:1031–1036.

19. Nakanishi, H., McLean, J., Wood, C., and Burnstock, C., 1969, The role of sympathetic nerves in control of the nonpregnant and pregnant human uterus, *J. Reprod. Med.* **2**:20–33.

20. Thorbert, G., Alm, P., Bjorkland, A., Owman, C., and Sjoberg, N., 1979, Adrenergic innervation of the human uterus: Disappearance of the transmitter and transmitter-forming enzymes during pregnancy, *Am. J. Obstet. Gynecol.* **135**:223–226.

21. Anderson, S. G., Still, J. G., and Greiss, F. C., 1977, Differential reactivity of the gravid uterine vasculatures: Effects of norepinephrine, *Am. J. Obstet. Gynecol.* **129**:293–298.

22. Assali, N. S., Nuwayhid, B., Brinkman, C. R. III, Tabsh, K., Erkkola, R., and Ushioda, E., 1981, Autonomic control of the pelvis circulation: *In vivo* and *in vitro* studies in pregnant and nonpregnant sheep, *Am. J. Obstet. Gynecol.* **141**:873–884.

23. Brinkman, C. R. III, Erkkola, R., Nuwayhid, B., and Assali, N., 1982, Adrenergic vasoconstrictor receptors in the uterine vascular bed, in: *Uterine and Placental Blood Flow* (A. H. Moawad and M. D. Lindheimer, eds.), Masson USA, New York, pp. 113–117.

24. Nuwayhid, B., Erkkola, R., Tabsh, K., Ushiodu, E., Murad, S., Brinkman, C. R. III, and Assali, N. S., 1982, Beta-adrenergic responses in the uterine vascular bed, in: *Uterine and Placental Blood Flow* (A. H. Moawad and M. D. Lindheimer, eds.), Masson USA, New York, pp. 119–126.

25. Brennen, S. C., McLaughlin, M. K., and Chez, R. A., 1977, Effects of prolonged infusion of beta-adrenergic agonists on uterine and umbilical blood flow in pregnant sheep, *Am. J. Obstet. Gynecol.* **128**:709–715.

26. Albright, G. A., Jouppila, R., Hollman, A. I., Jouppila, P., Vierola, H., and Koivula, A., 1981, Epinephrine does not alter human intervillous blood flow during epidural anesthesia, *Anesthesiology* **54**:131–135.

27. Brinkman, C. R. III, Nuwayhid, B., and Assali, N. S., 1975, Renal hypertension and pregnancy in the sheep. I. Behavior of uteroplacental vasomotor tone during mild hypertension, *Am. J. Obstet. Gynecol.* **121**:931–937.

28. Brinkman, C. R. III, 1981, Experimental renal hypertension in pregnant sheep. II. Response to a one-kidney model, *Am. J. Obstet. Gynecol.* **141**:895–904.

29. Clark, K. E., Durnwald, M., and Austin, J. E., 1982, A model for studying chronic reduction in uterine blood flow in pregnant sheep, *Am. J. Physiol.* **242**:H297–301.

30. Jansson, I., 1969, Forearm and myometrial blood flow in toxemia of pregnancy studied by venous occlusion pletysmography and [133]xenon clearance, *Acta Obstet. Gynecol. Scand. [Suppl.]* **8**:33–56.

31. Kauppila, A., Koskinen, M., Puolakka, J., Tuimala, R., and Kuikka, J., 1980, Decreased intervillous and unchanged myometrial blood flow in supine recumbency, *Obstet. Gynecol.* **55**:203–205.

32. Lunell, N. O., Lewander, R., Mamoun, I., Nylund, L., Sarby, S., and Thornstrom, S., 1984, Uteroplacental blood flow in pregnancy induced hypertension, *Scand. J. Clin. Lab. Invest. [Suppl.]* **169**:28–35.

33. Assali, N. S., Dasgupta, K., Kolin, A., and Holms, L., 1958, Measurement of uterine blood flow and uterine metabolism V. Changes during spontaneous and induced labor in unanesthetized pregnant sheep and dogs, *Am. J. Physiol.* **195**:614–620.

34. Assali, N. S., Holm, L., and Parker, H., 1961, Regional blood flow and vascular resistance in response to oxytocin in the pregnant sheep and dogs, *J. Appl. Physiol.* **16**:1087–1092.

35. Ramsey, E. M., Corner, G. W., and Donner, M. W., 1963, Serial and cineradioangiographic visualization of maternal circulation in the primate (hemochorial) placenta, *Am. J. Obstet. Gynecol.* **86**:213–225.

36. Borell, V., Fernstrom, I., Ohlson, L., and Wiqvist, N., 1965, Influence of uterine contractions on the uteroplacental blood flow at term, *Am. J. Obstet. Gynecol.* **93**:44–57.

37. Brotanek, V., Hendricks, C., and Yoshida, T., 1969, Changes in uterine blood flow during uterine contractions, *Am. J. Obstet. Gynecol.* **103**:1108–1116.

38. Janbu, T., Koss, K. S., Nesheim, B. I., and Wesche, J., 1985, Blood velocities in the uterine artery in humans during labor, *Acta Physiol. Scand.* **124**:153–161.

39. Janbu, T., and Nesheim, B. I., 1987, Uterine artery blood velocities during contractions in pregnancy and labor related to intrauterine pressure, *Br. J. Obstet. Gynecol.* **94**:1150–1155.

40. Resnick, R., Killam, A. P., Battaglia, F. C., Makowski, E. L., and Meschia, G., 1974, The stimulation of uterine blood flow by various estrogens, *Endocrinology* **64**:1192–1196.

41. Rosenfeld, C. R., Morris, F. H., Jr., Battaglia, F. C., Makowski, E. L., and Neschia, G., 1976, Effect of estradiol-17β on blood flows to reproductive and nonreproductive tissues in pregnant ewes, *Am. J. Obstet. Gynecol.* **124**:618–629.

42. Nuwayhid, B., Brinkman, C. R. III, Woods, J. R., Jr., Martinek, H., and Assali, N. S., 1975, Effects of estrogens on systemic and regional circulations in normal and renal hypertensive sheep, *Am. J. Obstet. Gynecol.* **123**:495–504.

43. Woods, J. R., Jr., Brinkman, C. R. III, Tyner, J., Martinek, H., and Assali, N. S., 1976, Action of histamine and its receptor blockers on uterine circulation in sheep, *Proc. Soc. Exp. Biol. Med.* **151**:811–816.

44. Killam, A. P., Rosenfeld, C. R., Battaglia, F. C., Makowski, E. L., and Meschia, G., 1973, Effect of estrogens on the uterine blood flow of oophorectomized ewes, *Am. J. Obstet. Gynecol.* **115**:1045–1052.

45. Fritz, M. A., Stanczyk, F. Z., and Novy, M. J., 1986, Maternal estradiol response to alterations in uteroplacental blood flow, *Am. J. Obstet. Gynecol.* **155**:1317–1325.

46. Caton, D., Abrams, R. M., Clapp, J. F., and Barron, D. H., 1974, The effect of exogenous progesterone on the rate of blood flow of the uterus of ovariectomized sheep, *O. J. Exp. Physiol.* **59**:225–231.

47. Greiss, F. C., and Anderson, S. G., 1970, Effect of ovarian hormones on the uterine vascular bed, *Am. J. Obstet. Gynecol.* **107**:829–836.

48. Resnik, R., Brink, G. W., and Plumer, M. H., 1977, The effect of progesterone on estrogen-induced uterine blood flow, *Am. J. Obstet. Gynecol.* **128**:251–254.

49. Ryan, M. J., Clark, K. E., VanOrden, D. E., Farley, D. F., Edvinsson, L., Sjoberg, N. O., VanOrden, L. S. III, and Brody, M. J., 1974, Role of prostaglandins in estrogen-induced uterine hyperemia, *Prostaglandins* **5**:257–268.

50. Blatchley, F. R., and Poyser, N. L., 1974, The effects of estrogen and progesterone on the release of prostaglandins from the uterus of the ovariectomized guinea pig, *J. Reprod. Fertil.* **40**:205–209.

51. Williams, K. I., Dembinska-Kiec., A., Zmuda, A., and Gryglewski, R. J., 1978, Prostacyclin formation by myometrial and decidual fractions of the pregnant rat uterus, *Prostaglandins* **15**:343–349.

52. Venuto, R. C., O'Dorisio, T., Stein, J. H., and Ferris, T. F., 1975, Uterine prostaglandin E secretion and uterine blood flow in the pregnant rabbit, *J. Clin. Invest.* **55**:193–197.

53. Goodman, R. P., Killam, A. P., Brash, A. R., and Branch, R. A., 1982, Prostacyclin production during pregnancy comparison of production during normal pregnancy and pregnancy complicated by hypertension, *Am. J. Obstet. Gynecol.* **142**:817–822.

54. Jogee, M., Myatt, L., and Elder, M. G., 1983, Decreased prostacyclin production by placental cells in culture from pregnancies complicated by fetal growth retardation, *Br. J. Obstet. Gynecol.* **90**:247–250.

55. Buckberg, G. D., Juck, J. C., Payne, D. B., Hoffman, J. I. E., Archie, J. P., and Fixler, D. E., 1971, Some source of error in measuring regional blood flow with radioactive microspheres, *J. Appl. Physiol.* **31**:598–604.

56. Landauer, M., Phernetton, T. M., Parisi, V. M., Clark, K. E., and Rankin, J. H. G., 1985, Ovine placental vascular response to the local application of prostacyclin, *Am. J. Obstet. Gynecol.* **151**:460–464.

57. Clark, K. E., Stys, G. H., and Austin, J. E., 1982, PGI_2 and PGD_2 vasodilatory effects on the uterine vascular bed, in: *Uterine and Placental Blood Flow* (A. H. Moawad and M. D. Lindheimer, eds.), Masson USA, New York, pp. 153–159.

58. Speroff, L. N., Haning, R. V., and Levin, R. M., 1977, The effect of angiotensin II and indomethacin on uterine artery blood flow in pregnant monkeys, *Obstet. Gynecol.* **50:**611–614.

59. Rankin, J., Bersenbrugge, A., Anderson, D., and Phernetton, T., 1979, Ovine placental vascular responses to indomethacin, *Am. J. Physiol.* **236:**H61–H72.

60. Naden, R. P., Iliya, C. A., Arant, B. S., Jr., Gant, N. F., and Rosenfeld, C. R., 1985, Hemodynamic effects of indomethacin in chronically instrumented pregnant sheep, *Am. J. Obstet. Gynecol.* **151:**484–493.

61. Lunell, N. O., Garby, B., Lewander, B., and Nuland, L., 1979, Comparison of uteroplacental blood flow in normal and in intrauterine growth-retarded pregnancy, *Gynecol. Obstet. Invest.* **10:**106–118.

62. Wolfson, M. R., and Blake, K. C. H., 1975, Isotopic method using xenon-133 for assessing placental blood flow and for detecting light for date babies, *South Afr. Med. J.* **49:**117–119.

63. Campbell, S., Griffin, D. R., Pearce, J. M., Diaz Recasens, J., Cohen-Overbeck, T. E., Willson, K., and Teague, M. J., 1983, New Doppler technique for assessing uteroplacental blood flow, *Lancet* **1:**675–677.

64. Suonio, S., Simpanen, A. L., Olkkonem, H., and Haring, P., 1976, Effect of the lateral recumbent position compared with the supine and upright positions on placental blood flow in normal late pregnancy, *Ann. Clin. Res.* **8:**22–26.

65. Zeek, P. M., and Assali, N. S., 1950, Vascular changes in the decidua associated with eclamptogenic toxemia of pregnancy, *Am. J. Clin. Pathol.* **20:**1099–1109.

66. Browne, J. C. M., and Veall, N., 1953, The maternal placental blood flow in normotensive and hypertensive women, *J. Obstet. Gynecol. Br. Emp.* **60:**141–147.

67. Dixon, H. G., Browne, J. C. M., and Davey, D. A., 1963, Choriodecidual and myometrial blood flow, *Lancet* **1:**369–373.

68. Fleischer, A., Schulman, H., Farmakides, G., Bracero, L., Grunfeld, L., Rochelson, B., and Koenigsberg, M., 1986, Uterine artery doppler velocimetry in pregnant women with hypertension, *Am. J. Obstet. Gynecol.* **154:**806–813.

69. Lowe, C. R., 1959, Effects of mothers smoking on birthweight of their children, *Br. Med. J.* **2:**673–676.

70. Murphy, J. F., Drumm, J. E., Daly, L., and Mulcahy, R., 1980, The effect of maternal cigarette smoking on fetal birthweight and on growth of fetal biparietal diameter, *Br. J. Obstet. Gynecol.* **87:**457–466.

71. Miller, H. C., Hassanein, K., and Hensleigh, P. A., 1976, Fetal growth retardation in relation to maternal smoking and weight gain in pregnancy, *Am. J. Obstet. Gynecol.* **125:**55–60.

72. Davies, D. P., Gray, O. P., Ellwood, P. C., and Abernethy, M., 1976, Cigarette smoking in pregnancy: Associations with maternal weight gain and fetal growth, *Lancet* **1:**385–387.

73. Kirschbaum, T. H., Dilts, P. V., and Brinkman, C. R. III, 1970, Some acute effects of smoking in sheep and their fetuses, *Obstet. Gynecol.* **35:**527–536.

74. Resnik, R., Brink, G. W., and Wilkes, M., 1979, Catecholamine-mediated reduction in uterine blood flow after nicotine infusion in the pregnant ewe, *J. Clin. Invest.* **63:**1133–1136.

75. Lehtovirta, P., and Forss, M., 1978, The acute effect of smoking on intervillous blood flow in the placenta, *Br. J. Obstet. Gynecol.* **85:**729–731.

76. Lehtovirta, P., and Forss, M., 1980, The acute effect of smoking on uteroplacental blood flow in normotensive and hypertensive pregnancy, *Int. J. Gynaecol. Obstet.* **18:**208–211.

77. Lichtenfeld, P. J., Rubin, D. B., and Feldman, R. S., 1984, Subarachnoid hemorrhage precipitated by cocaine snorting, *Arch. Neurol.* **41:**223–224.

78. Acker, D., Sachs, B. P., Tracey, K. J., and Wise, W. E., 1983, Abruptio placenta associated with cocaine use, *Am. J. Obstet. Gynecol.* **146:**220–221.

79. Woods, J. R., Jr., Plessinger, M. A., and Clark, K. E., 1987, Effect of cocaine on uterine blood flow and fetal oxygenation, *J.A.M.A.* **257:**957–961.

80. Chandler, K. D., and Bell, A. W., 1981, Effect of maternal exercise on fetal and maternal respiration and nutrient metabolism in the pregnant ewe, *J. Dev. Physiol.* **3:**161–176.

81. Lotgering, F. K., Gilbert, R. D., and Longo, L. D., 1983, Exercise response in pregnant sheep: Oxygen consumption, uterine blood flow and blood volume, *J. Appl. Physiol.* **55:**834–841.

82. Hohimer, A. R., Bissonnette, J. M., Metcalfe, J., and McKlan, T. A., 1984, Effect of exercise on uterine blood flow in the pregnant pygmy goat, *Am. J. Physiol.* **246:**H 207–212.

83. Steegers, E. A. P., Buunk, G., Brinkhorst, R. A., Jongsma, H. W., Wijn, P. F. F., and Hein, P. R., 1988, The influence of maternal exercise on the uteroplacental vascular bed resistance and the fetal heart rate during normal pregnancy, *Eur. J. Obstet. Gynecol. Reprod. Biol.* **27:**21–26.

85. Artal, R., Rutheford, S., Romem, Y., Kammula, R. K., Dorey, F. J., and Wiswell, R. A., 1986, Fetal heart rate response to maternal exercise, *Am. J. Obstet. Gynecol.* **155:**729–733.

86. Clapp, J. F., and Dickstein, S., 1984, Endurance exercise and pregnancy outcome, *Med. Sci. Sports Exercise* **16:**556–562.

19

Effects of Obstetric Analgesia and Anesthesia on Uterine Activity and Uteroplacental Blood Flow

Kenneth A. Conklin

1. Introduction

Obstetric analgesia and anesthesia may alter uterine activity and/or uterine blood flow by a direct effect of the drug on the myometrium or uterine vasculature or indirectly by alteration of maternal physiology. A change in uterine activity can be objectively documented by a change in myometrial tone or in the frequency and intensity of uterine contractions or inferred by alteration of the progress of labor, i.e., the rate of cervical dilatation or the duration of labor. Anesthetic effects on uterine blood flow can be documented by direct (e.g., flow probe) or indirect (e.g., isotope washout) measurement of uterine blood flow or inferred if the fetal heart rate pattern or fetal blood gases change in a manner characteristic of uteroplacental insufficiency (i.e., reduced uterine blood flow). Additional information can be obtained by investigating the direct effects of an analgesic or anesthetic agent *in vitro* using myometrial or uterine arterial tissue preparations.

The *in vivo* effects of obstetric analgesia and anesthesia, as mentioned above, may occur through maternal physiological changes in addition to the direct effects of the pharmacological agents used. In fact, for many agents and techniques the physiological changes entirely account for the changes in uterine activity and uterine blood flow that are observed. Those physiological parameters of importance are discussed individually.

Kenneth A. Conklin • Department of Anesthesiology, School of Medicine, University of California Los Angeles, Los Angeles, California 90024.

1.1. Pressure–Flow Relationship of the Uterine Vascular Bed

Uterine blood flow (UBF) may be defined as follows:

$$UBF = UPP/R = (UAP - UVP)/(R_i + R_e)$$

where uterine perfusion pressure (UPP) is uterine arterial pressure (UAP) minus uterine venous pressure (UVP), R_i is intrinsic vascular resistance (vascular smooth muscle tone), and R_e is extrinsic vascular resistance (intrauterine pressure).[1] The pressure–flow relationship is nearly linear (as measured in sheep), with the pressure intercept at about 25 mm Hg.[2] Vascular resistance is minimal, vasodilatation cannot occur, and the uterus is unable to maintain blood flow when perfusion pressure declines.[1,2] Thus, a drop in uterine arterial pressure or an increase in uterine venous pressure will decrease uterine blood flow; this has been demonstrated experimentally in sheep.[3] In humans it occurs as a result of compression of the aorta and inferior vena cava by the gravid uterus when the patient assumes the supine position (aortocaval compression,[4,5] which increases uterine venous pressure[6,7] and decreases uterine arterial pressure[8,9]). An increase in extrinsic vascular resistance also reduces uterine blood flow, which has been demonstrated in animal models[10–13] and in humans[4,14–16] during uterine contractions. The reduction of uterine blood flow is proportional to the increase in extrinsic vascular resistance,[10,13–16] although the increase in uterine venous pressure[17] and decrease in uterine arterial pressure[4] during contractions also contribute. Additionally, an increase in intrinsic vascular resistance also reduces blood flow (see Section 1.2).

The effect of reducing uterine blood flow by changes in the pressure–flow relationship is reflected in the well-being of the fetus and neonate. For example, fetal distress, as evidenced by changes in fetal heart rate, occurs when maternal systolic blood pressure drops below 70 to 100 mm Hg secondary to regional anesthesia or hemorrhage.[18–21] However, the effect of hemorrhage on uterine blood flow is, in part, related to elevated intrinsic vascular resistance from increased sympathetic nervous system activity.[22] The reduced uterine blood flow caused by aortocaval compression can also result in signs of fetal distress[23,24] and neonatal depression.[25–27] Furthermore, uterine contractions reduce uterine blood flow and fetal oxygen availability secondary to increased extrinsic vascular resistance.[28] Finally, reduced uterine blood flow in the supine position[29] or resulting from hypotension associated with anesthesia[30] impairs uterine activity, which is likely caused by a decrease in myometrial oxygenation.

1.2. Adrenergic Influences on Uterine Blood Flow and Uterine Activity

The uterine vasculature responds to physiological and pharmacological influences in a manner consistent with a vascular bed lacking sympathetic nervous system innervation (i.e., postganglionic sympathetic nervous system nerve terminals), possessing α-adrenergic receptors, but lacking β_2-adrenergic receptors (Table I).

This is supported by the following data. The pressor activity of ephedrine results primarily from release of norepinephrine from sympathetic nervous system postganglionic nerve terminals,[31] but it does not directly affect α receptors. When hypotension occurs, administration of ephedrine increases uterine blood flow by elevation of uterine perfusion pressure and does not cause uterine vasoconstriction or reduce uterine blood flow.[32–35] In contrast to the effect of ephedrine, the direct-acting α-adrenergic agonists methoxamine and phenylephrine increase uterine vascular resistance. Despite increasing uterine perfusion pressure, they reduce uterine blood flow in the normotensive animal and do not elevate the reduced uterine blood flow (and, in fact, may further decrease uterine blood flow) that occurs with hypotension.[32,33,36,37] Conflicting results have been observed with β-adrenergic agonists. In sheep, isoproterenol did not alter uterine blood flow.[38,39] In monkeys metaproterenol, a relatively selective β_2 agonist, appeared to cause vasodilation when studied angiographically.[40] However, several studies have demonstrated, by direct measurement, reduced uterine blood flow in monkeys[41] and sheep[42–45] with a variety of β_2 agonists. These latter results, however, are likely explained by systemic vasodilation resulting in decreased uterine arterial pressure, with minimal vasodilation occurring in the uterine vascular bed. Reflex sympathetic nervous system stimulation in response to the drop in blood pressure, resulting in elevated catecholamine blood levels and uterine vasoconstriction (see below), may also contribute to the decline in uterine blood flow.

The catecholamines epinephrine and norepinephrine possess α-adrenergic activity and produce uterine vasoconstriction that can be demonstrated *in vitro*, using isolated uterine arteries,[46] and *in vivo*.[2,38,39,45,47–52] Bolus intravenous injection of 0.1 to 1.0 μg/kg of epinephrine in sheep and guinea pigs reduces uterine blood flow for 3 to 4 min.[47,48] Continuous infusion of 0.1 to 1.25 μg/kg per min of epinephrine or norepinephrine in pregnant sheep reduces uterine blood flow by up to 60%,[45,49–52] with the response being proportional to the dose administered. The reduction of uterine blood flow that results from infusion of epinephrine and norepinephrine can produce severe asphyxia of the fetus.[53]

The myometrium possesses α- and β_2-adrenergic receptors (Table I). Stimulation of α receptors by infusion of norepinephrine increases uterine tone and the

Table I. Innervation and Adrenergic Receptors of the Uterine Vasculature and Myometrium

	Vasculature	Myometrium	Effect of stimulation
α receptors	Present	Present	Vasoconstriction, uterine stimulation
β_1 receptors	Absent	Absent	—
β_2 receptors	Absent	Present	Uterine relaxation
SNS[a] innervation	Absent	—[b]	—

[a] Sympathetic nervous system.
[b] Evidence for or against innervation is inconclusive.

frequency and intensity of uterine contractions, although some incoordination of uterine contractions may occur.[54-58] The elevated uterine activity induced by norepinephrine is blocked by α-adrenergic blocking agents.[58] Bolus intravenous injection of epinephrine in doses of 30 to 600 μg also increases uterine activity by α-adrenergic receptor stimulation.[59-61] However, infusion of low-dose epinephrine (0.5 to 20 μg/min: approximately 0.01 to 0.35 μg/kg per min) reduces uterine tone and the frequency and intensity of uterine contractions by stimulation of β₂-adrenergic receptors.[54-56,60-63] Since epinephrine-induced relaxation of the rat uterus is associated with elevated adenylate cyclase activity and adenosine 3′,5′-monophosphate (cAMP) levels, the action of epinephrine may be mediated through cAMP.[64,65]

In vitro results using myometrial samples from gravid uteri demonstrate stimulation of myometrial activity with epinephrine and norepinephrine,[66-69] but with concentrations (0.01 to 10 μg/ml) that are far in excess of those achieved with infusion of reasonable doses *in vitro*.[54,70,71] The effect of epinephrine and norepinephrine *in vitro* could be antagonized by α-adrenergic blockade, and although this did not result in depression of myometrial activity with epinephrine as might be expected, isoproterenol (a β₂ agonist) did depress activity, an effect that could be antagonized with the β blocker propranolol.[69]

1.3. The Influence of Pregnancy and Labor on Sympathetic Nervous System Activity

In adult humans and animals at rest, the sympathetic nervous system exerts a minor role in the maintenance of systemic arterial pressure. During pregnancy the contribution of the sympathetic nervous system to the maintenance of vasomotor tone, particularly in the capacitance system, increases progressively and reaches a peak near term.[72-74] Thus, pharmacological sympathectomy by ganglionic blockade or high spinal anesthesia produces minimal blood pressure changes in nonpregnant adult humans (less than a 10% decrease), whereas individuals with normal term pregnancies experience nearly a 50% drop in blood pressure.[72] The postpartum patient responds to sympathetic nervous system blockade as does the nonpregnant individual.[72] Despite the increased reliance of the parturient for blood pressure maintenance by the sympathetic nervous system, most studies demonstrate little change in catecholamine blood levels at term pregnancy[75-77] compared to the nonpregnant state.[78-83]

Plasma catecholamines, however, increase during maternal stress, including the stress of labor. Although epinephrine and norepinephrine blood levels are little changed from rest during the latent phase of the first stage of labor, they rise progressively during the active phase of the first stage and the second stage of labor and peak at the time of delivery.[77,84-88] Most studies demonstrate that epinephrine levels rise to a greater extent than norepinephrine levels,[86-88] which is consistent with the observation that emotional stress elevates epinephrine primarily, whereas physical exertion is more related to increases in norepinephrine.[89] The consequences of elevated catecholamines during labor or periods of maternal stress have been demonstrated in numerous studies. In sheep, maternal stress results in a significant elevation of the norepinephrine blood level and a concomitant reduction of uterine blood flow.[90] In rhesus monkeys and baboons, maternal stress reduces uterine blood flow, decreases fetal arterial oxygenation, and results in fetal heart rate changes consistent with fetal distress.[91-93] Maternal stress in humans results in similar fetal heart rate changes[88,94,95] as well as reduced uterine activity and increased duration of labor, effects that correlate with plasma epinephrine levels.[87,88]

1.4. Neural Reflexes and Uterine Activity

Ferguson demonstrated in postpartum rabbits that dilatation of the cervix, uterus, or vagina augmented uterine activity by reflex liberation of oxytocin.[96,97] Although controversial as a neurohumoral feedback mechanism for labor in humans, elevation of oxytocin starting with the second stage of labor[99-101] (see Chapter 11) is effectively blocked by interrupting the visceral afferent pathways.[97,98] A neural reflex in humans that does not involve oxytocin release has also been suggested.[102]

1.5. Influence of Maternal Respiratory and Acid–Base Status on Uterine Blood Flow

The method of maternal respiration is critical for understanding the following discussion. Spontaneous respiration causes a negative intrathoracic pressure, enhancing venous return to the heart, which increases cardiac output. Conversely, mechanical ventilation causes a positive intrathoracic pressure, which reduces venous return and cardiac output and may result in a decrease of arterial pressure.

In several animal species controlled maternal hyperventilation[103-105] or metabolic alkalosis produced by bicarbonate infusion[106,107] reduces uterine blood flow. The effects of hyperventilation can be attributed to the mechanical effects of positive-pressure ventilation on cardiac output[103,108] and to an increase in uterine vascular resistance that occurs with hypocapnia.[109] Fetal oxygenation or acid–base status is compromised by maternal alkalemia induced by bicarbonate infusion[107,110] and when mechanical hyperventilation is associated with maternal hypocapnia[103,104,110-113] but not when hyperventilation is induced without changes in maternal $P\text{CO}_2$ or pH.[110,111] Hypocapnia contributes to deterioration of the fetus by reducing umbilical blood flow and increasing the affinity of maternal hemoglobin for oxygen[113] in addition to elevating intrinsic vascular resistance. In humans, spontaneous hyperventilation results in little or no adverse effect on fetal oxygenation or acid–base status,[114-117] although maternal hypocapnia associated with positive-pressure ventilation does adversely effect these parameters[113,117-119] and the condition of the newborn.[118,119]

Studies of the effects of hypercapnia suggest that elevation of $P\text{CO}_2$ above normal increases sympathetic nervous system activity and that carbon dioxide itself causes vasodilatation of the uterine vascular bed, which is consistent with the observation that hypocapnia increases intrinsic vascular resistance. Thus, increasing maternal P_{CO_2} from 30 to 60 mm Hg in unanesthetized spontaneously breathing sheep elevates maternal arterial pressure.[120] Uterine blood flow increases since uterine perfusion pressure increases while intrinsic vascular resistance is unchanged; the effect of elevated sympathetic nervous system activity is opposed by the direct effects of carbon dioxide on the uterine vasculature. Above a P_{CO_2} of 60 mm Hg, intrinsic vascular resistance increases and uterine blood flow decreases because of the marked elevation of sympathetic nervous system activity. In mechanically ventilated sheep,[103] increasing maternal P_{CO_2} from 17 to 36 mm Hg also increases uterine blood flow, reflecting the direct effects of carbon dioxide. However, in contrast to the above results, an increase from 36 to 64 mm Hg reduces uterine blood flow, reflecting a further increase in sympathetic nervous system activity and reduced cardiac output from mechanical ventilation.

The effects of hypoxia on uterine blood flow are unclear. Studies have shown that spontaneous ventilation in sheep with 10% or 15% oxygen results in minimal or no change in uterine blood flow.[121–123] However, sheep ventilated with 6% and 12% oxygen exhibited elevation of uterine vascular resistance and reduction of uterine blood flow, although positive-pressure ventilation may have contributed to these results.[124]

2. Local Anesthetics and Regional Analgesia

Local anesthetic agents can directly affect uterine activity and uterine blood flow after (unintentional) intravascular injection or absorption from their site of injection or indirectly because of the consequences of regional (spinal or epidural) anesthesia or analgesia.

2.1. Direct Effects of Local Anesthetics on Uterine Activity and Uterine Blood Flow

Isolated uterine artery segments from gravid human uteri contract when exposed to local anesthetics.[125–127] The response is dose dependent, with maximal contraction being observed at 1 mg/ml of lidocaine.[125] Nearly all samples responded to lidocaine or mepivacaine concentrations above 50 μg/ml,[125–127] but few responded to concentrations of these agents at 20 μg/ml or less.[125] The response of arterial segments from gravid uteri is greater than that of segments from nongravid uteri.[127]

In vitro studies using myometrial strips from gravid uteri demonstrate that resting tone increases in a dose-dependent manner with lidocaine, bupivacaine, and chloroprocaine. However, little effect is observed at local anesthetic concentrations below 50 to 100 μg/ml,[128–130] blood levels far in excess of those (10 to 20 μg/ml) that produce systemic toxicity. Bupivacaine, lidocaine, and chloroprocaine, at or above concentrations of approximately 10, 25, and 50 µg/ml, respectively, also depress uterine contractions, which is primarily a reduction of contraction amplitude with only a small reduction of contraction frequency.[126-128] Although the mechanism to explain elevation of resting tone is unclear, the inhibitory action of local anesthetics on uterine activity likely results from antagonism of calcium entry into the myometrial cells or impairment of release of the ion from an intracellular storage site.[131,132]

In vivo studies also demonstrate effects of local anesthetics on uterine activity and uterine blood flow. Bolus intravenous injection of lidocaine, 20 to 80 mg,[133] mepivacaine, 20 to 80 mg,[133] bupivacaine, 10 to 40 mg,[134] or chloroprocaine, 60 to 240 mg,[135] in the gravid ewe decreases uterine blood flow and increases uterine tone. Intravenous infusion of lidocaine, chloroprocaine, or bupivacaine in gravid ewes also reduced uterine blood flow, with an ED_{25} for these agents of 19.5, 11.5, and 7 µg/ml, respectively.[136] The effect of lidocaine, chloroprocaine, and bupivacaine on myometrial tone was much less than the effect on uterine blood flow.[133-136] In humans receiving oxytocin during labor, lidocaine infusion (for 20 min) resulting in blood levels of 4 to 7 µg/ml augmented uterine activity while the drug was being administered but was followed by a period (approximately 30 min) of depressed activity when the infusion was discontinued.[137] In these studies, however, the concentrations of local anesthetics needed to affect uterine activity and uterine blood flow exceed the usual local anesthetic blood levels obtained during epidural blocks for labor and delivery (maximum of 4 µg/ml).[138,139] When lidocaine is infused so that blood levels of 0.8 to 4.0 µg/ml are attained in gravid ewes, uterine blood flow and uterine activity are unaffected.[138] Thus, well-conducted regional analgesia in the parturient would not be expected to affect uterine blood flow or uterine activity through a direct effect of the local anesthetic that is absorbed from its site of injection.

2.2. Effects of Local Anesthetics Used for Pain Relief on Uterine Activity and Uterine Blood Flow

2.2.1. Regional Analgesia and Uterine Activity

Conflicting results have been reported regarding the effects of epidural (caudal or lumbar) analgesia on uterine activity, cervical dilatation, and the progress of labor. Many early reports of the use of caudal analgesia state that progress was accelerated for the first stage of labor,[139-145] for both the first and second stages of labor,[146] or simply that progress, overall, was improved.[147-149] Caudal analgesia has also been reported to correct incoordinate uterine contractions and improve the progress of labor in situations of prolonged or dysfunctional labor.[150-152] More recent investigations report that use of lumbar epidural analgesia for labor also improves uterine activity,[153,154] accelerates the progress of labor,[154-159] and regularizes uterine contractions in cases of incoordinate uterine activity or prolonged dysfunctional labor.[154,157-163]

Positive effects of epidural analgesia on uterine activity are most likely related to the effect of the block on the epinephrine blood level. Inhibition of uterine activity occurs with infusion of epinephrine in a dose as low as 1 μg/min,[60] which results in a blood level of epinephrine of approximately 200 pg/ml.[70] Epinephrine levels of this magnitude or higher are commonly reported during labor.[76,164–166] Since epidural analgesia reduces the epinephrine blood level[77,86,164–168] by alleviating pain and denervating the adrenal glands (the source of circulating epinephrine),[86,165–167] improvement of uterine activity would be expected following institution of the block and relief of the inhibitory effect of epinephrine[54–56,60–63] on the myometrium. Although several studies have observed no change in uterine activity or in the progress of labor with caudal[169–173] or epidural[174–183] analgesia, the absence of a positive effect in most of these investigations can be explained by a variety of mechanism, as discussed below.

In contrast to the above studies, epidural analgesia has also been reported to inhibit uterine activity[184–187] or impair the progress of labor.[188,189] However, numerous investigators report that inhibition of uterine activity, when it occurs, is only a transient effect of epidural analgesia,[30,154,163,190–198] which is followed by return to control uterine activity,[30,163,196–198] improved uterine activity compared to control,[154,193,194] or acceleration of the progress of labor.[159] Additionally, several mechanisms can account for both a prolonged negative effect of epidural analgesia on uterine activity or the progress of labor as well as transient inhibition of uterine activity:

1. The transient decrease in uterine activity occurs immediately after injection of the local anesthetic and likely before conduction blockade has occurred.[30,192,194] Vasicka and Kretchmer[30] suggested that mechanical distension of the epidural space stimulates the spinal cord, which results in inhibition of uterine activity, possibly by a neural or neurohumoral reflex. Support for this theory is the observation by these authors that spinal anesthesia that does not distend the epidural space does not result in a transient inhibition of uterine activity.

2. Epinephrine may be added to local anesthetics to delay their absorption and prolong their effects. Epidural analgesia and anesthesia using these solutions results in systemic β_2-adrenergic effects from absorption of the epinephrine,[199,200] including inhibition of uterine activity[153,193,196,197,201] and prolongation of labor.[191,197,202,203] The transient inhibition of uterine activity reported in studies in which all or nearly all patients received epinephrine[163,192,198] is likely explained by this mechanism. Additionally, authors reporting no change in the progress of labor after epidural analgesia might have observed improved progress had epinephrine been omitted from the local anesthetic solution.[171,175,177]

3. A decrease in maternal blood pressure following regional analgesia or anesthesia decreases uterine activity secondary to reduction of uterine blood

flow.[30,190,204] Many studies document reduced blood pressure in a significant number of patients, which may have contributed to a transient decrease in uterine activity[159,192,195,198,201] or to the lack of improvement in the progress of labor[169,177] or uterine activity[172] following an epidural block. Hypotension may also account for reports of prolonged labor or impaired uterine activity in studies that did not report blood pressure measurements.[154,184—189,191,193,196,197]

4. In numerous studies patients labored in the supine position.[169,180,185,187,194,196,198] Therefore, aortocaval compression and reduction of uterine blood flow[4,5] may explain the observed transient[194,196,198] or prolonged[185,187] reduction of uterine activity following initiation of epidural analgesia and the observations that the progress of labor was unaltered by the block.[169,180] Several studies fail to report the position of the patient after initiation of regional analgesia. Thus, aortocaval compression may have contributed to impairment[30,184,188—192,195,197,201] or lack of improvement[173,195,197,201] of uterine activity or the progress of labor in these studies.

5. Initiation of epidural analgesia during the latent phase of the first stage of labor reduces uterine activity[163] and slows the progress of labor,[170,204—208] an effect attributed to interruption of a neural or neurohumoral reflex[96—102] that occurs with institution of the block.[98,178] This mechanism likely contributed to the reduced uterine activity reported in many studies in which analgesia was begun early in labor[30,184,185,192,196,198,201] and possibly in studies in which the time of initiation of analgesia was not stated.[186,187,189—191,195]

6. Excessive sedation impairs the progress of labor, especially when administered in the latent phase.[188,204—207] Thus, the concomitant administration of sedatives and/or narcotics in several studies may have accounted for the lack of a positive response to the epidural block.[169—171,177,179,189]

7. Finally, many studies report prolongation of the second stage of labor by epidural analgesia.[140—142,147,187,208] This effect, which is not related to impaired uterine activity, is attributed to two factors. First, relaxation of the pelvic floor (the levator ani muscles) interferes with internal rotation and flexion of the fetal head[142,146,162,177,189,208] and results in a higher than normal incidence of transverse and occiput—posterior presentations, which delays delivery. Since a conduction block of motor nerves to skeletal muscle requires a high dose of local anesthetic, this problem arises with caudal analgesia,[141,142,146,149,189,208] which requires a large local anesthetic dose to produce analgesia, and when excessive local anesthetic doses are used for lumbar epidural analgesia.[155,162,176,177,209] The use of appropriate amounts of local anesthetic for epidural analgesia, however, can prevent the occurrence of malrotation.[174,180] The second factor that may prolong the second stage of labor is interference with the reflex urge to bear down,

which occurs when perineal analgesia is produced by a caudal[142] or lumbar epidural[162,187] block. The resulting decrease in the parturient's voluntary effort can also delay delivery.

To summarize, uterine activity and the progress of labor will likely improve (although possibly after a transient decrease in uterine activity) following an epidural block that is appropriately administered (Table II); i.e., the local anesthetic should not contain epinephrine, hypotension from sympathetic nervous system blockade should be prevented by administration of an appropriate amount of intravenous balanced salt solution, aortocaval compression should be prevented by maintaining the lateral position or by using a hip wedge, analgesia should not be administered during the latent phase of the first stage of labor, excessive sedation should be avoided, and pelvic floor relaxation should be prevented by avoiding caudal analgesia (which is rarely used in current obstetric anesthesia practice) and using appropriate local anesthetic doses for epidural analgesia.

2.2.2. Regional Analgesia/Anesthesia and Uterine Blood Flow

Epidural (or spinal) analgesia or anesthesia that results in maternal hypotension produces a decrease in uterine blood flow as a consequence of a decrease in uterine perfusion pressure.[21,34,35,37,210–213] This can result in impaired fetal or neonatal well-being as evidenced by deterioration of the fetal heart rate pattern,[18,19,21,153,177,190,193,214–219] fetal blood gas status,[34,217–220] or umbilical cord blood gas status measured at the time of delivery.[221] Detrimental effects may in large part be related to supine positioning of the parturient in many of these studies.[18,19,177,190,193,214,215,218,219,221]

In contrast to the above reports, well-conducted regional analgesia or anesthesia, without hypotension or supine positioning, does not have a detrimental effect on uterine blood flow,[211,213,222–224] fetal heart rate,[163,212,216,225,226] or fetal acid–base or blood gas status.[157,163,214,219,224,226] In fact, many studies document an increase in uterine blood flow following epidural analgesia,[211,212,227–229] especially in parturients with pregnancy-induced hypertension,[230–234] as well as improvement in fetal acid–base status.[193,225,235,236] The improvements noted in these studies have been attributed to relief of vasospasm of the uteroplacental circulation

Table II. Peridural Analgesia Technique to Avoid Impairment of Uterine Activity or Progress of Labor

Omit epinephrine from the local anesthetic solution
Prevent hypotension by administering fluids
Avoid aortocaval compression
Do not administer analgesia during the latent phase of labor
Avoid excessive sedation
Avoid pelvic floor relaxation

caused by sympathetic nervous system blockade[227,228,230,231,234] and/or indirect reduction of sympathetic nervous system activity by relief of pain.[227] Since, as noted above, the uteroplacental vascular bed does not appear to be innervated by the sympathetic nervous system, reduction of intrinsic vascular resistance would have to be accounted for by the reduction of epinephrine and norepinephrine blood levels thats occur with regional analgesia and anesthesia.[77,84–86,164–168]

In sheep, uterine blood flow is reduced by approximately 20% with infusion of epinephrine or norepinephrine at the rate of 0.1 μg/kg per min,[45,49] which would result in blood levels (in humans) of at least 1000 and 3000 pg/ml, respectively.[70,71] Although the highest blood levels reported for epinephrine and norepinephrine during labor are approximately 600 and 2000 pg/ml, respectively,[165] these levels would likely reduce uterine blood flow for the following reasons: (1) interspecies differences exist with respect to the sensitivity of the uteroplacental vascular bed to catecholamines, and primates appear to be more sensitive than sheep[237]; (2) during labor, both epinephrine and norepinephrine are elevated, whereas in the sheep experiments either epinephrine or norepinephrine, but not both, was infused; and (3) blood levels of catecholamines that result from stress in primates reduce uterine blood flow and impair fetal well-being.[91–93] Therefore, well-managed regional analgesia in which hypotension and aortocaval compression are prevented can result in improved uterine blood flow and fetal well-being, especially in circumstances in which uteroplacental perfusion is compromised such as in pregnancy-induced hypertension. Finally, it should be noted that addition of epinephrine (100 μg) to the local anesthetic solution for epidural analgesia or anesthesia does not reduce uterine blood flow in sheep[238] or in humans.[222] However, this is not unexpected, since the addition of epinephrine raises the blood level by only 210 pg/ml without altering the norepinephrine blood level.[238]

3. General Anesthesia and Sedation

General anesthesia or sedation may alter uterine activity and/or uterine blood flow by mechanisms that are independent of the direct effects of the pharmacological agents on the myometrium or uterine vasculature. For example, inappropriately placing the parturient in the supine position during labor or an operative procedure may reduce uterine blood flow secondary to aortocaval compression.[4,5] This can result in impaired uterine activity[29] and in a higher incidence of neonatal acidosis and birth asphyxia.[23–27] Mechanical ventilation and alteration of maternal respiratory gases may also affect uterine blood flow. Mechanical hyperventilation during general anesthesia reduces uterine blood flow[103–105] and results in adverse effects on fetal oxygenation and acid–base status when associated with hypocapnia.[103,104,110–113,117–119] Maternal hypoventilation during general anesthesia or with excessive sedation may result in hypoxia and hypercapnia, either of which may reduce uterine blood flow.[120,124] Laryngoscopy, intubation, and surgical or other noxious stimuli enhance sympathetic nervous system activity and result in elevated catecholamine blood levels, which can reduce uterine blood flow[239-244] and compromise the fetus.[88,90-95,239-244] These effects can be ameliorated by enhancing the anesthetic depth[92,239] or providing maternal sedation.[91,93] Reducing the epinephrine blood level during labor by providing parenteral analgesia or sedation may also improve uterine activity,[185] although excessive medication during the latent phase of labor impairs cervical dilatation and the progress of labor.[188,204-207] Finally, an anesthetic agent, analgesic, or sedative may alter uterine activity or uterine blood flow by a direct effect on the myometrium or uterine vasculature.

3.1. Inhalation Anesthetics

In the central nervous system the gaseous and volatile inhalation anesthetics alter neuronal activity and interrupt synaptic transmission within many regions of the brain and depress transmission through the spinal cord. The potency of these general anesthetics correlates with their solubility in phospholipid bilayers.[245-247] Lipid solubility of these agents using olive oil as the model solvent yields similar results.[247] The current theory to explain this relationship is that anesthetic dissolves in the lipid component of cellular membranes, causing an increase in volume (volume expansion theory), with anesthesia occurring when the absorption of anesthetic expands the membrane volume beyond a critical amount (critical volume hypothesis).[248] Anesthetic dissolved in membranes may exert pressure on ionic channels, inhibiting ion transport and thus blocking axonal conduction and synaptic transmission, and may disrupt protein function within the membrane. Enhanced mobility of membrane components on exposure to anesthetics (fluidization theory) may also profoundly change membrane function.

Numerous other mechanisms have been proposed to explain anesthesia. Several theories propose impairment of central nervous system energy production and/or energy utilization, such as uncoupling of oxidative phosphorylation or inhibition of electron transport.[248] It has also been postulated that specific receptors for these anesthetics exist or that they interact with other receptor such as opioid or acetylcholine receptors.[248] Although these latter actions of anesthetics enhance our understanding of the many systemic effects of these agents, they likely have little relevance to the mechanism of anesthesia.

The halogenated anesthetics halothane, enflurane, and isoflurane relax the uterus by a direct effect on the myometrium. Halothane increases adenylate cyclase activity and cAMP levels in rat uterine muscle.[249] Thus, uterine relaxation by volatile anesthetics and β_2-adrenergic agonists (e.g., epinephrine[64,65]) may be through a common final pathway. However, the effects of halothane are not antagonized by β-adrenergic blocking agents, indicating that the anesthetic action on the uterus is not mediated through β-adrenergic receptors. Alternatively, inhalation anesthetics may relax the uterus by reducing free calcium ions in the cytoplasm of myometrial cells, a mechanism proposed for relaxation of tracheal smooth muscle exposed to halothane.[250] Additionally, inhalation anesthetics may cause uterine relaxation by an action at receptors. Halothane has been demonstrated to block α-adrenergic binding and the uterine contractile response.[251]

In vitro exposure of isolated uterine muscle from rats[252,253] and humans[254–258] to halothane, enflurane, and isoflurane produces a dose-related reduction of resting tone and of frequency and peak developed tension of uterine contractions, with equipotent anesthetic concentrations producing equivalent effects on the myometrium.[254] The effects are observed at anesthetic concentrations of 0.35 MAC (minimum alveolar concentration) and above, 1.0 MAC being the alveolar concentration of anesthetic at 1 atm pressure that produces immobility in 50% of those patients (or experimental animals) exposed to a noxious stimulus. (MAC is the standard measure of anesthetic potency for gaseous or volatile anesthetics and is analogous to the ED_{50} for drugs administered by other routes.) Cyclopropane has a smaller effect, and nitrous oxide has no effect on the uterus *in vitro*,[257] although evaluation of the latter agent is limited by its low potency (100% nitrous oxide equals approximately 0.95 MAC).

Clinically, the frequency and intensity of uterine contractions are reduced by halothane, enflurane, and isoflurane in concentrations of 0.35 to 0.5 MAC and above.[258–262] Fluroxene[263] and ether[30] also inhibit uterine activity. Above 1 MAC, marked uterine relaxation occurs, which can result in a substantial increase in blood loss at the time of vaginal delivery (although general anesthesia is not recommended for routine vaginal delivery),[258,261,264] cesarean section,[265] or therapeutic abortion.[266,267] Utilization of 0.5 to 0.75 MAC of one of these agents also increases blood loss during therapeutic abortion,[266,267] but not when administered during cesarean section for the short period of time from induction of anesthesia to delivery of the infant.[268–271] These agents, in concentrations of 0.75 MAC and above, also inhibit the response of the uterus to oxytocin[30,260,272,273] and the ergot alkaloids.[265,273] Finally, consistent with *in vitro* studies, cyclopropane has a smaller effect on uterine activity than the above agents,[30] and nitrous oxide has no clinical effect on the uterus.[30,258,259]

The effects of inhalation anesthetics on uterine blood flow depend on the concentration administered. General anesthesia for cesarean section, when performed with nitrous oxide as the only inhalation anesthetic, results in progressive deterioration of fetal acid–base status as the induction-to-delivery time increases.[274–276] Since this technique provides minimal anesthesia, compromise of the fetus likely results from elevated sympathetic nervous system activity and reduced uterine blood flow from noxious stimuli associated with anesthesia and surgery.[239–241,244] The addition of approximately 0.6 MAC of halothane or enflurane to the nitrous oxide technique enhances the depth of anesthesia and improves uterine blood flow because the sympathetic nervous system response to stimuli is reduced.[239] Attenuation of α_2-adrenergically mediated vasoconstriction by halothane would also be expected to contribute to improvement of uterine blood flow.[277]

For cesarean section, the addition of a potent inhalation anesthetic in low dose

(0.5 to 0.75 MAC) also results in improved neonatal (umbilical cord) acid–base status,[278–290] or acid–base status that is considered normal.[268–270,281–283] Higher concentrations of halothane, isoflurane, and methoxyflurane have been shown to cause uterine vasodilatation.[284,285] Thus, although concentrations of 1.0 to 1.5 MAC of a halogenated anesthetic reduce maternal cardiac output, maternal blood pressure, and uterine perfusion pressure, little or no effect on uterine blood flow or fetal well-being is observed.[284–286] High anesthetic concentrations (2 MAC), however, substantially reduce maternal cardiac output and blood pressure, which markedly decreases uterine perfusion pressure and uterine blood flow, resulting in fetal hypoxia and acidosis.[264,284–287] Although anesthetic concentrations of this magnitude are used to relax the uterus if intrauterine manipulation of the fetus is necessary at the time of delivery, the duration of administration is usually short and does not adversely affect the fetus.

3.2. Barbiturates

In vitro exposure to very high concentrations of barbiturates depresses the activity of isolated uterine muscle from rabbits, guinea pigs, and humans.[288,289] Prolongation of labor is observed in rabbits given very high doses of pentobarbital (i.e., equivalent to 2 g in humans).[290] However, administration of barbiturates in usual clinical doses or even reasonably high doses (e.g., 1 g of amobarbital in divided oral doses) does not affect uterine activity or progress when given in the active phase of labor in humans.[291–295] When barbiturates are administered in the latent phase, uterine activity may be depressed.[291] Intravenous administration of anesthetic doses of thiopental, thiamylal, or methohexital also does not depress uterine activity.[296] Some investigators have noted improvement in the progress of labor with oral administration of thiopental in divided doses totaling up to 1.3 g.[297,298] Improvement of uterine activity has also been noted when pentobarbital, in usual clinical doses, is administered to a parturient suffering considerable pain or excitement.[291] Barbiturates suppress sympathetic nervous system activity by acting on the cerebral cortex[299] and sympathetic ganglia[300] and also diminish epinephrine release from the adrenal medulla.[301,302] The improvement in progress of labor and uterine activity reported in these studies is most likely attributable to reduction of maternal catecholamine blood levels and relief of the inhibitory effect of epinephrine on the myometrium.[54–56,60–63]

Administration of intravenous anesthetic induction doses of thiopental, thiamylal, and methohexital transiently reduces uterine blood flow secondary to reduction of maternal arterial pressure and uterine perfusion pressure.[296,303–305] These agents do not appear to have a direct effect on intrinsic vascular resistance. Although the fetus is not severely compromised by the reduction in uterine blood flow, a certain degree of fetal hypoxia and acidosis does develop with administration of these drugs. However, sedative and anesthetic doses of pentobarbital, administered intravenously, prevent or correct the impairment of uterine blood flow and fetal distress that occurs in pregnant monkeys under stressful conditions.[91–93] These effects are likely attributable to barbiturate suppression of sympathetic nervous system activity.[299−302]

3.3. Ketamine

The frequency and intensity of uterine contractions and base-line uterine tone are generally increased following intravenous administration of ketamine, although the magnitude of the effects depends on the clinical setting or experimental model. During labor, 1.0 mg/kg of ketamine increases base-line uterine tone and the frequency, but not the intensity, of uterine contractions,[306] whereas doses of 1.5 and 2.2 mg/kg increase base-line tone and may result in uterine tetany.[307] Immediately post-partum the uterus responds to ketamine, 0.6 to 1.8 mg/kg, with an increase in the intensity of uterine contractions without a change in their frequency.[308] During the first and second trimesters of pregnancy (measured during therapeutic abortion), ketamine, 0.275 to 2.2 mg/kg, causes contractions or increases their frequency and intensity when they are present and increases base-line tone.[309−311] However, during elective cesarean section, when the uterus is quiescent, ketamine, 2 mg/kg, was without effect on uterine tone.[311] In the pregnant ewe, ketamine, 0.7 to 5.0 mg/kg, increases uterine tone and the frequency and intensity of uterine contractions.[296,312,313]

Ketamine elevates maternal cardiac output and arterial pressure when it is administered in doses of 0.7 to 5.0 mg/kg to pregnant ewes that are not in labor.[312−314] Since this effect is independent of the sympathetic nervous system, and catecholamine blood levels do not increase,[312,314] intrinsic vascular resistance is not elevated, and uterine blood flow increases secondary to the elevated uterine perfusion pressure (with ketamine doses of 1.8 to 5.0 mg/kg).[313,314] However, when administered to the ewe in labor, ketamine, 0.9 to 5.0 mg/kg, reduces uterine blood flow.[296] This most likely occurs because uterine tone and contraction frequency and intensity increase to a greater degree in this situation. In rabbits, uterine blood flow was unaltered by ketamine, 2 mg/kg, but decreased with a dose of 4 mg/kg.[306] Although the effects of ketamine on uterine blood flow in humans have not been studied, the administration of 1.0 and 2.0 mg/kg shortly before vaginal delivery[315,316] or for induction of anesthesia for elective cesarean section[317,318] does not adversely affect neonatal acid–base status. However, administration of ketamine in doses of 1.5 mg/kg and above during labor can result in fetal tachycardia and deterioration of fetal acid–base status.[307] The poorer fetal/neonatal response when ketamine is administered during labor, as compared to shortly before delivery, is likely attributable to the extended period of time between drug injection and delivery; i.e., impairment of uterine blood flow by ketamine can be tolerated for a short period but not a long period of time.

To summarize, the contracting uterus (i.e., during labor) is more sensitive than the quiescent uterus to the uterotonic effects of ketamine. Although ketamine elevates uterine perfusion pressure (an effect that appears to be independent of the sympathetic nervous system) and may enhance uterine blood flow, it can also reduce uterine blood flow by increasing uterine tone and uterine activity (i.e., extrinsic vascular resistance). Usual clinical doses of ketamine are 0.75 to 1.0 mg/kg for induction of general anesthesia for cesarean section or 0.2 mg/kg intermittently, up to a total dose of 1.0 mg/kg over 30 min, to provide analgesia during labor. However, administration of 2 mg/kg or less shortly before vaginal delivery or for induction of general anesthesia for cesarean section appears to be safe for the fetus and neonate, although a safe dose during labor cannot be determined from the available data.

3.4. Narcotics

Studies of the effects of morphine and meperidine on isolated uterine muscle have yielded variable results. These narcotics, in moderate concentrations, have minimal effects on uterine activity or uterine tone of myometrial samples from humans or other primates, although high concentrations of meperidine do elicit a depressant effect.[289,319,320] Myometrial samples from guinea pigs, cats, rabbits, sheep, dogs, and mice, however, respond to these drugs in high concentrations with an increase in uterine activity or uterine tone.[320–323] Similarly, intravenous injection of high-dose morphine or meperidine increases uterine tone in dogs, cats, and rabbits,[321,324] although subcutaneous injection of very high-dose morphine in dogs and rabbits decreases uterine tone and the frequency and intensity of uterine contractions.[325] The latter results may be related to hypotension produced by the high dose of morphine (up to 10 mg/kg) administered.

Administration of narcotics, especially in excessive amounts, during the latent phase of labor can inhibit uterine activity and slow the progress of labor.[188,204,326,327] The mechanism of this action is unclear. Numerous studies, however, demonstrate that uterine activity generally improves or labor is shortened after narcotic administration during spontaneous[185,328–332] or induced labor.[333] Similar results have been reported with the use of a narcotic plus promethazine[331,332,334–340] or hydroxyzine.[341] The beneficial effects have been attributed to elimination of anxiety and stress by providing analgesia and sedation, which reduces sympathetic nervous system activity and relieves the inhibitory effect of epinephrine on the uterus.[185,331,333] This mechanism likely explains the improved uterine activity when morphine or meperidine is administered to patients with incoordinate uterine activity.[342] Other studies report little or no effect[343–349] or a negative effect[350,351] of narcotics or narcotics plus a sedative–tranquilizer on uterine activity or progress of labor, but the time of drug administration was either during the latent phase or not stated. Finally, little information is available concerning the effects of narcotics on uterine blood flow, although fentanyl does not alter uterine blood flow in pregnant sheep.[352]

Narcotics injected into the epidural or subarachnoid space produce prolonged analgesia (18 to 24 hr) by interacting with opioid receptors in the substantia gelatinosa of the spinal cord. Epidural administration of a normal clinical dose (5 mg)[353] and a much larger dose (20 mg)[354] of morphine produced no significant changes in uterine blood flow or intraamniotic pressure in the pregnant ewe model (not in labor). Similarly, 50 and 100 μg of epidural fentanyl does not alter uterine blood flow or uterine activity in the pregnant ewe.[355]

3.5. Sedative Tranquilizers

Promethazine *in vitro* inhibits the spontaneous contractile activity of human myometrial strips from term pregnant and nonpregnant uteri.[356] The effect was proportional to the concentration of promethazine over a range of 5 to 20 μg/ml, and the drug also inhibited the uterine response to oxytocin. In contrast, promethazine used clinically, when administered alone, has been shown both to increase[332] and to decrease[351,356] uterine activity. In combination with meperidine, the drug has been observed to increase uterine activity[331,332] or to have no effect on uterine activity.[334,338] In the studies showing a decrease or no effect on uterine activity,[334,338,351,356] however, the drug or drugs were administered to some or all patients during the latent phase of labor, which may account for the results. Although promethazine has variable effects on uterine activity, most investigators have found that the drug in combination with meperidine[334-339] or morphine[340] shortens labor. The beneficial effect of promethazine on uterine activity and the progress of labor may be associated with reduction of sympathetic nervous system activity.[77,331] Two reports state that promethazine in combination with meperidine does not alter the course of labor,[343,349] although it is unclear at what stage of labor the drugs were given.

Hydroxyzine used alone,[345] in combination with meperidine,[343,345] and compared to placebo in patients who have received meperidine[357-359] has been found to be without effect on the duration of labor. In one study, however, the investigators felt that the combination of hydroxyzine and meperidine shortened labor.[341] Objective data for hydroxyzine are scant, with only one study demonstrating that uterine activity, although not decreasing after drug administration, did not increase to the degree expected with advancing labor.[351]

Diazepam, in concentrations of 1 to 33 μg/ml, reduces basal tone and the frequency and intensity of spontaneous contractions of isolated human myometrium.[360,361] In contrast, tocographic studies have shown no consistent effect of diazepam. A dose of 5 to 20 mg administered intravenously during labor has no effect[362] or increases[363] uterine tone and has either no effect on uterine activity[363] or reduces uterine activity by reducing either the frequency[362] or the intensity[364] of uterine contractions. A dose of 10 to 20 mg of diazepam administered intramuscularly does not alter uterine activity.[365] Most investigators who evaluated the effect of diazepam on the duration of labor found the drug to produce no significant effect,[366-369] although one study reported a shortening of labor[370] and another found the first stage of labor to be unaffected but the second stage to be prolonged.[371] In the latter study, prolongation of the second stage of labor was attributed to lack of desire of the parturient to push.[371] Finally, although one study reported deterioration of fetal acid–base status following intravenous administration of 20 mg of diazepam to patients during labor,[372] another study found no such adverse effect of the drug.[364] An investigation in pregnant ewes supports the latter results in that diazepam doses of 0.5 mg/kg and below do not affect uterine blood flow, and doses as high as 1 to 2 mg/kg reduce uterine blood flow by only 8% to 10% without affecting fetal acid–base status.[373] However, although sedatives administered to the parturient during labor are safe for the fetus, the neonate may be adversely affected by these drugs, which readily cross the placenta. For example, diazepam can result in respiratory depression, hypotonia, reluctance to feed, and impaired thermoregulation in the newborn infant.[374]

3.6. Neuromuscular Blocking Agents

Decamethonium, *d*-tubocurarine, and metocurine have, with few exceptions, been shown to have no effect on uterine activity,[375–380] uterine tone,[381] the progress of labor,[378] or the uterine response to uterotonic agents.[378–383] Although it has been suggested that *d*-tubocurarine reduces intraamniotic pressure[384] and the frequency and intensity of uterine contractions,[383] abdominal wall relaxation may explain the former results, and hypotension resulting from administration of a large dose of this relaxant (i.e., 30 mg[383]) may account for the latter results. It has also been stated that succinylcholine causes uterine relaxation.[385] However, in this study the drug was administered for difficult vaginal deliveries or retained placenta, and the beneficial effects were likely caused by relaxation of the perineal outlet and not the uterus. Other studies have demonstrated, subjectively[381] and objectively,[386] that succinylcholine does not affect uterine activity, uterine tone, or the uterine response to oxytocin.

3.7. Antihypertensive Agents

Antihypertensive agents are frequently used in the management of pregnancy-induced hypertension. Hydralazine and, more recently, labetalol have been shown to be effective for blood pressure control in these patients during labor and delivery. Sodium nitroprusside and nitroglycerin, both of which have a rapid onset and short duration of action, are useful for acute management of hypertensive crises and to control the hypertensive response to laryngoscopy and intubation during induction of general anesthesia. Most investigators (see below) have studied these drugs in the pregnant ewe made hypertensive by nephrectomy and renal artery constriction of the remaining kidney or by infusion of norepinephrine or phenylephrine. Since neither of these conditions duplicates the physiological state of pregnancy-induced hypertension, caution must be used in extrapolating these results to humans.

Hydralazine causes direct relaxation of arteriolar vascular smooth muscle by a mechanism that appears to involve activation of guanylate cyclase, which results in elevation of guanosine 3',5'-monophosphate (cGMP). Besides lowering arteriolar resistance, hydralazine has a positive inotropic and chronotropic effect on the heart (increasing cardiac output) and may reduce sympathetic nervous system activity by a central action.[387] This agent, and its derivative dihydralazine, effectively lowers blood pressure in the normotensive and the renal and pressor-induced hypertensive pregnant ewe[387,388] as well as in the parturient with pregnancy-induced hypertension.[389,390] In the normotensive ewe, hydralazine increases uterine blood flow despite the drop in uterine perfusion pressure, suggesting that this agent causes uteroplacental vasodilatation.[387] In both hypertensive ewe models, correcting the blood pressure to normal or control levels with hydralazine also increased uterine blood flow, although it still remained below that of the normotensive or control state.[387,388] In humans with pregnancy-induced hypertension, intravenous hydralazine administration caused a reduction of total uterine blood flow and intervillous blood flow and an increase in uterine vascular resistance, although none of these changes were significant.[389,390] Labetalol, a combined α- and β-adrenergic antagonist, produces effects in pressor-induced hypertensive sheep[391] and patients with pregnancy-induced hypertension[392] that are similar to those of hydralazine.

In the normotensive pregnant ewe, nitroglycerin and sodium nitroprusside reduce maternal blood pressure and uterine blood flow and increase uterine vascular conductance (except for very high-dose sodium nitroprusside).[393,394] Both agents activate guanylate cyclase[395] and thus appear to act by a mechanism similar to that of hydralazine. Sodium nitroprusside produces comparable changes in the renal-hypertensive pregnant ewe.[393] In both phenylephrine- and norepinephrine-induced hypertension, uterine blood flow decreased between 35% and 60% with an increase in maternal blood pressure of 20% to 25%.[388,394,396,397] Sodium nitroprusside and nitroglycerin improved uterine blood flow in each of these studies. However, when the antihypertensive agent was administered in a dose that returned maternal blood pressure to control values, uterine blood flow was still 25% to 30% below control values. Thus, sodium nitroprusside and nitroglycerin are unable to compensate fully for the detrimental effect of α-adrenergic agonists on the uteroplacental vascular bed.

4. Summary

The effect of regional analgesia or anesthesia, general anesthesia, or parenteral medication on uterine blood flow is dependent on how they affect uterine perfusion pressure, extrinsic vascular resistance, and intrinsic vascular resistance. Uterine activity may also be affected by analgesic technique, either by a direct effect of a drug on the myometrium or indirectly by altering uterine blood flow or interfering with neural influences on the uterus. However, when used in an appropriate manner, neither analgesia nor anesthesia has any substantive adverse effect on uterine blood flow or uterine activity. In fact, improvement of these parameters may be observed if therapeutic intervention reduces sympathetic nervous system activity.

References

1. Greiss, F. C., Jr., 1966, Pressure–flow relationship in the gravid uterine vascular bed, *Am. J. Obstet. Gynecol.* **96**:41–47.
2. Ladner, C., Brinkman, C. R. III, Weston, P., and Assali, N. S., 1970, Dynamics of uterine circulation in pregnant and nonpregnant sheep, *Am. J. Physiol.* **218**:257–263.
3. Berman, W., Jr., Goodlin, R. C., Heymann, M. A., and Rudolph, A. M., 1976, Relationships between pressure and flow in the umbilical and uterine circulations of the sheep, *Circ. Res.* **38**:262–266.
4. Bieniarz, J., Crottogini, J. J., Curuchet, E., Romero-Salinas, G., Yoshida, T., Poseiro, J. J., and Caldeyro-Barcia, R., 1968, Aortocaval compression by the uterus in late human pregnancy: II. An arteriographic study, *Am. J. Obstet. Gynecol.* **100**:203–217.
5. Kauppila, A., Koskinen, M., Puolakka, J., Tuimala, R., and Kuikka, J., 1980, Decreased intervillous and unchanged myometrial blood flow in supine recumbency, *Obstet. Gynecol.* **55**:203–205.
6. Scott, D. B., and Kerr, M. G., 1963, Inferior vena caval pressure in late pregnancy, *J. Obstet. Gynaecol. Br. Commonw.* **70**:1044–1049.
7. Kerr, M. G., 1965, The mechanical effects of the gravid uterus in late pregnancy, *J. Obstet. Gynaecol. Br. Commonw.* **72**:513–529.
8. Bieniarz, J., Branda, L. A., Maqueda, E., Morozovsky, J., and Caldeyro-Barcia, R., 1968, Aortocaval compression by the uterus in late pregnancy: III. Unreliability of the sphygmomanometric method in estimating uterine artery pressure, *Am. J. Obstet. Gynecol.* **102**:1106–1115.
9. Eckstein, K. L., and Marx, G. F., 1974, Aortocaval compression and uterine displacement, *Anesthesiology* **40**:92–96.
10. Ramsey, E. M., Corner, G. W., Jr., and Donner, M. W., 1963, Serial and cineradioangiographic visualization of maternal circulation in the primate (hemochorial) placenta, *Am. J. Obstet. Gynecol.* **86**:213–225.
11. Assali, N. S., Holm, L., and Parker, H., 1961, Regional blood flow and vascular resistance in response to oxytocin in the pregnant sheep and dog, *J. Appl. Physiol.* **16**:1087–1092.
12. Lees, M. H., Hill, J. D., Ochsner, A. J. III, Thomas, C. L., and Novy, M. J., 1971, Maternal placental and myometrial blood flow of the rhesus monkey during uterine contractions, *Am. J. Obstet. Gynecol.* **110**:68–81.
13. Greiss, F. C., Jr., 1965, Effect of labor on uterine blood flow, *Am. J. Obstet. Gynecol.* **93**:917–923.
14. Borell, U., Fernstrom, I., Ohlson, L., and Wiqvist, N., 1964, Effect of uterine contractions on the human uteroplacental blood circulation: An angiographic study, *Am. J. Obstet. Gynecol.* **89**:881–890.
15. Borell, U., Fernstrom, I., Ohlson, L., and Wiqvist, N., 1965, Influence of uterine contractions on the uteroplacental blood flow at term, *Am. J. Obstet. Gynecol.* **93**:44–57.
16. Brotanek, V., Hendricks, C. H., and Yoshida, T., 1969, Changes in uterine blood flow during uterine contractions, *Am. J. Obstet. Gynecol.* **103**:1108–1116.
17. Hendricks, C. H., 1958, The hemodynamics of a uterine contraction, *Am. J. Obstet. Gynecol.* **76**:969–982.
18. Hon, E. H., Reid, B. L., and Hehre, F. W., 1960, The electronic evaluation of fetal heart rate: II. Changes with maternal hypotension, *Am. J. Obstet. Gynecol.* **79**:209–215.
19. Ebner, H., Barcohana, J., and Bartoshuk, A. K., 1960, Influence of postspinal hypotension on the fetal electrocardiogram, *Am. J. Obstet. Gynecol.* **80**:569–572.
20. Romney, S. L., Gabel, P. V., and Takeda, Y., 1963, Experimental hemorrhage in late pregnancy: Effects on maternal and fetal hemodynamics, *Am. J. Obstet. Gynecol.* **87**:636–649.
21. Lucas, W., Kirschbaum, T., and Assali, N. S., 1965, Spinal shock and fetal oxygenation, *Am. J. Obstet. Gynecol.* **95**:583–587.

22. Greiss, F. C., Jr., 1966, Uterine vascular response to hemorrhage during pregnancy: With observations on therapy, *Obstet. Gynecol.* **27**:549–554.

23. Abitbol, M. M., 1985, Supine position in labor and associated fetal heart rate changes, *Obstet. Gynecol.* **65**:481–486.

24. Humphrey, M. D., Chang, A., Wood, E. C., Morgan, S., and Hounslow, D., 1974, A decrease in fetal pH during the second stage of labour, when conducted in the dorsal position, *J. Obstet. Gynaecol. Br. Commonw.* **81**:600–602.

25. Datta, S., Alper, M. H., Ostheimer, G. W., Brown, W. U., Jr., and Weiss, J. B., 1979, Effects of maternal position on epidural anesthesia for cesarean section, acid–base status, and bupivacaine concentrations at delivery, *Anesthesiology* **50**:205–209.

26. Humphrey, M., Hounslow, D., Morgan, S., and Wood, C., 1973, The influence of maternal posture at birth on the fetus, *J. Obstet. Gynaecol. Br. Commonw.* **80**:1075–1080.

27. Crawford, J. S., Burton, M., and Davies, P., 1972, Time and lateral tilt at cesarean section, *Br. J. Anaesth.* **44**:477–484.

28. Misrahy, G. A., Beran, A. V., Spradley, J. F., and Garwood, V. P., 1960, Fetal brain oxygen, *Am. J. Physiol.* **199**:959–964.

29. Caldeyro-Barcia, R., Noriega-Guerra, L., Cibils, L. A., Alvarez, H., Poseiro, J. J., Pose, S. V., Sica-Blanco, Y., Mendez-Bauer, C., Fielitz, C., and Gonzalez-Panizza, V. H., 1960, Effect of position changes on the intensity and frequency of uterine contractions during labor, *Am. J. Obstet. Gynecol.* **80**:284–290.

30. Vasicka, A., and Kretchmer, H., 1961, Effect of conduction and inhalation anesthesia on uterine contractions: Experimental study of the influences of anesthesia on intra-amniotic pressures, *Am. J. Obstet. Gynecol.* **82**:600–611.

31. Conklin, K. A., and Murad, S. H. N., 1982, Pharmacology of drugs in obstetric anesthesia, *Semin. Anesth.* **1**:83–100.

32. Eng, M., Berges, P. U., Ueland, K., Bonica, J. J., and Parer, J. T., 1971, The effects of methoxamine and ephedrine in normotensive pregnant primates, *Anesthesiology* **35**:354–360.

33. Ralston, D. H., Shnider, S. M., and deLorimier, A. A., 1974, Effects of equipotent ephedrine, metaraminol, mephentermine, and methoxamine on uterine blood flow in the pregnant ewe, *Anesthesiology* **40**:354–370.

34. Eng, M., Berges, P. U., Parer, J. T., Bonica, J. J., and Ueland, K., 1973, Spinal anesthesia and ephedrine in pregnant monkeys, *Am. J. Obstet. Gynecol.* **115**:1095–1099.

35. Hollmen, A. I., Jouppila, R., Albright, G. A., Jouppila, P., Vierola, H., and Koivula, A., 1984, Intervillous blood flow during caesarean section with prophylactic ephedrine and epidural anaesthesia, *Acta Anaesthesiol. Scand.* **28**:396–400.

36. Greiss, F. C., and van Wilkes, D., 1964, Effects of sympathomimetic drugs and angiotensin on the uterine vascular bed, *Obstet. Gynecol.* **23**:925–930.

37. Greiss, F. C., Jr., and Crandell, D. L., 1965, Therapy for hypotension induced by spinal anesthesia during pregnancy, *J.A.M.A.* **191**:793–796.

38. Greiss, F. C., Jr., and Pick, J. R., Jr., 1964, The uterine vascular bed: Adrenergic receptors, *Obstet. Gynecol.* **23**:209–213.

39. Greiss, F. C., Jr., 1972, Differential reactivity of the myoendometrial and placental vasculatures: Adrenergic responses, *Am. J. Obstet. Gynecol.* **112**:20–30.

40. Wallenburg, H. C. S., Mazer, J., and Hutchinson, D. L., 1973, Effects of a beta-adrenergic agent (metaproterenol) on uteroplacental circulation: An angiographic study in the pregnant rhesus monkey, *Am. J. Obstet. Gynecol.* **117**:1067–1075.

41. Thiagarajah, S., Harbert, G. M., Jr., and Bourgeois, F. J., 1985, Magnesium sulfate and ritodrine hydrochloride: Systemic and uterine hemodynamic effects, *Am. J. Obstet. Gynecol.* **153**:666–674.

42. Ehrenkranz, R. A., Walker, A. M., Oakes, G. K., McLaughlin, M. K., and Chez, R. A., 1976, Effect of ritodrine infusion on uterine and umbilical blood flow in pregnant sheep, *Am. J. Obstet. Gynecol.* **126**:343–349.

43. Ehrenkranz, R. A., Hamilton, L. A., Jr., Brennan, S. C., Oakes, G. K., Walker, A. M., and Chez,

R. A., 1977, Effects of salbutamol and isoxsuprine on uterine and umbilical blood flow in pregnant sheep, *Am. J. Obstet. Gynecol.* **128**:287–293.

44. Brennan, S. C., McLaughlin, M. K., and Chez, R. A., 1977, Effects of prolonged infusion of β-adrenergic agonists on uterine and umbilical blood flow in pregnant sheep, *Am. J. Obstet. Gynecol.* **128**:709–715.

45. Tabsh, K., Nuwayhid, B., Erkkola, R., Zugaib, M., Lieb, S., Ushioda, E., Brinkman, C. R. III, and Assali, N. S., 1981, Hemodynamic responses of the pelvic vascular bed to vasoactive stimuli in pregnant sheep, *Biol. Neonate* **39**:52–60.

46. van Nimwegen, D., and Dyer, D. C., 1974, The action of vasopressors on isolated uterine arteries, *Am. J. Obstet. Gynecol.* **118**:1099–1103.

47. Hood, D. D., Dewan, D. M., and James, F. M. III, 1986, Maternal and fetal effects of epinephrine in gravid ewes, *Anesthesiology* **64**:610–613.

48. Chestnut, D. H., Weiner, C. P., Martin, J. G., Herrig, J. E., and Wang, J. P., 1986, Effect of intravenous epinephrine on uterine artery blood flow velocity in the pregnant guinea pig, *Anesthesiology* **65**:633–636.

49. Greiss, F. C., 1963, The uterine vascular bed: Effect of adrenergic stimulation, *Obstet. Gynecol.* **21**:295–301.

50. Clapp, J. F. III, 1979, Effect of epinephrine infusion on maternal and uterine oxygen uptake in the pregnant ewe, *Am. J. Obstet. Gynecol.* **133**:208–212.

51. Rosenfeld, C. R., and West, J., 1977, Circulatory response to systemic infusion of norepinephrine in the pregnant ewe, *Am. J. Obstet. Gynecol.* **127**:376–383.

52. Rosenfeld, C. R., Barton, M. D., and Meschia, G., 1976, Effects of epinephrine on distribution of blood flow in the pregnant ewe, *Am. J. Obstet. Gynecol.* **124**:156–163.

53. Adamsons, K., Mueller-Heubach, E., and Myers, R. E., 1971, Production of fetal asphyxia in the rhesus monkey by administration of catecholamines to the mother, *Am. J. Obstet. Gynecol.* **109**:248–262.

54. Zuspan, F. P., Cibils, L. A., and Pose, S. V., 1962, Myometrial and cardiovascular responses to alterations in plasma epinephrine and norepinephrine, *Am. J. Obstet. Gynecol.* **84**:841–851.

55. Garrett, W. J., 1954, The effects of adrenaline and noradrenaline on the intact human uterus in late pregnancy and labour, *J. Obstet. Gynaecol. Br. Commonw.* **61**:586–589.

56. Kaiser, I. H., 1950, The effect of epinephrine and norepinephrine on the contractions of the human uterus in labor, *Surg. Gynecol. Obstet.* **90**:649–654.

57. Cibils, L. A., Pose, S. V., and Zuspan, F. P., 1962, Effect of *l*-norepinephrine infusion on uterine contractility and cardiovascular system, *Am. J. Obstet. Gynecol.* **84**:307–317.

58. Althabe, O., Jr., Schwarcz, R. L., Jr., Sala, N. L., and Fisch, L., 1968, Effect of phentolamine methanesulfonate upon uterine contractility induced by *l*-norepinephrine in pregnancy, *Am. J. Obstet. Gynecol.* **101**:1083–1088.

59. Brown, W. E., and Wilder, V. M., 1943, The response of the human uterus to epinephrine, *Am. J. Obstet. Gynecol.* **45**:659–665.

60. Woodbury, R. A., and Abreu, B. E., 1944, Influence of epinephrine upon the human gravid uterus, *Am. J. Obstet. Gynecol.* **48**:706–708.

61. Kaiser, I. H., and Harris, J. S., 1950, The effect of adrenalin on the pregnant human uterus, *Am. J. Obstet. Gynecol.* **59**:775–784.

62. Pose, S. V., Cibils, L. A., and Zuspan, F. P., 1962, Effect of *l*-epinephrine infusion on uterine contractility and cardiovascular system, *Am. J. Obstet. Gynecol.* **84**:297–306.

63. Stroup, P. E., 1962, The influence of epinephrine on uterine contractility, *Am. J. Obstet. Gynecol.* **84**:595–601.

64. Triner, L., Overweg, N. I. A., and Nahas, G. G., 1970, Cyclic 3′,5′ AMP and uterine contractility, *Nature* **225**:282–283.

65. Triner, L., Vulliemoz, Y., and Verosky, M., 1977, The action of halothane on adenylate cyclase, *Mol. Pharmacol.* **13**:976–979.

66. Pinto, R. M., Lerner, U., Pontelli, H., and Rabow, W., 1968, Action of epinephrine and nor-

epinephrine on contractile activity of the three separate layers of the human uterus, *Am. J. Obstet. Gynecol.* **102**:333–339.

67. Adair, F. L., and Haugen, J. A., 1939, A study of suspended human uterine muscle strips *in vitro*, *Am. J. Obstet. Gynecol.* **37**:753–762.

68. Sandberg, F., Ingelman-Sundberg, A., Linsgren, L., and Ryden, G., 1958, *In vitro* studies of the motility of the human uterus: Part III. The effects of adrenaline, noradrenaline and acetylcholine on the spontaneous motility in different parts of the pregnant and non-pregnant uterus, *J. Obstet. Gynaecol. Br. Emp.* **65**:965–972.

69. Stander, R. W., and Barden, T. P., 1966, Adrenergic receptor activity of catecholamines in human gestational myometrium, *Obstet. Gynecol.* **28**:768–774.

70. Clutter, W. E., Bier, D. M., Shah, S. D., and Cryer, P. E., 1980, Epinephrine plasma metabolic clearance rates and physiologic thresholds for metabolic and hemodynamic actions in man, *J. Clin. Invest.* **66**:94–101.

71. Silverberg, A. B., Shah, S. D., Haymond, M. W., and Cryer, P. E., 1978, Norepinephrine: Hormone and neurotransmitter in man, *Am. J. Physiol.* **234**:E252–E256.

72. Assali, N. S., and Prystowsky, H., 1950, Studies on autonomic blockade. I. Comparison between the effects of tetraethylammonium chloride (TEAC) and high selective spinal anesthesia on blood pressure of normal and toxemic pregnancy, *J. Clin. Invest.* **29**:1354–1366.

73. Assali, N. S., and Prystowsky, H., 1950, Studies on autonomic blockade. II. Observations on the nature of blood pressure fall with high selective spinal anesthesia in pregnant women, *J. Clin. Invest.* **29**:1367–1375.

74. Tabsh, K., Rudelstorfer, R., Nuwayhid, B., and Assali, N. S., 1986, Circulatory responses to hypovolemia in the pregnant and nonpregnant sheep after pharmacologic sympathectomy, *Am. J. Obstet. Gynecol.* **154**:411–419.

75. Whittaker, P. G., Gerrard, J., and Lind, T., 1985, Catecholamine responses to changes in posture during human pregnancy, *Br. J. Obstet. Gynaecol.* **92**:586–592.

76. Jones, C. M. III, and Greiss, F. C., Jr., 1982, The effect of labor on maternal and fetal circulating catecholamines, *Am. J. Obstet. Gynecol.* **144**:149–153.

77. Lederman, R. P., McCann, D. S., Work, B., Jr., and Huber, M. J., 1977, Endogenous plasma epinephrine and norepinephrine in last-trimester pregnancy and labor, *Am. J. Obstet. Gynecol.* **129**:5–8.

78. Robertson, D., Johnson, G. A., Robertson, R. M., Nies, A. S., Shand, D. G., and Oates, J. A., 1979, Comparative assessment of stimuli that release neuronal and adrenomedullary catecholamines in man, *Circulation* **59**:637–643.

79. Lake, C. R., Ziegler, M. G., and Kopin, I. J., 1976, Use of plasma norepinephrine for evaluation of sympathetic neuronal function in man, *Life Sci.* **18**:1315–1326.

80. Engelman, K., and Portnoy, B., 1970, A sensitive double-isotope derivative assay for norepinephrine and epinephrine: Normal resting human plasma levels, *Circ. Res.* **26**:53–57.

81. Galbo, H., Holst, J. J., and Christensen, N. J., 1975, Glucagon and plasma catecholamine responses to graded and prolonged exercise in man, *J. Appl. Physiol.* **38**:70–76.

82. Vendsalu, A., 1960, Studies on adrenaline and noradrenaline in human plasma, *Acta Physiol. Scand. [Suppl.]* **173**:1–123.

83. Johnson, G. A., Peuler, J. D., and Baker, C. A., 1977, Plasma catecholamines in normotensive subjects, *Curr. Ther. Res.* **21**:898–908.

84. Jouppila, R., Puolakka, J., Kauppila, A., and Vuori, J., 1984, Maternal and umbilical cord plasma noradrenaline concentrations during labour with and without segmental extradural analgesia, and during caesarean section, *Br. J. Anaesth.* **56**:251–255.

85. Falconer, A. D., and Powles, A. B., 1982, Plasma noradrenaline levels during labour: Influence of elective lumbar epidural blockade, *Anaesthesia* **37**:416–420.

86. Grenman, S., Erkkola, R., Kanto, J., Scheinin, M., Viinamaki, O., and Lindberg, R., 1986, Epidural and paracervical blockades in obstetrics: Catecholamines, arginine vasopressin and analgesic effect, *Acta Obstet. Gynecol. Scand.* **65**:699–704.

87. Lederman, R. P., Lederman, E., Work, B. A., Jr., and McCann, D. S., 1978, The relationship of maternal anxiety, plasma catecholamines, and plasma cortisol to progress in labor, *Am. J. Obstet. Gynecol.* **132:**495–500.

88. Lederman, R. P., Lederman, E., Work, B., Jr., and McCann, D. S., 1985, Anxiety and epinephrine in multiparous women in labor: Relationship to duration of labor and fetal heart rate pattern, *Am. J. Obstet. Gynecol.* **153:**870–877.

89. Dimsdale, J. E., and Moss, J., 1980, Plasma catecholamines in stress and exercise, *J.A.M.A.* **243:**340–342.

90. Shnider, S. M., Wright, R. G., Levinson, G., Roizen, M. F., Wallis, K. L., Rolbin, S. H., and Craft, J. B., 1979, Uterine blood flow and plasma norepinephrine changes during maternal stress in the pregnant ewe, *Anesthesiology* **50:**524–527.

91. Morishima, H. O., Yeh, M. N., and James, L. S., 1979, Reduced uterine blood flow and fetal hypoxemia with acute maternal stress: Experimental observation in the pregnant baboon, *Am. J. Obstet. Gynecol.* **134:**270–275.

92. Myers, R. E., 1975, Maternal psychological stress and fetal asphyxia: A study in the monkey, *Am. J. Obstet. Gynecol.* **122:**47–59.

93. Morishima, H. O., Pedersen, H., and Finster, M., 1978, The influence of maternal psychological stress on the fetus, *Am. J. Obstet. Gynecol.* **131:**286–290.

94. Talbert, D. G., Benson, P., and Dewhurst, J., 1982, Fetal response to maternal anxiety: A factor in antepartum heart rate monitoring, *J. Obstet. Gynaecol.* **3:**34–38.

95. Lederman, E., Lederman, R. P., Work, B. A., Jr., and McCann, D. S., 1981, Maternal psychological and physiologic correlates of fetal–newborn health status, *Am. J. Obstet. Gynecol.* **139:**956–958.

96. Ferguson, J. K. W., 1941, A study of the motility of the intact uterus at term, *Surg. Gynecol. Obstet.* **73:**359–366.

97. Vasicka, A., Kumaresan, P., Han, G. S., and Kumaresan, M., 1978, Plasma oxytocin in initiation of labor, *Am. J. Obstet. Gynecol.* **130:**263–273.

98. Goodfellow, C. F., Hull, M. G. R., Swaab, D. F., Dogterom, J., and Buijs, R. M., 1983, Oxytocin deficiency at delivery with epidural analgesia, *Br. J. Obstet. Gynaecol.* **90:**214–219.

99. Coch, J. A., Brovetto, J., Cabot, H. M., Fielitz, C. A., and Caldeyro-Barcia, R., 1965, Oxytocin-equivalent activity in the plasma of women in labor and during the puerperium, *Am. J. Obstet. Gynecol.* **91:**10–17.

100. Leake, R. D., Weitzman, R. E., Glatz, T. H., and Fisher, D. A., 1979, Stimulation of oxytocin secretion in the human, *Clin. Res.* **27:**99A.

101. Leake, R. D., 1983, Oxytocin, in: *Initiation of Parturition: Prevention of Prematurity, Report of the 4th Ross Conference on Obstetric Research,* Ross Laboratories, Columbus, OH, pp. 43–51.

102. Fisch, L., Sala, N. L., and Schwarcz, R. L., 1964, Effect of cervical dilatation upon uterine contractility in pregnant women and its relation to oxytocin secretion, *Am. J. Obstet. Gynecol.* **90:**108–114.

103. Levinson, G., Shnider, S. M., deLorimier, A. A., and Steffenson, J. L., 1974, Effects of maternal hyperventilation on uterine blood flow and fetal oxygenation and acid–base status, *Anesthesiology* **40:**340–347.

104. Behrman, R. E., Parer, J. T., and Novy, M. J., 1967, Acute maternal respiratory alkalosis (hyperventilation) in the pregnant rhesus monkey, *Pediatr. Res.* **1:**354–363.

105. Leduc, B., 1970, The effect of acute hypocapnia on maternal placental blood flow in rabbits, *J. Physiol. (Lond.)* **210:**165P–166P.

106. Buss, D. D., Bisgard, G. E., Rawlings, C. A., and Rankin, J. H. G., 1975, Uteroplacental blood flow during alkalosis in the sheep, *Am. J. Physiol.* **228:**1497–1500.

107. Ralston, D. H., Shnider, S. M., and deLorimier, A. A., 1974, Uterine blood flow and fetal acid base changes after bicarbonate administration to the pregnant ewe, *Anesthesiology* **40:**348–353.

108. Scott, D. B., Lees, M. M., Davie, I. T., Slawson, K. B., and Kerr, M. G., 1969, Observations on cardiorespiratory function during caesarean section, *Br. J. Anaesth.* **41:**489–495.

109. Fuller, E. O., Galletti, P. M., Chou, H. Y., and Peirce, E. C. II, 1972, Effect of pCO_2 on the vascular resistance of the pregnant sheep uterus, *Physiologist* **15**:140.

110. Motoyama, E. K., Rivard, G., Acheson, F., and Cook, C. D., 1967, The effect of changes in maternal pH and PCO_2, on the PO_2 of fetal lambs, *Anesthesiology* **28**:891–903.

111. Motoyama, E. K., Rivard, G., Acheson, F., and Cook, C. D., 1966, Adverse effect of maternal hyperventilation on the foetus, *Lancet* **1**:286–288.

112. Morishima, H. O., Moya, F., Bossers, A. C., and Daniel, S. S., 1964, Adverse effects of maternal hypocapnea on the newborn guinea pig, *Am. J. Obstet. Gynecol.* **88**:524–529.

113. Cook, P. T., 1984, The influence on foetal outcome of maternal carbon dioxide tension at caesarean section under general anaesthesia, *Anaesth. Intens. Care* **12**:296–302.

114. Miller, F. C., Petrie, R. H., Arce, J. J., Paul, R. H., and Hon, E. H., 1974, Hyperventilation during labor, *Am. J. Obstet. Gynecol.* **120**:489–495.

115. Saling, E., and Ligdas, P., 1969, The effect on the fetus of maternal hyperventilation during labour, *J. Obstet. Gynaecol. Br. Commonw.* **76**:877–880.

116. Lumley, J., Renou, P., Newman, W., and Wood, C., 1969, Hyperventilation in obstetrics, *Am. J. Obstet. Gynecol.* **103**:847–855.

117. Low, J. A., Boston, R. W., and Cervenko, F. W., 1970, Effect of low maternal carbon dioxide tension on placental gas exchange, *Am. J. Obstet. Gynecol.* **106**:1032–1043.

118. Peng, A. T. C., Blancato, L. S., and Motoyama, E. K., 1972, Effect of maternal hypocapnia v. eucapnia on the foetus during caesarean section, *Br. J. Anaesth.* **44**:1173–1178.

119. Moya, F., Morishima, H. O., Shnider, S. M., and James, L. S., 1965, Influence of maternal hyperventilation on the newborn infant, *Am. J. Obstet. Gynecol.* **91**:76–84.

120. Walker, A. M., Oakes, G. K., Ehrenkranz, R., McLaughlin, M., and Chez, R. A., 1976, Effects of hypercapnia on uterine and umbilical circulations in conscious pregnant sheep, *J. Appl. Physiol.* **41**:727–733.

121. Greiss, F. G., Jr., Anderson, S. G., and King, L. C., 1972, Uterine vascular bed: Effects of acute hypoxia, *Am. J. Obstet. Gynecol.* **113**:1057–1064.

122. Assali, N. S., Holm, L. W., and Sehgal, N., 1962, Hemodynamic changes in fetal lamb in utero in response to asphyxia, hypoxia, and hypercapnia, *Circ. Res.* **11**:423–430.

123. Makowski, E. L., Hertz, R. H., and Meschia, G., 1973, Effects of acute maternal hypoxia and hyperoxia on the blood flow to the pregnant uterus, *Am. J. Obstet. Gynecol.* **115**:624–631.

124. Dilts, P. V., Jr., Brinkman, C. R. III, Kirschbaum, T. H., and Assali, N. S., 1969, Uterine and systemic hemodynamic interrelationships and their response to hypoxia, *Am. J. Obstet. Gynecol.* **103**:138–157.

125. Gibbs, C. P., and Noel, S. C., 1977, Response of arterial segments from gravid human uterus to multiple concentrations of lignocaine, *Br. J. Anaesth.* **49**:409–412.

126. Cibils, L. A., 1976, Response of human uterine arteries to local anesthetics, *Am. J. Obstet. Gynecol.* **126**:202–210.

127. Gibbs, C. P., and Noel, S. C., 1976, Human uterine artery responses to lidocaine, *Am. J. Obstet. Gynecol.* **126**:313–315.

128. Munson, E. S., and Embro, W. J., 1978, Lidocaine, monoethylglycinexylidide, and isolated human uterine muscle, *Anesthesiology* **48**:183–186.

129. Willdeck-Lund, G., and Nilsson, B. A., 1979, The effect of local anaesthetic agents on the contractility of human myometrium in late pregnancy, *Acta Anaesth. Scand.* **23**:78–88.

130. McGaughey, H. S., Jr., Corey, E. L., Eastwood, D., and Thornton, W. N., Jr., 1962, Effect of synthetic anesthetics on the spontaneous motility of human uterine muscle *in vitro, Obstet. Gynecol.* **19**:233–240.

131. Feinstein, M. B., 1966, Inhibition of contraction and calcium exchange-ability in rat uterus by local anesthetics, *J. Pharmacol. Exp. Ther.* **152**:516–524.

132. Feinstein, M. B., and Paimre, M., 1969, Pharmacological action of local anesthetics on excitation–contraction coupling in striated and smooth muscle, *Fed. Proc.* **28**:1643–1648.

133. Greiss, F. C., Jr., Still, J. G., and Anderson, S. G., 1976, Effects of local anesthetic agents on the uterine vasculatures and myometrium, *Am. J. Obstet. Gynecol.* **124**:889–899.

134. Craft, J. B., Jr., Co, E. G., Yonekura, M. L., Roizen, M. F., Mazel, P., Gilman, R. M., and Johnson, J. L., 1979, Intravenous injection of bupivacaine in the pregnant ewe, *Anesthesiology* **51:**S309.

135. Craft, J. B., Jr., Dao, S. D., Yonekura, M. L., Co, E. G., Mackinnon, D., Roizen, M. F., Mazel, P., Gilman, R. M., and Shokes, L. K., 1980, Bolus injection of chloroprocaine in the pregnant ewe, *Anesthesiology* **53:**S313.

136. Fishburne, J. I., Jr., Greiss, F. C., Jr., Hopkinson, R., and Rhyne, A. L., 1979, Responses of the gravid uterine vasculature to arterial levels of local anesthetic agents, *Am. J. Obstet. Gynecol.* **133:**753–761.

137. Evans, J. A., Chastain, G. M., and Phillips, J. M., 1969, The use of local anesthetic agents in obstetrics, *So. Med. J.* **62:**519–524.

138. Biehl, D., Shnider, S. M., Levinson, G., and Callender, K., 1977, The direct effects of circulating lidocaine on uterine blood flow and foetal well-being in the pregnant ewe, *Can. Anaesth. Soc. J.* **24:**445–451.

139. Conklin, K. A., 1987, Pharmacology of local anesthetics, *J. Am. Assoc. Nurse Anesthetists* **55:**36–44.

140. Mengert, W. F., 1944, Continuous caudal anesthesia with procaine hydrochloride in 240 obstetric patients, *Am. J. Obstet. Gynecol.* **48:**100–102.

141. Hingson, R. A., and Edwards, W. B., 1943, Continuous caudal analgesia: An analysis of the first ten thousand confinements thus managed with the report of the authors' first thousand cases, *J.A.M.A.* **123:**538–546.

142. Hodges, W. R., 1944, Continuous caudal analgesia in obstetrics: Three hundred cases, *J.A.M.A.* **125:**336–341.

143. Galley, A. H., 1949, Continuous caudal analgesia in obstetrics, *Anaesthesia* **4:**154–168.

144. Block, N., 1944, Further studies with continuous drip caudal anesthesia, *Am. J. Obstet. Gynecol.* **47:**331–334.

145. Harer, W. B., Jr., Gunther, R. E., and Stubblefield, C. T., 1963, Long-acting single injection caudal anesthesia: A report on 525 deliveries with mepivacaine, *Am. J. Obstet. Gynecol.* **87:**236–241.

146. Ellis, G. J., and Sheffrey, J. B., 1945, Continuous caudal anesthesia as an analgesic and therapeutic agent, *Anesth. Analg.* **24:**193–204.

147. McCormick, C. O., Huber, C. P., Spahr, J. F., and Gillespie, C. F., 1944, An experience with one hundred cases of continuous caudal analgesia, *Am. J. Obstet. Gynecol.* **47:**297–311.

148. Hallet, R. L., 1953, The conduct of labor and results with continuous caudal anesthesia, *Am. J. Obstet. Gynecol.* **66:**54–61.

149. Lewis, M. S., and Austin, R. B., 1950, Continuous caudal versus saddle-block anesthesia in obstetrics, *Am. J. Obstet. Gynecol.* **59:**1146–1152.

150. Johnson, G. T., 1954, Continuous caudal analgesia: Experiences in the management of disordered uterine function in labour, *Br. Med. J.* **1:**627–629.

151. Johnson, G. t., 1957, Prolonged labour. A clinical trial of continuous caudal analgesia, *Br. Med. J.* **2:**386–389.

152. Arthur, H. R., and Johnson, G. T., 1952, Continuous caudal anaesthesia in the management of cervical dystocia, *J. Obstet. Gynaecol. Br. Emp.* **59:**372–376.

153. Jouppila, P., Jouppila, R., Kaar, K., and Merila, M., 1977, Fetal heart rate patterns and uterine activity after segmental epidural analgesia, *Br. J. Obstet. Gynaecol.* **84:**481–486.

154. Climie, C. R., 1964, The place of continuous lumbar epidural analgesia in the management of abnormally prolonged labour, *Med. J. Aust.* **2:**447–450.

155. Cowles, G. T., 1965, Experiences with lumbar epidural block, *Obstet. Gynecol.* **26:**734–739.

156. Printz, J. L., and McMaster, R. H., 1972, Continuous monitoring of fetal heart rate and uterine contractions in patients under epidural anesthesia, *Anesth. Analg.* **51:**876–880.

157. Bleyaert, A., Soetens, M., Vaes, L., van Steenberge, A. L., and van der Donck, A., 1979, Bupivacaine, 0.125 per cent, in obstetric epidural analgesia: Experience in three thousand cases, *Anesthesiology* **51:**435–438.

158. Tunstall, M. E., 1960, High extradural block for cervical dystocia, *Br. J. Anaesth.* **32**:292–294.

159. Moir, D. D., and Willocks, J., 1966, Continuous epidural analgesia in inco-ordinate uterine action, *Acta Anaesth. Scand. [Suppl.]* **23**:144–153.

160. Maltau, J. M., and Andersen, H. T., 1975, Epidural anaesthesia as an alternative to caesarean section in the treatment of prolonged, exhaustive labour, *Acta Anaesth. Scand.* **19**:349–354.

161. Moir, D. D., and Willocks, J., 1967, Management of incoordinate uterine action under continuous epidural analgesia, *Br. Med. J.* **3**:396–398.

162. Raabe, N., and Belfrage, P., 1976, Lumbar epidural analgesia in labour: A clinical analysis, *Acta Obstet. Gynecol. Scand.* **55**:125–129.

163. Raabe, N., and Belfrage, P., 1976, Epidural analgesia in labour: IV. Influence on uterine activity and fetal heart rate, *Acta Obstet. Gynecol. Scand.* **55**:305–310.

164. Irestedt, L., Lagercrantz, H., Hjemdahl, P., Hagnevik, K., and Belfrage, P., 1982, Fetal and maternal plasma catecholamine levels at elective cesarean section under general or epidural anesthesia versus vaginal delivery, *Am. J. Obstet. Gynecol.* **142**:1004–1010.

165. Shnider, S. M., Abboud, T. K., Artal, R., Henriksen, E. H., Stefani, S. J., and Levinson, G., 1983, Maternal catecholamines decrease during labor after lumbar epidural anesthesia, *Am. J. Obstet. Gynecol.* **147**:13–15.

166. Abboud, T. K., Yanagi, T., Artal, R., Costandi, J., and Henriksen, E., 1985, Effect of epidural analgesia during labor on fetal plasma catecholamine release, *Regional Anesth.* **10**:170–174.

167. Neumark, J., Hammerle, A. F., and Biegelmayer, C., 1985, Effects of epidural analgesia on plasma catecholamines and cortisol in parturition, *Acta Anaesth. Scand.* **29**:555–559.

168. Jones, C. R., McCullouch, J., Butters, L., Hamilton, C. A., Rubin, P. C., and Reid, J. L., 1985, Plasma catecholamines and modes of delivery: The relation between catecholamine levels and *in-vitro* platelet aggregation and adrenoreceptor radioligand binding characteristics, *Br. J. Obstet. Gynaecol.* **92**:593–599.

169. Gunther, R. E., and Harer, W. b., Jr., 1965, Single injection caudal anesthesia: Report on 531 deliveries with 1.5 per cent mepivacaine, *Am. J. Obstet. Gynecol.* **92**:305–309.

170. Ritmiller, L. F., and Rippmann, E. T., 1957, Caudal analgesia in obstetrics: Report of thirteen years' experience, *Obstet. Gynecol.* **9**:25–28.

171. Brown, H. O., Thomson, J. M., and Fitzgerald, J. E., 1946, An analysis of 500 obstetrical cases with continuous caudal anesthesia using pontocaine, *Anesthesiology* **7**:355–374.

172. Cibils, L. A., and Spackman, T. J., 1962, Caudal analgesia in first-stage labor: Effect on uterine activity and the cardiovascular system, *Am. J. Obstet. Gynecol.* **84**:1042–1050.

173. Fernandez-Sepulveda, R., and Gomez-Rogers, C., 1967, Single-dose caudal anesthesia: Its effect on uterine contractility, *Am. J. Obstet. Gynecol.* **98**:847–853.

174. Jouppila, R., Jouppila, P., Karinen, J. M., and Hollmen, A., 1979, Segmental epidural analgesia in labour: Related to the progress of labour, fetal malposition and instrumental delivery, *Acta Obstet. Gynecol. Scand.* **58**:135–139.

175. Phillips, J. C., Hockberg, C. J., Petrakis, J. K., and van Winkle, J. D., 1977, Epidural analgesia and its effects on the "normal" progress of labor, *Am. J. Obstet. Gynecol.* **129**:316–322.

176. Studd, J. W. W., Crawford, J. S., Duignan, N. M., Rowbotham, C. J. F., and Hughes, A. O., 1980, The effect of lumbar epidural analgesia on the rate of cervical dilatation and the outcome of labour of spontaneous onset, *Br. J. Obstet. Gynaecol.* **87**:1015–1021.

177. Kandel, P. F., Spoerel, W. E., and Kinch, R. A. H., 1966, Continuous epidural analgesia for labour and delivery: Review of 1000 cases, *Can. Med. Assoc. J.* **95**:947–953.

178. Sala, N. L., Schwarcz, R. L., Jr., Althabe, O., Jr., Fisch, L., and Fuente, O., 1970, Effect of epidural anesthesia upon uterine contractility induced by artificial cervical dilatation in human pregnancy, *Am. J. Obstet. Gynecol.* **106**:26–29.

179. Henry, J. S., Jr., Kingston, M. B., and Maughan, G. B., 1967, The effect of epidural anesthesia on oxytocin-induced labor, *Am. J. Obstet. Gynecol.* **97**:350–359.

180. Gal, D., Choudhry, R., Ung, K. A., Abadir, A., and Tancer, M. L., 1979, Segmental epidural analgesia for labor and delivery, *Acta Obstet. Gynecol. Scand.* **58**:429–431.

181. Schellenberg, J. C., 1977, Uterine activity during lumbar epidural analgesia with bupivacaine, *Am. J. Obstet. Gynecol.* **127:**26–31.
182. Abboud, T. K., Sheik-ol-Eslam, A., Yanagi, T., Murakawa, K., Costandi, J., Zakarian, M., Hoffman, D., and Haroutunian, S., 1985, Safety and efficacy of epinephrine added to bupivacaine for lumbar epidural analgesia in obstetrics, *Anesth. Analg.* **64:**585–591.
183. Abboud, T. K., David, S., Nagappala, S., Costandi, J., Yanagi, T., Haroutunian, S., and Yeh, S. U., 1984, Maternal, fetal, and neonatal effects of lidocaine with and without epinephrine for epidural anesthesia in obstetrics, *Anesth. Analg.* **63:**973–979.
184. Alexander, J. A., and Franklin, R. R., 1966, Effects of caudal anesthesia on uterine activity, *Obstet. Gynecol.* **27:**436–441.
185. Filler, W. W., Jr., Hall, W. C., and Filler, N. W., 1967, Analgesia in obstetrics: The effect of analgesia on uterine contractility and fetal heart rate, *Am. J. Obstet. Gynecol.* **98:**832–846.
186. Bates, R. G., and Helm, C. W., 1985, Uterine activity in the second stage of labour and the effect of epidural analgesia, *Br. J. Obstet. Gynaecol.* **92:**1246–1250.
187. Johnson, W. L., Winter, W. W., Eng, M., Bonica, J. J., and Hunter, C. A., 1972, Effect of pudendal, spinal, and peridural block anesthesia on the second stage of labor, *Am. J. Obstet. Gynecol.* **113:**166–175.
188. Friedman, E. A., 1956, Labor in multipara: A graphicostatistical analysis, *Obstet. Gynecol.* **8:**691–703.
189. Nicodemus, R. E., Ritmiller, L. F., and Ledden, L. J., 1945, Continuous caudal analgesia in obstetrics on trial, *Am. J. Obstet. Gynecol.* **50:**312–318.
190. Vasicka, A., Hutchinson, H. T., Eng, M., and Allen, C. R., 1964, Spinal and epidural anesthesia, fetal and uterine response to acute hypo- and hypertension, *Am. J. Obstet. Gynecol.* **90:**800–810.
191. Epstein, B. S., Coakley, C. S., Barter, R. H., and Chamberlain, G., 1970, New developments in epidural anesthesia for obstetrics, *Am. J. Obstet. Gynecol.* **106:**996–1003.
192. Lowensohn, R. I., Paul, R. H., Fales, S., Yeh, S. Y., and Hon, E. H., 1974, Intrapartum epidural anesthesia: An evaluation of effects on uterine activity, *Obstet. Gynecol.* **44:**388–393.
193. Zador, G., and Nilsson, B. A., 1974, Low dose intermittent epidural anaesthesia with lidocaine for vaginal delivery: II. Influence on labour and foetal acid–base status, *Acta Obstet. Gynecol. Scand.* *[Suppl.]* **34:**17–30.
194. Tyack, A. J., Parsons, R. J., Millar, D. R., and Nicholas, A. D. G., 1973, Uterine activity and plasma bupivacaine levels after caudal epidural analgesia, *J. Obstet. Gynaecol. Br. Commonw.* **80:**896–901.
195. Nielsen, J. S., Spoerel, W. E., Keenleyside, H. B., Slater, P. E., and Clancy, P. R., 1962, Continuous epidural analgesia for labour and delivery, *Can. Anaesth. Soc. J.* **9:**143–152.
196. Willdeck-Lund, G., Lindmark, G., and Nilsson, B. A., 1979, Effect of segmental epidural analgesia upon the uterine activity with special reference to the use of different local anaesthetic agents, *Acta Anaesth. Scand.* **23:**519–528.
197. Craft, J. B., Jr., Epstein, B. S., and Coakley, C. S., 1972, Effect of lidocaine with epinephrine versus lidocaine (plain) on induced labor, *Anesth. Analg.* **51:**243–246.
198. Hollmen, A., Jouppila, R., Pihlajaniemi, R., Karvonen, P., and Sjostedt, E., 1977, Selective lumbar epidural block in labour: A clinical analysis, *Acta Anaesth. Scand.* **21:**174–181.
199. Bonica, J. J., Akamatsu, T. J., Berges, P. U., Morikawa, K., and Kennedy, W. F., Jr., 1971, Circulatory effects of peridural block: II. Effects of epinephrine, *Anesthesiology* **34:**514–522.
200. Kennedy, W. F., Jr., Bonica, J. J., Ward, R. J., Tolas, A. G., Martin, W. E., and Grinstein, A., 1966, Cardiorespiratory effects of epinephrine when used in regional anesthesia, *Acta Anaesth. Scand.* *[Suppl.]* **23:**320–333.
201. Matadial, L., and Cibils, L. A., 1976, The effect of epidural anesthesia on uterine activity and blood pressure, *Am. J. Obstet. Gynecol.* **125:**846–854.
202. Gunther, R. E., and Bauman, J., 1969, Obstetrical caudal anesthesia: I. A randomized study comparing 1% mepivacaine with 1% lidocaine plus epinephrine, *Anesthesiology* **31:**5–19.
203. Gunther, R. E., and Belville, J. W., 1972, Obstetrical caudal anesthesia: II. A randomized study

comparing 1 per cent mepivacaine with 1 per cent mepivacaine plus epinephrine, *Anesthesiology* **37**:288–298.

204. Friedman, E. A., 1955, Primigravid labor: A graphicostatistical analysis, *Obstet. Gynecol.* **6**:567–589.

205. Friedman, E. A., and Sachtleben, M. R., 1959, Caudal anesthesia: The factors that influence its effect on labor, *Obstet. Gynecol.* **13**:442–450.

206. Friedman, E. A., and Sachtleben, M. R., 1961, Dysfunctional labor: I. Prolonged latent phase in the nullipara, *Obstet. Gynecol.* **17**:135–148.

207. Friedman, E. A., and Sachtleben, M. R., 1961, Dysfunctional labor: II. Protracted active-phase dilatation in the nullipara, *Obstet. Gynecol.* **17**:566–578.

208. Siever, J. M., and Mousel, L. H., 1943, Continuous caudal anesthesia in three hundred unselected obstetric cases, *J.A.M.A.* **122**:424–426.

209. Hoult, I. J., MacLennan, A. H., and Carrie, L. E. S., 1977, Lumbar epidural analgesia in labour: Relation to fetal malposition and instrumental delivery, *Br. Med. J.* **1**:14–16.

210. James, F. M. III, Greiss, F. C., Jr., and Kemp, R. A., 1970, An evaluation of vasopressor therapy for maternal hypotension during spinal anesthesia, *Anesthesiology* **33**:25–34.

211. Brotanek, V., Vasicka, A., Santiago, A., and Brotanek, J. D., 1973, The influence of epidural anesthesia on uterine blood flow, *Obstet. Gynecol.* **42**:276–282.

212. Huovinen, K., Lehtovirta, P., Forss, M., Kivalo, I., and Teramo, K., 1979, Changes in placental intervillous blood flow measured by the [133]xenon method during lumbar epidural block for elective caesarean section, *Acta Anaesth. Scand.* **23**:529–533.

213. Jouppila, R., Jouppila, P., Kuikka, J., and Hollmen, A., 1978, Placental blood flow during caesarean section under lumbar extradural analgesia, *Br. J. Anaesth.* **50**:275–279.

214. Thomas, G., 1975, The aetiology, characteristics and diagnostic relevance of late deceleration patterns in routine obstetric practice, *Br. J. Obstet. Gynaecol.* **82**:121–125.

215. Schifrin, B. S., 1972, Fetal heart rate patterns following epidural anaesthesia and oxytocin infusion during labour, *J. Obstet. Gynaecol. Br. Commonw.* **79**:332–339.

216. Maltau, J. M., 1975, The frequency of fetal bradycardia during selective epidural anaesthesia, *Acta Obstet. Gynaecol. Scand.* **54**:357–361.

217. Shnider, S. M., de Lorimier, A. A., Holl, J. W., Chapler, F. K., and Morishima, H. O., 1968, Vasopressors in obstetrics: I. Correction of fetal acidosis with ephedrine during spinal hypotension, *Am. J. Obstet. Gynecol.* **102**:911–919.

218. Zilianti, M., Salazar, J. R., Aller, J., and Aguero, O., 1970, Fetal heart rate and pH of fetal capillary blood during epidural analgesia in labor, *Obstet. Gynecol.* **36**:881–886.

219. McDonald, J. S., Bjorkman, L. L., and Reed, E. C., 1974, Epidural analgesia for obstetrics: A maternal, fetal and neonatal study, *Am. J. Obstet. Gynecol.* **120**:1055–1065.

220. Shnider, S. M., deLorimier, A. A., Asling, J. H., and Morishima, H. O., 1970, Vasopressors in obstetrics: II. Fetal hazards of methoxamine administration during obstetric spinal anesthesia, *Am. J. Obstet. Gynecol.* **106**:680–686.

221. Antoine, C., and Young, B. K., 1982, Fetal lactic acidosis with epidural anesthesia, *Am. J. Obstet. Gynecol.* **142**:55–59.

222. Albright, G. A., Jouppila, R., Hollmen, A. I., Jouppila, P., Vierola, H., and Koivula, A., 1981, Epinephrine does not alter human intervillous blood flow during epidural anesthesia, *Anesthesiology* **54**:131–135.

223. Jouppila, R., Jouppila, P., Hollmen, A., and Kuikka, J., 1978, Effect of segmental extradural analgesia on placental blood flow during normal labour, *Br. J. Anaesth.* **50**:563–567.

224. Wallis, K. L., Shnider, S. M., Hicks, J. S., and Spivey, H. T., 1976, Epidural anesthesia in the normotensive pregnant ewe: Effects on uterine blood flow and fetal acid–base status, *Anesthesiology* **44**:481–487.

225. Noble, A. D., Craft, I. L., Bootes, J. A. H., Edwards, P. A., Thomas, D. J., and Mills, K. L. M., 1971, Continuous lumbar epidural analgesia using bupivacaine: A study of the fetus and newborn child, *J. Obstet. Gynaecol. Br. Commonw.* **78**:559–563.

226. Maresh, M., Choong, K. H., and Beard, R. W., 1983, Delayed pushing with lumbar epidural analgesia in labor, *Br. J. Obstet. Gynaecol.* **90:**623–627.
227. Hollmen, A. I., Jouppila, R., Jouppila, P., Koivula, A., and Vierola, H., 1982, Effect of extradural analgesia using bupivacaine and 2-chloroprocaine on intervillous blood flow during normal labour, *Br. J. Anaesth.* **54:**837–842.
228. Johnson, T., and Clayton, C. G., 1955, Studies in placental action during prolonged and dysfunctional labours using radioactive sodium, *J. Obstet. Gynaecol. Br. Emp.* **62:**513–522.
229. Giles, W. B., Lah, F. X., and Trudinger, B. J., 1987, The effect of epidural anaesthesia for caesarean section on maternal uterine and fetal umbilical artery blood flow velocity waveforms, *Br. J. Obstet. Gynaecol.* **94:**55–59.
230. Jouppila, R., Jouppila, P., Hollmen, A., and Koivula, A., 1979, Epidural analgesia and placental blood flow during labour in pregnancies complicated by hypertension, *Br. J. Obstet. Gynaecol.* **86:**969–972.
231. Jouppila, P., Jouppila, R., Hollmen, A., and Koivula, A., 1982, Lumbar epidural analgesia to improve intervillous blood flow during labor in severe preeclampsia, *Obstet. Gynecol.* **59:**158–161.
232. Philipp, K., Leodolter, S., and Neumark, J., 1980, The influence of continuous epidural block on uteroplacental blood-flow, in: *Pregnancy Hypertension* (J. Bonnar, I. MacGillivray, and M. Symonds, eds.), MTP Press, Lancaster, pp. 539–542.
233. Jouppila, P., Jouppila, R., Barinoff, T., and Koivula, A., 1984, Placental blood flow during caesarean section performed under subarachnoid blockade, *Br. J. Anaesth.* **56:**1379–1383.
234. Janisch, H., Leodolter, S., Neumark, J., and Philipp, K., 1978, Der Einfluss der kontinuierlichen Epiduralanasthesie auf die utero-plazentare Durchblutung, *Z. Geburtshilfe u Perinatol.* **182:**343–346.
235. Pearson, J. F., and Davies, P., 1974, The effect of continuous lumbar epidural analgesia upon fetal acid–base status during the first stage of labour, *J. Obstet. Gynaecol. Br. Commonw.* **81:**971–974.
236. Pearson, J. F., and Davies, P., 1974, The effect of continuous lumbar epidural analgesia upon fetal acid–base status during the second stage of labour, *J. Obstet. Gynaecol. Br. Commonw.* **81:**975–979.
237. Myers, R. E., and Myers, S. E., 1979, Use of sedative, analgesic, and anesthetic drugs during labor and delivery: Bane or boon? *Am. J. Obstet. Gynecol.* **133:**83–104.
238. de Rosayro, A. M., Nahrwold, M. L., and Hill, A. B., 1981, Cardiovascular effects of epidural epinephrine in the pregnant sheep, *Reg. Anesth.* **6:**4–7.
239. Shnider, S. M., Wright, R. G., Levinson, G., Roizen, M. F., Rolbin, S. H., Biehl, D., Johnson, J., and Jones, M., 1978, Plasma norepinephrine and uterine blood flow changes during endotracheal intubation and general anesthesia in the pregnant ewe, in: *Abstracts of Scientific Papers* (American Society of Anesthesiologists), pp. 115–116.
240. Shnider, S. M., Abboud, T., Levinson, G., Wright, R. G., Kim, S., Henriksen, E., Highes, S. C., Roizen, M. F., and Johnson, J., 1980, General anesthesia for cesarean section: Maternal and fetal norepinephrine levels and neonatal neurobehavioral status, *Anesthesiology* **53:**S302.
241. Jouppila, P., Kuikka, J., Jouppila, R., and Hollmen, A., 1979, Effect of induction of general anesthesia for cesarean section on intervillous blood flow, *Acta Obstet. Gynecol. Scand.* **58:**249–253.
242. Brown, F. F. III, Owens, W. D., Felts, J. A., Spitznagel, E. L., Jr., and Cryer, P. E., 1982, Plasma epinephrine and norepinephrine levels during anesthesia: Enflurane-N_2O-O_2 compared with fentanyl-N_2O-O_2, *Anesth. Analg.* **61:**366–370.
243. Russell, W. J., Morris, R. G., Frewin, D. B., and Drew, S. E., 1981, Changes in plasma catecholamine concentrations during endotracheal intubation, *Br. J. Anaesth.* **53:**837–839.
244. Palahniuk, R. J., and Cumming, M., 1977, Foetal deterioration following thiopentone–nitrous oxide anaesthesia in the pregnant ewe, *Can. Anaesth. Soc. J.* **24:**361–370.
245. Janoff, A. S., and Miller, K. W., 1982, A critical assessment of the lipid theories of general

anesthetic action, in: *Biological Membranes* (D. Chapman, ed.), Academic Press, London, pp. 417–476.

246. Janoff, A. S., Pringle, M. J., and Miller, R. W., 1981, Correlation of general anesthetic potency with solubility in membranes, *Biochim. Biophys. Acta* **649:**125–128.

247. Smith, R. A., Porter, E. G., and Miller, K. W., 1981, The solubility of anesthetic gases in lipid bilayers, *Biochim. Biophys. Acta* **645:**327–338.

248. Ueda, I., and Kamaya, H., 1984, Molecular mechanisms of anesthesia, *Anesth. Analg.* **63:**929–954.

249. Yang, J. C., Triner, L., Vulliemoz, Y., Verosky, M., and Ngai, S. H., 1973, Effects of halothane on the cyclic 3′,5′-adenosine monophosphate (cyclic AMP) system in rat uterine muscle, *Anesthesiology* **38:**244–250.

250. Korenaga, S., Takeda, K., and Ito, Y., 1984, Differential effects of halothane on airway nerves and muscles, *Anesthesiology* **60:**309–318.

251. Wikberg, J. E. S., Hede, A. R., and Lindahl, M., 1985, Effect of general anaesthetics and organic solvents on alpha₁-adrenoceptors in the myometrium, *Acta Pharmacol. Toxicol.* **57:**53–59.

252. Naftalin, N. J., Phear, W. P. C., and Goldberg, A. H., 1975, Halothane and isometric contractions of isolated pregnant rat myometrium, *Anesthesiology* **42:**458–463.

253. Naftalin, N. J., Phear, W. P. C., and Goldberg, A. H., 1976, Halothane and calcium interaction in isolated pregnant and postpartum rat myometrium, *Anesthesiology* **45:**31–38.

254. Munson, E. S., and Embro, W. J., 1977, Enflurane, isoflurane, and halothane on isolated human uterine muscle, *Anesthesiology* **46:**11–14.

255. Naftalin, N. J., McKay, D. M., Phear, W. P. C., and Goldberg, A. H., 1977, The effects of halothane on pregnant and nonpregnant human myometrium, *Anesthesiology* **46:**15–19.

256. McDonald-Gibson, W. J., 1969, The influence of halothane (Fluothane) on isolated human uterine muscle, *J. Obstet. Gynaecol. Br. Commonw.* **76:**362–365.

257. Munson, E. S., Maier, W. R., and Caton, D., 1969, Effects of halothane, cyclopropane and nitrous oxide on isolated human uterine muscle, *J. Obstet. Gynaecol. Br. Commonw.* **76:**27–33.

258. Miller, J. R., Stoelting, V. K., Stander, R. W., and Watring, W., 1966, *In vitro* and *in vivo* responses of the uterus to halothane anesthesia, *Anesth. Analg.* **45:**583–589.

259. Westmoreland, R. T., Evans, J. A., and Chastain, G. M., 1974, Obstetric use of enflurane (Ethrane), *South Med. J.* **67:**527–530.

260. Marx, G. F., Kim, Y. I., Lin, C. C., Halevy, S., and Schulman, H., 1978, Postpartum uterine pressures under halothane or enflurane anesthesia, *Obstet. Gynecol.* **51:**695–698.

261. Stoelting, V. K., 1964, Fluothane in obstetric anesthesia, *Anesth. Analg.* **43:**243–246.

262. Neumark, J., and Clark, R. B., 1977, Halothane for intrauterine resuscitation of the fetus, *Anesthesiol. Rev.* **4:**13–18.

263. Zargham, I., Leviss, S. R., and Marx, G. F., 1974, Uterine pressures during fluroxene anesthesia, *Anesth. Analg.* **53:**568–572.

264. Cosmi, E. V., and Marx, G. F., 1969, The effect of anesthesia on the acid–base status of the fetus, *Anesthesiology* **30:**238–242.

265. Russell, J. T., 1958, Halothane and caesarean section, *Anaesthesia* **13:**241–242.

266. Cullen, B. F., Margolis, A. J., and Eger, E. I. II, 1970, The effects of anesthesia and pulmonary ventilation on blood loss during elective therapeutic abortion, *Anesthesiology* **32:**108–113.

267. Dolan, W. M., Eger, E. I. II, and Margolis, A. J., 1972, Forane increases bleeding in therapeutic suction abortion, *Anesthesiology* **36:**96–97.

268. Coleman, A. J., and Downing, J. W., 1975, Enflurane anesthesia for cesarean section, *Anesthesiology* **43:**354–357.

269. Galbert, M. W., and Gardner, A. E., 1972, Use of halothane in a balanced technic for cesarean section, *Anesth. Analg.* **51:**701–704.

270. Abboud, T. K., Kim, S. H., Henriksen, E. H., Chen, T., Eisenman, R., Levinson, G., and Shnider, S. M., 1985, Comparative maternal and neonatal effects of halothane and enflurane for cesarean section, *Acta Anaesth. Scand.* **29:**663–668.

271. Moir, D. D., 1970, Anaesthesia for caesarean section: An evaluation of a method using low concentration of halothane and 50 per cent of oxygen, *Br. J. Anaesth.* **42**:136–142.

272. Embrey, M. P., Garrett, W. J., and Pryer, D. L., 1958, Inhibitory action of halothane on contractility of human pregnant uterus, *Lancet* **2**:1093–1094.

273. Albert, C. A., Anderson, G., Wallace, W., Henley, E. E., Winshel, A. W., and Albert, S. N., 1959, Fluothane for obstetric anesthesia, *Obstet. Gynecol.* **13**:282–284.

274. Lumley, J., Walker, A., Marum, J., and Wood, C., 1970, Time: An important variable at caesarean section, *J. Obstet. Gynaecol. Br. Commonw.* **77**:10–23.

275. Fothergill, R. J., Robertson, A., and Bond, R. A., 1971, Neonatal acidaemia related to procrastination at caesarean section, *J. Obstet. Gynaecol. Br. Commonw.* **78**:1010–1023.

276. Datta, S., Ostheimer, G. W., Weiss, J. B., Brown, W. U., Jr., and Alper, M. H., 1981, Neonatal effect of prolonged anesthetic induction for cesarean section, *Obstet. Gynecol.* **58**:331–335.

277. Larach, D. R., Schuler, H. G., Derr, J. A., Larach, M. G., Hensley, F. A., and Zellis, R., 1987, Halothane selectively attenuates alpha$_2$-adrenoceptor mediated vasoconstriction, *in vivo* and *in vitro*, *Anesthesiology* **66**:781–791.

278. Datta, S., Ostheimer, G. W., Naulty, J. S., Knapp, R. M., and Weiss, J. B., 1981, General anesthesia for cesarean section: Effects of halothane on maternal and fetal acid–base and lactic acid concentration, *Anesthesiology* **55**:A309.

279. Magno, R., Karlsson, K., Selstam, U., and Wickstrom, I., 1976, Anesthesia for cesarean section V: Effects of enflurane anesthesia on the respiratory adaptation of the newborn in elective cesarean section, *Acta Anaesth. Scand.* **20**:147–155.

280. Crawford, J. S., Burton, M., and Davies, P., 1973, Anaesthesia for section: Further refinements of a technique, *Br. J. Anaesth.* **45**:726–732.

281. Marx, G. F., and Mateo, C. V., 1971, Effects of different oxygen concentrations during general anaesthesia for elective caesarean section, *Can. Anaesth. Soc. J.* **18**:587–593.

282. Kangas, L., Erkkola, R., Kanto, J., and Mansikka, M., 1976, Halothane anaesthesia in caesarean section, *Acta Anaesth. Scand.* **20**:189–194.

283. Warren, T. M., Datta, S., Ostheimer, G. W., Naulty, J. S., Weiss, J. B., and Morrison, J. A., 1983, Comparison of the maternal and neonatal effects of halothane, enflurane, and isoflurane for cesarean delivery, *Anesth. Analg.* **62**:516–520.

284. Smith, J. B., Manning, F. A., and Palahniuk, R. J., 1975, Maternal and foetal effects of methoxyflurane anaesthesia in the pregnant ewe, *Can. Anaesth. Soc. J.* **22**:449–459.

285. Palahniuk, R. J., and Shnider, S. M., 1974, Maternal and fetal cardiovascular and acid–base changes during halothane and isoflurane anesthesia in the pregnant ewe, *Anesthesiology* **41**:462–472.

286. Eng, M., Berges, P. U., der Yuen, D., Bonica, J. J., and Ueland, K., 1976, A comparison of the effects of the inhalation of 4% and 8% fluroxene in the pregnant primate, *Acta Anaesth. Scand.* **20**:183–188.

287. Eng, M., Bonica, J. J., Akamatsu, T. J., Berges, P. U., der Yuen, D., and Ueland, K., 1975, Maternal and fetal responses to halothane in pregnant monkeys, *Acta Anaesth. Scand.* **19**:154–158.

288. Mulinos, M. G., 1936, Anesthetic properties of sodium-ethyl-pentyl, malonylthiourea, *Proc. Soc. Exp. Biol. Med.* **34**:506–507.

289. Talbert, L. M., McGaughey, H. S., Jr., Corey, E. L., and Thornton, W. N., Jr., 1958, Effects of anesthetic and sedative agents commonly employed in obstetric practice on isolated human uterine muscle, *Am. J. Obstet. Gynecol.* **75**:16–22.

290. Snyder, F. F., 1949, The influence of barbiturates in the relief of pain, in: *Obstetric Analgesia and Anesthesia*, W. B. Saunders, Philadelphia, pp. 277–298.

291. Reynolds, S. R. M., Harris, J. S., and Kaiser, I. H., 1954, The effects of anesthesia and analgesia on labor, in: *Clinical Measurement of Uterine Forces in Pregnancy and Labor*, Charles C. Thomas, Springfield, IL, pp. 251–264.

292. Dodek, S. M., 1934, External hysterographic studies of the effect of certain analgesics and anesthetics upon the parturient human uterus, *Anesth. Analg.* **13**:8.

293. Seear, T., 1967, Pentobarbital anesthesia in labor, *Am. J. Obstet. Gynecol.* **99**:952–959.
294. Campbell, C., Phillips, O. C., and Frazier, T. M., 1961, Analgesia during labor: A comparison of pentobarbital, meperidine, and morphine, *Obstet. Gynecol.* **17**:714–718.
295. Irving, F. C., Berman, S., and Nelson, H. B., 1934, The barbiturates and other hypnotics in labor, *Surg. Gynecol. Obstet.* **58**:1–11.
296. Cosmi, E. V., 1976, Drugs, anesthetics and the fetus, in: *Reviews in Perinatal Medicine*, Volume 1 (E. M. Scarpelli and E. V. Cosmi, eds.), University Park Press, Baltimore, pp. 191–253.
297. MacPhail, F. L., Gray, H. R. D., and Bourne, W., 1937, Pentothal sodium as a hypnotic in obstetrics, *Can. Med. Assoc. J.* **37**:471–474.
298. Bourne, W., and Pauly, A. J., 1939, Thiobarbiturates in obstetrics: Pentothal and thioethamyl, *Can. Med. Assoc. J.* **40**:437–440.
299. Domino, E. F., 1962, Sites of action of some central nervous depressants, *Annu. Rev. Pharmacol.* **2**:215–250.
300. Exley, K. A., 1954, Depression of autonomic ganglia by barbiturates, *Br. J. Pharmacol. Chemother.* **9**:170–181.
301. Maynert, E. W., and Levi, R., 1964, Stress-induced release of brain norepinephrine and its inhibition by drugs, *J. Pharmacol. Exp. Ther.* **143**:90–95.
302. Holmes, J. C., and Schneider, F. H., 1973, Pentobarbitone inhibition of catecholamine secretion, *Br. J. Pharmacol.* **49**:205–213.
303. Cosmi, E. V., Tonelli, F., Batianon, V., and Condorelli, S., 1973, Asfissia fetale da tiopentale sodico e da lidocaina, *Acta Anaesth. Italia* **24**:147.
304. Cosmi, E. V., Condorelli, S., and Scarpelli, E. M., 1974, Fetal asphyxia induced by sodium thiopental, thiamylal, and methohexital, in: *Fourth European Congress on Perinatal Medicine, Prague,* Abstract IV, 3/12.
305. Wolkoff, A. S., Bawden, J. W., Flowers, C. E., and McGee, J. A., 1965, The effects of anesthesia on the unborn fetus, *Am. J. Obstet. Gynecol.* **93**:311–320.
306. Dick, W., Borst, R., Fodor, L., Haug, H., Milewski, P., Schumann, R., and Traub, E., 1973, Ketamin in obstetrical anesthesia: Clinical and experimental results, *J. Perinat. Med.* **1**:252–262.
307. Little, B., Chang, T., Chucot, L., Dill, W. A., Enrile, L. L., Glazko, A. J., Jassani, M., Kretchmer, H., and Sweet, A. Y., 1972, Study of ketamine as an obstetric anesthetic agent, *Am. J. Obstet. Gynecol.* **113**:247–260.
308. Marx, G. F., Hwang, H. S., and Chandra, P., 1979, Postpartum uterine pressures with different doses of ketamine, *Anesthesiology* **50**:163–166.
309. Galloon, S., 1976, Ketamine for obstetric delivery, *Anesthesiology* **44**:522–524.
310. Galloon, S., 1973, Ketamine and the pregnant uterus, *Can. Anaesth. Soc. J.* **20**:141–145.
311. Oats, J. N., Vasey, D. P., and Waldron, B. A., 1979, Effects of ketamine on the pregnant uterus, *Br. J. Anaesth.* **51**:1163–1166.
312. Craft, J. B., Jr., Coaldrake, L. A., Yonekura, M. L., Dao, S. D., Co, E. G., Roizen, M. F., Mazel, P., Gilman, R., Shokes, L., and Trevor, A. J., 1983, Ketamine, catecholamines, and uterine tone in pregnant ewes, *Am. J. Obstet. Gynecol.* **146**:429–434.
313. Cosmi, E. V., 1977, Effetti della ketamina sulla madre e sul feto. Studio sperimentale e clinico, *Minerva Anesth.* **43**:397.
314. Levinson, G., Shnider, S. M., Gildea, J. E., and deLorimier, A. A., 1973, Maternal and foetal cardiovascular and acid–base changes during ketamine anaesthesia in pregnant ewes, *Br. J. Anaesth.* **45**:1111–1115.
315. Galbert, M. W., and Gardner, A. E., 1973, Ketamine for obstetrical anesthesia, *Anesth. Analg.* **52**:926–930.
316. Maduska, A. L., and Hajghassemali, M., 1978, Arterial blood gases in mothers and infants during ketamine anesthesia for vaginal delivery, *Anesth. Analg.* **57**:121–123.
317. Meer, F. M., Downing, J. W., and Coleman, A. J., 1973, An intravenous method of anaesthesia for caesarean section. Part II: Ketamine, *Br. J. Anaesth.* **45**:191–196.
318. Bernstein, K., Gisselsson, L., Jacobsson, L., and Ohrlander, S., 1985, Influence of two different

anaesthetic agents on the newborn and the correlation between foetal oxygenation and induction–delivery time in elective caesarean section, *Acta Anaesth. Scand.* **29**:157–160.

319. Scoggin, W. A., McGaughey, H. S., Jr., Johnson, W. L., and Thornton, W. N., Jr., 1963, Uterine contractility in primates: A comparative study, *Surg. Forum* **14**:384–385.

320. Bueno-Montano, M., 1966, Species variation in contractile response of excised mammalian uterine tissues, *Am. J. Obstet. Gynecol.* **94**:1062–1067.

321. Gruber, C. M., Hart, E. R., and Gruber, C. M., Jr., 1941, The pharmacology and toxicology of the ethyl ester of 1-methyl-4-phenyl-piperidine-4-carboxylic acid (Demerol), *J. Pharmacol. Exp. Ther.* **73**:319–334.

322. Gruber, C. M., Brundage, J. T., deNote, A., and Heiligman, R., 1936, A comparison of the actions of dilaudid hydrochloride and morphine sulphate upon segments of excised intestine and uterus, *J. Pharmacol. Exp. Ther.* **55**:430–434.

323. Barbour, H. G., and Copenhaver, N. H., 1915, The response of the surviving uterus to morphin and scopolamin, *J. Pharmacol. Exp. Ther.* **7**:529–539.

324. Barbour, H. G., 1915, Morphin and scopolamin action upon the intact uterus, *J. Pharmacol. Exp. Ther.* **7**:547–555.

325. Slaughter, D., and Gross, E. G., 1937, The action of morphine on the contractions of the non-pregnant uterus in unanesthetized dogs and rabbits, *J. Pharmacol. Exp. Ther.* **59**:350–357.

326. Friedman, E. A., 1965, Effects of drugs on uterine contractility, *Anesthesiology* **26**:409–422.

327. Bolton, R. N., and Benson, R. C., 1958, The use of promazine and meperidine in labor, *West. J. Surg.* **66**:253–257.

328. Lindgren, L., 1959, The influence of anaesthetics and analgesics on different types of labour, *Acta Anaesth. Scand. [Suppl.]* **2**:49–56.

329. Gilbert, G., and Dixon, A. B., 1943, Observations on Demerol as an obstetric analgesic, *Am. J. Obstet. Gynecol.* **45**:320–326.

330. Sica-Blanco, Y., Rozada, H., and Remedio, M. R., 1967, Effect of meperidine on uterine contractility during pregnancy and prelabor, *Am. J. Obstet. Gynecol.* **97**:1096–1100.

331. Ballas, S., Toaff, M. E., and Toaff, R., 1976, Effects of intravenous meperidine and meperidine with promethazine on uterine activity and fetal heart rate during labor, *Israel J. Med. Sci.* **12**:1141–1147.

332. Riffel, H. D., Nochimson, D. J., Paul, R. H., and Hon, E. H. G., 1973, Effects of meperidine and promethazine during labor, *Obstet. Gynecol.* **42**:738–745.

333. DeVoe, S. J., DeVoe, K., Jr., Rigsby, W. C., and McDaniels, B. A., 1969, Effect of meperidine on uterine contractility, *Am. J. Obstet. Gynecol.* **105**:1004–1007.

334. Gordon, L. E., and Ruffin, C. L., 1958, Promethazine as an adjunct to obstetrical analgesia and sedation, *Am. J. Obstet. Gynecol.* **76**:147–151.

335. Carroll, J. J., and Moir, R. S., 1958, Use of promethazine (Phenergan) hydrochloride in obstetrics, *J.A.M.A.* **168**:2218–2224.

336. Carroll, J. J., and Hudson, P. W., 1955, Chlorpromazine and promethazine in obstetrics, *Can. Anaesth. Soc. J.* **2**:340–346.

337. Fitzgerald, W. J., Garcia, R. R., and Cassidy, J. J., 1958, Clinical evaluation of the analgesic effect of phenergan in labor, *Obstet. Gynecol.* **12**:703–704.

338. Eckerling, B., Goldman, J. A., and Gans, B., 1959, The combined intravenous use of Pethidine, Phenergan and Lorfan for analgesia in obstetrics, *Obstet. Gynecol.* **14**:331–336.

339. Leazar, M. A., 1960, Evaluation of promethazine and promazine for sedation in labor, *West. J. Surg.* **68**:135–137.

340. Pannullo, J. N., and Cerone, D. M., 1960, Promethazine in obstetrics, *J. Med. Soc. NJ* **57**:65–68.

341. Semler, W. L., 1965, Use of hydroxyzine in management of labor, *Wisconsin Med. J.* **64**:389–390.

342. Jeffcoate, T. N. A., Baker, K., and Martin, R. H., 1952, Inefficient uterine action, *Surg. Gynecol. Obstet.* **95**:257–273.

343. Malkasian, G. D., Jr., Smith, R. A., and Decker, D. G., 1967, Comparison of hydroxyzine–

meperidine and promethazine–meperidine for analgesia during labor, *Obstet. Gynecol.* **30**:568–575.

344. Eskes, T. K. A. B., 1962, Effect of morphine upon uterine contractility in late pregnancy, *Am. J. Obstet. Gynecol.* **84**:281–289.

345. Zsigmond, E. K., and Patterson, R. L., 1967, Double-blind evaluation of hydroxyzine hydrochloride in obstetric anesthesia, *Anesth. Analg.* **46**:275–280.

346. Corbit, J. D., Jr., and First, S. E., 1961, Clinical comparison of phenazocine and meperidine in obstetric analgesia, *Obstet. Gynecol.* **18**:488–491.

347. Bourne, A. W., and Burn, J. H., 1930, Action on the human uterus of anaesthetics and other drugs commonly used in labour, *Br. Med. J.* **2**:87–89.

348. Olson, R. O., and Riva, H. L., 1964, Evaluation of phenazocine with meperidine as an analgesic agent during labor, by the double blind method, *Am. J. Obstet. Gynecol.* **88**:601–605.

349. Stewart, R. H., 1961, Phenothiazine derivatives in labor and delivery: A study of four drugs, *Obstet. Gynecol.* **17**:701–713.

350. Larks, S. D., Dasgupta, K., Morton, D. G., and Bellamy, A. W., 1959, Effects of oxytocin and analgesic drugs on the human electrohysterogram, *Obstet. Gynecol.* **13**:405–412.

351. Petrie, R. H., Wu, R., Miller, F. C., Sacks, D. A., Sugarman, R., Paul, R. H., and Hon, E. H., 1976, The effect of drugs on uterine activity, *Obstet. Gynecol.* **48**:431–435.

352. Craft, J. B., Jr., Coaldrake, L. A., Bolan, J. C., Mondino, M., Mazel, P., Gilman, R. M., Shokes, L. K., and Woolf, W. A., 1983, Placental passage and uterine effects of fentanyl, *Anesth. Analg.* **62**:894–898.

353. Craft, J. B., Jr., Bolan, J. C., Coaldrake, L. A., Mondino, M., Mazel, P., Gilman, R. M., Shokes, L. K., and Woolf, W. A., 1982, The maternal and fetal cardiovascular effects of epidural morphine in the sheep model, *Am. J. Obstet. Gynecol.* **142**:835–839.

354. Rosen, M. A., Hughes, S. C., Curtis, J. D., Norton, M., Levinson, G., and Shnider, S. M., 1982, Effects of epidural morphine on uterine blood flow and acid–base status in the pregnant ewe, *Anesthesiology* **57**:A383.

355. Craft, J. B., Jr., Robichaux, A. G., Kim, H. S., Thorpe, D. H., Mazel, P., Woolf, W. A., and Stolte, A., 1984, The maternal and fetal cardiovascular effects of epidural fentanyl in the sheep model, *Am. J. Obstet. Gynecol.* **148**:1098–1104.

356. Zourlas, P. A., 1964, *In vitro* and *in vivo* effects of promethazine hydrochloride on human uterine contractility, *Am. J. Obstet. Gynecol.* **90**:115–119.

357. Gordon, E. M., and Nelson, H. B., 1965, A double blind study of the effect of tranquilizers during labor, *Am. J. Obstet. Gynecol.* **92**:299–304.

358. Inmon, W. B., 1963, A study of the effect of hydroxyzine hydrochloride on labor and delivery, *Am. J. Obstet. Gynecol.* **86**:853–855.

359. Brelje, M. C., and Garcia-Bunuel, R., 1966, Meperidine–hydroxyzine in obstetric analgesia, *Obstet. Gynecol.* **27**:350–354.

360. Cavanagh, D., Albores, E. A., and Todd, J., 1966, Comparative effects of two benzodiazepine compounds on isolated human myometrium, *Am. J. Obstet. Gynecol.* **94**:6–13.

361. Landesman, R., and Wilson, K. H., 1965, Relaxant effect of diazepam on uterine muscle, *Obstet. Gynecol.* **26**:552–556.

362. Toaff, M. E., Hezroni, J., and Toaff, R., 1977, Effect of diazepam on uterine activity during labor, *Israel J. Med. Sci.* **13**:1007–1012.

363. Yeh, S. Y., Paul, R. H., Cordero, L., and Hon, E. H., 1974, A study of diazepam during labor, *Obstet. Gynecol.* **43**:363–373.

364. Scher, J., Hailey, D. M., and Beard, R. W., 1972, The effects of diazepam on the fetus, *J. Obstet. Gynaecol. Br. Commonw.* **79**:635–638.

365. Soiva, K., Castren, O., and Ruponen, S., 1964, Tocographic studies with Librium and Valium, *Ann. Chir. Gynaecol. Fenn.* **53**:141–146.

366. Lee, D. T., 1968, The effects of diazepam (Valium) on labour, *Can. Med. Assoc. J.* **98**:446–448.

367. Timonen, S., and Hagner, M., 1966, Benzodiazepines in an unselected obstetrical series, *Ann. Chir. Gynaecol. Fenn.* **55**:65–68.
368. Nisbet, R., Boulas, S. H., and Kantor, H. I., 1967, Diazepam (Valium) during labor, *Obstet. Gynecol.* **29**:726–729.
369. Bepko, F., Lowe, E., and Waxman, B., 1965, Relief of the emotional factor in labor with parenterally administered diazepam, *Obstet. Gynecol.* **26**:852–857.
370. Friedman, E. A., Niswander, K. R., and Sachtleben, M. R., 1969, Effect of diazepam on labor, *Obstet. Gynecol.* **34**:82–86.
371. Davies, J. M., and Rosen, M., 1977, Intramuscular diazepam in labour: A double-blind trial in multiparae, *Br. J. Anaesth.* **49**:601–604.
372. Suonio, S., Kauppila, A., Jouppila, P., and Linna, O., 1972, The effect of intravenous diazepam on fetal and maternal acid–base balance during and after delivery, *Ann. Chir. Gynaecol. Fenn.* **60**:52–55.
373. Mofid, M., Brinkman, C. R. III, and Assali, N. S., 1973, Effects of diazepam on uteroplacental and fetal hemodynamics and metabolism, *Obstet. Gynecol.* **41**:364–368.
374. Cree, J. E., Meyer, J., and Hailey, D. M., 1973, Diazepam in labor: Its metabolism an effect on the clinical condition and thermogenesis of the newborn, *Br. Med. J.* **4**:251–255.
375. Harroun, P., and Fisher, C. W., 1949, The physiological effects of curare: Its failure to pass the placental membrane or inhibit uterine contractions, *Surg. Gynecol. Obstet.* **89**:73–75.
376. Bourke, J. D., 1949, Curare with general anesthesia for forceps delivery, *Lancet* **2**:511–512.
377. Hartnett, L. J., and Freiheit, H. J., 1950, Curare as an adjunct in the conduct of labor: A preliminary report on its use in 200 cases, *South. Med. J.* **43**:277–281.
378. Weinberg, A., and Talisman, M. R., 1956, The use of long-acting curarine in obstetrics: A preliminary report, *Am. J. Obstet. Gynecol.* **71**:871–875.
379. Austin, B. R., and Mering, T. W., 1951, The use of muscle-relaxing drugs in complicated deliveries, *Am. J. Obstet. Gynecol.* **62**:143–149.
380. Squire, J. J., and Roberts, L. M., 1957, Curare and obstetric anesthesia, *Obstet. Gynecol.* **10**:56–59.
381. Dennis, J. W., and Carroll, J. J., 1954, The use of short acting muscle relaxing drugs in obstetrical anaesthesia, *Can. Anaesth. Soc. J.* **1**:82–86.
382. French, T. A., 1958, Repository tubocurarine in obstetrics, *Obstet. Gynecol.* **12**:550–555.
383. Pittinger, C. B., Morris, L. E., and Keettel, W. C., 1953, Vaginal deliveries during profound curarization, *Am. J. Obstet. Gynecol.* **65**:635–638.
384. Scurr, C. F., 1951, A comparative review of the relaxants, *Br. J. Anaesth.* **23**:103–116.
385. Alver, E. C., White, C. W., Jr., Weiss, J. B., and Heerdegen, D. K., 1962, An effect of succinylcholine on the uterus, *Am. J. Obstet. Gynecol.* **83**:795–799.
386. Wiqvist, N., and Wahlin, A., 1962, Effect of succinylcholine on uterine motility, *Acta Anaesth. Scand.* **6**:71–75.
387. Brinkman, C. R. III, and Assali, N. S., 1976, Uteroplacental hemodynamic response to antihypertensive drugs in hypertensive pregnant sheep, in: *Hypertension in Pregnancy* (M. D. Lindheimer, A. I. Katz, and F. P. Zuspan, eds.), John Wiley & Sons, New York, pp. 363–375.
388. Ring, G., Krames, E., Shnider, S. M., Wallis, K. L., and Levinson, G., 1977, Comparison of nitroprusside and hydralazine in hypertensive pregnant ewes, *Obstet. Gynecol.* **50**:598–602.
389. Lunell, N. O., Lewander, R., Nylund, L., Sarby, B., and Thornstrom, S., 1983, Acute effect of dihydralazine on uteroplacental blood flow in hypertension during pregnancy, *Gynecol. Obstet. Invest.* **16**:274–282.
390. Jouppila, P., Kirkinen, P., Koivula, A., and Ylikorkala, O., 1985, Effects of dihydralazine infusion on the fetoplacental blood flow and maternal prostanoids, *Obstet. Gynecol.* **65**:115–118.
391. Mandell, G., Dewan, D., Eisenach, J., Rose, J., and Thomas, K., 1986, Labetalol treatment of norepinephrine induced hypertension in the gravid ewe, *Anesthesiology* **65**:A395.
392. Lunell, N. O., Fredholm, B., Hjemdahl, P., Lewander, R., Nisell, H., Nylund, L., Persson, B., Sarby, B., and Wager, J., 1982, Labetalol, a combined alpha- and beta-blocker, in hypertension of pregnancy, *Acta Med. Scand. [Suppl.]* **665**:143–147.

393. Lieb, S. M., Zugaib, M., Nuwayhid, B., Tabsh, K., Erkkola, R., Ushioda, E., Brinkman, C. R. III, and Assali, N. S., 1981, Nitroprusside-induced hemodynamic alterations in normotensive and hypertensive pregnant sheep, *Am. J. Obstet. Gynecol.* **139:**925–931.
394. Wheeler, A. S., James, F. M. III, Meis, P. J., Rose, J. C., Fishburne, J. I., Dewan, D. M., Urban, R. B., and Greiss, F. C., Jr., 1980, Effects of nitroglycerin and nitroprusside on the uterine vasculature of gravid ewes, *Anesthesiology* **52:**390–394.
395. Rapaport, R. M., and Murad, F., 1983, Endothelium-dependent and nitrovasodilator-induced relaxation of vascular smooth muscle: Role for cyclic GMP, *J. Cyclic Nucleotide Protein Phosphorylation Res.* **9:**281–296.
396. Craft, J. B., Jr., Co, E. G., Yonekura, M. L., and Gilman, R. M., 1980, Nitroglycerin therapy for phenylephrine-induced hypertension in pregnant ewes, *Anesth. Analg.* **59:**494–499.
397. Ellis, S. C., Wheeler, A. S., James, F. M. III, Rose, J. C., Meis, P. J., Shihabi, Z., Greiss, F. C., Jr., and Urban, R. B., 1982, Fetal and maternal effects of sodium nitroprusside used to counteract hypertension in gravid ewes, *Am. J. Obstet. Gynecol.* **143:**766–770.

Appendix

Conversion factors
  Estradiol: 1 μg/ml $\approx$ 3.7 μM; mol. wt. 272
  Progesterone: 1 μg/ml $\approx$ 3.2 μM; mol. wt. 314
  Oxytocin: 1 μU/ml $\approx$ 2 pg/ml $\approx$ 2 pM; mol. wt. 1007
  Prostaglandin E_2: 1 μg/ml $\approx$ 2.8 μM; mol. wt. 352
  Prostaglandin $F_{2\alpha}$: 1 μg/ml $\approx$ 2.8 μM; mol. wt. 354
  Calcium ionophore A23187(Galimycin): 1 μg/ml $\approx$ 1.9 μM; mol. wt. 523
  Calcium ionophore X-537A (Lasalocid): 1 μg/ml $\approx$ 1.6 μM; mol. wt. 591
  Nifedipine: 1 μg/ml $\approx$ 2.9 μM; mol. wt. 346
  Verapamil: 1 μg/ml $\approx$ 2.2 μM; mol. wt. 455
  Inositol trisphosphate: 1 μg/ml $\approx$ 2.4 μM; mol. wt. 420

Phosphorylatable light chain of myosin	mol. wt.
From fast twitch skeletal muscle	18,000
From slow twitch skeletal muscle	19,000
From cardiac muscle	19,000
From smooth muscle	20,000

Cell dimensions
  Pregnant myometrial cell:　　　length 300–600 μm; diameter 5–10 μm

Hormone levels in fetal circulation at term
Cortisol	22–70 ng/ml
Dehydroepiandrosterone sulfate (DHEA-S)	1–2 μg/ml
Adrenocorticotrophic hormone (ACTH)	160 pg/ml

Human daily steroid hormone production near term
17β-Estradiol	15–20 mg
Estriol	50–100 mg
Progesterone	250–600 mg
Aldosterone	1–2 mg
Deoxycorticosterone	3–12 mg

Synonyms
  Phospholipase C = (poly)phosphoinositide phosphodiesterase =
    phosphoinositidase C
  Ca,Mg-ATPase = (Ca^{2+},Mg^{2+})ATPase = calcium transport ATPase =
    calcium pump

Gestational length
  Cow   270–280 days
  Dog   69 days
  Elephant   22 months
  Guinea pig   68 days
  Horse   333 days
  Human   267 days
  Rabbit   31–32 days
  Rat   21–23 days
  Sheep   145–149 days

Index

Page numbers in italics indicate figures.